THE COMPREHENSIVE GUIDE TO WORK INJURY MANAGEMENT

Edited by
SUSAN J. ISERNHAGEN, PT
President
Isernhagen and Associates, Inc.
Duluth, Minnesota

AN ASPEN PUBLICATION
Aspen Publishers, Inc.
Gaithersburg, Maryland
1995

Library of Congress Cataloging-in-Publication Data

The comprehensive guide to work injury management/
edited by Susan J. Isernhagen.
p. cm.
Includes bibliographical references and index.
ISBN 0-8342-0558-0
1. Wounds and injuries—Treatment. 2. Industrial safety.
3. Physical therapy.
I. Isernhagen, Susan J.
RD97.5.C66 1994
658.3'82—dc20
94-18431
CIP

The authors have made every effort to ensure the accuracy of the information herein, particularly with regard to product selection, drug selection, and dose. However, appropriate information sources should be consulted, especially for new or unfamiliar drugs or procedures. It is the responsibility of every practitioner to evaluate the appropriateness of a particular opinion in the context of actual clinical situations and with due consideration to new developments. The authors acknowledge the mention of specific products as examples of products in current clinical practice; however, this should not be construed as endorsements of the products. The authors and the publisher cannot be held responsible for any typographical or other errors found in this book.

Editorial Services: Jane Colilla

Library of Congress Catalog Card Number: 94-18431
ISBN: 0-8342-0558-0

Printed in the United States of America

1 2 3 4 5

Table of Contents

Contributors

Susan H. Abeln, BSPT, ARM
President
Strategic Healthcare Alternatives
San Clemente, California

Mark A. Anderson, MA, PT
Director of Industrial Consulting
The Saunders Group
Minneapolis, Minnesota

Stephen T. Anderson
President
ErgometRx
St. Paul, Minnesota

Barbara Baum, MPT, MS, PT
Coordinator of Occupational Rehabilitation Services
Institute for Occupational Rehabilitation
Fairview Hospital and Healthcare Services
Minneapolis, Minnesota

Laurence N. Benz, MPT, ESC, OCS
President
Comprehensive Injury Management Associates
South Kentucky Physical Therapy, PSC
Elizabethtown, Kentucky

Thomas A. Broderick, MIS
Executive Director
Construction Safety Council
Hillside, Illinois

Allan M. Brown, PT, ATC
Co-Owner/Treasurer
Granite Hill Physical Therapy
Brunswick, Maine

Stephen E. Campbell, PT
Therapy Dynamics, LP
Dallas, Texas

Sara B. Chmielewski, PT
Director of Occupational Medicine
Isernhagen Clinics, Inc.
Duluth, Minnesota

Robert B. Clift, PhD, PA
Psychologist
Minneapolis, Minnesota

David W. Clifton, Jr., PT
President/CEO
Disability Management Associates
Lester, Pennsylvania

Rebecca Cox, PT
Director of Physical Therapy
Productive Rehabilitation Institute of Dallas for Ergonomics
Dallas, Texas

Linda E. Darphin, PT
Clinical Director
Occupational Performance Center
A Division of Baton Rouge Physical Therapy
Baton Rouge, Louisiana

Carl DeRosa, PhD, PT
Associate Professor and Chairman
Physical Therapy Program
Northern Arizona University
Flagstaff, Arizona

Melanie T. Ellexson, MBA, OTR, FAOTA
Assistant Vice President/Executive Director
STEPS Industrial Rehabilitation Clinics
Schwab Rehabilitation Hospital
Chicago, Illinois

Pam Garcy, MA
Department of Psychiatry
University of Texas Southwestern Medical Center
Dallas, Texas

Gloria J. Gebhard, MA, PT
Manager
Health Care Policy Section
Minnesota Department of Health
Minneapolis, Minnesota

R. Gary Gray, PT
Midland Physical Therapy, PC
Midland, Texas
WorkSTEPS, Inc.
Arlington, Texas

Elizabeth J. Green, OTR/L
Industrial Consultant
Charlotte, North Carolina

Marie-Louise Hallmark
Specialist in Ergonomics
Clinic of Rheumatology and Institute of Physical Therapy
University Hospital of Zurich
Zurich, Switzerland

Dennis L. Hart, PhD, PT
Director of Research and Outcomes Management
Rehability Corporation
Great Falls, Virginia

Heinz O. Hofer, MD, MA
Specialist in Physical Medicine and Rehabilitation
Clinic of Rheumatology and Institute of Physical Therapy
University Hospital of Zurich
Zurich, Switzerland

Dennis D. Isernhagen, PT
Vice President
Isernhagen Work Systems
Duluth, Minnesota

Susan J. Isernhagen, PT
President
Isernhagen and Associates, Inc.
Duluth, Minnesota

Laurie J. Johnson, PT
Director of Clinical Performance
Isernhagen and Associates, Inc.
Duluth, Minnesota

Kristine F. Kerr, MS, PT
Research Coordinator
Isernhagen Clinics, Inc.
Duluth, Minnesota

Barbara L. Kornblau, JD, OTR, CIRS, CCM, DAAPM
Miami, Florida

Andrea O. Kramer, MEd
Technical Writer
Racine, Wisconsin

Barbara A. Larson, OTR
Director of Industrial Medicine Programs
Health Dimensions Rehabilitation
Cambridge, Minnesota

Gordon Leeman, MS
Medical Case Manager
Productive Rehabilitation Institute of Dallas for Ergonomics
Dallas, Texas

Frank H. Leone, MBA, MPH
President
Ryan Associates
Executive Director
National Association of Occupational Health Professionals
Santa Barbara, California

Elroy C. Lundblad, MS
Safety Manager
SC Johnson Wax
Racine, Wisconsin

Leonard N. Matheson, PhD, CVE
Diplomate, ABVE
Director of ERIC Human Performance Laboratory
Santa Ana, California
Adjunct Assistant Professor
Washington University School of Medicine Program in Occupational Therapy
St. Louis, Missouri

Tom G. Mayer, MD
Clinical Professor of Orthopedic Surgery
University of Texas Southwestern Medical Center
Medical Director
Productive Rehabilitation Institute of Dallas for Ergonomics
Dallas, Texas

Sharyn McGuire, OT
Rehabilitation Consultant
Work Dynamics
Occupational Rehabilitation Services
Fremantle and Sydney, Australia

Barbara McPhee, MPH, Dip Phty
Head of Ergonomics Unit
Worksafe Australia
Sydney, Australia

Margot Miller, PT
Director of National Programs
Isernhagen Work Systems
Duluth, Minnesota

Mary A. Mistal, PT
Mississippi Physical Therapy
Jackson, Mississippi

Matthew Monsein, MD
Medical Director
Chronic Pain Rehabilitation Program
Sister Kenny Institute
Minneapolis, Minnesota

Michael Oliveri, MD
Specialist in Physical Medicine and Rehabilitation
Head of the Unit for Ergonomics and Industrial Rehabilitation
Swiss National Accident Insurance Fund
Bellikon, Switzerland

Deena E. Pease
Corporate Manager
Workers' and Unemployment Compensation
Weyerhaeuser Company
Tacoma, Washington

Marlene J. Perkins, RN
Nurse Practitioner
SC Johnson Wax
Racine, Wisconsin

Lucinda A. Pfalzer, PhD, PT
Associate Professor
Interim Associate Director for Research and Post-Professional MPT Program
Physical Therapy Department
School of Health Professions and Studies
University of Michigan at Flint
Flint, Michigan

James A. Porterfield, MA, PT, ATC
Rehabilitation and Health Center
Crystal Clinic
Akron, Ohio

Amy Santana, OTR
Director of Occupational Therapy
Productive Rehabilitation Institute of Dallas for Ergonomics
Dallas, Texas

Robin Saunders, MS, PT
Chief Executive Officer
Saunders Therapy Centers
Edina, Minnesota

Karen Schultz-Johnson, MS, OTR, CVE, CHT, FAOTA
Owner/Director
Rocky Mountain Hand Therapy
Glenwood Springs, Vail, and Aspen, Colorado

Richard L. Smith, MS, PT
Director/Owner
Missoula Physical Therapy Center
Missoula, Montana

William D. Sommerness, JD
Sommerness Law Offices, PA
Duluth, Minnesota

Joseph J. Sweere, DC, DABCO, FICC
Professor and Chairperson
Department of Occupational Health
Northwestern College of Chiropractic
Bloomington, Minnesota

Stephen R. Vance, PT, CHT
Co-Owner/President
Granite Hill Physical Therapy
Brunswick, Maine

Daniel J. Vaught, RN, CRRN, CIRS
Certified Case Manager
Karr Rehabilitation Services, Inc.
Duluth, Minnesota

Ann M. Walker, OTR
Director of Clinical Performance
Isernhagen Work Systems
Duluth, Minnesota

Lynne A. White
Browning/White and Associates
San Francisco, California

Michelle Wiklund, RRA
Director of Quality Research
Isernhagen and Associates, Inc.
Duluth, Minnesota

Preface

Work Injury: Management and Prevention, published in 1988, set the stage for spirited dialogue and problem solving on occupational medicine and occupational rehabilitation topics. It has been very rewarding to receive comments as I have traveled throughout the world. The book, which was authored by experts from many disciplines, met its purpose. The information provided a base of knowledge for readers, inspired them to think of work injury management as a continuum, and sparked additional interest for specialty work participation and increasing the base of knowledge.

The authors of *Work Injury: Management and Prevention* are to be commended for the interesting way in which they wrote, the lucidity of their comments (in both science and art), and their ability to integrate their thoughts with other authors. The book produced universal concepts that have only grown stronger since its time of publication. Those concepts were:

- Work injury management and prevention is a multidisciplinary effort.
- All parties involved, including medical professionals, legal professionals, insurance and vocational experts, employers, and employees, share responsibility and opportunity for improving work injury management and prevention.
- Communication is essential among those involved in work injury management and prevention. It is not possible for one person, one discipline, or one portion of the workers' compensation system to make changes without cooperation and communication with all others.
- While perceived as complex, the basic tenets of work injury management and prevention actually are quite simple. Professionals need to use a scientific base, have a positive approach, avoid adversarial relationships, and dedicate themselves to work together in order to alleviate work injuries.

- Work injury management and prevention are interesting fields that bring an ability to integrate logic, compassion, science, and relationships to improve the workplace and the work so that workers and employers can meet their goals of production in our society.

If the authors of *Work Injury: Management and Prevention* did such fine work in establishing these parameters, why another book? The first book, authored in 1986 and 1987 and published in 1988, set forth the fundamental principles and philosophies of work injury management and prevention. Ideas and methods that work never go out of style. As needs grow, they only grow more important. But additional development in principles, philosophies, and methods has taken place since that time. This new book therefore has its foundation in the first but goes significantly further in demonstrating advancements by referencing data, scientific studies, outcomes, and additional professional expertise to expand on the principles stated above.

A wide scope of professionals agreed to author chapters in this second book, *The Comprehensive Guide to Work Injury Management*, and for their dedication and work I am very grateful. They have provided us with another step forward in work injury management and prevention. In this book, even more than in the first, it is clear that all the prevention and management factors, when taken as a whole, provide a continuum of care that benefits the worker and the employer, and allows medical, vocational, and legal professionals the opportunity to do even better work.

There is one major group to whom this book is dedicated. This dedication is unsaid by all of the authors, but it is very clear. We are all committed to the productive worker. Whether we are in Europe, the Far East, or North America, our goal is for the healthiest, most productive worker possible. This book is dedicated to those who allow the productive work of the world to take place.

Acknowledgments

I gratefully acknowledge all of the professionals who have contacted me in the past years to comment on the first book. As I have continued teaching and writing, the positive comments and constructive critiques of my colleagues have allowed me to develop what I hope and believe is the mainstream approach of work injury management and prevention. This would not have been possible without all of the support and assistance I have received.

I thank my daughters, Aura and Jessica, who through the years between the first and second book have themselves come into the "worker" role. They have been of much assistance both personally and professionally to encourage me to complete this work. I also thank my husband, Dennis, for both his professional and personal support and his contributions to this book.

Many of my colleagues have been of great support in the professional growth that allows this second book to exist. I particularly thank Leonard Matheson and Dennis Hart, who have provided an "intellectual incubator" for continued thought processes on these subjects. Their dedication to science, communication, and ethical conduct matched my own, and their support allowed me to bring together these thoughts in a far more professional manner than would have been possible without their help. In addition, my work colleagues Margot Miller, Laurie Johnson, Ann Walker, and Michelle Wiklund have provided daily input to allow the field of work injury management and prevention to progress. My colleagues with whom I have worked to refine ideas and methods have been of profound help: thank you to Linda Darphin, Bill Sommerness, Helene Fearon, Libby Green, Richard Smith, Mary Mistal, Larry Benz, Barbara Larson, Jennie Gallagher, and Ginnie Halling.

Last, as always, this book would not have been possible without those who make our words concrete and coordinate the processes. I thank Gail Ernst, Ruth MacGregor, and Jan Voltzke for their highly accurate work in bringing this book to completion. In addition, my thanks go to Aspen Publishers, which, as before, has done a professional service in assistance to this work.

Introduction

Work injury management and prevention have reached a level of interest, funding, and urgency that could not have been predicted in earlier years. The human dilemma of injury creates an ethical and moral need to prevent injuries and manage them expertly if they do occur. Today there is a reaffirmation of human values that press for safer and more effective integration of the worker and the workplace.

The study of work injury prevention and management is preceded by the existence of meaningful work for productive workers. If the proposed solution to work injuries can assist industries and human endeavors to be more prosperous and productive, then we will meet not only the human need of health and productivity but also the industrial need of financial stability and improvement in products, processes, and services.

A concern for worker health may be the driving force. The accelerating force, however, is the high cost of work injuries, which puts tremendous pressure on industries attempting to stay in business. Companies are faced with regional and international competition. The cost of work injuries and the resulting decrease in productivity can affect the industry's viability.

The pressure for work injury cost containment has become a major force not only in industry but also in medicine. The managed care era is here. When work injury management was primarily a fee-for-service system, not only were medical, vocational, and psychological professionals paid for what they charged, but the types of services were openly subject to provider discretion and selection. The managed care era, which purports to cut costs while improving medical care, now forces proof of effectiveness. Therefore, in both injury prevention, which is traditionally paid for by the employer, and injury management, which is traditionally paid for by a third-party payor, work injury programs must be defined, categorized, and measured.

As a result, this book provides not only a higher level of science and medical art expertise but also direction toward efficacy. In as many areas as possible, studies or references to outcome are provided. Therefore the authors describe a movement toward demonstration as well as discussion.

A CADRE OF EXPERTS SPEAKS

In formulating our governing ideas, we asked several professionals for their opinions. The following have responded and will be quoted as they discuss their perception of what is current in the direction of work injury management and prevention. The same professionals have contributed to the conclusion also. Those whose ideas have been quoted or paraphrased are:

- Gunnar B.J. Andersson, MD, PhD, Orthopaedic Surgeon, Chicago, Illinois
- Jean Brisson, Work Injury Coordinator, Duluth, Minnesota
- Glenn Carmen, Director Workers' Compensation, Weston Corporation, Toronto, Canada
- Don Chaffin, Ergonomist, Engineer, University of Michigan, Ann Arbor, Michigan
- Jed Downs, MD, Occupational Medicine Physician, Duluth, Minnesota
- Melanie T. Ellexson, OTR, Industrial Rehabilitation Specialist, Chicago, Illinois
- Dennis L. Hart, PhD, PT, Industrial Rehabilitation Specialist, Falls Church, Virginia
- Leonard Matheson, PhD, Psychologist, Industrial Rehabilitation Specialist, Santa Ana, California
- Vert Mooney, MD, Orthopedic Surgeon, San Diego
- Michael Oliveri, MD, University of Zurich, Zurich, Switzerland
- Marilyn G. Peterson, RN, Workers' Compensation Specialist, Duluth, Minnesota
- Mark A. Rothstein, Professor of Law, Health Law and Policy Institute, Houston, Texas
- Peter Towne, PT, Private Practice Physical Therapist, Hamilton, Ohio

These experts were asked to give their impression of what is current (both positive and negative) in the work injury management system. They were candid and pointed in their remarks. The following summarizes their responses.

Jed Downs began his comments with three points that characterize our professional challenges:

1. Each individual has intrinsic value regardless of how difficult or easy he or she is to work with.

2. In order to make existing systems function, all players need to communicate and cooperate.
3. A point that must be recognized by those not directing care is that the physician is not omniscient.

These comments set the stage for those of the other professionals.

Return to Work

Jed Downs: The primary patient management question that needs to be answered as soon as possible is, when will the injured worker be able to be returned to work and what will his or her capabilities be?

Problems in the System

Jed Downs: Some employers and most insurers look at the employee as a commodity and a cost to be controlled. This causes conflict as long as the depersonalization and apathy continue.

Unionized worksites need to assist with return to work for the injured worker. Trade demarcations and seniority rules frequently prevent employable rehabilitated workers from working and this results in residual disability and unemployment.

Jean Brisson: Education is essential in both industry and the medical community. Industries have to stop seeing themselves as victims of the system and establish aggressive programs of injury prevention and injury management for when injuries do occur. These programs have to become an integrated part of the corporate culture and be communicated through the ranks.

Many health care providers are put off by the paperwork demands and state regulations that require specific guidelines. Industry is put off by costs and lack of control. These frustrations often allow injured employees to manage and manipulate their own claims and care. These systems may have to be reformed first before strong remedial action can take place.

Leonard Matheson: We are not requiring copayment for services provided.

Mark Rothstein: At too many companies there is such an emphasis on cost containment that it has become the only end. The goal should be injury prevention and employee rehabilitation. Then the cost containment will take care of itself. Instead, some companies are still obsessed with dubious screening devices and disability management programs that attempt to get injured workers off the disability rolls purely as a cost-saving measure.

Discrimination based on back injuries is by far the number one cause of ADA complaints filed with the EEOC. This will lead to further research into this type of problem as it relates to work.

Glenn Carmen: Disability costs are twofold. There are the direct measurable financial costs, such as benefit payments and assessment rates, and there are the human costs of employee disability, which are extremely difficult or impossible to put a dollar value on.

Gunnar Andersson: Unfortunately, the workers' compensation system works against medical providers. It rewards people for disability rather than encouraging rehabilitation.

Don Chaffin: Our ergonomic and work parameters need to emphasize prevention through ergonomics rather than being reactive to injury.

Melanie Ellexson: One of the problems in our system is lack of basic education for students on industrial rehabilitation. Continuing education and basic education should both be improved so that professionals working in either prevention or management of the injured worker have the expertise and practical experience necessary. We need to promote well-designed programs for workplace safety and to have the resources to deliver the service.

Light Duty

Jed Downs: Light duty and return-to-work programs need to be closed-ended. The employer and employee need additional leverage to assist in resolution of cases. A time must be reached at which suitable alternative duty or regular duty permanent jobs are assigned or job search takes place, and adequate time must be allowed for rehabilitation.

Communication

Michael Oliveri: An intense interdisciplinary collaboration between all parties involved in rehabilitation is very important.

Jean Brisson: As liaisons between industry and the medical community, professionals need to provide education to both spheres (employers and employees). Successful programs will team with industrial safety specialists, ergonomists, or other providers.

Marilyn Peterson: With an understanding of the work injury management system before the accident, the worker will know what is expected with regard to complying with medical care and early return to work.

Glenn Carmen: Employers need to involve providers in the return-to-work process. The employer knows the workplace and the employees. As employers begin to work with medical providers, there should be an improvement in the system.

Peter Towne: Physical therapists have an obligation and responsibility to make the system better by extending themselves into communities and worksites. Medical providers should have the expertise in musculoskeletal and movement dysfunction. This information should be shared with those we can best serve at the worksite through cooperative and joint research activities.

Prevention

Jean Brisson: A danger for industry in being more proactive in prevention is that they are already overwhelmingly burdened by workers' compensation costs. But industries must look beyond themselves to commit time and dollars to prevention.

Marilyn Peterson: Implementation of safety at all work levels is imperative. Involving employee representatives from all work areas and all staff levels on the safety committee assures cooperation and practicality of our adopted ideas. Immunity needs to be ensured.

Melanie Ellexson: Therapists tend to combine education, engineering, and work design changes to bring about long-term results. By studying the problems they can avoid quick fixes and one-time solutions, which are less likely to have a lasting effect. Therefore therapists should be used to develop long-range plans that benefit both the company and the worker.

Ergonomics

Don Chaffin: Most engineers are responsible for specifying and designing their work environments. But only about 4 percent understand the ergonomics at fundamental levels. Thus there is a factor of continuing to design jobs that require a large proportion of the work population to perform hazardous exertions.

Few engineers understand how to design jobs and equipment to accommodate individuals with acknowledged impairment.

All people in the team process have seen management become enlightened enough to start stressing ergonomics and prioritizing necessary changes in the work.

We need more professionals with the breadth of education and experience needed to use ergonomics effectively . . . more education is needed.

The System

Dennis Hart: There is a growing sense of urgency to justify our clinical services because of the increasing number of politicians and accountants who feel it would be more financially expedient simply to pay the injured worker a structured settlement instead of paying for clinical rehabilitation services.

Mark Rothstein: An important issue is OSHA regulation of ergonomic conditions in the workplace. This will be formative for employers' responses.

Gunnar Andersson: The legal system at the present time is set, and we have to work within this framework to improve the process of work injury management.

Early Intervention

Vert Mooney: What we are doing wrong is delaying in initiating effective programs.

Leonard Matheson: What we are doing wrong is we're not identifying problem (costly) cases early.

Peter Towne: Early intervention to effect early resolution of tissue trauma is important to be blended with the educational process of reinjury education.

Gunnar Andersson: All parties seem to agree that early appropriate care is the key to prevention of long-term disability. Therefore we need to disseminate information about *what constitutes* early appropriate care.

Rehabilitation

Vert Mooney: What we are doing right at this time is focusing on the return to work and physical training.

Leonard Matheson: We are not being vocational rehabilitators. Too many provide palliative care and not enough emphasize function over pain and impairment.

Gunnar Andersson: The negative effects of inactivity and the positive effects of activity need to continue to be emphasized.

Functional Testing

Vert Mooney: We have a lack of common utilization of functional tests to define baseline deficits and to determine when function plateaus reach maximum medical benefits.

There is a lack of agreement on effective, functional, usable tests.

Dennis Hart: Clinicians who do not evaluate the functional progress of their patients to use the results of that evaluation in planning are not demonstrating quality clinical problem-solving skills. If there are no objective findings demonstrating the need for clinical services, those services should not be remunerated.

Treatment Outcomes

Dennis Hart: Outcome data and quantification are critical in measurement of programs that are effective.

Just as bad is the philosophy that a certain number of treatments should be remunerated for a specific type of clinical problem. We only have ourselves to blame if we have not studied our outcomes, and therefore most clinical services do not have scientific evidence to support their effectiveness or necessity.

Leonard Matheson: The efficacy of every intervention must be evaluated. A scientific basis to "what works and what doesn't work" is necessary. If we answer this question we can make policy recommendations.

Michael Oliveri: More appropriate study designs should be created for clinical outcome of rehabilitation programs.

Standards of Practice

Leonard Matheson: What we are doing right is becoming more scientific. We are developing standards of practice.

Motivation

Michael Oliveri: In order to improve our work injury management system, we have to consider today's social and economic situation. In many countries there isn't a workplace available for everybody anymore. Under these conditions we have to accept that we cannot bring every patient back to work. Our main goal lies with working with *patients who are motivated* to get back to work either for personal financial reasons or to defend their self-confidence and social recognition. The challenge is to find valid criteria to select suitable patients for comprehensive rehabilitation programs.

Criteria should include cooperation and motivation of the patient. More importance should be given to the patient's self-rating.

It is of interest to study to what extent functional status correlates to patients' awareness of their resources and quality of life.

CONCLUSION

These recognized experts in the field have consolidated some very important thoughts.

1. Professionals must aspire to measuring techniques, methods, and processes in relation to outcome. By understanding which medical or rehabilitation practices provide the best outcome, we can direct the best care for injured workers.
2. The laws and the workers' compensation system often create problems in trying to improve medical management. We will need to be proactive in influencing these laws. If medical providers don't set the medical parameters, the politicians will!

3. The employer may see the employee as a monetary statistic rather than a human being who works. All involved in work injury prevention and management will need to encourage the return to humanity in treating the injured worker. When we do that, the monetary problems may take care of themselves.
4. Early intervention and early return to work are going to be extremely important in this entire system. When professionals understand the workplace demand, they can effectively communicate and take proactive roles. Thus disabilities should be able to be avoided.
5. Activation and functional testing should be chosen over palliative measures. Both the motivation and physical parameters are important in providing self-actualization for return to work.

These work injury management philosophies and practice indicators provide pathways for us to follow. The points detailed in these introductory comments are comprehensively covered in this volume. The reader will find in-depth discussions and solutions of these specific problems, with direction on interventions and methods of changing directions when the wrong path has been chosen. The reader may use the book to enhance the safety of the worker in the workplace, the return of a healthy worker to the work force after injury, and the relationship between employee and employer, which solidifies industrial competence.

Part I

Work Injury Prevention

The impact of rising health care costs (both human and financial) reaffirms the position that prevention of those health care problems would be the best cost saver. While few would dispute the wisdom of this thought, there is still less attention paid to prevention of injury than to management of injured workers. This is due to lack of automatic funding. When a worker is injured, an insurance system immediately responds by payment for services needed. When prevention is identified as the method to solve the problem, employers must reach deep into their pockets in order to fund these programs.

Although laws promulgated by the world's various occupational health and safety departments have assisted us in understanding the importance of prevention, in most countries funding is not available directly for these prevention services. Therefore they require an employer system with a deep knowledge of the human and financial distress caused by work injury that will take the first strong step to evaluate the need for prevention services and fund those services.

The operative words in the last paragraph were "evaluate the need." Because of costs related to prevention, employers have sought or been victims of pressures for a "shotgun" approach to preventing work injury. Because prevention is not well understood, many remedies have surfaced over the past years to convince employers that a method or product brings "the magic answer." This lack of cohesiveness in approach is destructive because when an employer seeks a simplistic answer and then is not rewarded with positive outcomes of injury reduction, there is a natural tendency not to pursue other prevention measures.

The logical method of establishing prevention, as with all interventions, is thorough evaluation of the multifactorial elements: workplace, workers, work, employment milieu, and the relationships that employers, unions, and employees have with one another. This evaluation can help pinpoint areas where attention should be focused. In most cases, prevention is a multivariate need that requires not only the initial evaluation but also constant ongoing reevaluation of methods to improve health, safety, and production in the workplace.

Mark Anderson presents us with strong concepts regarding working within industry with ergonomics. Elroy Lundblad, Marlene Perkins, and Andrea Kramer then take us to the actual implementation of ergonomic principles through the ergonomic team. Working in industry also is Tom Broderick, who presents the reader with a unique and innovative method of viewing the safety director and his or her importance in the workplace.

A cornerstone component of prevention, job analysis, is presented by Susan Isernhagen. Margot Miller then explains how prework screening is developed in an objective and nondiscriminatory way based on the job demands. Gary Gray and Stephen Campbell present us with prework screening case studies to show the integration of several prevention techniques. Back schools, which form a cornerstone for spinal injury prevention, are explained in a powerful and clear manner by Lynne White. Mary Mistal then takes us through the development of specific industrial back programs from her broad experience.

The next chapters deal with identification and remediation of factors related to one of the newest epidemics, upper-extremity repetitive strain. Cindy Pfalzer and Barbara McPhee have cooperated to write an extensive and pointed compendium of information on carpal tunnel syndrome research. Karen Schultz-Johnson discusses practical issues, with excellent ideas on prevention and return-to-work guidelines for the potential hand patient in industry. Rounding out the prevention and management side is Larry Benz, who demystifies carpal tunnel syndrome diagnosis. He also discusses development of a workable internal industrial prevention program.

The summary chapter brings all of these concepts together as Barbara Baum takes us through her philosophies regarding the methods of employee involvement. All prevention measures fail if the employee does not "buy in" to the concepts and does not practice prevention measures during work. While the employer's role is initially important in fostering safety and health through positive relationships with the workers, it will ultimately be action on the prevention principles by employees that will count.

These chapters form a comprehensive whole as the authors in their own ways tell the reader how we can consider many types of prevention opportunities, integrate them, and receive feedback on their effectiveness.

Chapter 1

Ergonomics: Analyzing Work from a Physiological Perspective

Mark A. Anderson

Some toothbrushes, truck seats, and needle nose pliers all have one thing in common. They, along with a myriad of other products, are all being touted as having been "ergonomically designed."

It seems in the 1990s that the word *ergonomics* has become a part of the common vernacular. This chapter will discuss some of the reasons why this has occurred. It will look at the evolution of the practice of ergonomics and will outline the components of the practice and the involved fields of study. It will discuss ergonomic practitioners and their backgrounds and define a common set of goals and objectives obtainable through the application of ergonomic principles. Finally, it will pull together a set of practical application guidelines.

While a number of factors are responsible for the development of ergonomics, two predominant reasons are evident.

LANDING GEAR AND WING FLAPS

The entry of the United States into World War II brought about an incredible marshaling of military and civilian resources. By World War II, air power had advanced to the point where it provided a significant contribution. Military flight training schools were under fierce pressure to train as many pilots as possible and get them into the war as quickly as possible. A certain number of mishaps were expected, but it quickly became apparent that an inordinate number of crash landings was occurring. In fact, over 400 crashes occurred within a 22-month period.[1] An investigation into the situation revealed two very important facts: (1) the control lever for the landing gear was *right next to* that for the wing flaps; (2) the configuration of the handles was *identical*. The pilots, under the pressure of landing the aircraft, thought they were lowering the wing flaps but were actually re-

The author thanks Mark R. Stultz, MSIE, PT, for his comments on the manuscript, and Sarah P. Merrill for her administrative support.

tracting their landing gear! The result was quite predictable. The technological age was progressing rapidly with ever-increasing complexity.

MASS PRODUCTION

From the industrial perspective monumental changes had started to occur. In this context they can be summed up in two words—mass production. Industries throughout the country and around the world enhanced and accelerated the concepts of greater productivity through assembly line mass production practices. One worker, rather than having multiple job demand responsibilities, now performed the same smaller set of tasks over and over and over again. Frederick Taylor's theories of "scientific management" were widely put into practice.[2]

So two major influences were being exerted: (1) the ever-increasing complexity of the technologies used on a routine basis; and (2) the rush for greater productivity through mass production. It has become apparent that the outcome of these two influences has not been entirely desirable.[3] Complex technologies require sophisticated control, and operator error can be significant. The impact of the operator error varies widely. Misprogramming your VCR is not a big deal; operator error at the Three Mile Island nuclear power plant was.[4]

The reported incidence of cumulative trauma disorders has increased dramatically in the last decade.[5] Performing the same series of repetitive tasks in sedentary and perhaps awkward positions has been implicated. A new set of disciplines—collectively called *ergonomics*—that examines the person–machine–environment interface is evolving to address a new world of technology.

THE SCIENCE OF ERGONOMICS

Definition

The word *ergonomics* is derived from the Greek words *ergos* ("work") and *nomos* ("laws" or "study of"). Many definitions of ergonomics exist—for example, "(to) discuss and apply information about human abilities, limitations, and other characteristics to the design of tools, machines, system tasks, jobs, and environments for productive, safe, comfortable, and effective use"[6] and "to optimize the functioning of a system by adapting it to human capacities and needs."[7] In more operational and perhaps more practical terms, the practice of ergonomics incorporates the concepts of "Work smarter—not harder" and "Fit the work to the person—not the person to the work!"

Ergonomic Practitioners

Another approach to gain insight into the practice of ergonomics is to examine the backgrounds and applications of ergonomics practitioners. For example, the systems design engineer may feel that the ability to carry through with the design, development, and implementation of a specific work station, tool, or piece of equipment is paramount. The health care provider's major concern may rest more with the specific pathophysiological risk factors influenced by a particular job design or work process. The engineering psychologist may feel that how the machine operator interacts with the controls is of the utmost importance. And the management consultant's major concern may be the organizational structure of the organization. All of these issues are important; all of them need to be addressed; and all of them fit under the umbrella of ergonomics.

Each professional must recognize his or her own professional capabilities and limitations. He or she must ensure a match of "work demands" to skills. A true multidisciplinary team then exists and functions as a whole greater than its individual parts.

At the most practical level the "best ergonomist" is not represented in the distinguished list outlined above. Practically speaking, the best ergonomists are the people who actually do the work. From firsthand experience they know the problems and, in many cases, have a good sense of how to fix them.

ECONOMICS OF ERGONOMICS

Trying to put a precise figure on the costs associated with ergonomic issues in the workplace may very well be impossible. Ergonomic interventions are designed to control workplace injuries as well as to promote productivity. In other words, both cost-added and value-added components make up the equation.

An important point to consider concerns opinion regarding the actual extent of cumulative trauma disorders due to and arising out of the course of employment. For example, Hadler contends that the prevalence of work-related cumulative trauma disorders (CTDs), at least to the extent reported by the Bureau of Labor Statistics (BLS), is greatly exaggerated.[8] On the other end of the spectrum, results of other studies suggest significant underreporting.[9]

Incontrovertible, however, is the fact that CTDs are the fastest increasing reported category of work-related illnesses.[10] In 1989, the National Institute for Occupational Safety and Health (NIOSH) estimated that the exposed workforce may be a proportion as high as 25 percent of all workers.[11] The incidence rate of ergonomic disorders over the past 10 years has increased dramatically. The percentage of increase has been as high as 50 percent:

- 50 percent for supermarket cashiers
- 41 percent for meat packers
- 40 percent for newspaper employees
- 30 percent for specialty glass workers
- 20 percent for poultry workers[12]

The American Academy of Orthopedic Surgeons in 1984 estimated that overall CTDs cost $27 billion a year in lost earnings and medical costs.[13] Liberty Mutual reports since 1987 the costs of upper-extremity CTDs, as a percent of all of Liberty Mutual's compensation costs, have quadrupled.[14]

As this juncture, regardless of the actual etiology, the costs associated with ergonomic-related disorders are monumental. The successful application of ergonomic principles can be part of the overall injury prevention and management scheme.

THE ERGONOMIC PROCESS OVERVIEW

Process versus Programs

Ergonomics, in and of itself, is inert. The successful application of ergonomic principles occurs only when the organization's culture either can already accommodate the needed changes or can be modified appropriately. The issue can be summed up by comparing the program versus the process approach.

The *program cycle* starts when a "problem" emerges, perhaps identified by an increase in injury reports, subjective complaints, or an OSHA inspection. The status quo has been upset. Attention is directed toward the problem and a program is initiated, such as training or a new safety policy.

The index qualifier is seen to decrease; the "problem" has diminished. No problem . . . no need for a program. However, typically the problem eventually reemerges and the cycle continues.

A process approach is inherently different. A process continues on an ongoing basis; never ending, it is modified to effectively address changing concerns.

Successful ergonomic intervention strategies fall into the process cycle. The true causes or triggers of the problems are identified, feasible interventions are introduced, the outcomes are tracked and measured, and the process continues.

An example illustrates the concept: A company noticed an increase in reported low back injuries in the warehouse/distribution center. The manager felt a body mechanics training program was the answer. The training was provided and the participants practiced the desired techniques. A few minutes later, a tour of the center revealed not a single worker using the new technique. They continued to lift with the back bent–straight leg technique. The manager concluded that body mechanics training was a "waste of time."

Was training really the appropriate intervention here? Further analysis revealed that 60 percent of the loads lifted were stored at floor level on pallets. Frequency monitoring indicated that a lift occurred two to three times per minute on average. The "correct technique" required a squat lift every 20 to 30 seconds. Was this metabolically feasible?

Additional body mechanics training was not the answer; the true cause or trigger was the material stored on the floor. Effectively, the solution was workplace modification—raise the storage level.

Objectives

The primary objectives of ergonomic intervention are twofold and mutually inclusive: (1) enhance performance, and (2) control fatigue. To meet the objectives, goals can be established in three primary measurable areas: (1) quality, (2) production, and (3) safety. In an integrated fashion, these goals are an overall measure of effectiveness.

Categories

Three primary kinds of parameters can affect performance and fatigue: physical, psychological, and psychosocial. Metabolic and mechanical stress factors make up the physical parameter category. Influencing behavioral outcome to achieve the objectives encompasses the psychological category. And the psychosocial category examines the organizational relationships inherent in the implementation and management of the worker/workplace interface.

A Set of Factors

Within these three primary categories, a set of observable and measurable factors have been identified that influence performance and fatigue.

- worker physical fitness—poor/good
- force—dynamic/static, high/low
- position—awkward/sustained
- repetition—rate of activity/job demand
- duty cycle—work/recovery rate
- vibration—whole body/segmental
- friction—high/low coefficient of friction
- work practices—training, policy, and procedures
- environment—temperature, ventilation, illumination
- job satisfaction—labor/management relationships
- operator control—ease of use[15]

Interrelated, these factors can have either a negative or a positive impact on the objectives of enhancing performance while controlling fatigue in the physical, psychological, and psychosocial realms.

Identify the risk factors, investigate the true causes or triggers, design and implement interventions, track the outcome, and repeat the process—these are the steps to apply ergonomics.

SYSTEMS APPROACH

The Systems Approach coordinates the aspects of the effective application of ergonomics. The systems design concept, as defined by the *Human Factors Design Handbook*, is a "mission-oriented grouping of elements into an integrated, functional whole. The system typically includes a physical facility, equipment, furnishings and fixtures and involves a variety of people who use, operate or maintain it."[16]

A set of general principles of the Systems Approach includes:

- The system is adapted to the human.
- The system facilitates the highest performance level of which the operator is capable.
- The system optimizes the physical and mental stress imposed on the operator.
- The system provides personal satisfaction for the user in terms of use.
- The system and its components function to serve the human.
- The system recognizes individual variation in human capabilities and limitations.
- The system design will influence human behavior either positively or adversely.
- A system, by definition, does not exist in isolation.

Whether the ergonomist's involvement is upstream, at the initial design phase, or further downstream, as part of an equipment, workstation, or work process retrofit, these principles still apply.

Ergonomic intervention at the design stage has been readily determined to be the more cost-effective. It can be very expensive to make modifications after the fact. While relatively inexpensive retrofit changes sometimes are possible, they do not replace the need for good design from the outset.

In a sentence, adequate systems design ensures the highest level of safety, quality, and productivity possible.

Foundation of Knowledge

Among the professionals practicing ergonomics, a common basic foundation of knowledge is required to ensure an adequate level of communication. The reader may have in-depth knowledge in one or more of the disciplines noted, but also needs to have a general sense of the basic principles of the others and how they all fit together.

So the practicing ergonomist needs to assimilate the foundation of knowledge—the education—and practically apply it in real world situations. The goal is to achieve a "system" that enhances worker/workplace effectiveness.

Fields of Study

An appropriate systems design is at the core of ergonomics and, among others, primarily incorporates foundation principles from the fields of

- work physiology
- engineering psychology
- anthropometry
- occupational biomechanics
- organizational development/management
- pathology
- epidemiology

This ergonomic discussion will begin by examining the principles of work physiology—the capabilities and limitations of the metabolic system of the body. Then it will advance to how humans interact with the surrounding work environment—engineering psychology.

An important field is the study of the shape, size, and form of the human body—anthropometry. This leads us into occupational biomechanics—the study of the motion of and forces acting on the body. This is the basis of biomechanical models. One must examine the organizational development aspects relating to the relationships of labor and management to accomplish the organization's efforts. From a medical perspective, an understanding of the pathology, including the etiology, of occupational injury and illness is necessary.

Finally, the discipline of epidemiology—statistical analysis of the risk factors and outcomes within the work environment—will be addressed.

These fields of study coalesce to provide a comprehensive ergonomic approach. Of course each field involves in-depth study and is a specialty unto itself. Practitioners are strongly encouraged to investigate in depth those fields of their choice and need.

FIELDS OF STUDY

Work Physiology

Work physiology is the science that investigates how the body responds to the physical stress of work demands. To accomplish work, the body is able to take in nutrients and convert them into chemical energy and then ultimately into mechanical energy (e.g., muscular contraction) and heat. This complex process is called metabolism. The field of work physiology is quite broad; a number of excellent references are available to the reader.[17]

Metabolic Efficiency

The metabolic efficiency of the body can be determined based on energy expenditure relative to workload. Higher efficiency levels result in less energy expended at higher workloads as well as a greater maximal level of energy expenditure.

Metabolic efficiency can be measured in several ways. A common method is to measure the volume of oxygen taken into the lungs per unit of time at known workloads. This measure is related to the number of kilocalories per minute expended by the body to sustain the activity.

To illustrate: two individuals, one fit and one less fit, are tested using a stationary bicycle as the ergometer. Oxygen consumption is measured as a function of energy expenditure. At progressive intervals the workload on the ergometer is increased. The test is taken to the maximum capacity for each individual. Two typical findings emerge: (1) at comparable workloads the fit individual will expend less energy—higher level of efficiency; (2) at maximum capacity the fit individual will have a higher metabolic level—greater capacity.

Metabolic capacity and efficiency are influenced by aerobic capacity (heart and lung fitness) and muscular fitness. Metabolic fatigue effects can be reduced by improving the body's metabolic efficiency: that is, improving physical fitness levels.

Blood Flow

Glucose and oxygen are stored in relatively small amounts within the muscle tissue. Consequently, sustained performance requires a continuous flow of oxygen and energy-rich blood into the tissue in addition to removal of metabolic waste products. Increase in circulation in relation to workload can be significant. The following blood flow changes with increased workload have been reported by Scherrer:

muscle at rest	4 ml/min/100g muscle
moderate work	80 ml/min/100g muscle
heavy work	150 ml/min/100g muscle[18]

External Factors

A number of external factors have been shown to have an influence on blood flow to working tissues and consequently on metabolic fatigue.

Type of Muscular Effort. The type of muscular effort has been shown to have a profound impact on blood flow. Static effort—sustained contraction of the muscle—results in blood vessel compression due to internal muscle pressure. At contraction levels of 60 percent and greater of the maximum voluntary contraction of the muscle, blood flow ceases.[19] The muscle depends on the quite limited initial reserves stored internally. Waste products accumulate, and only short-duration contractions are possible. Static efforts at high (maximal) force levels have been shown to be sustained for ten seconds or less, moderate effort for one minute and less, and slight effort (one-third of maximum) for less than four minutes. The guidelines are based on work conducted by Monad and Rohmert as summarized by Grandjean.[20]

Dynamic muscle contraction is the alternating contracting and then relaxing of muscle groups to perform tasks. Less research has been done examining dynamic muscle contraction responses compared to static contractions. What studies have been done show that the overall energy expenditure requirements of the working muscle are determined by the level of muscle contraction and the duty cycle (ratio of contraction/relaxation). Higher muscular force levels can be achieved and maintained with dynamic muscle contraction than with static contractions.[21]

An interesting study designed to reveal differences in blood flow as a result of a passive rest or an active rest following repeated upper-extremity work demands was conducted. The repeated work demands did result in decreased radial artery blood flow. However, the active rest period resulted in greater improvement in radial artery blood flow than the passive rest periods.[22]

In terms of enhancing performance and controlling fatigue, dynamic muscle contraction is an improvement over static muscle contractions; however, metabolic fatigue still can result, depending on the contraction level and duty cycle.

Position—Sustained/Awkward. A common perception held by those in industry is that lifting boxes all day certainly is tiring but that sedentary positions—sitting or standing—don't require any work! Ask someone in the military which he or she would rather do: stand for one hour at attention with full dress and 70 pounds of equipment, or go on a one-mile march (even double-time) with the same full dress and equipment. Metabolic fatigue certainly occurs as the result of sustained position. Blood flow—both volume and rate of flow—decreases. Pooling of fluid in the extremities also occurs. The body's tissues require ongoing nutrition even at low or minimal activity levels. Also mechanical fatigue is an issue.

The position of the body when sedentary is important to consider. Maintained awkward positions result in muscular contractions to maintain the position, and potential decrease in blood flow due to internal impingement or external contact stress.

What about neutral body postures? Office chair designers have attempted to define the quintessential office chair—and failed. Positioned in the most ideal seated postures, subjects will start to "fidget" in less than fifteen minutes. Ideal positions become less than ideal with time. Movement is an essential component in enhancing performance and controlling fatigue.

The principles of work physiology can be practically applied to effectively influence the body's physiological response to the physical stressors of work demands.

Engineering Psychology

To turn on a light at the wall switch, you typically push the switch up; to turn it off, you push it down. Why has the switch been designed to operate this way? The principles of engineering psychology are involved. Engineering psychology is closely tied to experimental psychology. The intent is to make use of the laws of behavior to accommodate the limits of human performance but also to enhance human performance.[23]

People equate "on" with "up"—if you want a light on, you push the switch up. This is a deeply ingrained behavior. The author once lived in a house where the kitchen light switch had been wired upside down: push the switch down to turn on the light. This created momentary confusion each time the switch was used—perhaps not a big issue, but an ongoing irritant.

A crucial aspect of a good systems design involves applying engineering psychology principles. As technologies become more complex, the system may overload human information-processing capabilities. So engineering psychology involves designing systems with these information-processing capabilities and limitations in mind.

- Jet fighter pilots have been reported to actually shut down some of their aircraft's information-gathering systems. They simply cannot handle the volume of data.
- A typical phone number with the area code is ten digits long—too long for most people to remember it long enough to dial it.

Accurate information processing is also based on reasonable expectations based on previous experiences.

- The horn on a 1983 Ford LTD station wagon is activated by depressing the turn signal stalk. Personal experience reveals it does absolutely no good to pound on the center of the steering wheel to warn the driver of the car that is about to back into you.
- A private pilot flying under and through low clouds narrowly averted hitting a large commercial airliner. He later described it as pure luck. His first response was to slam on the brakes with both feet. In the stress of the moment, an old habit was preeminent. He knew that cars and airplanes do not work the same—slamming on the brakes at 1,000 feet does not have the same effect as on the ground. But the old behavior took precedence.

Predicting system performance is the essence of engineering psychology. Several models have been developed that describe how this occurs.[24]

Donald Norman, in *The Psychology of Everyday Things,* outlines the principles of design in a practical manner:

1. Design for good visibility. Make it visually apparent what the control is for. Many people never learn how to program their VCRs or fully use the features of their telephones. The controls, by themselves, are not visually apparent.
2. Apply what Norman calls the principles of mapping. Make clear the relationship between two things—between controls and their movements and the results in the real world. Make use of physical analogies and cultural standards.
   - to steer a car to the right, turn the wheel to the right
   - an indicator moving up means an increase in volume
   - an indicator moving from left to right means an increase in volume
   - if a number of indicators are monitored to be in the safe range, have all of the indicators be in the same position in the safe range, so that one that moves out of the safe range is quickly identified
3. Provide feedback—return information to the user about the outcome. The problem becomes significant when more features are available but less feedback is provided. Without good feedback, how do you really know you have correctly programmed your VCR or your phone or even your alarm clock?[25]

Norman sums it up well:

> If an error is possible, someone will make it. The designer must assume that all possible errors will occur and design so as to minimize the chance of the error in the first place or its effect once it gets made. Errors should be easy to detect, they should have minimal consequences and, if possible, their effects should be reversible.[26]

Anthropometry

Door Heights, Kitchen Counters, and Stair Steps

An architect is designing a house. Countless design decisions have to be made. How high should the doors be? How wide? How high should the kitchen counter be? How high and deep should the step on the stairway be? Fortunately for the architect, standards exist for each of these questions. The standards are based on anthropometry. The word *anthropometry* is derived from two Greek words:

anthro ("man") and *metrein* ("to measure"). Anthropometry is the study of the physical dimensions—size, shape, and weight—of the human body.[27] Anthropometric principles are applied across the full spectrum of the practice of ergonomics:

1. design standards
   - machine guards
   - reaches
   - heights
   - handle configuration
   - general work station design
2. development of biomechanical models

Data Tables

The basis for anthropometry has been the careful measurement of the length, volume, and weight of body part segments. From this, measurement tables have been generated that calculate

- segment length
- center of rotation of the segment
- segment mass
- distribution of the body mass
- center of mass location
- inertial properties of the segment[28]

The outcome is a set of statistical data that describes the human size and form. Often the data are described in terms of means and standard deviations. Percentiles are also calculated.

Returning to the architect designing the house, let's address the questions of how high the door openings should be. One could look for the tallest person and make sure the door opening accommodates that person. But the tables show that only a very few people are actually that height. One could look for the average height individual in the data set. The 50th-percentile height indicates that half of the population is shorter and half is taller. But then this would accommodate only half of the population.

The convention adopted is a trade-off between allowing as many people as possible to pass through the door without hitting their heads and the efficient and economical use of building materials.

The 95th-percentile man is chosen to establish the standard opening height. Only 5 percent of the population of that data set would have to make an accommodation; 95 percent would not. A reasonable trade-off has been accomplished.

Let's examine the other end of the continuum. A workstation is being designed, including the layout of parts and materials. How far away, at maximum, should the supplies be placed and still be in reasonable reach? In a fixed workstation design one needs to look at the reach envelope of the smallest individual. The adopted convention is to design for the fifth-percentile woman.

In reality the goal is to build in reach flexibility to accommodate both ends of the reach envelope. A 95th-percentile man feels quite cramped at the 5th-percentile woman's reach.

So by applying the principles of anthropometry as part of the overall systems design, both objectives of enhancing human performance by controlling fatigue are met.

Occupational Biomechanics

According to Frankel and Nordin, "Biomechanics uses laws of physics and engineering concepts to describe motion undergone by the various body segments and the forces acting on those body parts during normal daily activities."[29] As defined by Chaffin and Andersson, occupational biomechanics focuses on the "study of the physical interaction of workers with their tools, machines and materials so as to enhance the worker's performance while minimizing the risk of future musculoskeletal disorders."[30]

In this chapter, the scope of occupational biomechanics will be limited to a discussion of the measurement and analysis of human movement that fall into the scope of kinesiology. Kinesiology has two major subparts: *kinematics*—motion of the body or of major body segments independent of the forces that result in movement—and *kinetics*—a description of the forces that cause movement. The principles of kinesiology provide the basis for one of the prime applications of occupational biomechanics—the development of quantitative biomechanical models.

Biomechanical Models

Biomechanical models allow us to address a number of purposes:

1. provide design data based on body postures, forces, and movements
2. allow a simplified representation of a complex system
3. predict the impact on the structures of the body based on the forces and motions imposed
4. allow the testing of hypotheses without the risk of injury[31]

Biomechanical models can be quite complex. For our purposes, the development of biomechanical models can be illustrated by examining the following case study (see Figure 1–1).[32]

As part of the assembly process, the operator is required to hold a 10-kg (22-lb) part at arm's length in front of the body. Does the design optimize human performance? As part of the answer, we can calculate the force level acting on the worker's elbow. What are the rotational moments and forces?

This is a problem in kinetics: measurement of the external forces acting on the elbow to impart motion. The analysis is static in nature: the worker has to hold the part.

Step 1. Determine the amount of force acting on the body part.

If dropped, the part will accelerate downward due to the gravitational force. The weight of the part is equal to the mass (*M*) measured in kilograms (kg) times the gravitational acceleration (meters per second^2, m – s^{-2})

$$\text{Weight} = Mg$$
$$\text{So weight} = 10 \text{ kg} \times 9.8 \text{ m} - \text{s}^{-2} = 98\text{N}$$
$$(\text{N} = \text{Newtons})$$

Step 2. Determine the weight-vector characteristics.

The weight is a force with four characteristics: (1) magnitude, (2) direction, (3) line of action, and (4) point of application. In our example:

1. Magnitude = 98N
2. Direction = downward
3. Line of action = vertical
4. Point of application = center of part's mass

Step 3. Determine the force acting on each hand.

We assume that the part is loaded symmetrically in the hands. Because the object is not moving, a state of static equilibrium is achieved: all external forces acting on the part add to zero. The hand reactive force *R* must equal the weight force, but be opposite in direction.

$$\Sigma \text{ forces} = 0$$
$$-98\text{N} + 2R \text{ hand} = 0$$
$$R \text{ hand} = \frac{98\text{N}}{2} = 49\text{N}$$

So 49N of force are acting on each hand.

Step 4. Estimate the elbow forces and moments.

A planar static analysis—forces acting in a single plane—is used. We make the following assumptions:

- Forearm and hand are one body segment.
- Length, weight, and center of mass of the segment represent an average male.

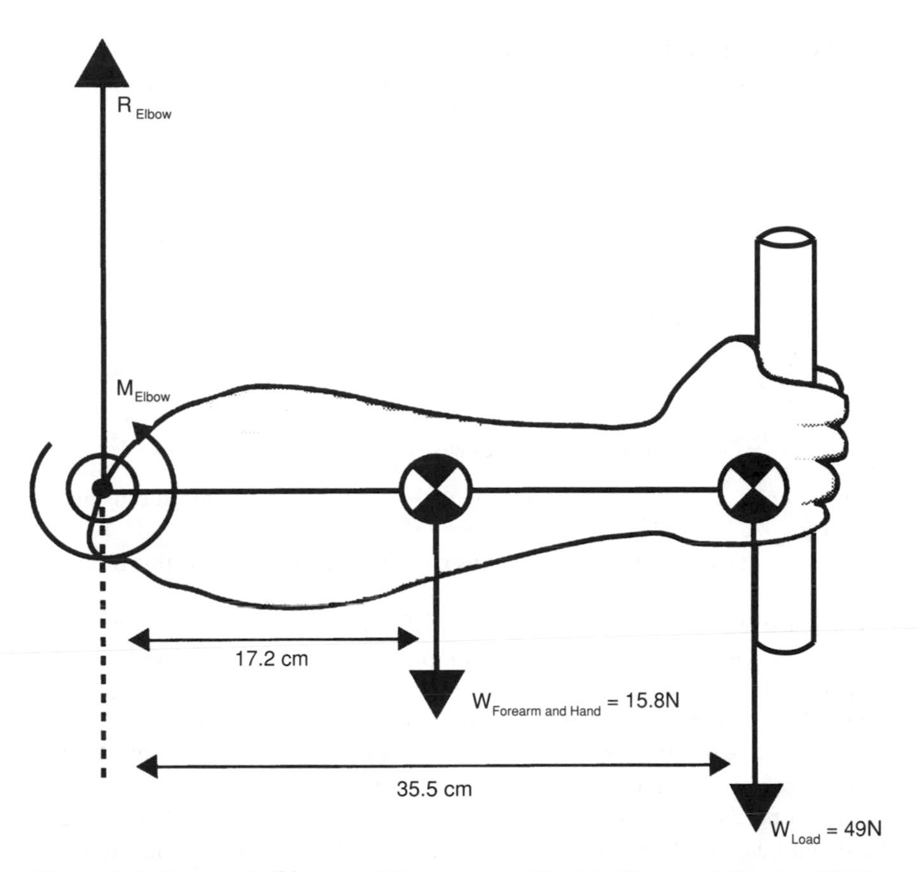

Figure 1–1 Free Body Diagram of Forearm and Hand in Horizontal Position Holding Load

---

- Line of action of the weight acts through the center of the mass of the hand—about the center of a power grip.

The reactive force of the elbow, R_E, counteracts the combined weight forces of the load (–49N) and the forearm-hand segment (–15.8N). The sum of the forces equals 0.

$$\Sigma \text{ Forces} = 0$$
$$-49\text{N} - 15.8\text{N} + R \text{ elbow} = 0$$
$$R \text{ elbow} = 64.8\text{N (upward)}$$

The body segment and load weights act at a distance from the elbow reactive force. In doing so they create a moment—a tendency to rotate. A moment is the force times the perpendicular distance from its line of action to the point of rotation. In this example with the forearm held horizontal, the moment arm is 17.2 cm (6.771 in) (based on the average male anthropometry).

Moments are vectors, so magnitude as well as directions about the point of rotation must be considered. The combined moments of the weight of the part must be counteracted by an equal but opposite direction reactive moment at the elbow (M_E). The second condition of equilibrium is the sum of the moments = 0.

$$\Sigma \text{ Moments} = 0$$
$$17.2 \text{ cm } (-15.8\text{N}) + 35.5 \text{ cm } (-49\text{N}) + M_E = 0$$
$$(-271.8\text{N cm}) + (-1739.5\text{N cm}) + M_E = 0$$
$$M_E = 2011.3\text{N cm}$$
or
$$M_E = 20.113 \text{ N}M \text{ (counterclockwise)}$$

So now we have determined the reactive forces at the elbow and the resulting moments.

Of course, a reasonable question at this time is, "Of what value is this information?"

The reactive forces reveal the magnitude of the tensile forces in the soft tissue holding the joint together as well as the compressive and shearing forces acting on the joint surfaces. The reactive moment reflects the strength (energy requirement) of the elbow flexors to maintain the posture or to initiate movement.

Consider this issue: what if the forearm-hand segment were at an angle other than horizontal? The moment at the elbow would decrease until at the vertical position it became zero. Should the design be modified as the reach is reduced? We could quantify the reduction in force and moment levels resulting from the modification and consequently evaluate the merits of the modification.

More complex models we could develop include

- a two-body-segment static model
- a model of nonparallel forces—for example, pushing or pulling a load
- a model that reflected the internal forces generated
- a model based on multiple limbs
- a model based on three dimensions, still static in nature

At an even higher level, we could introduce movement into the modeling practice. The direction of motion of the segment as well as the components of velocity and acceleration would have to be factored in.

This level of modeling complexity is beyond the scope of this chapter. The reader is encouraged to pursue other texts. However, it is possible for the practicing ergonomist to make practical use of the information obtained from the models. At this point, hardware and software is commercially available.

NIOSH Work Practices Guide for Manual Lifting

At a basic level, the *NIOSH Work Practices Guide for Manual Lifting* may be used.[33] For example, in those cases when assistance is required in choosing between two apparently equal design choices or when additional information is required to make decisions regarding prioritization of services, the *NIOSH Guide* can provide some cost-effective guidance or verification.

Recommended Weight Limit. The *NIOSH Guide* was introduced in 1981. Its initial equation, which resulted in the action limit and the maximal permissible limit, was derived from four primary bodies of knowledge: epidemiology, biomechanics, psychophysics, and physiology. The equation has been recently (1991) modified to compensate for two factors that were not accounted for in the original equation, namely hand-to-container coupling and asymmetry of the lift.[34] Some of the remaining factors—horizontal distance, vertical location, travel distance, and frequency of lift—have been modified as well. The equation, in its current state, is a calculation of recommended weight limit (RWL) and is determined as follows:

$$RWL = LC \times HM \times VM \times DM \times FM \times AM \times CM$$

where

	U.S. Customary Value
LC = Load Constant	51 lbs
HM = Horizontal Multiplier	$(10/H)$
VM = Vertical Multiplier	$1 - (.0075\,\lvert V - 30\rvert)$
DM = Distance Multiplier	$.82 + (1.8/D)$
AM = Asymmetric Multiplier	$1 - (.0032A)$
FM = Frequency Multiplier	(see Table 1–1)
CM = Coupling Multiplier	(see Exhibit 1–1)

H = Horizontal location of hands from midpoint between the ankles. Measured at the origin and the destination of the lift.

V = Vertical location of the hands from the floor. Measured at the origin and destination of the lift.

D = Vertical travel distance between the origin and the destination of the lift.

A = Angle of asymmetry—angular displacement of the load from the sagittal plane. Measured at the origin and destination of the lift (degrees).

F = Average frequency rate of lifting measured in lifts/min.

Lifting Index. The lifting index (LI) provides a simple estimate of the hazard of an overexertion injury for a manual lifting job.

$$LI = \frac{L}{RWL}$$

where

L = Load weight of the object lifted
RWL = Recommended weight limit

The lifting index provides the following recommendations:

1.0 or less . . . . no changes recommended
1.0 to 3.0 . . . . consider administrative controls
3.0 or greater . . . . consider engineering controls

NIOSH Guide Case Study. A manual material handling job demand has been identified as a significant factor in a series of reported back injuries. The *NIOSH Guide* can be used to analyze the job demand and provide rationale for improvements. The job involves palletizing boxes coming off a conveyor. The necessary measurements were taken for both the initiation and the termination of the lift. The weight of the box is 35 pounds.

Table 1–1 Frequency Multiplier Table (FM)

Frequency	*Work Duration (Continuous)*					
	≤ 8 hours		*≤ 2 hours*		*≤ 1 hour*	
Lifts/Min.	*V < 30*	*V > 30*	*V < 30*	*V > 30*	*V < 30*	*V > 30*
0.2	0.85	0.85	0.95	0.95	1.00	1.00
0.5	0.81	0.81	0.92	0.92	0.97	0.97
1	0.75	0.75	0.88	0.88	0.94	0.94
2	0.65	0.65	0.84	0.84	0.91	0.91
3	0.55	0.55	0.79	0.79	0.88	0.88
4	0.45	0.45	0.72	0.72	0.84	0.84
5	0.35	0.35	0.60	0.60	0.80	0.80
6	0.27	0.27	0.50	0.50	0.75	0.75
7	0.22	0.22	0.42	0.42	0.70	0.70
8	0.18	0.18	0.35	0.35	0.60	0.60
9	0.00	0.15	0.30	0.30	0.52	0.52
10	0.00	0.13	0.26	0.26	0.45	0.45
11	0.00	0.00	0.00	0.23	0.41	0.41
12	0.00	0.00	0.00	0.21	0.37	0.37
13	0.00	0.00	0.00	0.00	0.00	0.34
14	0.00	0.00	0.00	0.00	0.00	0.31
15	0.00	0.00	0.00	0.00	0.00	0.28
>15	0.00	0.00	0.00	0.00	0.00	0.00

Exhibit 1–1 Hand-to-Container Coupling Classification

GOOD	FAIR	POOR
1. For containers of optimal design (some boxes, crates, etc.), handles or hand-hold cutouts of optimal design 2. For loose parts or irregular objects that are not usually containerized (castings, stock, supply materials, etc.), a comfortable grip, easily wrapped around the object	1. For containers of optimal design, handles or hand-hold cutouts of less than optimal design 2. For containers of optimal design with no handles or hand-hold cutouts, or for loose parts or irregular objects, a grip in which the hand can be flexed about 90 degrees	1. Containers of less-than-optimal design with no handles or hand-hold cutouts; loose parts or irregular objects that are bulky or hard to handle

1. An optimal handle design has 1.9- to 3.8-cm (.75- to 1.5-in) diameter, >11.5-cm (4.5-in) length, 5-cm (2-in) clearance, cylindrical shape, and a smooth, nonslip surface.
2. An optimal hand-hold cutout has >3.8-cm (3-in) height, >11.5-cm (4.5-in) length, semioval shape, >5-cm (2-in) clearance, cylindrical shape, and a smooth, nonslip surface.
3. An optimal container design has <40-cm (16-in) frontal length, <30-cm (12-in) height, and a smooth nonslip surface.
4. A worker should be capable of clamping the fingers at nearly 90 degrees under the container, such as required when lifting a cardboard box from the floor.
5. A less-than-optimal container has a frontal length >40 cm (16 in), height >30 cm (12 in), rough or slippery surface, sharp edges, asymmetric center of mass, or unstable contents, or requires gloves.

Couplings	*V < 30 in*	*V ≥ 30 in*
Good	1.00	1.00
Fair	0.95	1.00
Poor	0.90	0.90

Calculate the recommended weight limit (RWL) and lifting index (LI):

Initiation of Lift:

Factor		Multiplier	
H	10 in	HM	10/10
V	30 in	VM	$1-(.0075\|30-30\|)$
D	24 in	DM	$0.82 + (1.8/24)$
F	1/min, ≤ 8 hour	FM	0.75 (from Table 1–1)
A	10°	AM	$1-(.0032 \times 10°)$
C	good	CM	1.00 (from Exhibit 1–1)

RWL: (51)(1)(1)(0.895)(0.75)(0.968)(1.0) = 33.1 pounds
LI: 35/33.1 = 1.06

Termination of Lift:

Factor		Multiplier	
H	20 in	HM	10/20
V	6 in	VM	1– (.0075|6 – 30|)
D	24 in	DM	0.82 + (1.8/24)
F	1/min, ≤ 8 hour	FM	0.75 (from Table 1–1)
A	30°	AM	1 – (.0032 × 30°)
C	good	CM	1.00 (from Exhibit 1–1)

RWL: (51)(0.5)(0.82)(0.895)(0.75)(0.904)(1.0) = 12.7 pounds
LI: 35/12.7 = 2.8

Case Study Interpretation. Calculations obtained from the initiation of the lift reveal a RWL of 33.1 pounds and an LI of 1.06. This part of the analysis demonstrates that minimal risk is evident.

However, at the termination of the lift the RWL is calculated to be 12.7 pounds and the LI is approaching 3 at 2.8. This suggests that at a minimum, administrative controls—such as job rotation or preplacement screening—would be advised. With an LI this close to 3, engineering controls could be considered.

The multipliers provide good guidance for what types of changes make sense. The horizontal multiplier at 0.50 all by itself reduces the RWL to 25.5 pounds. The frequency multiplier at 0.75 holds the second greatest importance.

An engineering control that reduced the horizontal distance factor combined with an administrative control that decreased the frequency could significantly reduce the risk of injury.

Organizational Development/Management

If the principles of work physiology, engineering, psychology, anthropometry, and occupational biomechanics are the "hardware," the principles of management are the "software." Increasingly, the traditional concepts of ergonomics are broadening to include the management principles of organizational development. Issues included are labor/management relationships, supervision given and received, peer interaction, corporate philosophies, and management style—in other words, all of those tangible and intangible factors that make up the "culture" of the organization.

Management styles have evolved. Lerman and Turner[35] note that while management goes back to ancient times it became a specialized field in the early twentieth century.

Classical Management

Classical Management is viewed as a systematic approach first developed in the early 1900s. Based on the work of Frederick W. Taylor,[36] Max Weber, Henri Fayol, and others, it stressed "scientific efficiency." The "science of management" was based on the premise that the primary motivating factor for workers was money.

Classical Management did result in increases in productivity, but now we recognize that it came at the expense of harmonious relationships between labor and management. The explosive growth of unions of the time in part was the consequence of the use and misuse of this approach.[37]

Human Relations Management

In the 1930s the Human Relations Management approach emerged.[38] Based a great deal on research conducted at the Hawthorne Works of the Western Electric Company, this approach explored the contributions of personality and cognitive functions of work.

Their conclusions included the notions that workers wanted more than just money from their work, and that effective management involved social as well as technical attributes.

Modern Systems and Contingency Management

In the 1990s, Modern Systems and Contingency Management considers the organization as a total system. Complex interactions are continuously occurring both inside and outside the organization.[39] Situations change, and the management response needs to allow the flexibility to respond to the changes. Lerman and Turner's answer to "What is the best management system?" is a resounding "It depends!" At this point there is significant support for the major influence of concepts of organizational development on productivity, quality, and safety in the workplace.

Ergonomics and Management

Management principles bring the principles of ergonomics to life. In overview, management principles can be divided into four broad categories.

1. Policies and Procedures
2. Ergonomic/Safety Teams
3. Claims Management
4. Medical Management[40]

In-depth discussions of each of these categories is beyond the scope of this chapter. In overview, management's commitment to, involvement in, and facilitation of the ergonomic process are critical to its success. Significant evidence exists that a management team that sends the message "we care" has major impact on controlling workplace injuries and illnesses.[41]

Pathology

A general understanding of the disease processes associated with occupational injuries and illnesses is an important part of the process. *Cumulative trauma disorders,* in the United States, has become the umbrella term for a wide range of neuromusculoskeletal diagnoses. More information regarding these issues is found elsewhere.

Epidemiology

A company of 1,000 employees involved in a light manufacturing industry has determined they are experiencing an incidence rate of 20 reports of carpal tunnel syndrome per every 200,000 work hours. Should the company be concerned?

A company has ten departments and can afford to invest a total of $20,000 for ergonomic interventions. Should they invest $2,000 per department or $20,000 all in one?

The field of epidemiology can help provide some answers. Epidemiology is defined in *Dorland's Illustrated Medical Dictionary* as "the science concerned with the study of factors determining and influencing the frequency and distribution of disease, injury and other health-related events and their causes in a defined human population for the purpose of establishing programs to prevent and control their development and spread."[42] The factors discussed include the set of factors previously outlined. The frequency and distribution of the health-related events can be determined through the interpretation of reactive statistical measures such as incidence rate, severity rate, and relative risk. Other data sources include workforce questionnaires, surveys, and interviews.

The study of this information can

- establish an injury and illness baseline against which future interventions can be measured
- provide guidance in resource allocation
- allow comparison of a particular company to industry-wide statistics
- provide a means by which the workforce can provide input and enhance communication

ERGONOMIC ANALYSIS AND APPLIED INTERVENTIONS

A universal scheme to conduct ergonomic analysis does not exist. Depending on a wide variety of possible variables, the specific ergonomic analysis needs to be tailored to the specific job demands in question.

Having said this, however, it is true that a good deal of commonness exists across job demands.

The ergonomic objectives of enhancing performance and controlling fatigue by positively influencing quality, production, and safety are uniform across the board.

Generically, the following set of factors can have an impact on physical, psychological, and psychosocial parameters:

- worker physical fitness—poor/good
- force—dynamic/static, high/low
- position—awkward/sustained
- repetition—rate of activity/job demand
- duty cycle—work/recovery rate
- vibration—whole body/segmental
- friction—high/low coefficient of friction
- work practices—training, policy and procedures
- environment—temperature, ventilation, illumination
- job satisfaction—labor/management relationships
- operator control—ease of use

Performing the Analysis

In an ergonomic job hazard analysis, the job is essentially evaluated by using the following sequence:

- Gain clearance from the worker and supervisor.
- Interview the area supervisor responsible for the workers performing the task.
- Select an average but experienced operator to observe and interview.
- Document the task via videotape and slides to enable later and more detailed analysis.
- If the task is repetitive, identify the cycle time of the activity and examine a minimum of two to four cycles when feasible.
- Identify and analyze the risk factors associated with each of the elements or tasks.
- Sketch the work area, including dimensions and distances in the diagram.
- Obtain appropriate weights and measures.
- Compare experienced faster and injury-free workers with those who are inexperienced, fatigued, uncomfortable, or complaining of pain or injury. Determine if there are differences in work technique among these groups of workers.[43]

In the study and practice of ergonomics it is readily apparent that no one right way exists to successfully intervene. All of the issues discussed so far are not independent; they interact. In response to this idiom, ergonomic interventions must also be interrelated. That is, if you intervene regarding a particular risk factor, you also influence the others. It does not work to make changes in isolation! With this in mind, the following analysis and interventions strategies are outlined.

Factors Analyzed

Force

The level of force involved in performing a particular job task is an important determinant in ergonomic analysis. Generally speaking, the goal is to identify why a particular force is imposed on the musculoskeletal system and then intervene to reduce the force level.

Generic guidelines have been developed to assist in analyzing and designing or redesigning the force requirement.[44] Essentially the guidelines will promote the reduction of static and also dynamic muscle contractions.

To reduce *static muscle contractions,*

- reduce or eliminate the need to grasp/hold objects
- limit static efforts:

  high effort level .... less than 10 seconds

  moderate effort ..... less than 60 seconds

  slight effort .... less than 4 minutes

To reduce *dynamic muscle contractions,*

- limit ongoing dynamic effort to 30 percent or less of maximum
- if less than five minutes, limit up to 50 percent of maximum

Force As a Function of Weight. Force levels are a function of the weight of the tools, containers, boxes, parts, carts, and so forth; whether lifted, carried, pushed, or pulled, the force required to move or manipulate the object directly impacts stress on the body. Using biomechanical models and/or the *NIOSH Guide*, it is possible to quantify this stress.

Many intervention strategies exist to control force levels related to the weight of the load:

1. Make use of mechanical devices (hoists, lifts, etc.) to eliminate manual lifting.
2. Slide rather than lift the weight.
3. Eliminate the effect of gravity by counterbalancing the weight—a method commonly used with tools.

4. Remove physical barriers, thereby reducing the horizontal distance (long lever arm).
5. Relocate storage heights, with heavier objects stored between mid-thigh and waist height.
6. Work with vendors to provide material either in smaller unit weights (e.g., 50 pounds rather than 100) or in bulk requiring handling with mechanical means.
7. Provide adjustable height surface (e.g., scissor tables) to maintain desired height of material.
8. Reposition the worker to provide greater mechanical advantage: for example, use body weight rather than musculoskeletal strength.
9. Reposition the work material: for example, brings parts and tools within reach envelope; place bin on a bin tipper or provide side drop-down bins.

Force As a Function of Friction. The coefficient of friction can have a major impact on controlling force levels. In some cases a functional increase is desired, in other cases a decrease:

Increase friction between hand and object by

1. using rubberized coating on the object (e.g., tool handle)
2. cleaning the object of lubricants
3. providing appropriate nonslip gloves
4. maintaining normal skin moisture; dry skin has about two-thirds the coefficient of friction of moist skin

Decrease friction between object and surface by

1. lining storage shelves with decreased friction liners (e.g., Teflon sheets)
2. being aware that spray-on products will reduce friction but may cause a toxic substance problem
3. using roller conveyor systems to transport materials
4. maintaining the quality of floor conditions to eliminate cracks and general deterioration
5. using appropriate type and size of casters/wheels as original equipment or retrofit, depending on floor type

Force As a Function of Grasp. Whether using tools or handling boxes, grasp has a major influence on controlling force levels.

1. A power grasp makes use of larger, more powerful muscles than does a pinch grasp; typically a maximal pinch is only 20 percent of a maximal power grasp.

2. Facilitate the use of power grasps by adjusting coupling (refer to the *NIOSH Guide*).[45]
3. Provide grasp spans of 1.5 to 2 inches as an ideal; spans greater or less result in less than desirable mechanical advantage.

Force As a Function of Glove Use. Gloves are commonly seen in work environments. The type and fit of the glove should reflect the purpose of the glove.

1. Is the glove truly necessary? Generally a gloved hand is able to produce a maximum of 25 percent to 30 percent less force than the ungloved hand.
2. A "one size fits all" policy is not desirable. A too-small glove increases the force required to overcome the resistance of the glove. A glove that is too large does hinder dexterity due to sloppiness of fit.
3. The glove material should reflect the material handled (cotton, leather, rubberized, etc.).[46]

Force As a Function of Component Fit. A poor fit of components during an assembly process may force the assembler to "bang in" the component using the hand or other body part as a hammer. Coordinated effort with the vendor—in house or off site—can assure the needed quality.

The type of fastener used may be at issue. Options include use of riveting, spot welding, and use of TORX or Starburst fastening systems rather than slotted fasteners.

Force As a Function of Tools. A switch from manual hand tools to power tools can reduce force levels. However, power tools bring their own set of issues including vibration and torque reaction force. Torque reaction occurs when a fastener reaches the end of its travel, transferring the torque to the tool and operator. Clutches and torque reaction bars can be employed. Also newer tools make use of pulse technology rather than impact and significantly reduce force requirements.

Handle size should be monitored to provide optimum power grasps. Triggering configuration should spread the force over a large area, not concentrate it in a smaller area.

Preventive maintenance is critical to ensure proper operation of the tool based on manufacturer specifications.

Providing sharp bits, blades, and abrasives significantly reduces the force required to use the manual or power tool.

Force As a Function of Work Flow and Rate. The factors of work rate and flow affect the impact of force on the musculoskeletal system. The duty cycle of the job demand will determine the force dose and exposure. Reducing either the dose (level of force) or the exposure (duration of the force) is desirable. Exposure can be reduced through administrative controls, including job rotation and job enlargement.

Position

To maintain adequate levels of performance, the human body requires adequate movement and recovery throughout work activity. Any jobs or activities that tend to force the worker out of an ergonomic neutral position or result in sustained awkward or even desirable positions should be evaluated. As much as possible the goal is to have the body in a neutral posture.

Undesirable positions include

- Prolonged or repeated non-neutral spinal positions

  Non-neutral spinal positions include bending the head, neck, and trunk forward, backward, or to the side with or without twisting. As individuals move away from the neutral posture, excessive compression and stress may be placed upon the discs, ligaments, or muscles resulting in fatigue, discomfort, or micro-trauma.

- Wrist deviations from neutral greater than 15 degrees

  The neutral position at the wrist can be demonstrated by making a tight fist. This angle is approximately 10 to 15 degrees of extension in most people and is the position of power in the wrist. This posture enables maximum force production while maintaining space within the carpal tunnel. As the wrist moves away from this neutral position the finger flexor tendons increase their contact against the carpal ligament or bones of the wrist. This increased contact may result in inflammation, and the pressure within the carpal tunnel may increase.

- Forearm rotation

  When the forearm is rotated toward the extremes of supination (palm up) and pronation (palm down), and combined with deviations of the wrist from neutral, there is a great degree of stress placed over the site where the forearm muscles attach to the elbow. When high force levels are combined with these postures, there is even more potential for trauma.

- Elbows sustained above mid-chest height

  Elbows positioned above mid-chest height place additional stress on the shoulder by forcing the muscles to maintain prolonged contractions. In addition to this inefficient use of energy, these positions also tend to increase pressure within the shoulder joint and to cause a reduction in blood flow to the tendons in the shoulder.

- Reaching frequently behind the body or above the shoulders

  These positions also tend to increase pressure within the shoulder joint while stretching many of the shoulder tendons and muscles. These postures also place the muscles at poor length when considering muscular length-tension relationships.

Modifying Work Positions. In evaluating the direction of force required at a particular workstation, it may be appropriate to alter the position or posture in which people have typically worked.

For example, if frequent or relatively heavy lifting is required or if significant downward forces are required, standing positions may be more appropriate than sitting positions. When light assembly or precision work is being performed, sitting workstations may be desired.

In some cases sit/stand workstations may provide a viable option. These provide for postural variability.

Adding adjustable-height workstations and lift tables to a work area allows for increased postural variety for the worker but also allows accommodation for variation in body stature between workers. Workstations can then also be easily adjusted for different-sized products. Turn tables can be used in many cases to bring parts closer to the worker, thereby reducing the need for sustained or forward reaches. These are particularly helpful when the worker needs to access the other side of the pallet.

Even in relatively well-designed ergonomic workstations, individuals who are required to work in one posture, even if it is "ideally correct," are exposed to static stresses. Evaluation of the workplace should include an assessment of how often that individual has the opportunity to break out of that posture to perform other movements or tasks.

Positioning Guidelines. A number of positioning guidelines are available to assist in the design/redesign process.[47] Guidelines are just that—guidelines; they need to be applied judiciously with a good understanding of all of the factors involved.

1. Work Height
   - Standing:

     A general standing work surface height is in the range of 34.6 to 36.6 in for women and 37.4 to 39.4 in for men. More precisely, work surface height should be related to the type of work.

     — hard work: 2 to 4 in below the elbow

     — delicate/precision work: 2 to 4 in above the elbow

     — heavy work: 6 to 15 in below the elbow

     An inclined work surface of 10° to 20° above horizontal is also desirable as possible.
   - Sitting:

     An office desk height without keyboard is in the range of 27.5 to 29 in for women and 29 to 31 in for men. For a keyboard, deduct 2 to 3 in. Suggested leg room is 27 in wide and 27 in high. An inclined work surface of 10° to 20° above horizontal as possible is recommended.

2. Seating

   Seating options have improved dramatically in the last decade. General guidelines include

   - Seat pan
     - — height adjustable from 15 to 21 in
     - — tilt 5° backward to 5° forward
     - — 16 to 18 in across
     - — 15 to 16.5 in from front to back
     - — waterfall front edge
     - — nonslip, permeable material
     - — swivel on base
   - Back support/rest
     - — lumbar pad of 2 in depth
     - — lumbar pad adjustment 4 to 8 in above seat pan
     - — incline adjustment of 105° to 10° to seat pan, 110° to 120° to horizontal
   - Five-point base with castors or glides
   - Foot rest as needed

Tool Use and Postures. Frequently workers use tools that have been specially designed for another purpose. This is often found when using pistol grip and in-line tools.

An in-line power tool should be used when there is need for a vertical drive that occurs between the waist and elbow height. Pistol grip tools should be considered when used on horizontal surfaces at waist height or for vertical surfaces between elbow and shoulder height.

Tool manufacturers have made major strides in the past decade in the design of ergonomically approved tools. These include bent handle pliers, ergonomic knives, reduced vibration power tools, and so on.

Rapid Machine Pacing in an Assembly Task. Production workers performing machine-paced tasks are frequently required to maintain a work rate above that which they can comfortably perform. In many cases, workers find themselves working ahead of the line in an attempt to create a buffer for fear that they may fall behind. Other workers find themselves working behind the line because they can't keep up the pace.

Both of these situations require the individual to work in positions other than directly in front of himself or herself, promoting awkward postures.

Workstation layout regarding placement of tools, parts, or materials can be examined to promote a reasonable reach envelope. A desired reach envelope is laid out horizontally within a 45-degree arc from midline to each side of the body.

Amount of forward reach also needs to be considered, recognizing that items stored above or below shoulder height need to be closer to the worker than those stored at shoulder height.

The most frequently used materials, tools, and controls should be placed at the optimal positions within the reach envelope. Rotating jigs and turn tables may be used so that the part or material can be brought close to the person when required.

Repetition Rate

Rates are frequently controlled by several factors, including machine pacing, incentive pay schemes, and production quotas. Regardless of the amount of perceived flexibility for changing these factors, it is important to realize that high repetition rates multiply the effects of all of the risk factors. In any case, it is important to document task cycle time and to determine the repetition rate per hour or per day.

High repetition rates tend to increase stress levels and force exertion levels and have the tendency, particularly in a machine-paced assembly task, to encourage workers to work ahead of the line.

Repetition needs to be considered from two perspectives. These include both static tasks, requiring "repetition" of sustained postures, and tasks that are truly highly repetitive. There are risks associated with both types of tasks.

Mechanical Aids. Repetition rates can frequently be reduced through the use of mechanical aids. For example, introducing power tools in the workplace may reduce the frequency and duration of forceful contractions and awkward and sustained positions.

Worker Rotation. In many cases, a feasible alternative to reducing repetition rates is the use of worker rotation. This reduces the overall exposure of the worker to particular types of repetition. In implementing worker rotation systems, work methods need to be analyzed and the required motions for each body part should be identified. After determining that the designs are acceptable or that redesign is not feasible, workers should be cross-trained to perform each job within the rotation schedule.

In addition to involving the workers in identifying a successful rotation strategy, it is also important to assure that the jobs into which workers rotate involve significantly different physical job demands. In some cases job rotation schemes rotate individuals through different jobs, but the actual physical demands are very similar from position to position.

Pay System. The type of pay schedule may be directly related to the pace at which an individual works. Incentive, piece-rate systems can actually drive workers to perform at higher paces than are considered desirable.

Add Staff. In some situations, it may be worthwhile to consider the addition of staff members. The cost per unit may increase based on this protocol, but the effect on production, quality, employee morale, and the risk of overall cumulative trauma must be weighed against this additional cost.

Reduce Line Speed. In some situations it may be possible to reduce the speed of the line or to change from a machine-paced to a self-paced work environment.

Contact Stresses

Sharp Edge or Weight Bearing. In evaluating the type and severity of contact stress, look for any part of the body that is in contact with a sharp edge or that maintains weight-bearing support upon any surface for a long period of time.

Tool handle size and shape should be examined for prominences that may promote increased pressure over any point or the grasping surface of the hand. Tools should be evaluated regarding the amount of localized pressure they produce in the palm of the hand. Finger contours on handles or triggering devices on tools may also produce unnecessary stress on the digits.

The size and shape of any machine guards should be examined in terms of potential for contact stress as well. Because these guards are positioned over parts that are likely to involve hand-intensive tasks, there is a great deal of potential for the hands to frequently contact the machine guard. Any sharp edges or sustained pressure on the guard should be evaluated and documented.

A frequently used intervention strategy for controlling contact stress is rounding work surface edges that come in contact with the worker. Another is to round the contours of the tool handles and trigger switches. Tools that require continuous or intermittent pressure on the fingers, palm, base of the wrist, forearm, and elbow should be avoided. Use of self-opening tools such as pliers and scissors that are spring loaded is recommended as possible. This reduces the contact stress required when opening the tool. When contact stress itself cannot be avoided, the goal is to distribute the pressure over a larger area as possible by increasing the contact surface.

Use of the Hands for Pounding. When the hands are used as hammers, the butt of the hand and the fleshy part of the fist are vulnerable to local nerve and soft tissue trauma. Using the hands in this manner increases the likelihood of local inflammation, which may cause unnecessary scarring and eventual reduction in blood flow to the nerves and other soft tissues.

Inappropriate techniques and work processes are frequently the culprit regarding contact stresses. Encourage workers to be aware of potential problem areas such as pressure over vulnerable areas of the body where nerves and blood vessels are close to the surface.

Standing and Sitting. Two areas of the body that are frequently not evaluated relevant to contact stress are the feet of people who stand all day and the buttocks/thighs of those who sit all day.

Chairs can be specifically evaluated by observing the front of the seat pan and the position of the back rest and evaluating the potential for pressure behind the knee or the back of the thigh caused by the edge of the seat pan.

Floor surfaces can affect the comfort of workers who are required to stand for the larger percentage of the day, particularly when there is limited potential for movement. Concrete, steel grates, uneven or vibrating floor surfaces may increase foot, leg, or spinal fatigue and discomfort and can also affect concentration and product quality. Antifatigue mats or shock-absorbing shoe inserts can improve comfort levels.

Vibration

Whole body vibration is frequently encountered in tasks such as truck driving and fork lift driving. Vibration of this type is suspected of weakening and disrupting soft tissue structures such as tendons and ligaments, thereby increasing the potential for cumulative trauma injury, especially as it relates to the spine.[48]

Segmental vibration is typically found in tasks that require the use of abrasive wheels and grinders, lathes, and power hand tools. Vibration from these sources has been shown to decrease sensitivity in the hand, resulting in an unnecessary increase in local muscle contractions.[49]

Like force, posture, repetition, and contact stress, vibration is frequently associated with other risk factors. When evaluating vibration, it is important to assess the duration of the exposure, the exposure patterns during the day, and the force levels and postures assumed during the vibration exposure.

The fastener types used with various power drivers and nut runners may also play a role in vibration exposure. Certain fasteners, because of the means by which they engage the power tool, may drive more easily, resulting in reduced exposure to vibration, sustained or high force levels, poor postures, and contact stresses.

For example, hex head screws have been shown to drive faster and with less muscular effort than Phillips screws, and Phillips screws with less time and effort than slotted screws. In some cases rivets, welding, or adhesives may replace the need for these types of fasteners.[50]

Source Control. When possible, it makes the most sense to control vibration at the source. This is true whether it is segmental or whole body vibration. To accomplish this, power tools should be maintained and balanced on a regular basis. This keeps vibration to a minimum. To control whole body vibration, floor quality as well as vehicle seat ability to attenuate vibration should be examined. Repair work or even replacing vehicle seats may be necessary to reduce exposure to whole body vibration.

Path Control. In many situations it may not be possible to control vibrations at the source. In this situation, the path of the vibration can be obstructed and the vibration dampened. Several vibration attenuation covers are available that can be attached directly to tools. Another option involves the wearing of gloves with padded palms. However, be aware that when you add these coverings the effective handle diameter increases and tool control and grip strength may be adversely affected.

Altering Abrasives or Tool Speed. Increasing the revolutions per minute at which the tool turns frequently helps to reduce the amplitude of vibration. General exposure to segmental vibrations may also be reduced by assuring that quick-cutting abrasives are readily available in grinding and sanding operations.

Work Environment

Cold. Cold environments, tools, or pneumatic tool exhaust may bring about a reduction in tissue sensitivity, manual dexterity, and grip strength. As sensitivity decreases, the amount of force exerted to perform a task increases and the individual performs more work than what is necessary.

Adequate personal protective equipment and appropriate worker rotation (in/out of cold environment) are effective. Directing tool exhaust away from the user is important.

Heat. Warm environments result in an obvious increase in metabolic demand. They may also affect an individual's ability to grasp tools and parts and to manipulate controls due to the effect of perspiration on grasp. As perspiration increases, the friction between the hand and the tool decreases and higher force levels are again required to maintain the integrity of the grasp. Warm and humid environments may also result in the fogging of eye protection, again complicating effective task completion. Adequate ventilation and clothing as well as worker rotation are effective control agents.

Lighting. The amount, quality, and direction of light needs to be evaluated. In many cases in today's office buildings, illumination levels are approximately 25 to 30 percent greater than desirable. In many situations it is advisable to decrease the amount of general overhead light and to bring in specific task lighting that can be effectively used in selected areas. Also consider the overall quality and level of the light in relation to the color and reflectivity of the walls, floors, and ceilings.

Glare is a commonly observed problem in office environments, where it is apparent on VDT screens. Glare can frequently be reduced or completely abolished by tilting or positioning the monitor away from the light source. In some cases overhead light diffusers can be used to help ensure that light rays are directed down and perpendicular to the work area. Window blinds can be used to redirect light.

Glare hoods or screens may be used over computer monitors to also reduce the intensity of glare. If glare screens are to be used, there is a great need to determine the quality of the glare screen. If a low-grade screen is used, character resolution is degraded, with resulting impairment of visual acuity.

Underillumination may facilitate forward bending of the trunk and head as individuals attempt to get closer to the item they are viewing. Task lighting can be effective to focus illumination where desired and at the same time control glare.

Noise. The appropriate noise level within an environment is a function of the communication requirements of the environment. Even in areas where there is no danger of hearing loss, the overall noise level can affect an individual's ability to concentrate on the task and can increase muscular tension.

Noise is essentially another form of vibration; the intervention strategies are similar to those for the control of vibration.

Controlling the noise at its source is always the best possible solution. For example, replacing noisy dot-matrix printers with laser printers can be effective in office environments.

If it is not possible to control the source of the noise, noise can also be controlled by controlling its path. This can be accomplished through the use of acoustical sound barriers, enclosures, and sound-absorbing tiles and carpet.

Organization, Job, and Skill Factors

In addition to the physical factors present in the workplace, organizational and psychological factors also play a role in increasing stress, muscular tension, and discomfort. It is becoming increasingly apparent that there are strong psychological issues related to reducing injuries in the workplace.

Bigos et al. determined that measures of psychological attributes were much stronger predictors of injury and illness rates than those factors of physical performance, such as strength and flexibility.[51]

The person responsible for job redesign needs to take these factors into consideration to make the environment more satisfying, less stressful, and less likely to bring about cumulative trauma disorders. Organization, job, and skill factors are all important components in improving workplace safety.

1. Organizational Factors
    - Unwanted overtime
    - Shift work (particularly that which involves rotating job schedules)
    - Incentive- or quota-based pay structures
    - Poor or inappropriate supervisory or coworker relationships
    - Perceived stability and growth potential within the job
2. Job Factors
    - Workload—overload or underload
    - Work pace—self-paced, incentive paced, or machine paced

- Work pressure—due to unrelenting workload
- Pressure to continue producing due to the interdependence of tasks, and performed the effects that not producing may have on coworker output or quality

3. Skill Requirement
   - Monotony due to lack of variety in task performance
   - Perception of significance or identity in the tasks performed
   - Sense of not being a part of the whole while performing a very small task as part of a larger process

CONCLUSION

The skillful application of ergonomic principles to analyze and design or redesign work involves the talents of many people: workers, supervisors, managers, safety professionals, health care providers, engineers, human resource personnel . . . the list goes on.

The objectives of enhancing performance and controlling fatigue can be met in the areas of quality, production, and safety. The physical, psychological, and psychosocial parameters of work can be positively influenced for the benefit of all.

NOTES

1. P. Fitts and R. Jones, *Analysis of Factors Contributing to 460 "Pilot Error" Experiences in Operating Aircraft Controls* (Dayton, Oh.: Aero Medical Laboratory, Wright Patterson Air Force Base, 1947).
2. F. Taylor, *Scientific Management* (New York: Harper & Row, 1947).
3. V. Putz-Anderson, ed., *Cumulative Trauma Disorders: A Manual for Musculoskeletal Diseases of the Upper Limbs* (New York: Taylor & Francis, 1988).
4. T. Rubinstein and A. Mason, The Accident That Shouldn't Have Happened: An Analysis of Three Mile Island, *IFEE Spectrum* (November 1979):33–57.
5. U.S. Dept. of Labor, Bureau of Labor Statistics, *Occupational Injuries and Illnesses in the United States by Industry, 1989* (Bureau of Labor Statistics Bulletin no. 2379, April).
6. A. Chapanis, Some Reflections on Progress, in *Proceedings of the Human Factors Society 29th Annual Meeting* (Santa Monica, Calif.: Human Factors Society, 1985), 1.
7. E. Grandjean, *Fitting the Task to the Man,* 9th ed. (New York, Taylor & Francis, 1988), ix.
8. N. Hadler, *Occupational Musculoskeletal Disorders* (New York: Raven Press, 1993).
9. L. Fine and B. Silverstein, Detection of Cumulative Trauma Disorders of Upper Extremities in the Workplace, *Journal of Occupational Medicine* 28 (1986):674.
10. U.S. Dept. of Labor, *Occupational Injuries.*
11. U.S Congress, House Committee on Government Operations, Subcommittee on Employment and

Housing, Hearing on Dramatic Rise in Repetitive Motion Injuries and OSHA's Response, June 6, 1989. Testimony of Dr. Lawrence J. Fine, Director of the Division of Surveillance, Hazard Evaluation, and Field Studies, U.S. Dept. of Health and Human Services (Washington, D.C.: National Institute for Occupational Safety and Health, 1989).

12. U.S. Dept. of Labor, *Occupational Injuries.*
13. M. Fletcher, Cumulative Trauma Disorders: Repetitive Motion Cases Cost Billions Annually, *Business Insurance* 24 (1990):36.
14. G. Brogmus and R. Marko, Cumulative Trauma Disorders of the Upper Extremity: The Magnitude of the Problem in U.S. Industry, in *Human Factors Design for Manufacturing and Process Planning,* ed. International Ergonomics Association (Santa Monica, Calif.: Human Factors Society, 1990), 49–59.
15. Grandjean, *Fitting the Task;* T. Armstrong, Cumulative Trauma Disorders of the Upper Limb and Identification of Work-Related Factors, in *Occupational Disorders of the Upper Extremity,* ed. L. Millender et al. (New York: Churchill Livingstone), 19–46.
16. W. Woodson et al., *Human Factors Design Handbook,* 2nd ed. (New York: McGraw-Hill Publishing Co., 1986), 2.
17. P. Astrand and K. Rodahl, *Textbook of Work Physiology,* 3rd ed. (New York: McGraw-Hill Publishing Co., 1986).
18. Grandjean, *Fitting the Task,* 8.
19. Ibid., 9.
20. Ibid., 9–10.
21. Ibid., 7.
22. P. Hansford et al., Blood Flow Changes at Wrist in Manual Workers after Preventive Interventions, *Journal of Hand Surgery* 2A (1986):505–508.
23. C. Wickens, *Engineering Psychology and Human Performance* (Glenview, Ill.: Scott, Foresman and Co., 1984), 3.
24. Ibid., 8–17.
25. D. Norman, *The Psychology of Everyday Things* (New York: Basic Books, Inc., 1988).
26. Ibid., 141.
27. D. Chaffin and G. Andersson, *Occupational Biomechanics* (New York: John Wiley & Sons, 1984), 53.
28. D. Winter, *Biomechanics of Human Movement* (New York: John Wiley & Sons, 1979), 47–64.
29. V. Frankel and M. Nordin, *Basic Biomechanics of the Skeletal System* (Philadelphia: Lea & Febiger, 1980), ix.
30. Chaffin and Andersson, *Occupational Biomechanics,* 1–2.
31. Ibid., 147–148.
32. Ibid., 148–155.
33. U.S. Dept. of Health and Human Services, *Work Practices Guide for Manual Lifting* (Cincinnati: U.S. Dept. of Commerce, National Technical Information Service, 1981).
34. T. Waters et al., Revised NIOSH Equation for the Design and Evaluation of Manual Lifting Tasks, *Ergonomics* 36 (1993):749–776.
35. P. Lerman and J. Turner, *One Day MBA* (Englewood Cliffs, N.J.: Prentice Hall, 1990).
36. Taylor, *Scientific Management.*

37. M. Smith and P. Sainfort, A Balance Theory of Job Design for Stress Reduction, *Int.* *Journal of Industrial Ergonomics* 4 (1989):49–67.
38. E. Mayo, *The Social Problems of an Industrial Civilization* (Andover, Mass.: Andover Press, 1945).
39. Smith and Sainfort, Balance Theory.
40. M. Anderson et al., *Reducing Injuries in the Workplace* (Minneapolis: Saunders Group, 1993).
41. S. Fitzaler and R. Berger, Chelsea Back Program: One Year Later, *Occupational Health and Safety* (1983):52–54.
42. *Dorland's Illustrated Medical Dictionary,* 27th ed. (Philadelphia: W.B. Saunders Co., 1988).
43. Anderson et al., *Reducing Injuries.*
44. Grandjean, *Fitting the Task,* 8–9.
45. Waters et al., Revised NIOSH Equation.
46. M. Sanders and E. McCormick, *Human Factors in Engineering and Design* (New York: McGraw-Hill Publishing Co., 1987), 317–319.
47. Grandjean, *Fitting the Task;* S. Rodgers and E. Eggleton, *Ergonomic Design for People at Work,* Vol. 1 (New York: Van Nostrand Reinhold, 1983) and Vol. 2 (New York: Van Nostrand Reinhold, 1986).
48. Sanders and McCormick, *Human Factors.*
49. P. Pelmear et al., *Hand-Arm Vibration: A Comprehensive Guide for Occupational Health Professionals* (New York: Van Nostrand Reinhold, 1992), 105–121.
50. T. Cederquist and M. Lindberg, Screwdrivers and Their Use from a Swedish Construction Industry Perspective, *Applied Ergonomics* 24 (1993):154–155.
51. S. Bigos et al., Back Injuries in Industry: A Retrospective Study. II. Injury Factors, *Spine* 11 (1986):246–251.

Chapter 2

Formation and Workings of an Ergonomics Team: A Case Study

Elroy C. Lundblad, Marlene J. Perkins, and Andrea O. Kramer

THE SAFETY/MEDICAL DILEMMA

The SC Johnson Wax Ergonomics Committee evolved from what could be construed as an ongoing disagreement between medical and safety personnel. Despite the mutual goal to provide a safe and healthful working environment, differing professional perspectives led to a persistent dilemma in the 1980s. This philosophical difference initially stymied development of a working relationship to mutually address ergonomic issues.

The key question was: Does a problem exist? Interpretation of data yielded opposing viewpoints, even though the data being examined by each department were the same. How could that be if, indeed, all data gathered concerned the same workers at the same facility?

The answer lay not so much in a difference of opinion as in definitions driven by unique professional obligations. The Safety and Medical Departments were each responsible to ensure that record keeping was in compliance with the Occupational Safety and Health Administration (OSHA) standards. Safety did not view individual areas of injuries as sufficiently significant in number to warrant an ergonomics program. Other safety issues took higher priority due to higher incidence and obligatory regulations. Ergonomics issues did not warrant immediate attention.

The Medical Department, on the other hand, was viewing the identical situation from a holistic perspective. Driven by medical definition, the department felt that the need for ergonomics could be derived by analyzing injury data related to ex-

---

This study is taken from the experience at SC Johnson Wax, headquartered in Racine, Wisconsin. Founded in 1886 as a manufacturer of parquet flooring, it is today one of the world's leading producers of specialty products for home, personal care, and insect control, and a leading supplier of products and services for commercial, industrial, and institutional facilities. SC Johnson Wax and its subsidiaries employ more than 13,000 people worldwide, about 2,700 of whom are based in Wisconsin.

tended use of soft tissue. As such, investigation was warranted when more than one such documented case was noted in the same work area or evolved from doing the same task. Witnessing multiple cases of strains and sprains constituted an ergonomically based concern for the Medical Department, despite the fact that the number of incidents related to a particular body part was insufficient to gain the immediate focus of the Safety Department. The Medical Department approach was to effect a cure before the problems grew in number and proportion. The underlying philosophy was that this could only be accomplished if the situation causing stress was arrested.

DEVELOPING A UNITED FRONT

Dialogue between the two groups continued into the 1990s as new industry trends became evident. OSHA changed its terminology regarding injuries and illnesses. Citations were becoming more commonplace. Enforcement meant greater dollar impact for violators. Workers' compensation costs continued to rise. The Safety Department broadened its previous definition of what it considered an ergonomics issue.

At this point, both the Safety and Medical Departments recognized that common needs existed. They had, after all, educated each other and had reached an understanding regarding each other's roles. They had analyzed injuries and illnesses to determine ergonomic impact on industry in general, and their company in particular, over a number of years. Together they decided that mutual benefit could be realized by pursuit of an ongoing working relationship to address ergonomic issues. The time was right for further action.

Together the two departments adopted a proactive, preventive mode, one aimed at averting all pain and suffering while reducing company costs.

The Medical Department sent a delegation to the Executive Committee to pursue involvement at a higher level within the company. As a result, two company representatives attended a national seminar to further expand knowledge and understanding of ergonomics. Their recommendation to the Vice President of Manufacturing was to develop a framework for a company-wide program using the seminar information and the OSHA meat industry regulation as a reference. This recommendation was approved.

THE TEAM STARTS HERE

The Safety and Medical Departments turned their attention to further fact finding at this point. They wanted to design a research-based program, soundly grounded in ergonomic principles but sufficiently customized to meet the company's unique needs. Books and articles were read from which pertinent information was gathered. Conversations were conducted with numerous companies of similar size and venue who shared similar concerns. Hearing what had and hadn't worked for others was useful to the planning process. Success stories served as inspiration and provided a basis from which to lay the needed groundwork. Perspective on failed attempts saved valuable time and effort.

The two departments then prepared a program prototype. It included a mission statement and a set of preliminary goals for submission to upper management.

SC Johnson Wax Ergonomics Mission Statement

The Ergonomics Committee of SC Johnson Wax actively commits to development of a strong ergonomics program as a means to effectuate the best possible health of workers.

To achieve that goal, the Ergonomics Committee will:

- Define and oversee a proactive, company-wide ergonomics program and measure its performance
- Write policies and procedures necessary to support the program
- Develop educational materials and make them available for company-wide use
- Through the provision of training and technical resources, ensure incorporation of sound ergonomic principles in the design of workstations
- Serve as a resource to safety committees and the Medical Department

Objectives

- To prevent potential ergonomic disorders in the workplace
- To reduce CTD incidences and their severity
- To return injured employees to productive status
- To reduce workers' compensation costs

Strategies and Responsibilities

The Ergonomics Committee will:

1. Serve as a resource to operational safety committees and the Medical Department

2. Ensure, through the provision of training and technical resources, the incorporation of sound ergonomic principles in the design of workstations
   - Develop written policy/procedure

- Define and oversee a proactive, company-wide ergonomics program and measure its performance
- Develop and make available ergonomics educational materials for company-wide use

3. Recognize and identify the potential for ergonomics disorders in the workplace
   - Develop an audit survey form
   - Maintain, at minimum, compliance with regulations

This document reflected the significance of an ergonomics program to employees, the Safety Department, the Medical Department, and management. In short, it highlighted the potential implications for the entire company. It was a model of the uniformity sought company-wide.

In order to accomplish these objectives and fulfill these responsibilities, the committee would need to be a repository of SC Johnson Wax ergonomics expertise. By selecting representatives from various disciplines, the Ergonomics Committee could serve to reflect and complement the team-management concept already adopted by the company. The groundwork completed by the Safety and Medical Departments could serve as a model for the type of accomplishments that can be achieved through teamwork.

GAINING NECESSARY APPROVAL

The next step for the two departments was to share their ideas and concerns relative to ergonomics with upper management. Not only was empathy for individual pain and suffering conveyed, but statistics relative to incidents and cost containment were prepared and presented. At SC Johnson Wax, back, shoulder, and carpal tunnel complaints were heard at the Medical Center. The group used national studies to compute the potential surgical costs associated with each case type. Then they were able to present potential cost savings based on those statistics.

Finally, they prepared an action plan, outlining the main components of the program they envisioned. The purpose of this step was twofold: to substantiate the program request with specifics and to assist with cost determination.

Action Plan: Proposed Program

- Develop a communication plan for employees and contractors with consultant.
- Schedule educational training sessions for Ergonomics Committee members, engineers, technicians, safety committee members, and managers with consultant.

- Increase the awareness of all untrained employees via internal sessions with operational trainers.
- Conduct facility-wide audit with the aid of a consultant.
- Oversee ongoing area audits.
- Formulate recommendations for corrective actions and compliance review.
- Measure and evaluate program, comparing medical and workers' compensation records with those of past history, monitoring attendance at educational programs and documenting/implementing recommendations.

Presented with this information, management was immediately receptive to the proposal. The request was viewed as both an effective means of minimizing human suffering and a cost-effective action. Approval was granted to convene an Ergonomics Committee and proceed further. Financial backing was provided. This enabled planning and implementation of a sound program, one inclusive of the utilization of a consultant who could devote time-dedicated professionalism to the program.

INTEGRATING WITH EXISTING INFRASTRUCTURE

Having received a strong endorsement from the Vice President of Manufacturing to proceed, the first step toward accomplishing their many goals was formation of the Ergonomics Committee. Constitution was fundamental to future success, given the company's recent adoption of the team management concept and its division of various operations into separate businesses. A single committee was needed to support both office and manufacturing environments and to present all areas and types of work. Members would need to collaborate, so it was necessary for the core group to be a workable size. It needed to be large enough to represent all areas and types of expertise, but small enough to be result oriented.

Potential committee members were selected to include resources important to implementing ergonomics adaptations and representatives from impacted areas. Included were personnel from engineering, office, technical services, research and development, purchasing, management, industrial hygiene, line operations, and warehouse, in addition to Safety and Medical Affairs. Personal contact was made with selected individuals to determine degree of interest and commitment to a grassroots ergonomics effort. A roster of enthusiastic members was finalized. Representing so many divisions would enable contributions from numerous perspectives.

UP AND RUNNING

With the selection process complete, it was time to take the team to task, launching them on a solution-producing expedition. The group convened in the spring of 1992.

At the initial meeting, introductions were made to meld and blend the group. Discussion regarding the group's purpose was instigated. The Safety and Medical committee members, who had worked so diligently toward this day, contributed enthusiastic remarks regarding the task at hand and initiated dialogue concerning the importance of the committee's work at present and to the company's future. They shared their expectations and visions and pointed out current trends within the company.

Then together committee members examined and reviewed the mission statement and other documents as generated by the Safety/Medical group. This provided a starting point for determining the company's resources and needs. With minor modifications, the committee approved the objectives and accepted the responsibilities outlined.

Defining Ergonomics within the Company

At a subsequent meeting, a definition of ergonomics was generated. Then all of the statements of explanation and intent were consolidated into a single document and presented to management. Its adoption was the first step toward establishment of consistent policy throughout the company.

Next the committee began work on the proposed program, creating a timeline and assuming specific responsibilities for portions of the individual projects. Some persons assumed responsibility for screening training materials. Discussion led to creation of a mini-audit form. Using it as a guideline, committee members circulated to identify improper workplace designs that could lead to ergonomically related OSHA recordable cases.

Hiring a Consultant

Next the timeline called for the hiring of a consultant to assist with larger scale program development conceived by the committee. Four professionals were interviewed before one was selected. The successful candidate was equally comfortable in office and manufacturing environments, had extensive background in conducting audits, and was experienced in group presentations. Each of these qualifications was necessary to effective implementation of the program the committee had designed.

Upon hiring, the consultant requested and received various items for examination. Pictures of furnishings and videos of workplace settings in the manufacturing plant were sent for preview. Documents describing the company philosophy and materials already generated by the committee were also forwarded in preparation for on-site visits.

The consultant's first visit was scheduled for a two-day period. A meeting with the Ergonomics Committee was first on the agenda. During this time, the committee verbalized its visions for the consultant who would help to actualize the dream. The consultant then outlined recommended steps in pursuit of these goals.

Following this meeting, a walk-through audit was completed. The consultant conversed with many employees along the way, stopping to ask questions about work habits, comfort levels, and schedules. Extensive notes were taken while looking and listening. Pictures were taken for use at later workshops. Workers' compensation records were reviewed. The product of this analysis was a series of recommendations, later submitted to the Ergonomics Committee for examination, further analysis, and action.

At this point, the majority of employees were still unaware of what ergonomics was or how it could impact each and every business within the company. So education was addressed next. With their consultant, the committee planned a three-day training workshop. Because they wanted to ensure a professional image, they designed personal invitations that served to introduce the concept of ergonomics, its importance to individual and company well-being, and the SC Johnson Wax committee members behind the initiative.

Ergonomics Inservices

As the teaching agenda was developed, careful consideration was given to which persons should be invited to what sessions. The group wanted to ensure awareness on the part of key persons within the company. They knew that this was fundamental to the program's growth and success. Other persons would need more in-depth knowledge to effectively perform their assigned tasks.

With that focus in mind, the greatest number of personnel was invited to attend the first session, a lecture highlighting ergonomics issues and principles. Subsequent sessions involved fewer people and were geared toward the technical analysis and evaluation of ergonomics issues and the generation of solutions.

Actual company scenarios, noted during the walk-through audit, were utilized during these roundtable sessions. Brainstorming generated multiple ideas for modification of existing situations. Paradigms fell away as participants explored potential solutions. Use of this problem-solving technique promoted high interest and involvement from audience participants. Attendees included engineers, technicians, operations safety committee members, safety trainers, managers/coaches/

coordinators, medical staff, and field personnel managers. Discussions were documented for future use.

The Committee Work Begins

With the successful workshop behind them, the committee turned its focus from discussion to making actual changes in the workplace. To pinpoint existing or potential problems, the committee created a survey and issued it to each operation. Its purpose was to gather safety or operational team perceptions of design problems and preventable injuries. Included on the survey were issues of bending/lifting, repetitive motion, temperature, light, posture, and workstation design that were applicable to all workers. In the manufacturing setting, issues related to tool grip, vibration, rotation, push/pull, and equipment were also examined.

The committee discussed the findings with each area or operation. It offered itself as a resource to the area safety committees as each worked toward implementation of sound ergonomics principles in the design of workstations and tasks.

NOTABLE CHANGES

Over the next few months, the committee worked to implement meaningful procedural changes that would help to effectuate positive changes in the ergonomics arena.

Medical Records and Tracking

The Medical Center redesigned the documentation forms it was using for persons seeking medical treatment. It instigated a new policy whereby each ergonomically related problem or injury is referred to the employee's department for remediation. Then special training or recommendations can be provided to personnel in areas of ergonomics incidence. The department manager and director also receive letters regarding ergonomics issues in their areas, including dollar amounts associated with a problem.

This procedure represents a more proactive approach since this form is issued *before* a situation becomes an incident. Such documentation provides statistics enabling tracking. Coupled with incident statistics maintained by the Safety Department, it can minimize the incidence of soft tissue injuries.

The Medical Department now conducts more frequent on-site reviews. By leaving the Medical Center to visit worksites, staff members are able to make recommendations and become a resource for prevention. Other committee members can be called in for consultation. Such attention demonstrates personal interest on the part of the company and enables problem solving by all the players involved.

Safety Department Reporting

Streamlining the reporting process inspired the Safety Department to spearhead design of a new Incident Report form and database. The form serves to record all injury, illness, fire, spill, and near-miss incidents. One portion functions as an incident investigation form for use by operational safety committees. Potential causal factors for numerous situations include ergonomically based issues.

Ongoing Education

Ongoing education was a need discerned by the committee. Videos were purchased and inservices were planned and made available to individual business operations. For example, back strains, sprains, and other injuries were prevalent in certain areas of the facility, so an in-house program was developed and presented to those employees in affected areas. It focused on proper lifting techniques and addressed investigation of back injuries.

The Ergonomics Office Model

Many persons expressed discomfort associated with their office workstations, so an office furnishing systems representative presented a program detailing current research related to ergonomics. Personnel in attendance were also updated on current regulatory issues, health implications associated with ergonomics problems, and methods of solving issues.

As a result of the brainstorming so crucial to group process, an office model was conceived for installation in the Medical Center. It will serve as a training center to educate employees and managers about office equipment. Samples of desks, chairs, adjustable keyboards, copy holders, foot rests, wrist supports, and other office equipment will be available. Personnel will be able to assess individual needs by experimentation with, and examination of, equipment available in house or in the marketplace. This will offer remedies to ergonomically related discomforts. It will also represent a cost savings opportunity since departmental staff wanting to modify existing equipment or purchase new workstation components can make wiser purchases.

A VISION FOR THE FUTURE

Pride in past accomplishments does not preclude committee members from continually looking to the future.

Long term, one goal is for the committee to become self-directed, rather than driven by the Safety or Medical Departments. Such a group would operate in a problem-solving mode, with members discussing issues as they arise, sharing findings, and offering testimonials. The committee could serve as a solution center. Membership would be rotated among personnel with similar background and knowledge who could contribute their expertise to the company's benefit and welfare.

The committee intends to remain flexible in response to individual or area needs and, through increased training and awareness, to promote consistency throughout the company. One concern to which the committee is attending is sensitive presentation and selection of materials, tools, and equipment. With so much available, attention must remain focused on meeting the particular and unique needs of SC Johnson Wax and its management system.

With that goal in mind, a SC Johnson Wax Ergonomics Manual is being produced as a reference guide for employees. The committee determined that its purpose should be to:

1. serve all workers as a ready reference
2. provide necessary information to the various operational sites
3. provide a framework for analyzing worksites
4. serve as documentation to OSHA of the SC Johnson Wax Ergonomics Program components

It should be written for use by on-line, self-directed work teams. Its contents should be targeted for both office and production areas, and should specify what to look for and why. It should include illustrations.

Having conferred with their consultant, the following topics were selected for inclusion as chapters in the manual:

1. Management, Committee, and Employee Involvement
2. Worksite Analysis, including review and analysis of injury records, ergonomics profiles, analysis hazards, and the company's annual survey
3. Hazard Prevention and Control, including guidelines for workstation design, tool and handle design, workstation methods, work practice controls, personal protective equipment, and administrative controls
4. Medical Management
5. Training and Education

Audits will continue. Worksite analysis will be utilized to monitor workstations, tool and equipment design, and work practice controls. Program success will be measured as statistics become available.

Much progress has been witnessed at SC Johnson Wax since the days of the safety/medical dilemma that stymied proactive solutions to long standing ergonomics issues. A vision of expanded awareness related to ergonomics factors, increased worker comfort, and greater productivity drives the committee forward. Having successfully passed through the embryonic stage of development, the SC Johnson Wax ergonomics program continues to grow toward maturity.

Chapter 3

The Role of the Safety Director in Injury Prevention

Thomas A. Broderick

HISTORICAL PERSPECTIVE

The relationship between the work environment and the health and safety of the worker has long been recognized. In antiquity, certain endeavors, such as mining and smelting of minerals, were known as the "dangerous trades." They were so dangerous, in fact, that slaves were enlisted to perform much of the work.

As early as the fourth century B.C., Hippocrates described the hazard of lead toxicity to miners and others who worked with it. The hazards of handling sulfur and zinc were noted by Pliny the Elder in Rome in the first century A.D. He described the first crude attempt at respiratory protection when workers in the "dusty trades" wore dried animal bladders over their faces. In the next century, Galen wrote in some detail on the pathology of lead poisoning.

Throughout the Middle Ages we find references to the deleterious effects of the work environment on the health of workers. Unfortunately, the life span was shortened by many nonoccupational factors such as tuberculosis, plagues, and other diseases.

King Rothari of Lombardi published a seventh-century compilation of statutes relating to the compensation of the victims of injuries sustained in altercations.

The Nordic King Canute described a system of compensating individuals for specific physical losses during the eleventh century. He set forth the principle of placing higher values on those body parts deemed more essential to normal function than others.

In 1473, Ulrich Ellenbog wrote the first publication devoted solely to occupational injury and disease. An Austrian physician, Ellenbog studied the toxic effects of heavy metals and the hazards faced by gold miners. Furthermore, he described hygiene and other preventive techniques.

Bernardo Ramazzini's *DeMorbis Artificum Diatribor* (The Diseases of Workers), published in 1700, was the first comprehensive treatise on occupational medicine. His description of the toxicological effects of materials found in the workplace and the ergonomic stressors inherent in many industrial processes earned him the moniker "father of industrial medicine." Unfortunately, many of the prevention strategies that he described were ignored for a few centuries. Perhaps his most significant contribution, however, is the addition of the question, "What is your occupation?" to the others posed by medical practitioners.

The Industrial Revolution wrought havoc on the health and safety of workers. Factories replaced the cottage industries. Animal and water power were replaced with steam and electricity. Children were recruited to work in the factories as cheap sources of labor. The 1900 census revealed that nearly two million children worked in the factories, mines, and quarries and at other hazardous workplaces. They were favored for many factory jobs because their size allowed them to crawl into machines to make adjustments and perform lubrication.

The human carnage that accompanied the technological advances near the turn of the century was so conspicuous that few doubted that reform would come. And come it did. Theodore Roosevelt recognized that the worker should not be required to sue his employer for loss of wages and payment of medical bills after receiving an occupational injury. In 1908, he saw the first workers' compensation act pass the legislature. Although it only covered federal employees, it served as a model for the states. By 1911, Wisconsin became the first state to develop a workers' compensation system. Two other states followed that year, and by 1920 42 states had workers' compensation systems.

Parallel with the enactment of workers' compensation statutes came the development of the modern occupational safety movement. The relationship between the cost of accidents, with resultant worker injuries and production loss, and the profitability of a corporation was not lost on astute industry magnates. The railroad and steel industries pioneered safety programs that contained many elements still vital to the success of current programs.

Seeking to standardize and share safety technologies, a group of business leaders, insurance executives, public officials, and early safety practitioners gathered in Milwaukee in 1912 for the first Cooperative Safety Congress. A second meeting was scheduled the following year in New York. It was at that meeting in 1913 that the National Council for Industrial Safety was created. The name was later changed to the National Safety Council to reflect an even broader mission involving accident prevention in the home, schools, public transportation, and so forth.

As delegates gathered at these early meetings, it became clear that a number of rules, regulations, standards, and codes had been developed by various corporations, engineering and manufacturing societies, and governmental departments. In 1920, the American Engineering Standards Committee developed the National Safety Code Program. This organization, known now as the American National Standards Institute (ANSI), designed a framework for the assembly of industrial safety standards. ANSI continues today to revise old standards and work through a consensus process to develop new ones. Each standard is developed by a committee made up of representatives from all facets of the industry that will be affected by the standard. Members bring their diverse talent and experience to the table so that the concerns of the trade associations, labor unions, suppliers, engineers, safety professionals, and others are reflected in the development process.

Most early safety regulation came at the state level. One of the first real attempts to legislate safety nationally came in 1936 with the Walsh-Healey Public Contracts Act. Any corporation awarded a federal contract of $10,000 or more was required by the act to maintain a safe and healthful workplace. Violation of applicable state health codes, factory inspection laws, or other safety regulations could result in "blacklisting" and loss of other federal contracts.

The Walsh-Healey Act and the Longshoremen's and Harbor Workers' Compensation Act were among the few examples of federal safety and health regulation until the 1960s. Then a flurry of social legislation saw the passage of several acts that regulated workplace safety. These included the Contract Workers and Safety Standards Act (Construction Safety Act), Federal Coal Mine Health and Safety Act, Service Contract Act, and several others.

In 1969, the nation experienced 14,200 workplace deaths (a rate of 55 per day) and 2,200,000 workplace injuries (about 8,500 a day). Clearly these losses provided the moral and economic impetus to construct more far-reaching safety legislation—legislation that would protect millions of working men and women not covered by the existing requirements that applied only to workers in special industries or those engaged in federal contract work. The Occupational Safety and Health Act of 1970 was introduced in an attempt to fill the gap in federally mandated safety regulation. After much discussion and debate, the act passed and was signed into law by President Nixon. By April 28, 1971, the act was fully effective and has remained essentially unchanged since that date.

Although legislative reform to strengthen the scope and requirements of the act has been discussed after each major industrial accident since its inception, several incidents in the late 1980s and early 1990s brought reform closer to reality. A construction accident in Bridgeport, Connecticut, claimed 28 workers in 1987. An explosion at a Texas petroleum refinery killed 23 workers. A North Carolina poultry-processing plant fire took 25 lives in 1991 as employees attempted to escape but found exit doors chained. These tragedies have underscored the urgency for Occupational Safety and Health Act (OSHA) reform among proponents of a stronger federal role in workplace safety and health.

Accompanying the historical mileposts in occupational safety have been a plethora of safety management philosophies. Dan Peterson summarizes these according to "eras" when each was in vogue, beginning with the inspection era in the early 1900s and continuing through the unsafe act and condition era, the industrial hygiene era, the noise era, the safety management era, the OSHA era, the accountability era, to today's human era, marked with an emphasis on human behavior.[1] Many corporate and plant safety programs cling to certain vestiges from early eras, while in other aspects evolving through emphasis shifts. An example is seen where a plant develops a Total Quality Management (TQM) program and begins to develop a behavior-based safety program, yet retains an accident investigation and reporting system clearly designed to place blame. This incongruity may not defeat the program but can hinder its development to full potential.

WORKPLACE CULTURE

To frame our discussion of how to develop a safety program, a review of social dynamics in the occupational setting will be helpful. Later in this chapter we will examine the relationship between the emphasis on quality management programs common in industry today and a state-of-the-art behavior-based safety approach. First, let's take a look inside of a typical U.S. workplace. It could be a paper mill or auto factory, retail store or construction site. We are going to examine the Sulo Snowshoe Company. This third-generation, family-owned snowshoe manufacturing company has 37 people working in three departments: sales, production, and maintenance.

The four-person sales force is made up of two men and two women. When they are seen in a group, they could actually pass for the U.S. Olympic Snowshoe Team. They are fit and trim and seem to radiate a wholesome "outdoorsy" glow. It's not surprising that when not on the road selling Sulo Snowshoes or attending winter sport trade shows, they get together socially to enjoy various outdoor sports.

The production department numbers 24 employees, including 3 in purchasing and 3 in shipping and receiving. The production staff still has a few old-timers who were hired by Sulo himself. The snowshoe makers are a proud lot. Several have had the trade handed down from fathers and grandfathers. Others have served long apprenticeships under the watchful eyes of the older craftsmen. To the uninitiated, the production floor can seem imposing. The workers have their own language since each part of the snowshoe has its own name. Specialized tools are used to craft each shoe. The production process is very physical: the frame is twisted into shape and the leather pulled through it, forming a sinewy cobweb. The binder installers often chide the "framer" for not being able to keep up. True, there are machines that could make the framing process somewhat easier and perhaps a bit faster, but such technology has not been favored at Sulo's. Handcrafting is the order of the day.

After work, the framers and binders often stop at the Sportsman to relax, toss back a cold one, and talk about the day's events. Rudy, one of the old-timers, captures the imagination of the younger workers with his stories about how production swelled during the Korean Conflict to meet the demands of the troops up around the 38th Parallel. "Cripes, we had almost three hunnert guys on production back den," he tells them. Then, at the urging of one of the binders, he recounts for the umpteenth time how he lost his thumb "slashing hide to string some shoes." Rudy proudly displays the nub (which was once his right thumb) like a badge of courage.

The maintenance department keeps Sulo's snowshoe factory humming along. Bill is the maintenance supervisor. He and his crew seem to have a sixth sense and have often shown up to fix a machine within minutes of a malfunction—even before production squawks. And squawk they do—enough, in fact, so that a mechanic was recently heard to utter something about those "stuck-up primadonnas in production." After all, without the maintenance department, the snowshoe factory would have fallen apart long ago.

Of course, Sulo Snowshoe Company is fictitious. Did you notice, however, that the description of the relationships seemed familiar? Every workplace is a societal microcosm, a culture with a unique identity. This culture has two major ingredients: ideologies and cultural forms.

At Sulo Snowshoe factory, the production staff had a strongly held belief that snowshoes should be crafted with the same skill and care with which they have been for centuries. This ideology was shared by the other departments, and all found it a collective source of pride and identity. Think about workplaces you are familiar with. Are there ideologies that have evolved among the workers?

Cultural forms are the second intrinsic feature of a culture. These may take the form of slang or jargon, myths and legends, taboos and gestures. A classic example of a cultural form is seen in a ceremony performed by high steel ironworkers who erect the steel skeleton of a building. When the last piece of steel is placed on the highest point of a building, a "topping out" ceremony is conducted. All ironworkers on the construction project gather to watch and cheer as the final piece, festooned with an American flag and an evergreen tree, is lifted into place and secured by two ironworker "connectors." This custom originated when early Nordic home builders would place a small evergreen on the final log that was hoisted into place to complete a lodge (see Figure 3–1).

Did we see any of these cultural forms at Sulo Snowshoe Company? The stories about the factory's effort to supply snowshoes during the Korean Conflict and Rudy losing his thumb on the job are both examples of cultural forms. The jargon used by the craftsmen to describe tools, materials, and procedures is another. Think about real workplace environments you are familiar with. Can you think of cultural forms specific to certain industries or professions?

To make things even more interesting and complex, occupational subcultures exist within the culture of many workplaces. The employees at Sulo Snowshoe

Figure 3–1 Ironworkers Perform a "Topping Out" Ceremony

Company collectively held a cultural identity. Whether in sales, production, or maintenance, all employees had a common loyalty and bond. This ethnocentric behavior is perhaps the most common occupational cultural ideology. However, we also observed that segments of Sulo's workforce displayed an "us versus them" attitude. The maintenance workers seemed to feel that the production staff had an aura of superiority. Furthermore, the maintenance workers felt that they were the ones who held the place together. We saw examples of this subculture pattern emerge with each of the groups at Sulo, where production tradesworkers possessed their own slang, practices, and stories. Think about your own work environment. Are you a member of an occupational subculture?

So what does workplace culture have to do with safety and health? It has plenty to do with it. Unfortunately, safety and health professionals often overlook the occupational culture in their zeal to implement a new program or change an existing one. Medical professionals sometimes fail to consider the occupational culture in treating occupational injuries and attempting to reintegrate the employee into the work environment.

Let's first examine the relationship between the safety and health practitioner and workplace cultures/subcultures. Many well-intentioned young safety managers feel compelled, upon taking a new position, to immediately set about implementing new programs and procedures at the plant and perhaps scrapping much of the existing program. Soon after, a fair number of them consider a career change. Why? Changing the safety program shakes certain cultural ideologies deeply rooted in the organizational structure. From the CEO to the worker on the line, concerns may be expressed such as "We can't do that, we'll go broke!" or "It's impossible to do the job that way, it's been done this way for 80 years!"

Certain cultural forms within an occupational subculture may actually be unnecessarily hazardous. To accept, in fact embrace, certain dangerous tasks or procedures serves two purposes for the subculture. First, the appearance of risk glorifies the status of those in the subculture whose occupation has adopted the "risky ritual." Second, it discourages those from outside of the subculture from attempting to perform tasks claimed by the subculture as "our work." High steel ironworkers present a classic example of an occupational subculture that displays both behaviors. It is possible for a worker to perform steel erection at any height while being secured from falling. Ironworkers seldom do perform the work while secured, however. This risk-taking behavior sets them apart from other trades and provides a measure of esteem. It also discourages those who have not gone through the rites of passage in the ironworking trade from risking a fatal fall. It is interesting to note that these cultural forms can be very influential in establishing relationships with external entities. That ironworkers should be able to work at heights without protection is acceptable practice within the construction industry, and that practice has been validated by OSHA. Whereas OSHA limits the distance most workers may fall in the workplace to less than 6 feet, ironworkers performing structural steel erection may fall up to 25 feet without protection.

Does this mean that we should acquiesce and allow these cultural factors to dictate? Absolutely not. But developing a keen ability to identify and understand occupational cultures/subcultures can give the practitioner a significant advantage when attempting to implement change. It should be noted that some hard-line behaviorists take exception to paying much attention to predisposing cultural factors and favor effecting immediate behavioral changes, then allowing worker ideologies to eventually adapt to the "new reality." While this approach has some philosophical merit, thrusting externally driven change on a population may wreak such havoc on the stability of the workplace that even the most enlightened manager may back down and return to the original system.

The occupational medical practitioner can deliver more effective care by understanding the cultural factors surrounding the employee before injury, during recovery, and in the return-to-work process. The most successful approach to work injury management draws on a strong patient-provider trust relationship supported by a team approach wherein the influence of the worker's occupational subculture becomes a positive force in hastening the recovery process and facilitating early and successful work reentry.

BUILDING A SAFETY PROGRAM

What is a successful safety program? One is tempted to respond, "one that allows an organization to operate for a long time without anyone getting hurt." This seems like a reasonable answer and probably the one most frequently given as a response to the question. In fact, most safety programs are built around that premise. The degree of the program's success is inversely proportionate to the number of accidents experienced. If you investigate accidents and identify the causes, you can make the necessary changes to obviate a recurrence.

Unfortunately, this system is reactive and measures failure. Accidents must present themselves so the workplace can learn from them and ultimately effect intervention strategies. This is not to say that a safety program should not do these things: accident investigation and thorough follow-up are valid measures. They should not, however, drive the program.

Observing behaviors, selecting safe ones, and deleting unsafe ones is a superior approach. We are then building a program based on a proactive process. All tasks in the workplace can be analyzed and broken down to a series of actions. If each of the actions in a series has a nil possibility for resulting in error, the task may be performed safely. By determining, then observing and quantifying safe behaviors, we can reduce risk.

The word *accident* is one that has nearly disappeared from the parlance of safety experts. It connotes a freak occurrence, one that could be neither predicted nor prevented. Very few incidents in the occupational setting fit that criterion. As seen in Figure 3–2, a particular unsafe behavior may be repeated many times before it results in a near-miss occurrence. Similarly, the near-miss is a predictor that statistically will occur repeatedly prior to resulting in an accident. As we remove the statistical basis for predicting accidents by removing the unsafe behaviors, we enhance the likelihood of having an injury-free workplace.

Sounds great, but how do we get there?

A safety program is not something you can buy and have. It is an integral part of the fabric of the workplace. There is no secret formula or one-size-fits-all solution. As we have seen, occupational cultures and subcultures are powerful. The ideologies and cultural forms existing in a particular workplace should be congruent with the goals and objectives set forth in the safety program. Rewarding safe behavior and disregarding unsafe behavior can be done successfully in any occupational environment. This eventually will create the necessary adjustment in ideologies and cultural forms to embrace an accident-free environment. It takes hard work and patience, but the end result will be a permanent and positive shift toward a safety culture.

Some occupational settings, such as construction, make behavior-based safety programming more difficult. In that particular field we often see workers who

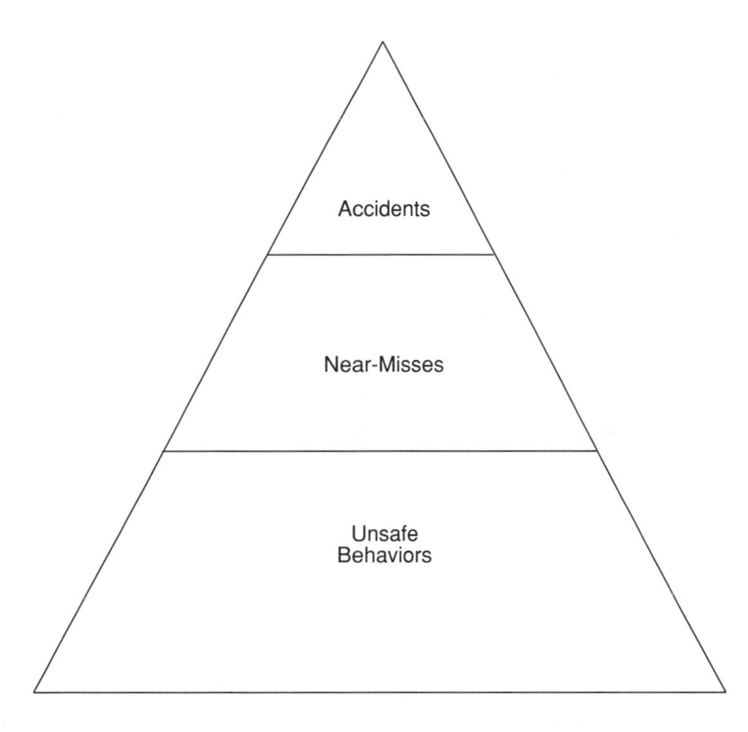

Figure 3–2 Model for Relation of Accident Incidence to Incidence of Near-Miss and Unsafe Behaviors

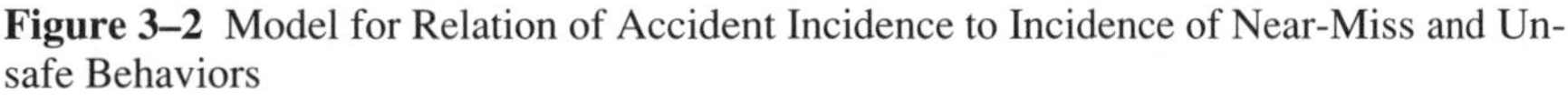

possess varying skill and experience levels meet for a short period of time to perform high-hazard tasks. Obviously, the dynamic nature of this type of work presents some obstacles to behavior modification techniques, which are better suited to the more stable environment of a factory or process industry.

Given this philosophical framework, let's examine some of thc componcnts of a modern safety program. It must be emphasized that a successful safety program may not contain all of these elements and that not all safety programs that contain all of these elements will be accident-free.

Management Commitment

If the entity that ultimately makes the decisions about the operation of a facility is totally and unequivocally committed to excellence in safety, the battle is nearly won. This commitment may (1) reflect practical concerns for reducing losses, (2) reflect a moral obligation to provide a safe work environment, or (3) reflect each to varying degrees. Most programs are driven by the latter.

Management's commitment to providing a safe and healthy work environment should be outlined in a clearly worded policy statement. This statement provides a foundation for the safety program by documenting the will of the highest authority in the organizational structure.

The structure of the management hierarchy should allow for line accountability in both directions. Executive management should accept responsibility for safety and communicate that sentiment often. Lower management echelons should expect to quickly access upper management with safety and health issues. Each level of the line management hierarchy must have a clearly defined role in the safety process.

The safety manager position is not normally, nor should it ever be, a line function. This is a staff role. The position of this job on the organizational chart does speak to the importance that executive management places on it. If the safety director reports to the president of the corporation or to the plant manager, the staff safety function is given access to all segments of the respective organization. This is a departure from the archaic system whereby the safety department is tucked under the personnel department's wing.

The organization should have a written safety plan/program. This does not mean a set of rigidly enforced "dos" and "don'ts." It is a living document that outlines the various components of the safety program. Ideally, safe operating procedures will be incorporated into the written procedures that should be in place for each task or position in the workplace. The safety program should contain a mechanism for reviewing these procedures, revising them if necessary, and developing new ones as technologies change or new ones are introduced. Other safety program elements may include specialized procedures such as a respiratory protection plan, an emergency action plan, inspection criteria, safety training requirements, and all other pragmatic features. To be useful and used, the written program should be lean on verbiage but comprehensive in scope. Keep it simple.

Finally, management commitment should extend throughout the organization. A good method for accomplishing this is through the establishment of a safety committee(s). In this setting workers should be on even ground with management. Such committees provide a window into some of the cultural/subcultural compartments that exist in the workplace, including those inhabited by managers and supervisors.

Management Accountability

The real trick to making a safety program keep on track once the commitment from top management has been secured is to keep all echelons of the management hierarchy interested and involved. Posters and contests may play a minor role in this effort, but they cannot ensure that the push for production will not overshadow safety procedures. Each manager and supervisor must have clearly defined safety responsibilities, with a system in place for evaluating this aspect of individual performance.

In general, accountability may be categorized as positive or punitive. Each type plays an important role in keeping a sharp edge on the safety program.

Managers may be expected to discuss safety in each sales and production meeting, conduct safety tours of their respective areas, give positive strokes to supervisors for safety-related activities, and create motivational activities. Similarly, supervisors are expected to conduct safety meetings, perform inspections and investigations, and give one-on-one positive strokes to acknowledge safe behavior displayed by individual employees. The safety specialist should examine what contributions each level of the management system can make and create a system that allows them to be quantified. The information will become a part of employees' personnel records and be considered a significant part of the performance appraisal process. If the top salary increases and bonuses are available only to those managers and supervisors who have a high safety score as well as a good production record, a powerful message affirming management's safety commitment will permeate the organization.

The second approach to accountability quantifies negative behaviors and their outcomes, often using accident statistics and dollar losses as yardsticks. Charging losses to individual divisions or departments is a tactic that sends a powerful message. A belief held by many managers is that accident costs are a cost of production for which the company has insurance. The cost of the insurance is usually not related back through the accounting process as a part of production overhead. By charging back accident costs to the appropriate business unit, managers are able to see the positive effects that good safety performance will have, especially in the loss-sensitive world of workers' compensation coverage.

Fixing accountability for safety within the organization is critical to the success of a safety program. It imparts ownership in the program. It cements the principle that safe production is the desired state rather than production with safety as an afterthought.

Personnel Policies

Demonstrating a company's commitment to safety should begin even before an employee is hired. The occupational ecology is most conducive to safety and health when jobs fit the workers both physically and psychologically. Each employer should carefully determine what criteria should be used to screen, interview, and hire new employees. This may be done in concert with medical professionals, attorneys, personnel consultants, and so on. It is important to preserve the dignity of prospective employees. Sometimes we think in terms of employers favoring people with jobs rather than appreciating the fact that prospective employees are offering us a significant share of their adult lives to give service to the employer.

The Americans with Disabilities Act has modified the way many firms conduct preemployment screenings. It forces employers to recognize that a worker can find a good psychological fit with a good job if the employer adapts the process to create a good physical fit.

Some developments have been made in the area of profiling prospective or new employees for attitudes and behaviors that are associated with risk taking. A correlation is drawn between the results of this analysis and subsequent accident frequency rates. Caution must be taken when using such a tool, however. A person may have many positive and compensating qualities. If an indication of risk-taking attitudes is the person's outstanding negative quality, a little coaching from supervisors and settling into a culture or subculture where a positive safety attitude is an important ideology will probably lead him or her to embrace that attitude also.

Training

The OSHA citation most frequently issued after completion of a workplace fatality investigation relates to the lack or inadequacy of the employer's training program. Employees must know how to do their jobs and do them safely. To ensure that this happens, each job must be thoroughly analyzed and specific safe operating procedures developed. Employees must be thoroughly familiar with these procedures from both the conceptual and the physical perspectives.

Training should be considered a process rather than an occurrence. Beginning with a thorough orientation upon hiring, proceeding with job- and task-specific training, hazardous operation training, specialized technical training, quality training, and so forth, the process will continue throughout each employee's work life. Each time an employee performs a new task, specific skill and safety instruction must be provided. Supervisors should closely support and coach employees who are adjusting to a new task and worker-machine fit.

Of particular concern is the area of supervisor and management training. Inject a significant component of safety emphasis into teaching employees about the responsibilities they assume while progressing within the supervisory/management hierarchy. After all, if they are to be held accountable for workplace safety, they must have the proper tools and training specific to each new duty.

Audits and Inspections

The first line of defense from accidents and illness for many employers is in the purchasing process. Hopefully, those who have authority to order supplies and equipment will do so with employee safety and health in mind. Some firms route requisitions through their safety department. This allows the safety department to advise the requisitioner and purchasing department of any concerns. Whether relating to hazardous materials, machine guarding, specific training requirements, personal protective equipment, or a host of other potential concerns, the safety department has the ability to provide input prior to the purchase.

To facilitate this process, the safety department should develop an audit form that may be used to scrutinize all new materials and equipment purchased. Wherever possible, safer and healthier products should be specified.

Each workplace should undergo regular and thorough inspections. Often the safety director overlooks the fact that both production and maintenance people do this inspection continually. Production people look for quality defects and for adjustments that will enhance productivity. Maintenance people look for symptoms of problems so they may be prevented or repairs scheduled to minimize downtime. Make safety considerations an integral part of these inspections.

Provide a framework for auditing for safe behaviors as well as unsafe ones. Many systems have been developed for reviewing potential hazards and safety procedures, including safety sampling, critical incident recall, fault tree analysis, and job safety observations and analysis. In addition to these audits with generic application, many employers have developed their own inspection and audit systems specific to particular operations.

Participation in the audit/inspection process should be a part of each supervisor's and manager's job description. His or her actual participation level should be considered in the performance review as an indicator of accountability.

The audit function should not stop here, though. Workers in teams or as individuals should be trained to observe and quantify safe and unsafe behaviors.

Record Keeping and Performance Analysis

As stated earlier, a safety program based on the analysis of accident data is one that accepts the axiom, "accidents will happen." The safety program is therefore reactive and tends to lose focus between accidents. This has been the source of frustration for many safety directors who see their program reach a plateau but never really seem to get beyond it. This speaks to the subject of basing the safety program on analyzing positive performance data rather than relying on injuries. Training accomplishments, inspections and audits, positive strokes for safe behaviors, safety innovations, and safety meetings are all positive, quantifiable contributions to the safety programs. We would hope that using these as evaluative criteria will give us a larger statistical base than an analysis of injury records.

Keeping injury and illness data is an important function, however. Management needs a good system for the collection and analysis of injury data, including associated direct and indirect costs.

A number of firms have developed computer programs for the compilation and analysis of accident and injury data. Others purchase commercially available software for this purpose. Some programs generate OSHA-required reports such as the OSHA 101–Supplemental Injury Report and the OSHA 200–Log and Summary of Occupational Injuries and Illnesses. Some perform statistical analysis of data, and many spot trends or "hot spots." Many safety professionals depend on paper records and a pocket calculator to perform the same analysis.

Insurance carriers possess a treasure trove of data that many safety professionals never see. The data may never reach the insured except for a periodic loss run that the accounting/financial people file in a drawer. The safety director must have a thorough working knowledge of the employer's insurance portfolio and maintain a dialogue with both claims and loss control representatives assigned to the account.

Loss control departments use claims data to focus their evaluation of each insured's risks. These reports and analyses may well be available to the insured's safety director. The insurance carrier's loss control department may have a number of other helpful tools such as training programs and materials, and safety signs and posters. Loss control consultants typically make periodic tours of their insured's facilities and create a report that is submitted back to the facility manager. A copy of this report is also provided to the carrier's underwriting department and is used as an evaluative tool to assess the risk at the subject facility. Consistently negative reports may effect an increase in premium or result in policy cancellation or failure to renew.

Claims departments are generally less generous with information, which is why it is a good idea for the safety director to forge a good relationship with the claims department. A growing number of companies have worked with their insurance carrier's claims department to develop systems to charge workers' compensation claims to the operating unit or department where the work injuries and/or illnesses have occurred. This can usually be done by proper coding of the first report of injury.

Quality and Safety

Those readers familiar with the work of W. Edwards Deming, Philip Crosby, and J.M. Juran in the area of quality improvement through Statistical Process Control (SPC) will have found much of the earlier discussion about behavior-based safety programs familiar. Workplaces that have adopted total quality management (TQM) programs find that the same technology applied to accident prevention is very effective.

For decades, American factories have relied on inspection of the finished product as the trigger for adjustment of the production process to obviate defects. This system is predicated on a system designed to produce errors. Quality gurus advocate building quality emphasis and assessment into each step of the production process, beginning with product design and the specification and procurement of raw materials. Variables that affect quality are identified throughout the process. Each step is then performed in the optimal fashion. Attention to upstream detail prevents downstream defects.

TQM typically involves participative groups that analyze and improve the processes. These quality improvement "teams" and steering committees empower employees to constantly scrutinize the process for areas that can be improved.

These features of the TQM system should sound familiar. When talking about developing a behavior-based safety approach, we were really applying the TQM principles to safety in strategies where the upstream behavior was the focal point, rather than the downstream effect—accidents. We also discussed empowering employees through the formation of safety committees to analyze the work process for safe and unsafe behaviors, then observe work in progress to quantify these. Through communication and training, the unsafe behaviors were replaced with work methods that rewarded employees with safer and often easier jobs and methods.

Readers interested in improving workplace safety and health in their own plant or those with which they work should first inventory the culture as it exists. Is there a TQM system in place now? Is the safety program based on "putting out fires" and the reactive approach based on injury data? Is a TQM system working hand in glove with a behavior-based safety program? This writer, of course, advocates the latter.

If neither a TQM program nor a behavior-based safety program is in place, an opportunity exists to develop both simultaneously. Many excellent resources are available to accomplish this task, and representative examples of these resources are listed in the section of this chapter titled "Additional Resources." As with successful implementation of any cultural change, these programs must be carefully developed and introduced in a fashion that solicits a high degree of involvement from all segments of the workplace supported by the strong and visible executive participation.

Many U.S. firms now have TQM programs but have antiquated health and safety programs. These organizations should study the concept of behavior-based safety programming and develop a plan for adopting this methodology. The cultural adaptation effected through the adoption of the TQM program will make this transition easier from both the practical and the social standpoints.

It must be stressed that each workplace is unique. There are no "one-size-fits-all" programs that may be taken off the shelf and successfully installed. Management must perceive that the desired future state is consistent with the overall business plan and fiscal goals of the organization. Workers must feel that the goal of the changes is to remove the element of risk from their relationship with the other workers and machinery in the workplace.

SAFETY DIRECTOR ROLE

The last decade has seen a significant increase in the scope of the safety and health professional's role in the corporate fabric. This person's responsibilities may include environmental compliance, security, fire prevention, wellness programs, medical and occupational injury management programs, and other duties. The increased scope of responsibilities carries a commensurately expanded requirement for education in each of these areas.

A number of universities have developed curricula to address this challenge. A growing number of safety, health, and environmental degree programs are available at both the undergraduate and graduate level. The American Society of Safety Engineers has identified over 100 such programs (see Additional Resources section for more information). In addition to university-based educational programs, professional and trade associations, nonprofit organizations, consultants, and labor organizations have also developed safety, health, and environmental education and training programs.

Safety practitioners may be kept abreast of new information and technologies. A number of periodicals that address health and safety issues are in circulation. A list of some of this resource literature is found in the Additional Resources section.

The safety director will probably be responsible for the move toward a behavior-based safety program. That person will work closely with executive management and the quality assurance director to ensure that an error- and accident-free culture exists. The programs must be treated as ongoing processes that will be continually evaluated and tweaked. The process must be dynamic and fresh. The safety director will use observations and communication skills to accomplish that end. A review of the works of Dan Peterson, the National Safety Council, and Krause, Hidley, and Hobson will provide the safety director with invaluable information on developing and maintaining a state-of-the art safety and health program.[2]

The safety director will also be required to keep abreast of new safety, health, and environmental standards and regulations. This task becomes more difficult as regulations increase in complexity and scope, and more essential as penalties for noncompliance become more costly. The use of criminal prosecutions for safety and environmental crimes has increased and will become more extensive. The threat of jail is real and provides good rationale for the development and maintenance of a safety-first-and-foremost corporate culture.

Although injury and illness prevention is the safety director's primary function, work injury management has become a critical secondary responsibility. Although some safety departments are still driven by the medical or personnel director, this practice is a throwback that reflects reactive rather than proactive safety programming.

Postinjury concern for workers should at least equal preinjury concern. Safety programs driven by analysis of accident data often build incentive/reward systems based on employees working for predetermined periods without lost-time injuries. Workers in this type of environment often receive a "double-whammy" when injured on the job. The initial and most obvious blow is the injury itself. The secondary and perhaps most medically and socially damaging blow is the postinjury treatment that this individual receives from his or her own employer and coworkers. The worker's injury may have caused a crew or even the whole plant to lose out on a prize or award. Accident investigations may be conducted to fix blame. This type of environment will make it difficult and unattractive for an employee to return to work.

The safety director should take an active role in developing a proactive safety program and an equally proactive return-to-work program. The latter begins by selecting the best medical care available for injured employees. The care providers will become partners on a team that will treat injured workers with a high degree of respect and concern.

Communication will be stressed as the safety director brings the work reentry team together. Injured workers need to know what to expect in terms of medical care and payment of wages if the injury involves lost time. Explaining to the worker the treatment process and workers' compensation benefits will set the tone for a positive relationship between the employee and the work reentry team.

Communication is also vital between all members of the team. Who is on the team? The safety director or designee will likely be the key coordinator. That person will involve the injured person's supervisor and coworkers, plant medical staff, outside providers such as physicians, nurses and therapists, insurance personnel, and sometimes family members of the injured worker.

A return-to-work plan with well-defined roles and goals will be helpful to facilitate a successful recovery and return to work. All members of the team should have a copy of the plan, including the employee. Active involvement of the injured worker's supervisor and department head will ensure that suitable work is ready for the employee as soon as a release is possible. By working as a team, the supervisor will be able to collaborate with medical personnel. Medical decisions are then based on the actual conditions that the employee will face in the workplace. The supervisor will also understand the employee's physical limitations and recovery process.

The safety director will ensure this process functions smoothly and communications remain fluid. This may involve periodic telephone conference calls or meetings. Using this approach will keep the employee informed and comfortable with the knowledge that the employer is concerned and working in his or her best interest to effect a successful return to work. Employers should remember that if they don't provide the worker with prompt and competent medical care and information about workers' compensation benefits, there are plaintiff attorneys who will be happy to help the injured employee.

CONCLUSION

Proactive safety and health programs focus on identifying and measuring safe behaviors in the workplace. Occupational cultures and subcultures exist in each workplace and exert a strong influence on worker behavior. Behavior-based safety programming ultimately shifts the occupational culture to engender a safety culture, often replacing unsafe behaviors with those conducive to employee well-being.

Behavior-based safety programming and total quality management (TQM) share many key elements. They fit nicely to create a defect-free, injury-free environment.

The safety director plays a key role in designing and supporting the behavior-based safety program. This staff position is also responsible for technical assistance to live management.

ADDITIONAL RESOURCES

Organizations

American Society of Safety Engineers
800 East Oakton Street
Des Plaines, IL 60018
(708) 692-4121

American Industrial Hygiene Association
2700 Prosperity Avenue, Suite #250
Fairfax, VA 22031
(703) 849-8888

National Fire Protection Association
Batterymarch Park
Quincy, MA 02269
(800) 344-3555

National Safety Council
1121 Spring Lake Drive
Itasca, IL 60143
(708) 775-2282

National Safety Management Society
12 Pickens Lane
Weaverville, NC 28787
(800) 321-2910
OSHA Training Institute
1555 Times Drive
Des Plaines, IL 60018
(708) 297-4810

Periodicals

Occupational Hazards
Penton Publishing Inc.
1100 Superior Avenue
Cleveland, OH 44114
(216) 696-7000
Occupational Health and Safety
P.O. Box 2573
Waco, TX 76702
(817) 776-9000
Occupational Health and Safety Canada
1450 Don Mills Road
Don Mills, Ontario M3B 2X7
(416) 445-6641
Professional Safety
American Society of Safety Engineers
1800 E. Oakton Street
Des Plaines, IL 60018
(708) 692-4121

NOTES

1. D. Peterson, *Safety Management: A Human Approach*, 2nd Ed. (Goshen, N.Y.: Aloray, Inc., 1988).
2. Peterson, *Safety Management;* National Safety Council, *Accident Prevention Manual for Industrial Operations*, 9th Ed. (Itasca, Ill., 1988); National Safety Council, *Fundamentals of Industrial Hygiene,* 3rd Ed, (Itasca, Ill., 1988); National Safety Council, *Introduction to Occupational Health and Safety* (Itasca, Ill., 1986); T.R. Kraus et al., *The Behavior-Based Safety Process* (New York: Van Nostrand Reinhold, 1990).

Chapter 4

Job Analysis

Susan J. Isernhagen

Job analysis is a process by which an individual evaluates a job (and its specific components) in order to make a definitive statement about that job—its risks, its requirements, and its productivity—in order to facilitate administrative planning. A job analysis is typically done (1) internally by employees such as safety directors, risk managers, occupational health nurses, and ergonomists (when they are available in industry), and/or (2) externally through the use of a consultant.

The type of outside consultant used depends upon the type of job analysis required. For example, a professional with an ergonomics background may do the production-related ergonomic job analysis. A physical therapist, occupational therapist, or physician may perform job analysis in relationship to injured workers or prevention of injury. An engineer may evaluate the plant layout and production design. A safety consultant may do job analysis in relationship to Occupational Safety and Health Administration (OSHA) and governmental standards. An industrial hygienist may analyze air and environmental and sensory factors. A vocational consultant may do job analysis in order to define vocational aspects of work for hiring or internal placement. Job analysis is also used in order to develop functional job descriptions that lead to prework screening.

Job analyses may be classified as either quantitative or qualitative. The quantitative job analysis involves measurement of such parameters as weights, distances, forces, repetitions, speed, productivity, and timing (MTM). It can be done by a trained technician who has the skills and tools available to do the measurements. Quantitative job analysis is often done in order to measure productivity or risk. Numbers can be put into formulas and formulas can indicate places for change of work methods. Quantitative measurement is very important as a basis for job analysis.

Qualitative job analysis, on the other hand, requires a professional with evaluation and synthesis skills. This professional will need a strong background in his or her field as well as experience, specific objectives, and a specific type of research information on which to base recommendations. The analysis requires professional judgment, which first utilizes measurements from quantitative analysis but goes beyond these by imparting logic, order, evaluation, or recommendations to the data analyzed. An industry will always desire qualitative information in addition to quantitative information, since the multifaceted needs of industry require an analysis of all the factors involved.

Job analysis has its basis in many professions. Exhibit 4–1 shows some of the areas addressed by the science of core job analysis specialists. The coordination of the specialties is an ongoing process in ergonomic job analysis. Even if a specialist utilizes primarily the scientific approach of his or her own professional group, it is important to recognize there are other ways to analyze. The most comprehensive analyst will utilize as many specialties as possible.

ASPECTS OF JOB ANALYSIS

Job analysts utilize a variety of options and structures. They include

- Ergonomics from both a safety and injury prevention point of view
- Repetitive strain guidelines: OSHA, national, and state regulations
- Manual material-handling procedures and policies
- Time measurement methods and formulas
- Americans with Disabilities Act and other nondiscrimination practices
- Risk management guidelines

These blend with the sciences in Exhibit 4–1.

SETTING UP A JOB ANALYSIS

The following procedure facilitates both a logical process and documentation. Both employer and consultant can utilize these steps and plans as a letter of agreement. Following all three steps will also maximize the usefulness of the analysis.

Step 1: Goals

The purpose for the specific job analysis should be as well defined as possible before the process begins. This should include a statement of desired outcome defined by goals. The employer usually initiates the process with a request. The internal or external consultant reiterates this in writing, setting specific goals. The employee (or payor) verifies these as accurate.

Exhibit 4–1 Approaches to Job Analysis

A. Production Approach
   1. Core routines
   2. Episodic work routines
   3. Seasonal work pressure
   4. Consumer-dependent pressures
   5. Task rotation
   6. Job rotation
B. Medical Approach
   1. Risk and stress factors
   2. Safety and healthful work/work practices
   3. Analysis of injury prevalence and seriousness
   4. Human function parameters
C. Engineering Approach
   1. Mathematical-formula oriented
   2. Logical
   3. Worksite design emphasis
D. Biomechanical Approach
   1. Both medical and engineering aspects in combination
   2. Duration/endurance/fatigue
   3. Positional tolerance
   4. Physical demands
E. Psychological Approach
   1. Rating of perceived effort, psychophysical demands
   2. Comparison of physical stress to mental stress
   3. Job satisfaction
   4. Work culture
F. Physiological
   1. Oxygen consumption/aerobic demands
   2. Energy expenditure
   3. Sensory environmental impact (cold, noise)
G. Functional Approach
   1. Functional components/factors of work
   2. Coordination
   3. Kinesiological evaluation of work motions
   4. Comparison of functional abilities of workers with physical demands of job

Step 2: Plan

The plan for doing the job analysis should also be established in writing prior to the implementation. If an outside consultant is used, the plan should also include information such as hours or days of analysis, costs, and the type of report to be generated. The plan should be approved in writing by the employer so that all parties understand their responsibilities and authorities during the job analysis.

It is helpful to acknowledge the components of job accomplishment. The first is the worker's being present to do the job. The second comprises the actions, motions, and functional tasks of accomplishing the job. The third comprises the material, setting, and tools that allow the worker to accomplish the work. In job analysis all three must be documented and considered. Exhibit 4–2 details the most important parts of worker, work, and worksite.

Exhibit 4–2 The Three Components of Work Analysis

I. Worker
- A. Gender variables
- B. Age variables
    1. Musculoskeletal system
    2. Cardio respiratory system
    3. Neurological-sensory system
- C. Research on combination of age and gender
- D. Anthropometric data
- E. Skill level
- F. Preexisting conditions
- G. Boeing studies
- H. Conclusions

II. Work
- A. Forces
- B. Angles
- C. Speech
- D. Repetitions
- E. Rest breaks
- F. Rodgers Maximum Voluntary Contraction Chart
- G. Borg Scale
- H. Stress level
- I. Boredom level
- J. Conclusions

III. Worksite
- A. Workstation
    1. Work area
    2. Seating
    3. Stand/kneel
- B. Objects of work
    1. Materials to be handled
    2. Objects to be used/manipulated
    3. Controls
    4. Tools
    5. People
- C. Environment
    1. Lighting
    2. Temperature
    3. Noise
    4. Other

Step 3: Evaluation

The actual evaluation component will be a function of the agreement discussed in Step 2. If the consultant has carte blanche for the analysis, additional time and money can be spent during the process. In most cases, however, there will be an established time period and fee structure. Therefore the evaluation should be kept within the framework of the agreement.

The evaluation will always be done on site. Methods that facilitate accuracy are

- The employees and supervisors should participate in a short introduction to the process of job analysis and the reason for the specific analysis prior to its initiation.
- There should be an agreement between employer, supervisor, and employees that the consultant can interact with the employees during their work. This may at times interfere slightly in productivity, but the ability for the consultant to interface with the workers is critical.
- There may be a need for the consultant to actually "try" the job being analyzed. This should also be discussed ahead of time and permission granted. Any productivity or safety criteria, however, should be respected if the consultant is to learn portions of the job by actually doing them.
- If pictures (slides, pictures, or videotape) are to be used during the evaluation, this should be preannounced. It should also be understood that the workers will go about their usual work unless specifically asked to stop so that a specific area can be filmed. The videotape or pictures should reflect the actual work and should be as clear as possible. In most cases, the pictures or film will be used only internally, and therefore no releases should be necessary. It is courteous, however, to ask the worker who will be filmed to grant permission prior to the filming or picture taking. If the film or picture will be used outside the industry for any reason, releases should be obtained not only from the employees but also from the employer.
- Measurements should be taken during the job analysis. If the actual work makes the measurement difficult, the consultant should come back when the work has been shut down to get completely accurate measurements. Some factors that should be analyzed and measured are listed in Exhibit 4–3.
- Additional information may be gained from the employees and supervisors in the form of questionnaires. These questionnaires can be given either before or after the evaluation, depending on which would be more feasible. This gives credibility and depth to the actual consultant evaluation.
- Recording and measurement tools are listed in Exhibit 4–4.

Exhibit 4–3 Factors of Job Analysis

Standing	Kneeling
Walking	Crouching
Sitting	Crawling
Reclining	Reaching
Lifting	Handling
Carrying	Fingering
Pushing	Feeling
Pulling	Talking
Climbing	Hearing
Maneuvering	Seeing
Balancing	Tasting
Bending	Smelling
Stooping	

Exhibit 4–4 Recording/Measurement Tools

1. VIDEO CAMERA: Choose the type best suited for analysis. The simplest model can be used for recording and playback. More sophisticated technology allows freeze-frame, transfer to computer pictures, and computer analysis options.
2. TAPE MEASURE: Both cloth and metal types have their own advantages; consider both.
3. SCALE: Weighing objects is preferred to estimating.
4. GONIOMETERS: Measure relationships/angles of work and force.
5. INCLINOMETERS: Very accurate for single plane, gravity changed positions.
6. STOP WATCHES: Both single measurement and sequence timing.
7. DYNAMOMETERS—PUSH/PULL, GRIP PINCH: For both direct measurement and psychophysical replication.
8. THERMOMETERS: For both air and direct contact.
9. CLIPBOARD: Note holder/writing surface.
10. TAPE RECORDER: Dictaphone type.
11. STILL CAMERA: Slides and pictures are both helpful for report demonstration and education.

Report/Conference

A written report should be prepared unless there is a preagreement that only verbal information is needed. The written report should follow the same outline as the structure of the job analysis itself—for example:

- Goals
- Plan
- Evaluation process

- Results
- Recommendations
- Provision for follow-up and outcome measurement
- Specific recommendations (both qualitative and quantitative)

For each recommendation there should be a statement of how the recommendation could be implemented, any equipment needed and where it can be obtained, methods that could be changed specifically, and the overall projected costs of the recommendation. Table 4–1 shows three formats for presenting information gained in one job analysis: ordering recommendations by priority, by timeline, or by cost. The format will be determined by the goals of the industry.

SAMPLE JOB ANALYSIS REPORT STRUCTURE: ARROWHEAD ASSEMBLY PLANT

Purpose

Management of the Arrowhead Assembly Plant recognizes the benefits of an educational program to reduce cumulative trauma of assembly line workers. The purpose of this job analysis was to define the physical demands of five different assembly jobs, evaluate them for stress factors relating to upper-extremity cumulative trauma, and prepare an injury prevention education program for the workers.

The analysis would indicate whether the stress factors were similar enough that the same education program could be given to all workers or whether individualized education programs would be needed for the five different line worker groups.

Method

An initial evaluation was done on September 19, 1993, of the five jobs on the assembly line. They were

1. Assembly Line Placer
2. Container Filler Operator
3. Container Sealer Operator
4. Container Labeler
5. Quality Control

The evaluation was to be done by Isernhagen and Associates, Inc., with an anticipated time of ten hours on site.

The report was planned to be presented to a committee of management, supervisors, and one representative from each line job. The information regarding the

Table 4–1 Three Formats for Presenting Job Analysis Recommendations

Ordered by Priority

Recommendation	*Timeline*	*Cost*
1. Educate workers in manual material-handling ergonomics.	0–6 months	$2,000
2. Change storage patterns (heavy items on shelf 2).	0–6 months	$0
3. Purchase one additional forklift.	12–24 months	$12,000
4. Purchase stock in 50 lb boxes (eliminate 80 lbs).	6–12 months	$0
5. Use back belts with loading/unloading activities.	0–6 months	$3,000

Ordered by Timeline

Timeline	*Recommendation*	*Cost*
1. 0–6 months	Educate workers in manual material-handling ergonomics.	$2,000
2. 0–6 months	Change storage patterns (heavy items on shelf 2).	$0
3. 0–6 months	Use back belts with loading/unloading activities.	$3,000
4. 6–12 months	Purchase orders in 50 lb boxes (eliminate 80 lbs).	$0
5. 12–24 months	Purchase one additional forklift.	$12,000

Ordered by Cost

Cost	*Recommendation*	*Timeline*
1. $0	Change storage patterns (heavy items on shelf 2).	0–6 months
2. $0	Purchase orders in 50 lb boxes (eliminate 80 lbs).	6–12 months
3. $2,000	Educate workers in manual material-handling ergonomics.	0–6 months
4. $3,000	Use back belts with loading/unloading activities	0–6 months
5. $12,000	Purchase one additional forklift.	12–24 months

analysis was designed to be discussed by the committee prior to the education program being developed.

Evaluation

Evaluation of the plant and the five positions revealed the following:

1. The most complex job of the five is Quality Control (Position 5). Because of malfunction in some of the current machinery, there is a higher-than-normal num-

ber of defective containers reaching the quality site. To remove those containers that do not meet quality standards, the operator must reach into the assembly line, pick out the container, inspect the part, and then place the container either back on the line or in a "scrap" bin for future recycling.

Six months prior, the quality control officer indicated that the scrap rate ratio was 1:96 (1 percent). Since current problems with the machinery have arisen, the scrap rate ratio has been as high as 1:14 and only as low as 1:54 (2–7 percent). This has increased both physical and mental stress on the operators since they are working at a much more rapid pace than the initial design required. They also are doing more repetitive reaching motions with the upper extremities. At times, they grasp two containers rather than one, an action that also puts additional stress on the structures.

This cumulative trauma stress level may be temporary and this should be evaluated by the production engineer.

2. The most repetitive job was that of Container Labeler (Position 4). Because of the configuration of the labels and the container, each container must be touched and adjusted after the automatic labeler machine applies the label. While stresses are not high for individual motions or forces, they are nevertheless highly repetitive. At this time, stretch breaks are not taken, and the only opportunities for rest are when the line shuts down, coffee break, and lunch break.

3. Regarding education program development, three of the jobs (Positions 1, 2, 3) are very similar in cumulative trauma stress factors. The workers from these three positions may be grouped in a joint education training session.

The other two jobs have different stress factors and workers will need to be educated individually. In the Quality Control position, the productivity portions of ergonomics will need to be evaluated as well as the cumulative trauma stress factors.

4. In evaluating the lost time and injury records it was found that only Position 5 (Quality Control) showed an increase in numbers of cumulative trauma injuries. Positions 1 through 4 have had a scattered pattern and a low incidence of injuries. However, cumulative trauma stress factors are present in Positions 1 through 4, and there are physical concerns of workers and supervisors. Education is identified as a primary need and requested option.

The injury records of Position 5 indicate that because of the increase in scrap rate and its resultant increase in work, this station is where the repetitive strains have appeared to be most frequent and also severe.

Proposed Action

Because issues other than educational program development have been discovered in Position 5, a verbal report will be given to the supervisory committee on the analysis. Educational recommendations for Positions 1 through 4 are presented here:

1. Regarding the educational session parameters for Positions 1, 2, and 3: a preliminary injury prevention presentation will be given to the supervisors and worker representatives at the committee meeting. A proposed method of worker education will be discussed. The proposal will indicate that two one-hour sessions be given to this work group. The first should emphasize cumulative trauma and how each individual can take responsibility for preventing it. The second should revolve around ergonomic changes identified by the group during the educational session. Included would be education in stretches that will be done on a regular basis throughout the day. Actual practice of the stretches will be included.

2. For Position 4, a modified version of the educational program will be given. In this case there will be increased attention to the low back and neck, since the positioning for this job includes spine stresses as well as upper extremity repetitive use. In addition, the job has an opportunity for modifications for positioning of the worker at the job. In the education program, there will be facilitation of work groups presenting ideas on improved ergonomic posturing during the work cycle.

The second session for this group will be an evaluation of the changes in work station design discussed and implemented by the workers after Session 1. The second session will not only develop a final plan for the workstation design but also include stretch breaks. In this way, the education session will be a problem-solving opportunity where the workers take responsibility for their own ergonomic changes.

3. For Position 5, the quality control group, there will be discussion with the supervisor and worker committee prior to the initiation of an education program. There appear to be some internal productivity and machine issues that must be addressed prior to education program development. Therefore the education program for the quality control group should be delayed until final changes are made in the process preceding quality control so that the scrap rate is lower. At that time a reevaluation will identify the type of educational classes needed.

FUNCTIONAL JOB DESCRIPTIONS (FJDs)

An outcome of job analysis can, may, and in many cases, should be a functional job description. Most of the essential information will have been obtained. The derivation of the additional components will be well worth the employer's cost and the consultant's time. Also, in many instances, the development of the functional job description will be the sole purpose of the job analysis.

FJDs are utilized by employers as a legal-ethical base for

- development of hiring parameters and qualifications

- description of job for new or transferring employees
- basis for determining/developing modified work or reasonable accommodations
- basis for development of functional, job-related prework screens

Exhibit 4–5 is an example of an FJD.

Exhibit 4–5 Functional Job Description

FACILITY:

JOB TITLE: Lineman

JOB OBJECTIVE: Performs tasks related to the installation, replacement, maintenance, trouble shooting, and repair of both lines and equipment for underground and overhead electrical transmission/distribution systems. Determines and gathers appropriate equipment and material required for the assigned job and loads it on truck or other vehicle. Uses hoists and other equipment to load heavy material. Drives or rides in vehicle to worksite. Unloads, lifts, and carries equipment and material to work area, walks over a variety of terrains. Inspects, removes, installs lines and equipment by climbing poles, towers, and structures or may work in underground vaults, manholes, and trenches. Replaces or installs equipment, performing necessary bolting, clamping, wiring, drilling, hoisting, and hauling away old equipment and material. Digs pole and anchor holes, using both hydraulic and hand tools. Climbs and cuts down trees using axes, power saws, and brushhook to clear brush and trees. Operates a variety of heavy equipment, backhoe, dozer blade, brush clipper, boom truck, all terrain vehicle, and snowmobile. Job is performed primarily outdoors in all weather conditions and at any time, day or night.

ESSENTIAL JOB FUNCTIONS:

A. Repair both lines and equipment for underground and overhead electrical transmission/distribution systems.
B. Install, inspect, and remove lines and equipment.
C. Communicate by spoken words to other workers, supervisors, and general public.
D. Read, interpret, and record information regarding work assignments, supplies, and equipment required.
E. Sort, gather, load, and unload equipment and material of various weights and dimensions on and off a truck or other vehicle.
F. Drive and perform functions of various vehicles and equipment.
G. Climb poles, towers, and structures.
H. Work in awkward positions restricted by safety belts and climbers.
I. Reach overhead for prolonged periods while operating equipment.
J. Grip, pinch, hold, and manipulate objects of various dimensions and weights with one or both hands for prolonged periods.

EQUIPMENT USED TO PERFORM JOB:

A. Clothing (safety toe boots, hard hat, safety glasses, and leather gloves)

continues

Exhibit 4–5 (continued)

B. Heavy equipment (line truck with hydraulic equipment, mechanical digger, trencher, backhoe, cable plow, dozer blade, Digger-Derrick trucks, pole unloader, aerial bucket trucks, off-road vehicle, snowmobile, and pickup truck)
C. Climbers and safety belt
D. Hand equipment and tools (axe, brushhook, power drill, hand drill, power saw, visemeter, ammeter, megger, phasing device, cant hook, pole pick, and cable splicing tools)
E. Block and tackle, hand line
F. Cable locating equipment
G. Power front breaker
H. Power capstan hoist
I. Hot sticks (fiber glass rods of various lengths)

SIGNIFICANT WORKSITE MEASUREMENTS:

A. Truckbed 4 ft high
B. Climbs poles 40–60 ft high
C. 5 ft from pole to insulator, 5½ ft to center of crossarm, 2 poles
D. 8½ in circumference poles for hot high line work

CRITICAL DEMANDS OF JOB:

- Rarely: 1–5% of the time in an 8-hour work day
- Occasionally: 6–33% of the time in an 8-hour work day
- Frequently: 34–66% of the time in an 8-hour work day
- Constantly: 67–100% of the time in an 8-hour work day

A. *Standing:* Constantly. On various types of surfaces (concrete, earth, blacktop). Performs numerous tasks in this position for up to 3-hour intervals.
B. *Walking:* Frequently. On various types of surfaces (smooth, uneven, concrete, earth, blacktop). Up to 3–5 miles per 8-hour day.
C. *Lifting:* Frequently. Different types of lifts are required, including a 120 lb pump lifted from ground to truck (48 in), usually a two-person lift with each person lifting 60 lbs (tools, equipment, and material stored in truck); up to 130 lbs from ground to waist height (32–38 in), usually a two-person lift at 65 lbs per person (material and equipment loaded and unloaded from truck or trailer); and up to 50 lbs from waist to overhead (positioning or replacing equipment).
D. *Carrying:* Frequently. Single-arm carry and two-arm front carry, up to 50 lbs, for a distance of up to 50 ft. Various types of equipment and material.
E. *Climbing:* Frequently. Performed on stairs up to 30 steps per interval; poles and steel structures up to 60 ft. Usually carrying equipment with a tool belt weighing up to 20 lbs.
F. *Pushing/Pulling:* Frequently. Up to 75 lbs of pull force when hoisting material up from ground to the person on pole. Hand-over-hand up to 20 repetitions. Up to 160 lbs pull force on secondary cables and tighten guy wires.
G. *Crouching/Kneeling:* Occasionally. Up to 5 minutes in one position while installing or removing equipment.
H. *Balancing:* Constantly. Walking on uneven surfaces, stepping over material and obstructions, climbing poles and stairs, and working on steel structures. Also performed in icy conditions.

continues

Exhibit 4–5 (continued)

I. *Reaching:* Frequently. Required to reach overhead and at shoulder height while repairing or replacing equipment. Must be able to maintain position for up to 5 minutes.
J. *Grasping/Handling:* Frequently. Required to grasp firmly (up to 100 lbs of force) numerous objects in various diameters and weights (with gloves). Use of hand tools requiring firm grasp (up to 100 lbs of force) to cut and clamp heavy multiple-wire cable. Fine hand dexterity to manipulate bolts, connections, and small materials. (This function can be modified by using different types of tools to cut multiple-wire cable.)
K. *Static Forward Bending:* Occasionally. May assume this position for up to 5 minutes when installing, repairing, or using equipment.
L. *Trunk Rotating:* Frequently. Rotation is performed to both right and left, primarily in a standing position. Extreme rotation is necessary to assume awkward positions while working on the pole.
M. *Static Neck Extending:* Occasionally. Static prolonged neck extension while working on the ground hoisting material to person on pole, or while working overhead.

NON-ESSENTIAL JOB FUNCTIONS:

Sitting, except when riding in truck to reach worksite or transport equipment.

POSSIBLE JOB MODIFICATIONS/ACCOMMODATIONS:

Using cable cutters with one or two hands, depending on grip strength.

RETURN TO WORK AND FUNCTIONAL JOB DESCRIPTIONS

When a worker is being matched to a job, particularly after illness or work injury, the FJD can be directly compared to the worker's Functional Capacity Evaluation (FCE) (see Table 4–2).

CONCLUSION

Job analysis has its roots in many professions. Its purposes and goals are varied. However, job analysis is a logical, structured process that is serving industry in many ways. The specialist embarking upon job analysis will succeed if the following principles are utilized:

- A professional background in the sciences on which at least one form of job analysis was developed
- Respect for other professionals and points of view for complementary types of job analysis (often leading to a cooperative effort)
- Structured measurement and evaluation method developed from a working plan, which is in turn derived from a goal-purpose statement
- Outcome presented by a report and interactive conference/meeting

Table 4–2 Capacity Assessment and Functional Job Demands Comparisons

Functional Job Description	*Functional Capacity Results*	*Capacity/ Demands Match?*
1. Receives phone order for supplies and fills them.	Functions: Sits for 30-minute periods. Sitting tolerance normal.	Yes
Hand coordination tasks include limited writing and filing.	Hand coordination test indicates limitation of 60%, but able to do limited writing.	Yes
Order filling includes reaching into shelves and boxing light items.	Reaching upwards needs to be limited to 60 in.	Yes, with limitation of shelf height of 60 in.
2. Does routine maintenance tasks, sweeping, hosing, painting, cleaning. Requires		
a) ability to sustain a grasp on long-handled tool;	Cannot sustain grasp longer than 2 seconds.	
b) ability to move upper extremities freely.	Has limitation in motion in all joints of upper extremities.	
3. Delivers stock orders and messages.		
a) Operates truck between plants (5–20 miles).	Cannot sustain grasp—this severely limits use of steering wheel.	No
b) Uses dolly for delivery between buildings (500–2500 ft).	Can push dolly with handle modification.	Yes
4. Manually loads and unloads trucks and dollies. Lifts loads of medium dimensions up to 50 lbs from heights of 2–65 in. Carrying distance is 2–10 ft. Up to 30 repetitions 4 times daily.	Ability to lift up to 75 lbs when leverage under the container can be used and the weight can rest on his forearms rather than be borne by grip strength. Height must be 12–60 in due to adapted lifting technique and loss of shoulder flexion.	Yes, if height is 12–60 in.

continues

Table 4–2 (continued)

Functional Job Description	*Functional Capacity Results*	*Capacity/ Demands Match?*
5. Uses word processor and adding machine for reports. Word processor requires dexterity and sensation with ability to do quick repetitions.	Hand coordination limited due to 50% loss of sensation and motion in the wrists, finger, and thumbs. Cannot type functionally.	No
Adding machine requires same, but can be modified to one-finger use, which is slower.	Cannot use full-finger machine method but can use one-fingered method on adding machine.	Yes, if modified to one-fingered approach
Sitting tolerance.	Normal sitting tolerance.	Yes

Result: Functional evaluation matched with the functional job description and allowed full-time return to work for this employee with modified work tasks.

Job analysis utilizes principles of ergonomics (see Chapter 1). It can be compliant with nondiscrimination laws by providing employers with functional job descriptions. It is the precursor of other interventions such as injury prevention education, prework screening, ergonomic adaptations, and development of modified work (alternative duty) and return-to-work programs. An accurate job analysis will be an industrial cornerstone, giving ultimate direction toward both productivity and worker safety.

Chapter 5

Functional Prework Screening

Margot Miller

Although screening is relatively new in the work arena, it is an accepted and traditional method of determining an applicant's suitability in a variety of settings. An athlete, for example, would never be considered for a specific position on the team without a tryout. The athlete understands that he or she will first need to demonstrate the skills and ability to play the position. Likewise, an individual applying for a secretarial position knows that he or she will need to demonstrate competency in typing and shorthand prior to being hired to perform such duties. And similarly, a singer expects to audition for a desired part.

Worker screening is really no different than screening the athlete, the secretary, or the singer. The purpose, determining if an applicant has the physical abilities, skill level, aptitude, dexterity, and so forth to perform the job duties in question, remains the same. And just like the female singer who is denied the leading role in "Phantom of the Opera" because she does not have the necessary soprano voice range, the worker applicant who cannot perform the essential functions of a job should be denied employment unless modifications are possible. Certainly this is an oversimplification of the problem facing employers and potential workers today. Nevertheless, the key is an appropriate test, exam, measure, audition, try-out—a screen—to determine if an applicant is able to fulfill the essential job requirements. This is the essence of functional prework screening—nondiscriminatory testing of an applicant's ability to perform the essential functions or critical demands of a job.

As we search for new prevention models, the most effective prevention model combines prework screening, ergonomic work analysis, and injury prevention education with safe work practices at the center.[1] This model illustrates the interrelationship of these three factors. For example, if the worksite is poorly designed, hiring fit individuals will not help to reduce work injury. If physically capable individuals are hired but they choose to use unsafe work methods, safety will not be assured. And education in the face of work that is too physically difficult will not prove to be effective in reducing work injury. Thus it is evident that reducing work injury will require an integration of ergonomic analysis, prework screening to match a potential employee to the work, and targeted injury prevention education with safe work practices at the center.[2] In this context, prework screening is viewed as one component of a work injury prevention program.

This chapter will give the reader an understanding of the different types of screens, the impact of the Americans with Disabilities Act (ADA), the steps involved in developing a screen, and an understanding of liability issues with screening. In addition, a sample prework screen developed from a functional job description will be reviewed.

TYPES OF SCREENS

Preplacement versus Preemployment Screens

Preemployment screens are just as their name indicates, those used to judge an applicant's suitability for a job *prior to employment or offer of employment.* Typical information would include pertinent education and training, degrees, certifications, and work experience. The screen is merely a qualifying process that assists the employer in determining the best candidate for a specific job.

Preplacement screens, on the other hand, are those performed *after a job offer has been made* to an applicant. The rationale is that a candidate should not be subjected to medical screening unless a job offer has been made. Additionally, employers are prevented from selecting only the strongest and most fit candidates. Only applicants passing the screen are placed. Preplacement screens assist the process of matching the worker's abilities to a specific job.

Functional versus Health Screens

Health screening involves any medical testing that would indicate that an individual has a health condition that could potentially be linked to risk of injury on the job—for example, obesity, even if obesity has not necessarily been shown to be a health risk on the job. High blood pressure, poor vision, and diabetes are other examples. It is critical that these be identified as health status indicators, indicators not necessarily related to work injury or ability. Certainly a qualified candidate for a job, a candidate who also happens to have diabetes, should have the same opportunity for employment as a qualified individual without diabetes. Two questions should be answered: Is the applicant able to do the job? And if so, can he or she do the job safely? Only when the health condition, diabetes in this case, interferes with the applicant's ability to do the job or interferes with safety on the job should the applicant be denied employment.

All too often a relationship is inferred between health status indicators such as obesity, high blood pressure, and poor posture and increased risk of work injury. Certainly caution must be taken to avoid putting health status indicators into health risk categories unless the potential risks have been objectified and validated. Even if a health status factor is found to be linked with "risk," it will not be enough evidence in itself to prevent a qualified applicant from being placed on the job.

Conversely, functional screening is testing to determine if the applicant is able to perform the job-related functions and, if so, to perform the functions safely. Agility tests and fit-for-duty evaluations are two such screens. In both cases, all applicants for a particular job must be made to go through the functional screen to ensure that functional competence exists. It is also necessary to ensure that a qualified person with a disability has as fair a chance for the job as the qualified person without a disability. It is the responsibility of the employer to demonstrate that any functional screens being used are job related and consistent with business necessity.

IMPACT OF THE AMERICANS WITH DISABILITIES ACT (ADA)

Intent of the ADA

The intent of the ADA regarding employment is really very simple—to make sure that qualified persons with a disability are given as fair a chance for employment as qualified persons without a disability. It has been said to be the most far-reaching comprehensive civil rights legislation since the 1960s.[3] The ADA clearly states that the disabled applicant must first be qualified for the job in question just as the nondisabled applicant must be qualified for the job. What the ADA does not do is give persons with a disability who are *not* qualified for a job special privileges. Rather, it gives persons with a disability who are qualified for a job equal opportunity to be considered for the job. The impact on screening utilized during the employment process is that such screening is now mandated to be nondiscriminatory. This is very good news for those who have experienced discrimination in the workplace due to a perceived or real disability.

Effect on Hiring Practices

Prior to a job offer, verbal screening regarding the physical demands of the job can occur. A list of the critical functions of the job is given to the applicant; the applicant is asked, "Can you do these physical requirements of the job?" The applicant answers whether he or she could perform the physical requirements listed. If the applicant reveals that he or she has a disability and may possibly have trouble with some of the requirements, then the employer may ask how he or she could do the job with modifications or accommodations. At this point the employer is merely trying to determine if the applicant can perform the job duties.

At the time an applicant is offered a job, the employer can request that the applicant go through a functional screen. Four scenarios may result from the screen:

1. The qualified applicant may not have the physical capability to perform the job—without reasonable accommodation for the nondisabled and with or without reasonable accommodations for the person with a disability. Employment is denied since the functional screen documented the applicant's inability to perform the essential functions of the job.
2. The qualified applicant may have a physical problem that could put him or her at severe risk to himself/herself or others. This is called "direct threat." It must be remembered, however, that identifying the applicant at risk must be done on the basis of objective evidence that such a direct threat exists, rather than based purely on supposition or weak risk factor evidence. Employment may be denied if evidence of direct threat is supported with fact.
3. The qualified applicant with a disability is unable to perform some of the physical requirements of the job in the "traditional" manner, but with reasonable accommodations can perform the physical requirements. The applicant must be considered for employment along with qualified nondisabled applicants. Since the ADA is not a "hire the handicapped" bill, the qualified person with a disability does not get preference over the qualified person without a disability. However, the ADA does mandate fair and equal consideration for employment for qualified individuals who just happen to have a disability as well.
4. The applicant performs all items on the functional screen safely and competently. The hiring and placement process then proceeds and is completed.

DEVELOPING A FUNCTIONAL PREWORK SCREEN

Defining the Purpose

It is critical for the employer and medical professionals to understand the purpose and intent of a functional prework screen. Employers must remember that prework screening is just one component of work injury prevention and that it cannot be used to "screen out" individuals. In the past, for example, isometric strength testing was used to screen out those who were believed to be at risk for back injury. However, such testing has been shown not to correlate with dynamic testing and therefore should not be considered job related.[4] Employers will need to incorporate other components to effectively reduce work injuries, including worker education, ergonomic intervention to make the workplace safer, and early intervention policies. Therefore employers and medical professionals need to be aware of what screening can and cannot do from the start as well as understand that screening will not eliminate all work injuries. Rather, the purpose of screening is to test an applicant on the physical requirements of the job to determine if he or she is capable of performing the functions safely. The screening process enables the employer to match the worker to the job, thereby decreasing the likelihood for work injury. In this context, prework screening plays a significant role in injury prevention.

Need for a Functional Job Description

A functional prework screen should never be developed without first having an accurate functional job description. Whether the job description is written by the employer or by an outside consultant, it is the employer's responsibility to determine its accuracy. For the prework screen to be an accurate representation of the essential functions of the job, the functional job description must accurately reflect the job functions. The Equal Employment Oppurtunity Commission (EEOC) regulations define essential functions as "the fundamental job duties of the employment position the individual with a disability holds or desires."[5] Employers must know what employees actually do in their jobs, including both "essential functions" and "marginal functions." A function can be considered essential for any of several reasons, including but not limited to the following:

- It is the reason the position exists.
- Only a limited number of employees can perform the function.
- The job cannot be performed adequately without this function.

"Marginal functions" are those that are not essential to the job: for example, they could be done by someone else, or they are not the duties for which the job was designed. In the case of a cashier in a local grocery, when business is slow, the cashier may stock shelves; however, in an identification of the essential functions, stocking shelves would be considered a marginal function.

Identifying the Critical Demands

The critical demands of a job are the essential functions of a job that require physical attributes and place stress on the individual as they are performed (e.g., lifting and carrying tasks). To help identify the critical demands from an injury perspective, one can review the workers' compensation files, review the OSHA

200 logs, and interview employees. Employees have a wealth of information regarding what they do every day. They know which functions are critical to their position, they will describe what is difficult on the job and where injuries have occurred, and they can explain individual modifications they have implemented to make their job or portions of their job easier. The final step in identifying the critical demands is for the employer and employee representative to sign off on the list of critical demands.

Determining the Use of Safe Work Practices

The consultant performing the job analysis, writing the functional job description, and identifying the critical demands should take note of whether safe work practices are used in the workplace. The functional tasks that are critical to the job and part of the functional screen must be tasks that are able to be performed safely in the workplace. First, it may be that by implementing ergonomic changes the task becomes safe, thereby eliminating the need for screening. For example, ergonomic changes may result in workers using a mechanical assist rather than lifting a 100-pound object from floor level. The task thus becomes safe for everyone, and screening this particular function is no longer necessary.

Second, the critical job tasks that are included in the screen must be performed in the same way in the workplace—that is, safely—or screening is irrelevant. For example, the screen may determine that an individual can safely lift 45-pound bags from a 12-inch raised surface to waist level but is unable to safely perform a full squat. If the product is stacked on 12-inch pallets, the individual will be able to perform this function safely on the job. However, if in actuality the product is stacked on the floor, the employee will be unable to perform the lift safely. Therefore safe work practices should be established up front so that they can become relevant in the screening process.

Performing a Musculoskeletal/Physical Assessment

A history and musculoskeletal/physical assessment is necessary prior to actual functional testing to determine that there are no medical or physical contraindications to testing. This information cannot be used to determine if the individual can do the job; rather, it provides a guideline to ensure a safe functional screen. Such information may include range of motion, strength, flexibility, blood pressure, and heart rate. It is important to note that this information is confidential and must not be given to the employer until after the individual is hired. Even then, the information must be kept in confidential files.

The purpose of identifying the physical attributes of a potential worker is to hire only those candidates who can physically do the job. Questions or tests that do not relate to the job must be eliminated to prevent the person from being labeled "disabled" or "high risk" by the employer, and more importantly having a negative influence on the person's chance for hire. The key is stressing ability rather than disability.

Selecting the Functional Components

It has been established that the functional activities included in the prework screen are determined by the critical demands of the job. It is essential that the critical demands be adequately measured; however, attention needs to be given to the cost involved for the employer as well as the practicality of the testing designed. Screening activities may include the following components: strength, endurance, aerobic capacity, coordination, and balance. As an example, a critical demand may involve lifting 45-pound bags of product from floor to overhead. The strength component would then simulate this task. If the 45-pound bags had to be handled frequently throughout the day, repetitions would be built into the screen to achieve endurance. The most critical step is making sure that the functional tasks chosen for the screen simulate the critical demands. If they do not, the results of the screen will be useless for determining if the candidate has the ability to perform the critical demands of the job.

Safety Standards during Testing

The importance of the kinesiological approach in determining an individual's functional capacities is evident when developing a functional prework screen. Inherent in the kinesiological approach is that the evaluator determines functional levels, including whether an activity was performed safely and whether maximum effort was given. The kinesiophysical approach stresses safety. The evaluator determines the use of proper body positioning, pacing, and coordination and identifies areas of weakness and/or dysfunction. Furthermore, the evaluator interacts with the worker candidate to ensure that safety is maintained and to further establish that he or she understands the testing procedure. The result of the screen is an objective outcome statement of the candidate's ability to perform the critical demands of the job using safe protocols.

The responsibility for ensuring that safe performance is used during the functional screen is the evaluator's. If the individual's performance becomes unsafe, the test is stopped and the evaluator documents the objective findings. The potential for injury certainly exists during the screening process when safety is not required or monitored. Requiring safe performance during screening is up to the evaluator. However, the question remains whether safety is a valid pass/fail issue in screening. If the employer enforces safe work practices at the workplace and has safe practices written into policies and procedures, then it is valid for the prework screen to require safe performance. On the other hand, if no safety regulations are enforced at the workplace, it is inconsistent to require safety during the screen. Ultimately, however, if an applicant is injured during a screen, the evaluator is liable if negligence occurred in allowing unsafe performance during the screen.

Establishing Outcome Criteria

In the development process, the employer should be involved in establishing the level of performance that will be required to indicate adequate ability to perform the critical job demands. The evaluator needs to know the minimum requirements for the screen as well to determine outcome criteria. All of the criteria for hiring should be documented before testing. It is important to note that the employer "owns" the job, validates the job description, and is hiring the worker. Therefore the responsibility for hiring criteria is with the employer.

The actual paperwork screen outcome statement should include the following:

- Critical demands of the job
- Minimum criteria for each critical demand
- Description of whether critical demand was met
- Recommendations to employer regarding test results[6]

The evaluator is responsible for scoring the worker; however, the scoring is based on the criteria established by the employer.

SETTING UP A PREWORK SCREEN

Identify the Critical Demands from a Functional Job Analysis (FJA)

The first step in setting up a prework screen is to obtain the critical demands from a functional job analysis. (Refer to the functional job description in Exhibit 5–1.) The critical demands are evaluated for the following physical requirements: strength, endurance, positional tolerances, ambulation, aerobic capacity, coordination, balance, and specific tool use.

Select Functional Activities

The second step in setting up a prework screen is to select the functional activities to include in the screen. It should be noted that the activities must be performed in the same manner during the screen as they are at the workplace. Refer to the prework screen in Exhibit 5–2.

Exhibit 5–1 Functional Job Description: Machine Operator

Job Title: Machinist

Critical Physical Demands:

- Lifting: Floor to 36 in up to 75 lbs: floor to 60 in up to 50 lbs
- Carrying: Single hand up to 50 lbs for 200 ft or less
- Front carry: Up to 75 lbs for 50 ft or less
- Crouching/kneeling for up to 5 minutes
- Overhead reaching for up to 3 minutes
- Static forward bending from standing, crouching, or kneeling
- Trunk rotation from standing, sitting, or kneeling
- Standing/walking up to 6 hours

Evaluate Job Relatedness of Functional Activities

The third step in setting up a prework screen is to evaluate the job relatedness of the selected functional activities. An effective method to determine job relevance is to have management, supervisors, and workers perform the screen and score their performance. Have them review both the test and the scoring. Make sure the outcome criteria are appropriate. Once the screen and scoring are finalized, have the employer sign off on it.

Establish Screening Protocol

1. *Consent form:* ADA does not appear to consider this a necessity. However, one can be helpful. A consent form is used to inform the applicant of the screening process. It is an indication by the applicant that he or she is not restricted by a physician from doing the activities included in the screen. The applicant is agreeing to participate in the screen and have the results released to the company.
2. *Explanation to applicant:* The applicant may already be familiar with the essential functions of the job. If not, these should be explained.
3. *History:* The history should identify any contraindications to testing such as blood pressure or heart problems, back instability, or medical precautions given by a physician.
4. *Musculoskeletal/physical assessment:* The musculoskeletal or physical assessment for a prework screening is a brief assessment of the areas of the body that will be stressed during the screen.
5. *Functional activities:* The applicant performs the actual activities selected for the screening. Safety should be monitored. Standardized protocols should be written for each screen developed to ensure consistency of testing.

Exhibit 5–2 Prework Screen: Machine Operator

Job Title: Machinist

Functional Activities	*Critical Demands*	*Testing*	*Met/Not Met*
1. LIFTING			
Floor to 36 in, 5 in 1 minute	75 lbs	25 lbs, 50 lbs, 75 lbs	____/____
36–60 in, 5 in 1 minute	50 lbs	25 lbs, 50 lbs	____/____
2. CARRYING			
Single hand, 200 ft right	50 lbs right	25 lbs right, 50 lbs right	____/____
Single hand, 200 ft left	50 lbs left	25 lbs left, 50 lbs left	____/____
Front, 50 ft	75 lbs	50, 75 lbs	____/____
3. CROUCHING/KNEELING			
Crouch/kneel	5 minutes	____ minutes	____/____
4. OVERHEAD REACH			
Overhead reach 12 in above crown level	3 minutes	____ minutes	____/____
5. FORWARD BEND			
With repetitive reach 30 in from waist			
Standing	2 minutes	____ minutes	____/____
6. TRUNK ROTATION			
Right	30 reps	____ reps	____/____
Left	30 reps	____ reps	____/____

COMMENTS:

Physical abilities (do/do not) match the critical physical demands of the job description. Modifications to the job or changes in the applicant's physical abilities (would/would not) be necessary in order for this individual to perform these tasks.

EVALUATOR ____________________ DATE ________

6. *Scoring:* The applicant's performance is rated using the criteria established by the employer. The applicant is informed; however, the evaluator should avoid making hiring interpretations.
7. *Release of results:* Disseminate results to parties agreed upon during the screening agreement process.

LIABILITY ISSUES WITH PREWORK SCREENING

Liability issues are important considerations when doing prework screening. The following is presented to give the reader insight into pertinent legal issues with screening.

State Law

Prior to offering screening services, state laws regarding evaluations should be reviewed. Specifically, determine if physician referral is necessary when performing this type of service.

Informed Consent

Informed consent is not technically required when performing screens; however, to help applicants understand the screening process and to facilitate compliance, informed consent forms are suggested. The applicant signs the form prior to participating in the screen. The purpose of the screen is identified for the applicant as well as what information will be collected. This also allows the applicant the opportunity to refuse to participate if he or she so desires.

Disclosure of Health Risks

If any health risks are identified during the screening process, it is the responsibility of the evaluator to inform the applicant and the physician. Some may question whether the employer should be informed as well, particularly if the findings have an impact on future abilities or influence the safety of the applicant or others. It then seems prudent that the evaluator inform the employer as well.

Nonfunctional Test Disclosure

According to the EEOC guidelines, information discovered at the time of screening that is not job related, such as information from the musculoskeletal assessment or history, cannot be used by the employer in making hiring decisions.

Safety in Testing

If an applicant is injured in the screening process, both the evaluator and the employer may be liable. The bottom line is that if safety is required and monitored during screening, then the concern for applicant injury during screening is eliminated.

CONCLUSION

Prework screening is a systematic process of measuring an applicant's ability to perform the critical demands of a job. The outcome criteria established by the employer will identify the minimal requirements of performance in order that an applicant be considered for hire. In summary, it is important for evaluators to

- Know the reason for the screen so that the screening process can achieve the goals.
- Develop the screen from a functional job description.
- Develop a screen that accurately reflects the essential functions/critical demands of the job.
- Validate the screen with the employer.
- Maintain safe principles.
- Document all steps in setting up the prework screen.
- Follow up the screening process to determine appropriateness and accuracy of the screen.

NOTES

1. S. Isernhagen, Ergonomic Basics, *Orthopaedic Physical Therapy Clinics of North America* 1, no. 1 (1992):23–36.
2. S. Isernhagen, There Is No Magic Answer . . . But There Are Effective Methods, *Orthopaedic Practice* 2, no. 4 (1990):13–14, 21.
3. M. Lotitio et al., *The Americans with Disabilities Act: Making the ADA Work for You* (San Francisco: Milt Wright Associates, Inc., 1992).
4. J. Rosencrantz et al., *Relationship between Dynamic and Static Lifting Capacity,* University of Iowa, Iowa City, Physical Therapy Graduate Program Dissertation, 1991.
5. U.S. Equal Employment Opportunity Commission, *Americans with Disabilities Act: Technical Assistance Manual* (Washington, D.C., January 1992), B-19.
6. H. Fearon, Prework Screening, in *Orthopaedic Physical Therapy Clinics of North America*, ed. S.J. Isernhagen (Philadelphia: W.B. Saunders, 1992), 37–47.

Chapter 6

Case Studies: Prework Screening

R. Gary Gray and Stephen E. Campbell

The extent to which low back injuries directly and indirectly affect health care providers, employers, and employees is immeasurable. In an effort to reduce the impact of back disorders on industry, and simultaneously enhance their own referrals and revenue sources, health care providers have introduced countless programs deemed capable of having a positive effect. A multitude of programs, some valid and others unsound, have been developed to assess potential performance capabilities of employees, in addition to programs that attempt to determine the possibility of future risk. Included in these programs are physical examinations, medical histories, radiographic studies, strength testing, job-specific tests, and education programs. The enactment of the Americans with Disabilities Act has affected the means by which many of these programs can be used and has also facilitated a strong interest in the need for case studies and validation requirements. The importance of content validity as it relates to measuring skills and abilities has come to the forefront as health care providers network with industries.

CASE STUDY 1

The following case study reflects the impact that a program centered upon prework screens has had on an employer in a manual-labor-intensive industry with a history of an extremely high rate and cost of back injuries. At the commencement of these programs, the company described was being so negatively impacted by back injuries that its existence as a solvent and viable employer was severely threatened. The program initiated was designed to provide services prior to the hiring of a potential employee (prework) while also allowing for the comprehensive follow-up of an injured worker (postemployment). Although it is difficult, if not impossible, to quantify and categorize the direct fiscal impact of each area of service provided, it is felt that the prework programs, including job-specific testing and functional matching, were significant contributors to the successful reduction in the overall costs of back injuries.

The components of the program mentioned above were categorized into prework programs and postemployment programs. The prework intervention included

- Job site analysis (including ergonomic design/redesign, job description formation, and formation of job-specific tests for each labor position)
- Preplacement evaluation (including cardiovascular profile, musculoskeletal evaluation, joint integrity testing, static/dynamic/isokinetic strength testing)
- Job-specific testing
- Job matching
- Training in safe and effective performance of job tasks

The implemented components of the postemployment portion of the program included the following:

- Proper medical treatment and rehabilitation of the injured worker (initially includes comparison to the individual's prehire data, determination of the possibility of symptom magnification, and return to light duty as soon as possible)
- Periodic review of safety training (including back schools and instruction in material handling)
- Periodic retesting to assure job match and reveal significant changes in the employee's functional capabilities

Much of the success of the program for this particular company can be credited to the job-specific testing component of the prework screen program. The job-specific component was developed during an initial job site analysis. At that time, each position of labor that had the potential to be screened was thoroughly analyzed, and a determination was made regarding the essential and most physically demanding tasks and requirements of each labor position. With most materials and equipment acquired directly from the job location, the selected task or tasks were simulated as closely as possible in a clinical setting, thereby providing the basis of a job-specific test. Each prospective employee performed the job-specific test designed in accordance with his or her particular job. Prior to testing, the prospective employees were allowed to suggest any reasonable accommodation that they might require. Each person was tested following the same specifically designed protocol for his or her particular labor position, and the scoring system was the same for each position, regardless of sex, age, or other physical influences. The company was then notified as to the success or failure of each applicant in accordance with the designed protocol. On occasion, and at the company's discretion, employees who failed to successfully and safely perform the job-specific components at one particular labor position were allowed to attempt the tasks for another labor position, usually at a lower physical demand capacity level. Following successful completion, the prospective employee was trained in proper body mechanics and material handling techniques related to his or her specific labor position.

The prework programs in combination with the postemployment suggestions were implemented in late 1989 by the company, an oil field contractor with approximately 400 employees. Data collected revealed a significant reduction in the number and cost of back injuries (see Figure 6–1). An indirect but extremely important and significant statistic was the reduction of the employer's incident and turnover rates. The incident rate is acquired from a formula in which the number of claims is multiplied times 200,000 and divided by the total number of person-hours worked. In 1989, the company's incident rate was 22.0 with an employee turnover of 265 percent. In 1992 the incident rate was 9.7 with an employee turnover of 35 percent, even though the number of person-hours worked increased 10.7 percent. Meanwhile, the company's mod rate, on which workers' compensa-

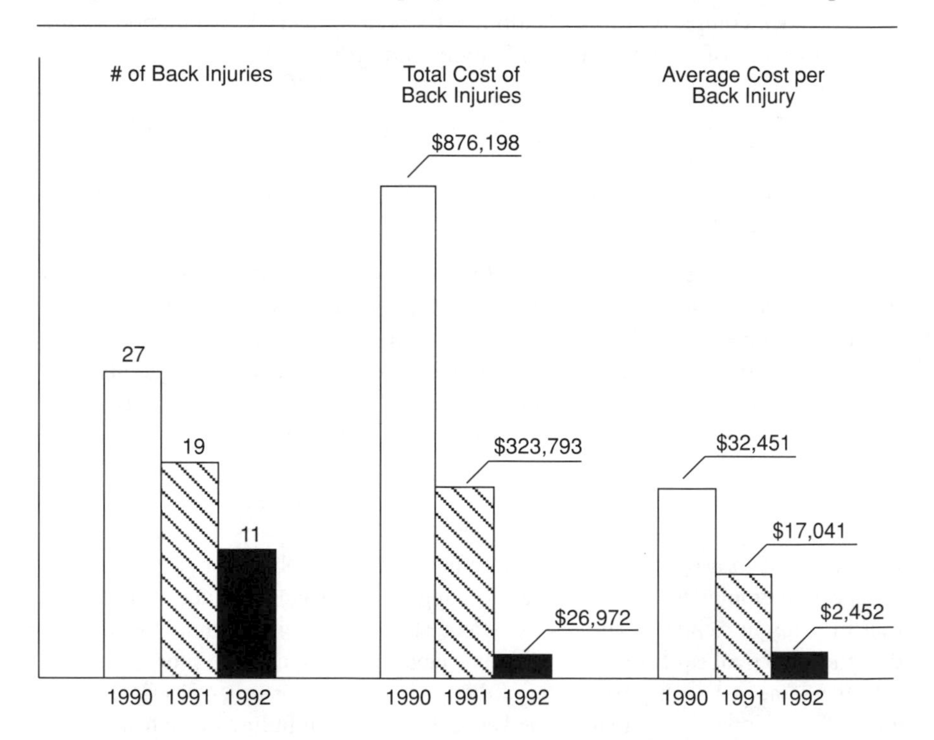

Figure 6–1 Back Injury Statistics for Case Study 1 Company, 1990–1992

tion premiums are based, has shown significant reduction in the past three years. Mod rate reduction has occurred as follows for the years of 1990, 1991, and 1992 respectively: 1.8. 0.97, 0.69.

CASE STUDY 2

The second case study involves a 310-employee, small manufacturing business in Texas. The labor force is generally nonskilled, with the majority of training being obtained while on the job. Forty percent of the work force is female, and physical demand levels range from sedentary to heavy. The method detailed in Case Study 1 was utilized here also. It was implemented at the beginning of the second quarter of 1991. In this case, body mechanics, back care, and material handling instructions were already in place and were continued with minor revisions prior to the implementation of preplacement screening.

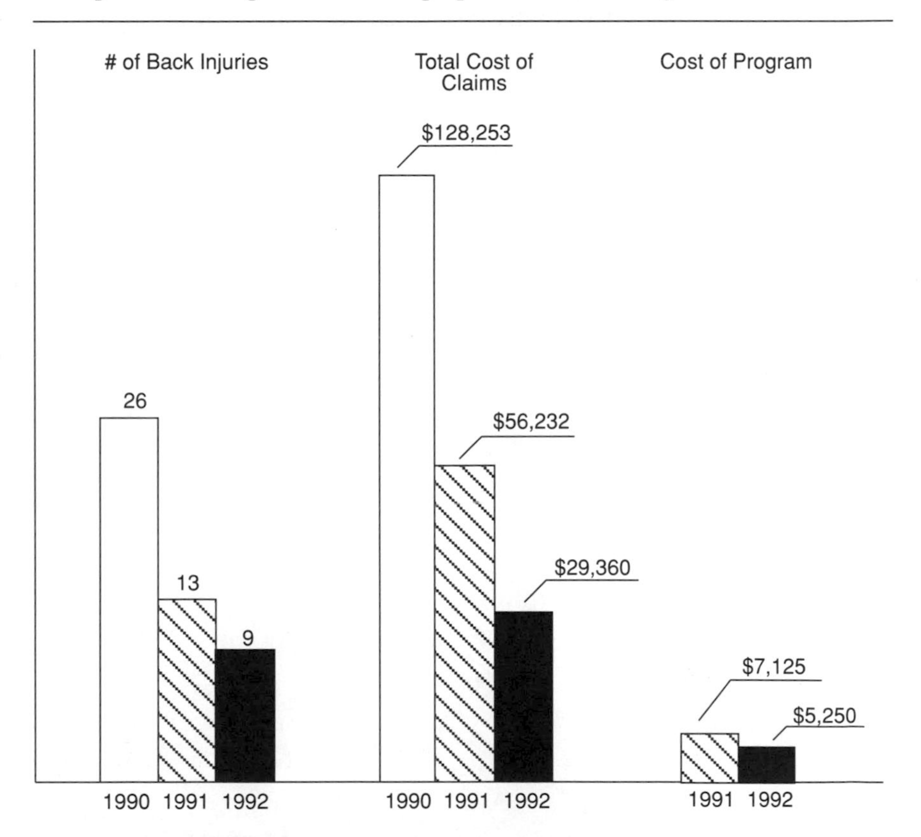

Figure 6–2 Back Injury Statistics for Case Study 2 Company, 1990–1992

Information on all claims (including back injuries) was recorded. Note that figures for 1990 demonstrate the need for this program, intervention began in 1991, and the results are evident in 1992 (see Figure 6–2).

FOCUS ON THE SOLUTIONS

Health care providers can make a significant impact in both the direct and indirect costs of back disorders to industry. A comprehensive approach that encompasses both prework and postemployment testing is the most effective means of providing successful intervention. An important portion of this revolves around job-specific testing and job matching. It is imperative that the health care provider, when working with industry, follow legal and ethical guidelines while administering safe and effective programs with the full realization of possible technical limitations. The importance of validation and case studies will assist in further facilitating the positive impact that health care providers can have on reducing the cost of back injuries to industry.

Chapter 7

Back School

Lynne A. White

Discussion is rampant about "the changing health care environment" of the 1990s. However, the last of the changes are probably years away. If one has been paying attention, it is clear that health care has been changing for the last 20 years. Changes are constant in the areas of clinical care, diagnostics, technology, medications, prevention, and rehabilitation. Changes have been ongoing for years within the areas of workers' compensation, federal and state health care programs, payment schedules, medical insurance, and the laws of health care.

Changes in the realm of prevention and patient education have also been ongoing, but probably not fast enough. During the same 20-year period of time, preventative awareness and patient education, particularly in the areas of heart disease, cancer, and AIDS, have greatly advanced. Information is being widely dispersed through school programs, all forms of media, and celebrity fundraising. Numerous books reach the best-seller list on each of these diseases and related subjects. These diseases and their associated health problems cost society billions of dollars and immeasurable human suffering. People also die from these diseases.

Back pain does not usually lead to death, but it still continues to cost society billions of dollars and immeasurable human suffering. Change in the field of patient education for those with back pain has been going on for 50 years. But even today, many patients are not getting the education they need in order to understand enough about their condition so that they can take responsibility for it.

LET'S TALK

They say talk is cheap, yet there is seemingly not enough of it going on in medical offices. Communication between patient and clinician is the glue that can bind the two together through the health care experience.

When asked to define the perfect caregiver, the patient does not immediately focus on the medical degree or specialty of the caregiver. What patients want is (1) more dialogue, (2) more health information, and (3) more involvement.[1] Translated, that means communication: more talk and more information.

When asked to define the perfect patient, the clinician would probably not start off the list with "industrial injury, off work for over six months, and physically unfit." What clinicians want from patients is (1) more attention and concentration, (2) more willingness to follow recommendations, and (3) more interest in accepting responsibility for condition. Translated, that, too, means communication: more listening and more responsiveness.

Communication is obviously a two-way path: caregivers and patients need both to listen and to be heard. Intelligent, creative, cost-effective patient education programs can offer an organized solution and offset many information gaps, misunderstandings, and poor communication skills. The results could be outrageously satisfying and a simple solution for even the not-so-perfect clinician and not-so-perfect patient.

A basic back school program, or any patient education program for that matter, should get the required medical outcome. It should take less time, cost less, and, most importantly, be less frustrating for all parties concerned. Now that's a change.

HISTORY OF BACK SCHOOL PROGRAMS

The use of education as a tool for the relief of back pain, and ultimately its prevention, is still sporadic and inconsistent, if not absent, in most medical settings other than major spine centers and the scattered offices of an enlightened few. In most offices, patient education directly associated with relieving and preventing back pain usually comes in the form of a handout. Unfortunately, time constraints and cost usually preclude a physician from taking the time to personally deliver a well-organized, structured, entertaining, provocative education program that assists the patient with understanding and managing his or her condition.

There have been, however, pioneers in various fields of spinal medicine that found that their patients responded extremely well, if not craved, education, and that education actually enhanced their total treatment plan. Over the past 40 years these few have persisted, and many of them still persist today, with patient, public, and medical education. If back pain is to be truly controllable and preventable[2] in the future, after 40 long years, one hopes that the future is now.

The 1950s

Formalized education programs for back pain sufferers did not begin to surface until the early 1950s, when Hans Kraus, MD, began writing about the treatment and prevention of low back pain caused by muscle deficiency.[3] Kraus's simple, systematic program of exercise, stressing relaxation and stretching of tight muscles and strengthening of weak muscles, was designed to improve physical fitness and relieve discomfort associated with back problems.

The 1960s

It was not until over a decade later, in 1966, that Harry Fahrni, MD, published *Backache Relieved.* His work was probably the first to attribute poor postural habits to a cause of back pain. He realized that back pain was a controllable condition and thus developed many techniques for resting the spine and using positioning and back health education to control back pain. He was one of the first to understand and demonstrate the differences between the ground-dwelling and industrialized cultures, and theorized that in ground-dwelling cultures individuals did not develop degenerative disc disease and painful spines until they were 50 years old, whereas in industrialized cultures individuals developed back pain in their early to mid-thirties. He wrote about these concepts[4] and trained physical therapists to deliver the message to his patients.

Another approach was found in the program at Rancho Los Amigos Hospital in California developed by Vert Mooney, MD, in the early 1960s.[5] Treatment was primarily behavioral, focusing on reinforcement of healthy behavior and essentially ignoring sick behavior. The program at this hospital pioneered some of the basic philosophies that are common components of pain programs as we know them today.

At about the same time as Dr. Fahrni was organizing his concepts, the Back School (the first use of this term) was first organized in 1969 at Danderyd Hospital, near Stockholm. The four-session educational program soon replaced most of the treatment modalities previously used in the physiotherapy department for patients with low back pain. Marianne Zachrisson Forssell, RPT, and her colleagues also developed a sound-slide program covering the back school information so that patients could refer to the material repeatedly if needed.[6] Attendance in a back school became commonplace for back pain sufferers who sought medical care in Sweden.

The 1970s: The Flexion versus Extension Decade

Ten years later Forssell wrote about the Back School in *Spine*, concisely describing the content of the education program.[7] The first lesson covered the general aspects of back disorders, including the occurrence of back pain and who might get it. There was information about anatomy and an explanation of how the back works. Treatment methods were discussed, but time was stated to be the great healer. During all teaching sessions, patients would lie down in the psoas position: that is, in flexion or with the pelvis titled forward, on their backs, with their knees bent over a specially made rest. The patients were advised to rest frequently in that position at home. The second visit covered the concepts of intradiscal pressure and intra-abdominal pressure, which were explained and demonstrated using the findings of Armstrong.[8] Patients were taught methods of creating a muscle corset by means of physical training of the abdominal muscles. This exercise was the only home exercise that patients were encouraged to do, preferably every day. During the third visit, patients were given lessons on carrying and lifting techniques. Simple activities of daily living were also covered. In the fourth and final lesson, the importance and necessity of physical activity were stressed, with emphasis on self-responsibility. Patients were told that there was no reason to believe that they had to avoid physical activities because of back pain. The therapists stressed that psychological and physical improvement would come with exercise. Patients were given a written summary of the back school concepts when they completed the course.

Zachrisson-Forssell visited the United States in the mid-1970s in an attempt to teach other therapists and physicians how to integrate the program into their practices. While her product was excellent, the marketing failed. The concept was considered new in America and was relatively unproved. The process required to deliver the information to patients no doubt seemed time consuming and labor intensive. Unfortunately, probably the most important single reason why this marvelously simple, educational program was not excitedly received and quickly implemented by clinicians at that time was that no one knew how to bill or get paid for the service.

It was not until 1977, when the major work was completed at the Volvo factory in Sweden by Berquist-Ullman and Larsson and published,[9] that many professionals were convinced that education could be effective in decreasing the incidence of low back pain and the lost time and dollars spent for industrial low back injuries.

In early 1976, the California Back School (CBS) was started in San Francisco by Arthur H. White, MD, A. William Mattmiller, RPT, and this author. There were two major guidelines from Arthur White: "Keep it simple and inexpensive" and "Don't confuse the educational process: No hands on, no heat, no ice—just education." During the next four years there occurred a teaming together of different disciplines, including orthopedic surgery, physical therapy, and business. Using the Fahrni and Swedish models, CBS experimented with written materials, sound-slide shows, and one-on-one training by the physical therapists in order to deliver the basics of spinal anatomy, body mechanics for home and work activities (using the concept of pelvic tilt for all patients), and personalized exercise programs for each patient. CBS's goal was to make the information interesting and challenging for the patient and figure out a way to get paid by third-party payers for the therapists' time for evaluating and training the patients.

The California Back School's obstacle course tested the patient's knowledge of body mechanics while walking, standing, sitting, lying down, bending forward, reaching, leaning sideways, twisting, pushing and pulling, and going over and under objects.[10] After patients had completed the three visits of training, they were again tested on the obstacle course in order to determine if they could use the information in real life situations, such as getting groceries out of the trunk of a car, carrying them into the house, and putting them away on shelves of various heights. Patients appeared to retain the information when given verbal instruction in a short, lecture format, reinforced with a short audiovisual show, and when they practiced the activities on the obstacle course and were given a written handout to take home. A review of 300 patients revealed that 89 percent sought no further medical treatment, 95 percent were able to resume normal activities, and 64 percent reported no significant change in their lifestyles.

During the 1970s in New Zealand, a physiotherapist, Robin McKenzie, was assessing and attempting to prove his findings from treating back pain patients in his clinic. According to McKenzie, there were three predisposing factors to low back pain: poor sitting posture, loss of extension, and frequency of flexion.[11] He recommended exercises and postural instructions that restored or maintained lumbar lordosis and suggested that, according to the studies by Armstrong[12] and Shah,[13] the nucleus pulposus migrated forward in lumbar extension and backward in lumbar flexion. McKenzie stressed self-treatment with extension exercises.[14] What physical therapists were not able to gather from Zachrisson-Forssell earlier, they gathered from Robin McKenzie. McKenzie's courses and educational materials popularized his extension concepts rapidly throughout the United States. Physical therapists became disciple-like and were known as McKenzie therapists, much to the chagrin of physicians still accustomed to the Williams' flexion routine.

The 1980s: Back Schools Spread in North America

By the end of the 1970s, the expanded team at the California Back School developed a two-day training seminar that included an extensive how-to manual and an audiovisual package with four sound-slide shows and a trainer's guide for use by clinicians for their patients.[15] By 1982, the group had trained approximately 1,500 professionals over a three-year period, and this, along with McKenzie's seminars, began a rapid spread of patient education programs for back health care.

In Canada, Hamilton Hall, MD, developed the Canadian Back Units, and by the early 1980s, his group had an estimated annual enrollment of 2,000 patients. Hall's program emphasized an attitude change, with patients assuming responsibility for their own health status. This attitude change was accomplished through group education, given by a variety of clinical specialists, about spinal anatomy, body mechanics, flexion exercises, pain and stress management, and relaxation techniques. The research from this group reported high rates of symptom reduction and patient satisfaction.[16]

In addition to articles and books by the developers of these various back school models, a profusion of articles appeared describing a variety of back school programs and concepts,[17] comparing outpatient with inpatient programs,[18] comparing one method with another,[19] and showing the effects of back school education in the industrial setting.[20]

During the second half of the 1980s, the use of exercise, education, and aggressive conservative care[21] expanded the basic three- or four-visit back school program into multidisciplinary, multivisit, intensive, and often expensive options.

While the core information on anatomy, the degenerative process, first aid, and basic body mechanics has remained about the same, the exercise-based approach, called stabilization training, was developed from a number of known theories by a variety of specialists.[22] Specialists in Northern California[23] began yet another wave of intense interest by physicians and physical therapists[24] in stabilization training as a tool in training patients with back and neck pain.

Stabilization training is now an integral part of a patient's education at most major spine centers, especially at the facilities of the concept originators and developers in Northern California and at those of their colleagues elsewhere in the United States. The training is basically one-on-one—one patient with one physical therapist or trainer. The training emphasizes methods for limiting and controlling lumbar movement, teaching the patient to keep his or her spine in a neutral, pain-free position while carrying on all activities of daily living, including recreational activities. Patient training is quite precise and definitely customized to the patient's diagnosis, abilities, and future physical requirements.

The expansion of patient education in the field of spinal medicine has also brought a national growth of large, thoughtfully designed rehabilitation facilities that include gyms with state-of-the-art aerobic and weight-training equipment, computerized back-testing machines, and life-sized work simulation areas used to teach injured workers how to do their job safely. Staffs include physicians, physical therapists, occupational therapists, manual therapists, psychologists, dietitians, social workers, and vocational rehabilitation counselors all working together to educate, train, strengthen, and cajole patients back to active, meaningful lives, and especially back to work. Patients began spending entire days, and even months, in these facilities being evaluated, exercised, and work-hardened,[25] mainly in groups.

Back pain and lost work days were costing industry billions of dollars throughout the United States,[26] if not in most industrialized nations. The term *return to work* became the mandate and battle cry from industry and the workers' compensation insurance carriers to the medical community at large dealing with back pain patients. While a program's expense had always been a major issue, it seemed to take a lesser role in importance as these rehabilitation facilities flourished. The predominant criterion for program success, or lack of it, in the mind of the payer was often measured by a facility's consistent ability to get the injured worker back on the job, not by the expense of the program—a feature that certainly made the prospect of opening a facility of this type financially more attractive than in the 1960s.

The 1990s: Let's Get Organized

Basic back school information, let alone personalized stabilization training for patients, is still not sufficiently widespread. Our consulting office has calls every day from individuals around the world trying to find facilities that offer this type of training in their community. The referral list of facilities throughout the United States that offer back school and stabilization training numbers only about 400. Many were members of the original North American Back School Association formed in 1984. When one considers the commonly quoted statistic that 80 percent of the population will have back pain at some time sufficiently serious to seek medical attention, there simply are not enough openings in the few spine centers and individual clinics around the country to even begin to handle the training required. So while the back school training concept of the 1970s became popularized and had a rapid growth in the 1980s, it has hardly even begun to touch the needs of huge numbers of back pain patients who are currently having sophisticated diagnostic tests, prolonged treatment, and even surgery, but who are still going untrained. Today one has only to sit in a spine center office for a short time to see the large numbers of patients who have had back pain for years, experienced a wide variety of therapeutic procedures, and gone through multiple surgeries, yet have never had any personalized back school education.

If the sluggish spread and use of patient education for the control of back pain are not at enough of an impasse in the 1990s, the minimal use of the same information for prevention programs in industry and for the general public is appalling. We have only begun to become aware of the costs and effects of cumulative trauma now that it has become a recognized, compensable distress. Ergonomic expertise for redesigning the workplace has been around for decades, and industry is still extremely slow to respond to the recommendations. Laws have had to be written and implemented setting forth ergonomic standards, guidelines, and strategies in order to protect the worker and regulate the employer. The process of getting government and industry attention about the necessity of industrial back injury prevention programs has been unfathomably slow and painful.

All of us are affected by the cost of back pain and health care in general. Premiums for workers' compensation insurance have risen steadily over the past 20 years. The rise in health care costs has also been staggering, and fees for service are being challenged by third-party payers on a daily basis. If a family member has had a medical problem or if we have become ill ourselves, we become acutely aware of the cost of medical care firsthand. Medical malpractice premiums have soared, and unless one is brave enough to go bare, we can feel that right in our overhead pocketbook.

The relatively few dedicated educators in the spinal medicine field actively using aggressive conservative care, and mainly education, as an important adjunct to clinical care, cannot carry the torch alone. The entire medical community dealing with patients with back pain must start an educational war on the epidemic—in the clinic, in industry, and in the public domain. Controlling and preventing back pain is everyone's responsibility. The extremely simple answers on how to do that have been around for 40 years. The age of patient education and back school is really just beginning, and must explode in the 1990s. All should get organized and demand that it happen.[27]

THE BASICS

At the very least, there is a core of information that can be given to most patients, regardless of their working diagnosis, over a very short period of time. The main purpose is to increase the patient's ability for self-care and functional capability. Function and independence are the goals. Most patients set their goal as pain reduction, and it is up to the clinician to help them set the correct priorities.

Most core back school programs include some basic spinal anatomy, basic body mechanics for activities of daily living, and a home or gym exercise program. Information can be given in a group setting within a two-week period of time. Three or four sessions of two hours each should prove sufficient.[28]

If general and family practitioners are to become the gatekeepers of most all patients at the onset of a medical problem, then they will also be the caregivers of patients with an initial onset of back pain. In a managed care system will there be time and dollars available to give a patient the basics of back health care? Will physical therapy of any kind be allowed for these patients? Will we return to medication, bed rest, and a tear-off sheet of exercises? In the past these back pain patients probably found their way to the chiropractor's or orthopedic surgeon's offices. Chiropractors have long been giving back school education and are well aware of the benefits. Orthopedic surgeons, physiatrists, and other specialists seeing numerous patients with spinal problems traditionally have sent these patients to educated physical therapists for back school and exercise training. What happens in the future is yet to be seen, but if we have just two hours of physical therapy available, perhaps the following lesson plan would at least give the simplest form of back school education.

1. Back School Basics:
   - Information on anatomy and where the pain is coming from
   - Posture practice
   - Information on pain relief and self treatment
   - Positioning practice for basic activities of daily living
   - Home exercises
2. Advanced Back School:
   - Neutral spine concept
   - Explicit, strong information on the value of trunk strength
   - Home exercises for stabilization, strength, and flexibility
   - Home exercise practice

For therapists who have been giving comprehensive back school education spread out over a luxurious time frame while patients made slow, methodical progress, the challenge will be to look at the lesson plans currently used and simply tighten them up. Clinicians may have the feeling that they will be on *fast forward*, and the fact is they will be.

STABILIZATION TRAINING: SOPHISTICATED BACK SCHOOL

While basic, core back school information is essential to each patient's understanding of his or her condition, stabilization training is a process that can take considerable time and effort—for both the patient and the physical therapist or trainer. Abdominal strengthening is the cornerstone of the stabilization program. This exercise approach to back rehabilitation combines the basics of back school, proprioceptive neuromuscular facilitation, sports medicine, manual therapy, and exercise. Numerous continuing education programs are available.[29]

It is tremendously helpful to the physical therapist and the patient if the patient's physician understands the stabilization concept and consistently motivates the patient. Using diagrams on a chalkboard or even examining table paper can help patients understand their specific problem and exactly why each step of the stabilization process is necessary in the rehabilitation process. If patients truly believe that their physician wants them to reach a certain level of stabilization training, it can make the process much more rewarding and effective. This training is hard work, and if patients think they can avoid the hard work by complaining to the physician about how hard the physical therapist is on them, there are definitely going to be less-than-perfect results. The physician should team with the physical therapist in order to get patient buy-in.[30]

Patients are taught specific stabilization exercises divided into four categories of difficulty. Patients are progressed in these categories as skill and strength levels advance. Each level increases in complexity, strength required, and reflexes. Most therapists use a large gym ball to accomplish some of the more difficult maneuvers. While each therapist might have his or her customized regime for each patient, the following example of a lesson plan might be used as a guide (see Table 7–1).

Specific exercises are introduced for each level of training. As the patient becomes stronger and more coordinated, the level is advanced. This training is quite precise, and goals are set according to the needs and lifestyle of each patient.[31]

Stabilization training is a self-help and management program that challenges patients to their highest level of capability. The concepts of this training can also be integrated in most back injury prevention programs.

NECK SCHOOL

Cervicothoracic stabilization training is an obvious extension to any back school and stabilization training program. Most patients with neck pain or thoracic pain are treated with cervical collars, traction, ultrasound, electric stimulation, soft tissues massage, and/or joint mobilization. These are all forms of hands-on therapy that are considered important in the pain-relieving phase of care. However, if communication, prevention, and self-responsibility are the name of the game in the "changing health care environment," a strong education program for these patients could maximize function and help prevent further injury. Basic neck school stresses postural reeducation, mobility, flexibility, and stabilization during more strenuous activities.[32]

THE FUTURE OF BACK SCHOOL

As health care reform aggressively goes into action, everyone involved in patient care will probably have yet another set of guidelines to follow. There will probably be clear direction about who will be seeing what patients, for how long, and for how much compensation. Physical therapy presumably will remain an important adjunct to medical care. While in the past most rehabilitation and training could conceivably go on for many months in order to get a patient properly educated and strengthened, it is most probable that all physical therapy visits will be greatly limited. If this does in fact turn out to be the case, it would be important to begin planning now how this back school education and training can be reduced to something acceptable to the payor and still be extremely beneficial to the patient. Every health care provider should be planning a strategy now that will set his or her clinic apart for efficacy, costs, and outcomes. Physicians and therapists who regularly see patients with back pain would not want to treat these patients without the added help of a formalized back school program, but they may have to in the future.

Table 7–1 Stabilization Progression Levels: Trunk Exercises

Level I Beginning

Supine	*Sidelying*	*Quadruped*	*Prone*
Abdominal bracing Pelvic clock Opposite hand-knee push Supported dying bug (alternate arm/leg lift) Short arc bridging Isolated free weight training Theraband exercises	Hip abduction	Four-point rock Four-point arm/leg lift	Gluteal sets Short arc upper/lower half extensions

Level II Intermediate

Supine	*Sidelying*	*Quadruped*	*Prone*	*Standing*
Partial/diagonal curls Unsupported dying bug Single straight-leg lowering/air bike One-legged bridge Bridging with leg lift/extension	Hip adduction Bilateral leg lift Fire hydrant	Four-point reciprocal arm/leg Fire hydrant	Upper/lower half extensions	Free weight training/isolated weight equipment Theraband resistance training

Level III Advanced

Supine	*Prone*	*Weight Training (Increase Endurance and Strength)*
Unsupported dying bug (add wrist and ankle weights)	Full ROM back extensions over ball (combined with sets of alternate arm lifts)	Free weights Weight machines

continues

Table 7–1 (continued)

Level III Advanced (continued)

Supine	*Prone*	*Weight Training (Increase Endurance and Strength)*
Partial and diagonal curls (on incline or with chest weight) Bilateral straight-leg lowering Ball walk/ tremble point Ball bridging	Bilateral leg extension over ball (combined with sets of alternate leg flutter) Prone dying bug (alternate arm leg lift)	

Level IV Sport and Job-Specific Training

Ball Exercises (Increase Lever Arm/Balance Difficulty)	*Equipment*	*Weight Training*
Bridge Push-up Partial/diagonal curls with medicine ball	Standing chair (hip flexion with knee extension) Roman chair (back extension)	Combine pulley (advanced job/sport-specific weight training)

Source: R. Robinson, The New Back School Prescription: Stabilization Training Part I, Spine, *State of the Art Reviews*, Vol. 5, No. 3, pp. 17–31, 1991.

It has been stated that the new health care reform will include "prevention," although it has not been clearly defined for what diseases or conditions. If Arthur White, MD, past president of the North American Spine Society, is correct, back schooling will begin before any form of formal education. It will be an informal part of a child's education at home. During the formative years of life, as the child's spine is maturing, parents will observe, encourage, and reward postures and body mechanics that will be commonly understood to be of long-range benefit. A child will not grow up with the misconception that he has one God-given posture that he is stuck with for his entire life. He will know that he is responsible for and has control of his own posture. He will grow to understand which muscles are required to maintain a healthy back. Individuals will no sooner blame their work for back pain than they would blame their work for tooth decay. Not only will back school be taught as the worker enters the workforce, but workers will automatically be tested and monitored for their ability to do any job safely without injury. If back pain should occur in a working person, the entire system will support the individual in his treatment and recovery. Even though the individual injured worker takes responsibility for letting his guard down, the employer and the insurance industry take responsibility for not watching closely enough and allowing the worker to get injured. All systems, therefore, go to work and share the cost of immediate accurate diagnosis, correction of the condition, and alteration of the job or future expected activities of the injured worker.[33]

While in our worst nightmare we do not want back school education regressing to the tear-sheet exercises of the 1950s, everyone currently involved in the care of patients with back pain will simply have to get more efficient and yet not jeopardize quality care. We should probably all be considering the following: If I am only given two or three hours over a one-month period of time to treat and train a patient, what information and techniques am I going to cram into that precious time? The results of rethinking all of the ideas of so many dedicated clinicians over the past 40 years could turn out to be an even better educational experience for patients suffering from back pain. Change always presents challenges, and this is most definitely one of them.

NOTES

1. S.W. Brown and A.P. Morley, *Marketing Strategies for Physicians* (Oradell, N.J.: Medical Economics Books, 1985).
2. A. White et al., *Back School and Other Conservative Approaches to Low Back Pain* (St. Louis: C.V. Mosby, 1983), 1–180.
3. H. Kraus, Diagnosis of Low Back Pain, *General Practitioner* 5(1952).35–39; H. Kraus, Prevention of Low Back Pain, *Journal of the Association for Physical Medicine and Rehabilitation* 6(1952):12–15.
4. W. Fahrni, *Backache and Primal Posture* (Vancouver, B.C.: Musqueam Publishers, Ltd., 1976); W. Fahrni, *Backache, Assessment and Treatment* (Vancouver, B.C.: Musqueam Publishers, Ltd., 1976), 1–89.
5. V. Mooney, Alternative Approaches for the Patient beyond the Help of Surgery, *Orthopaedic Clinics of North America* 6(1975):331–334.
6. M. Zachrisson, *The Low Back Pain School*, Danderyd Hospital, Danderyd, Sweden, Sound and slide program (4 lessons), 1970.
7. M. Forsell, M. The Back School, *Spine* 6 (1981):104–106.
8. J. Armstrong, *Lumbar Disc Lesions* (London: E&R Livingstone, Ltd., 1958).
9. M. Berquist-Ullman and U. Larsson, Acute Low Back Pain in Industry, *Acta Orthopedica Scandinavica* 170(1977):1–117.
10. White et al., *Back School*, 1–180.
11. R. McKenzie, *The Lumbar Spine: Mechanical Diagnosis and Therapy* (Waikanae, New Zealand: Spinal Publications, 1981).

12. Armstrong, *Lumbar Disc Lesions*.
13. J. Shah et al., The Distribution of Surface Strain in the Cadaveric Lumbar Spine, *Journal of Bone and Joint Surgery* 60B(1978):246–251.
14. McKenzie, *The Lumbar Spine*.
15. A. White et al., *The Back School: An Audiovisual Team Approach to Low Back Pain* (St. Louis: C.V. Mosby, 1984).
16. H. Hall, The Canadian Back Education Units, *Physiotherapy* 66(1980):115–117.
17. E. Attix and M. Tate, Low Back School: A Conservative Method for the Treatment of Low Back Pain, *Journal of Mississippi State Medical Association* 20(1979):4–9; C. Dutro and L. Wheeler, Back School and Chiropractic Practice, *Journal of Manipulative Physical Therapy* 9(1986):209–211; J. Fisk et al., Back Schools: Past, Present and Future, *Clinical Orthopedics* 179 (1983):18–23; C. Hayne, Back Schools and Total Back-Care Programs: A Review, *Physiotherapy* 70(1984):14–17; H. Hurri, The Swedish Back School in Chronic Low Back Pain. Part I. Benefits, *Scandinavian Journal of Rehabilitative Medicine* 21(1989):33–40; H. Hurri, The Swedish Back School in Chronic Low Back Pain. Part II. Factors Predicting the Outcome, *Scandinavian Journal of Rehabilitative Medicine* 21(1989):41–44; S. Linton and K. Kamwendo, Low Back Schools: A Critical Review, *Physical Therapy* 67(1987):1375–1383; J. Moffett et al., A Controlled, Prospective Study To Evaluate the Effectiveness of a Back School in Relief of Chronic Low Back Pain, *Spine* 11(1986):120–122; R. Pawlicki et al., The Low Back School: A New Palliative Approach to Low Back Pain, *West Virginia Medical Journal* 78(1982):249–251.
18. D. Bartorelli, Low Back Pain: A Team Approach, *Journal of Neurosurgical Nursing* 15(1983):41–44; L. Caruso et al., The Management of Work-Related Back Pain, *American Journal of Occupational Therapy* 41(1987):112–117; K. Harkappa et al., A Controlled Study on the Outcome of Inpatient and Outpatient Treatment of Low Back Pain. Part I. Pain, Disability, Compliance and Reported Treatment Benefits Three Months after Treatment, *Scandinavian Journal of Rehabilitative Medicine* 21(1989):81–89.
19. R. Stankovic and O. Johnell, Conservative Treatment of Acute Low-Back Pain. A Prospective Randomized Trial: McKenzie Method of Treatment versus Patient Education in "Mini Back School," *Spine* 15(1990):120–123.
20. S. Bigos and M. Battie, Acute Care To Prevent Back Disability: Ten Years of Progress, *Clinical Orthopedics* 221(1987):121–130; J. McElligot et al., Low Back Injury in Industry: The Value of a Recovery Program, *Connecticut Medicine* 53(1989):711–715; A Morris and J. Randolph, Back Rehabilitation Programs Speed Recovery of Injured Workers, *Occupational Health and Safety* 53(1984):64–68.
21. J.A. Saal and J.S. Saal, Nonoperative Treatment of Herniated Lumbar Intervertebral Disc with Radiculopathy: An Outcome Study, *Spine* 14(1989):431–437.
22. G. Johnson and V. Saliba, Post-graduate courses in orthopedic and neurological manual therapy and exercise training, offered by the Institute of Physical Art, 45 Tappan Road, San Anselmo, Calif. 94960; D. Morgan, Concepts in Functional Training and Postural Stabilization for the Low Back Injured, *Topics in Acute Care Trauma Rehabilitation* 2(1988):8–17; D. Morgan, et al., Course in education in manual therapy: Training the patient with low back dysfunction, offered by Folsom Physical Therapy, 115 Natoma St., Folsom, Calif. 95630.
23. J.A. Saal, General Principles and Guidelines for Rehabilitation of the Injured Athlete, *Physical Medicine and Rehabilitation: State of the Art Reviews* 1(1987):523–536; Saal and Saal, Nonoperative Treatment.
24. T. Sweeney et al., Cervicothoracic Muscular Stabilization, *Physical Medicine and Rehabilitation: State of the Art Reviews* 4(1990):335–359.

25. R. Lichter et al., Treatment of Chronic Low Back Pain: A Community Based Comprehensive Return-to-Work Physical Rehabilitation Program, *Clinical Orthopedics* 190(1984):115–123; L. Matheson, *Work Capacity Evaluation: Systematic Approach to Industrial Rehabilitation* (Anaheim, Calif.: ERIC, 1986); T. Mayer et al., A Prospective Two-Year Study of Functional Restoration in Industrial Low Back Injury: An Objective Assessment Procedure. *JAMA* 258(1989):1763–1767.
26. M. Liang and A. Komaraoff, Roentgenograms in Primary Care Patients with Acute Low Back Pain: A Cost-Effectiveness Analysis, *Archives of Internal Medicine* 142(1982):1108–1112.
27. L.A. White, The Evolution of Back School, Spine. *State of the Art Reviews* 5, no. 3(1991):325–332.
28. L. Martin, Back Basics: General Information for Back School Participants, Spine. *State of the Art Reviews* 5, no. 3(1991):333–340.
29. B.J. Headley, *The "Play-Ball" Exercise Program* (St. Paul, Minn.: Pain Resources, Ltd., 1990); Johnson and Saliba, Post-graduate courses; Morgan et al., Education in manual therapy.
30. J.A. Saal, The New Back School Prescription: Stabilization Training Part II, Spine. *State of the Art Reviews* 5, no. 3(1991):33–42.
31. San Francisco Spine Institute, *Dynamic Lumbar Stabilization Program*, audiovisual program available from 1850 Sullivan Avenue, Daly City, Calif. 94015, 1989.
32. Sweeney et al., Cervicothoracic Muscular Stabilization; T. Sweeney, Spine, *State of the Art Reviews* 5(1991):367–377.
33. White, The Evolution of Back School, Spine, 506.

Chapter 8

Establishing an Industrial Prevention Program

Mary A. Mistal

PREVENTION NEEDS

It has been estimated that in any industrial society, 10 percent to 15 percent of the adult population annually have some work disability as a result of back pain.[1] This results in a substantial impact on employers, with 1.3 billion lost person-days per year in the United States as a consequence of back pain.[2]

The escalating prevalence and devastating cost of back injuries to industry have prompted many to take proactive measures to control and reduce their losses. One of these measures is back care education programs administered in a group situation to the employees. The programs are based on available scientific knowledge of the physiology and mechanics of the spinal structures and their relationship to daily activities. The aim of these programs is to help individuals assume responsibility for their pain by providing a better understanding of the problem.[3]

The following concepts, when utilized by prevention specialists, will assist in methods of controlling the overwhelming losses from back injuries. The primary emphasis of program development should be the employees' self-responsibility for maintaining a healthy, fully functioning body in order to remain productive in all lift aspects.

PREVENTION PROGRAMS

Consulting Process

A consult for developing an effective back injury prevention program can be described as having four phases.

Phase I—Initial Contact

The initial contact with a particular industry is usually made with upper management and/or the safety director. In this preparatory phase, it is important to discover the specific needs of the company concerning the control of back injuries. One should ask questions concerning

- The number of employees
- Basic demographics of the employee population (i.e., age and gender)
- The presence of shift work or flextime and the availability, the shift times, and the number of hours per work shift
- The incidence rate of back injuries over the past year, differentiating new and recurrence rates
- The most frequently reported cause of a back injury
- Whether certain workers and/or job tasks produce a higher frequency of injuries
- The number of lost worker days in the past year
- The cost of lost productivity to the company during the previous year
- The previous annual cost of medical intervention
- The previous annual cost of legal intervention
- Past interventions (educational, ergonomic considerations, employee suggestions) and their success or failure
- The type of workers' compensation under which the company is protected (self-insured, governmental jurisdiction, state fund, or private)
- The company's basic philosophy and commitment to safety
- The purpose of the company (production versus service)
- The work environment (i.e., temperature, lighting, fumes, etc.)
- The postures and positions assumed most frequently
- Whether special clothing is required
- The tools and equipment necessary for job tasks
- The company's expectations that relate to the prevention consultation
- The potential benefits of intervention to the company and employees

This information is the foundation upon which the program is built. It provides a general overview of the company and insight into management's commitment to the program. Specific trouble areas can be identified in preparation of the job task observation phase. The appropriate services can then be determined. It must be stressed to management during this phase that no single program aspect will solve the complete problem. For any intervention to be successful, a constant dedication from all levels of management must be inherent. A cooperative and collaborative relationship between management and the consultant can then be established through the recognition of the responsibilities of each party.

Preliminary discussion of the type of services to be provided can occur at this point. The consultant can outline the basics of the educational problem and emphasize that customization for this specific company is imperative. This customization results from observation and photographing of job tasks within the company, particularly those of high risk. It should be explained that the photographs and/or videotaping are for the express purpose of analysis of the job tasks and utilization as visual aids in the instructional presentation.

Beyond an instructional program, other services can be offered. These may include, but are not limited to, prework exercises, stress reduction and management, early intervention following injury, changes in ergonomic factors, prework screening, wellness programs, and functional job specific testing.

Phase II—Observation

In this stage, the consultant walks through the facility to observe and document the postures, positions, body mechanics, essential job tasks, and working environment. Special attention is given to six work factors reported by Anderson[4] to be associated with sickness or absence resulting from low back pain:

1. Physically heavy work
2. Maintenance of a static position for prolonged periods of time, including prolonged sitting or lumbar flexion
3. Frequent heavy lifting or unexpected lifting
4. Spinal bending and twisting
5. Repetitive work
6. Exposure to cyclic loading and vibration, such as while driving or operating vibrating equipment

When using a camera, appropriate lighting is necessary. Static documentation of all aspects of the job can be provided through slides or photographs. Videotaping can be advantageous due to its ability to capture the dynamics of work. In either case, it is important to record as many proper biomechanical techniques and postures as possible. This lends greater credibility to the instruction program given as part of the prevention package.

Additional information is accumulated during this phase. The types of tools and equipment necessary for work are analyzed. The lighting, temperature, humidity, noise factors, clothing required, and the ability of the workers to freely assume required postures are documented. Measurements of horizontal and/or vertical distances, weights to be lifted, and push/pull forces can also be made. Suggested equipment for the consultant includes

1. 35 mm camera or video camera and film

2. Stop watch for repetitive and/or static activities
3. Tape measure
4. Goniometer, if needed, to measure work angles
5. Written notations material
6. Push/pull gauge

Once the walk-through has been completed, the next stage can begin.

Phase III—Program Development

The key tenets incorporated into the instructional program are

1. Basic anatomy of the spine, emphasizing the interrelationship of the spinal structures
2. Proper body mechanics related to the job-specific activities
3. Exercises including the essential concepts of flexibility, strengthening, and endurance
4. Self-care techniques
5. Information on intradiscal pressures as the foundation for the application of appropriate biomechanical techniques[5]

Colorful and interesting slides are used for the anatomy section. Individual illustrations of the spinal components are helpful to demonstrate the integrity of each and how they interrelate. Instruction is kept brief, precise, and geared toward the audience. Emphasis is placed on the cumulative nature of back pain and the effects of work and recreational activities on the development of pathology. Specificity of symptoms is not included to avoid the power of suggestion. The significance of employee self-responsibility is introduced at this point. Various medical treatments are mentioned, ranging from the most conservative to the most radical, with the underlying message that most back pain subsides on its own. However, back pain can become a lifetime problem with little guarantee of a cure if not understood and handled appropriately.

The body mechanics portion involves the fundamentals of safety, pacing, proper positioning, and use of the lower-extremity joints and musculature to assist with bending and/or lifting activities. Stabilization of the spine through abdominal muscle contraction is also stressed.

The photographs (slides) taken during the observation phase are included. These are to facilitate attention from the audience and reinforce the basic tenets initially introduced. At this time active audience participation is used to underscore the importance of spinal stabilization and incorporation of proper lifting, bending, and twisting. This can be done employing the props of job-specific tools, equipment, and weights. Static postures are demonstrated and practiced as well. Additionally, relevant problem solving takes place as the audience and consultant interact to ensure that the techniques being practiced can be assimilated into the work environment.

The essential exercise concepts of flexibility, strengthening, and endurance are promoted. The flexibility component is customized to the tasks and postures assumed in the workplace. This is done to acquaint the workers with the idea of prework warm-up exercises and the necessity of frequent breaks throughout the shift to assist in controlling stress and fatigue to the muscular and ligamentous structures at greatest risk. Other stretching exercises are also presented to use at home to assist with relaxation and self-care.

The exercises for strengthening include simple isometric abdominal contraction to enhance the concept of spinal stabilization and lower-extremity strengthening. Again, active audience participation is employed for emphasis. Supplementary exercises for strengthening the major muscle groups needed for job tasks are presented.

The importance of endurance exercise activity is highlighted. Presentation of the components required for aerobic exercise and its benefits is made. A brief discussion of aerobic physiology is conducted to enhance comprehension of the beneficial aspects related to reduced fatigue, greater energy level, and cardiovascular improvement. Comparisons between aerobic and anaerobic activities are presented.

Self-care aids are recommended. These include proper positions for resting, activities of personal care, activities of daily living, sitting, use of heat or ice for local relief, and when to seek medical attention. Emphasis on self-responsibility and self-care is reiterated.

The conclusion of the presentation reviews the high points of intradiscal pressures, the principles of proper body mechanics, exercise, and self-care methods. Any methods concerning the program or personal back issues from the workers complete the program.

Follow-up seminars are conducted in an abbreviated format. The sessions usually occur every three to six months. The original basic principles and troubleshooting are presented to confirm that comprehension, application, and assimilation has occurred.

Written materials can be prepared to augment the didactic instruction. These materials can be effective in any form from a single page to a 3 × 5 card or a booklet format designed for the company. The necessity for these materials should be discussed with management to ascertain the appropriateness and who is responsible for their production.

Phase IV—Follow-up and Supportive Services

In this final phase, ancillary consultation consists of providing follow-up presentations and other supportive services. The presentations, as mentioned earlier, are a condensed form of the original program. The main features are a review of intradiscal pressures, job specific body mechanics, and exercise. Enhanced interaction between the workers and the consultant is encouraged to collaboratively solve any problems that may have been identified in the intervening time. Creativity from the consultant is needed for these sessions to promote interest and heighten awareness.

Management Role

Since the success of the intervention largely depends on the proactive commitment of all levels of management, communication with line supervisors and middle and upper management is imperative. Worker compliance with the principles of proper body mechanics occurs from active and diligent observation by the line supervisors. Middle and upper management can assist with the monitoring of the success of the program through injury-reporting processes, loss run figures, and productivity changes. Communication with all levels of management can determine where inconsistencies in compliance exist, explore what changes can be made to reduce the noncompliance, and offer alternative solutions.

Exercise

Once the door has been opened through the educational program, suggestions for additional services can be offered. The prework warm-up exercise program has been previously introduced. It can be emphasized that such a program is an integral part to the total picture. The research validates the efficacy of such programs in enhancing productivity, morale, and injury prevention.

Stress Control

Stress reduction and management for all employees can be promoted. In the overall scheme, if stress is appropriately managed, other intervening factors, such as fatigue, can be reduced and subsequently result in a reduction of back injuries. General principles of stress management can be formulated and presented with unique factors inherent within the company to be underscored as benefits or detriments. Progressive relaxation and implementation of hourly stretch breaks can be proposed as effective job site stress reducers.

Early Care

Early intervention for injury is yet another service that can be provided. This may consist of on-site medical intervention, off-site contracting with preferred providers, and/or case management recommendations. The emphasis is on the reduction of loss time with expeditious, suitable, and aggressive care.

Ergonomics

Modification in ergonomic factors for certain job tasks and sites can be suggested. These modifications can be as simple as changing a seat height or as complex as the complete redesign of the setting. It is important to talk with the workers since they are the experts in their job and have the best understanding of how to make their job easier and more efficient. It is also important to bear cost-efficiency in mind. A company accepts or rejects recommendations according to the bottom line of how much will be saved or produced. For those work tasks with the highest propensity for injury, ergonomic modifications may prove to be the most cost-effective in the reduction of the number of reported injuries. Listening to the employees who perform the tasks and to their recommendations concerning improvement is the most effective method of understanding the hazards involved and the solutions to those hazards.

Prework Screening

For those companies that are hiring new employees on a regular basis, prework screening can be offered as an effective method of appropriately matching the worker to the job.[6] Simulated job activities are performed to determine the correlation of the prospective employee's physical ability with the demands of the job. Ingenuity and flexibility are key factors in the performance of prework screening, in addition to a thorough understanding of the work tasks to be tested to ensure reliability and validity of the testing process.

Wellness

Wellness programs have come of age in recent years. Many incorporate the concepts already presented in the back injury prevention programs and are expanded with additional health and wellness information. Depending on which illnesses result in the most lost time for the company, and on the extent of the company's interest in promoting the health and wellness of its employees, an appropriate program can be designed. These programs can include identification of lifestyle behaviors contributing to cardiovascular diseases, obesity, and poor conditioning. Through this identification, an educational program can be created to promote awareness of the available opportunities for self-improvement and the appropriate utilization of medical intervention.

Return-to-Work Testing

To assist in expeditiously returning the injured worker to work, functional job specific testing can be offered. This testing entails determination of the essential functions of the work tasks; work heights, distances, repetitiveness, and pacing; worker positions and postures; and job simulation. A thorough investigation of the work tasks is necessary to ascertain these facts, thus ensuring accuracy and validity of the test results. Parameters of testing are established, with emphasis on worker safety and employment of proper body mechanics for the specific work tasks involved. Testing should be performed once the injured employee's medical stabilization has been established. A general musculoskeletal examination is conducted initially to rule out any secondary problems that may exist or that may interfere with testing. Functional testing occurs after the preliminary physical examination. The test items are designed in accordance with the physical demands of the job. The worker is encouraged to perform each task to his or her safe physical maximum, employing proper body mechanics. Comparison of the test results with the physical demands of the job is then made to objectively correlate the worker to the job. Job site modifications may be indicated and recommended as a result of the functional testing.

CONCLUSION

A complete industrial injury prevention program should be designed to fit the unique needs of the industry. Particular aspects of the program are founded on the specific job tasks, the injury rate and causes, the worker population, and the potential suggestions for solving the presenting problems. A "total program" approach assists the management of a particular industry to choose the most beneficial and cost-effective method.

The best education program can end up being totally ineffective if it is not delivered in a meaningful manner to the audience. The education components require creativity, ingenuity, a sound scientific foundation, and humor. These elements, along with communication skills, can make even the most basic information lively and meaningful.

Overall, stressing the workers' self-responsibility in injury prevention is imperative. Without this and a firm commitment from management, no interventions can be efficacious in reducing the physical, monetary, and production losses incurred from an injury. Enhancing workers' awareness of their responsibility in maintaining safe work habits, performing proper body mechanics at all times, and altering detrimental lifestyle habits can positively affect their work and home life.

NOTES

1. L.N. Matheson, *Low Back Injuries in Industry* (Washington, D.C.: Forum Medicum, Inc., 1989).
2. L. Harris and Associates, *Nuprin Pain Report* (New York, 1985).
3. J.A. Klober Moffett et al., A Controlled, Prospective Study to Evaluate the Effectiveness of a Back School in the Relief of Chronic Low Back Pain, *Spine* 11, no. 2 (1986):120–122.
4. G.B.J. Anderson, Epidemiologic Aspects of Low Back Pain in Industry, *Spine* 6, no. 1 (1981):53–60.
5. A. Nachemson, In Vivo Measurements of Intradiscal Pressures, *Journal of Bone and Joint Surgery* 46A (1964):1077–1092.
6. H. Fearon, PreWork Screening, in *Orthopedic Physical Therapy Clinics of North America*, ed. S.J. Isernhagen (Philadelphia: W.B. Saunders, 1992), 37–47.

Chapter 9

Carpal Tunnel Syndrome Research

Lucinda A. Pfalzer and Barbara McPhee

Carpal tunnel syndrome (CTS) is one of a number of cumulative trauma disorders and is a significant cause of work-related impairment, disability, and compensation in hand-intensive tasks in business and industry.[1] The term *cumulative trauma disorder* (CTD) is used in the scientific and occupational literature in the United States to indicate a group of work-related musculoskeletal and peripheral nerve disorders associated with highly repetitive and/or forceful activity. Several other umbrella terms are used in the literature from other countries, such as *occupational overuse syndrome (OOS)*, *repetition strain injury (RSI)*, *repetitive motion disorder*, and *occupational cervicobrachial disorder (OCD)*.

Cumulative trauma disorders including CTS occur more frequently in persons who work at specific occupations or tasks than in those who do not (the general population), indicating that work factors contribute to their development.[2] Gerald F. Scannell, Assistant Secretary of Labor, Occupational Health and Safety Administration (OSHA), stated in 1989 that virtually every workplace in the United States has the potential to cause cumulative trauma disorders.[3] Epidemiological data in industrialized nations indicate that upper-extremity cumulative trauma disorders are increasing, although overall incidence is unknown. In the United States, over half of all work-related illnesses reported are CTDs (Figures 9–1 to 9–3).[4] As such, they, along with CTS, probably also are classified as work-related injuries under the categories of sprains/strains and overexertion/physical stress in Australia and the United States. Therefore the real prevalence and incidence are difficult to determine.

The U.S. Bureau of Labor Statistics (BLS) 1988 Annual Survey of Occupational Illnesses and Injuries reported a 25 percent increase from 1987 in new cases of occupational illness among workers in private industry.[5] Cumulative trauma disorders excluding back injuries accounted for more than 80 percent of this increase, to which CTS appears to be a major contributor.[6] Occurrence of CTS in the

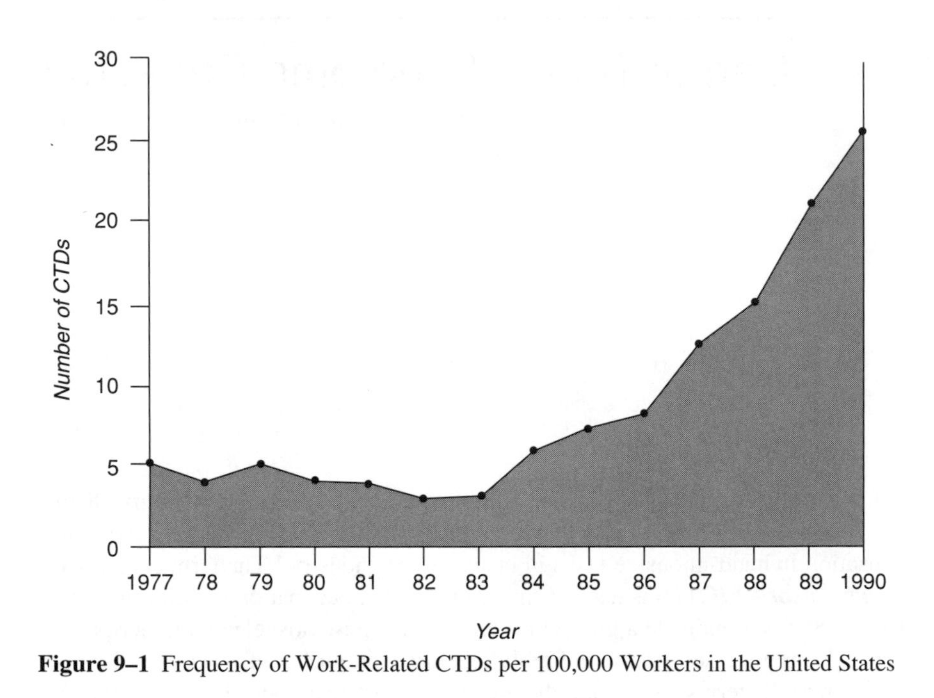

Figure 9–1 Frequency of Work-Related CTDs per 100,000 Workers in the United States

general U.S. population in 1982 was estimated at 1 percent, and as many as 10 percent of adults had occasional symptoms.[7] In 1988 Stevens et al., in a study of persons from Rochester, Minnesota, beginning with data collected from 1976 onward, found a low rate of occurrence in an adult population, with the peak age of incidence in the 60s, and many more females affected than males.[8] Franklin et al., from data collected from 1984 to 1988 on workers in Washington State, found a much higher frequency, a younger peak age, and almost equal rates between males (1.0) and females (1.2).[9] De Krom et al. reported a prevalence of 9.2 percent in the adult female population and 0.6 percent in the adult male population in the Netherlands, with 5.8 percent of the female CTS cases previously unreported.[10] Workers with CTS appear to be younger than the general population of people with CTS, and unlike the general CTS population, the worker population with CTS shows an almost equal distribution between males and females.[11]

The issues to be addressed when dealing with CTS are not simple. How the increasing prevalence and incidence are to be reversed will be the challenge that employers, policymakers, employees, and researchers must face. In doing so a range of theoretical and practical problems must be solved.

Some of the issues important in addressing this problem are

1. How to identify workers at risk and stratify them accordingly into groups of low, moderate, and high risk of developing cumulative trauma disorders,

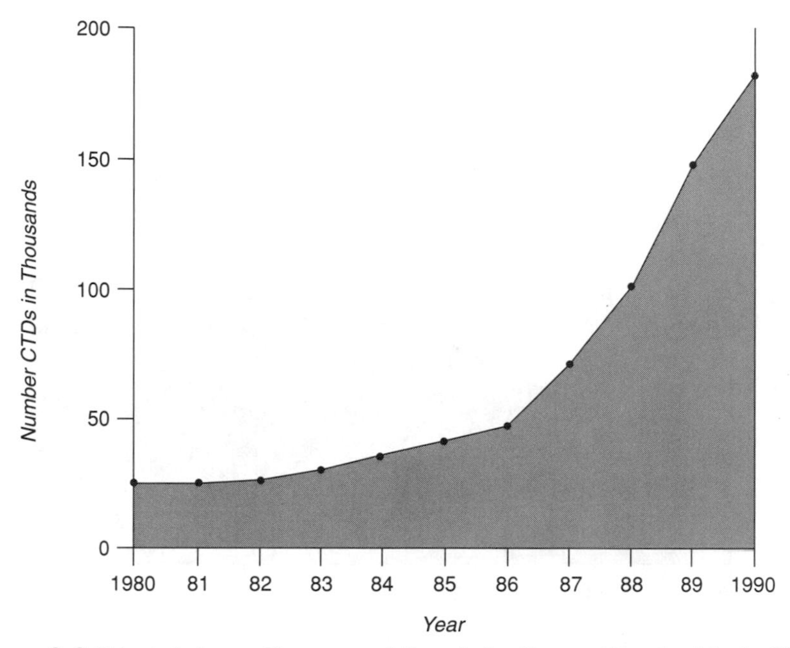

Figure 9–2 Private Industry: Frequency of Cumulative Trauma Disorders* in the United States

*Manufacturing accounted for about 160,000 of 185,400 cases.

including CTS. Which risk factors can be used to identify and stratify workers at risk? What are the reliability and validity of measurement tools used to screen for risk and to diagnose CTS?

2. How to demonstrate the efficacy of prevention. Are there intervention trials that demonstrate efficacy of primary, secondary, and tertiary prevention programs, and what are optimal prevention programs or their important components?
3. How to identify the biomechanical and morphological changes that cause CTS. What pathophysiology leads to development of pressure on the median nerve? What are the mechanisms of the sensory disturbances that are often the early symptoms of CTS?

This chapter will review selected relevant research about the problem in order to address the following issues:

- pathophysiology theories of CTS
- sources of information in this area that clinicians may access through a public, university, or hospital library and government agencies (see Appendix 9–A)

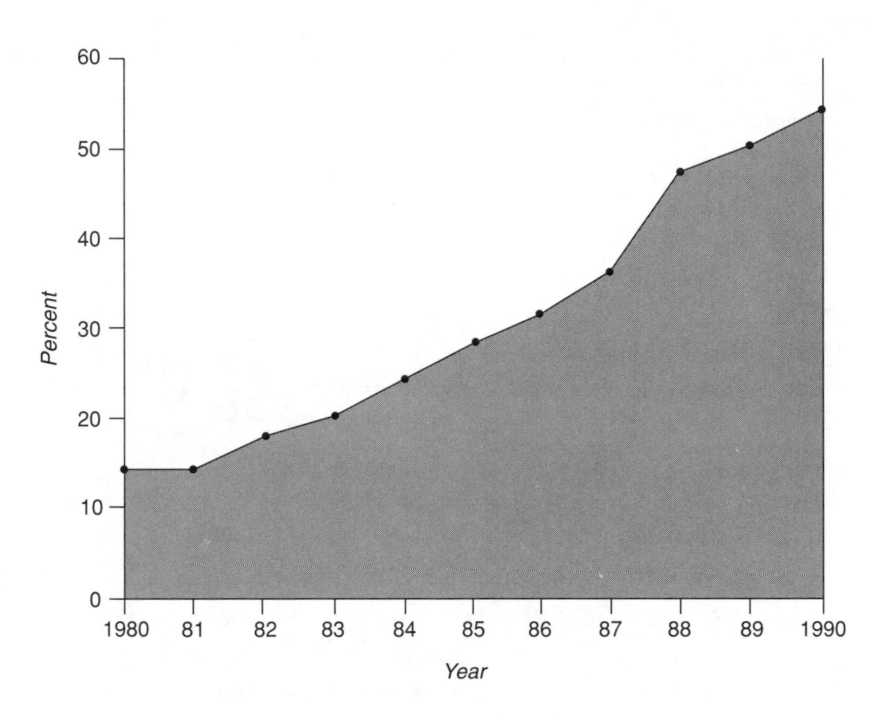

Figure 9–3 Private Industry: Frequency of Cumulative Trauma Disorders* in the United States

*Manufacturing accounted for about 160,000 of 185,400 cases.

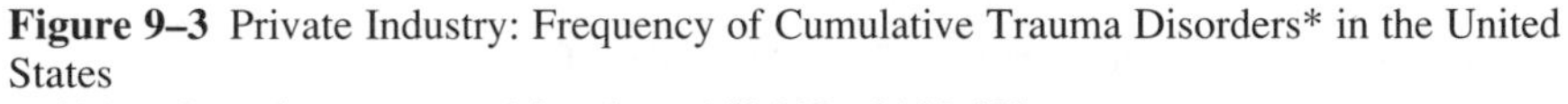

- epidemiological data (see Appendix 9–B)
- measurement issues in health outcomes, occupational exposure factors, and methodological problems (see Appendix 9–C)
- current interventions, including prevention strategies in the United States and Australia, diagnosis, and treatment (see Appendix 9–D)

PATHOGENESIS OF CTS

Signs and Symptoms

CTS is a combination of signs and symptoms characterized in its early stages by (1) diffuse aching, (2) pain, and (3) paresthesia. Paresthesia includes nocturnal paresthesia along the sensory distribution of the median nerve in the hand and fingers, which may refer into the wrist and forearm and occasionally up the entire upper extremity, and diminished sensory nerve latency along the median nerve distribution in the hand compared to the forearm.[12] In addition to the early signs and symptoms, the later stage of CTS is characterized by

- thenar and intrinsic muscle weakness in the index and middle fingers, with reduced grip and pinch strength
- thenar muscle atrophy
- diminished median motor nerve latency and conduction velocity
- possibly a positive Tinel's sign or Phalen's test
- occasionally other pseudomotor signs in the hand

The symptoms begin as episodic events that are relieved by rest and aggravated by use. Eventually rest brings minimal relief of the symptoms, with shorter exposure to work factors aggravating the symptoms.

Mechanisms of Injury

Peripheral neuropathies present with a wide spectrum of sensory disturbances that appear to be dependent on pathophysiological changes in axoplasmic flow and impulse propagation relative to alterations in sodium, potassium, and calcium channels in the nerve fiber.[13] CTS was recognized as a clinical problem in the medical literature in 1895 but was not fully described as a peripheral mononeuropathy, median nerve entrapment syndrome until 1947.[14] It may be acute, subacute, or chronic. A peripheral mononeuropathy such as CTS may be of unknown etiology, secondary to trauma or a systemic disease, or secondary to work factors such as repetitive work activities with inadequate work-rest ratios and task variation, and/or use of excessive force.[15] Work can aggravate CTS associated with other factors.

Many theories are proposed as to the cause(s) of pressure on the median nerve at the wrist leading to nerve compression and the resultant symptoms of CTS. These suggest that either biomechanical factors or vascular ischemia result in median nerve compression at the wrist. Arterial blood supply in the distal forearm to the median nerve is by small arterioles from the ulnar and radial arteries, forming a plexus. Elevated pressure can occlude blood flow, resulting in ischemia to the distal median nerve. Fibrosis restricting nerve gliding in the carpal tunnel is an example of a potential morphological cause of median nerve compression in the wrist. One potential biochemical cause of compression in the carpal tunnel is that proximal nerve compression or inflammation along the nerve course, such as in the thoracic outlet, results in a change in nerve metabolism or reduction in axoplasmic flow, increasing the risk of distal injury at the level of the wrist. The biochemical and morphological tissue changes need clarification.

Peripheral Nerve Anatomy

A peripheral nerve is a complex structure consisting of nerve fibers (axon with attached motor or sensory ending and cell body), connective tissue, and blood vessels. These three tissues comprising the nerve have different responses to trauma.

Nerve Fibers

Sensory type A fibers are large myelinated fibers relaying impulses associated with touch, pressure, cold, skeletal muscle tension, and joint position. These type A fibers and end organs alter their response to stimuli in both symptomatic and nonsymptomatic arms in persons with CTDs, changing the sensory information sent to the central nervous system (CNS) for processing.[16]

Connective Tissue

Three successive layers of connective tissue surround the nerve fibers, protecting their continuity. The endoneurium is the innermost layer, the perineurium is the middle layer, and the epineurium is the outermost layer. These layers of connective tissue are essential since the nerve fiber is easily damaged by stretching and compression. Damage or trauma to the nerve fiber can result in edema and an increased endoneurial fluid pressure.[17] This pressure can affect the microcirculation of the nerve fiber and nerve function.[18]

Microcirculation

Capillaries lie longitudinal to each nerve fascicle, obliquely entering the perineurium. Lundborg believes that with a rise in tissue pressure inside the nerve fascicles, these capillaries close like valves.[19] This may explain why small elevations in pressure are associated with reduced intrafascicular blood flow.[20]

Nerve Biomechanics

Maximal tensile loads that the median nerve can sustain prior to failure are in the range of 70 to 220 newtons.[21] Severe intraneural damage is produced by tensile loads long before the nerve fails. Debate over the elastic and biomechanical properties of nerves as composite tissues continues.[22] In various normal nerves, the elastic limit during tensile loading is about 20 percent of maximal elongation, with complete structural failure at about 30 percent of maximal elongation.[23] Beel et al. demonstrated that injured nerves have increased stiffness and decreased elasticity, altering their mechanical properties.[24] Tensile injuries are usually associated with acute trauma, such as brachial plexus injuries. During this type of injury, nerve fibers rupture before the endoneurium and the perineurium rupture.[25] Rydevik et al. demonstrated, in a rabbit tibial nerve that was transected, sutured, and stretched, that continuous impairment of capillary flow occurred at 15 percent of maximal elongation.[26] However, functional changes are minimal or nonexistent in slow, gradual tensile loading of a peripheral nerve.

Compression loading of a peripheral nerve can produce signs and symptoms such as paresthesia, pain, and muscle weakness.[27] The critical pressure levels that lead to these functional changes are identified when a nerve is compressed locally at one segment. The mechanical factors that appear important in determining the extent of the structural nerve injury (disturbances of intraneural blood flow, axonal transport, and nerve function) are the pressure level, type of compression, and duration of compression.[28] Lundborg et al. found that at 30 mm Hg of local compression, nerve function was altered, and that if the compression was maintained for 4 to 6 hours permanent nerve damage occurred.[29] The functional changes appear due to impaired blood flow in the compressed nerve.[30] Corresponding compressive pressures were reported in the carpal tunnel near the median nerve in patients with CTS. Control subjects had an average pressure of 2 mm Hg.[31] At about the same compressive load (30 mm Hg), axonal transport is altered. Upton and McComas state that such blockage in axonal transport may cause axons to be susceptible to additional compression distally, and they have called this the "double-crush syndrome."[32]

Longstanding intermittent compression of the nerve between about 30 to 80 mm Hg may induce intraneural edema and scar formation.[33] At the top end of this range of pressure (about 80 mm Hg of local compression), intraneural blood flow is blocked as ischemia occurs. Rydevik et al. found that even after two hours of compression, blood flow is restored with relief of the compression.[34] Rydevik and Norborg demonstrated that higher pressures (200–400 mm Hg) for shorter duration induced structural nerve damage with rapid loss of function and partial recovery.[35] Ochoa et al. demonstrated the edge effect of compression loading of large-diameter nerve fibers.[36] A lesion is induced at both edges of the compressed nerve segment, while the center of the compressed segment where the pressure is highest is mostly unaffected.

Rydevik and Lundborg demonstrated that intraneural blood vessels are also damaged at the edges of the compression.[37] This appears to be a consequence of the pressure gradient and shear displacement or deformation of the nerve, which are largest at the edges of the compression. Functionally this leads to motor and type A sensory fibers being at greater risk of damage, while smaller diameter fibers (C fibers) are spared.[38] At the wrist the median nerve has discharged most of its motor fibers and is largely composed of sensory fibers (type A and C) from the hand. As with other peripheral neuropathies, differential fiber loss does not explain the variety of sensory disturbances found.[39]

A recent case control study by Helme et al. of patients with RSI found altered sensory responses in the patient group, with modulation of both peripheral and central pain pathways.[40] This supports the premise that pathophysiological changes occur in the peripheral nerve along with changes in central processing of sensory feedback. The pathophysiological processes of altered sensory feedback from damaged peripheral nerves such as the median nerve in CTS and its contribution to chronic pain require further exploration of the morphological mechanisms. Two possible mechanisms that may explain the sensory disturbances are an insertion of abnormal nerve impulses back to the CNS or an amplification or distortion of normally generated nerve impulses back to the CNS.[41]

The basic mechanisms (biomechanical, neuromuscular, and neurovascular) of creating pressure on the median nerve also require further investigation.[42] While there is general agreement that compression on the median nerve leads to a rise in pressure with resultant interference with nerve homeostasis, conduction, and blood flow, research must encompass more than the end result of the pathophysiology of pressure on the nerve. It must also examine the outcome of the compression, altered sensory feedback to the central nervous system, and its effect on the CNS and motor and pain behavior.

Conditions and Disorders Coexisting with CTS

CTS can be occupationally and nonoccupationally induced. Nonoccupational CTS tends to present with the classical symptoms and signs previously described. It is occasionally associated with past or current history of systemic illnesses and acute trauma to the region (Exhibit 9–1). An estimated 20 to 30 percent of adults with CTS have related systemic diseases. On the other hand, occupational CTS initially may present with more generalized symptoms and signs unless acute trauma is the cause. Several studies in different populations of workers have associated CTS with overuse.[43] Work-related factors associated with occupational CTS include high repetition, high force (both static and dynamic), and awkward postures used to perform the work tasks. Vocational and avocational tasks requiring these hand-intensive efforts can contribute to risk or aggravate an existing case of CTS when performed over a long enough period of time.[44]

Work-related CTS occurs frequently in association with several musculoskeletal and neuromuscular disorders, such as flexor pollicis longus tenosynovitis and double crush syndrome, and systemic diseases, such as diabetes mellitis. Associated conditions include tenosynovitis and tendonitis, which are two of the most frequently occurring musculoskeletal disorders, along with other conditions such as ulnar nerve entrapment, basal joint arthritis of the thumb, lateral epicondylitis of the elbow, and brachial plexus injury.[45] Exhibit 9–1 lists the disorders, conditions, and diseases reported as associated with CTS that may increase the risk of developing CTS. Murray-Leslie and Wright found that 33 percent of the CTS cases also had lateral epicondylitis of the elbow as compared to only 7 percent of a control group.[46] One case study has found that 30 to 40 percent of persons with CTS also have ulnar nerve entrapment.[47] It has been reported that many workers complained of symptoms elsewhere in the upper quarter before or while they had

Exhibit 9–1 Factors Associated with Carpal Tunnel Syndrome Reported in the Literature

I. Anatomy
   A. Decreased size of carpal tunnel
      1. Abnormalities of the carpal bones
      2. Thickened transverse carpal ligament
      3. Acromegaly/myxedema
      4. Congenital anatomical abnormalities
   B. Increased contents of canal
      1. Trauma
         a. Colles fracture
         b. Volkmann ischemic contracture
         c. Distal radius and ulna fracture callus
         d. Post-traumatic osteophytes
         e. Distal radial epiphysis fracture
         f. Carpal bone with dislocation
         g. Pseudoarthrosis of the scaphoid
         h. Degenerative joint disease
      2. Neoplasm and other masses
         a. Benign/cysts and malignant tumor of bone
         b. Neuroma/nerve sheath tumor
         c. Lipoma
         d. Multiple myeloma
         e. Tuberculous granuloma
         f. Ganglions
         g. Metastatic tumor
         h. Neurofibromatosis
      3. Abnormal vasculature
         a. Persistent median artery (thrombosed or patent)
         b. Permanent shunt for renal dialysis
         c. Hematoma (hemophilia, anticoagulation therapy)
         d. Raynaud's phenomenon
         e. Other circulatory disturbances such as blood dyscrasia
      4. Abnormal muscle bellies
         a. Flexor digitorum sublimis
         b. Lumbricales
      5. Hypertrophic synovium
   C. Variation in the median nerve
      1. High division of nerve in forearm
      2. Variation in thenar innervation
      3. Accessory branch at distal portion of nerve
      4. Accessory branch proximal to tunnel
II. Physiology
   A. Neuropathic conditions
      1. Diabetes
      2. Alcoholism
      3. Proximal lesion of median nerve (double crush syndrome)
      4. Polyneuritis

continues

Exhibit 9–1 (continued)

- B. Inflammatory/autoimmune conditions
  1. Tenosynovitis (repeated flexion and rheumatoid arthritis)
  2. Rheumatoid arthritis and polymyalgia
  3. Infection such as tuberculosis
  4. Gout and pseudogout
  5. I° and II° amyloidosis
  6. Calcific tendinitis
  7. Suppurative tendinitis
  8. Dermatomyositis
  9. Lupus erythematosus
  10. Psoriasis
  11. Sarcoidosis
- C. Alternations of fluid balance (hormonal/metabolic)
  1. Pregnancy
  2. Eclampsia
  3. Myxedema
  4. Long-term hemodialysis/renal failure
  5. Horizontal position and muscle relaxation with possible prolonged wrist flexion or extension (sleep)
  6. Raynaud's disease
  7. Obesity
  8. Menopause
  9. Oral contraceptives

III. Position and use of the upper extremity and wrist
- A. Repetitive wrist flexion/extension
- B. Static and dynamic loading—repetitive forceful squeezing and release or pinching of a tool, or repetitive forceful torsion of a tool
- C. Finger motion with the wrist extended—constrained posture
  1. Typing
  2. Playing many musical instruments, etc.
  3. Use of pinch grips
- D. Mechanical stress concentrations/contact trauma—direct pressure
  1. Weight bearing with the wrist extended
  2. Contacting base of palm, palmar surface of fingers, or sides of fingers
  3. Paraplegia
  4. Long-distance bicycling, etc.
- E. Vibration forces on the hand and wrist
- F. Immobilization with the wrist in flexion and ulnar deviation
  1. Casting after Colles fracture
  2. Awkward sleep positions
- G. Use of poor-fitting gloves

IV. Work organization and administration factors

V. Work satisfaction and stress factors

Source: Adapted from *Occupational Hand and Upper Extremity Injuries and Diseases* by M.L. Kasdan (Ed.), p. 343, with permission of Hanley and Belfus, Inc., © 1991.

symptoms of CTS.[48] Therefore it is more appropriate to place work-related CTS in a framework of upper-quarter cumulative trauma disorders. A past medical history or a current report of tenosynovitis or tendonitis, epicondylitis, or ulnar nerve entrapment appears to be a predictor for developing CTS.

In several epidemiological studies, the ratio of cases of musculoskeletal disorders in the wrist and hand to cases of CTS is on the order of 5:1 up to 8:1.[49] Reports of this ratio have been as high as 2:1 in a number of case reports.[50] CT scans and MRI techniques are being explored as diagnostic tools and to aid in determining risk. However, this technology is expensive and has had mixed results to date.[51]

CTS presenting primarily in association with numerous other conditions and diseases has been previously reviewed in Exhibit 9–1. Along with an inability to accurately predict risk of developing work-related CTS from work factors alone, it is reasonable to believe that multiple occupational and individual factors are involved in producing the nerve compression. Debate continues regarding the importance and relationships of various work-related and personal factors and the causes of CTS.[52] Although basic knowledge is still lacking, epidemiological studies and meta-analysis of the literature such as the analysis by Stock strongly indicate a causal relationship with occupation.[53] Predictors of disorders need further attention to enable early identification and stratification of risk so that optimal prevention strategies may be undertaken.

EPIDEMIOLOGY OF CUMULATIVE TRAUMA DISORDERS AND CTS

Frequency of Disorders

Workers' compensation statistics or other centralized compilations of statistical data are a useful and efficient way of monitoring the occurrence of a disorder in a population. In the Nordic countries specialized databases have been established to allow the incidence and prevalence of work-related disorders of the locomotor system to be accurately determined. Most other countries must rely on workers' compensation statistics. Unfortunately, workers' compensation data are less reliable and of limited value for four main reasons:

1. The limited scope of what constitutes a "case," due to the requirement of a minimum absence from work of five days
2. The various disease/illness and injury classifications into which locomotor disorders may be put
3. The lack of sensitive or specific diagnostic criteria by which a set of signs and symptoms may be deemed to be a "case"
4. A significant percentage of workers with signs and/or symptoms who never seek or are not eligible for compensation for a variety of reasons

Generally, given these barriers, there is a tendency to underreport and misclassify CTDs both in the United States and Australia.[54]

Despite these difficulties, occupational musculoskeletal and neuromotor disorders appear to pose an increasing health problem for workers in industrialized countries. The incidence and prevalence statistics used as indicators of the frequency of injury or illness show a rising percentage of CTDs over the last eight years in the United States (Figure 9–3).[55] Upper-extremity CTDs represent over half of all work-related illnesses reported and an estimated 25 to 35 percent of all occupational injuries reported as overexertion/repetitive motion (mostly strains and sprains) in workers' compensation statistics. In Australia, these work-related disorders, when classified as injuries, are reported as overexertion/physical stress (mostly strains and sprains) (Figures 9–4 and 9–5) and as OOS when classified as an illness.[56]

While the U.S. statistics indicate a growing number of CTD cases, there is evidence of significant underreporting in the frequency statistics.[57] This problem is compounded by how the statistics are kept in both the United States and Australia. The U.S. Bureau of Labor Statistics (BLS) and the Australian Bureau of Statistics use workers' compensation data to compile their statistics, with the likelihood of underreporting the frequency of cases.[58] Mild or early cases of CTDs are not counted in the workers' compensation statistics because these workers are not absent from work for a long enough period of time to be classified as a case (minimum of five days absence). Severe underreporting of cases of occupational CTS is

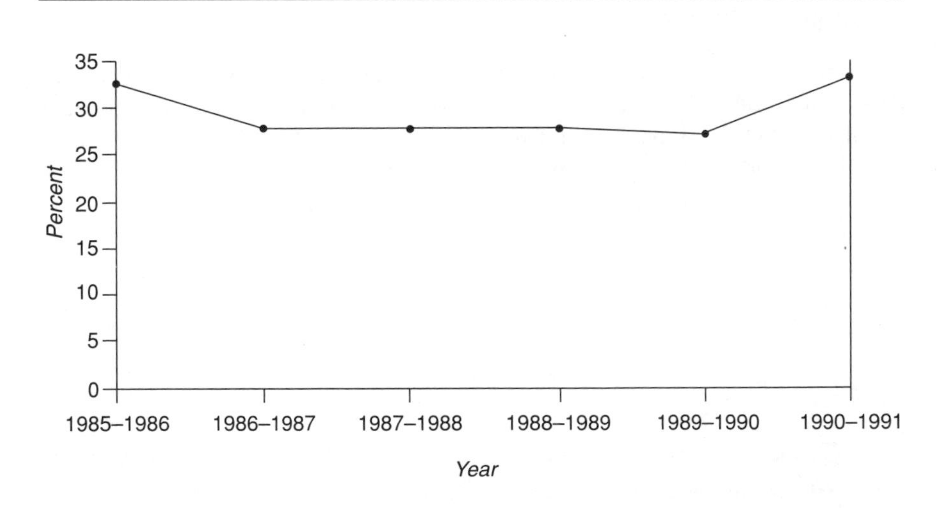

Figure 9–4 Work Injury: Percentage of Overall Accidents Due to Overexertion/Physical Stress in Australia

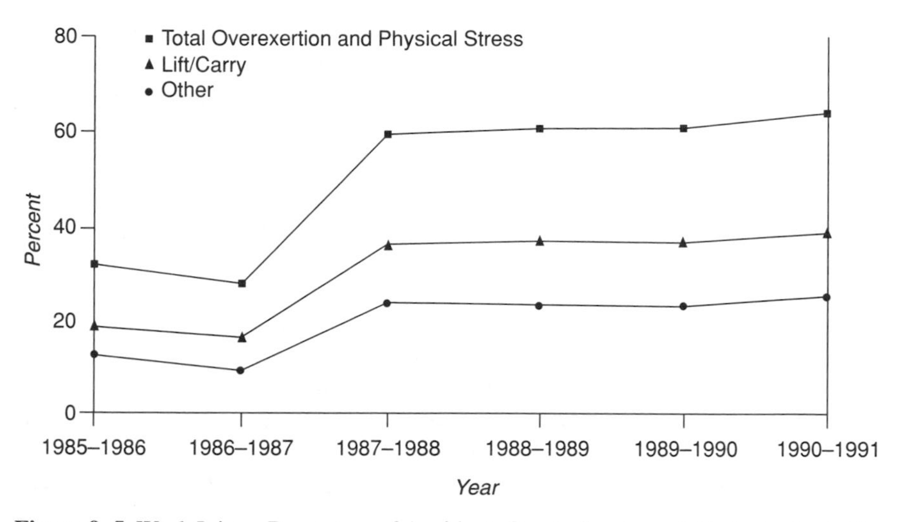

Figure 9–5 Work Injury: Percentage of Accidents Due to Sprains and Strains in Australia

documented in a California Department of Health Services survey specifically examining CTS.[59] In this study, 498 providers (30 percent response rate) reported caring for 7,214 cases of CTS during 1987 in Santa Clara County. Of these, the providers stated that nearly one-half (3,413 cases) of the cases were work related. Only 71 cases were reported through normal channels. A study conducted in two automobile plants identified four to ten times as many CTD cases through monitoring personal medical absences as through monitoring workers' compensation cases. A University of California study estimated that 130 percent more work-related cases of CTS occur than are reported.[60] This study found that only 60 percent of work-related injuries and 44 percent of work-related illnesses are reported.

Another problem with accurate frequency and severity data from workers' compensation sources is that many CTDs are reported as sprains or strains under work injury rather than as an occupational disease or illness, where they are more easily identifiable in the statistics. In Australia, the majority of CTDs may actually be found under the occupational injury category of "overexertion/physical stress" rather than under the occupational illness category of "OOS" (Workcover Statistics, Workcover Authority of NSW), and in the United States under the occupational injury category of "sprains and strains" rather than under the occupational illness category of "repetitive motion disorders." Approximately half of the sprains and strains reported in the state of New South Wales, Australia, from 1985 to the present are overexertion/physical stress work injuries (Figure 9–5).[61]

While there is an acknowledged problem of underreporting, the U.S. BLS data indicate a pattern of consistent increases since 1981 to the most recent data available, from 20 percent of all recorded illnesses (not including back injury, also considered a CTD in the United States) to approximately 60 percent of all recorded illnesses.[62] A 1991 report from the U.S. Department of Labor reported that CTDs rose from 115,000 cases in 1988 to 147,000 cases in 1989 and 185,400 cases in 1990 (Figure 9–2).[63] In 1988, CTDs accounted for some 50 percent of all reported work-related illnesses, in 1989 for some 52 percent, and in 1990 for some 56 percent (Figure 9–3), and they are expected to account for approximately 60 percent in 1991.[64] Tanaka et al. reported that CTDs were the second most frequently reported category of occupational illness after skin disease.[65] During a five-year period from 1980 to 1984, data from the Ohio Industrial Commission showed that the wrist was affected in almost half of all CTD claims, and 75 percent of these were tenosynovitis due to continuous motion.[66]

In 1987, manufacturing had about 20 percent of the private sector employment but slightly more than 33 percent of the total reported CTDs. By 1990, manufacturing accounted for about 160,000 (86 percent) of the 185,400 cases reported. Figure 9–6 shows the 1987 and 1988 frequencies and Figure 9–7 the 1991 frequency from BLS data in several industries.[67] The five manufacturing industries with the highest prevalence of occupational CTDs in 1989 were: (1) meatpacking,

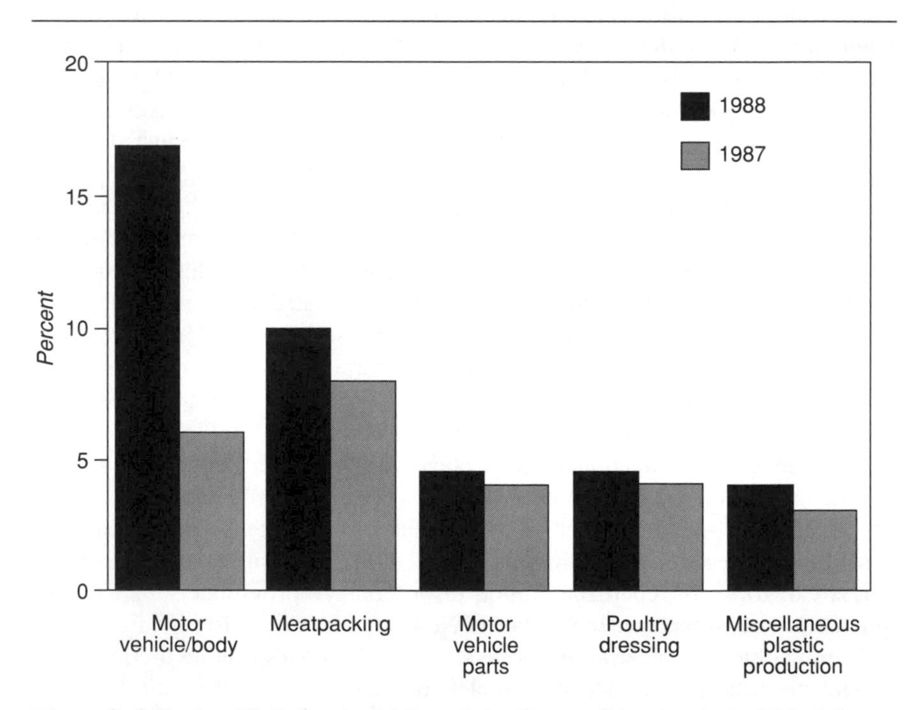

Figure 9–6 Hardest-Hit Industries of Cumulative Trauma Disorders in the United States

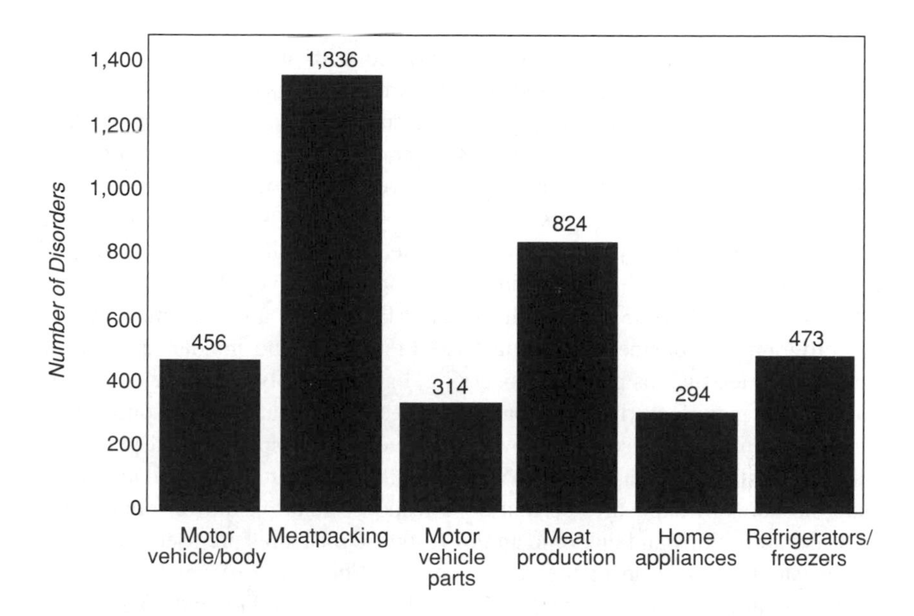

Figure 9–7 Work-Related Repetitive Motion Disorders per 100,000 Workers in Select Industries in the United States

(2) manufacturing household appliances, (3) rubber and plastics footwear, (4) office furniture and fixtures, and (5) motor vehicles and equipment. These injuries also appear to be on the rise in the white-collar workforce and are associated with VDT units/keyboard use.[68] Brogmas and Marko analyzed frequency and severity data from Liberty Mutual Insurance Company, the largest underwriter of workers' compensation in the United States. Their findings indicate that manufacturing continues to be overly represented in the area of CTDs (see Figure 9–2).[69] A recent report of the National Institute for Occupational Safety and Health (NIOSH) using 1981 to 1983 data estimated that in the United States 8 to 9 million workers are exposed to work-related risks associated with the development of these disorders.[70]

Generally in Australia routinely collected data on different aspects of occupational health and disease are available only from workers' compensation statistics. These have been of limited value for the reasons discussed earlier in this section. The difficulties have been compounded by major restructuring of the workers' compensation systems in most states and territories in the period 1987 to 1990. During this time a number of changes were made to the way compensation statistics were reported, collected, and analyzed. Also in 1990/1991 the basis on which cases were selected was changed. Therefore while statistics prior to 1987 can be compared year to year, they cannot be compared with statistics after that date (1990/1991 being the only fully modified set available at this point).

A more sensitive outcome or response measure than workers' compensation claims may be the use of self-reported symptoms from an employee survey.[71] However, one study found that symptoms of lower-back problems were more numerous among industrial workers when those symptoms were self-reported, as compared to when those symptoms were reported as a result of the completion of the Nordic questionnaire.[72] Self-administered employee surveys have been widely used by OSHA and NIOSH in Health Hazard Evaluation and Technical Assistance programs for business and industry.[73] Prevalence and incidence rates for several industries such as plastics processing, light assembly such as electronics, motor vehicle manufacturing, meatpacking, telecommunications, and publishing are a few examples. These data, however, are company-specific and generally should be considered as case reports. Several of the reports have limited and inadequate measures of exposures so that dose-response relationships are not available. Case definitions and outcome and/or response measures are used, but without consistent application of the same criteria. While this information can be helpful in identifying specific problems and risk factors for CTDs and CTS, cause-effect and dose-response relationships need further research. In addition, government statistics on frequency and severity of CTDs and CTS should be interpreted with caution. A review of issues relating to diagnostic criteria and screening can be found in this chapter under the Secondary Prevention section.

Given these problems, discussion in Australia of the occurrence of musculoskeletal disorders of the upper quarter associated with work and especially CTS is limited. Nevertheless, due to the so-called "epidemic" of repetition strain injuries (RSI, later renamed *occupational overuse injury* or OOS) that occurred in Australia in the mid-1980s, there was a heightened awareness in the community of the nature of the disorders and the range of work-related factors associated with their development. However, very little good-quality research was carried out at that time to further clarify the workplace issues and the nature and range of conditions being classified under the heading of RSI or OOS. Therefore information on the conditions is vague, and only broad categories of workers have been identified as being "at risk."

Statistics reveal that a number of main subgroups are represented in the epidemic: keyboard workers, especially in large financial organizations such as banks and insurance companies and in public service (federal, state, and municipal governments); food processing, especially in packaging; meat and poultry processing; electronics assembly; clothing manufacturing; construction; and light, medium, and heavy manufacturing industries. The growth in the number of cases of RSI (OOS) during the mid-1980s in Australia was greatest in the fastest growing area—keyboard work.[74] Occupationally induced CTS has received far less attention in Australia than in the United States, and it is very likely that many cases are misunderstood and misclassified as a result. In 1990/1991 in New South Wales, 154 new cases of CTS were compensated, and constituted 1.8 percent of the 8,463 new disease cases for that year. On the other hand, there were 6,883 new cases of sprains and strains of the limbs and 2,293 new cases of sprains and strains of the neck and shoulders, representing 16.5 percent and 5.1 percent respectively of the 41,235 new cases of injury during the same period.[75]

Severity of Disorders

In the United States in 1988, the actual cost of each CTD case ranged from $15,000 to $25,000.[76] More complicated or severe cases requiring extensive medical treatment such as surgery, compensation, and disability may cost up to $60,000.[77] The National Association of Rehabilitation Facilities (NARF) has reported that the average cost to treat CTS per case (direct medical cost) is approximately $3,500. But actual costs per case reported by Pinkham range between $15,000 and $25,000 when other direct and indirect costs such as workers' compensation are included.[78] Some workers receive additional compensation, which varies in amount depending on whether they are able to return to work. Disability claims received for CTS in the United States are often approximately $100,000.[79] In 1985, for example, $22,470 million was paid in workers' compensation costs, with $7,381 million for medical costs and $15,089 million for income benefits. A conservative estimate of total direct and indirect costs of all cumulative trauma disorders in the United States is at least $2.5 billion per year.

Risk Factors Associated with the Development of Disorders

Workplace Factors

A great deal of international literature about work-related CTS and other upper-extremity cumulative trauma disorders has been published, primarily in Australia, Europe and Scandinavia, Japan, and the United States. The majority of this literature is qualitative, increasing our general knowledge of the physical dimensions of the common cumulative trauma disorders (tendonitis, tenosynovitis, deQuervain's Syndrome, CTS, tension neck syndrome) and factors in the workplace that might give rise to these disorders. Several papers identify specific risk factors related to work. A risk factor is any attribute, experience, or exposure that increases the probability of occurrence of a disease or disorder, though it is not necessarily a causal factor.[80] By this definition, the type of work or occupation a person engages in can be a risk factor for CTS.

Many biomechanical and epidemiological studies have identified a range of occupational risk factors associated with the development of neck and upper extremity disorders. These include

1. Repetition and inappropriate work/rest cycles[81]
2. Forces[82]
   a) Static[83]
   b) Dynamic[84]
3. Extreme wrist postures or movements[85]
4. Use of poorly fitting gloves[86]
5. Exposure to cold[87]
6. Vibration[88]
7. Mechanical stress or direct pressure or trauma over the palm[89]
8. Work stress and job satisfaction[90]

An excellent meta-analysis of the literature by Stock highlighted the paucity of quantitative studies giving insight into risk factors and causal relationships in upper-extremity cumulative trauma disorders of the wrist, including CTS.[91] Nevertheless, the conclusions of the meta-analysis support causal relationships existing between workplace ergonomic factors of repetition and force and cumulative trauma disorders of the wrist, including CTS. Notwithstanding this evidence, some clinicians remain unconvinced that there are major causal links between these disorders and work factors.[92] There is a need for further quantitative studies for predicting risk and causal relationships through prospective and intervention studies.

Personal (Individual) Factors

Debate continues over which individual (personal) factors predict risk for CTS. They are reported in the literature as follows:

1. Age[93]
2. Gender[94]
3. Physiological attributes such as motor control
4. Job satisfaction and work stress[95]
5. Nonoccupational hand activities[96]
6. Pregnancy[97]
7. Physical activity[98]
8. Smoking[99]
9. Alcohol usage[100]
10. Systemic disease (rheumatism,[101] diabetes[102])
11. Previous repetitive work and job experience[103]
12. Other psychosocial factors such as stress[104]
13. Previous musculoskeletal disease or injury[105]

14. Oral contraceptives[106]
15. Increased weight/obesity[107]
16. Decreased height[108]
17. Increased body mass index (BMI)[109]
18. If female, s/p hysterectomy without oophorectomy[110] and, during the first year of menopause, history of dieting[111]
19. Wrist diameter ratio[112]
20. Hand dominance[113]
21. Family history of similar disorders[114]

Several of these factors are controversial, requiring confirmation from further research. Studies have suggested obesity or increased weight as a risk factor but have not found a statistically significant association.[115] De Krom's case control study did not find an association in females of the use of oral contraceptives, pregnancy, diabetes during pregnancy, menopause, age of onset for menarche or menopause, wrist fractures, or venous varicosities.[116] However, Dieck and Kelsey found an association between age at menopause in women and venous varicosities in men and CTS.[117] Rheumatism, thyroid disease, and diabetes were found not to be risk factors in men or women.[118] Bleeker found that external wrist circumference did not predict cross-sectional area of the carpal tunnel from CT scan.[119]

The Interaction of Work and Personal Factors in the Development of Signs and Symptoms

Empirical evidence indicates that increased risk of strain to any individual occurs when:

- new or excessive demands are placed on the worker who normally would have no difficulties coping with the job (e.g., temporary transfer to an unfamiliar job)
- the individual habitually works beyond his or her capacity either by direction or by personal choice
- personal, social, or environmental factors reduce the individual's tolerance to physical stress[120]

The relationship between physical workload and its effects on functional capacity and on the development and severity of symptoms appears to be modified by temporal factors, such as length of working day, periods worked without breaks, and the percentage of the working day spent doing repetitive activities in fixed postures. As well, personality, mood, the perception of load, work pressures, job satisfaction, and other personal factors may alter the individual's response to early signs of fatigue and discomfort.

The following factors need to be considered in the study of these disorders.

1. External load factors (task and workplace design and work organization) required by the task:
   - Number of movements
   - Static muscle work
   - Force
   - Work postures determined by equipment and furniture
   - Time worked without a break

   Usually these aspects are controlled by the employer, with little chance for the employee to influence change.
2. Factors that influence load but may vary between individuals:
   - Work postures adopted
   - Static muscle work used
   - Unnecessary force used
   - Number and duration of pauses taken
   - Speed and accuracy of movements[121]

   Many of these aspects can be modified through training in correct work techniques and rearrangements of work routines to accommodate individual differences.
3. Factors that alter the individual's response to a particular load (workplace, individual, and social factors):
   - Age
   - Gender
   - Physical capabilities
   - Environmental factors such as vibration, cold, noise, and other contaminants
   - Previous repetitive work and job experience
   - Psychosocial variables

   Most of these factors are beyond the control of either the worker or the employer but must be factored into the design of the work and workplace to accommodate difficulties that inevitably arise from them. Environmental factors and previous work experience may have a positive or negative effect on different individuals.

Occupational Groups at Risk

In epidemiological studies, categories of workers doing specific and demanding jobs such as welding,[122] butchery[123] and meatpacking,[124] telephony,[125] vehicle assembly work,[126] data entry,[127] electronics,[128] small press work,[129] light to heavy industry,[130] grocery checkers,[131] dental hygienists,[132] and sewing[133] have been identified as being at higher risk than the general population of developing symptoms of a variety of neck, shoulder, and upper-limb disorders. A review of epidemiological methods and occupational applications is found in Appendix 9–B.

Summary

Underreporting, misclassification, and overlapping categories of occupational injury and occupational illness lead to the conclusion that cumulative trauma disorders including CTS may comprise a greater percentage of occupational musculoskeletal injuries and illnesses than has previously been reported. The interrelationship of occupational and personal factors with the development of upper-extremity cumulative trauma disorders including CTS is not clear, and more study is necessary regarding what combinations of work and personal factors predict risk.

METHODOLOGICAL ISSUES IN THE STUDY OF CUMULATIVE TRAUMA DISORDERS AND CTS

Prevalence and Incidence Rates

Incidence rates represent the numbers of new cases (as defined) occurring in a specified group over a particular period of time. As such, incidence can be determined only through analytic studies such as randomized control trials, cohort studies, and case control studies. Incidence is a very powerful indicator of increasing or decreasing trends in the occurrence of a disorder.

Prevalence rate, on the other hand, is the number of cases in a specified group existing at any point in time. This might include new as well as recurrent cases. A higher prevalence rate might indicate increasing incidence and/or severity or chronicity of disorders that build up rather than improving and reverting to noncases. Prevalence may be determined through point or period prevalence studies (counting the number of cases at or during a specific period), through case studies, or retrospectively in case control studies. Prevalence indicates the numbers of cases, their severity, and how well they are being managed. More information on incidence and prevalence and how they may be calculated can be found in Appendix 9–B.

Measuring Occupational Exposure to Risk Factors

Along with determining outcomes (i.e., CTDs), measuring occupational exposure to workplace factors that might give rise to CTDs is possibly one of the most difficult areas to study. Not the least is the problem of which factors need to be measured. As well, the researcher must determine how such measurements might be made, recorded, and analyzed, and what level of exposure constitutes a risk for the development of disorders (over what period of time; how much repetition, force, or static muscle work is too much; individual susceptibility; what factors might magnify or ameliorate the impact of exposure). Often these are decided upon ad hoc when a significant proportion of the workers develops symptoms consistent with the outcome being studied, and therefore no uniformity is achieved across studies. In many cases measurement is limited by workplace restrictions, worker resistance or noncompliance, and the changes that might occur as the result of being studied (Hawthorne effect).

A variety of methods have been developed to study exposure, especially in epidemiology over the last 30 years. One of the more commonly used ones is to categorize workers solely on the basis of their jobs. These are usually specific and demanding and have been identified as giving rise to higher-than-expected levels of signs and symptoms related to musculoskeletal and neuromotor disorders. However, the exact elements of the jobs that are believed to lead to the symptoms are rarely identified.

A number of ways of estimating the effects of physical load on workers have been developed for ergonomics practice. Examples of more commonly used ones are OWAS and Posture Targeting for postural load analysis; Ergonomic Work Analysis (Finland) and AET (Arbeitswissenschaftliches Erhebungsverfahren zur Tätigkeitsanalyse) for general ergonomics analysis; and biomechanical load estimators such as the 2D and 3D models for estimating biomechanical load on the back and shoulders. However, none of these is designed specifically to study loads on the neck, shoulder, arm, and hand complex. A new method has been developed in Britain to quantify loads on the upper limbs called Rapid Upper Limb Assessment (RULA).[134] The major advantages of the system are that it is easy to learn, requiring minimal training, that it does not need special equipment, and that it has been designed to be a reliable tool for use by those without previous experience in observation. It provides a quick assesment of the postures of the neck, trunk, and upper limbs along with muscle function and external loads on the body. It can be used to assess large numbers of workers, with the derived scores indicating the level of loading experienced by the individual body parts.

Another approach is the psychophysical method. The psychophysical method of measuring load combines biomechanical and physiological stresses imposed by work into a single measure of perceived stress. It is commonly used in evaluating loads experienced by an individual during the execution of manual handling tasks but can be adapted for lighter, more repetitive work. The use of psychophysics in tests of work capacity generally allows the subject to be in control of several elements of the tasks that can be adjusted so that the work does not lead to undue fatigue or overexertion.[135]

This approach has some advantages in determining workload exposures for research purposes. It can be carried out on the job; it is relatively quick and inexpensive to undertake; it can give reproducible results that appear to be related to the development of symptoms of fatigue and musculoskeletal disorders. Its major disadvantage is that it is subjective and prone to the biases that subjects knowingly or unknowingly introduce.

The universal difficulty faced by all researchers studying humans is the issue of individual differences. When people are exposed to hazards in occupational situations, they react differently; this also can be said of their reactions to forceful and/or repetitive movements and prolonged fixed postures. An operation that is difficult and even damaging for one person may not constitute a risk for another. The higher the levels of physical stress, the greater will be the numbers who succumb to injury. Susceptibility to strain appears to be a continuum, with the highly susceptible at one end and the highly resilient at the other. If so, there is an argument for screening out susceptible individuals before permitting them to work at jobs known to cause symptoms, but this is not easy, nor is it usually acceptable. There must be some understanding of why some people are resilient and others are not, and tests used for screening must be highly reliable.

There appears to be some recent evidence that personal (individual) factors might influence this resilience in CTS. However, much more evidence is needed in the area of personal factors before they can be used to determine which individuals may be at greater risk of developing CTS and other cumulative trauma disorders as the result of their work. On the other hand, scientific evidence increasingly points to links between certain types of work and workplaces, and differences in individual work methods and the incidence of these disorders.[136]

Measuring Outcomes: Case Definition, Diagnosis, and Screening for CTS

Case Definition

A researcher may define a "case"—or outcome—in whatever way he or she pleases. Needless to say, there are difficulties in ensuring that case definitions are watertight and useable. Lengthy physical examinations and expensive tests may produce a marginally better result than, for instance, self-reported symptoms. However, training personnel, administering the examination and tests, and analyzing the results may not prove to be cost-effective. On another level the sensitivity, specificity, and validity of these procedures may be questionable, and the results may not truly reflect the prevalence of the outcome under examination.

The importance of standardized diagnostic criteria and reliable, valid tests for disorders such as CTS has been discussed elsewhere. Unless such criteria and tests are determined, universally agreed upon, and used, the issues discussed here will continue to be debated, and it is unlikely that any real progress will be made in either the prevention of the disorders or their case management.

Screening, Diagnosis, and Injury Surveillance

Screening workers for either predisposition or early signs of cumulative trauma disorders, particularly CTS, is a controversial area, not least because it may place the physical therapist or physician in an ethical dilemma. Preoffer screening rather than postoffer screening may unfairly discriminate against people applying for work if the tests have a low specificity and reliability. Also, if little or no attempt is made to reduce work risk factors, even the most resilient workers may succumb to injury.

Conversely, screening as a part of an injury surveillance program may be very effective. Nevertheless, it remains imperative to achieve the highest possible level of sensitivity, specificity, and validity for the tests and procedures used. During screening, workers are classified as to their likelihood of developing an illness or disorder in the future. Sensitivity, specificity, and predictive value are judged to determine which tests and measures are used to screen and classify workers. The calculation of sensitivity and specificity is usually based on chi-square analysis. Sensitivity, specificity, and predictive value are reviewed in Appendix 9–C.

The U.S. NIOSH SENSOR project to improve injury surveillance of CTS has generated data supporting the claim that there is significant underreporting of work-related CTS in the United States.[137] This project's case definition of CTS has also generated research on the sensitivity and predictive value of its clinical criteria and other clinical tests used to detect signs and symptoms of CTS. Katz et al. have reported a 38 percent misclassification rate using the SENSOR clinical criteria.[138] Electrophysiological tests were excluded from the case definitions in these studies so that they could be used as gold standards for determining cases from noncases. The NIOSH SENSOR case definition includes electrophysiological tests. De Krom et al. demonstrated low sensitivities and predictive values for several of the common special tests and clinical criteria used to screen for and diagnose CTS.[139] This indicates that the criteria used in the injury surveillance project are not adequate for use in diagnosis or screening of the workforce without inclusion of electrophysiological testing. Electrophysiological tests such as nerve conduction tests of sensory and motor latencies and nerve conduction have sensitivities and specificities ranging up to 90 to 95 percent depending on the test protocol used.[140] But problems remain with these testing procedures because of their cost and sophistication.

If we ignore those issues for the moment and examine the positive predictive value of the tests, the limitations in their usefulness are demonstrated.

Example 1. An example of the predictive value of a test is detection of CTS by a positive nerve conduction test battery (NCT). The test is 90 percent sensitive and

95 percent specific for CTS. Of 1,000 workers with wrist and hand pain in high-risk jobs, such as workers in a plastics processing plant with a frequency rate of 50 percent for all wrist hand complaints, 500 will have CTS. Ninety percent or 450 will have a positive NCT test (true positive). The 50 who have CTS and test negative are called false negatives. Of the 500 workers without CTS, 95 percent or 475 persons will have a negative NCT test, but 25 will have a positive test (false positive). The positive predictive value of the test is 450 divided by (450 + 25), or 95 percent. Ninety-five percent of those with a positive test actually have CTS (true positive), while the other 5 percent with a positive test are free of CTS (false positive). If these workers are considered for surgical treatment of their symptoms, 5 percent who undergo the surgery will not have CTS prior to surgery.

Example 2. If the NCT is applied to healthy job applicants under 30 years old without complaints of wrist and hand pain—a group in which the frequency rate of CTS is approximately 1 percent—it is much less useful. If 1,000 applicants are screened, of the 10 persons with CTS, 90 percent or 9 will have a positive NCT. Of the 990 persons without CTS, 95 percent or 941 will have a negative NCT, but 49 persons will have positive tests. Therefore the positive predictive value of the test is 9 divided by (9 + 49) or 16 percent. Only 16 percent of those with a positive test actually have CTS (true positive), while the other 94 percent with a positive test are free of CTS (false positive). If this test was used to screen job applicants, 84 percent of those denied employment because of a positive test would have been misclassified.

From a comparison of the predictive value of screening tests for CTS in the two preceding examples, it is reasonable to conclude that screening of low-risk applicants is not a feasible approach since outcome prediction will be low. However, routine screening surveillance of high-risk workers may offer good prediction of outcome.

Examples of other clinical tests are vibration screening,[141] current perception threshold screening,[142] and ridge detection tactility.[143] Muffly-Elsey et al. have proposed a clinical test battery for use in screening for upper-extremity CTDs.[144]

Risk Factors, Associations, and Cause-Effect Relationships in CTDs

In order to establish either an association with identified risk factors or, more importantly a cause-effect relationship, certain study design criteria must be met. Unfortunately, there has been much confusion and dispute about the contribution of different risk factors, both work and personal. This has come about because of inadequate study design and insufficient attention paid to important aspects of epidemiological research such as exposure measurement and outcome definition, bias, and confounding.

A major problem in the study of disorders such as CTS is that they can occur relatively frequently in some susceptible individuals in the normal course of life.[145] In 1984 Hernberg pointed out that apart from the inherent difficulties of all nonexperimental research, such as establishing a cause-effect relationship between two phenomena, there are problems in determining the contribution made by occupational factors (the "occupational etiologic fraction") to a work-related disease for which there are a variety of causes, both work and nonwork.[146] The role of the occupational factors in the development of a multifactorial disease may be small, and demonstrating the link becomes more difficult as the "fraction" decreases. Moreover, even when the occupational factors are significant, the magnitude of the effect may be influenced by other, nonwork factors that may need to be taken into account when studying certain complaints.

Study Design

Silverstein in her review of the literature of cumulative trauma disorders summarized four basic approaches in the identification of the cause-effect relationship of exposures and outcomes.[147] Type I studies use fairly rigorous definitions of health effects, often with precise diagnostic categories for such conditions as degenerative joint changes but much less precise diagnosis of nonarticular arthritic diseases. Measures of exposure are very imprecise (usually job classification only). On the whole, jobs chosen for study are physically heavy, and in some cases researchers have difficulty finding appropriate control populations because a high percentage of the community is engaged in the heavy work under study, such as coal mining.

Type II studies involve more precise task analysis and recording of work postures, usually with complex computerized video and analysis systems. Outcomes are determined by ratings of postural discomfort on a standardized form, the assumption being that pain and discomfort are precursors to disease and disability. These studies are useful in the planning and evaluation of job modifications.

Type III studies involve the use of existing centralized health records such as OSHA 200 logs, workers' compensation and insurance claims, and plant medical records as sources of data for analysis. The inherent problems with using such data were discussed in detail previously.

Type IV studies use clinical criteria to determine soft tissue health effects and more precise measures of exposure such as the detailed analysis of specific jobs.

Obviously Type IV studies, if well designed, are the most revealing and reliable, but they require considerable understanding of the nature of the disorders being studied, the range and type of exposure variables and possible confounders, and how to measure them.

The lack of well-designed prospective studies in the area of CTDs has led to a paucity of information on work-related causes of these disorders and their true incidence. Studies that are not prospective, such as case and point prevalence studies, are easier to design and conduct. However they do not establish cause-effect relationships because all factors are measured at the same point in time or cannot be established with certainty retrospectively. Therefore only significant associations between factors can be determined since it is not possible to determine which came first. While prevalence can be determined with these studies, incidence cannot.

In addition to these shortcomings, nonprospective studies miss severe cases such as workers who have left jobs because of health problems (healthy worker effect) and therefore underestimate associations between risk factors and cases.

Sampling and Bias in Different Worker Populations

Bias is any effect at any stage of investigation or inference that produces results that depart systematically from the true values. Sampling and other types of bias may affect the validity of both cross-sectional and longitudinal studies.

The most serious source of bias in studies of work-related musculoskeletal disorders is probably that of selection by health status. In many workplace-based studies, selection of "cases" (even when case definitions are tight, which is rare) has not taken into account the self-selection process that occurs in many semi-skilled and unskilled jobs. People who do not like the work they do for a range of reasons leave and find other employment. Similarly, if they find that the job causes them health problems, they may move on. This leads to what is referred to as the "healthy worker" effect: those workers who remain tend to be healthier than those who leave. There are two ways to avoid such distortions: use of controls and, in prospective studies, follow-up of subjects who leave to make sure their reason for leaving was not due to the problems being studied.

Biased samples also can be created if the case definition is too restrictive. This may lead to less serious or early cases being overlooked, thereby underestimating the association between the exposure variables and the outcome.

Self-reported symptoms may over- or underestimate prevalence. Depending on recall, which may in turn be related to whether the symptoms were considered important or work related, symptoms may or may not be recorded. When a questionnaire asks respondents about symptoms in the past year, it tends to overestimate cases. When a questionnaire asks only about current symptoms, it tends to underestimate the cases. Also, recall bias is an issue with self-report since symptomatic workers may be more aware of work-related risk factors than asymptomatic workers. This can lead to exaggerated associations between predictor factors and symptoms.

Data can be drawn from a number of sources, including clinical case series, occupational health center records, workers' compensation statistics, and research studies. The first three tend to underrepresent the prevalence of disorders in working populations because the primary purpose of the records is not for epidemiological investigations.

Confounding Factors

A confounding variable is one "that can cause or prevent the outcome of interest, is not an intermediate, and is not associated with the factor under investigation. Such a variable must be controlled in order to obtain an undistorted estimate of the effect of the study factor or risk."[148] In other words, a confounding variable distorts the true relationship between the study and outcome factors.

Silverstein identified a number of potential personal confounding factors in her prevalence study of CTDs with assembly workers in medium and heavy industry in the United States.[149] They were gender, age, prior injuries, chronic diseases, reproductive history, sports and hobbies, and previous jobs. As a matter of course, any study of CTDs should try to control for these factors. In Australia in the mid-1980s young and middle-aged women were overly represented as RSI cases. However, it was soon determined that the number of women undertaking highly repetitive work was proportionally greater. As well, most undertook household duties and child care in addition to their paid work, and these factors independently could have led to a range of symptoms.

On the other hand, Laubli, Grandjean, and Smith acknowledged the difficulties of selecting appropriate groups to control for work factors such as those brought about by rapid changes in office technology.[150] They argued it is often impossible to separate the effects of such changes from the study factors in question. Therefore comparisons of VDT and non-VDT users may be invalid if confounding factors are not taken into account.

Cannon et al. in their case control study did not report how they defined or measured work exposure factors.[151] Many of the early studies (1940s to 1960s) that reported an association between work factors and CTS used self-report of exposure variables and lacked controls.[152]

PREVENTION AND CONTROL OF WORK-RELATED CTDS AND CTS

Despite treatment advances and an increasing awareness of the need for well-managed rehabilitation programs for workers, the prevention of CTDs must be of the highest priority. Ultimately the onus is on health professionals, employers, and employees to seek safer ways of doing work where risks to health can be expected. In the developed world, many jobs in the 1990s have been identified as being associated with higher-than-expected frequencies of these disorders.

In 1950 an International Labor Organization/World Health Organization Joint Committee of Experts stated that occupational health should aim at the promotion and maintenance of the highest degree of physical, mental, and social well-being of workers in all occupations.[153] Rather than detracting from health, work should promote health. People's health should be improved by working. For many millions of workers this is still far from the case.

Prevention of work-related musculoskeletal disorders generally can be considered under three main headings:

1. *Primary prevention:* eliminating or minimizing risks to health or well-being
2. *Secondary prevention:* alleviating the symptoms of ill health or injury, minimizing residual disability, and eliminating or at least minimizing factors that may cause recurrence
3. *Tertiary prevention:* rehabilitating those with disabilities to as full function as possible and modifying the workplace to accommodate any residual disability (see Appendix 9-D for an overview of prevention)

The effective implementation and evaluation of preventive measures in the workplace may require a multidisciplinary approach involving ergonomics, occupational health, epidemiology, engineering, administration, and management. A prevention program will require cooperation, organization, and commitment, most particularly from senior management. It may be expensive in the short term because of the need to purchase new equipment or rearrange the work and/or the workplace, and it may temporarily reduce production. Often there is a reluctance on the part of management to accept short-term costs and organizational upheaval for the long-term benefit. Nevertheless, for successful long-term control of work-related musculoskeletal disorders and their associated costs, such an approach may be necessary.

Where ergonomics and other preventive strategies are not applied or are less than fully effective, problems may arise. Appropriate case management should aim at minimizing the severity of the disorders that do occur and returning individuals to work, modified as necessary, as soon as possible. Both Australian and New Zealand governments have issued Codes of Practice and Guidelines that aim at control of neck and upper-extremity disorders.[154]

Primary Prevention

A large body of literature on ergonomically designed tools, workstations, work tasks, or work organization to prevent cumulative trauma disorders is available. A number of case studies, reports, and reviews on attempts at primary prevention programs are found in the literature.[155] Outcomes of prospective primary prevention programs, however, are scarce. Most of the reports are case studies of a specific company or plant. Programs most frequently include ergonomic solutions for work design and environment, worker education and training, and exercise.

Ergonomics in the Workplace

Ergonomics is the matching of equipment, processes, and environments to people so that tasks and activities required of them are within their limitations but make the best use of their abilities. In the workplace the application of ergonomics aims to promote health, efficiency, and well-being in workers.

Hazard and Risk Identification. Investigation of ergonomic hazards (anything that has the potential to cause harm to a person) and attendent risks (factors that contribute to the occurrence of injury or loss from a hazard) may take many forms. One that has worked for ergonomists in Australia consists of four main components:

1. risk identification
2. risk assessment
3. risk control
4. control (solution) evaluation

Risk identification involves analysis of statistics and injury records, consultation with employees, and direct observations of the jobs or areas believed to have an increased risk of injury.

Risk assessment occurs when possible sources of injury have been identified. It involves determining which jobs are implicated and which workers are likely to be affected, the nature of the risks identified, how frequently they occur, and their severity.

Risk control involves the exploration of the ergonomics options for control. These include

- task design or redesign
- workplace layout modification
- modifying objects or equipment
- maintenance
- task-specific training

The cost, availability, and feasibility of control measures should be considered also. However, while solutions to identified problems may be put in place, evaluating whether they work is less likely. This involves revisiting worksites and in some way objectively determining the success or failure of the solutions. Often solutions need modification before they are effective, or they may have to be discarded and an alternative found.

Analysis of Workload Demands. To avoid mismatches between workers and their jobs, there must be some understanding of the demands of the work and the capacity of each worker to meet those demands. Measurement of workload and its effects on individuals and groups is one of the more challenging aspects of CTDs for therapists.

Work factors that are believed to contribute to these disorders fall into three main areas:

1. the task
2. the workplace, including the workstation and general work environment
3. the work organization

Several measurement techniques are now available to assist practitioners in quantifying workload factors. See Chapter 1 for further information.

Task Variation and Job Rotation

Task variation or multiskilling is highly desirable and can be achieved either through job enlargement, which requires careful job and task design to enable a number of different types of activities to be incorporated into a job description, or (less effectively) through job rotation. Job enlargement (enrichment) involves increasing and varying the job content either by adding tasks or by adding complexity to tasks. It is a much more acceptable alternative for providing variety, but requires careful planning and longer training periods.

In job or task rotation the job remains unchanged, but the worker gets more variety by moving from one component task to another. It is a ready way of spreading the load of particularly stressful jobs among a large group of employees, but it does have drawbacks. It works only where jobs are different enough to provide physical and mental variety, and many employees do not like rotating for a number of reasons, even when it is in their best interests to do so. Also, job rotation can mask the real causes of the problems and may only extend the period before problems eventually arise. Employees have to learn more skills and thus require more training and supervision.

Work Rates

Human performance varies between individuals and over time. Therefore work rates should aim to accommodate the physical and psychological capacities of all workers selected for the job. This is particularly important in machine-paced work.

Minimizing Aggravating Factors

Organizational difficulties of various kinds can arise in any enterprise. Mechanical and technical breakdowns and inefficiencies can have a disruptive effect on employees and usually involve periods of extra load to make up production or output. Poor quality control may require reworking for no additional productivity.

Therefore machine and equipment adjustment and maintenance are most important in the smooth and efficient operation of any system. Other organizational factors such as the need for overtime, shift work, peak loading, and bonus and other incentive schemes often require higher outputs than the employees can safely manage and should be avoided with careful planning.

Exercise Programs

Several reports suggest that exercise may be a useful preventive measure to reduce upper-extremity cumulative disorders.[156] Proposed mechanisms are increases in strength and endurance[157] or increased blood flow.[158] Silverstein et al. and Williams et al. reported equivocal results when only exercise intervention was used to determine if exercise alone could reduce the upper-extremity cumulative trauma disorders including CTS.[159] Silverstein reported slight increases in musculoskeletal discomfort in some workers. This may be a response to exercise training, and workers should be informed about safe exercise prescription (exercise intensity, duration, frequency, mode or type, and consideration of the worker's initial fitness level).

Guidelines have been issued by OSHA and the Department of Labor in 1989 to reduce these disorders in the meatpacking industry previously noted to have a high incidence of CTDs, including CTS.[160] The 1991 BLS statistics for meatpacking as compared to 1987 demonstrate a decrease in these disorders (Figures 9–6 and 9–7).[161]

Reports from light (Ethicon) and heavy (Mazda) manufacturing are examples of this case study approach.[162] Prospective evaluations of exercise and ergonomics programs were put in place and indicated success of the overall program.[163] Controls were not reported in these studies and which components of the prevention program were most effective was not examined.

Pause exercises (pause gymnastics), originally a Scandinavian concept, have gained acceptance in other countries.[164] They are rhythmic, free or set movements performed during the working day to help alleviate the effects of fixed work postures and repetitive movements. They usually include a series of full-range movements, sometimes done to music, and designed to meet the needs of particular working groups. Set movements should vary from time to time to avoid boredom, and should be performed moderately slowly and carefully to ensure maximum benefit. As work varies a great deal from time to time and between different groups of people, movements should be designed to take this into account.[165] They should be supervised initially and at regular intervals by a professional trained in anatomy and exercise physiology such as a physical therapist. Such a person can also ensure that movements are performed correctly, and where individuals have difficulties, these difficulties can be investigated and treated early.

Pause exercise programs aim to:

- Encourage changes of posture from those adopted for the majority of the work day
- Strengthen and stretch muscles that might be weak or tight
- Stimulate circulation and help lessen feelings of fatigue at the end of the working day

Lee et al. have provided a comprehensive review of the safety and effectiveness of a range of exercise programs for VDT/office workers.[166]

Work Pauses

Although there is little information on the benefits of work pauses, there is sufficient evidence to suggest that they are essential in certain tasks to avoid unnecessary fatigue.[167] They can be self-regulated or fixed and supervised, but to be effective their duration and frequency must be appropriate to levels of activity and fatigue. For example, at the end of a day or a week, more frequent, longer breaks may be required, and the system must be flexible to take different circumstances into account. Individually regulated breaks are the most desirable, but workers often have to be encouraged to pause from work even when they are tired. They must be positively discouraged from accumulating breaks.

Pause exercises and regulated work pauses are only temporary solutions in alleviating the effects of fixed, repetitive, or demanding work. In the long term, work should be designed to allow variation in tasks and movements and regular pauses throughout the day.

Worker Health Surveillance and Education Programs

Most jobs can be done in a variety of ways, and usually one way will be less stressful and fatiguing than others. It is important that the most efficient methods are identified for each job and that these methods only are taught to new employees and to those learning a new job or using new equipment. Even with training, however, employees may slip into inefficient practices, and these should be monitored and corrected in on-the-job supervision.

For the development of correct work techniques and postures, together with training and on-the-job supervision, supervisors should be consulted to help define correct and incorrect methods of work and to assist in initial training. Training, where possible, should be organized and run by a training officer or someone else skilled in teaching others. Education is an especially important aspect of ergonomics. If money, time, and expertise are used to produce an ergonomically sound workplace, then employees should understand why it has been so designed and how it can best be used.

Evaluation of Prevention Programs

As mentioned above, many publications dealing with the prevention, control, and management of neck and upper-limb disorders are now produced by government agencies,[168] universities,[169] and journals.[170] All deal with the identification of workloads that may be harmful; methods for measuring these workloads and/or assessing their impact; and control or prevention procedures suitable for different types of work. Monitoring of the effectiveness of programs is recommended in nearly all the documents, although criteria for monitoring are not discussed. Methods by which formal evaluation of quite extensive programs might be undertaken have not been addressed at all in any of the documents, and this seems to be a glaring oversight. When so many resources and so much time can be devoted to controlling these disorders, it would seem important to build some sort of evaluation into a prevention program, if only to help sell it to increasingly cost-conscious managers.

Some attempts have been made to evaluate the outcome of neck and upper-extremity disorder prevention programs, the most notable of which was conducted in Norway and included a cost-benefit analysis.[171] In 1975, an intervention study in a Norwegian electronics factory was initiated in response to an unusually high rate of sick leave in the preceding two years and increasing complaints of musculoskeletal disorders of the upper limb, neck, shoulder, and back.[172] This study attempted to discover retrospectively the reasons for the increasing rates of complaints and to evaluate the impact of ergonomic changes undertaken at the factory from 1975 onward.

It proved more difficult to argue that ergonomic changes directly led to a reduction in musculoskeletal disorders than to show that the old work situations contributed to the occurrence of the disorders. Nevertheless, there was evidence that the changes had a positive influence on health and were associated with decreasing complaints of symptoms. A study of assembly workers in Sweden demonstrated the effectiveness of instruction in correct work techniques to new workers in reducing the numbers of days lost through arm-neck-shoulder complaints.[173] This highlights an important area that has had little attention in the literature: the beneficial effects of training and education of workers, supervisors, and managers in what they can do for themselves to control musculoskeletal disorders arising from work.

A group of investigators in the United States attempted with some degree of success to establish an intervention program in a manufacturing industry.[174] Although the statistical analyses were never reported, complaints of disorders appeared to decrease, while the productivity in some jobs increased significantly. Changes included organizational rearrangements, such as the introduction of job rotation in selected areas and the provision of gloves to some workers, and engineering controls, such as the introduction of rotatable jigs, suspended tool retractors, and the redesign of components. Many of the researchers' recommendations were rejected as not being feasible, but some of the easier, less costly changes were made. Also some workers modified or redesigned their own tools, which helped decrease injuries and increase productivity. Generally, the changes that proved most successful were those where the front-line supervisor participated in and acted upon recommendations.

More recently in a commercial newsletter a U.S. company reported on a program that is applying the OSHA Guidelines for Meatpacking Plants to a baking company.[175] However, while it seems that the response to the program has been positive, no evaluation has been undertaken at this stage. A common feature of these intervention programs has been the interest and participation of occupational health professionals. They provide statistical and epidemiologic surveillance of the injuries and complaints, and thereby the mechanism by which the success of the program can be measured, however imprecisely.

The difficulties of measuring the effects of ergonomics change or any preventive health care program in the workplace are not insurmountable, but it is impossible to eliminate the influence of other factors that may alter the way people work or the way they perceive their work. In the case of musculoskeletal disorders, part of the problem is related to the "Hawthorne effect" (a change in performance of subjects merely because they are part of a study), part is related to the ubiquitous and ill-defined nature of the conditions being studied, and part is related to the numerous sources of bias and confounding variables that arise in the workplace. Nevertheless, as increasing numbers of these conditions are reported, there is an urgent need to demonstrate convincingly that certain preventive measures are effective.

Secondary Prevention

Early Detection of Symptoms

Phalen's and Tinel's tests demonstrate low to moderate sensitivity in detecting CTS in symptomatic workers. Results of Phalen's test sensitivity are reported as ranging from 66 percent to 75 percent, with specificities ranging from 36 to 80 percent.[176] Tinel's test sensitivity is reported as ranging from 44 up to 63 percent, with specificities ranging from 44 to 67 percent.[177] Many of these studies did not include controls[178] or confirm the presence of CTS neurophysiologically.[179]

The few studies that have included controls and neurophysiological confirmation of CTS[180] report sensitivities and specificities lower than previously reported values and conclude that these are not valid tests or significant indicators of CTS. If used as diagnostic criteria in a worker population with a prevalence of 15 percent, this level of sensitivity and specificity would lead to many false positive and false negative cases. Katz et al., when combining results of Phalen's and Tinel's tests, report a sensitivity of 46 percent and a specificity of 73 percent. When both tests combined were used to screen all workers at a plastics processing plant, sensitivity was 14 percent and specificity 96 percent.[181]

It may be concluded these tests are not accurate enough for utilization as screening or diagnostic tests except in very high-prevalence populations (greater than 50 percent prevalence in workers complaining of wrist and hand paresthesia and pain). These provocative tests should no longer be used for screening or diagnosis of CTS in secondary and tertiary prevention.

As new technologies develop and improve, new measures should be developed with improved sensitivities and predictive value. Research on outcomes of such screening is just beginning to emerge from the SENSOR project, which developed a case definition for CTS for use in ten states of the United States as part of a national effort to improve injury reporting.[182] Katz et al. reported preliminary results on the sensitivity, specificity, and predictive value of the SENSOR case definition for screening for CTS, demonstrating a 38 percent misclassification rate.[183] As previously described, risk factor identification for CTS has enjoyed more success. However, the predictive value and stratification of several proposed risk factors for workers remains unknown.

Body charts are a practical method of collecting information on symptoms of neck and upper-extremity disorders in the workplace. They are particularly useful in the absence of a specific diagnosis. The delineation of these conditions can be aided by the use of body charts that a worker or clinician may fill in.

Body charts have enabled researchers and those concerned with control of musculoskeletal disorders in the workplace to gain a clearer picture of symptom patterns and their prevalence without needing to categorize them as medical conditions. Further, the charts have enabled a systematic approach to prevention by identifying occupational groups with a high prevalence of symptoms in the neck and upper extremities and by pinpointing elements of their jobs that may be associated with symptom development.[184]

Intervention with Symptomatic Groups

Mismatches can occur when the demands of a job outstrip the worker's capacity to undertake the work safely. Therefore doctors and physical therapists must understand in some detail the demands of each task (workload measurement or assessment) and the capabilities of individual workers to perform those tasks (functional capacity assessments), and also to appreciate that these are likely to change over time.

Where particular work places unreasonable demands on workers, risks of injury can be reduced through better workplace layout or design, more adequate training, or more efficient work organization. Consideration will need to be given to modification of work to accommodate individuals with reduced physical capacity. As a last resort it may be necessary to advise people physically unsuited for certain work not to undertake that work but to seek a less demanding job. Workplace modification for the prevention of recurrence is dealt with extensively in the ergonomic and rehabilitation literature for return to work.[185]

Tertiary Prevention

Management and Rehabilitation of Workers with CTS

Recent research on outcomes of traditional approaches to the management of CTS has not been promising. CTS may be managed either conservatively or with surgical intervention. Repeated occurrences of CTS or failure of conservative treatment appear to be commonly accepted indications for surgical intervention.[186] The benefits of either approach need further controlled follow-up studies that use functional vocational outcome measures such as percent of cases who return to work, length of time to return to work, and percent requiring vocational retraining. Previous reports on surgical effectiveness are limited to physiological factors such as grip strength versus functional vocational or avocational outcomes over time.[187]

A recent follow-up study of open surgical release cases indicated that only a small number of patients returned to work. Return to work in laborers who did heavy work ranged from 8 to 12 weeks, while sedentary workers returned earlier.[188] Another recent study with 65 patients (75 hands) found that after endoscopic release, white-collar workers were able to return to work within 2 to 3 weeks and blue-collar workers were able to return in 2 to 8 weeks.[189]

One explanation for the poor surgical outcomes could be the previously mentioned criteria to select candidates for surgery. Many workers with severe late-stage cases already have incurred permanent damage. Another possibility is that some individuals are poor surgical candidates due to as yet unrecognized factors. Hagberg et al. found that postsurgical workers who had high rates of exposure to occupational vibration had poor results for symptom relief after surgery, with a risk rate of 18 of still having nocturnal paresthesia when compared to the low-exposure group.[190]

Savage et al. report a more positive outcome in a smaller sample of postsurgical vibration-exposed persons.[191] However, the amount of vibration exposure differed from that reported by Hagberg et al.[192] Bleeker et al. suggest that individuals identified with carpal tunnel stenosis by CT scan might benefit from early surgical intervention instead of waiting until after a course of failed conservative treatment.[193] An MRI study identified reasons for failed surgery as incomplete incision of the flexor retinaculum, excessive fat in the carpal tunnel, a persistent median artery, chronic median neuritis, and postoperative incisional neuromas.[194] Determination of risk based on carpal tunnel canal size from MRI and CT scan is controversial.[195] Arnoff et al. have identified several factors that predict poor return to work in patients with back injury, such as previous psychiatric history, history of substance abuse, and current legal involvement.[196] Similar research needs to be conducted with individuals with CTS to identify factors that predict poor return to work.

Reports of CTS release surgery having been performed on patients with peripheral polyneuropathy, cervical radiculopathy, thoracic outlet syndrome, brachial plexus injury, or forearm entrapment syndrome (pronator teres syndrome or anterior interosseous entrapment) of the median nerve should diminish with improved diagnosis and screening of proximal sites of neurovascular injury.[197] Eason et al. found cervical radiculopathy to be the most common cause for failed CTS surgery.[198] Parker and Imbus reported that in one plant over a two-year period, 110 CTD cases were referred to physicians for treatment, and 18 of these (16 percent) had CTS release surgery.[199] They reviewed the records of those 18 cases and concluded:

- Most had symptoms of shoulder and neck problems before or during the onset of hand symptoms.
- Usually cases with neurological symptoms did not present with classical symptoms of CTS, as symptoms were ill defined and involved.
- Electrodiagnostic studies were either essentially normal or slightly abnormal. No standard protocol was followed, and in some cases only the affected extremity was tested. In other cases only sensory nerve conduction tests were conducted.
- Most of those cases were operated on by well-qualified hand surgeons who were respected by their medical peers.

As risk stratification and diagnosis improve, better prediction of outcome and better selection of surgical candidates should ensue.

Those patients managed with conservative treatment have not fared much better on physiological factors such as relief of symptoms and functional problems such as return to work.[200] Corticosteroid injection into the carpal tunnel gives only temporary symptomatic relief in severe cases.[201] Relapses are common within 9 to 15 months or sooner if factors associated with injury such as work demands are not addressed.[202] Green found that injections provided relief of symptoms in many of the cases, and that in early to intermediate cases approximately half had permanent relief.[203] A problem with many of these papers is the lack of placebo or control groups.

One new treatment technique that appears promising is cold laser to the palmar wrist area. It is under investigation at General Motors in a double-blind trial by Goode et al. that also incorporates ergonomics intervention.[204] Case studies and small intervention trials are reported, with success on everything from heat modalities to electrical stimulation. However, these reports lack appropriate controls and have very small sample populations. NIOSH in 1985 issued a national strategy to deal with CTDs, including their position that heat modalities should not be used on acute cases of cumulative trauma disorders.[205]

Patients managed by work hardening also have limited success in return to work.[206] One explanation for this could be that by the time chronic, severe CTS cases enter work-hardening programs, permanent nerve damage has occurred.

Two conclusions can be drawn from these data on treatment: (1) alternative treatment approaches must be systematically investigated, and (2) more emphasis should be placed on prevention of the condition in the workplace.

CONCLUSION

Despite the problems of diagnosis, varying classification, and difficulties in preventing and treating many of the work-related neck and upper-extremity disorders, it is important to remember that many of these conditions are preventable. The complex interaction of work, personal, and social factors that may give rise to neck and upper-extremity complaints in workers means that ergonomics is an essential component of any such preventive program. Greater understanding of confounding factors, risk factors, and their predictive validity is needed before routine screening becomes a viable enterprise at the worksite. More emphasis on measuring functional outcomes and long-term follow-up of treatment (greater than three months) is needed to assess treatment effectiveness. Since long-term effects of treatment are not reported, emphasis should be on prevention and risk reduction.

NOTES

1. K. Cummings et al., Occupational Disease Surveillance: Carpal Tunnel Syndrome, *Morbidity and Mortality Weekly Report* 38, no. 28 (1989):485–489.
2. V. Putz-Anderson, ed., *Cumulative Trauma Disorders: A Manual for Musculoskeletal Diseases of the Upper Limbs* (New York: Taylor & Francis, 1988).
3. J.C. Bruening, Scannell Charts OSHA's Future Course, *Occupational Hazards* 51 (1989):27–30.
4. Bureau of Labor Statistics, *U.S. Department of Labor News*, no. 89-548 (1989):2–3; Bureau of Labor Statistics, *Occupational Injuries and Illnesses in the United States by Industry*, 1977 to 1989, Bureau of Labor Statistics Bulletin (Washington, D.C.: U.S. Dept. of Labor, 1990); G. Brogmus and R. Marko, Cumulative Trauma Disorders of the Upper Extremity: The Magnitude of the Problem in U.S. Industry, in *Human Factors in Design for Manufacturability and Process Planning* (Proceedings of the International Ergonomics Association, Honolulu, Hawaii, August 9–11, 1990), 49–59.
5. Bureau of Labor Statistics, *U.S. Department of Labor News*.
6. Ibid.

7. Bureau of Labor Statistics, *Occupational Injuries in the U.S. by Industry*, Bureau of Labor Statistics Bulletin no. 2130 (Washington, D.C.: Government Printing Office, 1982).
8. J.C. Stevens et al., Carpal Tunnel Syndrome in Rochester, Minnesota, 1961 to 1980, *Neurology* (Minneapolis) 38 (1988):134–138.
9. G.M. Franklin et al., Occupational Carpal Tunnel Syndrome in Washington State, 1984–1988, *American Journal of Public Health* 81 (1991):741–746.
10. M.C. de Krom et al., Risk Factors for Carpal Tunnel Syndrome, *American Journal of Epidemiology* 132 (1990):1102–1110.
11. Franklin et al., Occupational Carpal Tunnel Syndrome; Stevens et al., Carpal Tunnel Syndrome; de Krom et al., Risk Factors.
12. W.R. Brain et al., Spontaneous Compression of Both Median Nerves in the Carpal Tunnel, *Lancet* 1 (1947):277–282.
13. M. Devor, The Pathophysiology of Damaged Peripheral Nerves, in *Textbook of Pain*, ed. P. Walland and R. Melzak (New York: Churchill Livingstone, 1991), 63–81; J. Kimura, *Electrodiagnosis in Diseases of Nerve and Muscle: Principles and Practice*, 2nd ed. (Philadelphia: F.A. Davis, 1989).
14. Brain et al., Spontaneous Compression.
15. B.A. Silverstein et al., Hand Wrist Cumulative Trauma Disorders in Industry, *British Journal of Industrial Medicine* 43 (1986):779; B.A. Silverstein et al., Occupational Factors and Carpal Tunnel Syndrome, *American Journal of Industrial Medicine* 11 (1987):343; S.H. Rodgers, Recovery Time Needs for Repetitive Work, *Seminars in Occupational Medicine* 2, no. 1 (1987):19–24.
16. R.D. Helme et al., RSI Revisited: Evidence for Psychological and Physiological Differences from an Age, Sex, and Occupation Matched Control Group, *Australian and New Zealand Journal of Medicine* 22 (1992):23–29.
17. S. Sunderland, *Nerves and Nerve Injuries* (Edinburgh: Churchill Livingstone, 1991); G. Lundborg and L.B. Dahlin, Pathophysiology of Nerve Compression, *Hand Clinics* 8, no. 2 (1992):1–12.
18. Rydevik et al., Effects of Graded Compression on Intraneural Blood Flow: An in Vivo Study on Rabbit Tibial Nerve, *Journal of Hand Surgery* 6 (1981):3–12.
19. G. Lundborg, Structure and Function of the Intraneural Microvessels as Related to Trauma, Edema Formation, and Nerve Function, *Journal of Bone and Joint Surgery* 57A (1975):938–948.
20. G. Lundborg et al., Nerve Compression Injury and Increase in Endoneurial Fluid Pressure: A "Miniature Compartment Syndrome," *Journal of Neurology, Neurosurgery, and Psychiatry* 46 (1983):1119–1124.
21. Sunderland, *Nerves.*
22. Ibid.; J. Haftek, Stretch Injury of Peripheral Nerves: Acute Effects of Stretching on Rabbit Nerve, *Journal of Bone and Joint Surgery* 52B (1970):354.
23. Sunderland, *Nerves.*
24. J.A. Beel et al., Alterations in the Mechanical Properties of Peripheral Nerve Following Crush Injury, *Journal of Biomechanics* 17 (1984):185.
25. Sunderland, *Nerves.*
26. Rydevik et al., Effects.
27. Sunderland, *Nerves.*
28. R.J. Fowler et al., Recovery of Nerve Conduction after a Pneumatic Tourniquet: Observations on the Hind-Limb of the Baboon, *Journal of Neurology, Neurosurgery, and Psychiatry* 35 (1972):638–647; B. Rydevik and C. Norberg, Changes in Nerve Function and Nerve Fiber Structure Induced by Acute, Graded Compression, *Journal of Neurology, Neurosurgery, and Psychiatry* 43 (1980):1070.

29. G. Lundborg et al., Median Nerve Compression in the Carpal Tunnel: The Functional Response to Experimentally Induced Controlled Pressure, *Journal of Hand Surgery* 7 (1982):252–259.
30. Rydevik et al., Effects.
31. R.H. Gelberman et al., The Carpal Tunnel Syndrome: A Study of Carpal Canal Pressures, *Journal of Bone and Joint Surgery* 61A (1981):380–383.
32. R.M. Upton and A.J. McComas, The Double Crush in Nerve Entrapment Syndromes, *Lancet* 2 (1973):359.
33. B. Rydevik and G. Lundborg, Permeability of Intraneural Microvessels and Perineurium Following Acute, Graded Experimental Nerve Compression, *Scandinavian Journal of Plastic and Reconstructive Surgery* 11 (1977):179–187.
34. Rydevik et al., Effects.
35. Rydevik and Norborg, Changes.
36. J. Ochoa et al., Anatomical Changes in Peripheral Nerves Compressed by a Pneumatic Tourniquet, *Journal of Anatomy* 113 (1972):433–455.
37. Rydevik and Lundborg, Permeability.
38. Sunderland, *Nerves*.
39. J.W. Scadding, Peripheral Neuropathies, in *Textbook of Pain*, ed. P. Wall and R. Melzak (New York: Churchill Livingstone, 1991), 522–534.
40. Helme et al., RSI Revisited.
41. Silverstein et al., Occupational Factors.
42. B. Rydevik et al., The Biomechanics of Peripheral Nerves, in *Textbook of Pain*, ed. P. Wall and R. Melzak (New York: Churchill Livingstone, 1991), 75–87.
43. Stevens et al., Carpal Tunnel Syndrome; Silverstein et al., Hand Wrist; Silverstein et al., Occupational Factors; Rodgers, Recovery Time Needs; D.W. Badger, *Health Hazard Evaluation Report (HETA): Metalbestos Systems Inc.*, HETA #79-109-974 (Cincinnati: NIOSH, 1981); NIOSH, *Health Hazard Evaluation Report (HETA): Miller Electric Co.*, HETA #81-217-1086 (Cincinnati: NIOSH, 1982); R.L. Stephenson et al., *Health Hazard Evaluation Report (HETA): McGraw Edison Co.*, HETA #81-369-1591 and 81-466-1591 (Cincinnati: NIOSH, 1985); D.W. Badger, *Health Hazard Evaluation Report (HETA): KP Manufacturing Co.*, HETA #81-375-1277 (Cincinnati: NIOSH, 1983); P.L. Moody et al., *Health Hazard Evaluation Report (HETA): Donaldson Co.*, HETA #81-409-1290 (Cincinnati: NIOSH, 1983); NIOSH, *Health Hazard Evaluation Report (HETA): General Motors Corp.*, HETA #81-433-1452 (Cincinnati: NIOSH, 1984); NIOSH, *Health Hazard Evaluation Report (HETA): Pelton and Crane Co.*, HETA #83-233-1410 (Cincinnati: NIOSH, 1984); M.S. Crandall et al., *Health Hazard Evaluation Report (HETA): United Uniform Co.*, HETA #83-205-1702 (Cincinnati: NIOSH, 1986); J.M. Boiano and D.W. Badger, *Health Hazard Evaluation Report (HETA): Point Adams Packing Co.*, HETA #83-251-1685 (Cincinnati: NIOSH, 1986); P.J. Seligman et al., *Health Hazard Evaluation Report (HETA): Minneapolis Police Dept.*, HETA #84-417-1745 (Cincinnati: NIOSH, 1986); NIOSH, *Health Hazard Evaluation Report (HETA): AT&T Southern Bell and United Telephone*, HETA #85-452-1698 (Cincinnati: NIOSH, 1986); A.T. Nichting, *Health Hazard Evaluation Report (HETA): Devil's Lake Sioux Manufacturing Co.*, HETA #87-097-1820 (Cincinnati: NIOSH, 1987); J.M. Boiano et al., *Health Hazard Evaluation Report (HETA): ICI Americas, Inc. Indiana Army Ammunition Plant*, HETA #85-534-1855 (Cincinnati: NIOSH, 1987); A.T. Fidler et al., *Health Hazard Evaluation Report (HETA): Western Publishing Co.*, HETA #84-240-1902 (Cincinnati: NIOSH, 1988); F.D. Richardson et al., *Health Hazard Evaluation Report (HETA): Longmont Turkey Processors, Inc.*, HETA #86-505-1885 (Cincinnati: NIOSH, 1988); A.T. Fidler, *Health Hazard Evaluation*

Report (HETA): Standard Publishing Co., HETA #84-187-1966 (Cincinnati: NIOSH, 1989); T. Hales et al., *Health Hazard Evaluation Report (HETA): John Morrell & Co.*, HETA #88-150-1958 (Cincinnati: NIOSH, 1989); S. Tanaka and D. Habes, *Health Hazard Evaluation Report (HETA): Harvard Industries, Inc.—Anchor Swan Division*, HETA #87-428-2063 (Cincinnati: NIOSH, 1990); S. Burt et al., *Health Hazard Evaluation Report (HETA): Caldwell Manufacturing Co.*, HETA #88-361-2091 (Cincinnati: NIOSH, 1990); R.R. Hammel et al., *Health Hazard Evaluation Report (HETA): Bennett Industries*, HETA #89-146-2049 (Cincinnati: NIOSH, 1990); S. Burt et al., *Health Hazard Evaluation Report (HETA): Newsday, Inc.*, HETA #89-250-2046 (Cincinnati: NIOSH, 1990); T. Luopajarvi et al., Prevalence of Tenosynovitis and Other Injuries of the Upper Extremities in Repetitive Work, *Scandinavian Journal of Work Environment and Health* 5, no. 3 (suppl.) (1979):48–55; T.J. Armstrong et al., Investigation of Cumulative Trauma Disorders in a Poultry Processing Plant, *American Industrial Hygiene Association Journal* 43 (1982):103; B. Falck and P. Aarnio, Left-Sided Carpal Tunnel Syndrome in Butchers, *Scandinavian Journal of Work, Environment and Health* 9, no. 3 (1983):291–297; W.M. Keyserling et al., *Repetitive Trauma Disorders in the Garment Industry*, Dept. of Environmental Health Sciences, Harvard School of Public Health, Report #81-3220 (Boston: NIOSH, 1982); L. Punnett et al., Soft Tissue Disorders in the Upper Limbs of Female Garment Workers, *Scandinavian Journal of Work, Environment and Health* 11, no. 6 (1985):417–425; L.H. Morse, Repetitive Motion Musculoskeletal Problems in the Microelectronics Industry, *Occupational Med: State of the Art Reviews* 1, no. 1 (1986):167–174; V.R. Masear et al., An Industrial Cause of Carpal Tunnel Syndrome, *Journal of Hand Surgery* 11A (1986):222–227; J. Lecea, Reducing Repetitive Motions Injuries in Small Press Work: A Case Study, *Seminars in Occupational Medicine* 2, no. 1 (1987):69–70; B.A. Silverstein, The Prevalence of Upper Extremity Cumulative Trauma Disorders in Industry (Ph.D. diss., University of Michigan, 1985); R.H. Westgaard and A. Aarås, Postural Muscle Strain as a Causal Factor in the Development of Musculo-Skeletal Illnesses, *Applied Ergonomics* 15 (1984):162; A. Aarås and R.H. Westgaard, Further Studies of Postural Load and Musculoskeletal Injuries of Workers at an Electromechanical Assembly Plant, *Applied Ergonomics* 18 (1987):211; W. Margolis and J.F. Kraus, The Prevalence of Carpal Tunnel Syndrome Symptoms in Female Supermarket Checkers, *Journal of Occupational Medicine* 29 (1987):953–956; L. Punnett, Upper Extremity Musculoskeletal Disorders in Hospital Workers, *Journal of Hand Surgery*12A, no. 5, pt.2 (1987):858–862; R.G. Feldman et al., Risk Assessment in Electronic Assembly Workers: Carpal Tunnel Syndrome, *Journal of Hand Surgery* 12A, no. 5, pt. 2, (1987):849–855; P.A. Nathan et al., Relationship of Age and Sex to Sensory Conduction of the Median Nerve at the Carpal Tunnel and Association of Slowed Conduction with Symptoms, *Muscle and Nerve* 11 (1988):1149–1153; R.K. Sokas et al., Self-Reported Musculoskeletal Complaints among Garment Workers, *Journal of Industrial Medicine* 15, no. 2 (1989):197–206; G.M. Franklin et al., Occupational Carpal Tunnel Syndrome in Washington State, 1984–1987, *Neurology* (Minneapolis) 40 (1990):420; H. Morgenstern et al., A Cross-Sectional Study of Hand/Wrist Symptoms in Female Grocery Checkers, *American Journal of Industrial Medicine* 20, no. 2 (1991):209–218; S. Barnhart and L. Rosenstock, Carpal Tunnel Syndrome in Grocery Checkers: A Cluster of Work-Related Illness, *Western Journal of Medicine* 147 (1987):37–40; S. Barnhart et al., Carpal Tunnel Syndrome among Ski Manufacturing Workers, *Scandinavian Journal of Work, Environment, and Health* 17, no. 1 (1991):46–52; L.M. Finkel, The Effects of Repeated Mechanical Trauma in the Meat Industry, *American Journal of Industrial Medicine* 8 (1985):375–379; A. Moore et al., Quantifying Exposure in Occupational Manual Tasks with Cumulative Trauma Disorder Potential, *Ergonomics* 33, no. 12 (1991):1433–1453; R.R. McCormack et al., Prevalence of Tendonitis and Related Disorders of the Upper Extremity in a Manufacturing Workforce. *Journal of Rheumatology* 17 (1990):958–964; J.B. Osborn et al., Carpal Tunnel Syndrome among Minnesota Dental Hygienists, *Journal of Dental Hygiene* 64 (1990):79–85; T.J. Sluchak, Ergonomics: Origins, Focus, and Implementation Considerations, *American Association of Occupational Health*

Nurses Journal 40, no. 3 (1992):105–112; N.J. Barton et al., Occupational Causes of Disorders in the Upper Limb, *British Medical Journal* 304, no. 6822 (1992):309–311.

44. P. House et al., Use of Lifestyle Risk Ratios for CTS to Guide and Examine Workplace Health Programs, in *International Conference Physical Activity Fitness Health* (Toronto, Canada, Conference Proceedings, May 10–13, 1992).
45. Silverstein et al., Hand Wrist; Silverstein et al., Occupational Factors; Fowler et al., Recovery; House et al., Use of Lifestyle; L.C. Hurst et al., The Relationship of the Double Crush to Carpal Tunnel Syndrome, *Journal of Hand Surgery* 10B (1985):202–204; C.F. Murray-Leslie and V. Wright, Carpal Tunnel Syndrome, Humoral Epicondylitis, and the Cervical Spine: A Study of Clinical Dimensional Relations, *British Medical Journal* 1 (1976):1439–1442.
46. Murray-Leslie and Wright, Carpal Tunnel Syndrome.
47. J.G. Jones, Ulnar Tunnel Syndrome, *American Family Physician* 44, no. 2 (1991):497–502.
48. L.A. Pfalzer and B.J. Schmoll, Lifestyle Profile for Workers in a Plastics Processing Plant, *Cardiopulmonary Physical Therapy Journal* 3, no. 2 (1992):15–16; B. McPhee and D. Worth, Neck and Upper Extremity Pain in the Workplace, in *Physical Therapy of the Cervical and Thoracic Spine*, ed. R. Grant (New York: Churchill Livingstone, 1988), 310.
49. Silverstein et al., Hand Wrist; Silverstein et al., Occupational Factors; Badger, *Metalbestos;* NIOSH, *Miller Electric;* Badger, *KP Manufacturing;* Crandall et al., *United Uniform;* Boiano and Badger, *Point Adams;* Seligman et al., *Minneapolis;* Nichting, *Devil's Lake;* Fidler et al., *Western Publishing;* Tanaka and Habes, Harvard Industries; Armstrong et al., Investigation; Silverstein, Prevalence; Feldman et al., Risk Assessment; Morgenstern et al., Cross-Sectional Study; Osborn et al., Carpal Tunnel Syndrome; B.J. Mazerolle, Management of Repetitive Strain Injuries at an Electronic Plant (Proceedings of the Annual International Industrial Ergonomics and Safety Conference, Trends in Ergonomics/Human Factors IV, Miami, June, 1987), 1035–1041.
50. Silverstein et al., Occupational Factors; Stephenson et al., *McGraw Edison;* Moody et al., *Donaldson Co.;* Richardson et al., *Longmont Turkey;* Hales et al., *John Morrell;* Burt et al., *Caldwell;* Hammel et al., *Bennett;* Burt et al., *Newsday;* Falck and Aarnio, Left-Sided Carpal Tunnel; Keyserling et al., *Repetitive Trauma Disorders;* Punnett et al., Soft Tissue Disorders; Morse, Repetitive Motion; Punnett, Upper Extremity; Sokas et al., Self-Reported Musculoskeletal Complaints; Pfalzer and Schmoll, Lifestyle Profile.
51. M.L. Bleeker et al., Carpal Tunnel Syndrome: Role of Carpal Canal Size, *Neurology* (Minneapolis) 35 (1985):1599–1604; F.J. Winn, Jr., and D.J. Habes, Carpal Tunnel Area as a Risk Factor for Carpal Tunnel Syndrome, *Muscle and Nerve* 13, no. 3 (1990):254–258.
52. N.M. Hadler, Editorial: Cumulative Trauma Disorders, *Journal of Occupational Medicine* 32, no. 1 (1990):38–41; B.A. Silverstein, Editorial: Cumulative Trauma Disorders of the Upper Extremity: A Preventive Strategy Is Needed, *Journal of Occupational Medicine* 33, no. 5 (1991):642–644.
53. S.R. Stock, Workplace Ergonomic Factors and the Development of Musculoskeletal Disorders of the Neck and Upper Limbs: A Meta-Analysis, *American Journal of Industrial Medicine 19* (1991):1987–107.
54. Bureau of Labor Statistics, Annual Survey of Occupational Injuries and Illnesses, in *Handbook of Methods*, Washington, D.C.: Bureau of Labor Statistics Bulletin 2134-1, Chapter 17 (Washington, D.C.: Occupational Safety and Health Statistics, 1981), 122–134.
55. Bureau of Labor Statistics, *U.S. Dept. of Labor News*, no. 89-548 (1989):2–3; Bureau of Labor Statistics, *Occupational Injuries and Illnesses in the United States by Industry, 1977 to 1989.* (Washington, D.C.: Bureau of Labor Statistics Bulletin, U.S. Dept. of Labor, 1990).

56. Workers' compensation statistics (NSW, Australia: Workcover Authority), 1985–1986, 1986–1987, 1987–1988, 1988–1989, 1989–1990, 1990–1991.
57. Cummings et al., Occupational Disease Surveillance; S. Tanaka et al., Use of Worker's Compensation Claims Data for Surveillance of Cumulative Trauma Disorders, *Journal of Occupational Medicine* 30, no. 6 (1988):488–492; N.S. Seixas and K.D. Rosenman, Voluntary Reporting System for Occupational Disease: Pilot Project Evaluation. *Public Health Reports* 101 (1986):278–282.
58. Bureau of Labor Statistics, Annual Survey.
59. Cummings et al., Occupational Disease Surveillance.
60. S. McCurdy et al., Reporting of Occupational Injury and Illness in the Semiconductor Industry, *American Journal of Public Health* 81 (1991):85–89.
61. Workers' compensation statistics (NSW, Australia: Workcover Authority), 1985–1986, 1986–1987, 1987–1988, 1988–1989, 1989–1990, 1990–1991.
62. Bureau of Labor Statistics, *Occupational Injuries and Illnesses.*
63. Ibid.
64. Ibid.
65. Tanaka et al., Use of Worker's Compensation Claims Data.
66. Ibid.
67. Bureau of Labor Statistics, *U.S. Department of Labor News*; Bureau of Labor Statistics, *Occupational Illness and Injury.*
68. N.H. McAlister, Visual Display Terminals and Operator Morbidity, *Canadian Journal of Public Health* 78, no. 1 (1987):62–65; J. Melius et al., New Occupational Health Risks: Service Industries, an Example, *Proceedings of the Third Conference of the Joint U.S.-Finnish Science Symposium, October 22–24, 1986, Frankfort, Kentucky* (Cincinnati: NIOSH, 1986), 91–93.
69. G. Brogmus and R. Marko, Cumulative Trauma Disorders of the Upper Extremity.
70. U.S. Congress, Hearings on Dramatic Rise in Repetitive Motion Injuries and OSHA's Response, June 6, 1989, National Institute for Occupational Safety and Health, 101st Cong., 1st sess., testimony of Dr. Lawrence Fine, Director of the Division of Surveillance, Hazard Evaluation and Field Studies, U.S. Dept. of Health and Human Services.
71. J.N. Katz and C.R. Stirrat, A Self-Administered Hand Diagram for the Diagnosis of Carpal Tunnel Syndrome, *Journal of Hand Surgery* 15A (1990):360–363.
72. E. Holmstrom and U. Moritz, Low Back Pain: Correspondence between Questionnaire, Interview and Clinical Exam, *Scandinavian Journal of Rehabilitation Medicine* 23 (1991):119–125.
73. Badger, *Metalbestos;* NIOSH, *Miller Electric;* Stephenson et al., *McGraw Edison;* Badger, *KP Manufacturing;* Moody et al., *Donaldson;* NIOSH, *General Motors;* NIOSH, *Pelton and Crane;* Crandall et al., *United Uniform;* Boiano and Badger, *Point Adams;* Seligman et al., *Minneapolis Police Dept.;* NIOSH, *AT&T;* Nichting, *Devil's Lake;* Boiano et al., *ICI Americas;* Fidler et al., *Western Publishing;* Richardson et al., *Longmont Turkey;* Fidler, *Standard Publishing;* Hales et al., *John Morrell;* Tanaka and Habes, *Harvard Industries;* Burt et al., *Caldwell;* Hammel et al., *Bennett;* Burt et al., *Newsday.*
74. AGPS, *Repetition Strain Injury in the Australian Public Service,* Task Force Report (Canberra, 1985), 16; AGPS, *Worksafe Australia: Repetition Strain Injury (RSI): A Report and Model Code of Practice* (Canberra, 1986).
75. Workers' compensation statistics (NSW, Australia: Workcover Authority), 1990–1991.
76. J. Pinkham, CTS Impacts Thousands and Costs Are Skyrocketing, *Occupational Health and Safety* 57, no. 8 (1988a):6–7.

77. Ibid.
78. Ibid.
79. National Safety Council, *Accident Facts* (Washington, D.C.: Bureau of Labor Statistics, 1992).
80. J.M. Last, *Dictionary of Epidemiology* (New York: Oxford University Press, 1983), 29, 93.
81. Silverstein et al., Hand Wrist; Silverstein et al., Occupational Factors; Rodgers, Recovery Time; Luopajarvi et al., Prevalence of Tenosynovitis; Armstrong et al., Investigation; Morse, Repetitive Motion; Silverstein, Prevalence; T.J. Armstrong, Ergonomics and Cumulative Trauma Disorders, *Hand Clinics* 2, no. 3 (1986):553–565; T.J. Armstrong et al., Repetitive Trauma Disorders: Job Evaluation and Design, *Human Factors* 28, no. 3 (1986):325–336; T.J. Armstrong et al., Ergonomic Considerations in Wrist and Hand Tendonitis, *Journal of Hand Surgery* 12A, no. 5, pt. 2 (1987):830–837; Wieslander et al., Carpal Tunnel Syndrome (CTS) and Exposure to Vibration, Repetitive Wrist Movements, and Heavy Manual Work: A Case-Referent Study, *British Journal of Industrial Medicine* 46 (1989):43–47; R.W. Schoenmarklin, Biomechanical Analysis of Wrist Motion in Highly Repetitive Hand Intensive Industrial Jobs (Ph.D. diss., Ohio State University, 1991); R.W. Schoenmarklin and W.S. Marras, *Quantification of Wrist Motion in Highly Repetitive, Hand Intensive Industrial Jobs*, Final Report Grant No. 1 RO1 OH02621-01 and 02, RF Project 721720 (Washington, D.C.: NIOSH, 1991); T. Delgrosso and M.A. Boillat, Carpal Tunnel Syndrome: Role of Occupation, *International Archives of Occupational and Environmental Health* 63, no. 4 (1991):267–270; A. Freivalds, Pinching Forces Sustained during Lens Polishing Operations, *International Journal of Industrial Ergonomics* 6, no. 3 (1990):261–266; M.P. de Looze et al., Effects on Efficiency in Repetitive Lifting of Load and Frequency Combinations at a Constant Total Power Output, *European Journal of Applied Physiology* 65 (1992):469–474.
82. Silverstein et al., Hand Wrist; Silverstein et al., Occupational Factors; Silverstein, Prevalence; Westgaard and Aarås, Postural Muscle Strain; Aarås and Westgaard, Further Studies; Nathan et al., Relationship of Age; W.M. Keyserling et al., The Effectiveness of a Joint Labor-Management Program in Controlling Awkward Postures of the Trunk, Neck, and Shoulders: Results of a Field Study, *International Journal of Industrial Ergonomics* 11 (1993):51–65; V. Putz-Anderson and T.L. Galinsky, Psychophysically Determined Work Durations for Limiting Shoulder Girdle Fatigue from Elevated Manual Work, *International Journal of Industrial Ergonomics* 11 (1993):19–28; P.S. Helliwell et al., Work Related Upper Limb Disorder: The Relationship between Pain, Cumulative Load, Disability, and Psychological Factors, *Annals of the Rheumatic Diseases* 51, no. 12 (1992):1325–1329.
83. Armstrong et al., Investigation; Westgaard and Aarås, Postural Muscle Strain; Aarås and Westgaard, Further Studies; T.J. Armstrong and D.B. Chaffin, Some Biomechanical Aspects of the Carpal Tunnel, *Journal of Biomechanics* 12 (1979):567–570; Keyserling et al., Effectiveness; Putz-Anderson and Galinsky, Psychophysically Determined Work Durations; Helliwell et al., Work Related Upper Limb Disorder.
84. Armstrong et al., Ergonomic Considerations; Schoenmarklin, Biomechanical Analysis; Schoenmarklin and Marras, Quantification.
85. Rodgers, Recovery Time; Armstrong et al., Investigation; Falck and Aarnio, Left-Sided Carpal Tunnel; Masear et al., Industrial Cause; Margolis and Kraus, Prevalence; Armstrong et al., Repetitive Trauma Disorders; Armstrong et al., Ergonomic Considerations; Freivalds 1990; Keyserling et al., Effectiveness; Putz-Anderson and Galinsky, Psychophysically Determined Work Durations; G.S. Phalen, The Carpal Tunnel Syndrome, *Journal of Bone and Joint Surgery* 48A (1966):211–228; E.R. Tichauer, Some Aspects of Stresses on the Forearm and Hand in Industry, *Journal of Occupational Medicine* 8 (1966):67–71; E.R. Tichauer, *The Biomechanical Basis of Ergonomics: Anatomy Applied to the Design of Workstations* (New York: John Wiley &

Sons, 1978); L. Greenberg and D.B. Chaffin, *Workers and Their Tools* (Midland, Mich.: Pendell Publishing Co., 1977); T.J. Armstrong and D.B. Chaffin, Carpal Tunnel Syndrome and Selected Personal Attributes, *Journal of Occupational Medicine* 21, no. 7 (1979):481–486; E.J. McCormick and M.S. Sanders, *Human Factors in Engineering and Design*, 5th ed. (New York: McGraw-Hill, 1982); S. Konz, *Work Design: Industrial Ergonomics*, 2nd ed. (New York: John Wiley & Sons, 1983); T.J. Armstrong, *An Ergonomic Guide to Carpal Tunnel Syndrome* (Akron, Oh.: American Industrial Hygiene Association, 1983); T.J. Armstrong, Upper Extremity Posture: Definition, Measurement, and Control, in *The Ergonomics of Working Postures* (New York: Taylor & Francis, 1986), chap. 6; C.D. Browne et al., Occupational Repetition Injuries, *Medical Journal of Australia* 140, no. 6 (1984):329–332; R.C. Charash, *ErgoSPECS: Ergonomic Design Specifications Manual* (Minneapolis: Ergodynamics, Inc., 1989); T.M. Fraser, *The Worker at Work* (New York: Taylor & Francis, 1989); Eastman Kodak Company, *Ergonomic Design for People at Work*, 2 vols. (New York: Van Nostrand Reinhold, 1986); K.H.E. Kroemer, Cumulative Trauma Disorders: Their Recognition and Ergonomic Measures to Avoid Them, *Applied Ergonomics* 20, no. 4 (1989):274–280.

86. Armstrong et al., Investigation; Armstrong et al., Repetitive Trauma Disorders; McCormick and Sanders, *Human Factors;* Konz, *Work Design;* Y. Chen et al., Glove Size and Material Effects on Task Performance (Proceedings of the Human Factors Society 33rd Annual Meeting, Santa Monica, Ca., 1989), 708–712; D.J. Cochran et al., An Evaluation of Commercially Available Plastic Handled Knife Handles (Proceedings of the Human Factors Society 29th Annual Meeting, Santa Monica, Ca., 1985), 802–806

87. Armstrong et al., Repetitive Trauma Disorders.

88. Ibid.; M. Abbruzzese et al., A Comparative Electrophysiology and Histological Study of Sensory Conduction Velocity and Meissner Corpuscles of the Median Nerve in Pneumatic Tool Workers, *European Neurology* 16, no. 1–6 (1977)106–114; D.S. Chatterjee et al., Exploratory Electromyography in the Study of Vibration-Induced White Finger in Rock Drillers, *British Journal of Industrial Medicine* 39 (1982):89–97; M.H. Werner et al., Occupationally Acquired Vibratory Angioedema with Secondary Carpal Tunnel Syndrome, *Annals of Internal Medicine* 98 (1983):44–46; L. Ekenvall et al., Temperature and Vibration Thresholds in Vibration Syndrome, *British Journal of Industrial Medicine* 43 (1986):825–829; K. Koskimies et al., Carpal Tunnel Syndrome in Vibration Disease, *British Journal of Industrial Medicine* 47 (1990):411–416; M. Farkkila et al., Forestry Workers Exposed to Vibration: A Neurological Study, *British Journal of Industrial Medicine* 45 (1988):188–192; M. Bovenzi and A. Zandini, Occupational Musculoskeletal Disorders in the Neck and Upper Arms of Forestry Workers Exposed to Hand-Arm Vibration, *Ergonomics* 34, no. 5 (1991):547–562; D.E. Wasserman, Raynaud's Phenomenon As It Relates to Hand-Tool Vibration in the Workplace, *American Industrial Hygiene Association Journal* 46, no. 12 (1985):B10, B12, B14–18.

89. Armstrong et al., Repetitive Trauma Disorders; Wieslander et al., Carpal Tunnel Syndrome.

90. Luopajarvi et al., Prevalence of Tenosynovitis; L. Deves and R. Spillane, Occupational Health, Stress and Work Organization in Australia, *International Journal of Health Services*, 19 (1989):351–363; D. Muffly-Elsey and S. Finn-Wagner, Proposed Screening Tool for the Detection of Cumulative Trauma Disorders of the Upper Extremity, *Journal of Hand Surgery* 12A, no. 5, pt. 2 (1987):931–935; P. Leino, Symptoms of Stress Predict Musculoskeletal Disorders, *Journal of Epidemiology and Community Health* 43 (1989):293–300.

91. Stock, Workplace Ergonomic Factors.

92. Nathan et al., Relationship of Age; Hadler, Editorial; P.A. Nathan and R.C. Keniston, Carpal Tunnel Syndrome and Its Relation to General Physical Condition, *Occupational Diseases of the Hand* 9, no. 2 (1993):253–261; J.R. Schottland et al., Median Nerve Latencies in Poultry Pro-

cessing Workers: An Approach to Resolving the Role of Industrial "Cumulative Trauma" in the Development of Carpal Tunnel Syndrome, *Journal of Occupational Medicine* 33 (1991):627–631; N.M. Hadler, Is Carpal Tunnel Syndrome an Injury That Qualifies for Worker's Compensation Insurance? in *Clinical Concepts in Regional Musculoskeletal Illness*, ed. N.M. Hadler (New York: Grune & Stratton, 1987), 55–60.

93. Nathan and Keniston, Carpal Tunnel Syndrome; Schottland et al., Median Nerve Latencies.
94. Delgrosso and Boillat, 1991.
95. Luopajarvi et al., Prevalence of Tenosynovitis; Helliwell et al., Work Related Upper Limb Disorder; Leino, Symptoms of Stress; Nathan and Keniston, Carpal Tunnel Syndrome.
96. House et al., Use of Lifestyle Risk Ratios.
97. A.J. Voitk et al., Carpal Tunnel Syndrome in Pregnancy, *Canadian Medical Association Journal* 128 (1983):277–281.
98. Nathan and Keniston, Carpal Tunnel Syndrome.
99. Ibid.
100. Ibid.
101. M.A. Chamberlain et al., Carpal Tunnel Syndrome in Early Rheumatoid Arthritis, *Annals of Rheumatic Diseases* 29 (1970):149–152; H. Lang et al., Sensory Neuropathy in Rheumatoid Arthritis: An Electroneurographic Study, *Scandinavian Journal of Rheumatology* 10, no. 2 (1981):81–84.
102. Wieslander, Carpal Tunnel Syndrome; Keyserling et al., Effectiveness; D.W. Mulder et al., The Neuropathies Associated with Diabetic Mellitus: A Clinical Electromyographic Study of 103 Unselected Diabetic Patients, *Neurology* 11 (1961):275–284.
103. McCormack et al., Prevalence; Pfalzer and Schmoll, Lifestyle Profile; McPhee and Worth, Neck and Upper Extremity Pain.
104. Leino, Symptoms of Stress.
105. House et al., Use of Lifestyle Risk Ratios.
106. M.S. Sabour and H.E. Fadel, The Carpal Tunnel Syndrome—A New Complication Ascribed to the "Pill," *American Journal of Obstetrics and Gynecology* 107 (1971):1267; G.S. Dieck and J.L. Kelsey, An Epidemiological Study of Carpal Tunnel Syndrome in an Adult Female Population, *Preventive Medicine* 14 (1985):63–69.
107. Falck and Aarnio, Left-Sided Carpal Tunnel; Wieslander et al., Carpal Tunnel Syndrome; Nathan and Keniston, Carpal Tunnel Syndrome; Dieck and Kelsey, Epidemiological Study.
108. Dieck and Kelsey, Epidemiological Study; M.C. de Krom et al., Risk Factors for Carpal Tunnel Syndrome, *American Journal of Epidemiology* 132 (1990):1102–1110.
109. Nathan and Keniston, Carpal Tunnel Syndrome; de Krom et al., Risk Factors; M.P. Vessey et al., Epidemiology of Carpal Tunnel Syndrome in Women of Childbearing Age: Findings in a Large Cohort Study, *International Journal of Epidemiology* 19 (1990):655–659.
110. L.J. Cannon et al., Personal and Occupational Factors Associated with Carpal Tunnel Syndrome, *Journal of Occupational Medicine* 23 (1981):255–258; S.-E. Bjorkquist et al., Carpal Tunnel Syndrome in Ovariectomized Women, *Acta Obstetrica et Gynecologica Scandinavica* 56 (1977):127–130.
111. de Krom et al., Risk Factors.
112. Nathan and Keniston, Carpal Tunnel Syndrome; E.W. Johnson et al., Wrist Dimensions: Correction with Median Sensory Latencies, *Archives of Physical Medicine and Rehabilitation* 64 (1983):556–557; C. Gordon et al., Wrist Ratio Correlation with Carpal Tunnel Syndrome in Industry, *American Journal of Physical Medicine and Rehabilitation* 67 (1988):270–272.

113. Nathan and Keniston, Carpal Tunnel Syndrome; L. Reinstein, Hand Dominance in Carpal Tunnel Syndrome, *Medical Rehabilitation* 62, no. 5 (1981):202–203.
114. Delgrosso and Boillat, 1991.
115. Wieslander et al., Carpal Tunnel Syndrome; Nathan and Keniston, Carpal Tunnel Syndrome; de Krom et al., Risk Factors; Cannon et al., Personal and Occupational Factors; M.I. Kulick et al., Long-Term Analysis of Patients Having Surgical Treatment for Carpal Tunnel Syndrome, *Journal of Hand Surgery* 11A (1986):59–65.
116. Dieck and Kelsey, Epidemiological Study; de Krom et al., Risk Factors; Cannon et al., Personal and Occupational Factors.
117. Dieck and Kelsey, Epidemiological Study.
118. Wieslander et al., Carpal Tunnel Syndrome; de Krom et al., Risk Factors.
119. M.L. Bleeker, Medical Surveillance for Carpal Tunnel Syndrome in Workers, *Journal of Hand Surgery* 12A (1987):845–848.
120. B.J. McPhee, Musculoskeletal Complaints in Workers Engaged in Repetitive Work in Fixed Postures, in *Ergonomics: The Physiotherapist in the Workplace*, ed. M. Bullock (Edinburgh: Churchill Livingstone, 1990), 51.
121. Ibid.
122. P. Herberts and R. Kadefors, A Study of Painful Shoulders in Welders, *Acta Orthopedica Scandinavica* 47 (1976):381; P. Herberts et al., Shoulder Pain in Industry: An Epidemiological Study on Welders, *Acta Orthopedica Scandinavica* 52 (1981):299.
123. Falck and Aarnio, Left-Sided Carpal Tunnel.
124. Finkel, Effects.
125. D. Ferguson, Posture, Aching and Body Build in Telephonists, *Journal of Human Ergonomics* (Tokyo) 5 (1976):183.
126. A. Bjelle et al., Clinical and Ergonomic Factors in Prolonged Shoulder Pain among Industrial Workers, *British Journal of Industrial Medicine* 38 (1981):356.
127. J. Duncan and D. Ferguson, Keyboard Operating Posture and Symptoms in Operating, *Ergonomics* 17 (1974):651; R. Kukkonen et al., Prevention of Fatigue among Data Entry Operators, in *Ergonomics of Workstation Design*, ed. T.O. Kvalseth (London: Butterworths, 1983), 28.
128. Morse, Repetitive Motion; Feldman et al., Risk Assessment.
129. Lecea, Reducing Repetitive Motions.
130. Silverstein, Prevalence; Westgaard and Aarås, Postural Muscle Strain; Aarås and Westgaard, Further Studies; Franklin et al., Occupational Carpal Tunnel Syndrome; Barnhart et al., Carpal Tunnel Syndrome; McCormack et al., Prevalence; Barton et al., Occupational Causes.
131. Margolis and Kraus, Prevalence; Morgenstern et al., Cross-Sectional Study; Barnhart and Rosenstock, Carpal Tunnel Syndrome.
132. Osborn et al., Carpal Tunnel Syndrome.
133. Keyserling et al., Repetitive Trauma Disorders; Punnett et al., Soft Tissue Disorders; Nathan et al., Relationship of Age; L. Punnett et al., Soft Tissue Disorders in the Upper Limbs of Female Garment Workers, *Scandinavian Journal of Work, Environment and Health* 11 (1985):417.
134. L. McAtamney and E.N. Corlett, RULA: A Survey Method for the Investigation of Work-Related Upper Limb Disorders, *Applied Ergonomics* 24 (1993):91.
135. Putz-Anderson and Galinsky, Psychophysically Determined Work Durations.
136. McPhee, Musculoskeletal Complaints: Å. Kilbom et al., Disorders of the Cervicobrachial Region among Female Workers in the Electronics Industry, *International Journal of Industrial Ergonomics* 1 (1986):37.

137. Cummings et al., Occupational Disease Surveillance; T.D. Matte et al., The Selection and Definition of Targeted Work-Related Conditions for Surveillance under SENSOR, *American Journal of Public Health* 79, suppl. (Dec. 1989):21–25; J.N. Katz et al., Validation of a Surveillance Case Definition of Carpal Tunnel Syndrome, *American Journal of Public Health* 81 (1991):189–193.
138. Katz and Stirrat, Self-Administered Hand Diagram; Katz et al., Validation.
139. de Krom et al., Risk Factors.
140. Kimura, *Electrodiagnosis.*
141. T.C. Jetzer, Use of Vibration Testing in the Early Evaluation of Workers with Carpal Tunnel Syndrome, *Journal of Occupational Medicine* 33, no. 2 (1991):117–120.
142. J.J. Katims et al., Current Perception Threshold Screening for Carpal Tunnel Syndrome, *Archives of Environmental Health* 46, no. 4 (1991):207–212.
143. R.G. Radwin et al., Ridge Detection Deficits Associated with Carpal Tunnel Syndrome, *Journal of Occupational Medicine* 33 (1991):730–736.
144. D. Muffly-Elsey and S. Flinn-Wagner, Proposed Screening Tool for the Detection of Cumulative Trauma Disorders of the Upper Extremity, *Journal of Hand Surgery* 12A, no. 5 (1987):931–935.
145. Stevens et al., Carpal Tunnel Syndrome; de Krom et al., Risk Factors.
146. S. Hernberg, Work-Related Diseases—Some Problems in Study Design, *Scandinavian Journal of Work, Environment and Health* 10 (1984):367.
147. Silverstein, Prevalence.
148. Last, *Dictionary of Epidemiology.*
149. Silverstein, Prevalence.
150. T. Laubli et al., Visual Stress of Display Unit Operators, *Travail Humain* 43, no. 1 (1980):196; E. Grandjean, ed., *Ergonomics and Health in Modern Offices* (London: Taylor & Francis, 1984); M.J. Smith et al., An Investigation of Health Complaints and Job Stress in Video Display Operations, *Human Factors* 23 (1981):387–399.
151. Cannon et al., Personal and Occupational Factors.
152. Luopajarvi et al., Prevalence of Tenosynovitis; M.Q. Birbeck and T.C. Beer, Occupation in Relation to the Carpal Tunnel Syndrome, *Rheumatology Rehabilitation* 14 (1975):218–221; L. Hymovich and M. Lindholm, Hand, Wrist, and Forearm Injuries: The Result of Repetitive Motions, *Journal of Occupational Medicine* 8, no. 11 (1966):573–577; S. Rothfleisch and D. Sherman, Carpal Tunnel Syndrome: Biomechanical Aspects of Occupational Occurrence and Implications Regarding Surgical Management, *Orthopedic Review* 7, no. 6 (1978):107–109.
153. International Labor Organization/World Health Organization Joint Committee of Experts, Report of the Joint ILO/WHO Committee on Occupational Health, World Health Organization Technical Report Service, 1973.
154. Worksafe Australia, *National Code of Pracice for the Prevention and Management of Occupational Overuse Syndrome* (Canberra, 1990; first published 1986); Worksafe Australia, *Guidance Note for the Prevention of Occupational Overuse Syndrome in Keyboard Employment* (Canberra, 1989); Worksafe Australia, *Guidance Note for the Prevention of Occupational Overuse Syndrome in the Manufacturing Industry* (Canberra, 1992); Worksafe Australia, *National Standard and Code of Practice for Manual Handling (Occupational Overuse Syndrome)* (Canberra, 1993); New Zealand Department of Labour, *Occupational Overuse Syndrome: Guidelines for Prevention* (Wellington, 1991); New Zealand Department of Labour, *Occupational Overuse Syndrome: Treatment and Rehabilitation. A Practitioner's Guide* (Wellington, 1992).
155. Putz-Anderson, *Cumulative Trauma Disorders;* Lecea, Reducing Repetitive Motions; Aarås and Westgaard, Further Studies; Osborn et al., Carpal Tunnel Syndrome; McPhee and Worth, Neck and Upper Extremity Pain; Silverstein, Editorial; Stock, Workplace Ergonomic Factors;

McCormick and Sanders, *Human Factors;* Konz, *Work Design;* Armstrong, *Ergonomic Guide;* Eastman Kodak, *Ergonomic Design;* Kroemer, Cumulative Trauma Disorders; Kukkonen et al., Prevention; Punnett et al., Soft Tissue Disorders; I. Dimberg et al., Symptoms from the Neck and Upper Limb—An Epidemiological, Clinical, and Ergonomic Study Performed at Volvo, *Volvo Flygmotor*, Sweden, 1985; F. McKenzie et al., A Program for Control of Repetitive Trauma Disorders Associated with Hand Tool Operations in a Telecommunications Manufacturing Facility, *American Industrial Hygiene Association Journal* 46 (1985):674–678; J. Pinkham, Carpal Tunnel Syndrome Sufferers Find Relief with Ergonomic Designs, *Occupational Health and Safety* 57, no. 8 (1988):49–52; W. Karwowski and M.L. Kasdan, The Partnership of Ergonomics and Medical Intervention in Rehabilitation of Workers with Cumulative Trauma Disorders of the Hand, in *Ergonomics in Rehabilitation*, ed. A. Mital and W. Karwowski (Philadelphia: Taylor & Francis, 1988); H.L. Dortch and C.A. Trombly, The Effects of Education on Hand Use with Industrial Workers in Repetitive Jobs, *American Journal of Occupational Therapy* 44 (1990):777–782; I. Smolander et al., Policemen's Physical Fitness in Relation to the Frequency of Leisure-Time Physical Exercise, *International Archives of Occupational and Environmental Health* 54, no. 4 (1984):295–302; L.J. McCasland, Development of an Ergonomic Program—for the Meatpacking Industry, *American Association of Occupational Health Nurses Journal* 40, no. 3 (1992):138–142; B. King, Strategies to Combat Carpal Tunnel Syndrome. *Ed Welch on Workers' Compensation*, East Lansing, Mich. 1992; K. Sawyer, An On-Site Exercise Program to Prevent Carpal Tunnel Syndrome, *Professional Safety* 32, no. 5 (1987):17–20; P. Tadano, A Safety Prevention Program for VDT Operators: One Company's Approach, *Journal of Hand Therapy* 3, no. 2 (1990):64–71; Y.R. Lifshitz et al., The Effectiveness of an Ergonomics Program in Controlling Work Related Disorders in an Automotive Plant—A Case Study (Proceedings of the Human Factors Society, Santa Monica, Ca., 1993), 1986–1988; A.S. Kasdan and N.P. McElwain, Return to Work Programs Following Occupational Hand Injuries, *Occupational Medicine* 4 (1989):539–545; B.A. Silverstein et al., Can In-Plant Exercise Control Musculoskeletal Symptoms? *Journal of Occupational Medicine* 30 (1988):151–156; K.R. Gutekunst and M.T. Fogelman, An Ergonomics Program To Control Cumulative Trauma Disorders in a Manufacturing Environment (Proceedings of the Human Factors Society 32nd Annual Meeting, Anaheim, Ca., October 24–28, 1988), 656–659; T. Eisma, Ergonomic Design Protects Hands from Repetitive Motion, Stress, Vibration, *Occupational Health and Safety* 59 (April 1990):75–82; G. Lutz and T. Hansford, Cumulative Trauma Disorders Controls: The Ergonomics Program at Ethicon, Inc., *Journal of Hand Surgery* 12A, no. 5, pt. 2 (1987):863–869; T. Hansford et al., Blood Flow Changes at the Wrist in Manual Workers after Preventive Interventions, *Journal of Hand Surgery* 11A, no. 4 (1986):503–508; T.L. Williams, Exercise as a Prophylactic Device against Carpal Tunnel Syndrome (Proceedings of the Human Factors Society 33rd Annual Meeting, Santa Monica, Ca., 1989), 723–727; W. Laporte, The Influence of Gymnastic Pause upon Recovery Following Post Office Work, *Ergonomics* 9 (1966):501–506; U.S. Dept. of Labor, Occupational Safety and Health Administration and Bureau of Labor Statistics, *Ergonomics Program Management Guidelines for Meatpacking Plants* (1991); R.E. Thomas et al., Effects of Exercise on Carpal Tunnel Syndrome Symptoms, *Applied Ergonomics* 24, no. 2 (1993):101–108; K. Lee et al., A Review of Physical Exercises Recommended for VDT Operators, *Applied Ergonomics* 23, no. 6 (1992):387–408; S.A. Konz, and A. Mital, Carpal Tunnel Syndrome, *International Journal of Industrial Ergonomics* 5, no. 2 (1990):175–180; F. Gerr et al., Upper-Extremity Musculoskeletal Disorders of Occupational Origin, *Annual Review of Public Health* 12 (1991):543–566; S.J. Isernhagen, Principles of Prevention for Cumulative Trauma, *Occupational Medicine: State of the Art Reviews* 17, no. 1 (1992):147–153; D.M. Rempel et al., Work-Related Cumulative Trauma Disorders of the Upper Extremity, *JAMA* 267, no. 6 (1992):838–842; K. Parker and H. Imbus, *Cumulative Trauma Disorders—Current Issues and Ergonomic Solutions: A Systems*

Approach (Ann Arbor, Mich.: Lewis Publishers, 1992), 6; J. Winkel and R. Westgaard, Occupational and Individual Risk Factors for Shoulder-Neck Complaints. Part I—Guidelines for the Practitioner and Occupational and Individual Risk Factors for Shoulder-Neck Complaints. Part II—The Scientific Basis (Literature Review) for the Guide, *International Journal of Industrial Ergonomics* 10 (1992):79, 85; P.W. Buckle and D.A. Stubbs, Epidemiological Aspects of Musculoskeletal Disorders of the Shoulder and Upper Limbs, in *Contemporary Ergonomics*, ed. E.J. Lovesey (London: Taylor & Francis, 1990); M.L. Kasdan ed., *Occupational Hand and Upper Extremity Injuries and Diseases* (Philadelphia: Hanley & Belfus, Inc., Mosby Yearbook, 1991).

156. Lutz and Hansford, Cumulative Trauma Disorders; Hansford et al., Blood Flow Changes; Williams et al., Exercise; Laporte, Influence; Thomas et al., Effects of Exercise; Lee et al., Review.
157. Williams et al., Exercise; Laporte, Influence.
158. Lutz and Hansford, Cumulative Trauma Disorders; Hansford et al., Blood Flow Changes.
159. Silverstein et al., Can In-Plant Exercise; Williams, Exercise.
160. U.S. Dept. of Labor, *Ergonomics*.
161. Bureau of Labor Statistics, *U.S. Department of Labor News;* Bureau of Labor Statistics, *Occupational Injuries and Illnesses*.
162. Lutz and Hansford, Cumulative Trauma Disorders; Hansford et al., Blood Flow Changes; Williams, Exercise.
163. Eisma, Ergonomic Design; Lutz and Hansford, Cumulative Trauma Disorders.
164. McPhee and Worth, Neck and Upper Extremity Pain; A. Gore and D. Tasker, *Pause Gymnastics* (Sydney, Australia: CCH, 1986).
165. Gore and Tasker, *Pause Gymnastics*.
166. Lee et al., Review.
167. S. Pheasant, *Ergonomics, Work and Health* (Basingstoke: Macmillan Academic and Professional Ltd., 1991), 163–164.
168. *Worksafe Australia: National Code; Worksafe Australia: Guidance Note, Keyboard Employment; Worksafe Australia: Guidance Note: Manufacturing Industry; Worksafe Australia: National Standard;* New Zealand Dept. of Labour, *Occupational Overuse Syndrome: Guidelines;* New Zealand Dept. of Labour, *Occupational Overuse Syndrome: Treatment;* U.S. Dept. of Labor, *Ergonomics;* U.K. Health and Safety Executive, *Work Related Upper Limb Disorders: A Guide to Prevention* (London: HMSO, 1990).
169. L. McAtamney and E.N. Corlett, *Reducing the Risks of Work Related Upper Limb Disorders: A Guide and Methods* (Nottingham, U.K.: University of Nottingham, Institute for Occupational Ergonomics, 1992).
170. Winkel and Westgaard, Occupational and Individual Risk Factors.
171. S. Spilling et al., Cost Benefit Analysis of Work Environments: Investment at STK's Telephone Plant at Kongsvinger, in E.N. Corlett et al., eds., *The Ergonomics of Working Postures: Models, Methods, and Cases* (New York: Taylor & Francis, 1986).
172. Westgaard and Aarås, Postural Muscle Strain; Aarås and Westgaard, Further Studies; Spilling et al., Cost Benefit Analysis; R.H. Westgaard and A. Aarås, The Effect of Improved Workplace Design on the Development of Work-Related Musculoskeletal Illnesses, *Applied Ergonomics* 16 (1985):91; R.H. Westgaard and T. Janus, Individual and Work-Related Factors Associated with Symptoms of Musculoskeletal Complaints. II. Different Risk Factors among Sewing Machine Operators, *British Journal of Industrial Medicine* 49 (1992):154.
173. Kukkonen et al., Prevention of Fatigue.

174. J.D. McGlothlin et al., Can Job Changes Initiated by a Joint Labor-Management Task Force Reduce the Prevalence and Incidence of Cumulative Trauma Disorders of the Upper Extremity? (Proceedings of the 1984 International Conference on Occupational Ergonomics, Toronto, May 7–9, 1984), 336.

175. Konz, *Work Design.*

176. Katz and Stirrat, Self-Administered Hand Diagram; Phalen, Carpal Tunnel Syndrome; Katz et al., Validation; G.S. Phalen and J.L. Kendrick, Compression Neuropathy of the Median Nerve in the Carpal Tunnel, *JAMA* 164 (1957):524–530; D. Kendall, Aetiology, Diagnosis and Treatment of Paraesthesiae in the Hands, *British Medical Journal* 2 (1960):1633–1640; R.M. Szabo et al., Sensibility Testing in Patients with Carpal Tunnel Syndrome, *Journal of Bone and Joint Surgery* 66A (1984):60–64; H. Gellman et al., Carpal Tunnel Syndrome: An Evaluation of the Provocative Diagnostic Test, *Journal of Bone and Joint Surgery* 68A (1986):735–737; G.S. Phalen, Reflections on 31 Years' Experience with Carpal Tunnel Syndrome, *JAMA* 212 (1970):1365–1367.

177. Katz and Stirrat, Self-Administered Hand Diagram; Phalen, Carpal Tunnel Syndrome; Katz et al., Validation; Kendall, Aetiology; Gellman et al., Carpal Tunnel Syndrome; A.P. Bowles et al., Use of Tinel's Sign in CTS, *Annals of Neurology* 13 (1983):689–690; S.C. Loong, The Carpal Tunnel Syndrome: A Clinical and Electrophysiological Study of 250 Patients, *Proceedings of the Australian Association of Neurologists* 14 (1977):51–65; J.D. Stewart and A. Eisen, Tinel's Sign and the Carpal Tunnel Syndrome, *British Medical Journal* 2 (1978):1125–1126; S.S. Mossman and J.N. Blau, Tinel's Sign and the Carpal Tunnel Syndrome, *British Medical Journal* 294 (1987):680; H.J. Gelmers, The Significance of Tinel's Sign in the Diagnosis of Carpal Tunnel, *Acta Neurochirurgica* 49 (1979):255–258; P. Seror, Tinel's Sign in the Diagnosis of Carpal Tunnel Syndrome, *Journal of Hand Surgery* 12B (1987):364–365.

178. Phalen, Carpal Tunnel Syndrome; Kendall, Aetiology; Gellman et al., Carpal Tunnel Syndrome.

179. Phalen and Kendrick, Compression Neuropathy; Szabo et al., Sensibility Testing; Phalen, Reflections; Bowles et al., Use of Tinel's Sign; Loong, Carpal Tunnel Syndrome; Stewart and Eisen, Tinel's Sign; Mossman and Blau, Tinel's Sign.

180. Katz and Stirrat, Self-Administered Hand Diagram; Katz et al., Validation; W.E.M. Pryse-Phillips, Validation of a Diagnostic Sign in Carpal Tunnel Syndrome, *Journal of Neurology, Neurosurgery, and Psychiatry* 47 (1984):870–872; D.N. Golding et al., Clinical Tests for Carpal Tunnel Syndrome: An Evaluation, *British Journal of Rheumatology* 25 (1986):388–390; L. Heller et al., Evaluation of Tinel's and Phalen's Signs in Diagnosis of the Carpal Tunnel Syndrome, *European Neurology* 25 (1986):40–42; G. Eves et al., Clinical Tests for the Detection of CTS in Industry (Paper presented at the MPTA Fall Meeting, Kalamazoo, Mich., October, 1992).

181. Eves et al., Clinical Tests.

182. Matte et al., Selection.

183. Katz et al., Validation.

184. McPhee, Musculoskeletal Complaints; I. Kuorinka et al., Standardised Nordic Questionnaires for the Analysis of Musculoskeletal Symptoms, *Applied Ergonomics* 18 (1987):233.

185. Putz-Anderson, Cumulative Trauma Disorders; Lecea, Reducing Repetitive Motions; Feldman et al., Risk Assessment; Osborn et al., Carpal Tunnel Syndrome; McPhee and Worth, Neck and Upper Extremity Pain; Stock, Workplace Ergonomic Factors; Armstrong et al., Repetitive Trauma Disorders; Armstrong et al., Ergonomic Considerations; Wieslander et al., Carpal Tunnel Syndrome; Keyserling et al., Effectiveness; Putz-Anderson and Galinsky, Psychophysically Determined Work Durations; Helliwell et al., Work Related Upper Limb Disorders; Tichauer, Some Aspects of Stresses; Tichauer, *Biomechanical Basis;* Greenberg and Chaffin, *Workers;* Armstrong and Chaffin, Carpal Tunnel Syndrome; McCormick and Sanders, *Human Factors;*

Konz, *Work Design;* Armstrong, *Ergonomic Guide;* Armstrong, Upper Extremity Posture; Browne et al., Occupational Repetition Injuries; Charash, *ErgoSPECS;* Fraser, *Worker at Work;* Eastman Kodak, *Ergonomic Design;* Kroemer, Cumulative Trauma Disorders; Konz and Mital, Carpal Tunnel Syndrome; Isernhagen, Principles of Prevention; Parker and Imbus, *Cumulative Trauma Disorders;* Winkel and Westgaard, Occupational and Individual Risk Factors; Kasdan, *Occupational Hand and Upper Extremity Injuries.*

186. J.C.Y. Chow, Endoscopic Release of the Carpal Ligament: A New Technique for Carpal Tunnel Syndrome, *Arthroscopy* 5 (1989):19–25.
187. V.L. Young et al., Grip Strength before and after Carpal Tunnel Decompression, *Southern Medical Journal* 85 (1992):897–900.
188. R.R. Shenck, Current Treatment of Carpal Tunnel Syndrome: Survey Results (Paper presented at the 17th Meeting of the American Association of Hand Surgery, San Juan, Puerto Rico, November, 1987); R.R. Shenck, Carpal Tunnel Syndrome: The New "Industrial Epidemic," *American Association of Occupational Health Nurses Journal* 37, no. 6 (1989):226–231; K.H. Duncan et al., Treatment of Carpal Tunnel Syndrome by Members of the American Society for Surgery of the Hand: Results of a Questionnaire, *Journal of Hand Surgery* 12A (1987):384–391.
189. S.K. Brahme and D. Resnick, Magnetic Resonance Imaging of the Wrist, *Rheumatic Diseases Clinics* 17, no. 3 (1991):721–739.
190. M. Hagberg et al., Recovery from Symptoms after Carpal Tunnel Syndrome Surgery in Males in Relation to Vibration Exposure, *Journal of Hand Surgery* 16A (1991):66–71.
191. R. Savage et al., Carpal Tunnel Syndrome in Association with Vibration White Finger, *Journal of Hand Surgery* 15B (1990):100–103.
192. Hagberg et al., Recovery from Symptoms.
193. Bleeker et al., Carpal Tunnel Syndrome.
194. M. Mesgarzadeh et al., Carpal Tunnel: MR Imaging. Part II. Carpal Tunnel Syndrome, *Radiology* 171 (1989):749–754.
195. Bleeker et al., Carpal Tunnel Syndrome; Winn and Habes, Carpal Tunnel Area; Mesgarzadeh et al., Carpal Tunnel; W.D. Middleton et al., MR Imaging of the Carpal Tunnel: Normal Anatomy and Preliminary Findings in the Carpal Tunnel Syndrome, *American Journal of Radiology* 148 (1987):307–316; W. Jessurun et al., Anatomical Relations in the Carpal Tunnel: A Computed Tomographic Study, *Journal of Hand Surgery* 12B (1987):64–67.
196. G. Arnoff et al., Pain Treatment Programs: Do They Return Workers to the Work Place? *Occupational Medicine: State of the Art Reviews* 3 (1988):123–126.
197. B.T. Harter, Indications for Surgery in Work-Related Compression Neuropathies of the Upper Extremity, *Occupational Medicine* 4, no. 3 (1989):485–495.
198. S.Y. Eason et al., Carpal Tunnel Release: Analysis of Suboptimal Results, *Journal of Hand Surgery* 10B (1985):365–369.
199. Parker and Imbus, *Cumulative Trauma Disorders.*
200. Kasdan, *Occupational Hand and Upper Extremity Injuries;* Loong, Carpal Tunnel Syndrome; S.E. Mackinnon et al., Internal Neurolysis Fails to Improve the Results of Primary Carpal Tunnel Decompression, *Journal of Hand Surgery* 16A (1991):211–218; J. Newmark and F.H. Hochberg, Isolated Painless Manual Incoordination in 57 Musicians, *Journal of Neurology, Neurosurgery, and Psychiatry* 50 (1987):291; S.J. Kaplan et al., Predictive Factors in the Non-Surgical Treatment of Carpal Tunnel Syndrome, *Journal of Hand Surgery* 15B (1990):106–108; R.H. Gelberman et al., Carpal Tunnel Syndrome: Results of a Prospective Trial of Steroid Injection and Splinting, *Journal of Bone and Joint Surgery* 62A (1980):1181–1184; E.J. Groves and B.A.

Rider, A Comparison of Treatment Approaches Used after Carpal Tunnel Release Surgery, *American Journal of Occupational Therapy* 43, no. 6 (1989):398–402; V.L. Kruger et al., Carpal Tunnel Syndrome: Objective Measures and Splint Use, *Archives in Physiology and Medical Rehabilitation* 72, no. 7 (1991):517–520.

201. Gelberman et al., Carpal Tunnel Syndrome.
202. H.V. Goodman and J.B. Foster, Effect of Local Corticosteroid Injections on Median Nerve Conduction in Carpal Tunnel Syndrome, *Annals of Physical Medicine* 6 (1962):287–294.
203. D.P. Green, Diagnostic and Therapeutic Value of Carpal Tunnel Injection, *Journal of Hand Surgery* 9A (1984):850–854.
204. W. Goode et al., unpublished data from General Motors, Flint, Mich., 1991.
205. T.J. Pizatella et al., *The NIOSH Strategy for Reducing Musculoskeletal Injuries* (Morgantown, W.V.: U.S. Government Printing Office, 1985).
206. Birbeck and Beer, Occupation.

Appendix 9–A

Sources of Information on Cumulative Trauma Disorders and CTS

A. Government and other agency publications in the United States, Australia, and New Zealand are listed in Exhibit 9–A–1. These usually include population statistics and current guidelines and standards for record keeping, prevention, and treatment of cumulative trauma disorders. The U.S. Health Hazard Evaluation Technical Assistance Reports (HETA) include case studies from particular industries or businesses.

B. Selected books and monographs in occupational health are listed in Exhibit 9–A–2. These often contain a compilation or summary of current knowledge on CTDs.

C. Journals on occupational health in the United States, Australia, and New Zealand are listed in Exhibits 9–A–3, 9–A–4, and 9–A–5. These contain the original research relating to cumulative trauma disorders and may include some compilation and summary of original research. While there are some

---

Exhibit 9–A–1 Government and Other Agencies in the United States, Australia, and New Zealand

1. In the United States:
   a. Bureau of Labor Statistics, U.S. Dept. of Labor
   b. Occupational Safety and Health Administration, Centers for Disease Control, Public Health Service
   c. Individual state departments of labor and/or commerce such as Michigan Occupational Safety and Health Administration (MIOSHA)
   d. National Safety Council, Chicago
   e. Workers' Compensation Report
   f. California Work Injuries and Illness
   g. Corporations such as IBM's VDT user guidelines
   h. U.S. NIOSH Health Hazard Evaluation Reports, Cincinnati, Ohio, available from the National Technical Information Service (NTIS), U.S. Dept. of Commerce

2. In Australia:
   a. *Workcover News* (NSW)
   b. *Statistics Weekly* (NSW)
   c. Worksafe Australia (National Institute for Health and Occupational Safety)

seminal papers, many papers do not give more than a small insight into one aspect of the problem.

D. Conference proceedings on occupational health are listed in Exhibit 9–A–6. These publications contain some original research, reviews and overviews relating to cumulative trauma disorders.

E. Dissertation abstracts, Micro Media, Ann Arbor, Mich. These publications are dissertations that include original research relating to CTDs.

Exhibit 9–A–2 Books and Monographs in Occupational Health

1. Alexander, D.C., and B.M. Pulat. 1985. *Industrial ergonomics: A practitioner's guide.* Norcross, Ga.: Industrial Engineering and Management Press, Institute of Industrial Engineers.
2. Bullock, M.I. 1990. *Ergonomics: The physiotherapist in the workplace.* New York: Churchill Livingstone.
3. Chaffin, D.B., and G.B.J. Andersson. 1992. *Occupational biomechanics.* 2nd ed. New York: John Wiley & Sons.
4. Eastman Kodak Company, 1986. *Ergonomic design for people at work.* 2 vols. New York: Van Nostrand Reinhold.
5. Fraser, T.M. 1989. *The worker at work.* New York: Taylor & Francis.
6. Isernhagen, S., ed. 1988. *Work injury: Management and prevention.* Aspen Publishers, Inc.
7. Kasdan, M.L., ed. 1991. *Occupational hand and upper extremity injuries and diseases.* Philadelphia: Hanley & Belfus, Inc.
8. Mital, A., ed. 1989. *Advances in industrial ergonomics and safety.* New York: Taylor & Francis.
9. Pheasant, S. 1991. *Ergonomics, work and health.* New York: Macmillan.
10. Putz-Anderson, V., ed. 1988. *Cumulative trauma disorders: A manual for musculoskeletal diseases of the upper limbs.* New York: Taylor & Francis.
11. Taleisnik, J. 1985. *The wrist.* New York: Churchill Livingstone.
12. Grandjean, E. 1988. *Fitting the task to the man: A textbook of occupational ergonomics.* 4th ed. New York: Taylor & Francis.
13. Parker, K., and H. Imbus. 1992. Cumulative trauma disorders—Current issues and ergonomic solutions: A systems approach. Ann Arbor, Mich.: Lewis Publishers.

Exhibit 9–A–3 Journals in the United States on Occupational Health

1. American Industrial Hygiene Association Journal
2. American Journal of Epidemiology
3. American Journal of Industrial Medicine
4. American Journal of Physical Medicine
5. American Journal of Physiology
6. American Journal of Public Health
7. American Journal of Occupational Therapy
8. Archives of Physical Medicine and Rehabilitation
9. Applied Occupational and Environmental Hygiene

continues

Exhibit 9–A–3 (continued)

10. Clinical Biomechanics
11. Clinical Orthopedics and Related Research
12. Ergonomics
13. Hand Clinics
14. Human Factors
15. Journal of the American Medical Association
16. Journal of Applied Physiology
17. Journal of Biomechanics
18. Journal of Bone and Joint Surgery
19. Journal of Hand Surgery
20. Journal of Occupational and Organizational Psychology
21. Journal of Occupational Medicine
22. Journal of Orthopedic and Sports Physical Therapy
23. New England Journal of Medicine
24. Occupational Health and Safety
25. Occupational Medicine: State of the Art Reviews
26. Physical Therapy
27. Public Health Reports
28. Public Health Reviews
29. Rheumatology and Rehabilitation
30. Safety and Health
31. Yearbook of Occupational and Environmental Medicine

Exhibit 9–A–4 Journals in Australia and New Zealand on Occupational Health

1. Australian Journal of Physiotherapy
2. Australian Journal of Public Health
3. Health and Safety (Australian)
4. Journal of Occupational Health and Safety (Australia and New Zealand)
5. Medical Journal of Australia

Exhibit 9–A–5 International Journals on Occupational Health

1. Applied Ergonomics (U.K.)
2. Bulletin of WHO
3. British Medical Journal
4. British Journal of Industrial Medicine
5. Canadian Journal of Public Health
6. Canadian Journal of Occupational Safety
7. European Journal of Applied Physiology and Occupational Physiology
8. International Archives of Occupational and Environmental Health
9. International Journal of Epidemiology
10. International Journal of Human Computer Interaction

continues

Exhibit 9–A–5 (continued)

11. International Journal of Industrial Ergonomics
12. Industrial Health (Japan)
13. Lancet (U.K.)
14. Occupational Health (U.K.)
15. Occupational Health and Safety (Canadian)
16. Scandinavian Journal of Work, Environment and Health
17. Scandinavian Journal of Rehabilitation Medicine
18. Scandinavian Journal of Social Medicine

Exhibit 9–A–6 Conference Proceedings on Occupational Health

1. Annals of Proceedings of the Human Factors Society (1988, 32nd meeting, Anaheim, Calif.; 1989, 33rd meeting, Denver, Colo.; 1990, 34th meeting, Orlando, Fla. Available from the Human Factors Society in Santa Monica, Calif.
2. Proceedings of the 9th Annual Meeting of the American Society of Biomechanics (1985, Ann Arbor, Mich.)
3. Proceedings of the 12th Congress of the International Society of Biomechanics (1989, Los Angeles, Calif.)
4. Proceedings of the 9th International Conference on Production Research (1987, Cincinnati, Oh.)
5. Proceedings of the World Confederation of Physical Therapy (1988, Sydney; 1991, London)
6. Proceedings of the Annual International Industrial Ergonomics and Safety Conference, Trends in Ergonomics/Human Factors (begins with 1984, vol. 1, Utrecht, Netherlands)

Appendix 9–B

Epidemiological Research

FREQUENCY OF OCCURRENCE

Prevalence and Prevalence Rate

Prevalence statistics are helpful in identifying jobs where higher-than-expected rates of injury or illness occur.[1] Prevalence is the portion of a population that is experiencing an injury or illness at a specified time. If there are 300 employees with a specific job title and 30 are reported as having CTS, the prevalence for that group is 30/300 or 0.10 persons. A disadvantage with prevalence statistics is instability. They are a snapshot of the frequency of injury or illness and are dependent on the number and duration of new cases. Prevalence does not allow for inference about causal factors of the injury or illness. To examine causal factors, incidence statistics are preferred.

Incidence and Incidence Rate

The incidence of a disorder is the number of new cases during a specified period of time. It is referred to as *incidence rate* when related to exposure hours on a job. Incidence statistics are helpful in assessing the rate of change in new cases when plotted over time.[2] In the United States, typical incidence rates are calculated for approximately one year of work exposure.[3] This is the number of new claims per 200,000 hours of job exposure. The incidence rate is the number of new cases in a population for a specified unit of exposure time and is calculated as follows:[4]

$$\text{Incidence rate} = \frac{\text{New cases recorded on OSHA 200 logs} \times 200{,}000 \text{ worker-hours of exposure}}{\text{Hours worked}}$$

200,000 worker-hours = total number of worker-hours in a year for 100 full-time workers who work 50 × 40-hour weeks in a year

Hours worked = number of employees on job × exposure time

Exposure time = average number of hours per week spent on job × average number of weeks that worker on job worked during the year

Exposure time (for one year of work) = 40 hours/week × 50 week/year = 2,000 hours/year

If an increase in the incidence rate is reported, you should identify at what point in time the increase occurred and from this try to determine the associated factors. An example using the previous data from the plastics processing plant follows:

Incidence rate of CTS in machine operators (for that year)

$$= 30 \times 200{,}000 \text{ hours}$$

$$= 300 \times 2{,}000 \text{ hours}$$

$$\frac{6{,}000{,}000}{600{,}000} = 10 \text{ per } 200{,}000 \text{ hours}$$

Total number of machine operator employees = 300
New cases of CTS in these employees in one year = 30
Exposure time for one year = 2,000 hours/year

In this example the prevalence is counted as a period prevalence over one year so that prevalence rate and incidence rate are approximately the same. This is the case in many of the OSHA and BLS statistics in the United States.

A typical application of these data to attempt to determine causal relationships follows: An increased incidence rate of 10 for wrist cumulative trauma disorders is found for April, May, and June of 1993 as compared to 5 for January, February, and March. You review company work factors, including work organization and production factors, and find that production factors were unchanged. However, more employees worked overtime on a routine basis in March and April than in January and February. You then need to determine if this is a trend requiring intervention or if it is only a temporary condition. If it is a trend, such as employees continuing to work overtime in May and June, then intervention may be necessary. If overtime in May and June returned to January and February levels, then no intervention is needed at this point in time. However, it should be determined if this is an annual seasonal increase in workload. One solution to offset the increased seasonal workload is increasing the workforce, for example, by hiring additional temporary employees.

Incidence rate data may also be utilized to examine the effectiveness of an intervention or program to control or reduce CTDs such as CTS. If incidence rates are monitored before and after intervention, a reduction in incidence rate may indicate an effective program or intervention, although other unrelated factors may need to be taken into account.

Both prevalence and incidence are sensitive to the way cases and noncases are selected, classified, or determined, and to the type of outcome or response measure used to count the cases.[5] In CTS, as in most CTDs with a gradual onset, the screening and diagnostic criteria used to determine a case appear to lack sensitivity, specificity, and predictive value, with the notable exception of electrodiagnostic tests such as nerve conduction studies. See Appendix 9–C for further discussion of sensitivity, specificity, and predictive value of tests. The cost and patient suffering with electrodiagnostic testing are commonly cited as problems with widespread use for screening and diagnosis. These testing issues are further addressed in the main body of the chapter in the section Screening, Diagnosis, and Injury Surveillance.

Measuring Incidence and Prevalence in the Workplace

Workers' compensation case data underreport the frequency of the cases, since only severe cases where the person is off work for at least five days are counted as cases. Therefore workers' compensation data may not be a sensitive outcome or response measure if preventative programs are most effective in reducing mild to moderate cases where the worker is off work less than five days, as is the case with many cumulative trauma disorders. Another criterion used in workers' compensation is that the case must be related to a specific cause, and the gradual onset of cumulative trauma disorders often precludes meeting this criterion, leading to underreporting or misclassification. An outcome measure such as a visit to the industrial or medical clinic may still underreport the frequency of cases since persons with symptoms that occur at night may not relate those symptoms to their occupation and therefore may not report them.

Severity: Interpreting the Data

Severity statistics are used to measure costs to an employer associated with cumulative trauma disorders. As with incidence rates, severity can be monitored over time prior to and after a medical treatment or rehabilitation program as a measure of the effectiveness of control or treatment of cumulative trauma disorders. Severity statistics are affected by factors associated with recovery and rehabilitation and return to work after medical treatment or rehabilitation. Severity statistics provide little information on cause-effect relationships and primary prevention programs. Severity statistics also lack standardization; therefore severity at one plant may not be comparable to severity at another plant. Commonly used severity statistics are

1. Number of days lost per case of a given type of disorder such as CTS
2. Workers' compensation cost per case of a given type of disorder such as CTS
3. Number of days lost per employee per year
4. Workers' compensation per employee per year
5. Number of days lost per 100 employees per year
6. Workers' compensation per 100 employees per year

7. Number of days lost a year per 1,000,000 hours of work time (ANSI Severity Rate)[6]

In severity statistics, if the response or outcome measure used is workers' compensation cases or lost-time cases, as in frequency statistics, low severity statistics will result from only the severe cases being counted. If persons continue to work on light duty or with restricted activity or tasks, they are not reflected in the severity statistics, therefore not aiding in early detection and intervention. In addition, the time off work necessary for being counted as a case is variable over time and from one country to the next. A workers' compensation case was defined as a worker off work three or more days in Australia until 1987; since then a case is defined as a worker off work five or more days. This change in the case definition would tend to decrease the number of cases reported in the statistics.

NOTES

1. V. Putz-Anderson, ed., *Cumulative Trauma Disorders: A Manual for Musculoskeletal Diseases of the Upper Limbs* (New York: Taylor & Francis, 1988).
2. Ibid.
3. Bureau of Labor Statistics, *Occupational Injuries in the U.S. by Industry*, Bureau of Labor Statistics Bulletin no. 2130 (Washington, D.C.: Government Printing Office, 1982).
4. Bureau of Labor Statistics, *OSHA Procedure: Handbook of Methods* (Washington, D.C.: U.S. Government Printing Office, 1981).
5. Putz-Anderson, *Cumulative Trauma Disorders.*
6. American National Standard Institute, Methods for Recording and Measuring Work Injury Experience, Standard No. Z16.1 (New York: American National Standard, 1969).

Appendix 9–C

Measurement Issues

The relationships between screening measures and an illness or disorder are shown in Figure 9–C–1.

SENSITIVITY

Sensitivity is a calculation of the ability to correctly identify persons with a specific illness or disorder (percent of cases where a positive test result is a true positive case). It is calculated as a fraction or percent of all persons with an illness or disorder (denominator) who test positive (numerator) on that test or measure. The calculation for sensitivity is as follows:

$$\text{Sensitivity} = \frac{\text{true positives } (a)}{\text{truc positives + false negatives } (a{+}c)} \times 100\%$$

All persons with the illness or disorder = true positives + false negatives (a+c). A sensitivity of 90 percent indicates that 90 out of 100 people with a certain illness or disorder such as CTS will have a positive test (true positive), while 10 will have a negative test when in fact they have CTS (false negative).

SPECIFICITY

Specificity is a calculation of the ability to correctly identify persons who do not have a specific illness or disorder (percent of cases where a negative test result is a true negative case). This determines the accuracy of the test or measure. Specific-

	Positive Illness	Negative Illness	
Positive Tests	a = true positives	b = false positives	$a + b$
Negative Tests	c = false negatives	d = true negatives	$c + d$
	$a + c$	$b + d$	

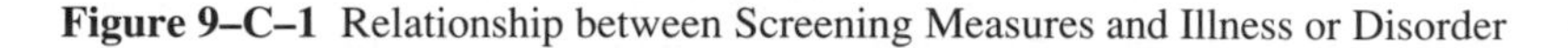

Figure 9–C–1 Relationship between Screening Measures and Illness or Disorder

ity is calculated as a fraction or percent of all persons without an illness or disorder (denominator) who test negative (numerator) on that test or measure. The calculation for specificity is as follows:

$$\text{Specificity} = \frac{\text{true negatives } (d)}{\text{true negatives + false positives } (d+b)} \times 100\%$$

All persons without the illness or disorder = true negatives + false positives (*d*+*b*). A specificity of 90 percent indicates that 90 out of 100 people who do not have a certain illness or disorder such as CTS will have a negative test (true negative) while 10 will have a positive test when in fact they do not have CTS (false positive).

PREDICTIVE VALUE

Predictive value of a test or measure is the probability that a test identifies persons with or without risk factors. Positive predictive value (PVpos) is the probability that a positive test identifies a person with risk factors (ability to identify a person at risk). Positive predictive value may be calculated as follows:

$$\text{PVpos} = \frac{\text{true positives } (a)}{\text{true positives + false positives } (a+b)} \times 100\%$$

Negative predictive value is the probability that a negative test identifies a person without risk factors (ability to identify a person without risk). Negative predictive value may be calculated as follows:

$$\text{PVneg} = \frac{\text{true negatives } (d)}{\text{true negatives + false negatives } (d+c)} \times 100\%$$

Predictive value is dependent on the frequency of an illness or disorder in a population. When prevalence of a disorder or disease is 50 percent or higher, there is little impact on predictive value (see Example 1 below). When prevalence of a disorder or disease is less than 10 percent, predictive value is severely affected as the ratio of true positives to false positives drops significantly (see Example 2 below).

Example 1. An example of the predictive value of a test is detection of CTS by a positive Phalen's test if the test is 60 percent sensitive and 90 percent specific for CTS and given to 1,000 workers with wrist and hand pain such as workers in a plastics processing plant who are in high-repetition jobs—a group in which the frequency of CTS is 50 percent for all wrist-hand complaints. Of the 500 persons with CTS, 60 percent or 300 will have a positive Phalen's test. Of the 500 persons without CTS, 90 percent or 450 persons will have a negative Phalen's test, but 50 will have a positive test. The positive predictive value of the test is 300 divided by (300+50), or 86 percent. This means that 86 percent of those with a positive test actually have CTS (true positive), while the other 14 percent with a positive test are free of CTS (false positive). If these persons were considered surgical candidates for treatment of their symptoms, 14 percent who underwent the surgery would have been free of CTS prior to surgery.

Example 2. What if the test is applied to healthy job applicants under age 30 without complaints of wrist and hand pain—a group in which the frequency of CTS is approximately 1 percent? If 1,000 applicants are screened, of the 10 persons with CTS, 60 percent or 6 will have a positive Phalen's test. Of the 990 persons without CTS, 90 percent or 891 will have a negative Phalen's test, but 99 persons will have positive tests. Therefore the positive predictive value of the test is 6 divided by (6+99) or 6 percent. Only 6 percent of those with a positive test actually have CTS (true positive), while the other 94 percent with a positive test are free of CTS (false positive). If this test was used to screen job applicants, 94 percent of those denied employment because of a positive test would have been misclassified.

Let us return to the example of the plastics workers and assume that the test sensitivity is actually 40 percent sensitive rather than 60 percent sensitive, and 90 percent specific for CTS. The test is given to 1,000 workers with wrist and hand pain, such as workers in a plastics-processing plant who are in high-repetition jobs—a group in which the frequency of CTS is 50 percent for all wrist-hand complaints. Of the 500 persons with CTS, 40 percent or 200 will have a positive Phalen's test. Of the 500 persons without CTS, 90 percent or 450 persons will have a negative Phalen's test, but 50 will have a positive test. The positive predictive value of the test is 200 divided by (200+50), or 80 percent. This means that 80 percent of those with a positive test actually have CTS (true positive) while the other 20 percent with a positive test are free of CTS (false positive). If these persons were considered surgical candidates for treatment of their symptoms, 20 percent who underwent the surgery would have been free of CTS prior to surgery.

If the test sensitivity in this same group of workers is actually 25 percent, of the 500 persons with CTS, 25 percent or 125 will have a positive Phalen's test. Of the 500 persons without CTS, 90 percent or 450 persons will have a negative Phalen's test, but 50 will have a positive test. The positive predictive value of the test is 125 divided by (125+50), or 71 percent. This means that 71 percent of those with a positive test actually have CTS (true positive), while the other 29 percent with a positive test are free of CTS (false positive). If these persons were considered surgical candidates for treatment of their symptoms, 29 percent who underwent the surgery would have been free of CTS prior to surgery.

Appendix 9–D

Prevention

Preventative medicine divides intervention into the categories of primary, secondary, and tertiary prevention (see Figure 9–D–1). Primary prevention attempts to prevent initiation of an illness or disorder. Primary prevention in the workplace includes such interventions as worker education and training in proper safety and work techniques, work environmental controls, and appropriate workplace design and organization.

Secondary prevention attempts to slow or halt progression of an illness or disorder. In the workplace, secondary prevention includes such interventions as screening and early identification of work- and non-work-related illness or disorders in a preclinical stage and the identification of the risk factors for the illness or disorder. At the preclinical stage, the disorder may be managed by modifying the work exposure—for example, by reducing exposure to tasks requiring high repetitions and/or forces and the interaction of the work exposure with other personal risk factors. In addition, treatment may be instituted to return the worker to full functional capacity.

Tertiary prevention in the workplace attempts to reduce work-related disability from an illness or disorder presenting with clinical impairments (signs and symptoms). Tertiary prevention includes reducing the work-related disability by workplace and job modification and accommodation and optimizing case management including medical treatment. This prevention model is shown in Figure 9–D–1.

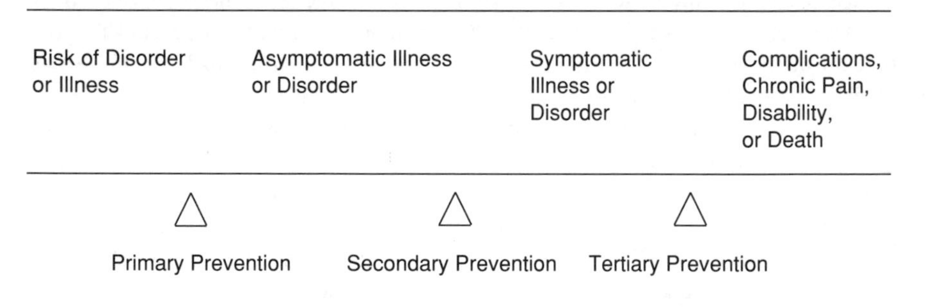

Figure 9–D–1 Course of Illness or Disorder and Points of Prevention

Chapter 10

Industrial Rehabilitation and the Hand Specialist

Karen Schultz-Johnson

THE SPECIAL ROLE OF THE HAND SPECIALIST

The hand specialist has a special contribution to make to industrial rehabilitation. According to Walker and Rehm, hand therapists "differ from other types of [testers] in their detailed understanding of pathologic anatomy and kinesiology, their understanding of ergonomic factors in task analysis [pertinent to the upper extremity] and their ability to examine the hand."[1] With its 27 bones, 37 muscle-tendon units, and three major nerves, the hand presents a challenge to even anatomy enthusiasts. A full understanding of the hand's complex biomechanics requires diligent focus.

The hand specialist participates in the three vital components of industrial rehabilitation: prevention, acute care therapy, and work-oriented or return-to-work programs.

Prevention encompasses

- ergonomics and safety consultation
- postoffer evaluation (previously preemployment screening)
- fitness evaluation

Each of these revolves around a comprehensive job analysis.

Acute care therapy involves

- immediate informal job analysis (IJA)
- wound care
- edema control

Note: Significant parts of the sections "The Upper Extremity Aspects of Lifting and Carrying," "Altering the Presentation of True Symptom Levels," and "Standardized Tests of Hand Function," will also appear in K.S. Schultz-Johnson, Functional Capacity Evaluation, in *Rehabilitation of the Hand,* 4th ed., ed. J.M. Hunter et al. (St Louis: C.V. Mosby, in press).

- range of motion exercises
- splinting
- desensitization
- scar modification
- programs to increase strength, dexterity, and endurance
- ADL evaluation and training, including recommendations for adaptive equipment
- assisting the physician and employer with identification of temporary modified work
- communication with the occupational health nurse

The *return-to-work aspect* of the program entails

- functional capacity evaluation (FCE)
- formal job analysis (JA)
- work simulation
- situational assessment
- employability-enhancing programs such as job modification, work hardening, work conditioning, and dominance retraining

PREVENTION

Prevention programs take many forms. Ergonomics and safety consultation, postoffer evaluation, and fitness evaluation all offer the employer a means to prevent injury to employees from occurring. The cost-effectiveness of prevention is documented;[2] it is more than just a fine idea.

Ergonomic and Safety Consultation

To accomplish the goal of reducing injury due to poor job, tool, and workstation design, the hand specialist provides an ergonomic consultation. While ergonomics addresses a major aspect of injury prevention, the safety consultation focuses upon spontaneous trauma such as lacerations and injury resulting from falls. Possessing a wide range of knowledge and skills, the consultant must be able to analyze the worker, the work, and the administration and/or management to identify problems. He or she must provide appropriate and cost-effective intervention to solve the problems in each of these areas.[3] A comprehensive knowledge of both the biomechanics and kinesiology of the upper extremity and the risk factors of work provides the basis for sound recommendations.

Ergonomic and safety consultation can be comprehensive or focused. It may involve industry-wide analysis and intervention or pinpoint a specific task or worker classification. The recommendations can be as sweeping as workstation redesign or as subtle as instruction for periodic exercise breaks (see Figure 10–1).

Postoffer Screening

According to the Americans with Disabilities Act (ADA), preemployment screenings are no longer legal. However, an employer can, once an offer of employment has been made, assess the worker to be sure that he or she can perform the essential functions of the job. Hand specialists can execute job-specific functional capacity evaluations to determine whether the worker has the physical ability to perform the job. They can also do sophisticated tests such as nerve conduction velocity and vibrometry (see Figure 10–2A and 10–2B) to screen for preexisting conditions that may place a worker at risk. When the tests reveal milder forms of pathology, the employer may choose to monitor the worker more carefully with repeat testing at consistent intervals. In severe cases of preexisting physical problems, employers may consider denying an applicant a job because of objective test results indicating that the worker appears at high risk for nerve compression syndromes. This denial would be based on the "direct threat" clause in the ADA, which states, in essence, that workers can be refused a job if it is found that they present a direct threat to themselves or others. However, the mere presence of early symptoms of a problem would be insufficient evidence to conclude that a "direct threat" to the worker exists. "Direct threat" can only be applied in the light of strong medical evidence. The hand specialist and physician will assist the employer's decision-making process.[4]

ACUTE CARE FOR WORK-RELATED UPPER-EXTREMITY INJURY

The therapist providing medical intervention to the upper extremity of an acutely injured person must take responsibility for gearing the program to return to work, even in the face of complex injury and open wounds. It is this work focus that sets the industrial clinic apart from any other setting. Instead of seeing the tip amputation, tendon repair, or nerve compression in a vacuum, the industrially oriented therapist sees a worker with an injury. The therapist sets goals and treatment plans in light of this perspective.

Informal Job Analysis

After the physical evaluation of the worker, the therapist undertakes the informal job analysis (IJA). The IJA acquaints those treating the worker with the required outcomes for the worker's return to the previous employment. The IJA approach elicits subjective information from the worker regarding the physical demands of the job. The therapist may need to verify critical aspects of the job with the employer. The IJA is a crucial part of the intake evaluation of an injured worker.

Figure 10–1 Exercise Card. *Source:* Copyright © Upper Extremity Technology, 1993.

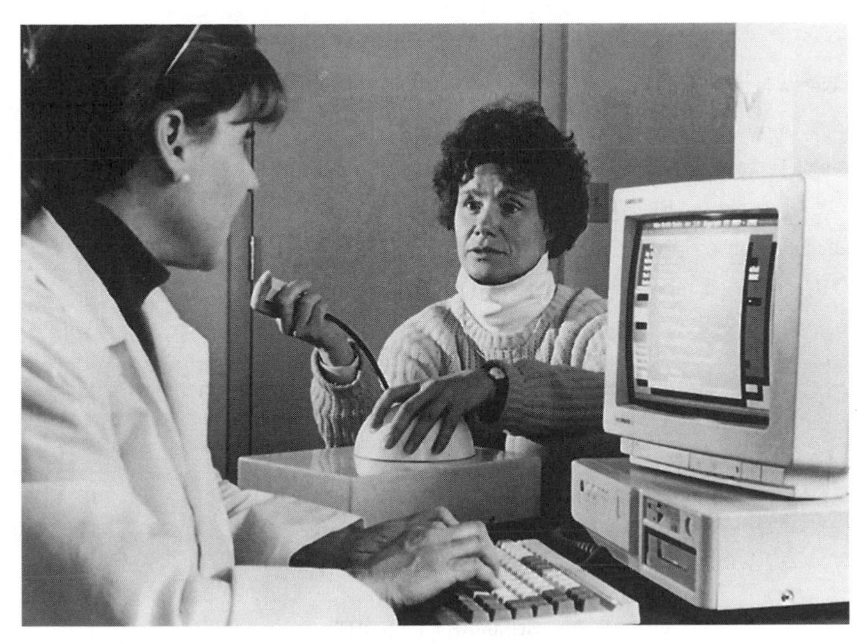

Figure 10–2A The Vibrotactile Tester Computerized Device for Measuring Skin Sensory Threshold. This instrument can aid in early detection at peripheral neuropathy—including carpal tunnel syndrome.

During the IJA process, the therapist creates a systematic description of the job in both U.S. Department of Labor terms and biomechanical terms. For example, to identify the "fingering" requirements of the job, the therapist asks the worker, "What tasks at your job involve using your fingers to pick things up, place them, hold them firmly? What size and shape are the objects? What are they made of? How much do they weigh?" Similar types of questions generate information on the various physical demands of the job. In addition to questions about the physical demands, the therapist seeks information about work postures, work schedules, task-handedness requirements, repetitive motions, tools and machines, gnosis and safety issues, and environmental conditions. If a job description is available from the employer, this may work as the core document for the IJA process. When the IJA does not produce an adequate picture of the task components and stresses, the hand specialist performs an on-site evaluation or obtains a videotape of the work.

Work Therapy and Work Conditioning

In contrast to *work hardening,* which requires tissue healing and homeostasis, *work therapy* can begin very early in the treatment process, before the injured limb

SUMMARY REPORT

Subject's Name: Susan Jones
Subject's Age: 38 Years Old
Date of Test: 09/16/93 Time: 10:35:15
Hand: RIGHT Finger: MIDDLE
Stimulus Routine: 50 Hz Vibration
NOTE: Subject reported numbness and tingling in right hand

			Subject's Threshold	
Test Name	*Normal Threshold Limit*	*Measured Threshold*	*Significantly Greater than Normal Limit*	*Percentage above Normal Limit*
50 Hz Vibration	14.0 μM	15.1 μM	YES	107%

SUMMARY REPORT

Subject's Name: Susan Jones
Subject's Age 38 Years Old
Date of Test: 09/16/93 Time: 10:43:15
Hand: LEFT Finger: MIDDLE
Stimulus Routine: 50 Hz Vibration
NOTE: No symptoms in left hand

			Subject's Threshold	
Test Name	*Normal Threshold Limit*	*Measured Threshold*	*Significantly Greater than Normal Limit*	*Percentage above Normal Limit*
50 Hz Vibration	14.0 μM	8.5 μM	NO	N/A

Figure 10–2B Reports Generated by the Vibrotactile Tester

has achieved tissue homeostasis. In conjunction with wound care, edema control, range of motion exercises, splinting, desensitization, scar modification, exercises to increase strength, dexterity, and endurance, and ADL evaluation, the therapy program includes work-oriented activities for rehabilitation. These activities may encourage range of motion, desensitize, and increase strength, endurance, and dexterity. Importantly, they help maintain the injured person's identity as a worker. Here too, work conditioning has an important role in maintaining general body fitness.

Splints and Braces

The hand specialist has another key contribution to make with the design and fabrication or the recommendation of upper-extremity splints and braces. Such splints and braces can[5]

- prevent deformity and maintain normal tissue length, balance, and excursion
- immobilize and stabilize
- protect healing structures
- correct deformity or dysfunction; reestablish normal tissue length, balance, and excursion
- control and modify scar formation
- substitute for dysfunctional tissue
- provide a forum for exercise

The hand specialist applies the functional splint or brace following an individual evaluation that reveals a medical need. Custom fitting the device to the specific worker, the therapist takes care to provide the most comfortable and function-compatible design. The practice of purchasing ready-made splints and applying these universally either prophylactically to the entire workforce or to the injured worker does not produce the desired results. Specific problems require specific splints that are chosen to match the size and function of the worker. Properly selected and applied, these devices can reduce the time of rehabilitation and can even facilitate return to work. The value of splints and braces to upper-extremity rehabilitation cannot be overstated.

Transition from Acute Care to Return to Work

The acute care therapist has an important role in helping the injured person return to some form of gainful employment as soon as it is feasible. Working with the occupational health nurse and treating physician, the therapist assists in identifying both appropriate temporary light duty and modifications or auxiliary aids that may facilitate early return to work and compliance with the ADA. The therapist may continue on with the worker into work hardening or may "hand off" to the work-hardening specialist. Often a hand specialist becomes involved in ergonomic and safety consultation with an employer as a natural sequelae to treating an individual worker. Thus the therapist equipped with the industrial perspective who offers medical care makes a key contribution to rehabilitation of the injured worker.

JOB ANALYSIS

Integral to all aspects of upper-extremity management is the upper-extremity-focused job analysis. The following section will focus upon aspects of the job analysis specific to upper-extremity function.

Task Taxonomies

As with any job analysis, the therapist first breaks the job down into specific tasks and then breaks each task into its component physical demand characteristics. In the upper extremity, the physical demand categories[6] are listed in Exhibit 10–1. Following these preliminary classifications, the therapist then makes the unique contribution of describing the task using a biomechanical perspective. According to Isernhagen, this entails describing movements, postures, and physical factors.[7]

Task Duration

To describe duration of each task and/or physical demand, the therapist may use general or specific terms. An industry standard in describing duration is shown in Exhibit 10–2. Precise timing of a particular task may be crucial to identifying risk

Exhibit 10–1 Upper Extremity Physical Demand Characteristics of Work

1. **FINGERING:** pinching or otherwise working with the fingers primarily (rather than the whole hand or arm as in handling)
2. **HANDLING:** seizing, holding, grasping, turning, or otherwise working with the hand or hands (fingering not involved)
3. **REACHING:** extending the hands and arms in any direction
4. **PUSHING:** exerting force upon an object so that the object moves away from the force
5. **PULLING:** exerting force upon an object so that the object moves towards the force (includes jerking)
6. **LIFTING:** raising or lowering an object from one level to another (including upward pulling)
7. **CARRYING:** transporting an object, usually holding it in the hands or arms or on the shoulder
8. **TWISTING/TORQUING (forearm/wrist):** exerting force upon an object in a circular manner via rotation at the forearm and wrist
9. **FEELING:** perceiving such attributes of objects and materials as size, shape, temperature, or texture, by touching with skin, particularly that of the fingertips
10. **CRAWLING:** moving about on the hands and knees or hands and feet
11. **CLIMBING:** ascending or descending ladders, stairs, scaffolding, ramps, poles, ropes, and the like using feet and legs and/or hands and arms

Exhibit 10–2 Industry Standard for Describing Task or Job Duration

Occasionally:	up to 33% of time at work
Frequently:	34–66% of time at work
Constantly:	67% or more of time at work

factors.[8] Importantly, the therapist considers the company's productivity standards along with its requirements for speed and quality.

Characteristics of Objects and Materials

In specifying the objects and materials used on the job, the therapist includes tools, machines, vehicles, parts, and substances. When considering the hand, one must delineate such object characteristics as size, shape, weight, texture, composition, and temperature.[9]

Prehension Patterns

A critical aspect of upper-extremity-focused job analysis is the prehension patterns (hand-object couplings in engineering terms) incorporated in each task. The literature presents several taxonomies of prehension patterns. Casanova and Grunert developed the most detailed analysis of prehension.[10] Kroemer's much simpler taxonomy (see Figure 10–3) communicates well with employers, rehabilitation professionals, and physicians.[11]

Task-Handedness

The determination of task-handedness has tremendous implications for the worker and for potential job or task modifications. A task may require one or two hands. If the task necessitates the use of two hands, the consultant must assess the skill level required of each hand. One hand may require a high level of skill while the other serves as an assist hand. In contrast, a task may require two skilled hands. Yet another task situation may require that a worker use a specific hand, for example, the right hand, to perform the task because of the unique configuration of the task or the design of a machine.

Safety from Trauma

The consultant screens for factors that jeopardize or protect the hand from sudden trauma. Checking for the presence of appropriate machine and tool guards, the

Hand-object couplings describe the geometric interaction between object surfaces and the anatomical sections of the hand.

Prehension occurs when the hand contains an object in the environment in a controlled manner.

Listed and pictured below are excerpts from Kroemer's article on hand-object couplings. This author has taken the liberty of renaming or adding names to Kroemer's couplings.

1. **FINGER TOUCH/STABILIZATION:** finger touches an object without prehending it 
2. **TIP PINCH:** thumb tip opposes one fingertip 
3. **PALMER PINCH:** thumb pad opposes the pads of two or more fingers 
4. **LATERAL/KEY/SIDE PINCH:** thumb pad opposes the radial side of a finger, often the index 
5. **THUMB-TWO FINGER GRIP/WRITING GRIP:** thumb and two fingers (often the index and middle) oppose each other near the tips 
6. **DISC GRIP:** thumb pad and the pads of three to four fingers oppose each other near the tips (object grasped does not touch palm) 
7. **PALM TOUCH/STABILIZATION:** some part of the inner surface of the hand touches the object without prehending it 
8. **HOOK GRIP:** one or several fingers hook(s) onto a ridge of a handle 
9. **GROSS GRASP/FINGER-PALM ENCLOSURE:** most or all of the palmer surface of the hand is in contact with the object while prehending it 
10. **CYLINDRICAL/POWER GRASP:** total palmer hand surface is grasping the object, which runs parallel to the knuckle and generally protrudes on one or both sides from the hand

Figure 10–3 Taxonomy of Prehension Patterns/Hand Object Couplings. *Source:* Adapted with permission from *Human Factors,* Vol. 28, No. 3, 1986.

therapist makes certain the workers know when and how to use them. It is crucial to note whether hands are used within the worker's line of vision or if the worker must use his or her hands using sensation as the primary guide. When hands are used outside the line of vision, the hand can be exposed to heat, moving machinery, or sharp edges. This exposure requires intact protective sensation and the ability to focus attention. Some tasks require the hand to be used within confined spaces. In such instances, impaired range of motion could have significant impact on the worker's ability to perform the job safely, or perform it at all.

General Ergonomic Issues

Many ergonomic issues affect the health of the upper extremity at the workplace. The key factors are listed below:

- repetitiveness of task[12]
- forcefulness; resistance engaged
- motion versus static positioning
- glove use[13]
- posture
- tool/human interface
- workplace characteristics
- worker position
- work schedule and pace

Repetitive and Static Motion

Task repetitiveness, especially when coupled with higher levels of resistance, has impact not only on the muscle, but also on the nerves, tendons, ligaments, and cartilage. Because the tendons of the hand often travel in tight sheaths and function around pulley-created angles, they are subject to shear and friction as force and repetition increase. Hand ligaments and cartilage bear high forces over small surface areas. With continued wear and tear from high repetition, especially in light of their poor vascularity, these structures fall prey to repetitive motion.[14] The sensitivity of nerves to pressure and decreases in blood supply makes these important structures vulnerable to the effects of repetitive motions as they cross the area of the joint.

The other end of the movement spectrum, static positioning, can be just as harmful. When workers must maintain a position, the muscles around the statically held joint must provide a long-duration contraction. Static muscle contractions reduce the blood flow as long as the contraction is maintained. Metabolites accumulate, and oxygen to the muscle can be quickly depleted.[15]

The role of synovial fluid chemical make-up in cumulative trauma disorders requires special attention. Synovial fluid is the body's natural lubrication system, and it is found around all moving body parts, including tendons and cartilage. Children seem to have a limitless tolerance to physical movement, no matter how repetitive. It is most common in the adult that inflammatory responses to activity occur. Thus it can be hypothesized that with age or hormonal changes, the viscosity of the synovial fluid changes. The synovial fluid can either lose viscosity and thus the ability to withstand high heat and friction,[16] or it can become too viscous and become a barrier to movement in itself.

Rohmert and Rodgers[17] have developed an algorithm to determine whether a task involves dangerous static contraction of muscle. To make the calculation requires information on the cycle time of the task (the time from the beginning of the task until the task begins again), the holding time of the task (the continuous muscle effort time), and the level of maximum voluntary contraction (MVC) of the target muscle. The graph in Figure 10–4 shows how these three factors interact. If the plotted point falls *under* the appropriate MVC line, then the contraction should be safe for the muscle. Thus repetition, force, and length of contraction often interact to cause or prevent worker injury.

Gloves

Offering many benefits, gloves can dampen vibration, reduce skin shear, and increase hand warmth, thereby increasing blood flow. However, gloves interfere with finger flexion and decrease sensory feedback. These factors cause the worker to contract muscles more forcefully, which leads to increased fatigue and harmful sequelae of static loading. Hertzberg[18] and Rodgers[19] studied the use of gloves on grip strength. They found that gloves can reduce grip strength up to 35 percent.

Posture

Posture of the upper extremity has great impact on performance. Studies have shown the importance of keeping the shoulder adducted as close to the body as possible.[20] Awkward position of the arm proximally can affect nerve function and blood supply distally. Ideally the forearm should be positioned in neutral to prevent stress to extrinsic flexors or extensors and to prevent static loading of the pronators and supinators. The Woody and Mathiowetz studies demonstrated that elbow and forearm position affects grip and pinch exertion levels.[21] Studying forearm posture during repeated, loaded-wrist flexion and extension, Hallbeck and Sheeley[22] found that forearm supination created the greatest force. To prevent

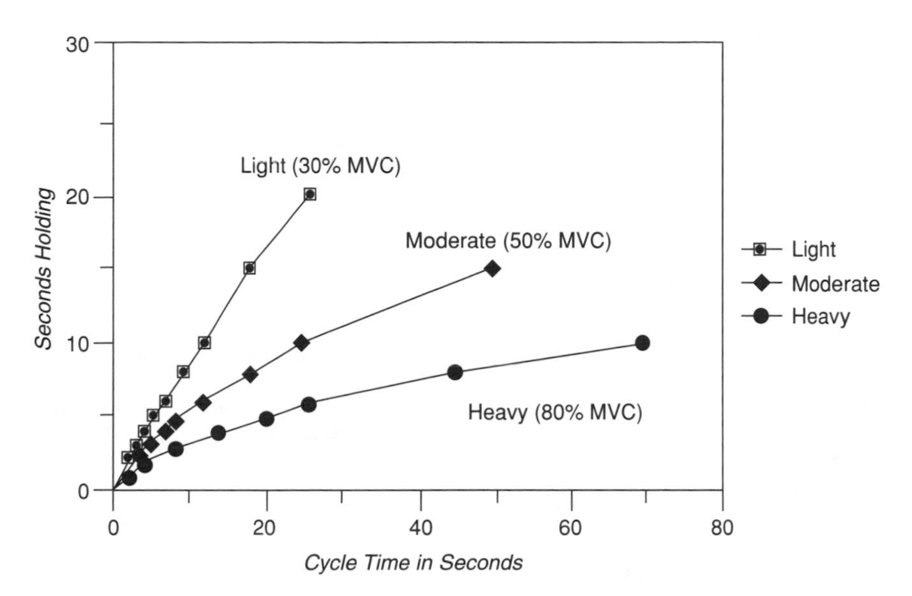

Figure 10–4 Static Work Recovery Curves Represent the Interaction between Task *Cycle Time, Holding Time,* and *Maximum Voluntary Contraction of Muscle* To Determine Risk Factors. *Source:* Adapted from Rohmert, W., Problems in Determining Rest Allowances. Part I: Use of Modern Methods To Evaluate Stress and Strain in Static Muscular Work, *Applied Ergonomics,* Vol. 4, pp. 91–95, with permission of Butterworth-Heinemann Ltd., © 1973.

awkward shoulder position, the work-tool-worker combination should allow the elbow to posture at approximately 90 degrees of flexion. Wrist deviation causes finger flexors and extensors to traverse extreme angles and makes them biomechanically inefficient, thus reducing grip and pinch strength[23] and potentially causing inflammation (see Table 10–1).

The Tool-Human Interface

The tool-human interface has profound effect on the health of the hand and upper extremity. The following principles summarize a review of literature:

- Keep the wrist maintained in neutral.[24]
- Use the appropriate muscle group for the tool.
- Use power grip for power and precision grip for precision.[25]
- Consider tool weight and use lighter tools when the task requires repetition.
- Align the center of gravity of the tool with that of hand.[26]

Table 10–1 The Effect of Wrist Angle on Grip Strength

Condition	*Percentage of Power Grip*
Power Grip—2-inch span	100
with 25° radial deviation	80
with 45° extension	75
with 45° ulnar deviation	75
with 45° flexion	60
with 65° flexion	45

Note: Based on data collected in student laboratories and industrial training courses, 1982–1985.

Source: Adapted from Rodgers, S., Recovery Time Needs for Repetitive Work, *Seminars in Occupational Medicine*, Vol. 2, No. 1, p. 20, with permission of Thieme Medical Publishers, Inc., © 1987.

- Use motor power over muscle power.
- Use job-specific tools.[27]
- Use tools that can be used by either hand.[28]
- Use tools that are specific to the population using them.
- Attend to tool activation—use spring-loaded hand tools and use triggers that activate with gross grasp versus single-finger flexion.
- Consider handle characteristics.

The anatomy and function of the hand determine optimum handle characteristics. Because the hand-tool interface is often the essence of the worker-tool interaction, it demands significant attention. The following list will outline the significant literature available on this topic.

- Tool handle *span* should not exceed the recommended maximum distance between the two edges of the handles of 3.5 inches for men and 3.124 inches for women.[29]
- Tool handle *curve* should not be greater than 0.5 inches over entire length.[30]
- Cylindrical handles should have a *diameter* between 1.5 and 2 inches—3.81–5.08 cm.[31]
- Minimum handle *length* for any tool is 4 inches (10.16 cm)[32]; for tools used with gloves add 0.5 inches (1.27 cm).[33]
- Shape of handle should allow the forces to be distributed over as large an area as possible; the pressure of the handle should be borne over fat pads of hand.[34]
- Flanged handle ends prevent the hand from sliding to dangerous parts of tools—hot or sharp.

- Optimal handle shape minimizes need for tight pinch or grasp (static loading).
- Handle texture should provide adequate friction coefficient to avoid fatigue.[35]
- The design of the grip surface should be compressible, nonconductive, and smooth.[36]
- The handle type—pistol grip versus cylindrical grip—interacts with the work orientation and should be chosen to ensure safe postures of the shoulder and wrist (Figure 10–5).
- Handles should absorb shock and dampen vibration[37] since percussion and vibration damage joints and cells and can cause vasoconstriction.

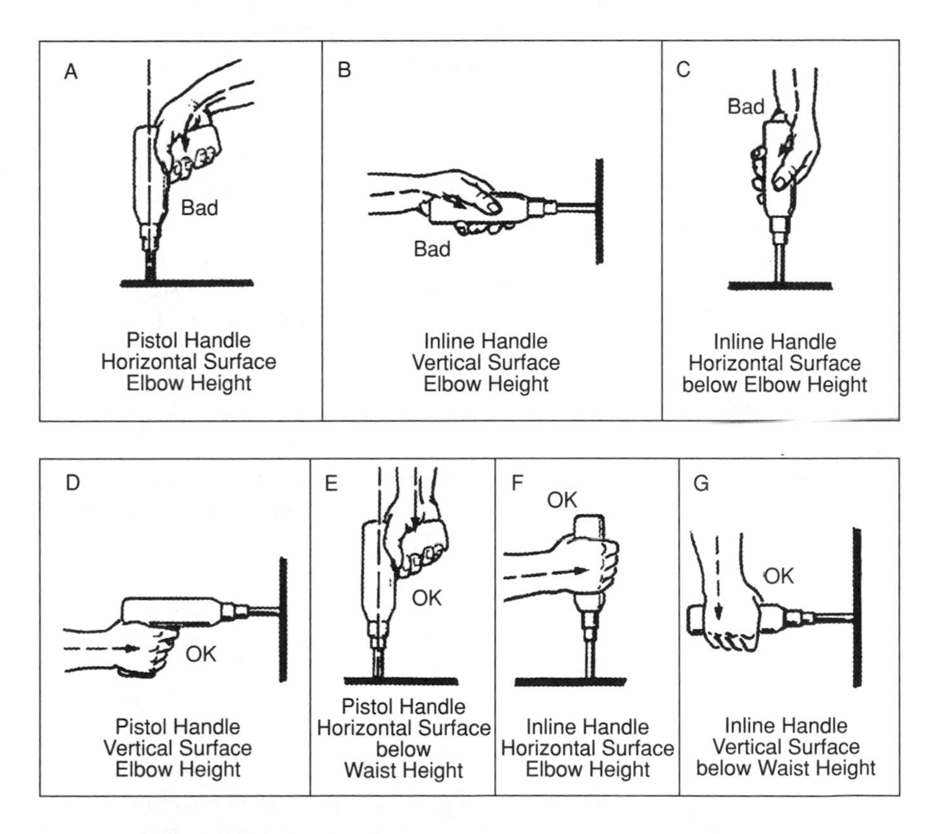

Figure 10–5 Relationship between Handle Design and Work Operation. *Source:* Reprinted from *Cumulative Trauma Disorders* by V. Putz-Anderson, p. 106, with permission of Taylor & Francis, © 1988.

Workplace Characteristics

Because the ambient temperature of the workplace affects blood flow, temperature becomes a critical factor of work. As temperature drops, the most distal aspects of the extremities experience vasoconstriction first. Thus the hand has primary vulnerability to cold. With synovial fluid viscosity and nerve and muscle nutrition reliant on the blood supply, it is easy to see why low temperature interacts with other factors such as repetition and static loading to create the scenario for cumulative trauma to occur.

When considering workplace characteristics and the upper extremity, one must consider work area and workstation height—the desk, assembly line, or bench. The bird's eye view should reveal that the design confines the work area to what can be conveniently reached with the sweep of the forearm with the upper arm in a natural position at the side of the trunk. The maximum area is what can be reached by extending the arm from the shoulder. These design constraints, in conjunction with proper workstation height, ensure minimum wear and tear on the shoulder and elbow.[38]

To design proper workstation height, one must relate the height of the worker to the height of the workstation as well as relate the height of the worker to the type of work being performed. Generally, optimum work height is achieved when the work surface allows the elbows to be at the sides of the body and flexed to 90 degrees. If the work surface is too high, the shoulder abducts and static loading results. Generally, precision work requires a higher work surface while heavier work necessitates a lower work surface.[39]

Worker Position

Worker position during work tasks has impact on posture and mechanical stresses. While it may seem out of context to focus upon the position of the trunk and legs during work, one must address the body as a whole, for "the arm bone's connected to the trunk bone." If the trunk lacks stability or optimal positioning, the upper-extremity musculature must produce a higher level of output to position the arm in space. As previously stated, poor proximal positioning will affect nerve conduction and blood supply to the extremity.

Job Analysis Summary

In the performance of a job analysis, the knowledge areas of anatomy, biomechanics, physiology, work practices, safety, and ergonomics come together to provide the basis for all industrial-medical intervention. The therapist utilizes expert observation and objective analytic techniques to describe and quantify each task. The job analysis is the core function of the industrially oriented therapist.

RETURN-TO-WORK PROGRAMS

The *return-to-work aspect* of the program entails functional capacity evaluation (FCE), job analysis (JA), work simulation, situational assessment, and employability-enhancing programs such as job modification, work hardening, work conditioning, and dominance retraining. Work hardening and work conditioning often straddle the acute care and return-to-work aspects of rehabilitation.

Functional Capacity Evaluation

A physical capacity evaluation can identify and/or clarify the skills and work tolerances available to a given worker. Such an evaluation can consist of a brief screening or a several-day evaluation, depending on the questions the evaluation needs to answer. When endurance and feasibility are major issues, the longer evaluation is generally recommended. When a question exists as to whether a worker has adequate physical tolerances and skills to begin or return to a *specific* job, a situational assessment or work simulation is usually the evaluation of choice. An FCE can determine candidacy for programs such as work hardening or dominance retraining. Maximizing the appropriate utilization of the programs, this initial screening is time- and cost-effective. The FCE serves many helpful purposes.

As with most FCEs, the upper-extremity FCE has five to six components,[40] depending on whether a target job is identified and a job analysis must be done.

1. initial interview
2. subjective effects of injury interview
3. physical evaluation
4. biomechanical analysis of targeted job
5. design and implementation of work activity testing, including standardized and nonstandardized testing
6. postactivity evaluation

Several aspects of the upper-extremity FCE warrant special focus. These are standardized testing, symptom magnification screening, and the upper-extremity aspects of lifting.

Standardized Tests of Hand Function

Standardized tests provide the evaluator with the opportunity to observe a specific physical demand. Because norms accompany these activity tests, they allow comparison of the tested individual to a larger group of people or workers. Because the tests require the evaluee to demonstrate function rather than talk about it, they will compare the client's *perceived* disability with his or her *actual* disability.

Standardized tests commonly administered as part of an upper-extremity FCE are: Purdue Pegboard, Minnesota Rate of Manipulation, Stromberg Dexterity Test, Crawford Small Parts Dexterity Test (Figure 10–6), Bennett Hand Tool Dexterity Test (Figure 10–7), Rosenbusch Test of Finger Dexterity (Figure 10–8), and the Jebsen Hand Function Test. In addition to these, some manufacturers have developed standardized testing systems such as Valpar (Figure 10–9) and JEVS.[41]

While the tests usually focus upon one or two physical demands, the evaluator should note the comprehensive physical demand requirements of each test. For example, the Purdue Pegboard focuses upon fingering and the ability to pick up, manipulate, and place small, light objects. However it also addresses fine coordination, dexterity, repetitive reaching between waist and chest level, speed, accuracy, prehension pattern availability and use, impact of sensation on function (feeling), endurance, and impact of pain on function. Cognitive and perceptual function as well as posture and balance will also affect the client's ability to perform the test. When the therapist has identified all of the physical demands of the standardized tests as well as the strength levels required for each demand, it then becomes straightforward to plug them into the evaluation plan.

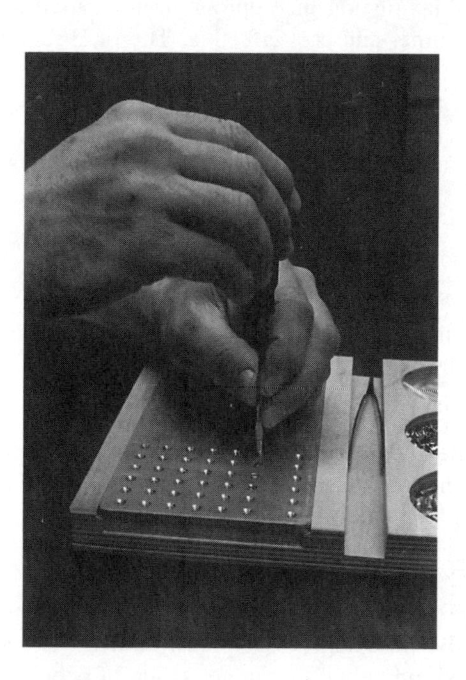

Figure 10–6 Crawford Small Parts Dexterity Test

Figure 10–7 Bennett Hand Tool Dexterity Test

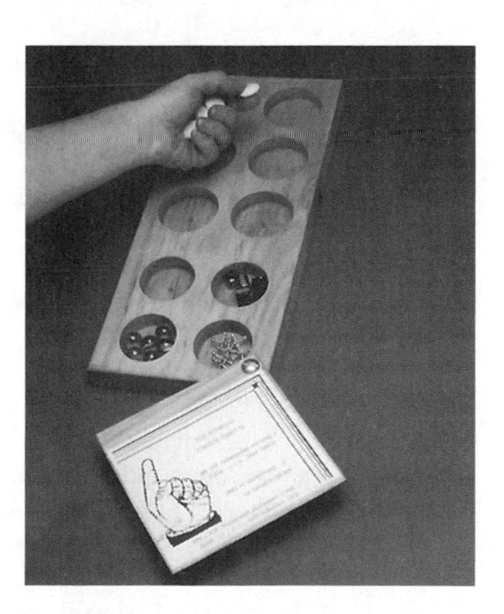

Figure 10–8 Rosenbusch Tests of Finger Dexterity

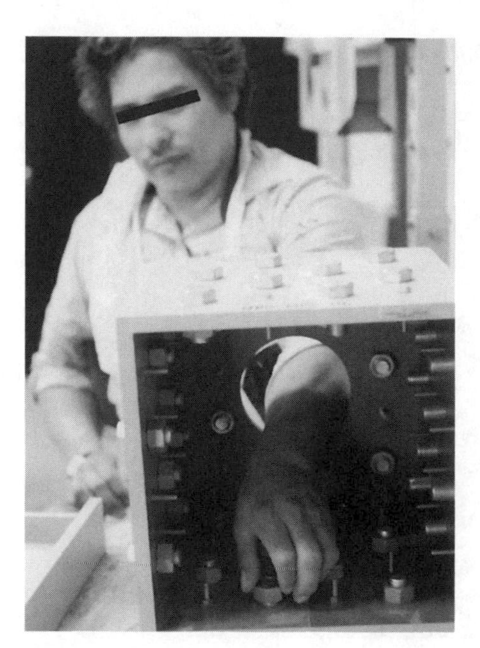

Figure 10–9 One of the Valpar Standardized Tests

Test norming can take several forms, and some tests include more than one form. Percentiles and stanines both offer a comparison between the evaluee and a group. Time-motion-measurement looks at the time a task should take according to industry standards. Because many of the commonly used standardized tests were developed several decades ago, the existing norms may be outdated. In addition, because the test may not have an appropriate norm group for an evaluee, the therapist must use care when interpreting the results of the test. Due to these flaws, it appears that the industry standard will soon become motion-time standards (MTS).

Farrell describes MTS as "a system to define work content by sequentially listing the motions performed by the worker."[42] MTS then assigns a span of time in which industry expects those actions to be completed. Lifecorp has developed computer software to convert raw scores from standardized tests to motion-time standards.[43] Because MTS has direct relevance to industry and a common language for action description across tests, in-clinic simulated tasks, and the workplace, therapists should look to MTS as the future of standardized testing.

In many clinics, standardized tests function both as work samples and as standardized tests. This is possible as long as the therapist acknowledges that the test cannot be used for *both* a work sample and a test with the same client. The tests were not normed with hours of practice as part of the standard. One must choose to use a test in one manner or the other.

Standardized tests form an integral part of the upper-extremity FCE. Fess states that "in an age of consumer awareness and accountability, it is no longer sufficient to rely on 'home-brewed,' non-validated evaluation tools."[44] The use of standardized and well-validated evaluation tools gives the clinician confidence in the accuracy of the data collected and the ability to compare these data with those of others in the clinic. However, one must also strongly consider the value of a standardized test as an opportunity to observe the client in a structured task rather than simply to generate a norm.

Altering the Presentation of True Symptom Levels

Symptom Magnifiers and Minimizers. Matheson coined the term "symptom magnification" to describe clients who use nonadaptive, static behavior to attempt to manipulate society with a display of symptoms.[45] In working with the industrially injured client, one will inevitably encounter a population of people who symptom magnify. Using the symptom magnification framework helps to identify successful interventions and minimizes useless conflict.

Knowledge of the types of symptom magnifiers facilitates the creation of an approach to working with the magnifying client. Matheson has thoroughly described three types of symptom magnifiers—the refugee, the identified patient, and the game player—and the description will not be reiterated here. However, for consideration, a fourth type of symptom magnifier, "the symptom misinterpreter"—will be presented.

Either because of difficulty processing the sensory and kinesthetic input or because of unrealistic belief systems about the way the body works, the symptom misinterpreter responds to the physical changes in the body in an extreme manner. Sometimes the nature of the problem can be revealed by asking the client, "What do you think is going on in your arm?" The sometimes astounding answers can vary from "I have cancer in it" to "My tendon is about to pop and roll up in my arm." While the reason for submaximal effort can then be understood, changing the beliefs of the client, and so the level of function, may not be possible.

In addressing the issue of symptom magnification, the evaluator should also keep in mind that the "symptom minimizer" also exists. This client keeps symptoms hidden so that he or she can rapidly return to normal activity or to avoid appearing "wimpy." Those who work in sports medicine see this frequently with athletes who wish to return to competition before possessing sufficient tissue integrity or resolution of problems. These two extremes of the continuum of clients—the symptom magnifier and symptom minimizer—challenge the therapist to objectively evaluate function and interpret the FCE results.

Identification of Symptom Magnification. When evaluating clients who have experienced *upper-extremity trauma,* the therapist has an edge in determining which clients are magnifying symptoms. First, the injury is visible, and second, it often results in somewhat predictable levels of function and dysfunction. Even clients with the often invisible cumulative trauma disorders should respond in a somewhat predictable manner to certain physical tests. Thus a thorough neuro-musculo-skeletal evaluation sheds much light on the behavior and subjective complaints of the evaluee.

The identification of symptom magnification rests heavily on the thoroughness of the neuro-musculo-skeletal evaluation. Therefore the therapist must have extensive knowledge of kinesiology, nerve function, and normal and anomalous anatomy. The therapist couples findings from this section of the evaluation with the comprehensive documentation during the interviews and activity section of the evaluation. During the neuro-musculo-skeletal evaluation as well as the other sections of the FCE, the therapist screens for behaviors, symptoms, and signs that are inconsistent with the patient's medical history.

While a review of the gamut of upper-extremity tests and techniques to identify symptom magnification is not possible within the confines of this chapter, an overview of screenings can be shared. The evaluator identifies any inconsistency in behavior. The client may use a tip pinch on one test and then state inability to do so on another test. The therapist may find discrepancies in perceived versus actual level of disability. During the interview, the client may claim inability to cut meat and then during the activity evaluation, demonstrate the ability to cut putty with a knife and fork. The client may demonstrate symptoms and signs that have no basis in anatomy, physiology, and kinesiology. One evaluee demonstrated this when he stated he had a "tremor"—an oscillating forearm pronation-supination movement—following a fall on an outstretched arm. Not only did this abnormal pattern of movement defy the laws of kinesiology, but when the evaluator stabilized the forearm to limit the pronation-supination movement, the client began flexing and extending his wrist. These behaviors clearly demonstrate symptom magnification since no physical injury or illness could produce the movement pattern described.

Researchers have produced several studies to investigate the use of grip testing in detecting symptom magnification.[46] These studies have given rise to two techniques employing the hydraulic dynamometer—the five-setting test and the rapid exchange grip (REG). Stokes reported that plotting force generated at each of the five dynamometer width settings generated a bell-shaped curve.[47]

Hildreth et al.[48] described the REG test, where subjects forcefully grip the dynamometer, which is rapidly alternated between the right and left hands. The maximum force generated during the REG is compared to that generated during a static grip test. The study found that a normal, motivated subject will produce less force during the REG than on the previous static grip testing, while the submaximal performer will produce more force during REG. The theory here is that on a test of static grip a subject can control grip force, whereas during REG the speed helps to eliminate conscious effort to perform submaximally.

Recent advances in evaluation documentation on computerized hand evaluation systems not only provide efficient recording of the five-setting test (with graph plotting to scan for the bell-shaped curve) (Figure 10–10) and the REG but also allow scanning of force-time curves generated during grip. While research has proven interpretation of force-time curves to be unreliable,[49] such curves can offer "a piece of the puzzle" when the specialist is trying to differentiate those generating maximum effort from those who are not (Figure 10–11).

Special Syndromes. Several "syndromes" demand special attention. One, *clenched fist syndrome*[50] or *nonfunctional co-contraction,* consists either of the forceful contraction of one group of muscles (usually flexors), or the simultaneous contraction of agonist and antagonist muscles, which results in inefficient motion or no motion (to be distinguished from functional co-contraction, which provides necessary stabilization of the upper extremity in space). It often manifests with forceful flexion of all of the fingers of the hand—hence the name *clenched fist syndrome*—but can manifest in a single digit or just the thumb. At times, the client clenches the fingers so tightly that the therapist cannot passively move them. The syndrome also manifests with slowly ratcheting movement of the fingers. A manual muscle test reveals that the client is contracting both the finger flexor and extensors simultaneously. Simmons and Vasile state that in clenched fist syndrome, no organic disease can be found and extension of the fingers is always possible under anesthesia. Psychiatrically, the clenched fist patient is classified as having severe anger and poor defenses. Prognosis for return to functional use of the hand is poor.

The recent clinical investigations into the phenomenon of *dystonia* complicate the interpretation of the findings of simultaneous contraction of agonists and antagonists. Jankovic and Schwartz write, "Although dystonic disorders were once considered to be of psychogenic origin, the weight of evidence now supports organic etiology. Physiologic studies have found impairment of normal motor control causing inappropriate co-activation of antagonistic muscles."[51, p.883–884] Thus these physicians advocate physical medicine intervention for people who display the symptoms described.

One of the most difficult problems an evaluator can encounter is the S-H-A-F-T Syndrome. In their 1978 article, Wallace and Fitzmorris describe the syndrome as

> a passive form of Munchausen's syndrome in which a patient submits to multiple surgical procedures to a part of the upper extremity. Apart from the obvious connotation, the title emphasizes that the patients are **S**ad, **H**ostile, **A**ngry, **F**rustrating, and **T**enacious [authors' caps and bold],

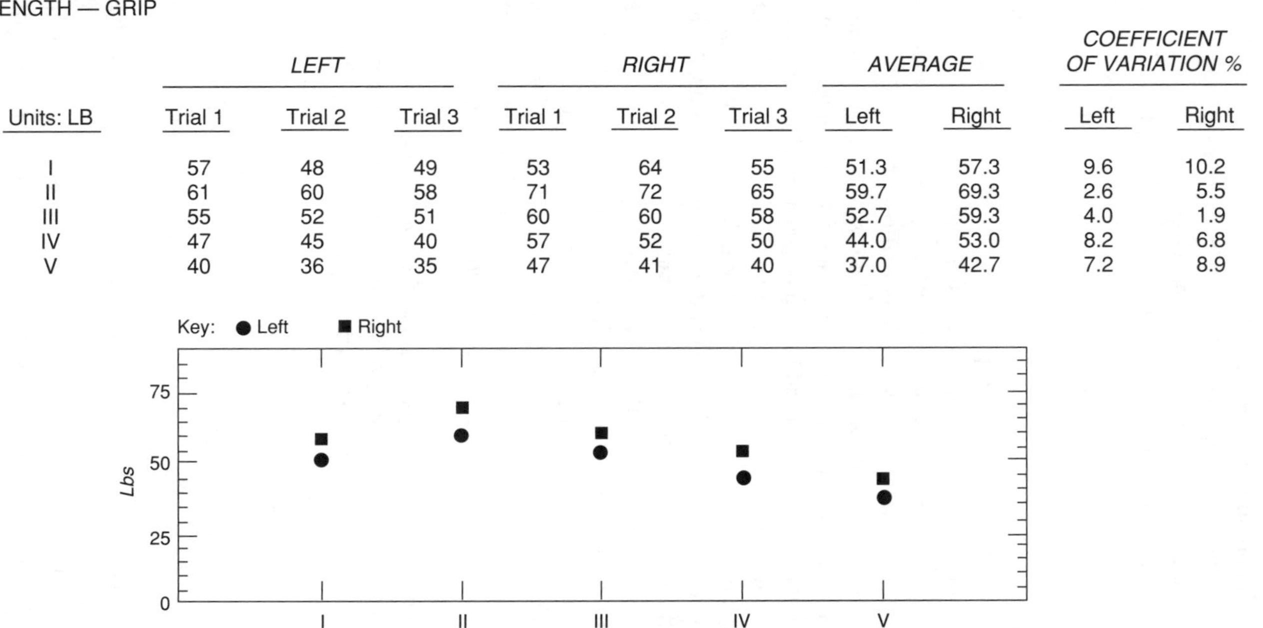

Rocky Mountain Hand Therapy

Name: Kate M.
ID Number: 123-45-6789
Examination Date: 9/18/92

STRENGTH — GRIP

	LEFT			*RIGHT*			*AVERAGE*		*COEFFICIENT OF VARIATION %*	
Units: LB	Trial 1	Trial 2	Trial 3	Trial 1	Trial 2	Trial 3	Left	Right	Left	Right
I	57	48	49	53	64	55	51.3	57.3	9.6	10.2
II	61	60	58	71	72	65	59.7	69.3	2.6	5.5
III	55	52	51	60	60	58	52.7	59.3	4.0	1.9
IV	47	45	40	57	52	50	44.0	53.0	8.2	6.8
V	40	36	35	47	41	40	37.0	42.7	7.2	8.9

Figure 10–10 Five Setting Dynamometer Test and Curve Print Out Done on the EVAL™ Computerized Upper Extremity Evaluation System by Greenleaf Medical

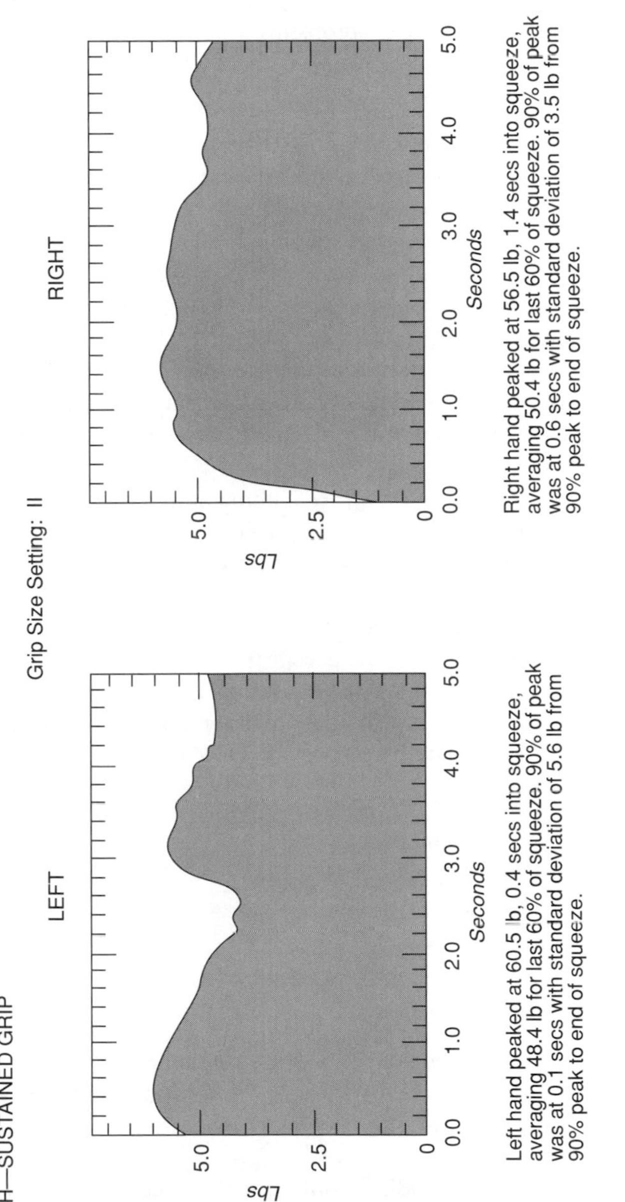

Figure 10–11 Force-Time Curve Generated on the EVAL™ Computerized Upper Extremity System by Greenleaf Medical

> with psychological problems similar to those of patients with Munchausen's disease—problems which are aimed at secondary gain of a psychological and/or financial nature.[52]

The article goes on to describe three patients who had a total of 25 operations as well as multiple injections of cortisone performed by 11 doctors in five different states. One patient, who according to the authors had no physical pathology, went on to have her arm amputated above the elbow.

The therapist performing FCEs must have an awareness of the more extreme forms of symptom magnification and the syndromes that mimic symptom magnification. Workers who spend significant time in the workers' compensation system sometimes do so because their problems fall into esoteric categories and no physician identifies a definitive diagnosis. Spending significant time with a worker during evaluation and treatment, a therapist can facilitate case resolution when he or she has awareness of the more obscure diagnostic categories and identifies objective findings that indicate specific diagnosis.

Submaximum Voluntary Effort. Submaximum voluntary effort can be conscious or unconscious. Several authors have written about screening techniques to identify when an evaluee fails to deliver an optimum effort. One approach, computation of coefficient of variation (COV), gives a percentage variation between multiple trials of a given static test. To accept the COV as a valid indicator of submaximum effort, one must accept the premise that repetitive trials within a brief span of time will be stable. While this may seem a reasonable assumption,[53] many factors besides symptom magnification can interfere with delivery of maximum effort.[54] These include but are not limited to

- misunderstanding instructions
- anxiety with test taking
- lack of familiarity with testing equipment
- very low endurance
- fear of reinjury
- fear of exacerbation of pain
- unidentified impairment

Matheson has suggested that COV testing "not directly involve an impaired component of the biomechanical system."[55, p.12] Yet most therapists who test the upper extremity for maximum effort look at the injured extremity for consistency of effort.

King designed the Validity Profile to look at maximum effort over several tests rather than just one.[56] This approach greatly enhances the possibility of identifying submaximum effort. King's work points out the importance of looking at multiple tests rather than just one test's COV before rendering an interpretation of submaximum effort. However, while this approach is the best we have so far, it will still not render 100 percent accuracy.

In summary, to identify symptom magnification, the therapist must first compare the evaluee's responses during the interview with observations during the neuro-musculo-skeletal evaluation and the activity evaluation. Next the therapist screens for the consistency and medical appropriateness of the client's behavior. Often, if the evaluator listens closely to the client, he or she can discover an alternate explanation for the ongoing problems. If, through this careful attention to the client, the therapist discovers the nature of the symptom magnification, he or she can facilitate case closure, saving everyone involved precious time and resources.

The Upper-Extremity Aspects of Lifting and Carrying

Because employment often revolves around the issue of lifting and carrying, these physical demands require special attention. While a plethora of information is readily available on the evaluation of lifting as it relates to the back-injured worker, literature, research, and data on evaluation of the upper extremity in lifting are much more scarce. In 1990, this author undertook a comprehensive review of literature and survey of practitioners regarding the upper-extremity factors in lifting. The following section summarizes this effort.[57]

Central to the lifting evaluation is the biomechanical analysis of the lifting task. Chaffin and Anderson group the factors that define the manual-materials-handling task or system.[58] These are

1. worker characteristics
2. material/container characteristics
3. task characteristics
4. work practices

These categories assist the evaluator in structuring a report of the many complex and interrelated aspects of the subject's lifting performance.

The interaction between multiple factors will affect the upper extremity during lifting and carrying. The critical factors listed below determine the upper extremity's response to the work demand placed upon it:

- prehension pattern
- direction of torque and force
- presence or absence of handles
- use of gloves
- arc of the lift
- whether the lift is one- or two-handed
- range of motion required at each joint

- duration
- handle characteristics, orientation, and position
- load characteristics
    1. shape
    2. size
    3. center of gravity of the load
        a. stable
        b. shifting
    4. weight

Prehension Patterns Used during Lifting. The relationship between prehension patterns and lifting and carrying has significant implication for a worker's ability to perform a task. This effect can involve the cardiovascular system as well as the total body biomechanics.[59] To assess the interaction between lifting and prehension patterns, therapist-evaluators may use interchangeable handles with the lifting assessment equipment. These handles attach to the load in a stationary or dynamic manner. The prehension pattern often changes during the course of a lift. The evaluator should investigate the reason for such changes. The lift may normally involve multiple prehension patterns, or the worker may use various patterns of prehension to compensate for weakness, hypersensitivity, or other problems.

In the case of the dynamic or moving handle, the handle attaches to the weight with a hook-type fastener that allows the weight to swing on the handle. The use of such dynamic handles will often preclude wrist deviation during lifting, since the handle moves on the weight rather than the wrist angulating. Thus the sequelae of wrist deviation during lifting will not become apparent unless the handle attaches statically to the container. Because many dysfunctions only become apparent in the presence of wrist deviation, the therapist must carefully consider the design of the lifting evaluation before stating conclusions regarding ability and disability.

Body Mechanics. Because the body performs lifting as a whole unit, lifting evaluation protocols should consider addressing comprehensive body mechanics. Even when the evaluee has no documented back, neck, or lower extremity involvement, the use of poor body mechanics during the assessment creates the potential for injury. The therapist should consider instructing the client in appropriate body mechanics for the lifting task. Some therapists will not initiate or continue an evaluation if the client cannot consistently demonstrate safe body mechanics.

Documenting Client Performance during the Lifting Evaluation. The form in Exhibit 10–3 has been used successfully during on-site visits to document the important characteristics of a lifting task. With this information, the therapist determines several other crucial task characteristics. What muscle groups engage the

Exhibit 10–3 Lifting Form Used during On-site Visits To Document the Important Characteristics of a Lifting Task

JOB SITE LIFTING EVALUATION WORKSHEET

Characteristic of Object Lifted:

Object lifted: _____

Dimensions of object: _____ × _____ × _____

Weight of object: _____ lbs

Texture of object: ____________________

Handles: ____________________ NONE

Center of gravity: Centered vs Off-center /// Stable vs Shifting

Hand-Object Coupling

Right Hand	Left Hand	Hand-Object Coupling Used
		Hook (fingers only)
		Hook (palm only)
		Gross grasp
		Palm touch stabilization (flat palm and fingers)
		Lumbrical (MPs at 90°, IPs extended)
		Lateral pinch
		Palmar pinch
		Forearm stabilization
		Other:

Characteristics of Lifting Task

Height at which lift starts: _____ in

Height at which lift ends: _____ in

Duration of lift: _____ sec / min

Repetition of lift: Repeats _____ × per min / hr / shift

Number of hands: 1 hand vs 2 hands

Does lift involve more than 1 person? Yes No #: _____

Biomechanics of Lifting Task

Equal weight distribution between right and left hands? Yes No

Skin pressure generated on what body part? __________

Does the lift require wrist deviation? Yes No

At what height? _____ in

What is maximum wrist deviation angle? _____°

continues

Exhibit 10–3 (continued)

Strength and ROM

	Muscle strength	Extreme ROM
Neck		
Shoulder	Right / Left	Right / Left
Elbow	Right / Left	Right / Left
Forearm	Right / Left	Right / Left
Wrist	Right / Left	Right / Left
Fingers	Right / Left	Right / Left
Thumb	Right / Left	Right / Left

Source: Copyright © Karen Schultz-Johnson, 1993.

load? What is the duration of the lift? What percent of maximum voluntary contraction of a given muscle group does the lifter use during the lift? Which joints require flexibility? Stability? How do these designations change as the lift progresses and the load moves from one height to another?

When a patient has difficulty performing a lifting task, the therapist must identify the reason. Often this consists of recognizing the "weak link" in the system, which could be any one of a combination of the following:

- inadequate muscle contraction
- inadequate tendon excursion
- lack of joint flexibility
- lack of joint stability
- decreased skin tolerance to pressure, texture, temperature, or shear
- inadequate physiological status to support effort (i.e., circulatory or respiratory insufficiency)
- impaired sensation
- impaired proprioception
- joint or soft tissue problems outside of the upper extremity
- pain experience
- poor motivation
- cognitive problems

The form in Exhibit 10–4 provides a format for documentation of client performance during the lifting evaluation. This form combines components of task analysis and many of the potential limiting factors listed above. It also notes maximum weight lifted to various heights on a one-time or repetitive basis. In addition to the quantity of weight lifted and number of repetitions, the therapist should note the quality of the task performance. The speed, control, and smoothness of the lift can reveal a great deal about the evaluee. A subject's ability to problem solve and spontaneously adapt significantly influences accomplishment of physically challenging tasks.

EMPLOYABILITY-ENHANCING PROGRAMS

When the FCE results indicate that a worker does not have the ability to return to work, the evaluator may suggest employability-enhancing programs such as work hardening, dominance retraining, prosthesis training, or modification of the worker, tool, or job. These programs require special expertise on the part of the therapist implementing them. When candidacy is carefully screened, these programs often result in returning the worker to the workplace.

Dominance Retraining

Following significant injury to a dominant hand or arm, a person must learn to use the previously nondominant arm at an improved level of function. The ability to effectively use the nondominant hand is often the crucial factor in returning a person to the workforce. In a concentrated program of work-related activities graded from gross to fine, and graded by developmental sequence of prehension patterns, a person can significantly improve nondominant hand skills. Focusing first upon skill and then upon speed, this program lasts approximately 8 to 12 weeks,[60] and clients usually participate five days per week. Tool modification frequently plays a role in this program. Dominance retraining programs often teach writing with a nondominant hand and can also offer one-handed typing. Importantly, the program addresses activities of daily living and incorporates them into the program. This "total immersion" technique has proven successful for the well-screened candidate.

Prosthesis Evaluation and Functional Training

A person who has undergone amputation of any part of the arm or hand may benefit from an evaluation to determine if a cosmetic or functional prosthesis will

Exhibit 10–4 Format for Documentation of Client Performance during the Lifting Evaluation

TWO-HANDED LIFTING EVALUATION WORKSHEET

Pre-test Screening

_____ History of back problems?

_____ History of knee problems?

_____ Able to demonstrate safe body mechanics before instruction?

_____ Able to demonstrate safe body mechanics after instruction?

Characteristics of Object Lifted:

Objected lifted: ______________________________

Dimensions of object: _____ × _____ × _____

Weight of object: _____ lbs

Texture of object: ______________________________

Handles: _____ NONE

Center of gravity: Centered vs Off-center /// Stable vs Shifting

Hand-Object Coupling

Right Hand	Left Hand	Hand-Object Coupling Used
		Hook (fingers only)
		Hook (palm only)
		Gross grasp
		Palm touch stabilization (flat palm and fingers)
		Lumbrical (MPs at 90°, IPs extended)
		Lateral pinch
		Palmar pinch
		Forearm stabilization
		Other:

Factors Affecting Lifting Ability

_____ Joint stability:

_____ Wrist deviation:

At what height? _____ in

At what wrist angle? _____°

_____ Sensation:

_____ Proprioception:

continues

Exhibit 10–4 (continued)

_____ Skin tolerance:

Location: ____

_____ Endurance:

After _____ lifts at varied / same weight

_____ Physiological status:

_____ Pain:

Location: ____

_____ Motivation:

_____ Non–upper extremity factors: ______________________________________

Factors Affecting Lifting Ability

Strength and ROM

	Muscle strength	Extreme ROM
Neck		
Shoulder	Right / Left	Right / Left
Elbow	Right / Left	Right / Left
Forearm	Right / Left	Right / Left
Wrist	Right / Left	Right / Left
Fingers	Right / Left	Right / Left
Thumb	Right / Left	Right / Left

Observations

_____ Unequal Left and Right weight distribution?

Majority of load with: Right vs Left

_____ Maintain control of load?

Lost control at _____ lbs and _____ in

Lost control at _____ lbs and _____ in

Lost control at _____ lbs and _____ in

Lost control at _____ lbs and _____ in

_____ Able to lift from floor level?

_____ Performance consistent with medical history?

_____ Non-functional co-contraction / Dystonia?

continues

Exhibit 10–4 (continued)

Arc of Lift

Indicate level at which lift initiated and maximum weight lifted to each level:

_____ to reach: _____ lbs

_____ to shoulder: _____ lbs

_____ to waist: _____ lbs

_____ to knuckle: _____ lbs

Arc of Lower

Indicate level at which lift ended and maximum weight lowered from each level:

Reach to _____: _____ lbs
Shoulder to _____: _____ lbs
Waist to _____: _____ lbs
Knuckle to _____: _____ lbs

Source: Copyright © Karen Schultz-Johnson, 1993.

enhance his or her employability. The therapist attempts to match the type of prosthetic appliance and the worker's needs and job goals. An FCE may reveal that the person who already has a prosthesis does not have adequate skills with the device to perform ADL or work tasks independently. Prosthesis training offers the opportunity to practice work-related skills and to explore tools and job modification.

Modifying the Worker, Tool, and Job

An FCE and job site visit may reveal that modification of the worker, tool, workstation, or job technique will facilitate return to work. Initially, the therapist decides what needs to be modified:

- client
- tool(s)
- workstation/workplace
- technique of performing task

Often multiple components integrate to form the problem and the modification must also be multifaceted—for example, replacing a standard tool with a ratchet tool (tool modification) to alter inappropriate body mechanics (change in technique).

Worker modification involves altering the worker's approach to the job. The therapist trains the worker in task-efficient approaches and energy conservation techniques. This includes teaching appropriate body mechanics and pacing as well as joint, nerve, and tendon protection techniques.

The therapist may identify the need for external supports or protectors. These items may

- protect the worker with external stimulation: gloves, tip protectors
- stabilize a joint: flexible or rigid splints
- position and support a joint to minimize static loading around a joint: slings, chair- or table-mounted arm rest, wrist rest

Implementing tool modification demands creativity, practicality, and a knowledge of tool design. The therapist may identify the need to change tool handle characteristics such as size, shape, texture, and density. A change in tool handle configuration may affect the hand and wrist position as it alters prehension patterns. To solve a return-to-work barrier, the therapist may recommend that the worker change from an energy-intensive tool to one less demanding—for example, changing from a manual to a power tool or from one with less leverage to one with greater leverage. Ratchet tools also conserve energy since they avoid the need to repetitively change grip. The therapist must carefully screen tools imparting vibration to be sure that this potentially harmful stimulus is minimized.

Changes to the workplace and the work task may have positive ramifications for the worker. A simple change in placement of objects at the workstation (e.g., the height and distance from the worker) may enable a worker to perform the job. Job rotation can relieve mechanical stresses. Altering the approach to one that is ergonomically and biomechanically sound may also provide the opportunity a worker needs to remain on the job.

Use of adaptive devices, identified in the ADA as "auxiliary aids," should not be overlooked. Straightforward, age-old solutions to problems may work very well, such as simple to complex writing aids. The therapist can explore the use of a universal cuff or adapted cutting board as well as the use of one-handed techniques with or without adaptive equipment.

Barriers to return to work can often be overcome with careful analysis and skillful intervention. For years, federally mandated workers' compensation has required employers to provide appropriate medical intervention after injury. The industrial community must bear in mind that the ADA now compels employers to provide reasonable accommodation to allow a worker to return to work. In this arena the necessary outcome is clear: return the worker to the workforce. With this outcome firmly in mind, the hand specialist participates as a powerful and competent ally in all phases of rehabilitation of the hand- or arm-injured worker.

NOTES

1. S. Walker and R. Rehm, *Work Capacity Evaluation for Patients with Soft Tissue Inflammatory Disorders* (audio tape) (Torrance, Calif.: SLMOT Discovery House, 1983).
2. J.L. Bly et al., Impact of Work Site Health Promotion on Health Care Costs and Utilization, *Journal of the American Medical Association* 256, no. 23 (1989).
3. National Interdisciplinary Committee on Health Ergonomics, recommendations (Paper in progress, Alexandria, Va., Ergonomics Rehabilitation Research Society, 1993).
4. B. Thompson, personal communication, October 1993; S. Isernhagen, personal communication; November 1993.
5. K.S. Schultz-Johnson, Splinting: A Problem Solving Approach, in *Current Concepts in Hand Rehabilitation,* ed. B. Stanley and S. Tribuzi (Philadelphia: F.A. Davis, 1992), 238–271.
6. U.S. Dept. of Labor, *Dictionary of Occupational Titles* (Washington, D.C.: U.S. Government Printing Office, 1992).
7. S. Isernhagen, Functional Job Descriptions, *Seminars in Occupational Medicine* 2, no. 1 (1987). 51–55.
8. S. Rodgers, Recovery Time Needs for Repetitive Work, *Seminars in Occupational Medicine* 2, no. 1 (1987):19–24.
9. K.S. Schultz-Johnson, *Upper Extremity Functional Capacity Evaluation,* seminar syllabus, 1992.
10. J.S. Casanova and B.K. Grunert, Adult Prehension: Patterns and Nomenclature for Pinches, *Journal of Hand Therapy* 2, no. 4 (1989):231–244.
11. K.H. Kroemer, Coupling the Hand with the Handle: An Improved Notation of Touch, Grip, and Grasp, *Human Factors* 3 (1986):337–339.
12. T. Armstrong, Ergonomics and Cumulative Trauma Disorders, in *Occupational Injuries,* ed. M.L. Kasdan (Philadelphia: W.B. Saunders, 1986), 553–565.
13. T. Hertzberg, Some Contributions of Applied Physical Anthropometry to Human Engineering, *Annals of the New York Academy of Sciences* 63, no. 4 (1955):616–629.
14. M.L. Rowe, The Diagnosis of Tendon and Tendon Sheath Injuries, *Seminars in Occupational Medicine* 2, no. 1 (1987):1–6.
15. W. Rohmert, Problems in Determining Rest Allowances, Part I: Use of Modern Methods to Evaluate Stress and Strain in Static Muscular Work, *Applied Ergonomics* 4 (1973): 91–95; A.R. Lind and G.W. McNichol, Circulatory Responses to Sustained Grip Contractions Performed during Exercise, Both Rhythmic and Static, *Journal of Physiology* 192 (1967):595–607.
16. Rowe, Diagnosis of Tendon and Tendon Sheath Injuries.
17. Rodgers, Recovery Time; Rohmert, Problems.
18. Hertzberg, Some Contributions.
19. Rodgers, Recovery Time.
20. B.D. Chaffin, Vocalized Muscle Fatigue, a Definition in Measurement, *Journal of Occupational Medicine* 15 (1973):346–354.
21. R. Woody and V. Mathiowetz, Effect of Forearm Position on Pinch Strength Measurements, *Journal of Hand Therapy* 1 (1988):124–126.
22. S.M. Hallbeck and G.A. Sheeley, Wrist Fatigue in Pronation and Supination for Dynamic Flexion and Extension: A Pilot Study, in *Advances in Industrial Ergonomics and Safety IV,* ed. S. Kumar (Philadelphia: Taylor & Francis, 1992).

23. Rodgers, Recovery Time; J.E. Fernandez and J.B. Dahalan, Effect of Deviated Wrist Posture on Pinch Strength for Females; in *Advances in Industrial Ergonomics and Safety IV,* ed. S. Kumar (Philadelphia: Taylor & Francis, 1992); E.R. Tischauer, *The Biomechanical Basis of Ergonomics* (New York: John Wiley & Sons, 1978), 41–43, 69–70; A.H. Kamal and E.J. Moore, Effects of Lateral Position/Glove Type on Peak Lateral Pinch Force, in *Advances in Industrial Ergonomics and Safety IV,* ed. S. Kumar (Philadelphia: Taylor & Francis, 1992).
24. Tischauer, *Biomechanical Basis.*
25. S. Konz, *Work Design: Industrial Ergonomics* (Ohio: Grid Publishing, Inc., 1983).
26. L. Greenberg and D. Chaffin, *Workers and Their Tools* (Midland, Mich.: Pendell Press, 1976).
27. Konz, *Work Design.*
28. Ibid.
29. T.M. Fraser, *Ergonomic Principles in the Design of Hand Tools,* Occupational Safety and Health Series, no. 50 (Geneva: International Labour Office, 1980).
30. Greenberg and Chaffin, *Workers.*
31. M. Ayoub and P. Lopresti, Determination of Optimum Size Cylindrical Handle by Use of Electromyography, *Ergonomics* 14, no. 4 (1971):509–518.
32. Fraser, Ergonomic Principles.
33. Human Factors Section, Health, Safety and Human Factors Laboratory, Eastman Kodak Company, *Ergonomic Design for People at Work I* (New York: Van Nostrand Reinhold, 1983).
34. S. Meagher, Design of Tools for Control of Cumulative Trauma Disorder in the Workplace (Paper presented at the Cumulative Trauma Symposium, Stanford, 1986); S. Meagher, Hand Tools: Cumulative Trauma Disorders Caused by Improper Use of Design Elements, in *Trends in Ergonomics/Human Factors III,* ed. W. Karwowski (New York: North Holland, 1986), 581–587.
35. S. Comaish and E. Bottoms, The Skin and Friction: Deviations from Amonton's Laws and Effects of Hydration and Lubrication, *British Journal of Dermatology* 84 (1971):37–43; P. Naylor, The Skin Surface and Reaction, *British Journal of Dermatology* 67 (1955):239–248.
36. M. Ayoub, *Ergonomics* (Los Angeles: University of Southern California, 1984).
37. S. Johnson, Ergonomic Tool Design (Paper presented at the American Society of Hand Therapists Annual Meeting, San Antonio, Tex., September 1988); A.J. Brammer, *Vibration Effects on the Hand and Arm in Industry* (New York: John Wiley & Sons, 1981).
38. E. Grandjean, *Fitting the Task to the Man,* 4th ed. (Philadelphia: Taylor & Francis, 1988).
39. Ibid.
40. K.S. Schultz-Johnson, Assessment of Upper Extremity Injured Persons' Return to Work Potential, *Journal of Hand Surgery* 2, no. 5, part 2 (1987):950–957.
41. Valpar test is obtainable from Valpar International, 2450 West Ruthrauff, Tucson, AZ 85705; JEVS is obtainable from JEVS, 1528 Walnut Street, Suite 1502, Philadelphia, PA 19102; Bennet Hand Tool Dexterity Test, Stromberg Dexterity Test, and Crawford Small Parts Dexterity Test are obtainable from Psychological Corporation, 7500 Old Oak Blvd., Cleveland, OH 44130; Purdue Pegboard is obtainable from JA Preston Corporation, 60 Page Road, Clifton, NJ 07012; Minnesota Rate of Manipulation Test is obtainable from American Guideline Service, Inc., Publisher's Building, Circle Pines, MN 55014; Rosenbusch Test of Finger Dexterity is obtainable from Upper Extremity Technology, 2001 Blake Avenue, Suite 2A, Glenwood Springs, CO 81601.
42. J.M. Farrell, Predetermined Motion-Time Standards in Rehabilitation: A Review, *Work* 3, no. 2 (1993):56–72.
43. Available from Lifecorp, P.O. Box 13853, Tucson, AZ 85732.

44. E.E. Fess, Documentation: Essential Elements of an Upper Extremity Assessment Battery, in *Rehabilitation of the Hand,* ed. J.M. Hunter et al. (St. Louis: C.V. Mosby, 1978), 49.

45. L.N. Matheson, Symptom Magnification Syndrome, in *Work Injury: Management and Prevention,* ed. S.J. Isernhagen (Gaithersburg, Md.: Aspen Publishers, Inc., 1988), 257–282.

46. S.N. Chengalur et al., Assessing Sincerity of Effort in Maximal Grip Strength Tests: Part II, *American Journal of Physical Medicine and Rehabilitation* 69 (1990):148–153; J.C. Gilbert and R.G.L. Knowlton, Simple Method to Determine Sincerity of Effort during a Maximal Isometric Test of Grip Strength, *American Journal of Physical Medicine* 62 (1983):135–144; D.H. Hildreth et al., Detection of Submaximal Effort by the Use of Rapid Exchange Grip, *Journal of Hand Surgery* 14A (1989):742–745; H.M. Stokes, The Seriously Uninjured Hand—Weakness of Grip, *Journal of Occupational Medicine* 25, no. 9 (1983): 683–684.

47. Stokes, Seriously Uninjured Hand.

48. Hildreth et al., Detection.

49. R.G. Hazard et al., Isokinetic Trunk and Lifting Strength Measurements: Variability as an Indicator of Effort, *Spine* 13 (1988):54–57.

50. B.P. Simmons and R.G. Vasile, The Clenched Fist Syndrome, *Journal of Hand Surgery* 5 (1980):420–424.

51. T. Jankovic and K.S. Schwartz, Use of Botulinum Toxin in the Treatment of Hand Dystonia, *Journal of Hand Surgery* 18A (1993):883–887.

52. R.F. Wallace and C.S. Fitzmorris, The S-H-A-F-T Syndrome in the Upper Extremity, *Journal of Hand Surgery* 3, no. 5 (1978):492.

53. Chengaler et al., Assessing Sincerity; G.A. Smith et al., Assessing Sincerity of Effort in Maximal Grip Strength Tests, *American Journal of Physical Medicine and Rehabilitation* 68 (1989):73–80.

54. K.H.E. Kroemer and W.S. Marras, Toward an Objective Assessment of the Maximum Voluntary Contraction Component in Routine Muscle Strength Measurements, *European Journal of Applied Physiology* 45 (1980):1–9.

55. L.H. Matheson, How Do You Know that He Tried His Best? The Reliability Crisis in Industrial Rehabilitation, *Industrial Rehabilitation Quarterly* 1, no. 1 (1988):1, 11–12.

56. J.W. King and B.H. Berryhill, Assessing Maximum Effort in Upper Extremity Functional Testing, *Work* 1, no. 3 (1991):65–76.

57. K.S. Schultz-Johnson, Upper Extremity Factors in the Evaluation of Lifting, *Journal of Hand Therapy* 3, no. 2 (1990):72–85.

58. D.B. Chaffin and G.B.J. Anderson, *Occupational Biomechanics* (New York: John Wiley & Sons, 1984).

59. S.H. Snook and C.H. Irvin, Maximum Acceptable Weight of Lift, *American Hygiene Association Journal* 28, no. 4 (1967):322–329; S.H. Snook, The Design of Manual Handling Tasks, *Ergonomics* 21, no. 12 (1978):963–985; Chaffin and Anderson, *Occupational Biomechanics.*

60. M. Kasch, personal communication, 1991.

Chapter 11

Carpal Tunnel Syndrome Measurement and Surveillance Management

Laurence N. Benz

The focus of this chapter is to detail methodology utilized in measuring carpal tunnel syndrome (CTS). Specifically, it addresses the use of nerve conduction studies in determining the extent of damage to the median nerve and the use of this information in treatment and prognosis of those patients with clinical symptoms of CTS. The current trends and techniques of CTS surveillance and management in industry will be evaluated with a conceptual framework presented along with a case study.

THE "GOLD STANDARD" IN CTS DIAGNOSIS

Nerve conduction studies have long been considered the "gold standard" for determining the diagnosis of CTS. Although there has been much recent scrutiny regarding the sensitivity, cost, and practicality of testing in industry, there has been little supportive evidence denying its utility and the value of its role. The first and most practical question should be whether electrodiagnosis (EMG/NCS) in CTS is actually reliable. Researchers and clinicians have become interested in EMG/NCS since noting several quandaries in CTS diagnosis. Fully realizing that EMG/NCS does not in and of itself "diagnose," what has been particularly perplexing is the high regard by orthopedic physicians for the value of information ascertained by the EMG/NCS but their low confidence in its sensitivity. Many traditionally cite that it is only 60 to 70 percent reliable and would often operate when patients had normal studies, even in circumstances where the economic incentive of operating did not exist (e.g., in the military). A detailed exploration of the literature revealed few if any studies to support such a lack of sensitivity. The resolution of this issue only occurred after various studies that utilized more than the conventional techniques documented discrepancies in sensitivity among various techniques used by those performing the EMG/NCS. As a result, a review of techniques will assist in fully understanding CTS measurement. This review is intended to point out those features and factors specifically related to nerve conduction testing that are relevant to carpal tunnel testing only. A more extensive review of electrophysiologic testing is available in many texts.

At some point everybody remembers studying the basic action potential (Figure 11–1, A) that is elicited with stimulation of a peripheral nerve. In nerve conduction studies, a nerve is stimulated artificially (through an electric stimulator) and the nerve's response is measured (Figure 11–1, A and B). Both motor and sensory branches of a nerve may be recorded. The evaluator is primarily trying to examine the two main components of a peripheral nerve, the myelin and axon (Figure 11–1, C). A peripheral nerve contains literally thousands of these nerve axons. The myelin is responsible for the speed of conduction of a nerve, and the axon is the nerve body on which a potential is transmitted in the conduction of an impulse. The important parameters measured in these studies are as follows:

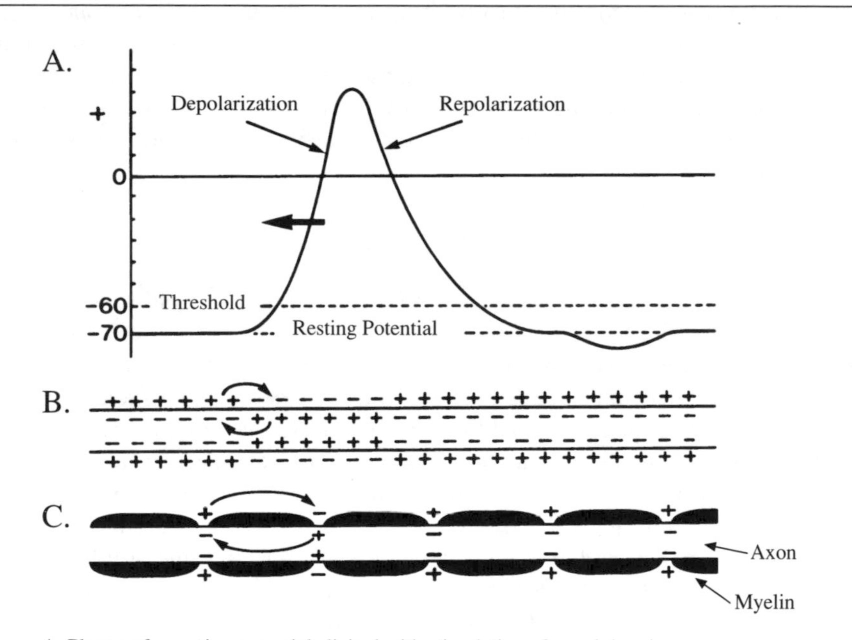

A. Phases of an action potential elicited with stimulation of a peripheral nerve.

B. Local circuit conduction. In a nerve conduction study this is measured after the nerve is stimulated artificially.

C. Mechanism of conduction in a myelinated fiber.

Figure 11–1 Action Potential. *Source:* Adapted from *Clinical Electromyography: Nerve Conduction Studies* by S.J. Oh, p. 7, with permission of Williams and Wilkins, © 1984.

1. *Latency* (see Figure 11–2): This is the time that an impulse takes to travel from point of electric stimulation to point of the action potential. In a motor nerve it includes the time necessary for the impulse to travel across an endplate to the muscle fibers themselves. There is no endplate for sensory nerves. The more myelin that a nerve axon contains, the more rapidly it will conduct. In a case of nerve compression like CTS, it is the myelin that is typically involved first. Classically, the latency has been examined most commonly in the determination of CTS. A prolonged latency of the median nerve at the wrist has been the hallmark of findings of CTS.
2. *Amplitude* (see Figure 11–2): The size of the evoked response refers to the number of available axons that conduct and are recorded with stimulation of the nerve. The amplitude is often times compared at various sites of stimulation along a nerve. This helps in determining if there is a compression and its extent.
3. *Conduction velocity* (see Figure 11–3): By stimulating a nerve at two sites, a determination of the velocity can be assessed. The latency of the distal site is subtracted from the proximal, and this difference is divided into the distance measured between the two sites. In CTS, the forearm conduction velocity is often calculated to rule out a compression in that segment.

MEDIAN NERVE COMPRESSION

Along with a limited review of the common parameters measured in a CTS evaluation, an understanding of the physiology of median nerve compression at

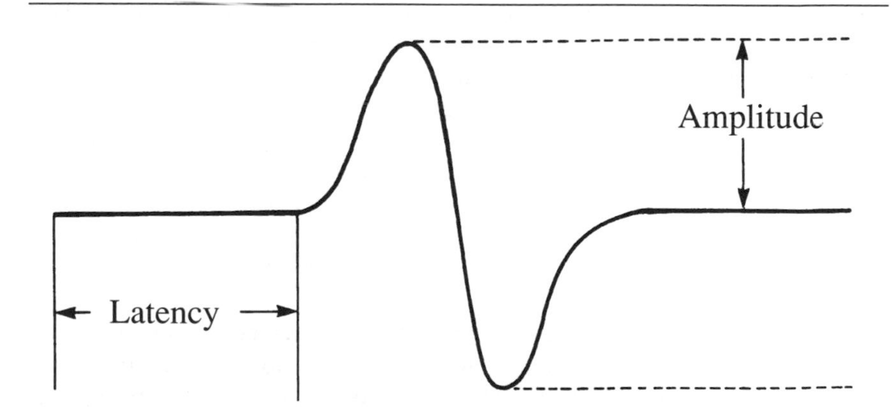

Figure 11–2 Parameters Measured in Electrodiagnosis of CTS. *Source:* Adapted from *Clinical Electromyography: Nerve Conduction Studies* by S.J. Oh, p. 51, with permission of Williams and Wilkins, © 1984.

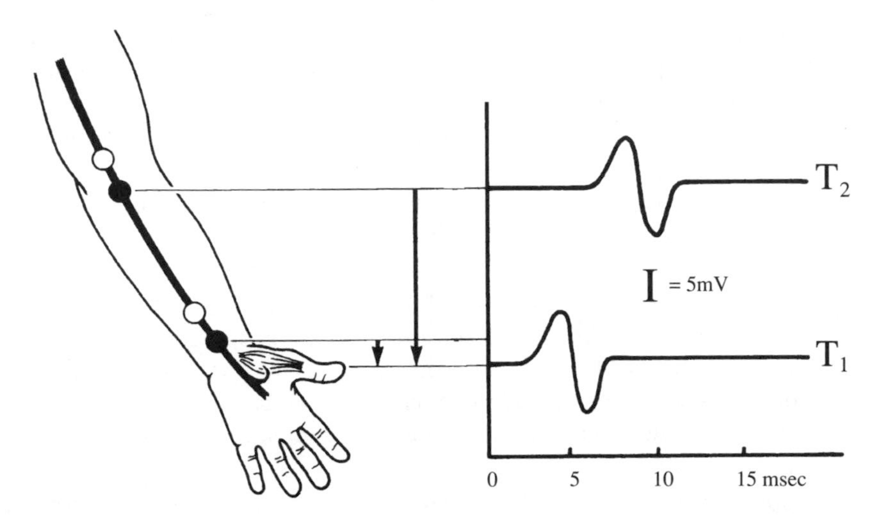

Figure 11–3 Conduction Velocity. By stimulating a nerve at two sites, a conduction velocity between the two sites can be calculated. *Source:* Adapted from *Clinical Electromyography: Nerve Conduction Studies* by S.J. Oh, p. 18, with permission of Williams and Wilkins, © 1984.

the wrist is in order. CTS begins as a neuropraxia. With compression on the nerve due to swelling or physical positioning of the transverse carpal ligament, the myelin of the median nerve begins to unravel. Because the sensory portion of the median nerve is located more superficially and in the peripheral portion of the nerve bundle, it is more vulnerable to compression. This results in slower conduction of these sensory branches, typically before the motor ones. Clinically, the patient experiences numbness and tingling. It often occurs first at night due to the wrist assuming a more flexed posture and pooling of swelling and inflammation at the wrist compartment. The prognosis can be excellent if the compressive source is taken away. This is the basis for success of treatment of CTS when we can recognize, identify, and treat the problem early in its course. The compressive source most often is continuous, repetitive activity of the wrist in a flexed posture in conjunction with force. At this stage of the problem, the myelin is the only involved segment of the nerve.

With continued progression and compression of the nerve, axontmesis occurs. This happens when actual nerve axons are damaged or destroyed. The prognosis is poor to unknown depending on the extent of the axonal damage. The axons must regenerate to heal, and the amount of time is dependent on the number of axons involved as well as whether the compression source has been removed or continues. Mild axon damage can heal in a timely manner if the compression source is taken away. Conversely, serious axon damage is almost always an indication for surgery. Only a properly performed nerve conduction study can ascertain this extent of axon involvement.

Often in communication with patients, companies, and occasionally physicians about these "stages" of CTS the author uses an illustrative example.

> Picture one-way four-lane highways on each side of a bridge under certain traffic conditions. The highway lanes are synonymous with the median nerve's axons, and the bridge is the carpal tunnel. A neuropraxia is a terrible traffic jam. The roads going on all sides of the bridge are normal, but the motor vehicles cannot get to the other side because the traffic is backed up. Some of the traffic lanes are moving a little faster than others, and eventually the vehicles do get to the other side. If we can get to the source of the traffic jam and alleviate its cause, the normal traffic flow can be restored.
>
> If the traffic was backed up because the four lanes have been reduced down to two due to actual road damage, this would be analogous to an axontmesis. Eventually the road damage would extend to both sides of the bridge. Only after these roads had been rebuilt (a lengthy process compared to a traffic jam!) could the normal traffic flow be continued.
>
> In CTS, the nerve's axons can eventually deteriorate along its entire course (before and after it enters the carpal tunnel). The role of nerve conduction studies is to differentiate this traffic jam from road damage. If they are performed properly, the assessment and prognosis can be accurately made. Conservative treatment versus indications for surgical intervention can in large part be based on the results of the nerve conduction study in conjunction with the clinical presentation.

NEW TECHNIQUES EVALUATED

Since the advent of clinical nerve conduction studies in the late 1940s, a number of typical and conventional studies have emerged that represent the basic techniques used in determining peripheral compression neuropathies such as CTS. A fundamental understanding of these techniques is imperative. Because of the refinement in the instrumentation used in conduction studies, many nonconventional techniques are currently used that have improved the sensitivity of these studies. However, the techniques vary among electromyographers such that extreme variability in sensitivity exists. Therefore a description of the conventional techniques is in order.

1. Study of the median motor branch (Figure 11–4). This is referred to as the distal median motor latency. This is the recording of the median motor

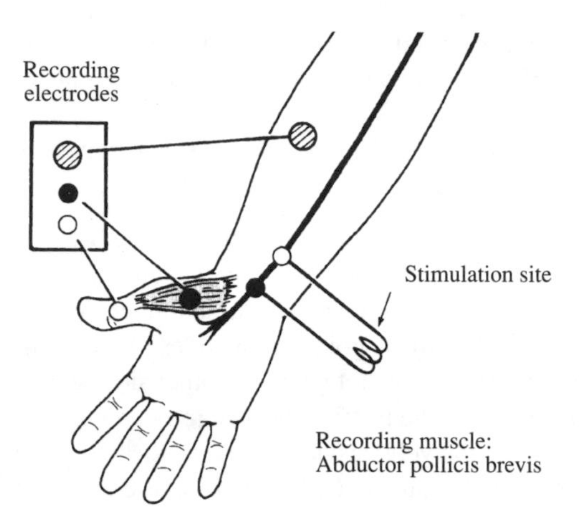

Figure 11–4 Median Motor Branch. The nerve is stimulated at a fixed distance (usually 8 cm) at the wrist above the carpal tunnel. *Source:* Adapted from *Clinical Electromyography: Nerve Conduction Studies* by S.J. Oh, p. 68, with permission of Williams and Wilkins, © 1984.

---

branch with surface electrodes attached to the abductor pollicis brevis of the mass of the thumb. The nerve is stimulated a fixed or standard distance (typically 8 cm) at the wrist above the carpal tunnel. This conduction time (latency) is less than 4.5 milliseconds in normals.

2. Study of the median sensory branch. The sensory branch can be measured in two ways, orthodromically (Figure 11–5) or antidromically (Figure 11–6). The orthodromic ("with nature") method involves stimulating the median sensory area (typically the second or third digit) while recording a standard distance (usually 14 cm) along the course of the median nerve at the wrist above the carpal ligament. The second method is antidromic ("against nature"). Recording occurs via electrodes wrapped around the median nerve digits (usually II or III), with the median nerve stimulated a fixed distance at the wrist above the carpal ligament. The usual conduction time is less than 3.7 milliseconds for both techniques. There are some variations of these techniques relative to distances and recording areas.

The conventional techniques that have traditionally been utilized have been evaluated by countless studies to adequately determine their relative sensitivity in determining carpal tunnel syndrome. Perhaps the most exhaustive evaluation was the Rochester study, which examined 1,016 hands of patients with clinical symptoms of CTS.[1] The median motor distal latency was seen as 37.5 percent sensitive,

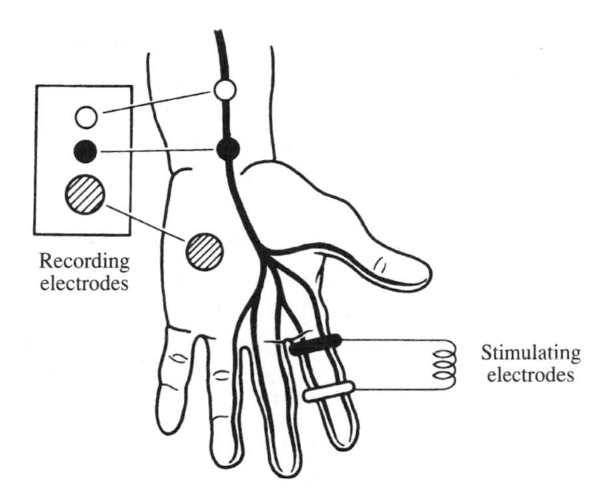

Figure 11–5 Orthodromic Median Sensory Stimulation. The sensory branch is stimulated while recording at the wrist proximal to the carpal tunnel (usually 14 cm). *Source:* Reprinted from *Clinical Electromyography: Nerve Conduction Studies* by S.J. Oh, p. 69, with permission of Williams and Wilkins, © 1984.

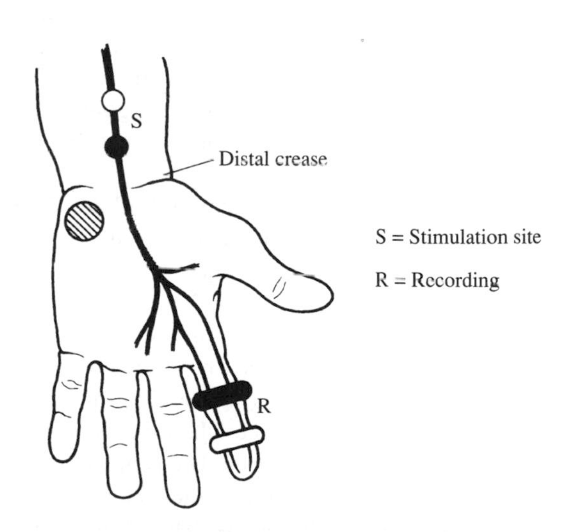

Figure 11–6 Antidromic Median Sensory Stimulation. The nerve is stimulated at the wrist above the carpal tunnel with recording electrodes around the median supplied digit (II or III) (usually 14 cm). *Source:* Adapted from *Clinical Electromyography: Nerve Conduction Studies* by S.J. Oh, p. 189, with permission of Williams and Wilkins, © 1984.

a percentage that could be increased to 50 percent if the conduction time was compared to the ulnar motor distal latency on the same hand. With comparison to the median nerve on the opposite hand, the sensitivity was increased to 51 percent—obviously, a poor rate with significant risks for false negatives if that technique is used alone.

The sensory studies were evaluated by both orthodromic and antidromic means. As expected, their sensitivity was better but not tremendously so, with only a 64 percent sensitivity rate. Greater sensitivity was detected when examining the amplitudes of these sensory responses. Antidromic amplitudes were fairly sensitive, but these values are seldom closely evaluated without their latency counterparts, so that when the latency values are normal the amplitudes are not typically considered.

The last test considered was the needle electromyographic examination (EMG). This test involves inserting an electrically sensitive needle into one of the thumb's muscles and monitoring the electrical potentials of that muscle at rest and with activation of that muscle. For the purposes of determining CTS, this test is particularly ineffective, with only 18 percent displaying positive results. This is mainly due to the injury process of CTS and should not be interpreted as meaning that EMG is not indicated or useful as an evaluative test. Many patients have concomitant pathologies that are better evaluated with EMG than with nerve conduction studies. Because electromyography and nerve conduction studies are extensions of the clinician's hand, the determination of whether to perform either or both tests is totally based on the patient's clinical findings. Most often both tests are performed to various extents. A reasonable argument can be made not to include EMG when isolated CTS without other clinical findings or suspicion is shown by nerve conduction studies.

With this general lack of sensitivity of traditional nerve conduction studies (at best 64 percent), a more innovative and certainly a more accurate technique was indicated. As electromyographic equipment became more refined in terms of noise filtering and amplification, a palmar technique developed. In relatively recent years (the 1980s), many electromyographers began recording from shorter segments of the median nerve than the conventional distances of 8 cm for motor and 14 cm for sensory. This has substantially increased the diagnostic utility of conduction studies, as evidenced by the Rochester study, which showed sensitivity of 91 percent. This sensitivity has been shown in many subsequent studies and is consistent for both orthodromic (Figure 11–7) and antidromic techniques (Figure 11–8) where the difference between the stimulation point and recording area is decreased to 7 or 8 cm.

There remains one consistent difference between orthodromic and antidromic techniques. That is, only the antidromic technique can stimulate the nerve before and after it departs the carpal tunnel. Although the sensitivity of the nerve conduc-

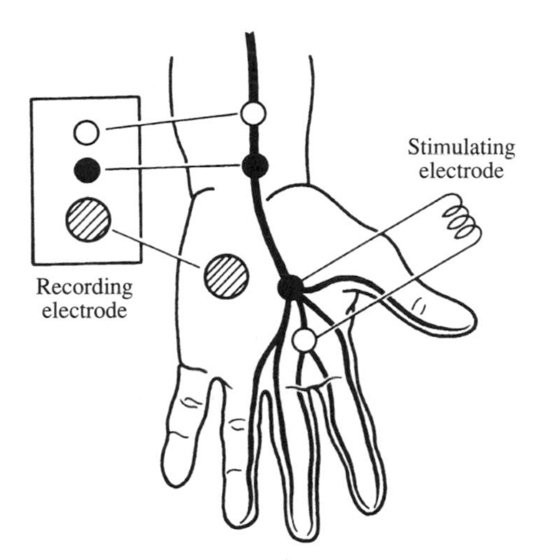

Figure 11–7 Palmar Orthodromic Median Sensory Stimulation. The nerve is stimulated at mid-palm with recording at the wrist (6–8 cm). *Source:* Reprinted from *Clinical Electromyography: Nerve Conduction Studies* by S.J. Oh, p. 69, with permission of Williams and Wilkins, © 1984.

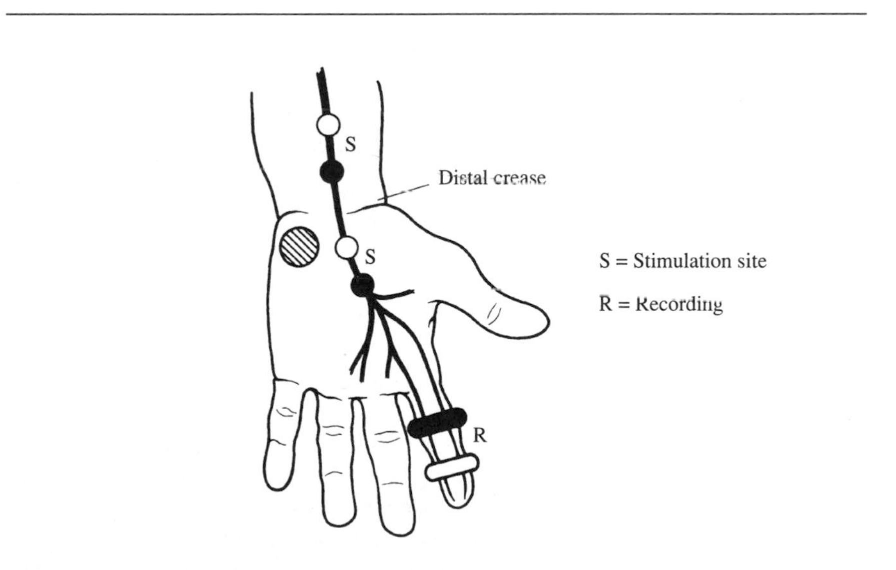

Figure 11–8 Antidromic Median Sensory Stimulation. The median sensory branch is stimulated and recorded before and after it leaves the carpal tunnel. *Source:* Adapted from *Clinical Electromyography: Nerve Conduction Studies* by S.J. Oh, p. 189, with permission of Williams and Wilkins, © 1984.

tion study increases with a short-segment orthodromic technique, both stimulation points go through the carpal tunnel and are recorded at the wrist, not allowing for measuring differences in conduction due to changes occurring at the carpal tunnel. The antidromic technique is the only method that allows clear visualization of the nerve and its status due to compression. Ulnar entrapments at the elbow as well as many other compression neuropathies are routinely identified by this type of technique. By identification of the difference between a traffic jam and whether the roads of the highway are intact, we can project the success of conservative treatment versus a true indication for surgery. Unfortunately, many practicing electromyographers do not incorporate any palmar technique and thus continue to have a high false-negative rate. Only through awareness that various techniques exist can one determine whether these more sensitive measurements were made. Furthermore, from a prognostic standpoint, only if the antidromic palmar technique was used can conclusions regarding the status of the nerve be made.

CASE EXAMPLES

The following are case examples of conduction studies performed on industrial clients that show how the antidromic palmar study helped determine the patient's prognosis and treatment.

Case 1: This patient is a 46-year-old sewing worker who has recently switched to a sideseam sew that requires continual repetitive motion of her wrist in conjunction with constant pinching of digits II and III with digit I. After six weeks of performing she started developing pain, numbness, and tingling primarily at night, and daytime swelling near her wrist. She has been referred for a nerve conduction study. When the conventional antidromic technique was used (stimulating at the wrist and recording at a 14-cm distance from digit III), the values in Figure 11–9, A, were recorded. A check against normal values for the latency and amplitude would show that this response was essentially normal. But a 7-cm antidromic technique was additionally performed (stimulating above the carpal tunnel and recording from digit III). The response is shown in Figure 11–9, B. Obviously, there is a conduction abnormality between the two points that accounts for the greater-than-50-percent increase in amplitude of the nerve after it departs the carpal tunnel (the roads on the other side of the bridge remain intact!). Unfortunately, many of these patients are either misdiagnosed or mistreated because only the conventional techniques were used in their nerve conduction study. This particular patient was treated very conservatively with mild job modification, night splinting, stretching exercises, pause breaks at work, and mandatory rest periods. Her symptoms resolved and she continues her present job tasks. A large number of these false-negative patients are treated with suspicion and omission until they

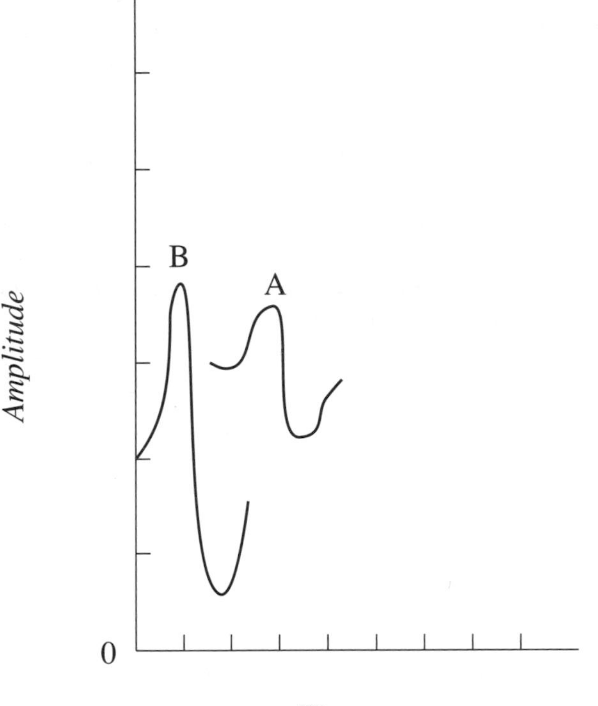

Figure 11–9 Nerve Conduction Results for Case Study 1. A: Conventional antidromic technique. B: Short-segment (7-cm) antidromic technique.

---

develop substantial problems later, at which point the nerve's axons are damaged and the patient undergoes surgical decompression. This case study demonstrates:

- If only conventional techniques are used, CTS misdiagnosis can occur.
- If erroneous evaluation was done, this symptomatic worker may be ignored and left without proper treatment, so that further damage to the nerve results and she becomes a surgical candidate.

Case 2. This patient is a 55-year-old keyboard operator who has developed symptoms consistent with CTS over a five-month period. She has been treated with splints, anti-inflammatory medications, and work restrictions. Her symptoms have failed to subside. She has been referred for a nerve conduction study that demonstrated the following printout of her antidromic sensory response from wrist to digit III (Figure 11–10, A). Her results show prolonged response with a decreased amplitude. The inclination is to suggest severe CTS. However, a palmar study was performed distal to the wrist (after the nerve had been through the carpal tunnel) (Figure 11–10, B) that showed a substantial increase in amplitude. This demonstrates evidence of a neuropraxia (bad traffic jam). Because the nerve's axons remain undamaged, conservative treatment is often effective. This patient received a series of two steroid injections into her carpal tunnel, some job modification with a resting splint, instruction in pause breaks, and mandatory rest at regular work intervals. Unfortunately, the vast numbers of these types of patients undergo surgery. Only the palmar study was able to differentiate the extent of injury and suggest a reasonably good prognosis.

Case 3. This patient is a 55-year-old tobacco farmer who develops parasthesias on a recurrent basis every fall when he begins "stripping" tobacco. On physical

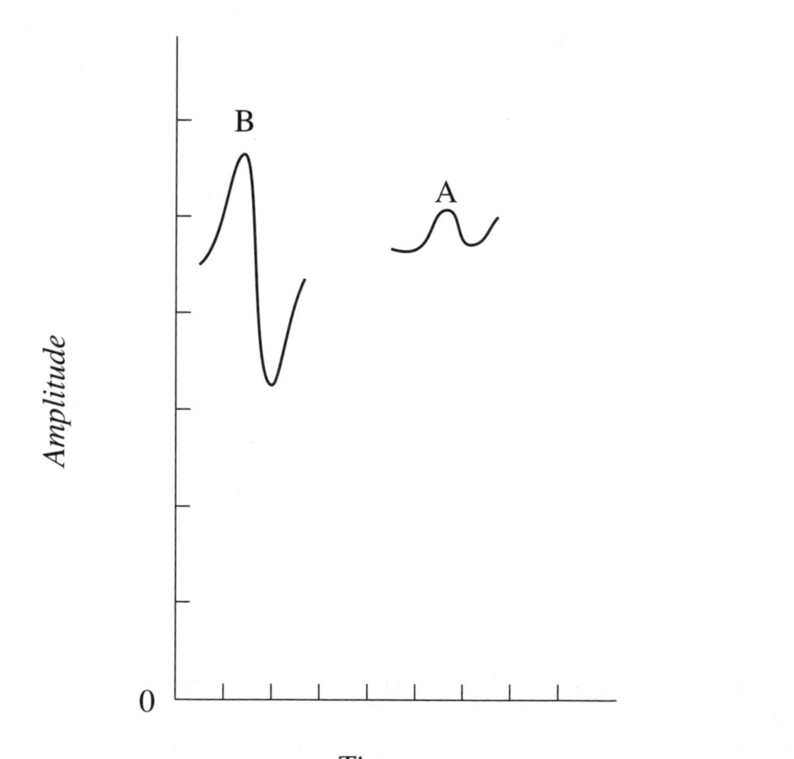

Figure 11–10 Nerve Conduction Results for Case Study 2. A: Antidromic technique, response from wrist to digit III. B: Palmar study.

exam he has obvious thenar atrophy. He is unresponsive to treatment and has been referred for EMG/NCS studies. The printout of the median sensory at the wrist (Figure 11–11, A) shows delayed conduction and loss of amplitude. In addition, his palmar study (Figure 11–11, B) demonstrates normal conduction but no significant change in amplitude between stimulation points. This patient has obvious axon damage and was a surgical candidate. A successful decompression has allowed him to continue his vocation to this day.

Case 4. This case involved a 48-year-old female small assembly worker. She is a known adult onset diabetic with several other medical complications. She complained of parasthesias to both hands in the median nerve distribution. The role of the EMG/NCS in her case was to differentiate between a diabetic neuropathy and carpal tunnel syndrome. Her distal median motor and sensory responses showed conduction slowing along the wrist and forearm. In addition, she showed a gener-

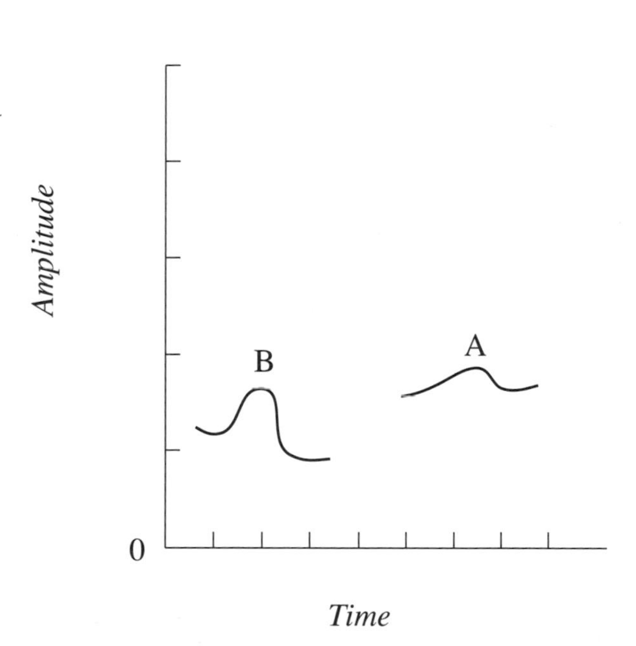

Figure 11–11 Nerve Conduction Results for Case Study 3. A: Antidromic technique, response from wrist to digit III. B: Palmar study.

alized slowing in all of her sensory nerves (upper and lower extremities). The slowing of her median nerves was of a much greater magnitude than that of her ulnar nerves. In addition, the palmar study showed slowing and amplitude changes consistent with a superimposed process (in Figure 11–12, A and B show the printout of the wrist and palm median sensory responses respectively). Many of these types of patients are excused as diabetic polyneuropathies without identification of a concomitant CTS. Many others are labeled as CTS without identification of the generalized diabetic polyneuropathy. Patients with diabetic neuropathies are more predisposed to compression neuropathies like CTS, and only a properly performed nerve conduction study in these cases will clearly differentiate. This particular patient had a surgical decompression with relief and was assigned to a different job in the factory that was nonrepetitive in nature.

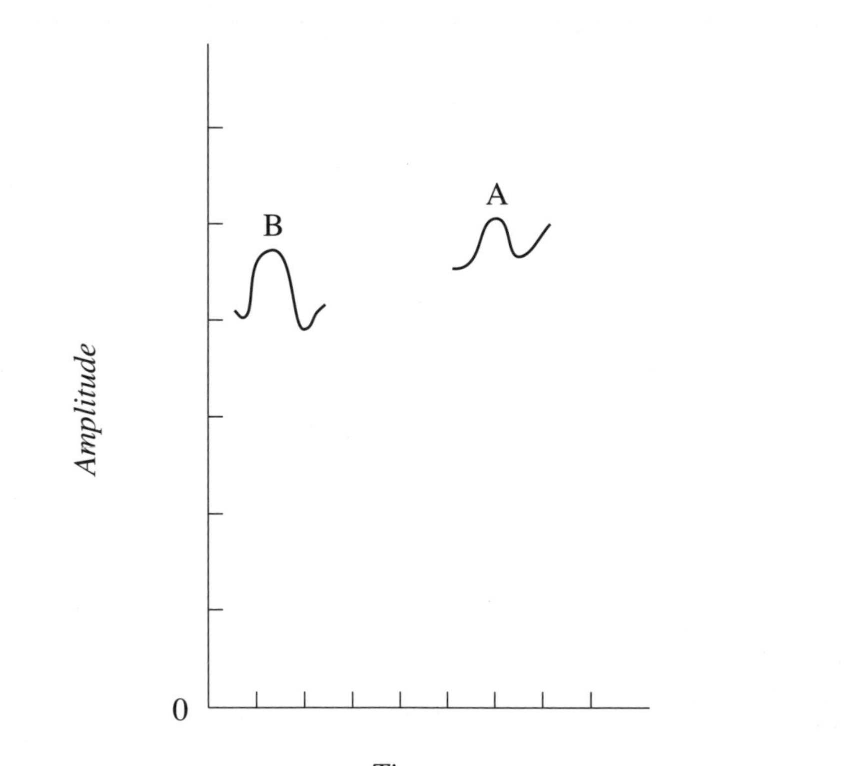

Figure 11–12 Nerve Conduction Results for Case Study 4. A: Antidromic technique, response from wrist to digit III. B: Palmar study.

CARPAL TUNNEL SCREENING AND SURVEILLANCE

Over the past few years, there have been more and more attempts to quantify an employee's or prospective employee's susceptibility to the risk of cumulative trauma, or more specifically to carpal tunnel syndrome in industry. Such screenings or tests supposedly would enable a company or supervisor to preclude hiring of a prospective employee, determine if a prospective employee had subclinical involvement at time of hire, correctly place an employee in a job classification that would match his or her susceptibility, provide baseline data if an employee developed symptoms associated with carpal tunnel, or provide the basis for determining the extent of monitoring necessary for the specific employee in order to prevent the occurrence of carpal tunnel syndrome. Although there are many instruments currently available on the marketplace today that purport to serve these purposes, most do not meet the basic comprehensive approach to work injury management, namely an analysis of the work, worker, and worksite.

Before discussion of the various means of screening and monitoring, one must examine the value of such screenings. If one particular device could give a specific value to determine an employee's susceptibility, would it be helpful? With the implementation of the Americans with Disabilities Act, this susceptibility could not certainly preclude hiring a prospective employee. Even if such a test demonstrated subclinical involvement, an employee who could still perform the essential job tasks with or without accommodation would still have to be hired, thus rendering the outcome of the screening test irrelevant. For the use of preplacement, such a test result would only be worthwhile if it could be shown that all other employees within a certain job classification made their preemployment or current test score worse by performing their job tasks over a period of time.

Although it would seem logical that such a numerical value taken prior to and after development of symptoms would serve as an appropriate baseline, no case law currently in existence demonstrates how this baseline would determine a work- or non-work-related cause of carpal tunnel or whether this baseline would have any true worth. Therefore the value of such a baseline figure is negligible, particularly in light of the potential errors from calibration and instrumentation.

The last and really the only significant worth of this screening device, then, would be to determine the extent to which an employee should be monitored for the development of carpal tunnel syndrome. Such a test value should be used only as a small portion of a comprehensive prevention program, and only if there is an accurate testing device available.

SCREENING PROCESS

Many of the current methods used in screening for carpal tunnel syndrome include the use of instruments that measure the median nerve velocity. These instruments attempt to be an extension of a nerve conduction study, which is part of a comprehensive clinical electrophysiologic examination. Unfortunately, due to various instrumentation and engineering pitfalls, these devices do not offer the reliability, sensitivity, or specificity necessary to make a true evaluation of the conductibility of the median nerve. The instrumentation errors include the lack of a differential preamplifier and the necessary filtering of low- and high-frequency signals. Furthermore, the lack of adjustment necessary to correct for temperature variation (the most vulnerable variable in measuring a nerve's conduction) and the measuring of only the motor portion of the nerve (the last branch involved in compression neuropathies) expand the inherent error of "nerve monitoring" through the use of these simple hand-held devices. Additionally, these instruments do not calculate the nerve conduction proximal and distal to its course through the carpal tunnel, thus giving no actual information on conductive changes occurring at each point. Lastly, the devices do not show a picture of the nerve's response.

Vibrometers are the second type of measuring device currently being marketed for screening of carpal tunnel syndrome. These instruments purportedly detect the perception threshold of the digits of the hand to vibration. The concept being sold is that the ability to perceive vibration precedes damage to the median nerve itself. For example, if you could detect these changes in vibration, you could, in theory, diagnose and prevent the problem before serious damage occurred. (One manufacturer claims that their instrument detects damage before it actually occurs!) This is an obvious impossibility when one considers that vibration is mediated along a particular nerve pathway that normally operates uninterrupted unless "real" damage has occurred. Although there are not as many potential instrumental errors as with nerve monitors, there are many more dependent variables. The patient subjectively responds to the perceived vibration level, creating many potential recording errors. Other limiting factors include the amount of callus formation on a patient's fingers and the always-present risks of reliability, repeatability, and room temperature. It is quite possible that these tests can be properly performed by a very trained and experienced medical technician in a laboratory in much the same way that licensed audiologists are trained in hearing tests, but not by paraprofessionals or nonmedical personnel in an industrial setting.

There is no known causal relationship between a person's current ability to perceive vibration and what will affect this ability in the future. There is also no evidence that certain perceived threshold ranges are more susceptible than others to the development of CTS. Doesn't it make more sense to educate workers on the initial symptoms and warning signs rather than to calculate the specific frequency of the high and low vibration levels of their second and third digits?

Although numerous studies have documented decreased sensitivity to vibration in patients diagnosed with CTS, there have been few independent studies showing the value of this testing in an industrial population. A study on Texas manufacturing plant workers that utilized a control group, a group receiving nerve monitoring, and a group receiving vibrometry testing failed to show any difference between the working group with symptoms and the control group with regard to 120-hz vibration threshold tests. Furthermore, the researchers concluded that high false-negative rates associated with nerve monitoring and vibration perception limit the usefulness of these methods in screening for CTS.[2]

Given these current testing limitations, the actual diagnosis of CTS through a comprehensive clinical evaluation to include the use of electroneuromyography is much less of a difficulty than trying to determine an employee's "risk" of developing CTS. For nonmedical professionals to ascertain a level of risk of an employee seems to be out of any scope of practice. Simplistic, inaccurate screening methods and equipment are incongruous with what our current medical knowledge dictates, particularly when only one or a few factors (motor nerve measurement or vibrometry threshold) are measured. Therefore a screen in the strictest sense cannot be comprehensive enough to deal with the complex "syndrome" aspect of carpal tunnel injury.

SURVEILLANCE AND SCREENING

An obvious and clear differentiation between a screen and diagnosis exists. Screens never diagnose even though manufacturers often claim they do. It is ludicrous to state that a hand-held "nerve monitor" or "vibrometer detector" aids in diagnosis when it is traditionally accepted that even a properly performed electromyogram/nerve conduction study (EMG/NCS) by a qualified medical practitioner cannot "diagnose." If an electromyogram is truly only an extension of the clinician's hands, how can these other instruments be utilized? It is questionable when manufacturers claim as a selling point that these instruments can be used by nonmedical personnel!

Even if a "perfect" machine or test were developed that identified 100 percent of those employees who would develop CTS, the information would be useless unless a comprehensive preventive approach were undertaken. The concept behind a screen should be to efficiently identify patients or employees through the use of a few tests or examinations that can accurately ascertain their current status or level of risk in the development of CTS or cumulative trauma.

With the understanding, then, that the worth of any screening examination is to determine the extent to which an employee should be monitored for the evolution of carpal tunnel, a comprehensive approach should be developed that incorporates the work, worker, and worksite. A proper program should provide an analysis of the known risk factors that pertain to these three variables. Exhibit 11–1 lists some of these essential elements.[3]

Exhibit 11–1 Risk Factors

Work risk factors
Repetition rate
Force of weight of workload
Posture
Body mechanics
Duration of movement
Coordination needed to perform tasks
Vibration or impact

Worker
Strength
Safe motion
Stability
Flexibility
Endurance

Worksite
Workstation dimensions
Sitting and standing equipment
Tools and product
Design
Light
Temperature

A comprehensive approach encompassing these factors should contain essential components of an effective overall management program for carpal tunnel and cumulative trauma. These components should be screening and risk assessment, prevention, intervention, and monitoring. Over the past several years a program has been developed containing all of these concepts. This program has been tested in industry and continues to be refined. Exhibit 11–2 shows the framework of these necessary components in a total surveillance program.

Screening and Risk Assessment

Although each component of this program is integral, none is more crucial than the classification of employees into risk categories. The results of the screening and its tests determine this risk level. This examination usually takes 30 to 45 minutes depending on the number of examiners (physical and occupational therapists) and how many stations are set up to facilitate a more efficient process.

Prior to the Americans with Disabilities Act, a part of this process was finding out the employee's past occupational and injury history. Because those with previous cumulative trauma in repetitive occupations are at a greater risk for recurrent or new injuries, this information was valuable in that risk determination equation. In the instances where a company has new employees who are integrated into this program, this portion of the screening is not included. However, the overwhelming majority of employees that have been tested in one program work at companies that, because of serious problems with carpal tunnel and upper extremity cumulative trauma, have just begun this program. An interview format is utilized for this portion.

A comprehensive but efficient evaluation is then administered. Provocative testing consisting of Phalen's test, Tinel's test, and conventional nerve conduction

Exhibit 11–2 Management Program Components

- I. Screening and risk assessment
  - A. Previous job and injury history
  - B. Evaluation
    1. Provocative testing
    2. Job task testing
    3. Employee knowledge and awareness testing
    4. Anthropometric testing
    5. Current work demands assessment
  - C. Predictive determination
    1. Low risk
    2. Moderate risk
    3. High risk
- II. Prevention
  - A. Stretching program
  - B. Job task rotation
  - C. Early identification of symptoms
  - D. Pause breaks
- III. Intervention
  - A. Acute injury management
  - B. Treatment protocols and algorithms
  - C. Treatment compliance assurance
- IV. Monitoring
  - A. Questionnaires
  - B. Follow-up monitoring
  - C. Retrospective analysis

studies of the median and ulnar nerve is recorded using a standardized approach and measuring scheme. The nerve conduction studies are performed by appropriately licensed medical personnel who utilize the palmar technique. Other testing includes sensation with pinprick, two-point discrimination, monofilament testing, and muscle testing with a dynamometer.

These tests were chosen based on their relative reliability and repeatability. Result determination is based on the outcome of each individual test times weighted values that incorporate the test's sensitivity and specificity as shown in valid research. These weighted values have evolved as this testing has increased, allowing for retrospective analysis. The nerve conduction study continues to be the "gold standard," and its results have been invaluable when compared to those of the same patient who has developed symptoms.

Other testing includes "anthropometric measurement," in which the employee's wrist dimensions are measured, and "job task testing" (specifically hand coordination tested with a standardized exam). An assessment of employee knowledge and awareness is made based on responses to questions concerning the early warning signs of carpal tunnel and cumulative trauma, familiarity with these basic injuries, understanding of the appropriate procedures that should take place when an injury occurs (when to report to supervisor, acute management, etc.), and understanding of the role of posture, body mechanics, and simple ergonomics at home and in the workplace.

The results of this examination determine the risk level of the employee. In conjunction with a cursory job demands assessment that incorporates the work and worksite, a final determination is calculated. The employee is placed in a

"Low," "Moderate," or "High" risk category. The categories have different degrees of monitoring and intervention.

If an employee is in a high-risk category, a more formal job demands assessment is undertaken, with ergonomic changes such as proper work height, appropriate chairs, and accessory equipment (ergonomic scissors, fitted wrist rests, etc.). This job demands assessment is also used to communicate with new employees and medical personnel the specifics of the employee's job. In addition, if the high-risk employee's evaluation score is particularly high, specifically on the nerve conduction study, a referral to the company's medical doctor is made for definitive diagnosis. Several employees in this program have been treated successfully based solely on this component's ability to identify employees who already are suffering from carpal tunnel syndrome. All employees are evaluated and treated in a nonadversarial manner.

Prevention

All employees are given an initial education class that stresses identification of early warning symptoms, instruction in pause breaks (brief interruptions of prescribed stretches throughout the workday), job task rotation, and specific instructions on what to do when an injury occurs—when and where to report, how it will be treated, etc. Moderate- and high-risk employees have this educational component repeated during the year in an abbreviated manner. In addition, questionnaires administered at risk-determined intervals (monthly, quarterly, or semiannually) prompt and test the employees on these same basic concepts.

The prevention component also includes a stretching program that all employees participate in prior to starting their work shift. Some companies also allow a second stretching to be done after the lunch break. There are seven stretches that have all major muscle groups involved but place most emphasis on the upper extremities. For muscle relaxation, increased circulation, and stress reduction benefits, this program has been very well documented for injury reduction even when used in isolation from the overall surveillance management program.[4]

Intervention

The intervention portion of this program utilizes treatment algorithms specifically targeting employee complaints (numbness, pain, tenderness, or swelling). It initiates treatment with predetermined time frames for reevaluation, referral, and expected outcome. This component is individually designed for each company since it may include specific alternative work and job rotation. This integral component appropriately treats the patients as well as containing cost. The company's physician and other potential treating practitioners are oriented to these protocols and algorithms, and they have input into the specifics of the treatment and time frames.

Lastly, continuous monitoring of these components is performed. Outcome analysis is calculated for incidence, lost workdays, costs, and outcomes. Retrospective analysis has shown that there is always an increase in risk of complaints with the initiation of this program. This should be interpreted positively since the overall incidence, lost workdays, and costs have always decreased. This initial increase in complaints eventually decreases as the employees' understanding, education, and compliance with the overall program progress. In the case study that follows, there was a 25 percent increase in complaints filed after initiation of the program. This lasted for the first three months, at which time complaints decreased to their preprogram levels. For the remainder of the first year there were varied fluctuations in complaint numbers, all related to the amount of productivity the factory demanded.

Case Study

The first industry to participate in this comprehensive method for surveillance was a large sewing company. Although the company as a whole had a high incidence of carpal tunnel syndrome and upper-extremity cumulative trauma, a particular work area that was responsible for a specific seam stitch was most noteworthy. This 45-employee department averaged three CTS cases every six weeks (eight per year), with two separate cumulative trauma cases (tendonitis, cervical soft tissue pain, or other upper-extremity dysfunction) every eight weeks. A case was counted only if more than two lost workdays occurred and the worker received medical treatment outside of the factory. There had also been one low back pain claim every three months. These averages were consistent over a prior two-year period and were matched relatively by the same department at a different factory in another geographic location. In an effort to curb these costly problems, the sewing company participated in the full program for that particular department only.

After the evaluation and predictive component was summarized, a structured job site analysis with appropriate ergonomic intervention was implemented for the entire department. Adjustable chairs and demonstration of appropriate work height dominated the emphasis of this component, as employees were taught hands-on techniques. The prevention program was also initiated at that time. The stretching portion was performed twice a day (at the beginning of the work shift and after lunch), with job task rotation and pause breaks mandated. Although the job task consisted of one main type of stitch, the employees were instructed to better mix the other components of the job—namely, stacking of material, transportation of items, and counting finished pieces. Early recognition of symptoms was taught to all employees of this department, who were also instructed in the proper protocol for filing a medical complaint. Intervention consisted of acute injury management (ice, over-the-counter anti-inflammatories, and range of motion exercises) coupled with expected outcomes and disposition of the employee dictated by a treatment algorithm. If carpal tunnel was suspected, nerve conduction studies were performed and compared to the initial screening for documented conduction differences. Monitoring consisted of questionnaires to the high-risk employees (32 out of the 45) on a monthly basis (completion before issuance of a paycheck) and to the moderate- and low-risk category at quarterly or semiannual intervals respectively.

Using the same definition for a documented case, follow-up statistics revealed a significant reduction in cases of carpal tunnel syndrome and upper-extremity cumulative trauma. At the end of one year, only two cases of carpal tunnel syndrome were documented (one that underwent surgery and was subsequently back at work and another that resulted in the patient seeking different employment). There was only one case of upper-extremity cumulative trauma (cervical myofascial pain) with significant lost work time. Although there were many other instances of complaints and medical treatment of conditions, none resulted in lost work time greater than two days. Analysis of the costs demonstrated medical cost savings of over 68 percent for that particular work area when compared to the previous year. Further analysis showed no statistically significant reduction in CTS and cumulative trauma cases at the other sewing factory that had a department performing this same type of stitch. The cost of implementing this program was less than the average cost of the medical treatment of one carpal tunnel case for that previous year.

CONCLUSION

CTS measurement and surveillance management can be integrated into any occupational medicine program. The financial savings obtained through proper identification and surveillance of CTS are enormous. The most current techniques utilized by electromyographers enable not only a more accurate assessment but a prognosis of CTS as well. The blending of this prognosis assessment with surveillance management will drastically affect the type of treatment and the expected outcome. The surveillance management program must be comprehensive by classifying the risk level of the worker and emphasizing early identification of symptoms, acute intervention, prevention, and continuous monitoring.

NOTES

1. J.C. Stevens, AAEE Minimonograph #26: The Electrodiagnosis of Carpal Tunnel Syndrome, *Muscle and Nerve* 10, no. 2 (1987):99–113.

2. K.A. Grant et al., Use of Motor Conduction Testing and Vibration Sensitivity Testing as Screening Tools for Carpal Tunnel Syndrome in Industry, *Journal of Hand Surgery* 17A, no. 1 (1992):71–76.
3. S. Isernhagen, *Cumulative Trauma Prevention: An Educational Program* (Duluth, Minn.: Isernhagen Work Systems, 1989).
4. L. Hebert, Body at Work, *Occupational Health and Safety* (October 1992):48–54.

SUGGESTED READING

American Association of Electrodiagnostic Medicine. 1993. Issues and opinions: Summary statement: Practice parameter for electrodiagnostic studies in carpal tunnel syndrome. *Muscle and Nerve* 16, no. 12:1390–1391.

Johnson, E.W. 1988. Carpal tunnel syndrome. In *Practical electromyography*, 2nd ed., ed. E.W. Johnson, 187–205. Baltimore: Williams & Wilkins.

Kimura, J. 1989. *Electrodiagnosis in disease of nerve and muscle,* 2nd ed. Philadelphia: F.A. Davis.

Nelson, R.M., and D.E. Nestor. 1991. Electrophysiological evaluation: An overview. In *Clinical electrotherapy*, 2nd ed., ed. R.M. Nelson and D.P. Currier, 331–360. Norwalk: Appleton & Lange.

Chapter 12

Working with the Workers: The "Buy In" to Maintenance of Wellness

Barbara Baum

A multitude of federal, state, and occupation-specific documents have been and continue to be written in an attempt to provide guidelines and standards that, when applied, will reduce the frequency and severity of work-related injuries. The prevention interventions included in such documents appear to reflect an overall need for today's workplace to be safe, fiscally viable, and attuned to the skills and abilities of diverse individuals within the working population. The various guidelines and standards are directly or indirectly reflective of similar goals:

1. increased health, wellness, and welfare of the workforce
2. maintenance of productivity standards
3. decreased insurance premiums and workers' compensation costs for employers
4. decreased spending on disability benefits and programs by federal and state agencies

In fact, comprehensive risk-management, injury/reinjury prevention programs that are well planned and implemented should not only achieve these goals but also result in lost-time containment, liability reduction, increased productivity, decreased workforce turnover, and increased employee morale. Economically and humanistically, injury prevention benefits employers, employees, and society as a whole.

As evidenced by the number of chapters in this text addressing the topic, injury prevention involves many components. For each component to interface or fit smoothly with the others, the primary need is for employers to have

1. well-written and systematically implemented policies and procedures related to workplace accident and injury reduction (i.e., a thorough and effective safety program)

2. well-written and systematically implemented policies and procedures related to thc employment of both nondisabled and disabled individuals, including application procedures, rates of pay, promotions, benefits, termination, and injury reporting/management
3, thorough, objective analysis of the physical requirements/demands, hazards, and essential functions of each job category
4. well-developed and systematically implemented education and training programs for all levels of employees regarding workplace policies, procedures, physical requirements/demands, and hazards

With regard to program philosophy, the key elements of workplace injury/reinjury prevention should include the belief that

1. Risk identification, assessment, and control are critical components of an injury prevention program.
2. Maintenance of wellness is a critical goal for both injured and uninjured employees.
3. Education at all levels is a prerequisite component of any program designed to maintain wellness by preventing injuries and illnesses.
4. Many work-related injuries can be prevented.
5. Safety is a condition of employment, is not optional, and requires the committed involvement of all employees.
6. The focus of training should be to teach the worker how to identify and reduce risks and how to apply general principles of safe body use in both work and nonwork settings. Safety as a life attitude leads to safe habits and decreased injury rate and severity.

Conceptually, then, injury/reinjury prevention programs begin with evaluation of the worker, the work to be performed, and the worksite. Prevention and treatment interventions provide workers with the education and other tools that enable appropriate assumption of self-responsibility for behaving safely at all times. The end goal is the maintenance of wellness for all workers, injured and noninjured (see Figure 12–1).

Functionally, the key elements of successful injury/reinjury prevention programs are

1. actively demonstrated management commitment and employee involvement
2. team effort and open, timely, effective communication involving management, engineering, medical, supervisory, and line personnel
3. actively demonstrated commitment by all employees to assume self-responsibility for behaving safely, not only at work but also at home and play

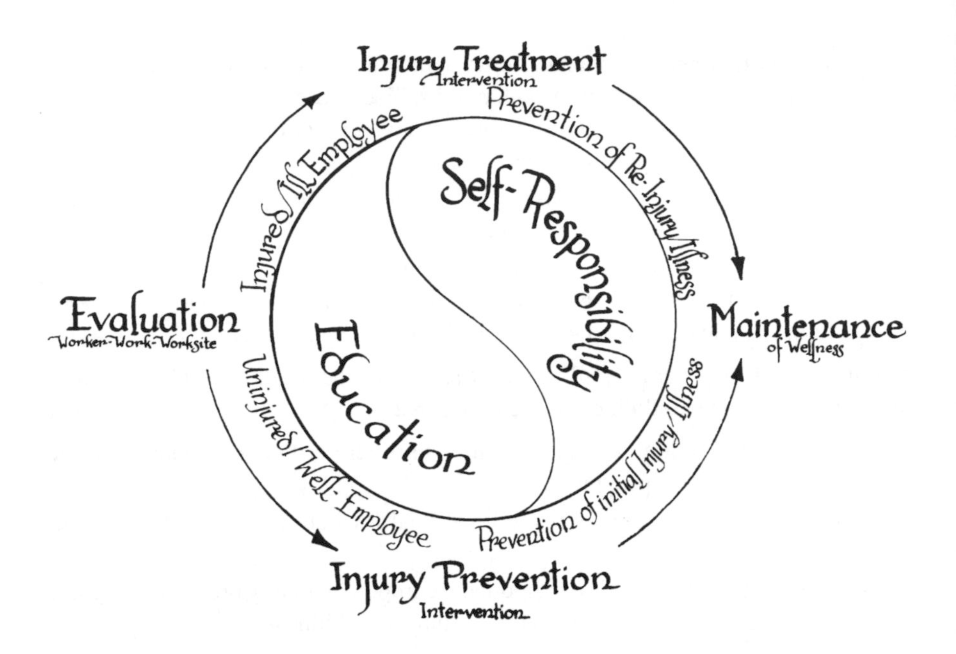

Figure 12–1 Goals of an Injury/Reinjury Prevention Program. *Source:* Developed by Barbara Baum, MS, PT, Coordinator, Occupational Rehabilitation Services, Institute for Occupational Rehabilitation, Fairview Hospital and Healthcare Services, Minneapolis, Minnesota. Not to be reproduced without permission.

The development and maintenance of a positive attitude about safety and the maintenance of wellness in the workplace are an evolutionary process. The process depends upon the "E's"—Evaluation, Ergonomics, Education, Equipment, Exercise, Emotions, Empowerment, and Enforcement. It begins with safety awareness, which is followed by program implementation and then is sustained through the maintenance of enthusiasm generated by positive behavior reinforcement and employee involvement and ownership. A prototype of this evolutionary process is the program used by a labor-management group in Minneapolis to address excessive work injuries. This group referred to their program as "PEER." It involved:

1. P = PROMOTE safety as an attitude and way of behaving.
2. E = EVALUATE the workers, work tasks, and worksite to identify work hazards, and set goals to reduce these hazards.
3. E = EDUCATE everyone at all levels as to his or her role and responsibility in worksite injury prevention, including participation in programs/processes designed to promote a safe work environment.

4. R = REVIEW workers, work tasks and worksite (including injury and accident investigation records) to determine the effectiveness of injury prevention measures and establish new goals.[1]

Through the "PEER" program, the labor-management group was able to successfully reduce their injury rate and severity significantly. The key to their success was the "buy-in" on the part of employees at all levels to the concept that injury prevention was a priority goal.

"Buy-in" to the concept of injury prevention as a priority goal must be accomplished first with those in positions of top management. Incentives for employers to develop and implement injury prevention programs include the need to address the direct and indirect economic costs of work-related injuries, employee health and wellness issues related to compliance with the Americans with Disabilities Act, methods of promoting and maintaining continuous quality improvement (CQI), and concerns related to compliance with state and OSHA safety guidelines and standards. Through an education and training process, persons in top management can be provided with information that will enable them to clearly identify the relationship between these areas of concern and the role that an effective injury prevention program can play in addressing these concerns. When provided with the appropriate information, they will recognize that injury prevention makes sound economic, legal, and humanistic sense in both the short and the long term for the employer.

Once top management "buy in" to the importance of an injury prevention program, they must take the steps to actively demonstrate their commitment to such a program. Such commitment is demonstrated by management's provision of organizational resources to address safety through

1. employee involvement in identifying potential hazards and potential solutions
2. implementation of effective solutions
3. development of a written program with specified goals
4. regular program review and evaluation with written report[2]

More specifically, management's commitment to safety can be demonstrated through provision of organizational resources to enable the implementation of a program that includes the following components:

1. worksite analysis—a thorough evaluation of workplace ergonomic hazards
2. hazard prevention and control through established procedures related to engineering controls, work practice controls, personal protective equipment, and administrative controls
3. medical management

4. training and education so that managers, supervisors, and employees come to understand workplace hazards associated with job/job tasks and the manner in which they are performed[3]

True commitment and personal concern by top management to injury prevention is evidenced when employee safety and health are at the same level of administrative importance as production, lines of authority and responsibility and resources are assigned to reflect this importance, and the written safety program is not only endorsed but advocated by those at the highest level of management. Active advocacy is demonstrated by the provision in writing, with employee involvement, of specific goals, plans, objectives, and implementation dates. It is further demonstrated by regular program review and evaluation of injury/illness trends, employee surveys, updating of goals, and written documentation of the effect of injury prevention interventions.[4]

Active demonstration by top management of their "buy-in" to the importance of injury prevention and employee involvement is the foundation for achieving "buy-in" by the workers to the importance of maintaining a state of wellness and their personal responsibility in accomplishing this goal. By taking the lead and providing the resources that will empower workers to take responsibility for their own health, top management sets the stage for the training, education, and effective teamwork that will enable safety and injury prevention to become an intrinsic component of the culture and daily life in the workplace. As safe behaviors are developed, nurtured, and reinforced in the work setting, individuals are likely to internalize these behaviors and subsequently apply them habitually whether they are at home, work, or play. Safety becomes a way of life.

Given that management commitment is the foundation for worker "buy-in" to the maintenance of wellness, then the cornerstone to this foundation is education. Indeed, just as management education precedes management commitment, so must education and training precede changes in worker behavior related to safety. Worker training and education directed toward change in safety behavior must be carefully designed if they are to be effective. In designing the training and education, consideration must be given to several factors:

1. problem identification that allows for delineation of the relationship between safety concerns and human behaviors related to these concerns
2. clarification of the factors contributing to those worker behaviors that result in unsafe acts
3. identification of those prevention interventions that will most likely influence worker behavior in such a manner as to motivate different, safer, and healthier behavior
4. postintervention evaluation to assess the degree to which worker behavior has, in fact, changed

Once these factors have been taken into consideration in the design phase, then the specifics of the education and training process can logically incorporate those principles delineated in OSHA guidelines related to this component of hazard prevention and control.

In describing education and training as one of four key components for hazard prevention and control, OSHA guidelines indicate that

1. Procedures for safe work should be developed that are understood and followed by all appropriate parties as a result of training, positive reinforcement, correction of unsafe performance, and, if necessary, enforcement through a clearly communicated disciplinary system.
2. Employees must actively participate in their own protection and such participation can be expected only if employees understand the need for personal protective equipment, know how to use it, and understand its benefits and limitations.
3. Procedures related to "preventive discipline" should be incorporated that will result in actions being taken to instill a sense of discipline before a problem arises. Such procedures/processes entail program activities that encourage employees to perform in the expected safe manner and that are evidenced in safety orientation, job safety training, recognition for proper work procedures, and safety observations made in routine walk-arounds.
4. When "preventive discipline" fails, unsafe behaviors should be corrected through reinstruction, persuasion, warnings, interviews, and, if necessary, penalties.

Actively demonstrated management commitment should result in the identification of general worksite and job-specific hazards and the establishment of safety rules and practices related to these hazards. These safety rules and practices, developed with employee involvement, then become standards that encourage employees to work safely. Employee education and training related to hazards and safety rules and practices become the vehicle for initiating workplace safety as a cultural norm and motivating workers to behave in a manner consistent with this cultural norm through an understanding and internalization of the role that the individual plays in assuming responsibility for safety in the workplace.

To be effective, education and training related to injury prevention in the workplace must encompass more than the "how to's" of performing job tasks safely. Indeed, to secure true worker "buy-in" to the importance of injury prevention not only to themselves but also to their families, their coworkers, and the company as a whole, education and training must aim to assure that each employee

1. knows how to and can do his or her job safely

2. has an understanding of the principles related to the maintenance of wellness and prevention of injury that are relevant to his or her individual work and lifestyle
3. has an understanding of, and an ability to identify, the many factors that can result in bodily injury and illness
4. assumes a role as an active participant in the prevention of injuries at work, home, and play
5. has an understanding of management's commitment and expectations of employees related to injury prevention

With these overall aims in mind, the specific objectives and content of the worker training and education can be determined based upon previously gathered data related to the identification of safety concerns, human behaviors related to these concerns, factors contributing to unsafe worker behaviors, and prevention interventions that will most likely influence an increase in safe worker behavior. Once the specific objectives and content have been determined, the next step is to design the education and training efforts in a way that will maximize worker "buy-in" to these efforts such that their attitude toward safety and their responsibility to behave safely will be enhanced. Given that workers are adults, the design of the education and training efforts must take into consideration the principles of adult learning and behavior change.

Education and training efforts are maximized when they are designed in such a way as to consider what learning is and those factors that affect adult learning. According to Webster, learning is "knowledge or skill acquired by instruction or study" or "modification of a behavioral tendency by experience (as exposure to conditioning)."[5] In other words, effective education and training efforts will enable the student worker to gain new or reinforce preexisting knowledge or skills, as well as provide the student with an experience that will facilitate the modification of preexisting behavior patterns. As adults, individuals are most likely to learn when they:

1. want to learn
2. are in a situation where they can "do" rather than just see or listen
3. are presented with an opportunity to solve realistic problems
4. can interpret new experiences on the basis of past experiences
5. are in an informal environment
6. are provided with a variety of teaching methods
7. are provided with guidance, not grades
8. are provided with feedback[6]

In the words of Confucius,

> "What I hear, I forget;
> What I see, I remember;
> What I do, I understand."

The challenge and opportunity of injury prevention education and training efforts is to design and implement learning experiences that will impart knowledge and skill in such a way that the "student workers" will be motivated and have the confidence to assume their responsibility of performing their job tasks in a safe and efficient manner. If true modification of their behavior has occurred, then the "student workers" will also have the knowledge, skill, and motivation to perform non-job-related tasks in a safe and efficient manner at home and play. In other words, the workers will have bought in to the importance of injury prevention and will have internalized the message that they have only one body, which is with them 24 hours a day and which no one except themselves can take care of, so that they must apply the ABCs × 2 principle of Always Being Careful twice by thinking of what they are going to do before actually performing a task—whether at home, work, or play (Figure 12–2).

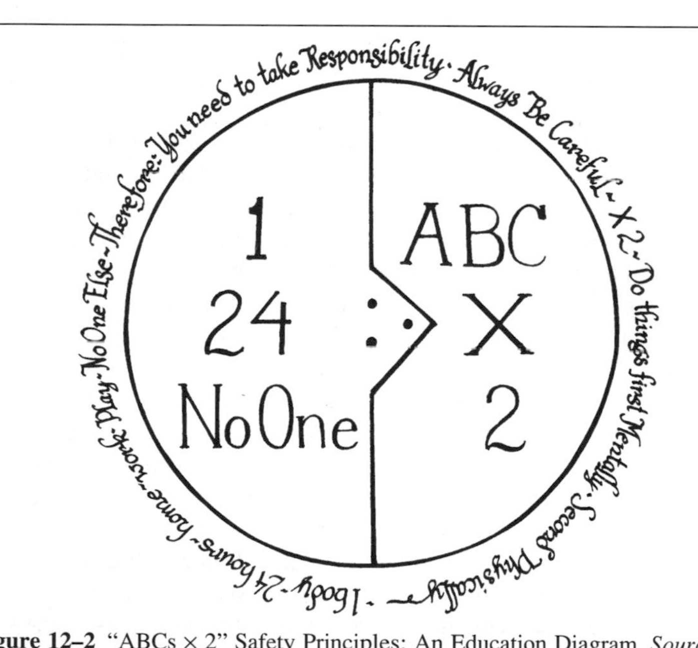

Figure 12–2 "ABCs × 2" Safety Principles: An Education Diagram. *Source:* Developed by Barbara Baum, MS, PT, Coordinator, Occupational Rehabilitation Services, Institute for Occupational Rehabilitation, Fairview Hospital and Healthcare Services, Minneapolis, Minnesota. Not to be reproduced without permission.

Implementation of the formal worker education and training program related to injury prevention is the culmination of a great deal of planning that begins with educating management and securing management "buy-in" to the importance of an injury prevention program and continues with gathering and analyzing injury and job hazard data; establishing objectives/goals for the injury prevention program in general and the education and training efforts specifically; designing the education and training content and learning experiences; selecting participants for, scheduling, and promoting the training sessions; and arranging for rooms, audiovisual aids, handouts, and so forth. Active involvement of management and other workers at all levels of authority and responsibility throughout this planning process is a key ingredient to the ultimate "buy-in" of the workers to the concept of job safety and injury prevention. Active involvement from the beginning of the process promotes the development of worker ownership in the process as well as demonstrating management's commitment to empowering the workers not only to participate in injury prevention activities but also to take seriously their role in assuming responsibility for behaving safely.

Effective implementation of the actual worker education and training sessions related to injury prevention requires thorough planning and use of presenters who have appropriate content and skill knowledge as well as participatory teaching experience and familiarity with the jobs performed by the participants. An introduction by management personnel at the beginning of each session and their participation during each session are frequently helpful in demonstrating the company's commitment to and willingness to work as a part of the "team" in developing and implementing injury prevention programs. The general content of these sessions should articulate and repeatedly emphasize that the key elements of injury prevention are team effort, management commitment, and individual self-responsibility and that all components are equally important.

The primary emphasis of the education and training sessions should be on the importance of the individual worker in preventing injuries and thus maintaining a state of well-being. To internalize the message that they must Always Be Careful, at home, work, and play, by putting their head/knowledge in gear before they put their body/skills in gear, workers must be given information regarding the anatomy and function of pertinent body parts and systems as well as techniques and tools that will enable them to plan for and perform functional activities in a manner that will minimize their risk for injury by protecting these body parts and systems. Concepts such as the "L's"—*L*ook for *L*ogic; Remember *L*evers, *L*oads and *L*imitations; Use your *L*egs for power and a firm base of support; and Keep your *L*ordosis or *L*ow back arch by tightening your abdominals and using the strut principle—should be taught, discussed, practiced, observed on video and slides, reviewed, and practiced again.

Internalization of the knowledge and skill learned in the education and training sessions will be demonstrated through their consistent application in the safe performance of work-related activities. This internalization of behaving safely should be preserved provided the worker receives positive reinforcement, such as positive written and verbal feedback from management and coworkers; requests to assist in safety training programs; an increased sense of physical well-being; incorporation by others of new ways of performing job tasks; and a personal level of awareness that safety has become an intrinsic component of one's values and thus the way one thinks and behaves twenty-four hours a day.

A successful injury prevention program requires management commitment and "buy-in" by workers at all levels of an organization. The development, implementation, and preservation of such a program do not happen overnight, but are an evolutionary process. The program requires the development of policies, procedures, and protocols that educate and empower all individuals to be involved and assure ongoing evaluation and reevaluation for program components. Ultimately, such a program should support an internal, ongoing process for assuring that safety and injury prevention become and remain integral components of the cultures and values of all workers not only at their workplace but also at home and when at play.

NOTES

1. J. Lennes, Commissioner of the Minnesota Department of Labor and Industry's Luncheon Speech (Presented at the Minnesota Safety Council Annual Meeting, 1992).
2. U.S. Department of Labor, Occupational Safety and Health Administration Safety and Health Program Management Guidelines: Issuance of Voluntary Guidelines; Notice, *Federal Register* 54, no. 16 (1989):3904–3916.
3. Ibid.
4. Ibid.
5. G. & C. Merriam Company, *Webster's New Collegiate Dictionary* (Springfield, Mass.: 1980), 649.
6. Burlington Northern Railroad, Safety Skills Workshop Training Manual (developed and printed for internal use by Burlington Northern Railroad, 1993).

Part II

Work Injury Management

The cost of workers' compensation claims continues to rise despite efforts to manage care. This is due to several factors. The first factor is lack of universal attention to effective methods. This is related to the second issue, which is the dearth of information regarding effective outcomes of medical management. The third factor is the slow realization of the medical community to provide cost-effective, outcome-effective care as the first priority rather than utilizing fee-for-service care to keep medical facilities open and profitable.

In the 1990s workers' compensation medical providers began to feel pressures caused by the crisis intervention practiced by insurance companies and workers' compensation systems. This intervention came in several forms of managed care, such as fee cutting, treatment limitations, institution of gatekeepers to the workers' compensation system, and in some cases denial of workers' compensation care. However, these front-end medical cost-cutting measures were confounded by escalating costs for disability payments, medical-legal cases, and overuse of expensive medical testing, especially in chronic cases.

What has happened as a result is a slow evolution of medical providers becoming more cognizant of the need to provide early and effective service, and to return people to work at a modified stage in order to prevent chronic disability. Employers have also made a management change in that most are willing to bring people back to work on modified or full duty. In addition, disability management has taken a turn away from the former adversarial dehumanizing process to more of a positive process, recognizing that for those with chronic disabilities, work is still a viable option and that for those for whom it cannot work, humane case resolution is another viable option. Therefore there have been many changes in the 1990s that have led to a better understanding of medical work injury management.

Steve Vance, a provider of on-site therapy, and Allan Brown begin the section by discussing an innovative early intervention model that combines early intervention with prevention of reinjury and reduction in chronic disability. Carl DeRosa and Jim Porterfield further explore aspects of one of the more expensive work problems, back injury. Their lucid extrapolation of important spinal treatment parameters helps us to understand the back injury dilemma. Robin Saunders discusses clinical early intervention that constitutes an enlightened approach to managing the injured worker and measuring effectiveness in early care.

Roles of different professions and different scientific protocols are then examined. Joseph Sweere, DC, gives a chiropractic point of view to the "team approach," a refreshing statement of concepts that dissolves some of the lines between traditional medical care and chiropractic.

Highly utilized programs such as functional capacity evaluation and work rehabilitation have rapidly progressed since their discovery in the early 1980s. Dennis Hart discusses tests and measurements regarding functional specialities. This chapter can be a cornerstone for further outcome research. The cardiopulmonary system, often overlooked in traditional medical work injury management, is clarified by Steve Anderson, who provides a piece of the puzzle formerly missing. Laurie Johnson and Susan Isernhagen discuss the kinesiophysical approach and new concepts important in functional capacity evaluation. Ann Walker provides case studies that demonstrate how good functional evaluation can minimize cost even in difficult cases.

Work rehabilitation is defined by four authors, experts in work conditioning/ work hardening. Linda Darphin defines the differences in the programs. Libby Green explores one of the most powerful parameters of work rehabilitation, the opportunity for work behavior to be modified. Richard Smith discusses the small clinic approach, where much work rehabilitation takes place today and where more will take place in the future. Barbara Larson completes this section with outcome studies to explore the satisfactory resolution of work rehabilitation cases. The concepts are then put into practice together by Sara Chmielewski and Dan Vaught, who examine the process of return to work after work rehabilitation.

The "technologizing"of science has had a tendency to lead practitioners away from the emphasis on the worker and toward focus on equipment. Dennis Isernhagen presents ideas and concepts based on current research that refocuses the emphasis on function. Leonard Matheson shows why the functional approach will continue to be the most important in worker self-actualization. Matt Monsein and Robert Clift take the psychological management of the worker one step further with their discussion of returning the chronic pain patient to productive work.

Since work rehabilitation and functional evaluation are often brought into the medical-legal system, the legal applicability of these principles is discussed. Bill Sommerness presents an important view regarding how clinicians, employers, and insurance companies must view the objectivity and admissibility of functional evaluations. Our understanding of work injury management is heightened through

Tom Mayer et al.'s chapter on defining work rehabilitation into logical sequences: the primary, secondary, and tertiary levels of intervention.

Adding important points of view are international authors who have diverse workers' compensation/health care systems. Michael Oliveri and Malou Hallmark provide an excellent overview of how work rehabilitation has been integrated into Switzerland, a country with a more "socialized medicine" context. Sharyn McGuire broadens the discussion in the use of functional restoration and evaluation in Australia.

The authors have made a concerted effort to provide us not only with the newest of materials but also with those principles that will stand the test of time. Accuracy, outcome, legality, and objectivity will always be hallmarks of the medical work injury rehabilitation process.

Chapter 13

On-Site Medical Care and Physical Therapy Impact

Stephen R. Vance and Allan M. Brown

The roles of physical therapists are expanding within the scope of health care. One arena that appears to have the elements of a strong foundation for successful physical therapy intervention is industrial medicine. Escalating health care costs related to workers' compensation and the inherent medical and indemnity benefits continue to pose major problems for this country's industries. As a result many companies, especially self-insured industries, are opting to gain more and better control over their claims and medical management by looking at alternative health care delivery systems. Preferred provider organizations (PPOs), health maintenance organizations (HMOs), and other managed care programs are coming to the forefront as viable options for the many floundering state-supported systems.

Physical therapists with strong and thorough evaluation skills and an understanding of pathophysiology, human kinematics, and function are ideally suited to treat the injured worker. Consequently therapists are becoming major players in the industrial arena. Clinical skills coupled with functional mind-sets tend to make therapists an extremely cost-effective and viable alternative to existing systems.

While therapeutic intervention with work-related injuries has been occurring for decades, a relatively new aspect has been the recent migration of therapists out of the hospitals and private clinics into on-site clinics housed within the industry itself: into the trenches, as it were. Along with this migration, therapists are bringing emphasis on new aspects of treatment that are active, functionally based, and related to the physical demands of the job. Possibly the most dramatic and effective role that on-site therapists fill is that involving early intervention in acute musculoskeletal injuries.

Work by Waddell, Fordyce, Matheson, and others indicates that often a person's perception of his or her disability does more to determine his or her disability than any treatment or objective testing. With that in mind, it becomes paramount to utilize an early intervention model that will have the best chance to prevent these illness perceptions from ever occurring. Early intervention promotes a sense of priority that tells injured workers that their injury is of concern to the employer. Not only does this process enhance credibility for the employer, it also creates a foundation for early recovery. It becomes readily apparent to injured workers that their employer shares a common concern for their welfare and is willing not only to accept responsibility for the injury but to provide first-rate care for the worker. The message being sent to the injured worker is one of cooperation and assistance and not an adversarial stance. The injured worker is now more likely to "buy into" the program and focus efforts on rehabilitation goals and return to work.

CONTROL

Early intervention in musculoskeletal injuries in the workplace affords employers control over what is happening to their employees. For example, in the state of Maine there is a very liberal workers' compensation system. Until recent changes in the workers' compensation laws, employees could obtain treatment from whomever they chose; and if they did not like what a particular health care provider was telling them or how their case was being managed, they could switch health care providers as often as they pleased. There was little that employers could do to control the type of treatment or service or the level of quality of service their employees were getting. By changing to on-site treatment, however, employers are allowed to gain significant control over the medical services being afforded employees as well as the quality of these services. This allows the employers to obtain services from health care providers who have a true interest and expertise in industrial injury management. Increased medical costs and disability are prevented by dealing with practitioners who understand the work environment and who work toward the maximum functional gains and safest return-to-work potential for both employee and employer.

The employer often has the opportunity to profile different services and providers in attempting to establish a working relationship. The employee usually does not have the luxury of interviewing health care providers to ensure that they are knowledgeable, credible, and competent. In some instances employees may feel that their "right to choose" has been stripped away by the employer. However, quality programs soon put to rest employees' fears of being controlled merely for control's sake by gaining the respect and trust of employees and convincing them that the main concern is their well-being.

In 1985, Bath Iron Works, a shipbuilding industry and the largest private employer in the state of Maine, was faced with the problem of passive treatment approaches for employee back injuries. The standard procedure for low back pain utilized by most practitioners at the time was to place the injured worker on two to three weeks of bed rest and to follow up with recheck appointments if the injured worker was not better. Standard X-rays and anti-inflammatories were often prescribed. During the three weeks out of work, the injured worker often would discuss his or her ailment with friends and family, and possibly seek out treatment from numerous physicians, chiropractors, or therapists. This often put the injured worker in a situation where he or she was receiving many different and conflicting messages concerning the nature of his or her ailment and the most appropriate way to handle his or her problem. These mixed approaches were often not in concert with each other. The differing expectations of the employer, medical department, family physician, and injured worker often established rather than eliminated obstacles in the way of a successful return to work.

By placing a physical therapist on site, Bath Iron Works Medical Department enhanced their system by gaining earlier and better control of the treatment approach of injured workers. In this model, the physical therapist receives patients from physician/nurse referral or vocational rehabilitation services. The therapist's role is to evaluate and initiate treatment if indicated. During the first intervention with injured employees, the therapist must demonstrate to patients that they are involved in a very credible, high-quality program that places their injury and recovery at a priority. This credible introduction establishes an invaluable rapport. If the therapist does his or her job well, patients will feel cared for and feel that their health is of concern to the employer. Satisfaction with care given early and compassionately often prevents doctor shopping and outside medical care.

Often in industrial rehabilitation, practitioners are faced with injured workers who have developed hostilities toward their employer. The psychological and emotional anxieties that accompany an injury are enhanced by delayed care. This often promotes a feeling of helplessness and hopelessness within the injured employee. The injured employee's perception of "victimization" occurs in as little as two to three weeks after the injury. In contrast, the patient's frustrations are minimized by early intervention. Less time is now spent defusing an adversarial situation. The main energies of both therapist and patient can now be directed toward the objective of restoring function to the patient as fully as possible, and facilitating return to work without compromising worker health and safety.

DIRECTION

The second contribution of on-site physical therapy to industry is direction. Direction can take many forms. The multitude of different treatment techniques by different disciplines makes it virtually impossible for injured workers to select the appropriate treatment for their particular ailment. Occupational injuries require treatment that is active in nature and that will lead toward independence from rather than dependence on the medical practitioners. For example, physical therapists employed in spinal cord centers will not help paraplegics back into a wheelchair from which they have fallen but will teach them how to get back into the wheelchair. Industrial physical therapists, likewise, must not disable injured workers with treatment that focuses on symptoms and keeps them out of the work environment, but must address the cause of injury and teach them how to cope within the work environment. Treatment for the injured worker must be functionally oriented, with the understanding that the worker may not always be pain-free. With that mind-set, it then becomes part of the physical therapist's role to help provide the direction of treatment so the employer and the employee recognize it as the best treatment serving the needs of the patient and maximizing his or her safe functional capabilities towards the goal of a safe and successful return to work.

AWARENESS

The on-site physical therapist is more familiar with work demands, specific injury management systems, the employees, and how to get things done within the industry than off-site practitioners are. On-site programs are more successful if the therapist not only provides treatment services but gains an understanding and a heightened awareness of the industry's needs. Information available and important for the on-site therapist to become familiar with includes

- seasonal work changes
- fluctuations in size of workforce
- referral patterns of in-house and community medical practitioners
- availability of transitional work
- management commitment and attitudes toward work injuries
- labor-management interaction
- corporate philosophy

MODELS OF PHYSICAL THERAPY INTERVENTION ON SITE

Primary Gatekeeper

As managed care programs develop, a "primary gatekeeper" is becoming the point of entry into the health care delivery system. The gatekeeper's responsibilities include directing the evaluation and treatment of patients to ensure that the most appropriate and cost-effective treatment is provided. Traditionally, the point of entry has been the physician.

In Case Study 1, an outdoor apparel mail order company, physical therapists have been designated as the primary gatekeeper.

Traditionally, physical therapists have received patients through referrals emanating from physicians, occupational health nurses, insurance companies, and chiropractors. In states where direct access exists, therapists are obligated to accept patients from any and all referral sources. For these therapists, a negative factor arises in that reimbursement policies from insurance carriers may require physician referral.

In Case Study 1, the physical therapy, gatekeeper model, the industry, which employed up to 5,000 individuals, had no on-site medical facility. After a pilot program in 1990 that integrated on-site physical therapy with ergonomics, a reduction in lost-time injuries of 65 percent and 58 percent was realized in the respective distribution and manufacturing divisions. The average duration of physical therapy treatment was 4.5 visits. Physical therapy was considered key in this successful approach.

This program has now expanded to include multiple on-site clinics at varying company locations. Injuries are first reported to coworkers who have been trained in soft tissue injury identification, and workers are then directed to the on-site therapist. From this point of entry the therapist decides to treat or not. If medical input is needed, the therapist makes the referral to the physician.

Credit is due to this industry and its employee health department for exploring a rather progressive and unique delivery system. Not only has the company facilitated early and effective physical therapy treatment for the worker, but the entire company benefits since lost time is not crippling productivity or financial stability.

Clinic PT/Acute Injury Evaluation/Management

In Case Study 2, Bath Iron Works (BIW), referrals are received from the in-house medical department as well as outside health care providers. A unique role for physical therapy at BIW is acute injury evaluation in the medical clinic. In the "clinic PT" role, the physical therapist sees musculoskeletal injuries that have been triaged by the clinical nurse. The objective is to promote early intervention and early mobilization, and to emphasize function rather than dysfunction. The "clinic PT" does not become the treating therapist. The "clinic PT" may determine if ongoing therapy is needed and refer the patient appropriately, or may decide that the problem is self-limiting and initiate minimal treatment through a home exercise program. If the problem is beyond the scope of therapy, the patient is referred to the in-house physicians for their expertise in further medical care or consultation. Indications exist that this role has helped to decrease the number of subsequent diagnostics and office visits off site and lost time.

EQUIPMENT

Equipping an on-site clinic depends primarily on space. In most industries space is at a premium. Therapists must realize that the industry exists to produce or manufacture a product and not to house on-site physical therapy. Therefore in most cases space allotted for on-site clinics is significantly limited. Exceptions, however, are two large paper mills in Maine that have provided excellent space and equipment, including a full line of weight-training equipment and modalities for on-site therapy. Both Boise-Cascade and International Paper offer their employees fully equipped and staffed programs highlighting their commitment to provide the best possible care to their employees.

Equipment should be chosen wisely. The modalities chosen should be the most effective in treatment of inflammation, swelling, and pain. All other equipment should be active and functionally oriented. A set of lifting boxes and a treadmill will achieve better results than a $40,000 isokinetic module in these settings. And, as always, the most effective equipment is carried into the clinic every day by the therapist. Hands and mind continue to be the best assets.

OUTCOMES DATA

Back Schools

Figures 13–1 and 13–2 illustrate effectiveness of preventative back schools conducted in BIW for Department 50 (shipfitters) and Department 11 (pipe covering). The back school was presented by the on-site therapist in a two-hour program of slides, lecture, practical demonstration, and written quizzes. The back school was designed and customized for BIW to teach proper body mechanics and the appropriate lifting and material handling techniques.

A second strong benefit was the opportunity it afforded the therapist to provide firsthand information about treatment approaches and the natural history of low back pain. By discussing topics such as bed rest, manipulation, exercise, diagnostic tests, and pain-free status versus pain management, much of the miscommunication and therefore inappropriate illness behavior could be prevented if any of the attendees did incur a back injury.

On-Site Treatment

Table 13–1 summarizes the impact of physical therapy at Bath Iron Works from 1986 to 1987. Aspects of early intervention, early mobilization, education, and functionally oriented treatment helped to keep workers safely and comfortably at work while receiving treatment for their injuries. The willingness of the industry

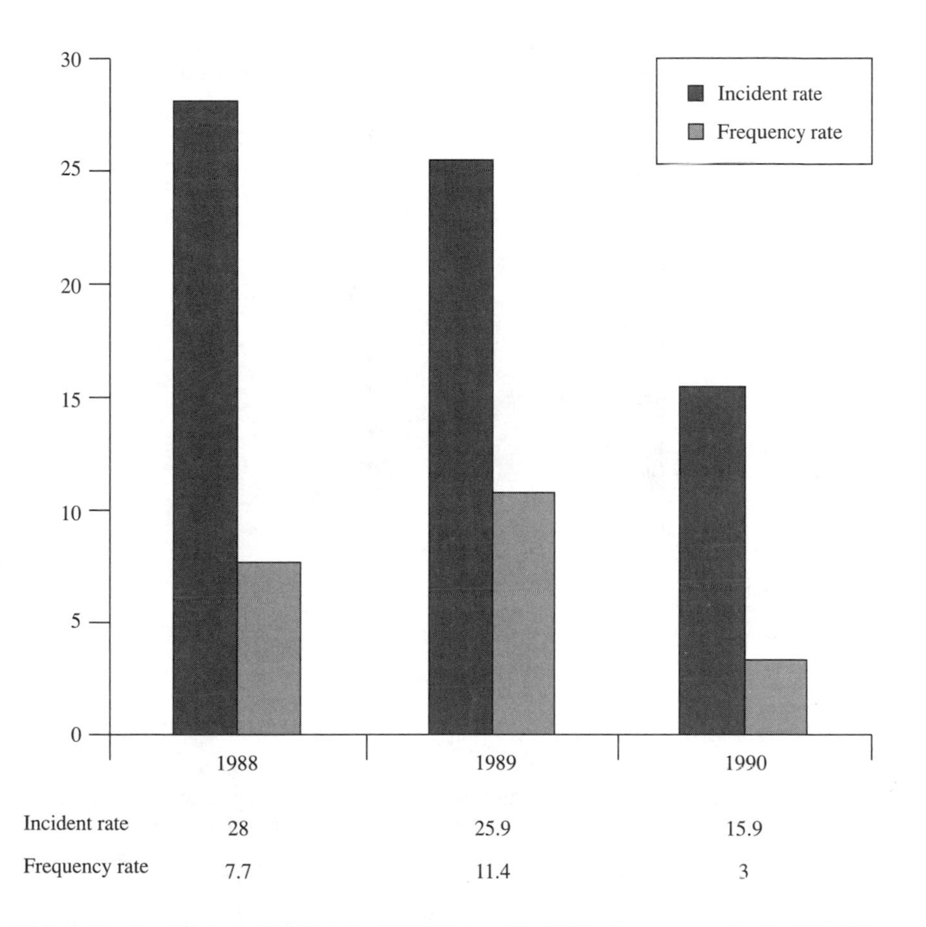

26 fewer lost-time injuries at $7,071 each – $18,232 cost of Back School = net cost reduction $165,614.

225 attendees in 1989 + 343 attendees in 1990 = 568 total. Data were collected in the first seven months of each calendar year.

Figure 13–1 Back Injuries at Bath Iron Works, Department 50, before and after Institution of Preventive Back School Program in 1988. Data courtesy of Bath Iron Works.

---

and its medical department to accommodate injured workers and provide transitional work also figured heavily in the success.

Significant comparison can be made from 1985 to 1986 in the time from date of injury to intervention. The early intervention correlates with less overall time in rehabilitation and an earlier successful return to work. In subsequent tracking of these programs over the last six years, a delayed treatment onset time is linked

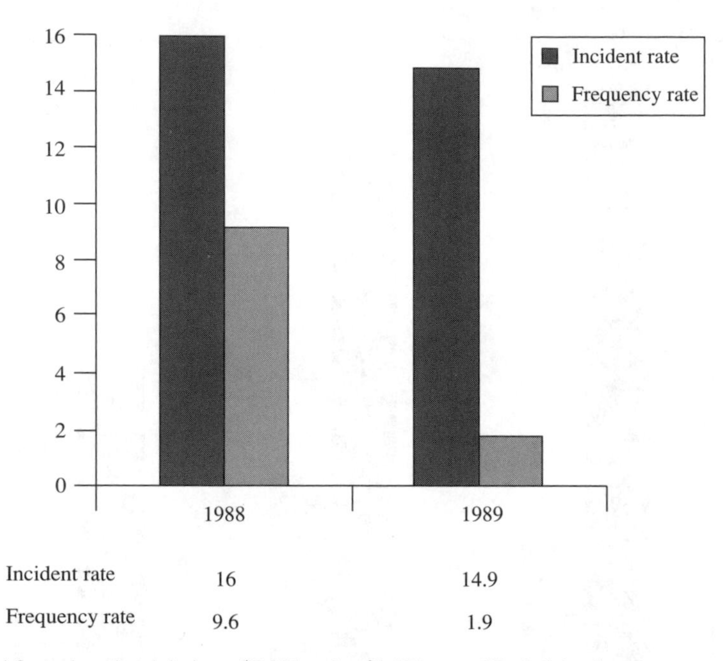

4 fewer lost-time injuries at $7,071 each – $1,536 cost of Back School = net cost reduction $26,748.

70% of employees attended Back School.

Figure 13–2 Back Injuries at Bath Iron Works, Department 11, before and after Institution of Preventive Back School Program in 1988. Data Courtesy of Bath Iron Works.

---

with an increase in the length of overall therapy duration. The year-end statistics for return-to-work success for 1992 are illustrated in Figure 13–3. It is important to note that over 70 percent of patients seen by physical therapy on site returned to their parent department. Thc average PT treatment duration for on-site involvement was 7.6 visits.

Other industries utilizing similar models have realized dramatic reductions of lost work time injuries. One paper industry (Boise-Cascade) recently reached 4.5 million safe person-hours without a lost-time injury. Injuries still occurred, but early intervention and treatment of employees on site and ergonomic changes tailored to the individual's needs kept employees from losing time.

STAFF/TEAM

Quality programs are the combination of vision, dedication, skill, opportunity, and hard work. They are only as good as the individuals involved. Many "fran-

Table 13–1 Work Injury Measures at Bath Iron Works before and after Institution of On-Site Physical Therapy Program in 1986. Data courtesy of Bath Iron Works.

	1985	*1986*	*1987 (first 6 months)*
Number of employees in physical therapy	75	182	141
Average intervention interval	2.5 months	1 month	3 days
Average duration of visit	1 hour	1 hour	½ hour
Average duration of treatment	3 months	2 months	2 weeks
Lost workdays per injury	31.1	16.5	10.9
Total number on compensation list	316	244	231
Total cost—medical and indemnity	$4,128,545	$935,318	$489,255
Total number of lost workdays	16,929	6,838	1,871
Total cost for physical therapy services	$144,750	$130,594	$159,584

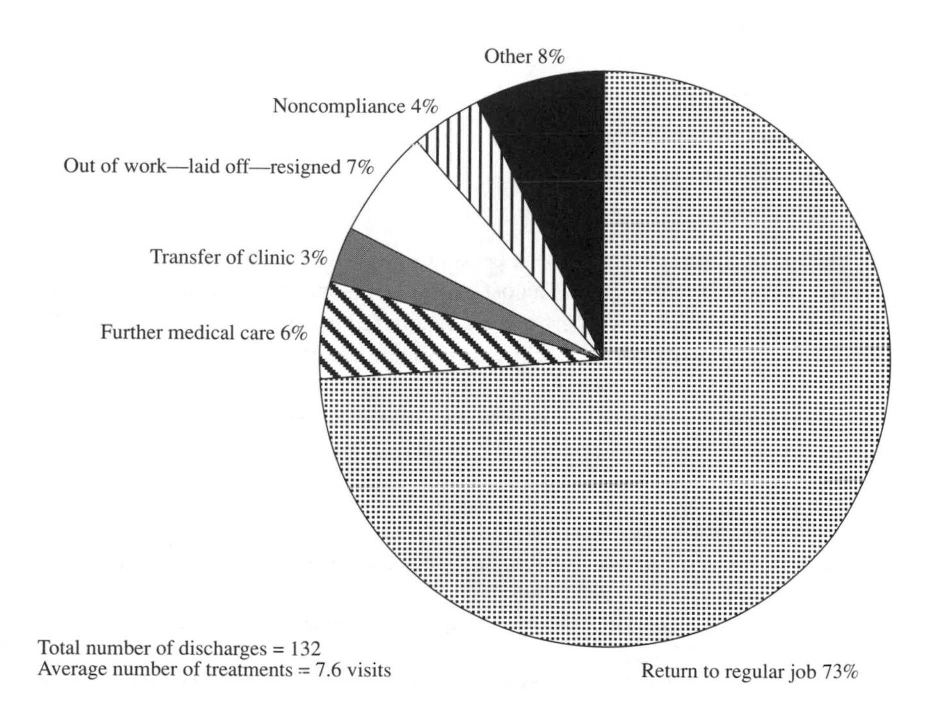

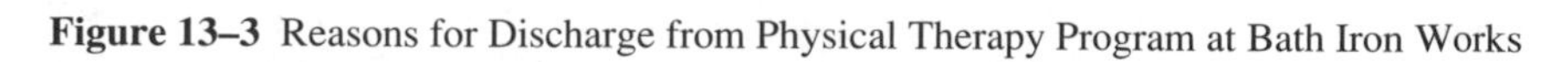

Figure 13–3 Reasons for Discharge from Physical Therapy Program at Bath Iron Works

chised" programs have failed due to the dilution of quality individuals. Case Studies 1 and 2 represent therapists that believe in the importance of their work more than the location of their work.

On-site clinics are not esthetically ideal. Lack of space, lack of equipment, and a sense of being "out of one's element" make industrial rehabilitation unappealing to many health care practitioners. When staffing on-site clinics, the therapists' professionalism, independence, autonomy, and drive are very important characteristics.

Team building is of utmost importance. The on-site physical therapy programs are but a part of the overall team approach to managing work-related injury. No therapist, however skilled, can come close to being as successful as a good team. Team members traditionally include other health and safety personnel as well as production supervisors and the injured worker. Without close collaboration among team members, all goals and objectives for work injury management would fall short of success.

The effectiveness of on-site physical therapy within a team approach has been clearly illustrated in the preceding examples. Elements of effective on-site programs include

1. team approach
2. functional outcomes
3. early intervention
4. active versus passive treatment
5. communication
6. education

The therapist within this framework becomes an intimate part of a program that provides highest levels of care in a cost-effective manner to allow maximum safe return to work for injured workers and an overall reduction in disability and medical costs industry-wide—truly a win-win situation for everyone involved.

REFERENCES

Eastman Kodak Company, Human Factors Section, *Ergonomic Design for People at Work,* Vol. 1 (New York; Van Nostrand Reinhold, 1983).

C.M. Bettencourt et al., Using Work Simulating to Treat Adults with Back Injuries, *American Journal of Occupational Therapy* 40 (1986):12–18.

S. Bigos et al., Back Injuries in Industry: A Retrospective Study II. Injury Factors, *Spine* 11 (1986):246–256.

A.R. Lind and G.W. McNicol, Circulatory Responses to Sustained Hand-Grip Contractions Performed during Other Exercise, Both Rhythmic and Static, *Journal of Physiology* (London) 192 (1967):595–607.

L.N. Matheson, Symptom Magnification Syndrome (Paper presented at National Rehabilitation Association meeting, Anchorage, Alaska, 1984).

T. Mayer and R. Gatchel, *Functional Restoration for Spinal Disorders: The Sports Medicine Approach to Low Back Pain* (Philadelphia: Lea & Febiger, in press).

T. Mayer et al., Objective Assessment of Spine Function Following Industrial Injury: A Prospective Study with Comparison Group and One-Year Follow-Up: Volvo Award in Clinical Sciences, *Spine* 10 (1985):482–493.

T. Mayer et al., A Prospective Two-Year Study of Functional Restoration in Industrial Low Back Injury: An Objective Assessment Procedure, *JAMA* 258 (1987):1763–1767.

A. Nachemson, Work for All—For Those with Low Back Pain as Well, *Clinical Orthopedics* 179 (1983):77–85.

C.E. Newman, Jr., Rehabilitation Can Be a Cost Containment Device, *Hospitals,* March 1979, 45–46.

S.H. Rodgers, Recovery Time Needs for Repetitive Work, *Seminars in Occupational Medicine* 2 (1987):19–24.

J. Troup et al., The Perception of Back Pain and the Role of Psychophysical Tests of Lifting Capacity: 1987 Volvo Award in Clinical Sciences, *Spine* 12 (1987):645–657.

M.J. Wang et al., Grip Strength Changes When Wearing Three Types of Gloves (Paper presented at HF Interface '87, Western New York Human Factors Society, Rochester, N.Y., May 1987).

Chapter 14

The Objectives of Treatment for Mechanical Low Back Pain

Carl DeRosa and James A. Porterfield

THE RISING COST OF HEALTH CARE

All industrialized societies are currently plagued with the rising cost of health care. As an example, health care accounted for 12 percent of the gross national product (GNP) in the United States in 1991, and the total cost of health care expenditures in the United States reached approximately $750 billion in 1991.[1] On its present course, the cost of health care could reach $1.5 trillion by the end of the decade. Furthermore, it is predicted that the United States will be using at least 15 percent of the GNP for health care by the year 2000.[2] Even more alarming is the fact that industry is currently spending nearly 30 percent of its profit margin on health care delivery.[3]

Although health care expenditures for the United States continue to increase, there are intensive efforts to reduce costs. Industry, third-party payors, and the government lead the charge in cost-containment efforts.

As the scope of practice environments has broadened, medical providers now are thrust directly under the purview of third-party payors, such as insurance companies, industrial commissions, and Medicare. All totaled, the cost factor has become much more than just a government, industry, and insurance company issue: the consumer is also exerting greater pressure to influence costs.

SOCIETAL COST OF LOW BACK PAIN

At the very center of rising health care costs is low back pain. With the exception of the common cold, back pain is the most common health problem in the United States. It is so common that Waddell has suggested that back pain is a universal, self-limiting condition that can be viewed as a normal human experience.[4] The National Center for Health Statistics notes that 14.3 percent of new patient visits to physicians are for complaints of low back pain.[5]

It is recognized that 80 percent of the population will experience back pain symptoms at some point during their lifetime and that sciatica, which can be defined simply as low back and leg pain, occurs in approximately 40 percent of all adults.[6] Sciatica, like low back pain, also has a natural history that favors recovery.[7] Thus both low back pain and sciatica not only have a fairly high incidence in society but also have a favorable prognosis.

Back pain accounts for 25 percent of all lost workdays in the United States and is responsible for at least 25 percent of all workers' compensation claims. According to Andersson, back pain is the most frequent cause of limitations to activity in people below the age of 45, the second most frequent reason for seeing a physician, the fifth most frequent reason for hospitalization, and the third-ranking reason for surgical procedures.[8]

Other health care practitioners besides physicians are typically sought out by patients seeking relief for their low back pain problem. Of the 100 million visits made to chiropractors each year, one-half are for complaints of low back pain.[9] Over five million office visits are made to physical therapy offices for low back pain each year.[10] Patients also seek help from osteopaths, naturopaths, acupuncturists, massage therapists, and a variety of other licensed and nonlicensed individuals.

DISABILITY AND ITS RELATIONSHIP TO LOW BACK PAIN

Although there is good evidence that back pain is a common occurrence and that this incidence has not changed appreciably over the years, it is the disability from low back pain that has risen dramatically over the past several years.[11] Disability is largely a subjective report by the patient stating his or her limited capacity for activities of daily living or employment. This limited capacity is essentially a comparison that takes into consideration the capacity for activities of daily living that an uninjured person of the same sex and age would have.[12] Understanding disability and its relationship to low back pain is critical: low back pain is the most common cause of the disability in those aged less than 45 years.[13]

Clearly one of the major difficulties in gaining control of the low back pain problem is the subjective nature of disability since varying interpretations of limited capacity occur. As a result, diverse opinions as to the extent and ramifications of a disability from low back pain often result. This wide range of opinions regarding disability is influenced not only by the view of the patient but also by views of others such as physicians, physical therapists, employers, lawyers, and insurance companies.

It is becoming increasingly clear that an understanding of disability and its ramifications is of paramount importance for clinicians involved in the treatment of low back pain. Without such knowledge, treatment becomes palliative at best and redundant and costly at its worst. It is not just the disability from low back pain that must be understood, but more importantly how this particular aspect fits into the complete scheme of low back disorders. A better perspective can be gained from the realization that according to various estimates, over 70 percent of the total costs of spinal disorders can be attributed to a minority of patients whose symptoms result in disability. Snook and Jensen observed that a breakdown of compensable work injuries revealed that one-third of the total cost was for medical care and two-thirds for disability payment, and that 90 percent of the total compensation costs in the United States were produced by 25 percent of those individuals suffering low back injuries.[14] Spengler et al. found that 10 percent of all workers' compensation claims accounted for 79 percent of the disability costs.[15] Furthermore, the Workers Compensation Research Institute has noted that only 25 percent of back injuries are responsible for 90 percent of the back injury costs.

An understanding of such epidemiological information is essential in order to appreciate the current medical trend to identify, as early as possible, potential chronic pain syndrome patients or those patients with a propensity toward disability. While early identification appears to be a daunting task, the literature clearly points out the necessity to develop such clinical skills. It has been suggested, for example, that four different dimensions of pain behavior might include marked distortion in ambulation or posture, a negative affect, facial or audible expressions of distress, and the avoidance of activity.[16] Early identification of these patients is essential because the objective of treatment for these patients becomes focused on the promotion of activity rather than the promotion of analgesia.

Despite the subjective nature of low back pain reports, and of disability in particular, it is becoming increasingly clear that all health professionals who deal with spinal disorders will be responsible for justifying their treatment based on measurable outcomes that can be appraised by both medical and economic criteria.[17] Physical therapists in particular can be well positioned to affect low back pain or disability since there is strong suggestion that early and aggressive rehabilitation strategies have the potential to reduce the duration of disability, which subsequently reduces societal cost.[18]

AGING AND THE INJURY PROCESS

A complete understanding of the basic sciences related to the low back and the implications of the aging and injury process is essential for the physical therapist and other clinicians treating patients with low back pain. However, among those whose education is replete with treatment technique skills, there is often a misconception that the myriad techniques and modalities being learned are potential "cures" for musculoskeletal problems. Failure to achieve "the cure" is typically ascribed to the wrong treatment technique or a poor choice in modalities. Little regard is given to the intent of treatment and the proper objective of treatment.

An understanding of the aging and injury process and the science of tissue healing and repair is perhaps more important than physical therapy technique. Clinicians often seek out more "technique courses" in search of the clinical skill that they perceive is somehow missing from their repertoire. Continuing education courses focused on techniques to treat the low back abound within physical therapy and other professions. The frustration to find the ultimate technique, coupled with a lack of understanding regarding the interplay of biopsychosocial factors, frustrates both the clinician and the patient.

This emphasis on technique has greatly contributed to the palliative approach traditionally used for the management of low back disorders. Instead of an emphasis on reactivation, attention has almost solely centered on pain relief. While such a focus may be indicated in the initial stages of injury, the realities of the aging and injury process dictate that the primary focus of treatment should be on education for self-management and the restoration of function. With the current level of knowledge regarding the degenerative process, the only rational treatment programs are "exercise programs that offer stabilization of the range (of spinal movement) by muscle activity and improved nutrition of the joint by mechanical activity."[19, p.438] Thus it is essential that the ultimate treatment goal be understood: reactivation of the patient. It will ultimately be the patient's own physical activity that maintains or improves his or her musculoskeletal health.

The diversity of treatment approaches toward low back pain was critically examined by the Quebec Workers' Health and Safety Commission.[20] Upon the request of the Quebec Workers' Health and Safety Commission, the Quebec Task Force on Spinal Disorders took a comprehensive look at the scientific evidence behind the assessment and management of activity-related spinal disorders. Of particular relevance was that this report was commissioned in part due to the increase in physical therapy treatments for low back disorders and the wide variation in types and duration of treatment from one institution to another for what appeared to be the same clinical entity.

This dilemma—the great diversity of assessment and treatment approaches—is extremely important to recognize. But a more fundamental question is raised. *Why* are there so many different schools of thought for the care of the spine? Certainly this is not the case with other areas of the musculoskeletal system. Are the rules of injury and healing for the musculoskeletal tissues "different" for the spine? Perhaps, considering the spine's intimate relationship with the central nervous system and the presence of a distinctive structure, the intervertebral disc, a case could be made for the uniqueness of the spine. However, it does not appear logical to conclude that gross dissimilarity should be the rule. It is the integration of the science with rational objectives of treatment for low back pain that comprises the future of treatment for this problem.

THE OBJECTIVES OF TREATMENT FOR MUSCULOSKELETAL DISORDERS

When the complete spectrum of nonsurgical treatment approaches for musculoskeletal disorders is carefully analyzed, each intervention is designed to achieve one or more of the following objectives (see Exhibit 14–1).

- Objective 1: *Modulating pain or promoting analgesia.* Patients typically enter the health care system because they are in pain and seeking relief. Therefore it is logical and ethical that this serve as an appropriate objective. The challenge, however, is determining whether pain modulation should be the treatment goal or whether it is simply a strategy to move the patient quickly and efficiently toward another objective.
- Objective 2: *Promoting active movement by the patient as quickly and safely as possible by clinically applying specific, nondestructive, and controlled forces that influence the neuromusculoskeletal system.* The effect of applying such controlled forces on the musculoskeletal tissues is most likely limited to
    1. *Altering fluid dynamics.* Stasis, venous congestion, swelling, altered pH, and edema are some of the ways in which the fluid milieu results in mechanical distortion and chemical insult to an injured region.
    2. *Increasing afferent input into the central nervous system.* A barrage of sensory stimuli into the central nervous system initiates several powerful neurophysiological reflexes: pain modulation via stimulation of the large-fiber afferents and alteration (increasing or decreasing) in the resting state of muscle tension. In many cases, these are desired clinical outcomes, but again they need to be considered in the context of the ultimate goal of reactivating the patient.

Exhibit 14–1 Objectives of Treatment

1. Pain modulation/promote analgesia
2. Introduce controlled, nondestructive forces in order to promote active movement by the patient
3. Enhance neuromuscular performance
4. Biomechanical counseling

3. *Modifying connective tissue.* With movement serving as a desired outcome, it can be concluded that the extensibility of connective tissue potentially influences at least the quantity and perhaps the quality of the movement. However, the question arises as to the force required to actually change the structure of collagenous tissue and whether it is realistically possible to apply adequate force over a period of time to elongate connective tissue in the typical clinical setting.

- Objective 3: *Enhancing neuromuscular performance.* There is intuitive as well as scientific evidence that training the neuromusculoskeletal system is of benefit for patients with musculoskeletal disorders. In a global sense, training refers to a planned disruption of current homeostasis in order to stimulate the necessary anatomical and physiological adaptations necessary to restore an improved homeostasis.[21] The adaptations can be as directly observable as change in muscle size, as measurable as an increase in muscle power or endurance, or as enigmatic as alteration in central nervous system activity. Despite the various means available to accomplish these objectives, the end result remains the same: enhanced neuromuscular performance.
- Objective 4: *Patient education in the form of biomechanical counseling.* Since management is primarily focused on activity-related, mechanical disorders of the musculoskeletal system, it is important that the therapist begin to teach the patient to take an active role in managing his or her musculoskeletal problem. An understanding of the pathomechanics of injury is important for the clinician to understand and then share with the patient, using common, understandable terminology, in an educational process that takes the form of biomechanical counseling. This is becoming increasingly important in today's changing health care system since the time period over which the clinician actually interacts directly with the patient is a minuscule portion of the time that the patient is actually awake and active.

INTEGRATING THE SCIENCE OF ANATOMY AND THE OBJECTIVES OF TREATMENT: SIMILARITIES OF TREATMENT TECHNIQUES

Perhaps one of the best means to integrate treatment approaches under each objective of treatment is to organize technique around the spinal anatomy. Such an organizational approach allows for similarities between different treatment techniques to be better appreciated. For the purposes of this chapter, the anatomy will be divided into the muscular system and associated fascia, the joints and related tissues, and the intervertebral discs and venous plexus. The reader is referred to other textbooks for a complete review of the functional anatomy of the low back.[22]

Muscles and Fascia

Numerous treatment techniques are purported to have an effect on the muscles and fascial system. These include topical application of modalities such as heat; cold; the various forms of electrical stimulation; manual and mechanical methods to palpate, shorten, stretch, or generally mobilize the tissue; and exercise to increase the functional capacities of the muscle, such as strength, power, endurance, coordination, and improved movement patterns. Based upon the new knowledge of low back anatomy and a clearer understanding of the epidemiology of the low back pain problem, these techniques can be applied to the four objectives of treatment (Tables 14–1 and 14–2).

Topical Modalities

It is questionable whether any of the superficial topical thermomodalities are indicated for the muscle system of the low back, since it does not appear likely that any topically applied thermomodality can have a direct effect on the muscle or fascial tissue owing to the depth of the muscle and significant fatty and connective tissue coverings. Any effect on the muscular system can only be on reflex connections from the skin receptors to the muscles that occur at the central nervous system level and as such serve a role in desensitizing the region. Modalities should not be used under the pretense of decreasing inflammation or warming the muscle system of the low back since this is not deemed effective through our understanding of the structural relationships in the low back.

Electrical Stimulation

Electrical stimulation, appropriately applied, appears to have the potential to initiate and propagate low-grade muscle contraction in selective although not necessarily exclusive areas: the superficial erector spinae, the deep erector spinae, and the multifidus. Electrode placement, however, must take into account the tendinous regions of these muscles versus their fleshy muscle bellies.

Table 14–1 Treatment Influences on Fascia

Anatomy	*Clinical Skill*	*Objective*
Thoracolumbar fascia	Exercise: latissimus dorsi	3
	Exercise: posterior spinal muscles	3
Hip fascia	Stretching	2
	Exercise: hip extensors	3

Table 14–2 Treatment Influences on Muscles

Anatomy	*Clinical Skill*	*Objective*
S. erector spinae	Exercise: strength, endurance, fiber type, power movement	3
	Palpation pressure	2
	Lengthening/shortening	2
	Electrical stimulation	1,2
D. erector spinae	Exercise: stabilization, postural considerations, movement specificity	3
	Palpation/pressure	2
	Lengthening/shortening	2
	Electrical stimulation	1,2
Multifidus	Exercise: strength, endurance, fiber type, movement specificity	3
	Palpation/pressure	2
	Muscle energy	2
	Electrical stimulation	1,2

The *reasons* to apply electrical stimulation can be placed under two of the objectives of treatment:

1. It modulates pain with a transcutaneous nerve stimulation (TENS) effect or elicits a low-grade muscle contraction, which actively utilizes a muscle-pumping action to minimize stasis and alter the chemical environment of the muscle.
2. It is a method of generating a controlled force via muscle contraction into the low back region in order to change the resting state of stiffness of the muscle. This change is accomplished by increasing the afferent input into the central nervous system, which potentially decreases the efferent output to the musculature.

The clinical question, however, is related to the indications for electrical stimulation in treating low back disorders. It is suggested here that a very specific group of patients is indicated for such treatment and that only for an even smaller group does the effect of electrical stimulation serve as a treatment goal. The patients for whom electric stimulation is indicated are those who are truly acutely injured. Modulation of pain and influence on fluid dynamics associated with tissue injury make electrical stimulation a useful modality. The acute injury, however, must be clearly defined as the state in which injury corresponds with the known time frames for soft tissue injury and repair. The other indication to be considered is whether the electrical stimulation modality allows for immediate reactivation of the patient.

Facilitating an Early Return to Movement

Objective 2 is that of facilitating the patient's early active movement by the application of specific, controlled, and nondestructive forces directly into the injured region. A variety of techniques that fall under this objective appear to be directed at the muscle and fascial system. These are discussed individually.

Palpation. This is one of the most common techniques performed in both evaluation and treatment. At its simplest it can be viewed as gradations of force application from light touch to very deep pressure. During the application of this pressure a combination of other forces such as shear or twisting can be applied.

In order for palpation techniques to begin to provide meaningful information to the clinician, an awareness of the fascia, tendon, and muscle belly locations must be evident, and it is proposed that such techniques be introduced with the study of anatomy. Relevant anatomy is as follows:

1. The superficial erector spinae muscle belly appears markedly lateral to the spinous processes and at the mid-lumbar region (Figure 14–1).
2. The deep erector spinae has a significant muscular attachment to the posterior superior iliac spine and in the region immediately lateral to the iliac crest (Figure 14–2). It is muscular rather than tendinous in this region.
3. Any palpation over the dorsal surface of the sacrum or immediately adjacent to the lower lumbar spinous processes is limited to the multifidus muscle belly since only the connective tissues of the erector spinae apo-

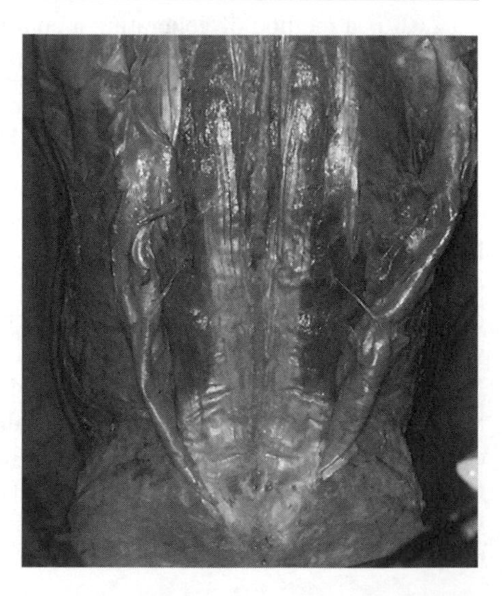

Figure 14–1 Posterior View of Superficial Erector Spinae with the Thoracolumbar Fascia Reflected. Note the extent of the erector spinae tendon (erector spinae aponeurosis) and how markedly lateral to the spinous processes it reaches. The actual muscle belly of the superficial erector spinae muscle begins in the midlumbar spine. *Source:* C. DeRosa, Integration of the Objectives of Treatment for Low Back Pain with Lumbar Spine Anatomy (PhD diss., Union Institute, Cincinnati, Ohio, 1993).

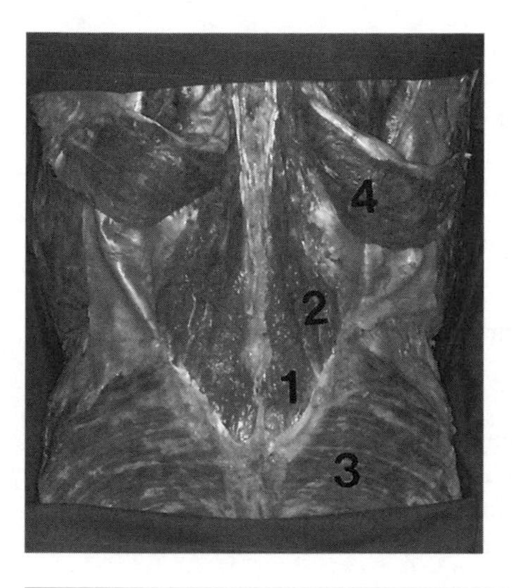

Figure 14–2 Posterior View of Deep Erector Spinae and Multifidus Muscles. 1) multifidus muscle, 2) deep erector spinae muscle, 3) gluteus maximus muscle, 4) reflected superficial erector spinae muscle and tendon. The superficial erector spinae tendons (4) have been reflected superiorly in order to expose the multifidus muscle (1) and the deep erector spinae muscle (2). Note that the deep erector spinae muscle is just lateral to the region of the posterior superior iliac spine, while the multifidus covers the dorsal surface of the sacrum. *Source:* C. DeRosa, Integration of the Objectives of Treatment for Low Back Pain with Lumbar Spine Anatomy (PhD diss., Union Institute, Cincinnati, Ohio, 1993).

neurosis and the posterior layer of the thoracolumbar fascia cover this muscle (Figure 14–2).

Pressure on Muscle. Pressure techniques typically fall under such treatment titles as massage, trigger point therapy, acupressure, rolfing, or transverse friction. Shortening of the muscle tissue takes tension away from hypothetically injured tissue and is exemplified by techniques such as postural positioning and strain/counterstrain. Stretching of muscle either alters the connective tissue matrix of the muscle or changes the resting tension of the muscle via stimulation of the golgi tendon organs or muscle spindles, which secondarily results in a decrease or increase in resting muscle stiffness.

Most likely there is very little difference in the overall clinical outcome of any of the pressure techniques. The techniques potentially modulate pain (Objective 1) because of the powerful stimulus of touch, but their primary influence is most likely via Objective 2. The techniques appear to have the potential to affect fluid dynamics as well as increase afferent input into the central nervous system. There does not appear to be any evidence that collagenous tissue is actually "broken down" with any of the techniques.

Shortening of Muscle

Techniques to shorten tissue present an interesting application of the muscle anatomy. The sequence of usage for clinical practice usually requires palpation of sensitive regions or putting the patient through different movement patterns to reproduce familiar pain, analyzing the response of the tissue to palpation, and then passively moving the segment or segments in order to shorten the tissue to reach the point in the range of motion in which the painful stimulus is decreased.

An approach of this sort is seen in treatment of other areas of the musculoskeletal system. The shoulder injured with a forced abduction-external rotation maneuver is often placed in a position of adduction and internal rotation to remove the tensile force from the injured tissue and alter the degree of surrounding muscle spasm. Likewise, a forward head posture is often assumed following whiplash injury in order to place the injured soft tissues at rest. A balance between shortening the tissue to promote healing and controlled motion to prevent adaptive shortening must occur, however.

These techniques can be placed under Objective 2. They have the ability to facilitate early activity by the patient. They cannot serve as a treatment endpoint but must be used as an adjunct to promote early activity by the patient. Integrating the technique with the anatomy and the clinical problem helps bring perspective to their application. Note that many of these techniques have excellent potential to be carried out by the patient if instructed properly.

Stretching of Muscle

Indications for stretching need to be carefully reevaluated. Stretching and mobilization techniques have been used largely on the basis of anecdotal reports. There is very little evidence that stretching and increased mobility decrease the potential for low back pain occurrences.[23] Indeed, when one looks at this technique using the same clinical rationale as that used for peripheral joint injuries, one must question the wisdom of aggressively mobilizing tissue to such an extent that mobility is increased. With most connective tissue injuries in the musculoskeletal system, such as the ligamentous injuries of the knee or ankle, or the capsular injuries of the shoulder, the emphasis is on restoration of stability, with less emphasis on increasing mobility. The spine should perhaps be viewed in much the same manner, especially since instability between spinal segments may be an important source of low back pain and a sequela to connective tissue injury. With an understanding of the three-dimensional aspects of the lumbar spinal anatomy, a more realistic view of tissue that can and cannot be stretched is gained. It is easy, for example, to view a picture of the iliotibial tract in a textbook and assume that it is a simple matter to stretch the tissue by adducting the hip. After one views the tissue, however, and realizes its extent, strength, and intimate association with the hip musculature, it is apparent that it would be extremely difficult to apply forces of sufficient magnitude to cause permanent elongation of the tissue.

The same might be said of the impressive array of connective tissue in the low back, notably the thoracolumbar fascia and the erector spinae aponeurosis. The strength of these tissues is such that they offer great resistance to any permanent deformation. Any changes in motion that are seen following stretching maneuvers are most likely due to altered efferent output to the muscle rather than permanent changes in the collagen framework.

As our understanding of collagen and neurophysiology has increased, the important future clinical question that needs to be asked regarding "limitation in motion" is whether the change is true tightness due to adaptive shortening or whether it is due to increased efferent output from the nervous system to the muscle. Tomorrow's clinician must answer this question first rather than assume stretching is indicated simply because the muscle seems tight. Increased efferent output to the muscle may in fact be the central nervous system's response to stabilize an injured spinal segment.

Tissue Mobilization

There are many different treatment techniques designed to increase tissue mobility. They include traction, mobilization, manipulation, myofascial techniques, soft tissue massage, contract-relax techniques, and muscle energy, to name only a few of the most common. Each one of these techniques meets the criteria listed under Objective 2 and therefore is indicated to facilitate early reactivation of the patient, but rarely does one of them serve as a treatment goal in and of itself. Stated more succinctly, patients move through a greater range of motion and often with less pain following the application of very specific forces such as those utilizing contract-relax techniques, the isometric contractions and subsequent stretch of muscle energy, or the rhythmic motion of mobilization. There is little evidence, however, that such techniques alter the long-term course of the low back problem. Therefore they should be taught as useful adjuncts for encouraging activity by the patient and need to be viewed in such a context. Patients often "feel better" following the application of such techniques, and the clinician of the past has accepted this as the final goal of treatment. As evidenced by the escalation of the low back pain problem as well as the cost, this approach was short-sighted and only served to perpetuate the low back pain problem. When combined with an emphasis in exercise (Objective 3) and self-management (Objective 4), however, the techniques may be indicated.

If symptoms of chronic pain syndrome are evident, then it is questionable whether such techniques have any indication. Chronic pain syndrome is here defined as that classification in which no direct relationship between application of physical stimulus and the pain response is evident. The patient is instead distinguished by such characteristics as anguish, depression, malaise, adoption of sick role, and discouragement.[24] The physical findings of examination are not proportionate to the pain response. To treat this population of patients with techniques that have the physiological effects described under Objective 2 does a disservice to their real need: restoring and improving function, and identifying measurable gains in their functional capacity. If we consider the possible anatomical and physiological effects of those treatment techniques that have been grouped under Objective 2, it is difficult to justify their use for chronic pain syndrome.

Physical Retraining

The last topic facilitating a more comprehensive understanding of the muscle and fascia system relates to the physical training and retraining of the muscular system (Objective 3). This objective is perhaps the most important reason to study the anatomy in the manner described: a true appreciation of both the gross and finer aspect of the muscles and their relationship to the lumbar and pelvic regions by analyzing their attachments, directions, cross-sectional area, and contribution in working with the surrounding hard tissue system (bone and articular cartilage) to attenuate trunk and ground forces.

Muscle insufficiency increases the stress on passive tissues of the spine since shock-absorbing capabilities of the muscles cannot be optimally utilized. Clinically, however, there is more to this statement than simply encouraging exercise: the physical therapist must know how to teach and direct exercise for the patient in pain since pain is the reason that the patient entered the medical system. It is unrealistic to wait until the patient is pain-free to begin an active exercise program. There is very little evidence that increasing physical activity makes any low back problem worse. It is not simply a matter of encouraging exercise for patients in the same manner that fitness programs are encouraged for the general population. This important difference in approach is the fundamental reason why muscle structure and function of this region must be clearly understood with the same breadth and depth as the anatomy of the extremities.

Physical training programs are clearly within the realm of Objective 3—enhancing neuromuscular performance. They can take many different approaches: resistance exercises (isokinetic, isoinertial, isometric), stabilization exercises (although it is unlikely that no movement occurs), Feldenkrais training, and work-hardening programs, to name only a few. The effect of each of these training programs is to train the neuromuscular system so that the muscles are better able to act as shock absorbers and to teach the patient patterns of movement that minimize physical stress to the injured tissues.

The anatomical perspective presented here provides support to the concept that the physical therapist must be aware of the broad influence of upper-extremity and hip muscles but also begin to pay much closer attention to the posterior spinal muscles. Historically, resistance exercises for the spinal extensors were avoided because of a concern that the contents of the intervertebral foramen might be compromised with such movements, or the injury to the apophyseal joint would occur. Any understanding of the biomechanics of the joints and foramen, as implied through the dissection approach described here, suggests that these muscles can be appropriately trained without compromising other tissues. As our knowledge of the anatomy and mechanics of the spinal extensors has increased, they have assumed a greater importance, and it appears that they will be viewed as the most important muscular consideration in the future.

Treatment Related to the Joints and Related Tissues

Many techniques are purported to have their effect at the joint, notably the various means of joint mobilization and manipulation. Integrating the clinical anatomy with these techniques allows a more critical analysis of these treatment approaches (Table 14–3).

Several important clinical questions are raised when the anatomy is detailed, such as

1. whether palpation of hypo- or hypermobility is within the realm of the tactile sense, owing to the small amount of intervertebral movement available
2. the asymmetry at adjacent levels, the asymmetry from side to side, and whether bony palpation to determine positional faults is a realistic evaluation tool

Lastly, but perhaps most importantly, realistic goals in conservative treatment are better visualized when age-related or degenerative changes of the joints are recognized.

Both manipulation and mobilization are techniques that fall under Objective 2. All three of the potential changes discussed previously under this objective (fluid dynamics, influence on the nervous system, and connective tissue changes) are

Table 14–3 Treatment Influences on Apophyseal Joints

Anatomy	*Clinical Skill*	*Objective*
Joint capsule	Muscle energy	2
	Contract/relax	
	Mobilization	
	Manipulation	
	Mobility testing	—
Articular cartilage	Manipulation	2
	Mobilization	
	Muscle energy	
	Active ROM	3
	Exercise: movement, cyclic compression, postural training	3
	Biomechanical counseling	4

potential effects of such treatment maneuvers. In keeping with the current emphasis of reactivation of the patient, they do not in themselves serve as treatment goals, but rather as adjuncts for eliciting early activity by the patient. Historically they have been taught as a "final common pathway" of treatment. There is little evidence that any long-term effect is gained with the exclusive use of such treatments.

Since mobilization and manipulation are common manual techniques utilized, their limitations need to be clearly understood by the clinician. Without recognition of these limitations, treatment via these approaches is no different than palliative-treatment-overutilizing modalities. Extensive, long-term treatment utilizing mobilization or manipulation discourages self-management. This is the basis of including such techniques under Objective 2.

The single most important reason to view these techniques in the context of the anatomy is to allow realistic treatment goals to be developed. It is clear, for example, that any force thought to be applied to the joints also falls on the adjacent tissues. It is unreasonable to assume that a singular force can be directed exclusively toward the joint.

Finally, three areas of the joint functional anatomy need to be recognized:

1. Compression of the anterior aspect of the joint occurs with anterior shear. Anterior shear occurs as a result of the lumbar lordosis and also during forward bending. Treatment strategies to unload the spine from such anterior shear appear to be based upon a logical rationale. However, long-term effects can only be accomplished via postural reeducation and training programs. They can only be temporarily altered by mobilization and manipulation. Exercise (Objective 3) has the most potential to influence anterior shear.
2. The normal weight-bearing relationship (compressive force) between the bone-disc interface and the apophyseal joints in the upright, standing posture is approximately 85 percent and 15 percent respectively.[25] A loss of hip extension capability limits the capacity to place the leg behind the body during the push-off phase of gait. This results in increased lumbar extension motion needed to place the lower extremity behind the body. There might be an association between loss of hip range of motion (adaptive changes in the hip capsule) and increased compression and subsequent degeneration of the facet articular cartilage in the lumbar spine.[26]
3. Articular cartilage is primarily nourished by the cyclic compression that occurs with movement. This strongly correlates with the intent of the techniques delineated under Objective 2—encouraging early active movement by the patient. While changes in articular cartilage cannot be reversed, the overall health of the articular cartilage is maintained with movement.

Treatments Regarding the Intervertebral Disc and Venous Plexus

The intervertebral disc is of extremely important clinical interest. First, the intervertebral disc itself is most likely a source of pain.[27] Second, the intervertebral disc injury or degeneration affects the adjacent tissues—notably the posterior longitudinal ligament, the dura mater, and the nerve root.

Postural Positioning and Traction

Postural positioning and traction are two of the more common techniques traditionally taught as methods to influence the intervertebral disc (Table 14–4). The original rationale for both of these techniques was based upon early studies regarding intervertebral disc pressure.[28] It is traditionally taught that a decrease in disc pressure subsequently relieves the mechanical stress on the posterior longitudinal ligament, dura mater, or nerve roots. However, inflammation, that is, an altered chemical environment within the spinal canal, is a sequela to mechanical insult of pain-sensitive tissues, and the inflammatory state may be a more important factor to consider than the mechanical stress of pressure.

Traction to the lumbar spine has minimal ability to significantly cause distraction at the bone-disc interface. However, when traction is viewed in the context of Objective 2, a more logical rationale can be suggested. Since muscle contraction of the spinal extensor muscles causes a significant rise in intervertebral disc pressure, the effect of traction may simply be to decrease the resting state of muscle tension through afferent input into the central nervous system, which secondarily lowers the intradiscal pressure.

Postural positioning and repeated motions such as extension or flexion active movements also fall under Objective 2 since motion affects fluid dynamics and introduces a barrage of afferent input into the central nervous system. This is a more logical scientific rationale, and eliminates pointless argument and questionable assumptions regarding mechanical influences on the disc and which tech-

Table 14–4 Treatment Influences on Intervertebral Disc and Vascular Tissue

Anatomy	*Clinical Skill*	*Objective*
Intervertebral disc	Postural positioning	2
	Traction	
	Reduce muscle contraction	
	Repetitive movement	
	Exercise	2,3
	Biomechanical counseling	4
Venous plexus	Active motion	2
	Mobilization	
	Repetitive movement	

nique is of more value. Currently, comparisons between techniques generate confusion (not only among clinicians but also among third-party payors) because claims are made that the techniques influence one particular tissue. As was pointed out in the discussion regarding mobilization and manipulation, it is unreasonable to assume that any technique can be isolated to one particular structure. Therefore when viewed in the context of Objective 2, the technique is placed in proper perspective, as an adjunct to facilitate early reactivation of the patient. The effects of such techniques are clearly seen to have specific indications but probably do not serve as treatment goals in and of themselves.

Active Repetitive Movement

Recognition of the consequences of inflammation within the spinal canal has become increasingly important. Active repetitive movement may serve the important function of minimizing fluid stasis and ultimately altering the chemical environment surrounding the tissues. This rationale further substantiates the current understanding that excessive rest is counterproductive in the management of low back disorders.[29] Excessive rest, that is, minimal movement, potentiates stasis and results in increased fluid buildup. The relationship of the venous plexus to the above-named tissues, namely the disc, posterior longitudinal ligament, dura mater, and nerve root, should be recognized. The extent of the venous network and its ramifications on structures located within the spinal canal must be known in order to fully understand the effects of venous congestion, tissue fluid buildup, and stasis (Figure 14–3).

Several investigators have speculated that engorgement of the epidural venous sinuses may increase external pressure on sensitized nerve roots in a stenotic spinal canal, especially in the presence of diminished right heart compliance.[30] Since the spinal canal is relatively inelastic, a stenotic canal encroaches upon the nerve roots. The epidural venous sinuses also occupy this same region of the canal, and cardiopulmonary resistance could result in engorgement of the venous channels (Figure 14–3). This pain is not typically elicited immediately after the patient reclines, but gradually increases with prolonged recumbency.[31]

Furthermore, Hoyland et al. utilized cadaver studies to show that intervertebral disc material could compress most of the venous channels passing through the inferior half of the intervertebral foramen. They further demonstrated that nerve root changes, as evidenced by long-standing inflammation, are associated with chronic venous congestion, even in the absence of any direct nerve compression via the disc.[32] In essence, compromise of the venous channels ultimately results in compromise to the nerve root.

The importance of spinal movement is recognized when intervertebral disc nutrition is studied with disc structure and function. As with articular cartilage, cyclic compression is of benefit in facilitating the transfer of fluid into and out of the

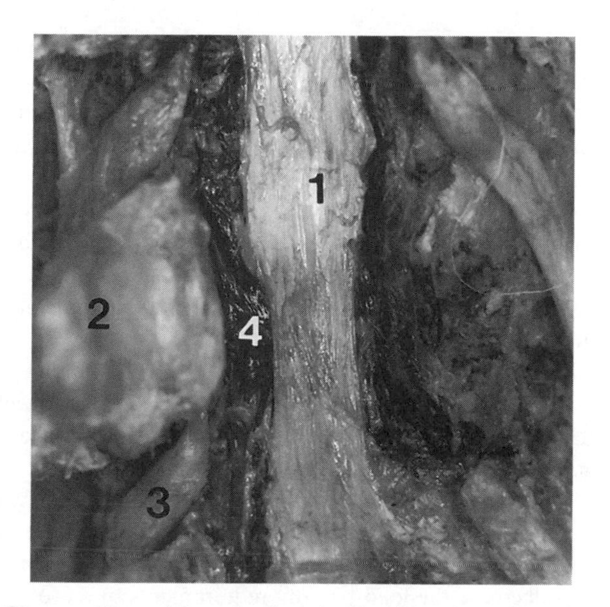

Figure 14–3 Close-up View of Posterior Longitudinal Ligament of Lumbar Canal. 1) posterior longitudinal ligament, 2) lumbar facet surface, 3) cut end of lumbar nerve root, 4) venous plexus of lumbar canal. The cauda equina has been removed and only the distal end of nerve root (3) at level of intervertebral foramen has been left in place. Note the rich venous network (4) intimately related to the posterior longitudinal ligament (1), underlying intervertebral disc, nerve root (3), and region of the apophyseal joint (2). *Source:* C. DeRosa, Integration of the Objectives of Treatment for Low Back Pain with Lumbar Spine Anatomy (PhD diss., Union Institute, Cincinnati, Ohio, 1993).

---

intervertebral disc. Movement helps maintain the health of the intervertebral disc, and the emphasis of clinical application should be methods available to the clinician to facilitate motion. The long-term outcome of nonsurgical management of disc disorders is actually quite favorable when compared with surgical intervention.[33] Therefore maintaining motion and muscular strength is important in order to provide the necessary mechanical stresses conducive to maximizing disc nutrition.

Setting Realistic Goals

Finally, one important aspect of the injury and degeneration process needs to be recognized in order to establish realistic, cost-effective treatment goals. This is especially true when one studies the anatomy of the intervertebral disc and the apophyseal joints in the low back. In the past, the treatment focus was on restoring the patient to a normal, preinjured state, that is, regaining 100 percent function.

This is an unrealistic goal for most significant musculoskeletal injuries. The philosophy of patient management must change to one accepting the fact that the patient has lost a hypothetical percentage (let us assume 20 percent for discussion purposes) of the function of his or her back due to injury or degenerative changes. Our goal as clinicians is perhaps not to try to restore the 20 percent but rather to teach the person how to manage his or her daily activities and occupational demands by maximizing the remaining 80 percent of function. This is the intent of Objective 4, and is a completely different clinical philosophy than currently used, but one that appears more realistic based upon the true anatomical changes that occur with injury and degeneration.

This further illustrates the importance of stressing both Objective 3 (enhancing neuromuscular performance) and Objective 4 (biomechanical counseling), which serve as the means by which individuals are taught patterns of movement to minimize mechanical stresses to injured tissues. Once articular cartilage or disc tissues are structurally altered, they lose their ability to attenuate forces with the same effectiveness as in the preinjured state. Although the injured tissue may repair, it rarely regenerates into tissue identical in structure or function. Increasing muscle sufficiency helps decrease the load-bearing requirements of these passive tissues of the spine. This makes exercise and training programs highly indicated for these problems. Enhancing neuromuscular performance becomes a treatment goal, while pain modulation is only indicated as an adjunct to strategies designed to enhance neuromuscular efficiency.

Many patients have exacerbation or reinjury of their unique low back disorder. This is most likely because they have previously injured tissues in the back, or early degenerative changes have occurred, and now that particular area is vulnerable for reinjury. Many activity-related low back problems feature this history of similar recurrent episodes. Patients with frequent exacerbations of their unique back problem are highly indicated for the objectives of enhancing neuromuscular performance and biomechanical counseling since those will be the primary means by which they will manage their low back problems. Table 14–5 summarizes the classifications of patients and matches them to the objectives of treatment.

FUTURE DIRECTIONS OF BACK INJURY MANAGEMENT IN THE INDUSTRIAL SETTING

An understanding of the epidemiology of the low back problem and the concern with the rising cost for care, coupled with advances in knowledge of the anatomy and biomechanics of the spine and the science of soft tissue healing, has resulted in many changes in regard to management of the low back problem in society, and particularly in industry. Numerous other additional influences, such as the psychological state of the individual, socioeconomic factors influencing his or her

Table 14–5 Matching the Objectives of Treatment to a Patient Classification System

	Patient Classification		
Objective	*Acute*	*Reinjury*	*Chronic*
Pain modulation/analgesia	XX	X	X
Application of controlled forces to enhance movement	XX	X	—
Enhance neuromuscular performance	—	XX	XX
Biomechanical counseling	X	XX	X

Note: XX = Strongly indicated and serves as treatment goal; X = occasionally indicated but does not serve as treatment goal; — = Not typically indicated or a treatment goal.

Source: Adapted from DeRosa, C.P. and Porterfield, J.A., A Physical Therapy Model for the Treatment of Low Back Pain, *Physical Therapy*, Vol. 72, pp. 261–272, with permission of the American Physical Therapy Association, © 1992.

perception of the problem, and satisfaction with the work environment, have only recently been given due recognition. Thus the model of care has expanded from the traditional physical sciences to also include the biological, psychological, and social sciences. This approach has resulted in more realistic intervention strategies for the management of the back problem in industry. This more global approach has also required that we view the problem not only from the perspective of the patient and the clinician, but also from the perspective of industry itself.

The key to the successful management of the indemnity claim is to recognize the extent of the problem that faces industry. The problem has three basic categories. The first is maintaining and controlling the workers' compensation claim from the initiation of the claim to the period of return to work. At the present time the injured employee often seeks care for an injury without the required information to determine if the treatment plan is appropriate. The industry usually does not have the internal communication system to capably manage the process.

The next aspect of the problem pertains to reduction of the indemnity claim. The current plan used in the state of Ohio serves as an example. The indemnity claim is a claim that is eight days in duration. In the state of Ohio the injured employee does not get compensated for the first seven lost days. Once the eighth lost day is reached, the injured employee becomes compensable. He or she gets compensated for days 8, 9, 10, 11, 12, 13, and 14. Once the 15th day is reached, the injured employee receives compensation for the first seven days of his or her injury. One can easily see the predicament created for both the industry and the patient.

Developing a network of medical services that meets the need of industry forms the basis to minimize the chance of reaching the eighth day of injury. Most musculoskeletal injuries respond well enough in the first week to return to at least a modified work level. If the industry can develop a system that can work in concert with the medical profession, then the industry can make great strides to reduce this aspect of the overall problem.

The last and certainly not the only problem is the escalating health care costs. The medical profession should realize the difficulty that industry is facing with the significant increase in overall business costs. The most frustrated group of all those involved in managing the injury claim is the industry, because it is responsible for the bills and perceives that it has no control.

Practitioners should provide services that will assist industry in developing a structured plan for program development geared toward shifting the responsibility of control from the medical profession to industry, and to develop programs that can address the problems.

Shifting the responsibility of control may seem like a radical thought, but recognizing the similarities between industry and medicine make this an attractive alternative. The similarities that exist between most industries and the medical system are quite evident. Understanding these similarities and being able to work them into a presentation that defines the particulars has proven to be a successful way to develop and enhance the interest of industry. It is our contention that both medicine and industry are attempting to improve their respective services and procedures for the injured worker by establishing a standardized evaluation and treatment model. This model has been discussed in this chapter and has been successfully marketed and implemented in the industrial setting.

Another similarity between medicine and industry is that both conduct a people-oriented business. Both continually strive to enhance internal and external communication systems that are required to assure quality and growth. Both medicine and industry recognize and fully understand the need for strategic and long-range planning, and both work to develop an environment that readily adapts to change. Change is necessary, and if the respective business is not set up to accommodate change, growth potential is hampered.

Both medicine and industry strongly advocate continuing education and staff development and training, and recognize that growth is a product of gradual perceptual change. With these similarities as the basis, they need to work together to develop and implement programs that will serve not just themselves but the public.

PROGRAM DEVELOPMENT: NETWORKING MEDICAL SERVICES AND THE ORGANIZATION

It appears that those in physical therapy, orthopedics, and family practice medicine who network themselves are the ones that are best aligned to succeed in this rapidly changing business of industrial medicine. The following represents the basis for such a program. It should be recognized that this is one of many plans available, but we are convinced that it is significant enough to be carefully scrutinized by our profession.

The plan has as its basis the concept of the physical therapist as the director of communications. Communication is with the patient, between the referral source and the rehabilitation counselor, with the industrial contact, and with the manager at the respective industry. The physical therapist is best positioned and qualified to carry out this task, and this represents a significant area of potential growth within the profession.

The first aspect of gaining control of the injury claim is to recognize who currently actually has control of the system—the patient. If a patient returns to the physician's office and the doctor suggests returning to work the next day, the patient simply replies that his back continues to give some difficulty and he does not think that he will be able to go back and do his job and may become reinjured. If the physician holds to the decision and sends the patient back to work, the patient is initially upset, and if he becomes reinjured (which is likely because from all accounts in the literature the unhappy employee is the most vulnerable), then the patient directs his anger toward the physician who returned him to work against his will. The patient/employee will most likely not return to that physician. Therefore the physician most likely makes the decision to keep the injured employee away from work for a predetermined time or until the patient is ready to return to work.

If patients do not like their work environment, then there is no real incentive to return. It is the successful program that recognizes the potential for flexibility on both the medical and the industrial side, with the well-being and respect of the patient held in the balance.

The key to the program is a multifaceted education program—education for management that explains the process, and for the supervisor who has first interaction with the patient and who keeps in contact with the injured employee while he or she is away from work. The patient education program should begin with information regarding the structure and function of the body, how the body breaks down with gradual consistent overuse and rapid impact, and how the body heals and responds to appropriate treatment. The patient education program should continue to include information that permits patients to make the decision as to whether they are receiving appropriate health care.

Patient education should allow the patient to answer the following important questions:

1. Do you have an understandable diagnosis?
2. Is there active progressive rehabilitation initiated on the day of injury or the next day?

3. Are there understandable short-term and long-term treatment goals that are reachable and realistic?
4. Do you feel as if you are an integral part of the treatment and goaling process?

If the answer to these questions is no, then the injured employee is most likely not receiving the appropriate health care that is geared to an early, safe return to work.

As the program continues to develop within the industrial setting, the clinician should review and if necessary assist in the development of an injured employee procedure policy. This procedure policy will vary from industry to industry, but as the network of medical services becomes defined, the clinician who can combine the industrial perspective with the medical network is the most likely to succeed.

The director of communications should then develop/improve/implement health education programs for the employees. These programs should have a global perspective, addressing such factors as fatigue, weight gain, fat in the diet, eating a variety of foods, providing proper hydration for the body, diminishing the smoking habit, obtaining proper rest behaviors, and exercise. The communication director should provide the necessary information to make wise decisions with regard to realizing positive outcomes.

It is beginning to be recognized that industrial involvement in the development of the medical staff is just as important as medical involvement in industrial education. Learning, growing, and succeeding together is the goal, and the physical therapist possesses the tools and the expertise to develop and implement such a program. Decreasing the incidence of low back pain and related industrial injuries is one avenue that should be pursued in the industrial setting, but the networking of medical services and the organization of communication and education within industry represents a definitive direction for the growth of our profession.

CONCLUSION

The current fractionization of the health care dollar for the various practitioners in the rehabilitation arena will not be tolerated by a system predicted to fall far short of the necessary dollars in the future. Profound changes are imminent. A careful analysis of some of the various rehabilitation participants—physical therapists, physicians, chiropractors, occupational therapists, athletic trainers, and exercise physiologists—reveals that significant overlap exists among the professions' treatment approaches and techniques. In the past, "turf" was protected via historical precedent and state law. The reality of monetary limitations available for rehabilitation in the future, however, invites speculation that an "intersection" of the various professions will occur, with a resultant blurring of the approaches utilized by the individual professions.

More than any other area of the musculoskeletal system, the low back has been the recipient of numerous treatment approaches. It is reasonable to conclude that part of the blame for escalation of the low back problem unwittingly falls on the health care system itself. Consider the bewilderment of the patient with an acute onset of simple low back pain: if the patient was seen by three different practitioners within a three-month period, he or she would most likely be given three different diagnoses and three different methods of management. What is most likely a self-limiting problem begins to elicit much greater concern in the patient's mind because of the perceived differences and uncertainty in management. Instead of helping to curtail the low back dilemma, health care professionals, with the best intentions, unknowingly contribute to its growth. This iatrogenicity of low back pain has only recently been increasingly recognized.

After careful review of the myriad treatment techniques, four objectives of treatment have been suggested as the categories under which treatment techniques could be placed. A focus on these objectives of treatment, rather than treatment techniques themselves, coupled with an understanding of the importance of a successful goal-setting process for both the patient and industry, offers the best hope for efficient management of the back pain problem in industry.

NOTES

1. American Physical Therapy Association, *APTA Delegates Handbook* (Alexandria, Va., 1992).
2. H. Aaron and W.B. Schwartz, Rationing Health Care: The Choice before Us, *Science* 247 (1990):418–422.
3. McManis Associates Inc., Technology 2000: A Window to the Future, *Health Care Competition Week,* 6 February (1989): 1–4; National Center for Health Statistics, *Physiotherapy Office Visits: National Ambulatory Medical Care Survey;* Vital and Health Statistics, no. 120, DHHS Publication #(PHS) 86-1250 (Hyattsville, Md.: Public Health Service, 1986).
4. G. Waddell, A New Clinical Model for the Treatment of Low Back Pain, *Spine* 12 (1987):632–644.
5. J. Frymoyer, Magnitude of the Problem, in *The Lumbar Spine,* ed. J.N. Weinstein and S.W. Wiesel (Philadelphia: W.B. Saunders, 1990), 33.
6. Waddell, New Clinical Model; J.W. Frymoyer et al., Risk Factors in Low Back Pain: An Epidemiological Study, *Journal of Bone and Joint Surgery* 65A (1983):213–218.
7. H.O. Svensson and G.B.J. Andersson, Low Back Pain in 40–47 Year Old Men: I. Frequency of Occurrence and Impact on Medical Services, *Scandinavian Journal of Rehabilitation Medicine* 14 (1982):47–53.
8. G.B.J. Andersson, The Epidemiology of Spinal Disorders, in *The Adult Spine,* ed. by J.W. Frymoyer (New York: Raven Press, 1991), 107.
9. J. McCulloch, Costs: Surgery vs. Chemonucleolysis. *Alternatives in Spinal Surgery* 2 (1985):3–4.
10. McManis Associates Inc., Technology 2000.
11. N.M. Hadler, The Language of Diagnosis, in *Clinical Concepts in Regional Musculoskeletal Illness,* ed. N.M. Hadler (Orlando: Grune & Stratton, 1987), 3–24.

12. G. Waddell and C.J. Main, Assessment of Severity in Lowback Disorders, *Spine* 9 (1984):204–208.
13. J.L. Kelsey, *Epidemiology of Musculoskeletal Disorders* (New York: Oxford University Press, 1982), 145–167.
14. S.H. Snook and R.C. Jensen, Cost, in *Occupational Low Back Pain,* ed. M.H. Pope et al. (New York: Praeger, 1984), 115–121.
15. D.M. Spengler et al., Back Injuries in Industry: A Retrospective Study. I. Overview and Cost Analysis, *Spine* 11 (1986):241–245.
16. D.C. Turk et al., An Empirical Examination of the "Pain Behavior" Construct, *Journal of Behavioral Medicine* 8 (1985):119–130.
17. P.M. Ellwood, Shattuck Lecture. Outcomes Management: A Technology of Patient Experience, *New England Journal of Medicine* 318 (1988):1549–1567.
18. W.L. Cats-Baril and J.W. Frymoyer, Identifying Patients at Risk of Becoming Disabled Due to Low Back Pain: The Vermont Rehabilitation Engineering Center Predictive Model, *Spine* 16 (1991):605–607.
19. V. Mooney, The Facet Syndrome, in *The Lumbar Spine*, ed. J.N. Weinstein and S.W. Wiesel (Philadelphia: W.B. Saunders, 1990), 438.
20. Quebec Task Force on Spinal Disorders, Scientific Approach to the Assessment and Management of Activity-Related Spinal Disorders, *Spine* 12, no. 7, suppl. (1987):S1–S59.
21. R.W. Fry et al., Overtraining in Athletes, *Sports Medicine* 12, no. 1 (1991):32–65.
22. J.A. Porterfield and C. DeRosa, *Mechanical Low Back Pain: Perspectives in Functional Anatomy* (Philadelphia: W.B. Saunders, 1991).
23. M. Battie et al., The Role of Spinal Flexibility in Back Pain Complaints within Industry: A Prospective Study, *Spine* 15 (1990):768–773.
24. Waddell, New Clinical Model.
25. M.A. Adams and W.C. Hutton, The Effect of Posture on the Role of the Apophyseal Joints in Resisting Intervertebral Compressive Forces, *Journal of Bone and Joint Surgery* (Britain) 62B (1980):358–362.
26. G. Mellin, Correlations of Hip Mobility with Degree of Back Pain and Lumbar Spinal Mobility in Chronic Low-Back Pain Patients, *Spine* 13 (1988):668–670.
27. S.R. Garfin and H.N. Herkowitz, Disc Disease—Does It Exist? in *The Lumbar Spine*, ed. J.N. Weinstein and S.W. Wiesel (Philadelphia: W.B. Saunders, 1990), 369–381.
28. A.L. Nachemson, The Load on Lumbar Discs in Different Positions of the Body, *Clinical Orthopedics* 45 (1966):107.
29. R.A. Deyo et al., How Many Days of Bedrest for Acute Low Back Pain? *New England Journal of Medicine* 315 (1986):1064–1070.
30. M.M. LaBan, "Vespers Curse" Night Pain—The Bane of Hypnos, *Archives of Physical Medicine and Rehabilitation* 65 (1984):501–504; M.M. LaBan and D.F. Wesolowski, Night Pain Associated with Diminished Cardiopulmonary Compliance, *American Journal of Physical Medicine and Rehabilitation* 67 (1988):155–160.
31. W.W. Parke, The Significance of Venous Return Impairment in Ischemic Radiculopathy and Myelopathy, *Orthopedic Clinics of North America* 22 (1991):213–221.
32. J.A. Hoyland et al., Intervertebral Foramen Venous Obstruction: A Cause of Periradicular Fibrosis, *Spine* 14 (1989):538–568.
33. H. Weber, Lumbar Disc Herniation: A Controlled Prospective Study with Ten Years Observation, *Spine* 8 (1983):131–140.

Chapter 15

Physical Therapy Early Intervention

Robin Saunders

THE PHILOSOPHICAL SHIFT IN TREATMENT: PASSIVE THERAPY IS OUT!

Since its beginnings in the 1920s, the physical therapy profession has changed rapidly. The role has changed drastically from that of a technician to that of a skilled evaluator who is qualified to determine appropriate treatment and rehabilitation for patients. Historically, though, there is a questionable track record when treating some musculoskeletal injuries, in particular back injuries. Along with other members of the medical profession, therapists have tended to focus on passive treatments such as medication, bed rest, and passive modalities aimed at pain relief.

In a very critical look at the literature supporting these treatments, the Quebec Task Force found that the variety of treatments widely used for low back pain have little scientific basis. This important study found that exercise and patient education are the treatments that consistently work in the long term.[1] Nachemson also criticizes most of the traditional forms of lower back pain treatment, instead advocating a more active treatment approach consisting of exercise.[2]

Deyo et al.'s study comparing two days and seven days bed rest for acute low back pain strongly suggests that the medical profession is generally too passive and careful with the acute back pain patient.[3] Their study found that patients with shorter periods of bed rest had better long-term results.

Despite the evidence described above, one still sees bed rest and passive modalities prescribed for several days or weeks for acute low back pain. In fact, some physicians inappropriately call this "conservative care," trying more invasive forms of treatment such as epidural steroid injections or surgery when "conservative care" fails. Actually, "conservative care" consisting of exercise, education, and functional training has not been tried at all. Perhaps if these things had been tried, the physician would have found that the more invasive treatments would not have been necessary.

The Quebec Task Force Study provides important information concerning the treatment of workers with low back pain. Its findings must be analyzed carefully, however, to avoid misinterpretation. For example, of all Quebec workers who were disabled with back pain in 1981, 76 percent were off work less than one month and accounted for only 7.6 percent of total workers' compensation expenses. On the other hand, 7.4 percent were off work longer than six months and accounted for 76 percent of all costs.[4] With this in mind, one might assume that we should concentrate our efforts on the small percentage of low back pain patients who have the complicated, chronic problems. After all, these few patients are the most costly to the system.

If we look at the problem a different way, however, we may find a better solution. Consider these questions: Do the patients with complicated back problems start out with minor, simple back problems? And do they become complicated problems because they were not treated aggressively when they were "simple" problems? In other words, could we prevent a minor back problem from becoming a chronic problem through early, effective treatment? We can if we direct our attention to patient education, exercise programs, and other rehabilitation techniques instead of concentrating only on the acutely painful episode.

Another fact that needs to be discussed is the recurrence rate in low back pain sufferers. The Quebec Study reported that approximately 87 percent of patients who suffer one episode of low back pain recover spontaneously in a short time, but that the recurrence rate for these patients is significantly greater than for those who have never suffered an episode.

Perhaps physical therapists and other medical practitioners need to shift their emphasis from treating a current episode via pain relief treatments to preventing the next episode, which occurs in many of these patients' cases. The recurrence rate for low back pain is so high because the painful episode is mistakenly treated as if it were a single-event injury. The fact is that most back injuries are a result of the cumulative effects of poor posture, faulty body mechanics, stressful living and working conditions, and a general decline of physical fitness. Repetitive and prolonged positions at work, poor ergonomic conditions, and attitudes about back injuries also contribute to their occurrence. Therefore most low back injuries, excluding those obviously caused by trauma, should be considered cumulative trauma injuries and be treated as such.

ACTIVE TREATMENT FOR CUMULATIVE TRAUMA INJURIES

Treatment that focuses on prevention is not a new concept in medicine. Consider how heart disease is currently treated. A patient may not be aware of an existing heart disorder until he or she experiences an episode such as angina, shortness of breath, or a cardiac arrest. Once the symptomatic episode is relieved, however, the factors leading to heart disease are still present unless aggressively managed. These factors may include poor diet, high stress level, high blood pressure, and obesity. The total management of a cardiac patient involves treating all of these factors, not just the symptomatic episode (Figure 15–1).

The medical community should treat back injuries and other cumulative trauma injuries in exactly the same fashion. In other words, the back pain episode should simply be a warning sign that a back problem is in existence. Then management should focus on managing and decreasing the risk factors associated with a back disorder (Figure 15–2).

Common sense tells us that aggressive rehabilitation addressing the causative factors of cumulative trauma places the individual at lower risk of reinjury. Certainly Lindstrom's study of Volvo workers in Sweden is convincing.[5] Workers who had been off work for eight weeks were assigned to either a control group or an activity group. The control group continued to treat with their own physicians. The other took part in an activation program consisting of functional capacity testing, a workplace visit, patient education, and an aggressive physical conditioning program. The results were impressive. On the average, the study group went back

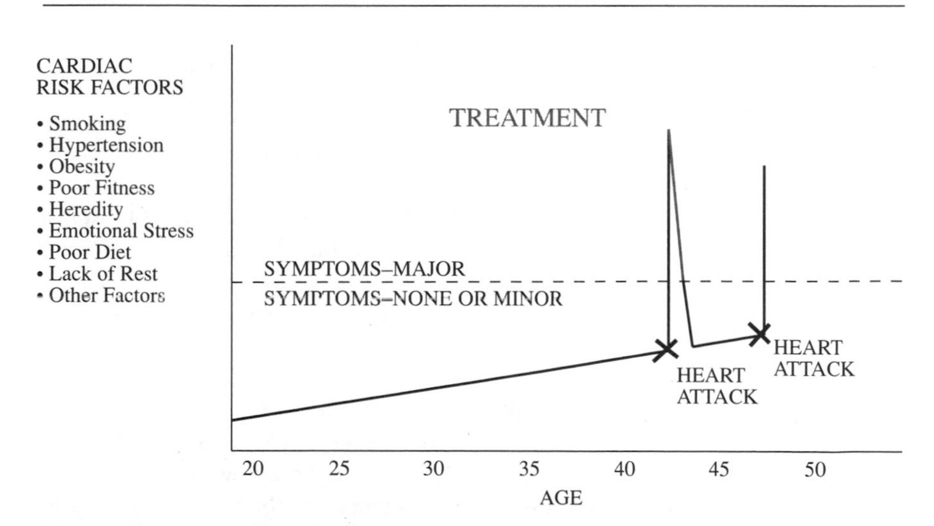

Figure 15–1 The Relation between Risk Factors and Heart Attacks. The list on the left represents risk factors for cardiac problems. The problems can be slowly worsening with time, but the individual has not yet experienced major symptoms. A heart attack occurs. Treatment for the acute heart attack may be successful, but if long-term management does not address the risk factors, the cardiac problem is still present and may continue to worsen, eventually leading to another heart attack.

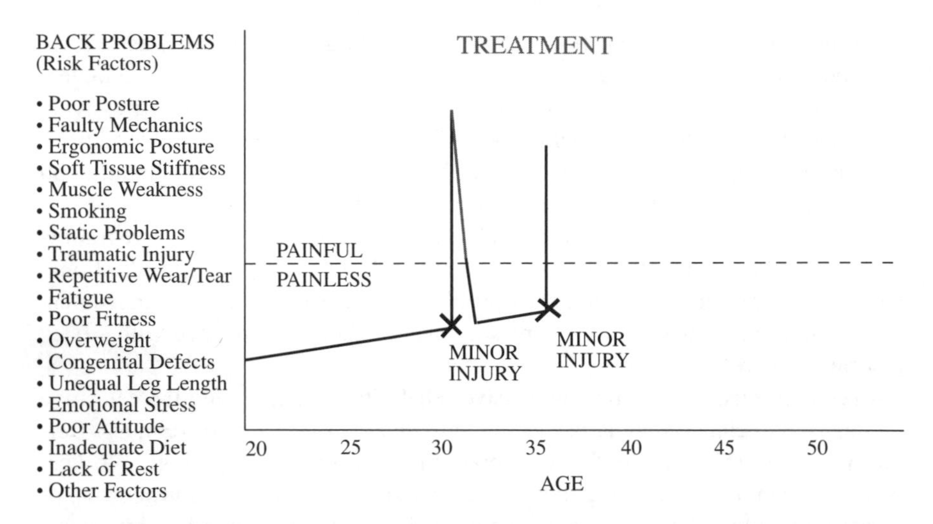

Figure 15–2 The Relation between Risk Factors and Back Problems. The list on the left represents risk factors for back problems. The problems can be slowly worsening with time, but the individual has not yet experienced major symptoms. A minor incident occurs that causes pain. Treatment directed toward pain relief effectively alleviates the pain. However, if the treatment does not address the risk factors, the back problem is still present and may continue to worsen, eventually leading to a major disorder. *Source:* Reprinted from *Evaluation, Treatment and Prevention of Musculoskeletal Disorders*, Vol. 1, 3rd ed., p. 375 by H.D. Saunders and R. Saunders, with permission of Educational Opportunities, © 1993.

to work five weeks sooner and had 7.5 weeks less loss time due to back pain over the next two years.

Treatment for an injured ankle, knee, or shoulder commonly consists of rest in a position of biomechanical neutrality for a period of time. Except in the case of a severe injury, complete rest for more than a day or two is rarely advised. In fact, patients with these conditions are well acquainted with the harmful effects of immobility ("stiffening up") and are usually open to the idea of stretching even if mild to moderate pain exists.

On the other hand, perhaps because of some of the misconceptions and fear surrounding back disorders, patients tend to be less willing to move through some of the pain that mild exercise causes in the subacute stages of low back injury. One of the roles of a physical therapist is to teach patients exactly what movements to avoid and what movements are most appropriate or safe. We should also teach the patient how much pain and what quality of pain is acceptable. We should instruct them to exercise noninvolved body parts to avoid the deconditioning that inevitably occurs when a patient is recuperating from an acute painful episode. We must teach them exactly what biomechanical neutrality is so that they do not make the common mistake of resting their backs by lying on a soft couch or in a La-Z-Boy recliner.

The patient who is pronounced "cured" after the painful episode has subsided has been done a disservice if he or she is returned to full activity without adequate rehabilitation. The effects of immobilization on soft tissue are well documented.[6] If a patient has been relatively inactive during the recuperating process, we should assume that deconditioning has taken place. Active exercise is necessary to restore the spinal structures to their preinjury condition. This must be a minimum requirement, since often the cause of the injury was preinjury imbalances in strength or flexibility. Patients who have a history of minor sprains and strains that "get better" but recur intermittently must be taught that each painful episode was not a distinct occurrence but an accumulation of a variety of risk factors. Without treating their *real* problems—poor posture, faulty body mechanics, stressful working and living conditions, and so forth—the patient will never be "cured."

Rehabilitation of back injuries is based on commonsense principles and the sound knowledge of normal physiology. Historically, we have learned our lesson when treating other areas of the body. For example, an elderly person with a hip fracture used to rest in bed longer than is currently recommended. We found that the complications of bed rest were serious, sometimes fatal. We found that by encouraging early movement, we actually caused the fracture to heal more quickly. The same is true in the case of postsurgical or postchildbirth patients. Early activation decreased the complications and improved the final outcome. We are finally learning this lesson with back patients as well.

EARLY INTERVENTION TECHNIQUES

Individual Exercise

A common mistake that many physical therapists make is giving patients the impression that their exercise program is standard, not unique. If a patient feels that he or she has been given the same exercise program that every other back patient has, he or she will not be motivated to perform it. Imagine going to the doctor and receiving the impression that the physician gave you the same drug prescription he or she gave everyone. You would quickly lose faith in that physician. It is no wonder then that patients respond poorly to standardized exercise programs and very well to individualized exercise.

Professionally drawn or preprinted exercise diagrams can be used, but only if it is clear to the patient that the choice of exercises was specific and individualized. In other words, if the patient does not have weak abdominal muscles, abdominal exercises should not be stressed. If the patient does not lack rotation flexibility, rotation exercises should not be stressed.

The skilled therapist will discover as much information as possible about the patient's "exercise ethics." In other words, how much has this patient been willing to exercise in the past? If the patient has a history of being fairly physically active and performing exercise on his or her own, that patient is more likely to follow through with the therapist's instructions. If, on the other hand, the patient has a history of being quite sedentary, he or she may need more encouragement and explanation regarding the beneficial effects and the absolute necessity of following through with the exercise program. For those patients who clearly do not like to exercise, the exercise program should be kept short and to the point. The patient must not be given any exercise that is not absolutely essential. Then the therapist should make it clear to the patient that the exercises have been minimized as much as possible, but that the patient is expected to perform these essential exercises.

The clinical use of special equipment that is unavailable to the patient elsewhere should be minimized. The therapist should try to use equipment such as free weights, mats, and elastic tubing, and should emphasize performing aerobic activities that can be continued at home, such as walking. Persistence, not high technology, is the key. Fancy equipment may encourage patients to keep their appointments, but it will not keep the patients exercising once they have been discharged to a home program. The therapist who uses such equipment may inadvertently be giving patients a message of dependence, and patients may feel that they are not capable of continuing the rehabilitation on their own at home.

Each patient varies in his or her response to exercise. Some patients with severe objective findings can actually progress quite rapidly. On the other hand, some patients with minimal objective findings have trouble performing the simplest exercise. The therapist should evaluate each case independently to find the right combination of exercises for each individual.

The therapist should pay close attention to signs and symptoms and use caution when progressing exercises in the patient who has peripheralization of symptoms (symptoms referring into the extremities). Often a change in the exercise technique or trying the same exercise in a different way (a hip flexor stretch performed while kneeling instead of supine, for example) makes the exercise more tolerable.

In some cases, it is appropriate to progress the patient's program in spite of symptom response. For example, performing a straight leg raise may cause increased peripheralization of symptoms in the case of an adherent nerve root or certain chronic conditions. The experienced therapist will consider the pathology and carefully weigh symptom response versus objective improvement to decide when and how aggressively to progress the exercises.

Manual Therapy and Modalities

The use of passive therapy, including manual therapy and modalities, has been deemphasized in this chapter. Obviously, however, there is a place for such treatment techniques. In the case of facet joint impingement, chronic stiffness, or acute spasm, manual therapy and/or modalities are very much indicated. However, from the very beginning the patient should be told that these treatment techniques are beneficial because they will help the patient to become more independent. Emphasis should be on the "real" long-term treatment, which, of course, is exercise and lifestyle change.

For most musculoskeletal injuries, one should see a positive response to manual therapy or modalities within three to six treatments. Continuing manual therapy or modalities for more than a month is rarely indicated, even if more active forms of treatment are taking place concurrently.

Exercise should begin as soon as possible, often as soon as the first day of treatment. If the patient's condition is acutely painful, the first exercise may be as simple as gentle range of motion in a short arc, or movement of an uninvolved body part.

Once exercise has begun, it is normal for patients to have good days and bad days. When the patient arrives at the clinic having a bad day, it is tempting for the therapist to provide electric stimulation or other modalities for pain relief. Therapists should be wary of sending a mixed message, however. It is sometimes better to first try to find an exercise or position that relieves the symptoms because this helps persuade the patient that self-management of the problem really is possible.

Functional Activities

Functional activities are important to include in the patient's exercise regime. The patient does not have to be enrolled in a work-hardening or work-conditioning program to practice lifting, bending, or other relevant activities. Physical therapists who treat sports injuries are well acquainted with the importance of functional outcomes. Functional activities such as throwing, kicking, and twisting are incorporated into sports rehabilitation programs, and the exact activities are individualized depending on the sport. Physical therapists who treat work-related injuries should treat patients the same way.

Very early on, therapists should be discussing with their patients the correct and incorrect ways to perform activities of daily living (Figure 15–3). We often forget that our patients have very busy lives and that such activities as doing laundry and grocery shopping cannot be postponed. We cannot postpone teaching patients how to do these activities safely.

WORK-HARDENING AND WORK-CONDITIONING PROGRAMS—WHERE DO THEY FIT IN?

It is not the purpose of this chapter to discuss work-hardening and work-conditioning programs in depth. This is covered in detail in other areas of this book.

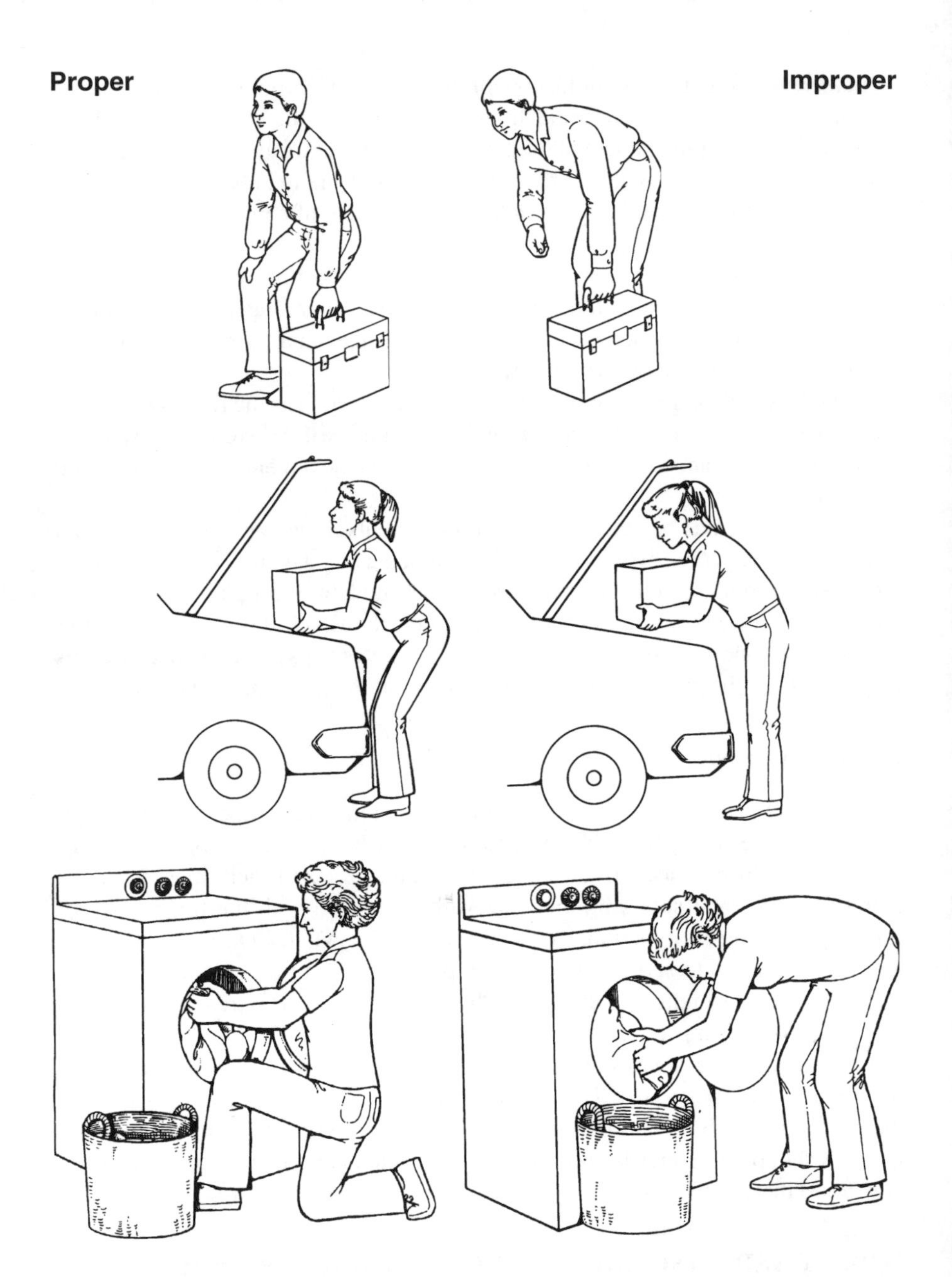

Figure 15–3 Examples of Proper and Improper Technique for Common Activities of Daily Living. *Source:* Reprinted from *Evaluation, Treatment and Prevention of Musculoskeletal Disorders*, Vol. 1, 3rd ed., p. 344, by H.D. Saunders and R. Saunders, with permission of Educational Opportunities, © 1993.

Work-hardening and work-conditioning programs are mentioned here briefly to discuss how to prevent the need for them. In other words, if we are treating our patients appropriately, aggressively, and early, the need for work hardening and work conditioning as a response to chronic condition should be minimal.

In work-hardening centers, one often sees patients who would not need these services if they had been treated more aggressively early on. Unfortunately, many of our patients relate a history of having received hot packs, ultrasound, and massage for several weeks or months, but when asked about their exercise program barely remember which exercises they were to have performed. Some patients have not even received exercise.

This is not to say that acute care can overcome all back problems and that work-hardening and work-conditioning programs will never be necessary. Even when good care is given from the very beginning, work-hardening and work-conditioning programs can be crucial for the complete rehabilitation of the patient. However, if appropriate early intervention takes place, the patients who eventually need work-hardening and work-conditioning programs should be fewer. This should be the goal.

ASSURING QUALITY

In the busy physical therapy practice, quality improvement sometimes gets neglected. The clinic or hospital's administrators and supervisors may support the concept of aggressive early intervention, but there is no system in place that ensures that it is actually happening. When the clinic is understaffed and therapists are overbooked, it is easy for them to slip into the habit of providing modalities rather than education or exercise. A good quality improvement program prevents this from happening by clearly defining goals and expectations, and continually monitoring how each therapist is performing relative to these goals.

Table 15–1 summarizes Saunders Therapy Centers' quality improvement (QI) philosophies for our orthopedic outpatient clinic. The objectives that have been chosen to monitor are those that represent minimum quality standards for the type of services provided and the market in which they are provided. Objectives are structured to emphasize the following:

1. good documentation and communication with referral sources
2. a philosophy that encourages active versus passive therapies
3. setting realistic functional goals
4. acceptable outcomes (meeting the goals)
5. satisfied customers (patients, referral sources, and payors)

In addition to measuring these objectives, statistics should be gathered relating to the average number of visits per patient referral and the average charge per visit. These data, along with the data generated from the quality improvement program,

Table 15–1 Objectives for a Quality Improvement Program in an Orthopedic, Outpatient Physical Therapy Clinic

Orthopedic Physical Therapy Quality Improvement Objectives

Objective	*Measure*	*Applicants*	*Time*	*Source*	*Goal*
Maximize timeliness of written communication	Percent of notes sent out per schedule	All patients	Quarterly	Flowsheet	85%
Maximize clear goal setting in initial evaluation and discharge notes	Percent of initial and discharge notes with adequate documentation of goal setting	Random sampling	Quarterly	Notes	90%
Maximize achievement of goals set at time of initial evaluation	Percent of charts displaying documentation of achievement of goals	Random sampling	Quarterly	Notes	85%
Provide adequate documentation supporting use of treatment modalities	Percent of charts with adequate documentation supporting use of modalities	Random sampling	Quarterly	Notes	70%
Maximize patient's feeling of knowledge and ability to manage own symptoms	Percent of patients indicating they learned to effectively manage own symptoms	All patients complete questionnaire	Quarterly	Questionnaire	90% 60% return rate required
Maximize patient satisfaction	Percent of patients indicating satisfaction	All patients complete questionnaire	Quarterly	Questionnaire	90% 60% return rate required
Maximize referral source satisfaction	Percent of referral sources indicating satisfaction	Sources complete questionnaire	Yearly	Questionnaire	90%

Note: The complete quality improvement program includes criteria for meeting each objective, and detailed instructions for data collection and report preparation.

Source: Reprinted with permission from Saunders Therapy Centers, *PA Employee Manual*, Minneapolis, Minn., 1992.

provide valuable information that can be used in a number of ways, including evaluating employee performance, marketing, and negotiating insurance contracts.

The objectives are relatively easy to measure, although there is some increased work required by the clerical and administrative staff. Results should be compiled quarterly. Clinics will undoubtedly realize the following benefits after initiating a formal quality improvement program.

1. *Marketing based on service and outcome:* When explaining philosophy and methods to potential referral sources, treatment centers can actually quote statistics rather than vague generalities. For example, a goal is that referring physicians will receive initial evaluation reports within six working days. Insurance claims adjusters need to know that over 90 percent of patients learn to self-manage their symptoms. Referral sources and payors never fail to be impressed with the dedication shown to quality improvement. QI programs alone may be the deciding factor in winning major HMO and workers' compensation insurance contracts.

2. *Improved productivity and retention of staff:* Job satisfaction is often linked directly to dedication to quality improvement. This manifests itself in turnover rates much lower than industry average.

3. *Objective feedback for the staff:* If QI programs are structured to truly measure what the staff determines is important, the quarterly QI reports provide a consistent, systematic way for the staff to determine how they are performing. Meeting most objectives requires a team effort. For example, in order for written communication to be timely, the therapist, the transcriptionist, and the secretary all have to do their parts. Many factors affect patient satisfaction—perceived outcome, the friendliness and professionalism of the staff, and the physical environment, to name a few. Therefore a positive quarterly report affects all staff members and makes the working atmosphere one of "pulling together" to achieve a common goal.

4. *Taking the guesswork out of management:* Managers who receive quarterly reports that are easy to read and truly reflect what is happening in the clinic on a daily basis are capable of more effective clinic administration. For example, at Saunders Therapy Centers, the management team never really knew how to measure quality prior to initiating our quality improvement program. The management style was one of "putting out fires" as they occurred. If negative feedback was not received, there was an assumption that everything was going well. When a complaint was received, there was a strong reaction, with an uneasy feeling about the solution. Did the solution really fix the problem, or were there other undiscovered problems? With the QI program in place, the management team and staff feel secure. We feel the objectives accurately measure our goals, and there is a system in place to measure performance relative to the objectives. Therefore there is not a question about quality.

5. *Better preparation for future changes.* The medical marketplace is changing rapidly. Increased managed care, increased government involvement, and tougher utilization review are realities that have to be faced now and in the future. Clinics must be confident that they will be able to pass the strictest quality improvement and cost-containment policies promulgated by managed care plans, governmental agencies, or utilization review companies in the future. The hard work is setting up internal data collection and quality management programs—if this is done, no one else will ask more from practitioners than they have already demanded from themselves.

CONCLUSION

The implementation of a quality early intervention program follows these principles:

1. Passive therapy should be used only as an adjunct to active therapy.
2. Treatment of cumulative trauma injuries should be similar to treatment of heart disease (prevention oriented).
3. Exercise and education are the keys to successful treatment of cumulative trauma injuries.
4. Exercise programs must be individualized for each patient.
5. Functional activities should be incorporated into each patient's program from the beginning.
6. Modalities and manual therapy play an important role in patient treatment, but care must be taken to avoid overutilization.
7. Appropriate early intervention should decrease the necessity of work-hardening and work-conditioning programs, since clients will return to work early.
8. The physical therapy practice must have a formal method of assuring quality.
9. An effective quality improvement program assists in nearly every aspect of clinical management, including performance evaluation, marketing, and negotiating.

Employers, insurance adjusters, medical providers, and rehabilitation consultants all play an important role in the early intervention process. Furthermore, we cannot neglect the most important member of the team—the injured worker. For early intervention to succeed, the employee must take an active part in the prevention process, report problems early, and follow through with medical advice. The employer must provide an atmosphere in which this can occur. The insurance adjuster must support and facilitate the employer's efforts and strive for a nonadversarial relationship when problems do occur. The medical provider must provide appropriate early treatment and communicate effectively with all involved parties. Rehabilitation consultants can bring the team together by facilitating communication and solving problems when they do arise.

Physical therapists possess the knowledge and ability to appropriately treat musculoskeletal cumulative trauma injuries and be essential members of the above team. The focus is on exercise and education. Patients need to know how to help themselves and decrease their dependence on passive treatments such as bedrest, medication, modalities, and surgical intervention. Experience has shown that early intervention is successful when team members work together to accomplish their mutual goals.

NOTES

1. Quebec Task Force on Spinal Disorders, Scientific Approach to the Assessment and Management of Activity Related Spinal Disorders, *Spine* 12, no. 7, suppl. (1987):S1–S59.
2. A. Nachemson, Newest Knowledge of Low Back Pain: A Critical Look, *Clinical Orthopaedics and Related Research* 279 (1992):8–20.
3. R. Deyo et al., How Many Days of Bed Rest for Acute Low Back Pain? *New England Journal of Medicine* 315 (1986):1064–1070.
4. Quebec Task Force, Scientific Approach.
5. L. Lindstrom et al., The Effect of Graded Activity on Patients with Subacute Low Back Pain: A Randomized Prospective Clinical Study with an Operant-Conditioning Behavioral Approach, *Physical Therapy* 72, no. 4 (1992):279–293.
6. V. Janda, Muscles, Central Nervous Motor Regulation and Back Problems, in *The Neurological Mechanisms in Manipulative Therapy*, ed. I.M. Korr (New York, Plenum Press, 1978); W.H. Kirkaldy-Willis, *Managing Low Back Pain*, 2nd ed. (New York: Churchill Livingston, 1988).

Chapter 16

The Role of Chiropractic in Return to Work

Joseph J. Sweere

The purpose of this chapter is to describe the role doctors of chiropractic (DCs) currently assume in returning workers to suitable gainful employment following workplace injury and illness. Chiropractic physicians function in the role of primary care, portal-of-entry caregivers for ill and injured workers and have done so throughout the profession's existence. Contemporary chiropractic care includes differential diagnosis, treatment, and physical rehabilitation of a wide variety of work-induced disorders. A significant amount of work-related injury involves the musculoskeletal system, particularly the spine, most often caused or precipitated by manual material handling, repetitive-motion-associated trauma, and vibration-related injuries.[1] The Bureau of Labor Statistics reports that approximately one million workers sustained back injuries in 1980, representing one of five injuries and illnesses in the workplace.[2] DCs' education, training, and clinical practice place significant emphasis on the diagnosis and clinical management of spinal and related neuromusculoskeletal disorders.[3]

THE SUBLUXATION COMPLEX

Subluxation is defined as an aberrant relationship between two adjacent articular structures that has functional or pathological sequelae, causing an alteration in the biomechanical and/or neurophysiological reflections of those articular structures, their proximal structures, and/or other body systems that may be directly or indirectly affected by them.[4] Spinal subluxation not only has a variety of physical and clinical manifestations but also is often visualized on neutral, static, erect-posture radiographs (Figures 16–1, 16–2, and 16–3).

Doctors of chiropractic, often in conjunction with various family practitioners and other specialists in allopathic and osteopathic medicine, utilize comprehensive standard health assessment procedures including history gathering, physical

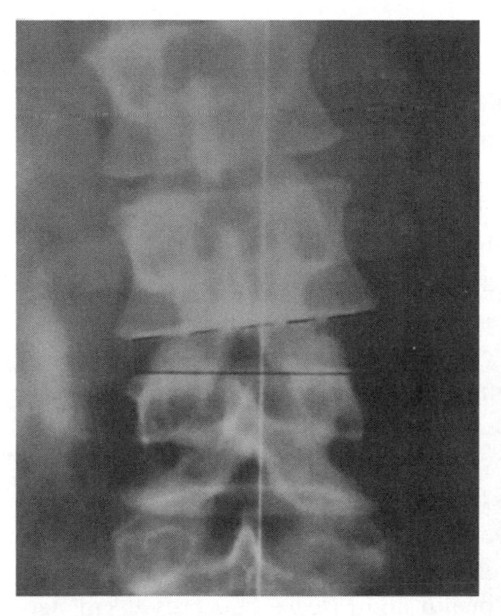

Figure 16–1 Lateral Flexion Subluxation of the 3rd Lumbar Vertebra

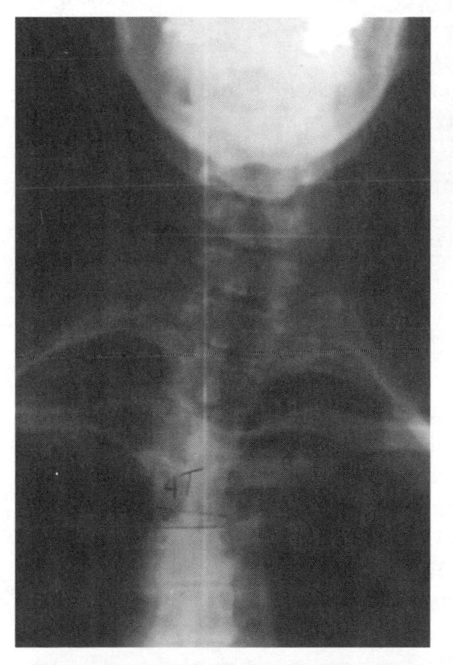

Figure 16–2 Lateral Flexion/Rotational Subluxation of the 4th Thoracic Vertebra

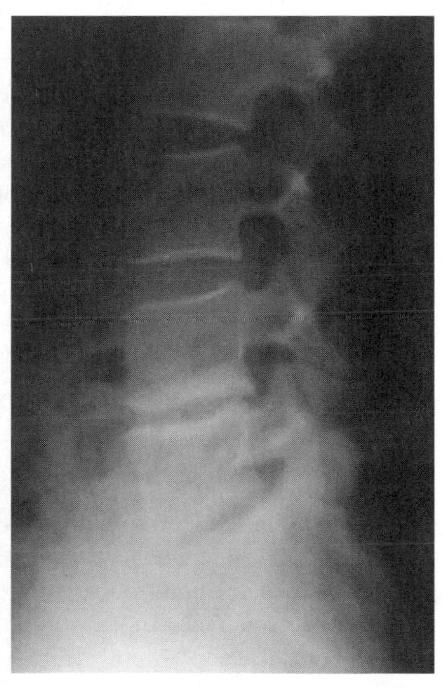

Figure 16–3 Posterior Subluxation of the 4th Lumbar Vertebra Coexistent with Degenerative Disc Disease

and laboratory examination, radiographic and diagnostic imaging, electroneurodiagnostic analysis, and an array of orthopedic and neurologic testing procedures common to all primary care health care disciplines.

In addition, and unique to their approach, chiropractors seek to identify and ameliorate the subluxation complex. Traditionally, the most common means of positively affecting the subluxation complex is through the use of a hands-on manual therapeutic procedure known as the chiropractic adjustment. Adjustment procedures utilize short-lever, high-velocity manual or mechanical forces applied to a specific body part for the purpose of reducing any of the hypothesized components of the subluxation complex. These components may include vertebral segment or other joint dysfunction, facilitated neural activity, altered somato-autonomic reflexes, joint surface disrelationship, nerve compression, cord compression, vertebrobasilar insufficiency, and neurodystrophy.[5]

Numerous other detrimental effects of the subluxation complex are theorized, and varying scientific documentation exists for their validity. In addition, chiropractic physicians place significant emphasis on psychosocial, stress reduction, nutritional, and lifestyle factors in health and disease and provide related counseling for their patients.

CHIROPRACTIC CARE FOR INJURED WORKERS

When a patient presents to a doctor of chiropractic with a history of what he or she believes to be a disorder related to work activity or work exposure, the chiropractor must objectively assess the patient's description of the condition. To legitimize the care within the workers' compensation system, establishing an accurate cause-effect relationship between the presenting symptomatology and the patient's history and the alleged work incident(s) is imperative.

Clinical History

The history must clearly link the injury sustained to the patient's current symptoms, condition, or alleged impairment, and must also describe an accurate and reasonable mechanism of injury or exposure. The history should describe what other care, if any, has been rendered, where, when, by whom, and the effect of the therapy. The history should also reveal any contributory or potentially contributory previous injuries, illnesses, or surgical procedures that may have had an effect on the causation, predisposition, or outcome of the current injury or healing process. A discussion of the potential of a similar condition in a family member may be relevant, particularly in gradual onset disorders. It is important to review and list any prescribed and over-the-counter medications currently being used by the patient for the current symptoms, as well as for concurrent and preexisting conditions. It is important to learn the exact duties and physical demands of the tasks the employee performs at work, and whether the patient has a part-time job in addition to his or her full-time employment. The history should provide a review of lifestyle considerations such as the use or abuse of alcohol, caffeine, tobacco products, and prescribed, over-the-counter, and recreational drugs.

Inquiry regarding the patient's nutritional status including vitamin and mineral supplementation is appropriate, particularly regarding nutrients important in the repair of soft tissue injury. Information regarding the patient's sleep habits, type of bedding, pillow(s), and preferred sleep postures may be important, particularly when dealing with lower spinal or neck disorders. Quality and style of work shoes may be important. Distance commuted to and from work can also have an effect on the healing time or causation or aggravation of neuromusculoskeletal injury, particularly of the lower spine. Recreational activities are reviewed as well as hobbies and interests, with particular focus on activities that are dependent on physical strength, agility, and endurance.

Previous workers' compensation injuries and illnesses are assessed, and information is sought regarding the potential for previous permanent partial or permanent total disability or impairment ratings. Inquiry regarding the patient's job satisfaction and relationships at home and at work can be important.

The Patient's Presenting Symptoms

Symptoms are the patient's subjective expression of the illness or effects of injury he or she is experiencing. They are helpful in arriving at a diagnosis; care must be exercised to learn their type, severity, frequency, duration, and location, and if any of the symptoms have changed from the time of injury. It is helpful to know what relieves and what aggravates the symptom(s), and which symptoms, if any, existed prior to the injury. In addition to the doctor's carefully listening to and listing the patient's description of his or her subjective symptoms, a generic symptoms survey questionnaire is generally used. The history questionnaire considers symptoms that may be manifesting from any of the body systems, including the cardiovascular, respiratory, gastrointestinal, endocrine, genitourinary, nervous system and brain, and the eyes, ears, nose, mouth, throat, and skin, as well as psychological symptoms. Patients are asked to describe symptoms in the past and the present, rating severity with single, double, triple, or quadruple "x"s. A section for women only and men only is also provided that seeks information regarding symptoms unique to each gender. Following industrial, automotive, and personal injuries, a detailed separate history questionnaire specifically designed to solicit information regarding neuromusculoskeletal injuries, disorders, and conditions is also used by many chiropractors. History of previous sprains, strains, contusions, crushing injuries, fractures, punctures, gunshot, stab wounds, burns, and lacerations is considered, and whether residual symptoms may be present and preexistent to the patient's current clinical presentation. A pain diagram is often used whereupon the patient designates the exact location(s) of his or her symptoms. Many chiropractors also utilize the Minnesota Multiphasic Personal Inventory (MMPI) to assist them in the evaluation of the patient's psychological status. This is particularly true when the patient's subjective description of his or her disorder is difficult to objectify with standard physical testing procedures. The evaluation is performed in consultation with a clinical psychologist.

The Chiropractic Physical Assessment

Following a careful review of the patient's history and presenting symptoms, the chiropractor, like any primary care physician, pursues a standard series of physical examination procedures. The extent of the evaluation is determined by the nature of the disorder. If the condition under evaluation is found to be localized to a particular body part, such as a sprain of an ankle, knee, wrist, or elbow, and if it had an acute traumatic onset with a clearly defined mechanism of injury, the examination may be appropriately focused on that body region. If, however, the patient presents with more generalized symptoms that involve multiple joints or body systems, to arrive at an appropriate diagnosis the examination may need to be comprehensive.

The Importance of Differential Diagnosis

The doctor must also be concerned about the potential that the neuromusculoskeletal complaints that the patient describes may be visceral in origin. Common visceral conditions that can refer pain to the lower portion of the spine, the pelvis, and occasionally the lower extremities are diseases of the prostate, bowel, appendix, kidney, bladder, uterus, and ovaries, including fallopian and ectopic pregnancies. Gastric ulcers, esophagitis, pancreatitis, and liver, gall bladder, and heart disease can manifest as middle and upper spinal pain, with pain referral patterns extending to the shoulders and upper extremities.

While most spinal pain is mechanical in origin, the doctor needs to also be alert for the potential that the pain may be arising from a destructive bone or joint pathology or a disease of the surrounding supportive soft tissues. Conditions such as osteomyelitis, rheumatoid arthritis, multiple myeloma, primary tumors, and secondary metastatic cancers from such organs as the uterus, prostate, bowel, liver, breasts, and lungs must all be ruled out.

Patients suffering the early stages of specific neurological disorders such as multiple sclerosis and Parkinson's disease, and brain and spinal cord tumors, both intradural and extradural, may present with various localized pains, as well as disturbances of gait and other motor functions, and with these symptoms may consult a doctor of chiropractic for what they believe is a work-related disorder. Under all circumstances, the physician must make a definitive differential diagnosis prior to the commencement of care. While manual therapy is considered safe to administer under most circumstances, there are times when it could accentuate or aggravate an already fragile situation. Copresenting symptoms such as fever, nausea, chills, sweating, malaise, excessive fatigue, blood found in the stools, urine, or sputum, chronic cough, prolonged diarrhea or vomiting, otherwise unexplained loss of weight or appetite, spotting or prolonged or excessive menstrual flow, skin changes, lymphadenopathy, night pain, pain that is not affected by position, loss of bowel or bladder sphincter control, or a history of previous cancer should alert the physician to search beyond the routine in patients who express neuromusculoskeletal complaints.

Fracture of bone is also an important consideration in the management of pain and dysfunction of the neuromusculoskeletal system. In conditions that have a traumatic origin, such as direct blows or falls, and those wherein the body has been subjected to notable compressive, rotational, or sheer forces, the diagnosis, when aided by appropriate standard radiographic imaging procedures, is generally quite simple. However, when there is no history of significant trauma, the possibility of a spontaneous, pathologically induced fracture still remains. Conditions such as advanced osteoporosis can result in a spontaneous fracture of the femoral neck or a compressive collapse of one or more vertebral bodies from even such a simple act as lifting a suitcase, coughing, or opening a window. Repetitious micro-trauma can also result in stress fractures, particularly of the lower extremities, even in young, otherwise healthy subjects. Since these have an insidious onset, it is important for the doctor to consider this possibility in individuals whose history warrants such investigation.

Primary vascular disease, such as an abdominal aortic aneurysm, can cause back pain; and advanced atherosclerosis of the distal aorta and common iliac vessels can produce vascular claudication of the lower extremities, which can resemble mechanical back pain with resultant sciatic neuralgia.

The Physical Examination

When other than the most routine condition presents itself to a physician for diagnosis and care, every reasonable measure must be taken to ascertain the patient's underlying condition. In the typical general chiropractic evaluation the following physical examination procedures are common:

- vital signs
- evaluation of the head, eyes, ears, nose, and throat
- evaluation of the chest and respiratory system

- evaluation of the heart and circulatory system
- evaluation of the abdomen and groin
- evaluation of the extremities
- evaluation of the skin
- neurological evaluation
- orthopedic/structural evaluation
- radiographic evaluation, as clinically indicated
- laboratory examination, as clinically indicated
- psychological assessment, as clinically indicated

Gynecological, breast, anal, and colorectal exams are conducted by the chiropractor when historically and clinically indicated, or the patient may be referred to another clinician for these specialized procedures.

In addition to the evaluation for spinal segmental subluxation, chiropractors pay special attention to local and regional structural faults that are known to result in shifts away from optimal axial loading. Some of these are congenital in nature and others are acquired as the result of trauma or disease. Many can be observed during the plumb-line evaluation, and others are discernable only on load-bearing radiographic studies. Included in the list of structurally related conditions of concern to the chiropractor are

- Scoliosis
- Leg length inequality
- Lordosis (hyperlordosis)
- Ramrod spine (flatback)
- Type I round back deformity (swayback)
- Type II round back deformity
- Juvenile kyphosis/Scheuermann's disease
- Cervical rib(s)
- Block vertebrae
- Deformed vertebrae
- Spondylolysis
- Spondylolisthesis
- Pedicogenic stenosis (congenital or acquired)
- Asymmetrical lumbar facets (tropism)
- Transitional lumbosacral vertebral segmentation (lumbarization of S-1 or sacralization of L-5)
- Extra or missing spinal segments

- Degenerative disc disease
- Pelvic subluxation/sacroiliac fixation, malposition
- Facet imbrication at the lumbosacral and cervical spine
- Pronation or supination of the ankle/arch mechanism
- Common biomechanical knee disorders
    1. Genu valgum
    2. Genu varum
    3. Flexion contracture
    4. Genu recurvatum
- Residuals of previous ankle, knee, or hip joint surgery
- Residuals of previous spinal surgery
- Spondylosis/hypertrophic changes
- Residuals of coxa vara or coxa plana
- Surgical arthrodesis of one or more of the ankle joints
- Deviation of the total body posture away from the gravity line in the anteropostero plane
- Deviation of the total body posture away from the gravity line in the lateral plane
- Presence of episacroiliac lipomas overlying the sacroiliac joints (ESIL's Nodes)
- Residuals of previous fractures that have left deformity
- Pes cavus or pes planus
- Corns, calluses, bunions, hammertoes, plantar warts
- Muscle tone
- Obesity
- Asthenia

The above conditions are considered a routine part of the general appraisal of the patient who presents to the chiropractor with a work-related injury or disorder. Presence or absence of the above conditions allows the clinician to assess whether any, or combinations thereof, have contributed to the predisposition, causation, or aggravation of the current condition, as well as determine the effect each, or combinations, may have on the patient's prognosis and healing time following injury.

The Diagnosis

Diagnoses rendered by chiropractors for work-related disorders are arrived at through an accurate history, the presenting symptoms, the mechanism of injury or tissue insult, and the results of the various examination procedures described above. Attention is also given to the acquisition and review of the previous medical records obtained from other care providers.

Common descriptive terms are used to render the diagnosis so that the workers' compensation insurance carrier and employer are able to gain an accurate portrayal of the patient's condition.

Diagnosis Sample # 1

Acute, moderate to severe, contusion, compression, hyperflexion subluxation sprain and strain of the upper thoracic and cervical spine with resultant

1. Grade I-II posterior midline protrusion of the intervertebral disc between C-5 and C-6
2. Moderate, continuous bilateral C-5, C-6 radiculitis
3. Mild to moderate recurrent suboccipital vascular syndrome

The above diagnosis was rendered for a 42-year-old male subject who was bent forward from the waist when a 55-pound bale of insulation fell on his head, neck, and upper spine from a distance of 12 feet above.

The words chosen describe the mechanism of injury, the tissues damaged, the extent of the damage, and the secondary effects that were exhibited by the patient.

The symptoms he presented with were exquisite diffuse upper spinal and neck pain; generalized grade III–IV spasm of the cervical and upper thoracic spinal musculature; inability to hold his head without the aid of external support; and pain, numbness, and weakness of his upper extremities, which extended from his neck to his thumb, fore, and middle fingers of both hands equally. He also described moderate to severe headache, dizziness, and occasional blurred vision. He denied any loss of consciousness and was able to recognize his surroundings immediately following the injury and continuously through the time of his initial examination, which took place within 45 minutes from the time of injury.

In the above diagnosis, the term *acute* was used because the mechanism of injury was sudden, violent, and traumatic. The phrase *moderate to severe* was used to describe the fact that he had significant loss of function but had some retained function, such as the ability to walk about and ride in a car for presentation at my clinic. He was, however, unable to drive the car because he had to hold his head with both hands to support and stabilize it, suggesting that the ligaments and other supportive tissues were torn. The terms *contusion*, *compression*, and *hyperflexion* all describe the nature and direction of the tissue trauma based upon the position his body was in when the bale fell onto him from above. The term *subluxation* was used to describe the joint surface disrelationship that resulted from the trauma. The term *sprain* in the above diagnosis describes the damage to the ligament, capsular, and cartilaginous tissues. The term *strain* describes the damage to the muscles, tendons, and fascia. The term *posterior midline* describes the direction in which the intervertebral discal material herniated (rather than anterior, antero-lateral, lateral, or postero-lateral). The term *Grade I–II disc protrusion* describes the herniation of the intervertebral disc material between the fifth and sixth cervical vertebrae into the central canal of the spinal column, impinging upon and displacing the C-5 and C-6 nerve roots (radicals) resulting in their inflammation (*radiculitis*). The term *suboccipital vascular syndrome* was used to describe the damage sustained to the blood vessels that supply the posterior portion of the brain, which resulted in the headaches and intermittent blurred vision.

Diagnosis Sample # 2

> Chronic, recurrent, mild to moderate unilateral right (preferred side) lateral epicondylitis with resultant effusion and loss of function, secondary to longstanding occupational overuse strain. (This is a cumulative trauma disorder secondary to repetitious micro-trauma.)

The above diagnosis was given for a 37-year-old female drill press operator who had been assigned to continuous duty on a poorly designed workstation for the previous seven years.

The word *chronic* is used to indicate that the problem had a gradual onset over a long period of time. The word *recurrent* is used to suggest that the symptoms would come and go and that this current episode was not the first time that similar symptoms were manifest. The adjectives *mild to moderate* are used to denote the severity, suggesting that the problem was not totally incapacitating but was a clear and present distraction and source of discomfort sufficient for her to seek professional intervention. The phrase *right (preferred side)* is used to inform the reader that the individual is right-handed and that therefore the problem is of greater concern from an occupational standpoint. It also suggests that the repetitious use of the right arm in the patient's work, on a daily basis, may have been a causative or contributing factor to the disorder, and that recovery may be delayed if the patient is expected to return to the poorly designed workstation that played at least a partial role in causing the disorder. The term *lateral epicondylitis* is used to describe the body part involved; in this case, the site is that of the radiohumeral bursa. The phrase *with resultant effusion and loss of function* suggests that the bursa has become swollen and painfully inflamed to the point of weakness and inability to use the arm. The phrase *secondary to longstanding occupational overuse strain* defines the problem as having had a gradual onset and as being work related. This last phrase affirms that the patient's disorder was occupationally induced and establishes it as a workers' compensation claim. The parenthesized statement "this is a cumulative trauma disorder secondary to repetitious micro-trauma" helps the reader to understand that the condition was not due to a contusion, blow, or other sudden wrenching sprain or strain, but rather to the effect of having repetitiously used the body part in a manner that produced a progressive breakdown of the tissues at the cellular level, gradually overcoming their ability to self-repair.

To arrive at a diagnosis physicians generally rely more on objective *signs* rather than subjective *symptoms*. The term *signs* refers to those physical indications of abnormality that the physician is able to perceive with his or her senses and that are measurable or can be quantified. Examples of signs are swelling, cyanosis, edema, pallor, and atrophy. Subjective *symptoms* are those words and statements the patient uses to describe the abnormal sensations he or she is experiencing. Examples of symptoms are "pain," "stiffness," "numbness," "burning," and "itching." Some symptoms may also be signs. For example, fever, spasm, and crepitus are often described by the patient, but can also be observed, measured, or heard by the physician examiner.

When patients have fears regarding their future work ability, it is common for them to exaggerate their symptoms. Discerning physicians must be sensitive about this reality and be able to accurately distinguish between unconscious embellishment of symptoms and indications of clear-cut malingering. True malingerers are individuals who consciously and willfully engage in deceptive behavior with a predetermined intention to defraud the workers' compensation system. An assortment of tests and procedures specifically designed to evaluate for symptom magnification can be utilized by the physician examiner to identify such behavior.[6]

Diagnostic and Analytical Considerations Unique to Chiropractic

In addition to standard orthopedic and neurological testing procedures, including traditional and advanced diagnostic imaging and clinical laboratory and electroneurodiagnostic procedures, doctors of chiropractic utilize a variety of noninvasive analytical devices and procedures that are more unique to their specialty. Plumb-line analysis, bilateral weight distribution scales, scoliometry, manual and computerized postural goniometry, thermosensitive instrumentation, and manual and computerized strength and range of motion evaluation procedures are examples. Chiropractors also commonly utilize load-bearing, erect-posture X-ray studies, which may include spinal radiographs obtained at the patient's maximum range of motion in flexion, extension, and lateral bending. When generalized structural imbalance is visualized by the physician, or when the patient presents with clinical indications of injury or structural imbalance at more than one region of the spine, such as pain and dysfunction within the lumbar spine and pelvis, as well as the thoracic and cervical spine, chiropractors often utilize 14" x 36" full-spine radiography. Recently, digital videofluoroscopy, which allows spinal intersegmental motion to be analyzed, has come into use within the chiropractic profession. This technology involves interfacing an imaging-processing computer with an X-ray image intensifier.

Standard health, pain, and function perception questionnaires, such as the McGill Pain Questionnaire, the Oswestry Disability Score, and various types of visual analog scales (pain diagrams) are also commonly utilized by chiropractors, particularly in complex cases where there is a contrast between the patient's subjective symptoms and the doctor's objective findings. Nutritional status questionnaires are also a common component of the chiropractor's approach to clinical diagnosis and case management. Finally, because of chiropractic's long tradition as a hands-on analytical and therapeutic specialty, a variety of specialized palpation methods are used by chiropractors that are unique to their field of study. Appropriate utilization of any of these procedures is left to the physician's discretion, based upon the history and clinical findings in each condition being evaluated.

Communicating with the Employer/Insurer

Upon diagnosing a patient's disorder as being work related, and accepting the responsibilities of a workers' compensation primary care provider, the doctor must provide the employer and the employer's insurer with timely and accurate information regarding the injured worker. Immediately following the patient's first visit, the clinician telephones the employer to inform them that their employee has consulted his or her office or clinic for evaluation and care that the patient believes is necessitated by the alleged work injury or illness. In most instances, particularly with larger firms, the doctor's discussion is with the employer's personnel director, who in many cases will be familiar with the matter since most often the employee will have informed his or her employer and a first report of injury or illness will already be filed with the employer's insurance carrier or self-insured risk management firm. During this initial telephone conversation, the doctor briefly reviews the history of the work-related incident, as described by the patient, with the personnel manager or employer representative. This affords the employer an opportunity to confirm or deny the patient's description of the incident, as well as to offer feedback and commentary regarding the circumstances of the accident that may further assist the physician in understanding the mechanism of injury or illness. Assuming the employer is acknowledging the employee's condition as being work related, and the appropriateness of the physician to proceed with care, the doctor communicates the following information to the employer in terms and language they can understand and relate to:

- the clinical impression and diagnosis
- temporary partial impairment, if any
- temporary total impairment, if any
- the treatment plan
- the prognosis, short and long term

- potential modifiers to the healing time, such as the patient's age, lifestyle, and concurrent health factors such as diabetes, obesity, heart disease, use of nicotine, previous similar injuries, and job satisfaction

In addition, during this initial discussion with the employer, the doctor will inquire about the company's return-to-work policy and whether it will make an effort to accommodate the worker's temporary restrictions from full duty, if necessary, and will encourage the employer to provide such modified employment as soon as the patient's recovery process suggests that it is appropriate to do so.

The Treatment Plan

For the majority of work-related neuromusculoskeletal injuries, the primary care that doctors of chiropractic provide consists of highly specific manual or mechanical adjustive procedures directed at reducing existing spinal subluxations and other related structural disrelationships. In addition, most doctors of chiropractic select from and use a wide variety of adjunctive therapies that are designed to render a positive effect on the repair of the soft tissue components of the structural disorder under care. Examples of some of the more common chiropractic therapies are as follows:

- myofascial "trigger point" release techniques
- massage
- muscle stripping
- transverse friction massage applied to tendons/joints
- ultrasound
- electrotherapy; sine-wave, low-volt, TENS, micro-currents
- diathermy
- infrared
- hydrotherapy, including whirlpool
- ice (cryotherapy)
- acupuncture/acupressure/meridian therapy
- reflex and subtle energy techniques
- applied kinesiology
- exercise therapy
- muscle-stretching techniques
- traction; sustained and intermittent
- ultraviolet therapy

Devices designed to temporarily assist ambulation are also commonly used by chiropractors during the course of patient care, including canes, crutches, walkers, and wheelchairs. In addition, the use of a variety of prosthetic and orthotic braces and supports for the ankle/arch mechanism, knee, groin, lower spine, upper spine, rib cage, clavicles, shoulder girdle, elbows, and wrists is common. Plaster casting, adhesive strapping, and elastic bandaging are also commonly employed.

Chiropractors also utilize a wide range of nutritional supplementation to facilitate healing and restoration of bone, muscle, and joint function. Examples are use of phosphorous-free calcium, magnesium, vitamins B, C, D, and E, proteolytic enzymes, and various specific homeopathic formulations. In all cases, the decision to incorporate nutritional supplementation is based on careful review of the patient's history, symptoms, and lifestyle. For example, if the patient is a smoker or uses tobacco in any form, vitamin C intake is generally advised. If the patient describes having a limited intake of dairy products and other sources of calcium and magnesium, or has an intolerance to dairy products, these minerals may be recommended.

When it is observed that stress plays a major role in the patient's condition, care may include biofeedback, hypnosis, and psychological counseling. Some chiropractors have adequate credentials to provide their patients these services directly; however, the majority refer their patients to appropriately trained professionals for care requirements that are outside their professional training or scope of practice.

SPECIFIC CLINICAL CASE MANAGEMENT

Lower Spinal Disorders

Spinal injuries are generally categorized by severity, ranging from fracture and dislocation, which can produce paralysis if the spinal cord is damaged, or even death if the brain stem is compromised, to simple muscle strain, which will recover without professional attention in 3 to 10 days.

Low back pain is among the most common disorders treated by chiropractors. Since most back pain is considered to be self-limiting, mechanical in origin, and, for the most part, non-life-threatening, chiropractors are called upon most often to treat noncomplicated lower spinal disorders. These would include such conditions as sprains, strains, spinal and pelvic subluxation syndromes, articular facet syndromes, intervertebral disc syndromes with and without sciatic nerve root involvement, degenerative joint disease, bursitis, hyperextension/hyperflexion (whiplash) syndrome, spinal stenosis, and any combination of the above disorders. The following represent examples of typical conditions with a description of the care rendered in each case.

Example # 1

Diagnosis. Acute hyperflexion rotary subluxation sprain and strain of the lumbosacral spine with associated Grade II protrusion of the intervertebral disc between L-5 and the sacrum with resultant moderate right sciatic nerve involvement.

Acute Phase. In this scenario, the patient, usually a person between ages 20 and 50, will present in a state of acute distress. The mechanism of injury described by patients is typically that of being bent forward from the waist, lifting an object while twisting their upper trunk and torso in relation to their feet. In some instances they will describe an acute, stabbing pain across the lumbosacral region of their lower back, occasionally accompanied by a popping or tearing sound. In many instances they are unable to continue in their work activity, and seek out professional attention within a day or so following such an episode. Some patients go to their homes and apply heat or cold to the injured areas hoping the pain and spasm will begin to subside. However, when the leg pain and numbness are paramount, they usually seek professional assistance quite early in the process of recovery.

Patients usually have great difficulty getting in and out of bed, their vehicle, chairs, and other furniture, and will carefully guard every change of position involving the muscles and joints of the lower spine and pelvis. In most cases, muscle spasm will cause their posture to be rigidly deviated either right or left of the midline in postero-lateral disc herniations and produce a forward bent presentation in central/midline discal protrusions and be most easily observed during the plumb-line evaluation. The patient will describe significant increase in pain upon coughing, sneezing, and straining at the stool. Linder's Sign, Soto-Hall, and the sitting Von-Bechterew's signs will all be positive, indicating nerve root involvement with exacerbation of the pain into the involved extremity.

The patient will describe pain and often paresthesia or partial anesthesia along the distribution of the S-1 dermatome, extending from the midline of the spine to the buttocks and hip, descending posteriorly along the thigh, into the calf and the outer aspect and little-toe side of the foot. There may be diminishment of the achilles tendon reflex; however, in some cases there may be a temporary exaggeration of the reflex. There is usually some unsteadiness of motor control of the affected limb, manifesting as an antalgic gait, slight to moderate foot drop, and guardedness during ascending or descending steps. With the patient recumbent or sitting, straight-leg-raising ability is notably diminished on the affected side. The patient is unable to raise both limbs simultaneously while back lying on the examination table, and a sit-up effort is also markedly stressful. In most instances, there is no disturbance of bowel and bladder sphincter control.

Radiological evaluation is generally negative for bone, joint, or soft tissue pathology, but varying stages of degenerative joint disease are common in the above diagnosis. Malposition subluxation of the fifth lumbar vertebra is commonly observed. Malposition fixation subluxation of the sacrum or innominate(s) (sacroiliac joint[s]) is also commonly found, coincidental with intervertebral disc herniations.

If neurological deficits are notably present, evidenced by such factors as more than a mild amount of foot drop, demonstrable weakness of the lower-extremity extensor muscles, total or near-total absence of the achilles tendon reflex response, significant loss of sensory perception upon pinwheel evaluation, or any indication of bowel or bladder sphincter control, the patient is referred for more advanced diagnostic imaging. Traditionally, computer-assisted tomography (CT scan) or magnetic resonance imaging (MRI) would provide the clinician with information regarding the physical status of the herniated discal material resulting in actual displacement of the exiting nerve root(s) and cauda equina. Whether the herniated discal material is contained within the confines of the annulus fibrosis or is totally prolapsed is also important in the diagnosis, as well as the prognosis regarding the potential for a satisfactory recovery under conservative, nonsurgical management. Fragmentation of the discal material can also be observed, and when present, generally suggests a more negative prognosis for recovery without surgical intervention. In addition, advanced imaging techniques can assist in ruling out the rare finding of a spinal cord tumor, either intra- or extradural in nature, a hematoma, a cyst, or another space-occupying lesion within the central neural canal or intervertebral foramen. The presence or absence of pediculostenosis, which is traditionally congenital in origin, resulting in a reduction in size of the central neural canal in either the sagital or coronal planes, can also be ruled on by these advanced diagnostic imaging techniques.

Laboratory evaluations of body excretions, including blood and urine, are generally negative. If there are indications of blood in the stool, fecal samples may be obtained and tested for occult blood.

In addition, one or more of the several structurally related conditions described earlier in this chapter may be found to exist coincidentally with this diagnosis. Conditions such as mild to moderate scoliosis, leg-length inequality, lordosis, flatback, Types I or II round back deformity, juvenile kyphosis, spondylolysis or spondylolisthesis, and facet asymmetry are commonly found. The role of each of these factors alone or in combination with discal herniation is not well established in the literature. However, it has been observed that they can have an important relationship to the vulnerability of the patient to the above diagnosis, and must also be considered in the patient's clinical management and prognosis.

Assuming that the above structurally related conditions are not present, or are only minimally present, the likelihood of the doctor of chiropractic being able to conservatively assist the patient in making a full or nearly total recovery from a grade II herniated disc in the lower spine, without the need for surgical intervention, is quite high. This also assumes good patient cooperation and understanding and encouragement from the patient's family, friends, and the employer/insurer.

Following the completion of a thorough history, symptoms presentation, and physical, orthopedic, neurologic, structural, X-ray, and necessary laboratory evaluations as described previously, the patient is briefed by the doctor regarding the positive and negative clinical findings, direct and indirect causative factors, the probable diagnosis, the treatment plan, including home care, and the prognosis.

During the acute phase of care, the doctor's primary therapeutic objective is to physically shift the offending herniated discal material away from the nerve root. This can usually be accomplished by carefully identifying and reducing the offending accompanying spinal or pelvic subluxation. In most cases of herniation of the L-5, S-1 intervertebral disc with accompanying S-l nerve root involvement, the accompanying subluxation is that of the fifth lumbar vertebra or the sacrum. In either case, the doctor chooses the most appropriate position for the patient to assume, and which of a variety of chiropractic table designs would be most appropriate for the most successful and specific manual reduction of the offending subluxation. When the patient is properly informed as to what to expect, he or she and the doctor are positioned properly, and optimal leverage, force, direction, and speed are incorporated, the adjustive manuever is well tolerated by the patient, with only minimal discomfort or protective muscle splinting experienced.

Upon successful completion of the adjustive reduction of the offending subluxation, the patient is assisted from the table and asked to walk about the treatment room. He or she is asked to approach a chair and attempt to assume a seated posture momentarily and then arise from the chair while the doctor observes for an indication of decreased muscle spasm, and improved function. If a significant reduction of a contained herniated disc has been accomplished by the adjustive thrust, it is common for the patient to immediately report improved function and less pain. If notable inflammation and swelling has accompanied the syndrome, and protective muscle splinting prevents the patient from being able to passively participate in thc maneuver, it is more difficult to attain more than a minimal reduction of the discal herniation on the initial visits. Physiological therapeutic measures are frequently incorporated to facilitate reduction of pain and muscle spasm during the acute phase and may involve use of electrical therapies such as sine wave, TENS, interferential current, or ultrasound. If indications of bleeding within the tissues are present, the patient is instructed in the home use of ice packs through the first 72 hours following injury. This is usually followed by the application of various forms of moist heat or infrared therapy in conjunction with gentle stretch, massage, and trigger point therapy techniques. Patients are advised to avoid all stressful physical activity that provokes pain, muscle spasm, and fatigue. They are required to avoid prolonged sitting and sleeping on their abdomen. Often, a flexible, wrap-around lumbosacral support belt or corset is used during the acute phase and may be used during sleep if it provides additional comfort. Naturally occurring anti-inflammatory nutrients derived from botanicals such as papaya and pineapple (bromalein) are used to assist the tissue healing response. Other appropriate nutritional support is given, proportionate to the patient's need as identified by history, symptoms, and clinical observation.

Traditionally, by applying these conservative procedures, in this author's experience, a notable reduction of the discal herniation can be obtained within 3 to 10 office visits extended over the first 10 days to three weeks of care. By this time, patients are usually able to sleep through the night, arise more comfortably from beds, furniture, and their automobile, and walk about without physical assistance or the need to use a cane or chair as a walker. There is a progressive reduction of the muscle spasm, and the patient has much less pain upon coughing, sneezing, and straining at the stool. Ranges of motion are improved upon each office visit, and the patient's posture resumes a more upright stance. During each visit, the same or modified specific manual adjustment is performed, as well as the reduction of other subluxations found that may exist in the spine, pelvis, or lower extremities that may have a bearing on the condition. Soft tissue rehabilitation techniques are used to facilitate a decrease of pain and spasm, reduction of adhesions, and increased joint mobility. Electrical stimulation of the musculature of the affected lower extremity is utilized to prevent the reaction of degeneration and resultant atrophy that can follow peripheral nerve injury.

Semiacute, Rehabilitation, and Recovery Phase. As the patient's gait and ranges of motion approach normal and the pain and paresthesia are lessened, the patient is instructed in a graduated program of specifically prescribed home exercises that include both stretching and strengthening maneuvers. Since patient compliance with physical rehabilitation efforts is critical for optimum outcomes, the use of a professional physical rehabilitation facility may be warranted in cases such as this and should commence within the first 30 days of care if long-term adhesions, muscle atrophy, and propensity to recurrence of injury are to be avoided.

Return to Work. Most patients suffering from a herniated disc must be restricted from work activities for the initial phase of their care. The time away from work will vary greatly with each individual. The patient's tissue healing response, the physical and postural demands of the patient's job, the patient's lifestyle, attitude, and general health, and the availability of work within physical restrictions will all affect the decision regarding exactly when he or she should attempt a return to work. In all cases, however, the doctor must be careful to encourage physical activities that will not aggravate or delay the healing process. It is also imperative that the doctor be sensitive to the possibility of associated emotional effects of such injuries, as situational depression commonly accompanies major illnesses and injury, and this can significantly affect the patient's perception of pain and disability and willingness to return to work.

Upon successful clinical management of the above syndrome, patients usually return to work within 30 to 60 days if their work activity is reasonably light, and within 60 to 120 days if the work activity requires significant physical demands. Professional chiropractic care for intervertebral disc syndromes traditionally takes place at daily intervals through the acute phase; and as progress is observed the patient is treated on a triweekly basis, then biweekly, weekly, and eventually bimonthly. Professional care is generally required throughout the healing time. It is prudent for the doctor to monitor patients' total recovery time following their return to work by office visits that are gradually extended to monthly intervals in order to ascertain full range of motion, tissue strength, joint mobility, and optimal structural stability.

Example # 2

Diagnosis. Acute moderate rotary facet syndrome of the lumbosacral spine with associated myositis.

Acute Phase. In this condition, the patient presents to the doctor of chiropractic with a history of a sudden-onset, acutely disabling episode of low back pain without the presence of sciatic nerve involvement. Frequently the patient describes the mechanism of injury as a twisting or hyperextension work activity such as that necessitated by shoveling wet snow, sand, or gravel. This rotational movement of the patient's torso and pelvis while the patient's feet are planted solidly is thought to produce excessive rotational and shear forces within the posterior facet joints of the lumbar spine, resulting in simple, noncomplicated subluxation.

Patients who have congenital or acquired facet tropism, wherein the facet joint facings are anatomically arranged asymmetrically, with one pair of inferior and superior facets facing coronally and the opposite pair sagitally, are known to be more vulnerable to this condition.

The traditional history, symptoms presentation, report of mechanism of injury, and physical, orthopedic, neurological, structural, radiographic, and related evaluations are accomplished in a fashion similar to the preceding case. In this situation, however, while the patient is in significant pain and is often almost totally temporarily disabled as the result of the condition, the clinical management and the prognosis are much different. In simple lumbosacral facet syndrome patients, there are no indications of nerve root or cauda equina involvement, and therefore no motor or sensory neurological signs are present. The offending spinal segment is usually easily identified by static and motion palpation, common orthopedic tests, and neutral, erect-posture static X-ray evaluation. The specific manual adjustive reduction of the subluxation is usually easy to accomplish, and the patient's response to the correction is usually immediate, with total restoration of function within days of the onset. In uncomplicated cases there are no residuals and time lost from work is minimal. Patients are instructed in appropriate preventive measures, including the need to avoid planting their feet during rotational work activities. The simple admonition given by the doctor, "when you move your nose, move your toes," is one that patients are more likely to remember and implement than lectures regarding the importance of maintaining a neutral, nonrotational relationship of the pelvis to the feet. Occasionally, when patients are not compliant with the above instruction, or their work activity produces repetitious or continuous rotational or hyperextension forces, this syndrome can become recurrent and chronic. Persons engaged in certain occupations, such as plumbers, electricians, sheet rock hangers, construction workers, and athletes involved in activities such as gymnastics, are thought to be more vulnerable to incurring a lumbosacral facet syndrome.

Repetitious Micro-Trauma Disorders of the Upper Extremity

Diagnostic Considerations

When patients present to doctors of chiropractic with a variety of conditions that involve such symptoms as pain, inflammation, paresthesia, weakness, and loss of function of the upper spine, neck, shoulder, arm, hand, and fingers, including carpal tunnel syndrome, arriving at an accurate diagnosis can be a significant challenge. While many cumulative trauma disorders can be diagnosed quite simply and without expensive or time-consuming specialized tests, combinations of the following diagnostic activities may be required in arriving at an accurate diagnosis and treatment plan in the appropriate clinical management of these disorders:

- a careful general health history, with emphasis on the patient's occupational history and potential exposure factors
- a historical review of the patient's previous neuromusculoskeletal injuries, conditions, and disorders
- a careful physical, orthopedic, and neurologic examination that may be both general and regional in nature
- a comprehensive structural/postural/plumb-line evaluation
- strength and flexibility testing
- the use of pain diagrams (visual analog scale)
- radiographic and diagnostic imaging, as follows:
    1. plain, neutral, erect-posture X-ray studies, which may include full-spine and pelvis and regional studies of the spine. Special studies such as flexion, extension, and oblique views of the spine may be necessary.

2. plain and special X-ray studies of the chest, rib cage, clavicle, shoulder, upper arm, elbow, forearm, wrist, and hand
3. computerized axial tomography (CT scans)
4. magnetic resonance imaging (MRI)
5. bone scans (as clinically indicated)

- laboratory evaluations (as clinically and historically indicated)
- thermography (as clinically indicated)
- electroneurodiagnostic testing, including electromyelography and somatosensory-evoked potentials (SSEP)
- special tests including vibrometry (loss of vibration sense) and neurometry (testing for alpha, beta, and gamma nerve fiber sensitivity)

Carpal Tunnel Syndrome

Since carpal tunnel syndrome (CTS) is among the most challenging and costly of the above conditions, the following information is offered as chiropractic management of this disorder.

Acute Phase. This phase involves the following treatments.

- *Work Restriction*—Restriction of all work duties and recreational activities that require repetitious flexion, extension, or ulnar or radial deviation of the involved wrist.
- *Splinting*—The use of a carefully fitted splint to restrict the above motions, used continuously, including during sleep, until the acute symptoms subside.
- *Ice*—Ice packs (cryotherapy) as needed for pain, swelling, and inflammation. If swelling or puffiness is present, elevation of the involved extremity supported by pillows may be beneficial.
- *Nonsteroidal Anti-inflammatories*—Naturally occurring anti-inflammatory nutritional agents such as bromelain are recommended. It is common that the patient has already been using nonsteroidal, over-the-counter agents such as buffered aspirin, ibuprofen, and acetaminophen, and these may be continued if benefits are noted.
- *Ultrasound*—Underwater ultrasound therapy, applied to the affected wrist as well as all related trigger points. The therapy is applied within the patient's pain-free tolerance and is used on alternating days throughout the acute phase of care.
- *Electrotherapy*—Application of high-volt, low-amperage therapy (TENS) and various forms of micro-currents to the involved musculature can be beneficial.
- *Muscle Stripping*—Manual, deep "stripping" of the involved musculature,

particularly of the flexor muscles whose tendons traverse through the carpal tunnel from the elbow to the palm of the hand, the pronator teres, and any and all musculature that may be splinting and compensating for the affected muscles throughout the arm, including the biceps and triceps. During muscle stripping, patients are supine (back-lying), with their arm positioned on the doctor's lap. Since a liquid-like hand lotion is used in this activity, the doctor or therapist spreads a Turkish towel over his or her lap to protect the clothing. Transverse friction massage may be incorporated; however, during the acute phase this may be too painful and potentially aggravate the condition. Therefore the stripping should be performed in the direction following the long axis of the muscle fibers, with particular attention given to any dense, granular, painful fibrocytic nodules (trigger points) discovered. These nodules may occur in the belly, origin, or insertion of the muscles. Intensity, depth, and duration for each treatment must be determined by the patient's pain threshold and tolerance. It is prudent to proceed somewhat cautiously during the initial phase of care since an exacerbation of the condition can occur if the doctor or therapist is overzealous.

- *Trigger Point Therapy*—Application of specific manual, electrical, or other mechanical therapeutic procedures to all existing active trigger points involved in the syndrome. An active trigger point is one that, when acted upon by mechanical forces, reproduces the patient's symptomatology. Examples of these forces would be the doctor's examining finger or thumb, elbow pressure, or externally applied energy such as ultrasound. In addition to trigger points and myofascial nodules that may be found in the arm proper, active trigger points important to the forearm, wrist, and hand can often be found with the upper spine, scapula, rib cage, and shoulder girdle musculature. This is the case because these muscles are involved in work and recreational activities of the upper extremity and are often overused as well. Of particular importance is a trigger point that can be found with high levels of predictability at the site of the origin of the infraspinatous muscle near the apex of the inferior/exterior border of the scapula. Invariably, when present, this trigger point reproduces the pain, numbness, heaviness, and distressful sensations the patient describes.
- *Stretching Exercises*—Gentle, within the patient's pain tolerance throughout the acute phase.
- *Specific Spinal and Related Joint Adjusting and Manipulative Therapy*—All subluxations throughout the pelvis and spine, with particular emphasis on concern for primary subluxation detection and correction, are incorporated. Effort is made to restore optimal axial loading and hydrostatic efficiency of the intervertebral discs involved in the subluxation complex. Consideration

for restoring optimal global postural balance and structural integrity of the entire human frame is emphasized in addition to local or regional care.

- *Nutritional Considerations*—Our approach includes a comprehensive review of the patient's nutritional status, with particular inquiry and observation for the need of the B vitamins, vitamins C, D, E, beta carotene, and the minerals calcium, magnesium, and potassium. When indicated, vitamin B6 is recommended. If the patient is obese, caloric intake is adjusted, along with a graduated aerobic exercise program. Adequate pure water intake is emphasized, with recommended daily ingestion of one-half ounce of water for each pound of body weight. More water is recommended if the patient is athletic or works in excessively warm environments.

Sub-Acute and Recovery Phase. Treatment in this phase consists of continued application of the above therapeutic strategies, but reduced in frequency from daily or triweekly office visits to biweekly, weekly, and bimonthly until full resolution of the disorder is attained. Resumption of work and recreational activities is permitted on a gradual basis, with significant education and recommendations regarding workstation ergonomics, use of hand tools, and lifestyle factors.

Prognosis. The above regimen of conservative care usually produces notable improvement within the first 10 days of professional management, with evidence of decreased pain and numbness and increased sleep and function. At this stage, return to work/recreational activities can be expected, with progressive improvement (both subjectively and objectively) continuously through complete resolution of the disorder. If no substantial improvement is manifest within the first 30 days of conservative care, or if deterioration seems apparent, the patient is referred for nerve conduction velocity (EMG) testing and potentially steroidal injection and/or surgical consultation.

Commentary. From the initial consultation through the end of a therapeutic trial of conservative management, the prudent clinician will render a somewhat guarded prognosis for a number of reasons. In most cases the patient presenting with carpal tunnel syndrome (or a related repetitious overuse syndrome) of the upper extremity will have had the symptoms complex developing gradually, and frequently intermittently, for many months prior to consulting a physician. During this phase the condition has usually not been disabling and therefore no professional attention may have been sought. Some patients will have consulted physicians who did not take the early phases of the condition seriously and dismissed the presenting complaints as insignificant, recommending noncorrective, nonpreventive care such as heat and aspirin. At times, this casual attitude is also taken by the occupational nurse, personnel manager, or safety director to whom the condition is first reported, with the hope that by ignoring or downscaling the importance of the condition, it might dissipate without professional care. In most instances, the patient continues to perform the activities that caused (or at least aggravated) the condition through the time of the gradual onset. Consequently, by the time the individual's condition is sufficiently severe to warrant serious clinical intervention, advanced, late-stage pathology is often present.

Doctor's/Employer's Recommendations for Workers Involved in Repetitious Use of the Upper Trunk, Neck, and Upper Extremities

- Stop smoking/chewing and all use/abuse of nicotine.
- Avoid gripping hand tools too tightly. Let the tool do the work; choose tools that fit your hand(s).
- Avoid unnecessary bending, stooping, twisting, reaching, and lifting.
- Alternate hands when/where possible during task performance.
- Drink adequate amounts of water. (About 1/2 ounce of pure water per pound of body weight daily is recommended, more if your work is in excessively warm conditions.)
- Avoid (minimize) caffeine products (coffee, tea, chocolate, many soft drinks, and nonprescription pain relievers).
- Avoid alcohol and illicit drug use/abuse.
- Use a good bed and mattress and a properly supporting pillow (ask your doctor).
- Avoid stomach sleeping and avoid sleeping with your hand(s) under your head or neck.
- If your job requires repetitious use of body parts, participate in diversionary activities and take frequent stretch breaks.
- Take advantage of all *adjustable* aspects of your workstation(s) to help create the best fit between you and your machine(s) and tools.
- Consider rotation of workstations if feasible.
- Avoid exposure to cold temperatures/drafts. Guard against cold with adequate clothing and protection. If you can avoid it, don't sit/stand near fans and vents.
- Consume a diet that assures adequate amounts of calcium, magnesium, potassium, and vitamins B, C, D, and E. Consult your doctor about which foods are rich sources of these nutrients. If you choose to supplement your diet with various vitamins and minerals, it is helpful to talk to your doctor about the recommended dosages of products available. Use of vitamin B6 (pyrodoxine) is felt by some to be helpful as a prevention for carpal tunnel and related nerve entrapment disorders. (Ask your doctor to help guide you in this decision.)

- Minimize use of birth control pills.
- Normalize your body weight.
- Avoid fluid retention by avoiding excessive sodium chloride (table salt and salty foods) and increasing potassium-rich foods such as bananas, oranges, orange juice, and squash.
- Avoid (minimize) stressful relationships and situations.
- *Immediately report* any early signs of upper-extremity overuse disorder to your supervisor and/or doctor—symptoms such as pain, weakness, stiffness, or numbness of your shoulder, arm, elbow, forearm, wrist, hand, or fingers.

THE CLINICAL EFFECTIVENESS AND COST-EFFECTIVENESS OF CHIROPRACTIC CARE—THE MANGA STUDY

In this age of health care cost containment, outcomes assessment, and quality assurance, concerned readers might ask questions such as: Does chiropractic work? How effective is this unique form of care? How does it compare with care rendered by other disciplines in health carc? Is it safe? How do the costs compare? Are patients satisfied with the care they receive from doctors of chiropractic?

These and other questions are perhaps best at least partially answered by citing a recent (August 1993) study by Pran Manga, PhD, and Associates, which was funded by the Ontario Ministry of Health. Dr. Manga is a professor and director of the Masters in Health Administration Program, University of Ottawa, and president of Pran Manga and Associates, Inc., Ottawa, Canada.

His study is entitled *The Effectiveness and Cost-Effectiveness of Chiropractic Management of Low-Back Pain.*[7] While the study examines only the question of chiropractic's effectiveness in the clinical management of lower spinal disorders, the findings suggest that similar outcomes might be achieved by its intervention in other neuromusculoskeletal disorders commonly diagnosed and treated by chiropractors. Dr. Manga's study, which represents perhaps the most objective, most comprehensive, and most critical review of the worldwide scientific literature on the subject, provides over 300 references in the bibliographic section of the work.

Regarding the question of the clinical and cost effectiveness, patient satisfaction, and safety of chiropractic care, Manga's Executive Summary of his report states in part as follows:

> On the evidence, particularly the most scientifically valid clinical studies, spinal manipulation applied by chiropractors is shown to be more effective than alternative treatments for low-back pain.[8]
>
> There is no clinical or case-control study that demonstrates or even implies that chiropractic spinal manipulation is unsafe in the treatment of low-back pain. Some medical treatments are equally safe, but others are

unsafe and generate iatrogenic complications for low-back pain patients. Our reading of the literature suggests that chiropractic manipulation is safer than medical management of low-back pain.[9]

There is good empirical evidence that patients are very satisfied with chiropractic management of low-back pain and considerably less satisfied with physician management.[10]

Since low-back pain is of such significant concern to workers' compensation, chiropractors should be engaged at a senior level by Workers' Compensation Boards to assess policy, procedures and treatment of workers with back injuries.[11]

Finally, the government should take all reasonable steps to actively encourage cooperation between providers, particularly the chiropractic, medical and physiotherapy professions.[12]

CONCLUSION

Chiropractic is a holistic approach to primary health care that maintains that work is therapeutic and emphasizes earliest possible return to the workplace following injury. Holistically oriented physicians respect all forms of healing and recognize optimal health as a manifestation of harmony between the body, mind, and spirit. Chiropractic is also a *participatory* system of health care wherein the doctor and the patient are considered equals, with patients involved in all major decisions regarding their care. Holistic health philosophy asserts that *the patient assumes the primary responsibility* for attaining and maintaining his or her own health.

A better understanding of the role that doctors of chiropractic can assume on the team of those responsible for the health and safety of workers everywhere will be of assistance to all who participate in work injury management.

NOTES

1. Centers for Disease Control, Leading Work-Related Diseases and Injuries—United States, *Morbidity and Mortality Weekly Reports* 32, no. 2 (1983):24–26, 32.
2. Bureau of Labor Statistics, *Back Injuries Associated with Lifting*, Bulletin 2144 (Washington, D.C.: U.S. Government Printing Office, August 1982), 1.
3. National Board of Chiropractic Examiners, *Job Analysis of Chiropractic* (Greeley, Colo., 1993), 3–6.
4. American Chiropractic Association, *Indexed Synopsis of Policies on Public Health and Related Matters*, C-22 (Arlington, Va., 1992–1993).

5. R. Leach, Chiropractic Theories, in *Chiropractic Family Practice*, Vol. 1, ed. J.J. Sweere (Gaithersburg, Md.: Aspen Publishers, Inc., 1992), 2–1 to 2–3.
6. R. Evans, Malingering and Psychogenic Rheumatism, in *Chiropractic Family Practice*, Vol. 1, Part 15, ed. J.J. Sweere (Gaithersburg, Md.: Aspen Publishers, Inc., 1992).
7. P. Manga et al., *A Study To Examine the Effectiveness and Cost-Effectiveness of Chiropractic Management of Low-Back Pain* (Ottawa, Canada: Pran Manga & Associates, 1993), 11–13.
8. Ibid., p. 11.
9. Ibid.
10. Ibid., p. 12.
11. Ibid., p. 13.
12. Ibid.

SUGGESTED READING

1. Haldeman, S., ed. 1992. *Principles and practice of chiropractic*. 2nd ed. Norwalk, Conn.: Appleton & Lange Publishers.
2. Sweere, J.J., ed. 1992. *Chiropractic family practice: A clinical manual*. Gaithersburg, Md.: Aspen Publishers.
3. Wardwell, W.I., ed. 1992. *Chiropractic: History and evolution of a new profession*. St. Louis, Mo.: Mosby-Year Book Publishers.
4. Plaugher, G., ed. 1993. *Textbook of clinical chiropractic: A specific biomechanical approach*. Baltimore, Md.: Williams & Wilkens Publishers.
5. Bergman, T., et al., eds. 1993. *Chiropractic technique*. New York: Churchill Livingstone Publishers, Inc.
6. Haldeman, S., et al., eds. 1993. *Guidelines for chiropractic quality assurance and practice parameters: Proceedings of the Mercy Center Consensus Conference*. Gaithersburg, Md.: Aspen Publishers, Inc.
7. Kelner, M., et al., eds. 1986. *Chiropractors—Do they help?* Ontario, Can.: Fitzhenry & Whiteside Publishers Ltd.

Chapter 17

Tests and Measurements in Returning Injured Workers to Work

Dennis L. Hart

WHY ARE MEASUREMENTS NECESSARY?

The expert in functional measurements must be able to perform tests and measurements on clients with work-related injuries for the purpose of quantifying and differentiating between physical impairment, disability, and functional abilities (and deficits). The measurements made by the clinician will affect the future of the client from a clinical, social, and economic perspective. Just as important is the increasing influence from managed care systems on all clinicians, which demands justification of clinical recommendations for initiation and continuation of treatment, including cessation of treatment. These measurements provide the assessment of objective necessity of care that is needed for justification of recommended treatment programs. Because many of the clients in industrial rehabilitation have the potential for being involved in litigation, functional experts must be prepared to justify their decisions in a court of law much more than most other clinicians in other clinical specialties.

WHY WE NEED OBJECTIVE NECESSITY OF CARE

In a previous publication, the philosophy of using documented "objective necessity of care" as the means of determining whether auditors will approve remuneration for clinical services was described.[1] In 1992, a study from the state of Washington described patient outcomes after lumbar spinal fusions and demonstrated the impact of not having reliable, valid scientific outcome studies and good documentation to justify the need for clinical services.[2] The purpose of that paper was to determine if lumbar spinal fusion improved the success rate of laminectomy for specific low back disorders. The authors synthesized information from 47 articles from the literature. In the absence of randomized clinical studies, a firm conclusion that fusion improves the outcome of laminectomy for spinal stenosis associated with degenerative spondylolisthesis could not be made. Because of study flaws and publication bias, the estimate of 68 percent satisfactory rate of patients with fusions was considered an overestimate. The lack of studies comparing the difference between outcomes from nonsurgical treatments and other surgical treatments, including no treatment, made it impossible to determine the extent to which the patient success rate at follow-up was indeed related to the surgical fusion. The analyses did not support the superiority of any fusion over other clinical outcomes. Finally, although higher rates of solid fusion were associated with better outcome, the "association between pseudarthrosis and symptoms in individual patients could not be determined."[3]

It is possible that this study will be used to formulate policies for payment (in this case, no payment) for lumbar spinal fusion in people receiving workers' compensation benefits. This would be a dangerous outcome of a lack of research supporting a specific treatment. Although the literature does not support specifically the use of lumbar fusion, there may be people who will benefit from such surgery. Unfortunately, because the medical community has not established a reliable and valid method of determining indications for fusion surgery, and is troubled by a poor association found between fusion solidity and pain relief,[4] remuneration for this type of treatment may be denied. If this were to happen, we, both clinicians and patients, could all suffer because of the ubiquitous lack of scientific justification for the great majority of our treatment procedures.

These points emphasize the need for all clinicians to justify clinical techniques through scientific means. In lieu of scientific studies that justify clinical services, clinicians need to improve the ability to provide documented "objective necessity of care"[5] for patient examinations, progress notes, and discharge summaries that are consistent with standardized patient treatment protocols. Without the scientific studies, documented objective necessity of care, or standardized treatment protocols, medical professionals may not only forfeit remuneration for a specific treatment but eliminate any payment for global groups of services in the future. Therefore, at a time when national health care reforms are being formulated, improvement in our objective clinical documentation and scientific efficacy studies is imperative. For those in private practice, that is, not a university environment, research opportunities become some of the best marketing dollars spent. Valid research that supports more than one treatment method for specific conditions allows clinicians to focus on effective treatment, making clinicians more efficient and better able to compete.

WHAT IS FUNCTION?

Before a discussion of measurements in industrial rehabilitation, a discussion about what should be measured is important.

Function by the simplest definition is the ability to perform work, leisure, and social activities. The term *function* encompasses our every activity, including bodily functions within our tissues and organs, and progressing to the highest of intellectual thought processes. The manner in which we control the function of our organs and systems allows us to manipulate our environment so we can maintain physiological consistency (homeostasis) to survive. Each bodily part, tissue, and so forth contributes to our ultimate reaction to and interaction with our environment, which is function.

Human level of function can be altered by many normal or abnormal activities in life such as normal aging or injury. When there is a reduction of the level of function of an individual, that individual has functional limitations. Functional limitations may result from physical impairments, such as anatomical pathology. Functional limitations also may be found in the absence of confirmable anatomical pathology (e.g., negative findings on magnetic resonance imaging). Additionally, it should be emphasized that many physical impairments do not manifest themselves in functional losses.[6]

Function does not necessarily relate to complaints of pain. For example, people with chronic low back pain almost universally report diminished levels of activity (or function) as a result of their pain, yet the actual levels of activity performed have not been well related to the subjective reports of pain.[7] In addition, people with regular, sometimes even daily, reports of pain (e.g., coal miners) frequently work without functional limitations even during the reported episodes of pain.[8] Through the study of new radiological techniques such as computed axial tomography or magnetic resonance imaging, a surprisingly high number of people who do not report functional deficits have significant radiological abnormalities.[9] Therefore function in some cases may be more related to the person's perception of the problem, the diagnosis of the problem, the availability of services to treat the perceived problem, or the learned perception of the method of treatment than to the actual pathology. The difference between the actual level of function and the perceived level of function needs to be differentiated. This is the purpose of functional capacity evaluations and work hardening.

WHAT IS A FUNCTIONAL MEASUREMENT?

When you quantify the ability to perform a task that is part of normal daily activity (perhaps including work), you have made a functional measurement. Since function can be exceedingly variable, functional measurements likewise can be varied. Some examples of functional measures are standing, sitting, lifting, carrying, steering, performing fine hand manipulations, and using power hand tools.

The more the functional task being measured simulates the actual task to be performed, the more valid the measurement.[10] The need for job simulations to be part of functional measurements is expected in standards and guidelines for industrial services.[11] Routine physical examinations including deep tendon reflexes and radiological findings do not allow direct correlation with the ability of the person to function safely. Therefore these routine physical examinations do not produce reliable or valid determinations of function in clients with work-related problems.

In the standards for industrial services,[12] the functional capacity evaluation and work-hardening program provide the procedures necessary to quantify functional abilities. The typical generic movements associated with the performance of work tasks include but are not limited to sitting, standing, balance, reaching horizontally and vertically, kneeling, stooping, crouching, lifting or carrying bilaterally or unilaterally, pushing, pulling, walking, climbing (ladders and stairs), fingering, feeling, and handling.

JOB SIMULATION

As the field of functional analyses progresses, job simulations will become the hallmark of functional measurements. For example, when the job requires the task of lifting, the task of lifting needs to be tested in the clinic to determine the ability of the individual to lift. Some experts believe that the closer the job simulations represent the actual task performed on the job, the more valid are the measurements of the function of lifting.[13]

As the need for job simulation increases in response to the Americans with Disabilities Act,[14] the reliability of these new measurements also should increase. In the case of lifting, for example, the literature demonstrates that isometric measurements of lifting forces can be reliably made.[15] There are many types of isometric testing devices. However, these are not job simulated, and therefore it is predicted that these devices, and others producing other nonfunctional measures, will become less prevalent in our clinics. New lifting devices and measures that do simulate the job tasks must be validated through research.

THE MEASUREMENT PROCESS IN INDUSTRIAL SERVICES

The American Physical Therapy Association confirms that in the most basic form, a measurement is a numeral assigned to an object, event, person, or class (category) according to a set of rules.[16] However, if that measurement is not reliable, the measurement cannot be valid. If the measurement is not valid, it should not be used to make clinical decisions.[17] The following aspects of measurement need specific attention as they relate to industrial services.

Reliability

Reliability of measurements is critical. There are several aspects of reliability that affect functional measurements.

Internal consistency relates to the extent to which items or elements contribute to a measurement or an interpretation of a set of measurements.[18] For example, if a client has a biomechanical problem, such as physical impairment in the form of reduced trunk forward bending, the impact of that loss of range of motion should be present whenever the client performs an activity that requires trunk forward bending. The degree of forward bending can be quantified several ways, for example, with an inclinometer[19] or a flexible ruler.[20] When the same client performs another activity requiring similar biomechanical trunk forward bending movements, the same or similar measurement tools should be used to quantify the second type of forward bending. The measurements should be the same. If the measurements are not the same, then one must ask why the measurements were different. Multiple measurements of similar physical impairment will provide a basis to assess internal consistency of testing procedures. This internal consistency, particularly if it occurs with multiple measurements of physical impairment and functional abilities, will improve the reliability and validity of overall testing procedures. The probability of valid interpretations of results also improves.

Intertester reliability is perhaps the most important form of reliability because it relates to the consistency or equivalence of measurements when more than one person makes the measurements. Intertester reliability indicates agreement or consistency of measurements taken by different examiners.[21] If several clinicians take the measurements and the results are different, poor intertester reliability is indicated. The results of the measurement should not be used by other clinicians to interpret patient measurements, particularly changes in patient function or physical impairment.

Intratester reliability, the second most important form of reliability, relates to the consistency of measurements when one person takes repeated measurements separated in time. Intratester reliability indicates stability (reliability) over time of the measurement.[22] Good intratester reliability allows one clinician to take repeated measurements and interpret changes in his or her patient. If the measurement has good intratester reliability but poor intertester reliability, the measurement has little clinical value. This is particularly important if the referral source is attempting to compare the progress of the same client between two different clinics. Every attempt should be made to use measurements that have good intratester and, more importantly, good intertester reliability.

There is increasing interest in reliability measures in medicine. The National Institute for Occupational Safety and Health (NIOSH) has devoted much effort to the assessment of reliable measures in people with low back pain.[23] These researchers used standards for statistical reliability that eliminated many measures that are common in the clinic, and their final grouping of measurements is not easily used in the clinic. A stronger assessment of clinically reliable and functional measurements is found in the article of Waddell et al.[24] These researchers used reliable measurements to form a simple and practical scale of physical impairment of people with chronic low back pain. But probably the most promising direction for the future is the approach by Delitto et al. concerning classification systems that can be used for testing clinical efficacy studies for treatments.[25]

There are several ways to improve the reliability of clinical procedures. First, attempt to use only clinical tests and measurements that have been shown to be reliable. Second, if you have the interest, test your clinical measures, possibly with the assistance of an experienced researcher. In any event, the clinician should know the level of reliability of clinical procedures used in the clinic. The clinician may find that some traditional measures are not reliable and may need to be abandoned.

Another way to improve the reliability of clinical procedures is to group several measures of one clinical issue, such as symptom magnification. Some of the measures or observations may be more reliable than others. Once the collection of measurements has been made, provide an ordinal scale of forced choices from which to choose the final rating. In this way, a collection of scores can be used to formulate an opinion that will have a better chance of being reliable.

Collecting your own measures of reliability in the field of industrial services is not a difficult task. Take the example of determining if your patients can provide reliable measures of function on your functional capacity evaluation. Most of us have developed our testing around the need to test agreement among scores of each test. We have the client perform the tests more than once. When you have more than one measure, you can test for reliability from your own data. You do not have to design a special research project. You may relinquish some internal validity because of your research design, but the measures still can be tested for reliability.

For example, in our practice, the data collection for functional testing of people with chronic pain syndromes was designed to test reliability of our measures while clinical tests were being conducted. Several measures have been tested, including isometric hand grasp (reliable, ICC = .94 left and ICC = .87 right),[26] isometric standing arm lift (reliable, ICC = .94),[27] and isometric push and pull (reliable, ICC = .96 for push and ICC = .95 for pull).[28] The tests for which it is harder to determine reliability are the dynamic tests.

Validity

Validity is a complex and on many occasions a confusing concept. There are many forms of validity. In a simplistic form, validity relates to the degree to which a useful (meaningful) interpretation can be inferred from a measurement.[29] In other words, if the measurement is valid, the measurement represents the degree to which you have measured what you believe you have measured.[30] A measurement must be reliable before it can be valid. Unreliable measurements cannot be valid.

The term *validity* commonly is misused in industrial rehabilitation. For example, apparently because of marketing and proprietary reasons, several sellers of "functional capacity systems" make claims that their products can assess "validity profiles" of clients or can determine if the effort of the client has been "valid, conditionally valid, invalid, or conditionally invalid." Some claim that the use of computer software can assess validity. Some take measurements of physiological responses, such as heart rate following short activities, and conclude whether the changes of the heart rates are related to expected or unexpected pain behaviors. A steady state for cardiac work is required to produce valid assessments of the physiological needs for changes in heart rate, which cannot be determined from short functional tests. Still others will take several findings from their tests and predict future treatment needs and expected outcomes following their treatment. Predictive validity studies have not yet been published for functional tasks. Until that time, the best measure of valid functional tests is whether the client did or did not perform a functional task.

From a research perspective, the user of these services must be wary of the marketing versus scientifically supported conclusions. For example, "validity profiles" may appear to be a reasonable approach for the assessment of clients, but there is no peer-reviewed published scientific justification for the use of the term *validity profile* as that term relates to functional testing. A measurement will be either valid or not. Does a collection of several different measurements provide a "profile" of validity? What is probably meant by these statements is that several different measurements are collected and the consistency of the measurements or expected findings is related and compared. This internal consistency of responses can allow the clinician to determine the degree to which the client compares with other similar clients. However, because of the higher variability of test scores in injured clients, probably the better comparison to use is the scores of a single subject.

If internal consistency can be confirmed for a set of functional tests, there should be concurrent validity of the measurements. This can and should be tested.

The terms *conditionally valid* and *conditionally invalid* are not common terms in the field of validity testing. A measurement is either valid or not valid. The placement of conditions on these terms offers little to the scientific assessment of the clinical procedures and therefore should be renamed, discarded, or ignored as marketing efforts, not science.

Computers can assist in the assessment of internal consistency of measurements, but the addition of a computer or computer software does not assure validity. The measurements must be tested for reliability and the appropriate form of validity. If the measurements are reliable and valid, the addition of the computer may save time for the clinician for the assessment of validity of the findings. Computers are not equivalent to validity!

Computers allow for the collection of immense sets of data. Rapid interpretation of these large data sets is facilitated by the computer. Possibly new measurements or decision-making techniques will be developed because of computers.

A good example from which to study the concept of validity is the comparison of lifting ability measured isometrically as compared to dynamically. Isometric[31] and "isoinertial"[32] measures of lifting abilities have been shown to be reliable, but when compared, there can be as much as two to three times the force generated isometrically as compared to dynamically.[33] Therefore which measure of lifting ability is "valid"?

Newton and Waddell reviewed the literature on the use of iso-machines to test the forces/torques from the movement of the trunk. Following the review of more than 100 articles published about these machines, they concluded that there "is inadequate scientific evidence to support the use of either isokinetic or isoinertial testing of trunk muscle strength in preemployment screening, routine clinical assessment, or medico-legal evaluation."[34, p.801] In a further study, Newton et al. reported that there were correlations between isokinetic lifting scores and both maximal isometric and psychophysical lift strengths (0.47–0.62) in both females and males, and patients and normal subjects.[35] Although the machines tested had good reliability and strong correlations between multiple measures of torque and peak force in dynamic lifting, the ability of the machines to discriminate between normal subjects and patients with low back pain was limited and little better than isometric measures or clinical measures of physical impairment. Furthermore, the measures of "consistency of effort" in isokinetic performance were unreliable and invalid. It is clear that the measurements of functional effort and the relation between different measurements need more study.

In reality, different measurements of lifting measure different aspects of force generation even though they may be related.[36] For example, an isometric measure from a constrained, nonfunctional position produces an unrealistically high measure (possibly unsafe for determination of ergonomically safe return-to-work restrictions) of dynamic functional lifting ability. The dynamic lifting, if performed in a manner that closely simulates the physical demands of the job, should be more predictive of the ability of the client to perform this type of lift at work. However, there are few data to confirm that conclusion. Probably the best articles to compare the predictive validity of dynamic lifting tests were conducted by Alpert et al. and Matheson et al.[37] In these papers, the authors demonstrated that there were relationships between several types of dynamic lifting methods—between isokinetic trunk strengths and lifting abilities—and there was even predictive validity of lifting capacity from dynamic lifting machine testing. The "gold standard" to which they were comparing the dynamic testing results was functional lifting, not machine-controlled lifting measures.

Therefore we can say that if a client performed the lifting activity in much the same way as he or she performed it at work (hopefully safely, but there are several interpretations of what is safe), the chances are good that he or she will be able to perform the same activity at work without undue risk of injury. If the client is released under the restrictions determined through dynamic assessments of job-simulated activities and work within his or her restrictions on the job, the rate of reinjury should not be different from the rate of injury of the normal population. Indeed, this was the conclusion of Mayer et al.[38]

Some have proposed using variability of maximal scores as an assessment of whether clients "tried their best." It is logical that if you try your best during a maximal exertion, the maximal scores should be similar. However, one must consider that clients with chronic low back pain syndromes can also produce reliable submaximal isometric efforts. Take the study of Hart et al.[39] A subject sample of 92 patients with chronic pain syndromes who were being tested for their functional abilities were asked to produce three submaximal (10 to 90 percent maximal isometric voluntary contractions) standing static arm lifts after completing three maximal arm lifts (maximal isometric voluntary contractions). The results confirm that even in a patient population of people with chronic pain syndromes receiving workers' compensation benefits, many of whom were involved in litigation, subjects could reliably produce submaximal efforts (see Table 17–1). This means that the absence of high variability is not conclusive of the absence of inappropriate symptom behavior. Likewise, the absence of high variability does not confirm that the clients produced their maximum effort. However, it implies that if there is high variability, there could be symptom magnification, inappropriate pain behavior, or a physical reason influencing the variability. Again, this set of data confirms the need for multiple tests to confirm consistency of efforts during functional testing.

To improve the development of valid tests in the industrial rehabilitation clinic, standardized testing protocols (Maryland standards) should be developed and used. The measures used should be tested for reliability, and the systematic analysis of the various forms of validity, including internal consistency of measurements, concurrent validity, and predictive validity of outcomes, should be made. This type of testing frequently will require professionals who have research backgrounds but should be performed in clinics with real clients. An excellent paradigm for testing that can be emulated in clinics is the work by Waddell et al.[40]

Subjective Measurement

Measurements typically fall between two points on a continuum from subjective to objective. The difference between subjective and objective measurements

Table 17–1 Difference in Variability between Maximum and Submaximum Static Arm Lifts

Subjects Performed Three Submaximal Lifts:
Mean ± *SD* = 39.3 ± 21 Lbs, 40.5 ± 22 Lbs, 42.5 ± 23 Lbs*

Source	*df*	*MS*	*F*
Subjects	91	1399.25	49.1**
Reps	2	242.2	8.5**
Residual	182	28.5	

Subjects Performed Three Maximal Lifts:
Mean ± *SD* = 73.5 ± 38 Lbs, 72.6 ± 37 Lbs, 72.9 ± 39 Lbs***

Source	*df*	*MS*	*F*
Subjects	91	4239.6	95.8**
Reps	2	18.8	0.4
Residual	182	44.2	

*ICC = .94, *N* = 92.
**$p<0.05$.
***ICC = .97, *N* = 92.

Source: D.L. Hart et al., "Difference in Variability between Maximal and Submaximal Static Arm Lifts," *Physical Therapy* 69 (1990) (abstract).

is the level of reliability.[41] Subjective measurements are affected by some aspect of the person obtaining the measurement[42] and have low reliability. In the clinical environment, there are many "extraneous variables" that can influence measurements of physical impairment and function and make the measurements less reliable. If the extraneous variables influence the clinician who is recording the measurement, the measurement will be unreliable and therefore invalid.

The most common extraneous variable affecting functional measurements is pain. Quantification of the intensity of pain is not reliable because the measure is influenced by many factors, most of which are not neuroanatomical in nature. If the intensity of pain is unreliable, then measures of pain intensity have limited value for making decisions about clinical treatment. Take, for example, the activity level of those with chronic low back pain. These people universally report lower activity levels. However, when observed, their subjective reports of pain are not closely related to their level of function.[43] If there is a poor relation between reports of pain and physical impairment in people with chronic low back pain,[44] if there is little evidence that early return to ergonomically safe work is detrimental,[45] and if, when they return, the rate of reinjury is no different from the normal population,[46] then the use of self-reported pain intensity when making decisions about returning to work has reduced validity in clients with chronic low back pain.

This is part of the justification for continuing safe functional testing in spite of continued chronic pain in clients with low back pain.

Other extraneous variables that should be considered include litigation, the specific attorney, physician, insurance carrier, adjuster, or therapist working with the client, the referral mechanism (referral for profit), and the local economy. If these factors are ignored, the analysis of your data may not be complete and could affect your results.

Objective Measurement

When the measurement is reliable, the measurement becomes objective. Objective measurements are influenced little by the person obtaining the measurement.[47] Clinical measurements cannot be entirely objective or subjective. Therefore each clinician should test the reliability of clinical measurements or use measurements that have been shown to be reliable.

Measurements made by machines will not always be as reliable as measurements made by clinical observations. Remember, the difference between objective and subjective is the level of reliability.[48] Therefore the assessment of the reliability of your clinical measures is paramount to your understanding of the level of objectivity of your clinical findings.

IMPACT OF SUBJECTIVE FUNCTIONAL INFORMATION ON REHABILITATION PLANNING

In discussing the importance of objective functional measures in the industrial rehabilitation process, the rehabilitation counselor synthesizes specific information to assist the client in making suitable occupational choices. Examples include vocational evaluations (made up standardized paper and pencil psychological tests and/or performance-based work samples with normative data), medical examinations (often performed by a physician with a specific medical specialty but typically not standardized), and a psychological evaluation (performed by a clinical or rehabilitation psychologist and often including a clinical interview and standardized tests, such as the Minnesota Multiphasic Personality Inventory) or, in the case of traumatic brain injury, a neuropsychological evaluation.

In the case of physical disability, the traditional approach, typically demanded from legal precedents, of most rehabilitation counselors has been to request information pertaining to the residual physical capacities of the injured worker from the treating physician or an independent medical examiner. Unfortunately, most physicians do not use standardized or objective techniques to measure functional capacities. The physical information traditionally comes in the form of "work restrictions," which more often than not liberally delineate what the patient *cannot* do. These restrictions are based on the physician's knowledge of the injured worker's diagnosis, prognosis, physical response to treatment, and subjective statements. For vocational rehabilitation planning, what injured workers *can do* is more important in the assessment of their occupational capabilities than what they *cannot do*.

All jobs have physical demands. This is not to say that all jobs are physically demanding. However, all jobs do require some degree of talking, hearing, seeing, walking, standing, sitting, reaching, lifting, carrying, feeling, and so forth. It is incumbent upon the rehabilitation counselor to help clients select occupations that are within ergonomically safe capabilities, physically and otherwise. In doing so, the counselor minimizes the chance of developing and implementing a plan that may result in an unsuccessful outcome or be unsafe for the client. If a rehabilitation plan fails to lead the client toward a positive outcome, the client may experience a great degree of frustration, lose faith in the ability of the counselor to assist in the case, lose motivation to make another attempt in a new direction, and internalize the failure and manifest maladaptive behaviors (see Chapter 28 on self-efficacy). Where many clients begin the vocational rehabilitation process with manifestations of psychosocial impact of the disability, the above scenario compounds the problems and may be devastating.

The counselor's reliance on subjective physical capacity information may adversely affect the occupational potential of the injured worker and the interests of a third-party payor. In situations where subjective physical information is used, the counselor will not be able to devise a feasible, goal-oriented and timely rehabilitation plan. Failure to develop a valid rehabilitation plan may result in a much greater cost to the third-party payor, society, and possibly the injured worker.

STANDARDIZATION OF THE MEASUREMENT PROCESS

To increase our ability to make the best, that is, the most reliable and valid, decisions for the benefit of the injured worker, decisions must be based on the best measurements possible. Unfortunately, when dealing with injured workers, not all measurements are straightforward. A standardized approach to the measurement process is needed to reduce the error involved in making any measurement, particularly in separating subjective from objective measurements.

The determination of the reliability of the observation or measurement and the adherence to standard rules of clinical observation are both important because so many clinical decisions arise from subjective clinical observations as compared to measurements made from simple instruments. If standardized rules are followed in the determinations of these observations, the likelihood of these observations becoming reliable improves. In the clinical management of the injured worker, where there are numerous extraneous variables that can influence clinical decisions, the reliability of the measurements or observations becomes crucial.

STANDARDS OF FUNCTIONAL TESTING

As with many new and different clinical techniques, functional testing, work conditioning, and work-hardening procedures have had to pass the test of time and credibility before gaining acceptance in our communities. Acceptance in some geographical areas was easier than others. There are two specific reasons why these industrial rehabilitation services have had difficulty being accepted in some areas. First, these services do not have the goal of "getting rid of the pain" in many clients. This can be perceived as not designed to "get the client better," which is contrary to the medical model of treatment. Therefore many clinicians and legal advocates have had difficulty accepting the goals of these services.

Second, these services were developed to quantify physical impairment, disability, and function as well as to improve functional abilities and vocational potential and decrease physical impairment and disability in spite of ongoing reports of pain. Therefore industrial services demand that the clinicians make their clinical decisions from "measurements." This philosophy is not consistent with many of the current philosophies of clinicians, who have been trained to continue to treat clients as long as painful syndromes continue, which may be long after the effectiveness of their treatment has ceased. Therefore the measurements of industrial rehabilitation services were contrary to tradition.

Finally, at this time in history, insurance carriers and governmental organizations are demanding justification for clinical services prior to payment for those services. Unfortunately, there is little literature to support payment for many clinical services. Therefore, in lieu of clinical research efficacy studies, governing bodies are requesting treatment standards, guidelines, and protocols with which clinicians must comply prior to being remunerated.

Hart et al. offer a set of guidelines for functional testing that can be used as an initial set of standards from which more thorough criteria can be developed.[49] The guidelines describe indications, contraindications, purposes, definitions, procedures, and so forth for testing function.

The Joint Committee of Industrial Services for the state of Maryland[50] has advanced these basic standards in conjunction with the standards from the American Physical Therapy Association[51] to provide a comprehensive set of standards for industrial services. The purpose of this committee was to respond to the request of the Industrial Commission of Maryland to establish treatment standards in lieu of research to justify such services. Similar standards are being requested by political bodies around the country for the purpose of allowing nonmedical personnel to judge the variety of clinical services from which they will make determinations of what to remunerate. This will become more common for all medical practices until valid outcome measures for our services are developed.

OUTCOME MEASURES IN INDUSTRIAL SERVICES

Clinicians are being pressed for outcome data that support their services. Insurance carriers, vocational rehabilitation counselors, employers, and physicians all request outcome data. The process of collecting good outcome data is difficult because of the problem of controlling extraneous variables and developing control groups in the clinic outside of the university research environment. Take, for example, the priority outcome measure: return to work and its corollary of length of time the client remains on the job following work hardening. These pieces of data are easy to collect by simply following the client over time with telephone or mail response questionnaires. Unfortunately, if you want the outcome data to reflect the effect of the industrial clinical services delivered—that is, work hardening—collection and analysis of the effect of many, if not all, of the extraneous variables that can affect the outcome must be made. Only when the effect of extraneous variables is controlled is there good external validity.

Some extraneous variables that influence outcome in the industrial rehabilitation setting are

1. employers who fire the client shortly after the client returns to work
2. attorneys who encourage more treatment with chronic pain syndromes or encourage litigation instead of return to work
3. physicians who will not release the client to work within measured functional abilities or who encourage continued treatment with chronic pain syndromes
4. therapists who encourage continued treatment with chronic pain syndromes
5. poor economic times that result in the loss of the previous job or returning to another job even with reasonable accommodations
6. employees who have not been valued employees in the past and therefore are not wanted back
7. financial incentives for referral to clinical services
8. poor client self-efficacy (see Chapter 28)

If outcome data are collected, these extraneous variables need to be measured to place the variables under statistical control. This will increase the validity of the outcome data and improve external validity of the outcome data project.

If clinicians only track return to work as their outcome measure of interest, their results should be considered very limited for external validity. In many respects, their results should be considered spurious because of the lack of control of extraneous variables. However, one can predict that when the above variables are controlled, the reported rates of return to work following functional training—for example, greater than 80 percent[52]—may, in reality, be higher.

Deyo has recommended that outcome measures for complaints that are common in industrial services, such as low back pain, should consist of symptoms, range of motion, strength, functional status, psychologic measures, and costs/utilization of services.[53] Of course, any measure should have good reliability prior to being used as an outcome measure. It should be emphasized that Deyo's use of functional measures relates to previous published questionnaires of perceived disability or functional ability, not measured functional abilities from functional capacity evaluation (FCE) tests. This may be because we do not have reliability or validity studies on FCEs. These studies need to be performed. The probability that the measured level of safe work (and limitations) through the use of an FCE will be one of the best outcome measures is good.

In any case, Deyo recommends and Waddell et al. confirm that we should measure the outcome variables of interest, not simply assume an association between such variables as function and physical impairment.[54] In a well-designed and controlled study, Waddell et al.[55] provide a scientific paradigm for all of us to emulate. They start with good operational definitions of physical impairment measures (supported by previous research).[56] These measures were tested for their reliability. If the reliability was not high enough, the researchers improved the standardization or skill of taking the measure until the reliability was adequate. Those tests that could not be improved in reliability were eliminated (a process most clinicians should accept!). The reliable tests whose incidence in the patient population was adequate were compared to behavioral signs.[57] Then measures were compared between normals and patients with chronic low back pain to determine if the measures could discriminate between patients and normals. The measures of physical impairment were compared to the level of perceived disability as determined through questionnaires. Finally, the physical impairment tests that were reliable, that discriminated between patients and normals, and that explained disability were grouped as a "Final Physical Impairment Scale" for people with chronic low back pain.

Waddell et al.'s paper is an excellent example of a well-designed research project on validity that is clinically practical. Through its process, one learns that in people with chronic low back pain, the Final Physical Impairment Scale consists of tests that are measures of current functional limitation rather than permanent anatomic or structural impairment. This fact casts strong doubt on the physical basis of permanent disability (the American Medical Association calls this "medical impairment") due to chronic low back pain.[58] From a perspective of measurement, we need to measure functional abilities, and we should be prepared to accept the strong possibility that in this population functional disability may not be related to physical impairment.

The need for further testing of reliability and validity of functional capacity measures is punctuated by the fact that the estimated functional capacities requested by and received from physicians commonly are invalid if not measured through simulated activities. For example, in one study by Frey and Netherton elderly patients were assessed on their ability to function around their home. The patients' assessments of their functional abilities at home were compared to nurses' assessments of those abilities after the nurses observed the patients at home. The physicians' assessments of their patients' abilities to function at home were also compared to those of the other two groups. There was poor agreement between the physicians' reports of function and either the patients' or the nurses' assessments (even though the nurses had observed patient function on site).[59]

The implications of these findings for injured workers receiving compensation benefits are twofold. First, to improve the validity of an assessment of function, one needs to assess the injured worker performing tasks during simulated work activities to improve the validity of the functional measurement. Second, if the functional tasks are not assessed during simulated functional testing, there will be little correlation between the findings during a typical clinical physical examination and the actual ability to function on the job.[60]

To better understand the need to measure simulated functional tasks, one simply needs to compare the functional "restrictions" reported by physicians on their patients to the results of measured functional capacity evaluations of patients in industrial clinics. There is typically no correlation. This is not surprising, with the conclusions of Waddell et al. confirming that there is no relation between measured physical impairment and patient-reported perceived disability and the results of Frey, confirming that simulated activities should be considered the new measure of "disability."[61]

CLASSIFICATION OF PATIENTS FOR CLINICAL EFFICACY STUDIES

Possibly the best approach to the future of reliable, valid, and practical clinical outcome studies is the work performed by those who classify patients into treatment groups according to clinical signs and symptoms. Delitto et al.[62] confirmed that clinicians could reliably classify patients into treatment groups by using a standardized clinical classification system based on signs and symptoms. The authors then tested the outcome of treating patients who had been classified by clinical movement signs and symptoms according to their treatment classification system, and demonstrated that patients had improved outcome if classified and treated according to the classification system.[63]

From these papers, those in industrial services can learn that they must classify clients according to classification systems that are yet to be developed. Systems that classify these clients must be tested for their reliability. Outcome or prescriptive validity studies must be performed to confirm if function and physical impairment are quantified, and then these variables must be improved with treatment. Only then will we have justification for our services.

In the field of identification and classification of patient behaviors in industrial rehabilitation, in order to be able to trust the validity of functional capacity measurements, the level of maladaptive behavior or symptom magnification tendencies must be interpreted. At this time, only initial descriptions of this behavior have been published.[64] These descriptions produce the foundation for the investigation into the classification of clients with behavior that may negatively influence their commitment to their own rehabilitation process (see Chapter 28). This area awaits further investigation.

HOW CAN YOUR CLINIC ESTABLISH ITS OWN RELIABILITY AND CREDIBILITY?

Most clinical professions, including physical and occupational therapy, have been forced to justify their clinical credibility. The following plan may be used in preparation for the defense of a clinic and its clinical procedures.

Step 1: Standardize All of the Clinical Procedures.

Clinicians cannot expect to confirm reliability of their clinical procedures and measurements if the procedures or measurements are not conducted in a standardized manner. Take, for example, the clinician who uses different criteria for the identification of a maximal lifting activity depending on whether a male or a female is being tested. Without standardized procedures for the lifting test regardless of whether the client is a male or female, the chance that the results of the test will be consistent or reliable is low. Measures of low reliability should not be used to make clinical decisions. If the reliability is low, improve the standardization and retest the reliability.

The astute clinician should write all procedures down and study them to make sure that what is written matches what is done clinically. All clinicians should participate. When a measurement process is identified that is different between clinicians, the procedure should be modified through consensus. If the standardized approach to the procedure is not agreeable to all, the procedure should be eliminated. The final product will be a set of procedures on which all of a clinic's clinicians can agree.

Step 2: Operationally Define All Observations/Measurements.

During the process of writing down procedures/measurements, the need to operationally define all procedures/measurements will be discovered. They should be defined, and communication enhanced among clinicians. If thoughts cannot be communicated internally in a clear and concise way, they will not be communicated well to outside parties. Communication is a two-way street: the clinicians' half of the communication process should be standardized and operationally defined.

The literature can help in this process. The operational definitions of others for whom you and the literature have respect may be used. For example, Waddell et al.[65] defined the tests used for the measurements of physical impairment and nonorganic physical signs for people with low back pain. Matheson defined the term *symptom magnification*.[66] The American Physical Therapy Association defined *work hardening* and *work conditioning*.[67] Using accepted operational definitions will save you time. However, one must not expect to find all procedures and measurements defined in the literature or by other clinicians. The clinician should be stimulated to define clinical procedures and measurements because many of the industrial rehabilitation procedures and measurements have yet to be standardized or defined.

Step 3: Test for the Reliability of Measurements.

Reliability is not difficult to determine. If it appears complicated, the clinician can call a research consultant to assist.

Obtaining and training on a new piece of equipment offers a good opportunity to test for reliability. Unless it is a piece of equipment that others have tested, it will require testing for reliability. Computerized motion analysis equipment may be useful in the litigious workers' compensation system, but attorneys may request reliability studies on your ability to convert a video image to a numerical assessment of a motion disorder of their client prior to their referring clients to be tested. In this case, your research efforts become a pathway to future referrals. If not, your pseudoresearch efforts could cause embarrassment and loss of credibility.

Step 4: Determine the Relation between Functional Measures and Physical Impairment/Disability.

Waddell et al. have eloquently and scientifically confirmed what many had predicted: there is at best a poor relation between physical impairment and disability of people with chronic low back pain.[68] The relation between physical impairment and function has yet to be determined. Therefore the true meaning of measurements and procedures must be further evaluated.

An easy way to start this assessment is to compare the results of two different measurement techniques or instruments. This will provide an assessment of concurrent validity. For example, it would seem logical that there would be a relation between the ability to climb ladders and the ability to ascend stairs—two functional measures. Likewise, if there is a limitation of shoulder flexion (physical impairment), and if the ability to lift an object over one's head (functional measure) requires shoulder flexion, the two measures should be related.

A good simple clinical research example of determination of relations between clinical observations is the paper by Tenhula et al.[69] In this paper, the authors confirmed the relation between clinical observations of lateral shift of the trunk and the reproduction of clinical symptoms by means of a simple research paradigm. Those interested in this type of work can use this project as a good example.

The relationships between procedures and measurements must be investigated. This allows interpretation of the internal consistency of testing procedures.

Step 5: Monitor Long-Term Outcome for Validity of Measurements.

Predictive validity is probably the most valued prize in all research. There is no well-designed published study that confirms whether clients return to the predicted level of physical activity because of functional testing or work hardening. If they do return to work within their measured functional levels, will that increase the level of future pathology? Mayer et al. reported that the rate of injury of their population of patients with chronic pain syndromes was no higher than that of the normal population if patients returned to work within their measured functional capacity.[70]

To improve external validity of outcome studies, the many extraneous variables that may affect the outcome of clients need to be collected. For example, the following can be included in outcome data collection at a minimum: demographic client information, treatment type, time in program, costs (direct and indirect), reason for progression and discharge in the program, physician, insurance carrier, adjuster, vocational rehabilitation company and counselor, industry, attorney (plaintiff and defense), changes in economic stability in area, and any other reason why the client may or may not return to work. It is better to collect more than less information for outcome data.

Step 6: A Classification System of Clinical Client Traits That Affect Physical Impairment, Disability, and Function Can Be Developed.

Since a confirmable diagnosis is still elusive for the majority of clients with low back pain syndromes,[71] classifying clients into groups can be used for directing treatment or functional programs. For example, medical diagnoses should not be considered as appropriate differentiaters from which to design treatment, since the probability of those diagnoses being reliable and valid is not good. When clinicians classify their clients into treatment groups for which they design specific treatments, the results of their treatments may be better[72] than the results of treatment protocols that are provided to patients that have not been classified.[73] A classification system for clients in industrial rehabilitation programs needs to be developed.

CONCLUSION

The level of subjectivity versus objectivity of measurement becomes a discussion of the reliability of measurement. Since clinical measurements cannot be truly 100 percent subjective or objective, and since objectivity and reliability are measured along a continuum,[74] the degree of reliability needs to be established for each measurement. In the absence of established levels of reliability, a standard set of rules of measurement must be followed for each measurement that will improve the chances of the measurement being reliable.

A subjective measurement has poor reliability and an objective measurement has high reliability. Measurements obtained from "high-tech" machines are not necessarily objective, and likewise measurements from "low-tech" observations/devices are not necessarily subjective.[75]

The classification of clients into treatment groups should positively influence the ability to collect quality outcome studies on the efficacy of industrial rehabilitation services. Only then will clinicians be able to justify their services.

NOTES

1. D.L. Hart, Standards in Tests and Measurements, *Orthopedic Physical Therapy Clinics* 1 (1992):75–82.
2. J.A. Turner et al., Patient Outcomes after Lumbar Spinal Fusions, *JAMA* 268 (1992):907–911.
3. Ibid., p. 910.
4. A.F. DePalma and R.H. Rothman, The Nature of Pseudarthrosis, *Clinics in Orthopaedics* 59 (1968):113–118; T.J. Flatley and H. Derderian, Closed Loop Instrumentation of the Lumbar Spine, *Clinics in Orthopaedics* 196 (1985):273–278.
5. Hart, Standards.
6. G. Waddell et al., Objective Clinical Evaluation of Physical Impairment in Chronic Low Back Pain, *Spine* 17, no. 6 (1992):617–628.
7. G. Waddell, A New Clinical Model for the Treatment of Low-Back Pain, *Spine* 12 (1987):632–644; G. Waddell et al., Nonorganic Physical Signs in Low-Back Pain, *Spine* 5 (1980):117–125; W.E. Fordyce et al., Pain Complaint-Exercise Performance Relationship in Chronic Pain, *Pain* 10 (1981):311–321.
8. D. Apts, personal communication.
9. S.W. Wiesel et al., A Study of Computer-Assisted Tomography. 1. The Incidence of Positive CAT Scans in an Asymptomatic Group of Patients, *Spine* 9 (1984):549–551.

10. W.D. Frey and D.R. Netherton, Functional Assessment: A New Measure for Disability, *Business and Health* (June 1988):30–34.
11. D.L. Hart et al., Guidelines for Functional Capacity Evaluation of People with Medical Conditions, *Journal of Orthopaedics and Sports Physical Therapy* 18, no. 6 (1993):682–686; D.L. Hart et al., *Standards for Personnel Performing Functional Capacity Evaluations, Work Conditioning and Work Hardening Programs* (Baltimore, Md.: State of Maryland, 1993).
12. Hart et al., Guidelines; Hart et al., *Standards;* P. Helm-Williams, P: Industrial Rehabilitation: Developing Guidelines, *PT Magazine*, March 1993, 65–72.
13. Frey and Netherton, Functional Assessment; S.J. Isernhagen, *Work Injury: Management and Prevention* (Gaithersburg, Md.: Aspen Publishers, Inc., 1988).
14. Americans with Disabilities Act, 42 U.S.C. §§ 12101-12213, 1990.
15. T.M. Khalil et al., Acceptable Maximum Effort (AME): A Psychophysical Measure of Strength in Back Pain Patients, *Spine* 12 (1987):372–376.
16. Task Force on Standards for Measurement in Physical Therapy, Standards for Tests and Measurements in Physical Therapy Practice, *Physical Therapy* 71 (1991):589–622.
17. Hart, Standards; Task Force, Standards.
18. Task Force, Standards.
19. T.G. Mayer et al., Use of Noninvasive Techniques for Quantification of Spinal Range-of-Motion in Normal Subjects and Chronic Low-Back Dysfunction Patients, *Spine* 9 (1984):588–595; American Medical Association, *Guides to the Evaluation of Permanent Impairment*, 3rd ed. rev. (Chicago, 1990), 86–101.
20. D.L. Hart and S. Rose, Reliability of Non-Invasive Measures of the Posture of the Trunk, *Journal of Orthopaedics and Sports Physical Therapy* 8 (1986):180–184.
21. Hart, Standards; Task Forces, Standards.
22. Task Force, Standards.
23. R.M. Nelson and D.E. Nestor, Standardization Assessment of Industrial Low-Back Injuries: Development of the NIOSH Low-Back Atlas, *Topics in Acute Care and Trauma Rehabilitation* 2 (1988):16–30; National Institute for Occupational Safety and Health (NIOSH), *National Institute for Occupational Safety and Health Low Back Atlas* (Morgantown, West Va.: U.S. Dept. of Health and Human Services, 1988).
24. Waddell et al., Objective Clinical Evaluation.
25. A. Delitto et al., Evidence for Use of an Extension-Mobilization Category in Acute Low Back Syndrome: A Prescriptive Validation Pilot Study, *Physical Therapy* 73 (1993):216–228; A. Delitto et al., Reliability of a Clinical Examination to Classify Patients with Low Back Syndrome, *Physical Therapy Practice* 1 (1992):1–9.
26. D.L. Hart and L.E. Shauf, Test-Retest Reliability of the Static Arm Lift and Hand Grasp Tests for Functional Capacity Evaluations, *Physical Therapy* 68 (1988):825 (abstract).
27. Ibid.
28. D.L. Hart et al., Test-Retest Reliability of the Static Push and Pull, *Physical Therapy* 68 (1988):824 (abstract).
29. Task Force, Standards.
30. Hart, Standards.
31. Khalil et al., Acceptable Maximum Effort; Hart and Shauf, Test-Retest Reliability; D. Chaffin et al., Preemployment Strength Testing, *Journal of Occupational Medicine* 20, no. 6 (1978):403–408.

32. T.G. Mayer et al., Progressive Isoinertial Lifting Evaluation. I. A Standardized Protocol and Normative Database, *Spine* 13 (1988):933–997; J. Alpert et al., The Reliability and Validity of Two New Tests of Maximum Lifting Capacity, *Spine* 1, no. 1 (1991):13–29; L. Matheson et al., Effect of Instructions on Isokinetic Trunk Strength Testing Variability, Reliability, Absolute Value, and Predictive Validity, *Spine* 17, no. 8 (1992):914–921.
33. D.L. Hart, unpublished data, and confirmed by L.N. Matheson, unpublished data, personal communication.
34. M. Newton and G. Waddell, Trunk Strength Testing with Iso-Machines. Part 1: Review of a Decade of Scientific Evidence, *Spine* 18, no. 7 (1993):801–811.
35. M. Newton et al., Trunk Strength Testing with Iso-Machines. Part 2: Experimental Evaluation of the Cybex II Back Testing System in Normal Subjects and Patients with Chronic Low Back Pain, *Spine* 18, no. 7 (1993):812–824.
36. Ibid.
37. Alpert et al., Reliability; Matheson et al., Effect of Instructions.
38. T.G. Mayer et al., A Prospective Two-Year Study of Functional Restoration in Industrial Low Back Injury, *JAMA* 258 (1987):1763–1767.
39. D.L. Hart et al., Difference in Variability between Maximal and Submaximal Static Arm Lifts, *Physical Therapy* 69 (1990) (abstract).
40. Waddell et al., Objective Clinical Evaluation.
41. J.M. Rothstein, On Defining Subjective and Objective Measurements, *Physical Therapy* 69 (1989):577–579.
42. Task Force, Standards.
43. W.E. Fordyce et al., Pain Complaint-Exercise Performance Relationship in Chronic Pain, *Pain* 10 (1981):311–321.
44. Waddell et al., Objective Clinical Evaluation.
45. Waddell, New Clinical Model.
46. Mayer et al., Prospective Two-Year Study.
47. Task Force, Standards.
48. Rothstein, On Defining.
49. Hart et al., Guidelines.
50. Hart et al., Standards.
51. Helm-Williams, Industrial Rehabilitation.
52. Mayer et al., Prospective Two-Year Study.
53. R. Deyo, Measuring the Functional Status of Patients with Low Back Pain, *Archives of Physical Medicine and Rehabilitation* 69 (1988):1044–1053.
54. Ibid.; Waddell et al., Objective Clinical Evaluation.
55. Waddell et al., Objective Clinical Evaluation.
56. G. Waddell et al., Clinical Evaluation of Disability in Back Pain, in *The Adult Spine: Principles and Practice*, ed. J.W. Frymoyer (New York: Raven Press, 1991), 155–167.
57. Waddell et al., Nonorganic Physical Signs.
58. American Medical Association, *Guides;* Waddell et al., Objective Clinical Evaluation.
59. Frey and Netherton, Functional Assessment.
60. Ibid.

61. Waddell et al., Objective Clinical Evaluation; Frey and Netherton, Functional Assessment.
62. Delitto et al., Reliability.
63. Delitto et al., Evidence.
64. L.N. Matheson, *Symptom Magnification Syndrome* (Anaheim, Calif.: Employment and Rehabilitation Institute of California, 1987); L.N. Matheson, Symptom Magnification Syndrome Structured Interview: Rationale and Procedure, *Journal of Occupational Rehabilitation* 1, no. 1 (1991):43–56.
65. Waddell et al., Nonorganic Physical Signs.
66. Matheson, *Symptom Magnification Syndrome;* Matheson, Symptom Magnification Syndrome.
67. Helm-Williams, Industrial Rehabilitation.
68. Waddell et al., Objective Clinical Evaluation.
69. J.A. Tenhula et al., Association between Direction of Lateral Lumbar Shift, Movement Tests, and Side of Symptoms in Patients with Low Back Pain Syndrome, *Physical Therapy* 70, no. 8 (1990):480–486.
70. Mayer et al., Prospective Two-Year Study.
71. W.O. Spitzer et al., Scientific Approach to the Assessment and Management of Activity-Related Spinal Disorders, *Spine* 12 (1987):S16–S21.
72. Delitto et al., Evidence.
73. Spitzer et al., Scientific Approach.
74. Task Force, Standards.
75. Rothstein, On Defining.

Chapter 18

Cardiopulmonary System: Implications for Work Evaluation and Rehabilitation

Stephen T. Anderson

This chapter discusses cardiopulmonary function measurements and how they can be applied to the process of returning the injured worker to his or her job. The saying "Use it or lose it" can be applied to the injured worker who is unable to perform even moderate aerobic exercise. The longer the worker remains impaired as a result of a physical injury, the greater will be the worker's reduction in cardiopulmonary function. Since each job in the workplace requires a specific amount of energy expenditure, the fact that a worker becomes injured in the first place gives rise to the question "What caused the injury?" Might the original cause of the injury be related to the worker's lack of functional aerobic capacity to meet the energy requirements of the job? And, once the physical injury is repaired, will the now-deconditioned worker have the necessary aerobic capacity to perform the job? This chapter provides a discussion of measurement and training techniques to be able to answer such questions conclusively.

Appreciation of the clinical and research value of measuring physiological function during conditions of stress has increased dramatically in recent years. The earliest application of stress testing was in the evaluation of the maximum performance potential of athletes. These same techniques were then applied to the study of individuals with circulatory and pulmonary diseases. With the advent of the microprocessor, the cost and complexity of performing stress tests were reduced, allowing for a more rapid and more complete evaluation of physiologic function as well as a simplification of the use of the testing systems. As a result, stress testing has become a major tool for use in evaluating the functional aerobic capacity of an individual and in forming and evaluating exercise training programs.

Historically, stress testing has had, as its main focus, the evaluation of "maximal" performance, most commonly using estimates of work based upon treadmill speed and grade. However, research has shown that estimating $\dot{V}O_2$ from maximal treadmill time can produce results that are statistically different from techniques utilizing direct measurement of $\dot{V}O_2$. In one study the 95 percent confidence limits for predicting $\dot{V}O_2$ based on treadmill time ranged nearly 6 METs.[1] In addition to providing poor reproducibility for the purpose of evaluating clinical interventions, treadmill time cannot provide the clinical information that is even more useful for evaluating patients' frequently submaximal ability to perform activities of daily living. With the increase in understanding of submaximal exercise physiology and the reduction in cost and complexity of stress-testing instrumentation, the very purpose and utilization of exercise stress testing as a clinical tool are evolving rapidly. To delineate the difference between the classical treadmill test and the evolving testing technology this chapter will use a separate terminology—Global Aerobic Testing and Evaluation (GATE)—to discuss this technology and its clinical applications.

Because GATE is noninvasive and easily tolerated by the patient, full clinical utilization is now possible to answer the following general questions:

- Does the patient have signs of disease?
- Does the patient require further investigation before participation in a rehabilitation program?
- What is the cause of the patient's symptoms?
- What is the patient's degree of impairment?
- How is the patient responding to therapy?

Similarly, because of the high degree of subject tolerance, GATE is being used increasingly to evaluate performance of a wide range of individuals, from those that are merely sedentary to professional athletes. Physiologic variables measured by the GATE technology provide objective data upon which an exercise training program can be formed and upon which, once implemented, the effectiveness of the exercise training program can be evaluated. GATE technology can also be used to answer the following specific questions:

- What is the subject's functional aerobic exercise capacity?
- What exercise intensity can the subject safely tolerate?
- At what exercise intensity can the subject obtain the greatest improvement in functional aerobic exercise capacity?
- Is the exercise training program producing positive results?

Answering the latter question is particularly significant in light of today's health care environment, with its increasing requirements for documentation of improvement resulting from the provision of therapy. It is also important to the patient. To understand just how important the provision of objective data is to a recovering patient, one simply needs to ask one of the many patients who have avoided heart transplantation and who have been rehabilitated instead by a combination of exercise and pharmaceutical therapy. The objective measurements discussed in this chapter become the "carrot" that provides the positive feedback needed to encourage the patient to comply with the exercise prescription and thus to recover. To an injured worker such feedback can mean a quicker return to the job and an increased awareness of injury and disease prevention.

PHYSIOLOGIC BASIS

For better appreciating the role of the GATE technology, a basic understanding of exercise physiology is important.

Skeletal Muscles

Physical activity significantly increases the demand for energy—as much as 10 to 20 times above resting values. The body's ability to respond to this increased demand for energy is a result of the skeletal muscles' ability to vary their metabolic rate. The variance is produced by a complex and closely integrated system made up of ventilation, diffusion from the alveoli to the pulmonary capillaries, circulation, and the diffusion gradient of O_2 from the blood to the muscle mitochondria. Wasserman[2] lists five factors that affect the oxygen supply to the cells with increased metabolic activity:

1. the partial pressure of O_2 in the arterial blood
2. the hemoglobin concentration and arterial O_2 content
3. the cardiac output
4. the distribution of perfusion to the tissues in need of O_2
5. the hemoglobin's affinity for O_2

The absolute intensity of the physical activity has a major effect on the transformation in metabolism that occurs during exercise. This is a consequence of (1) the structure of the nervous system and (2) the presence of two major types of fiber in skeletal muscle. Figure 18–1 illustrates how the body is set up to accommodate a wide range of physical demands through specific and segregated pathways of nerves, muscles, and fuels. The red, slow-twitch muscle fibers metabolize glucose aerobically into carbon dioxide and water with a release of large amounts of energy as ATP. The white, fast-twitch fibers break down glucose anaerobically with much lower ATP yield and the release of lactic acid. Since white muscle fibers are available for isometric and forceful contractions, they frequently are used at times when blood flow is inadequate or unavailable. Consequently they can be regarded

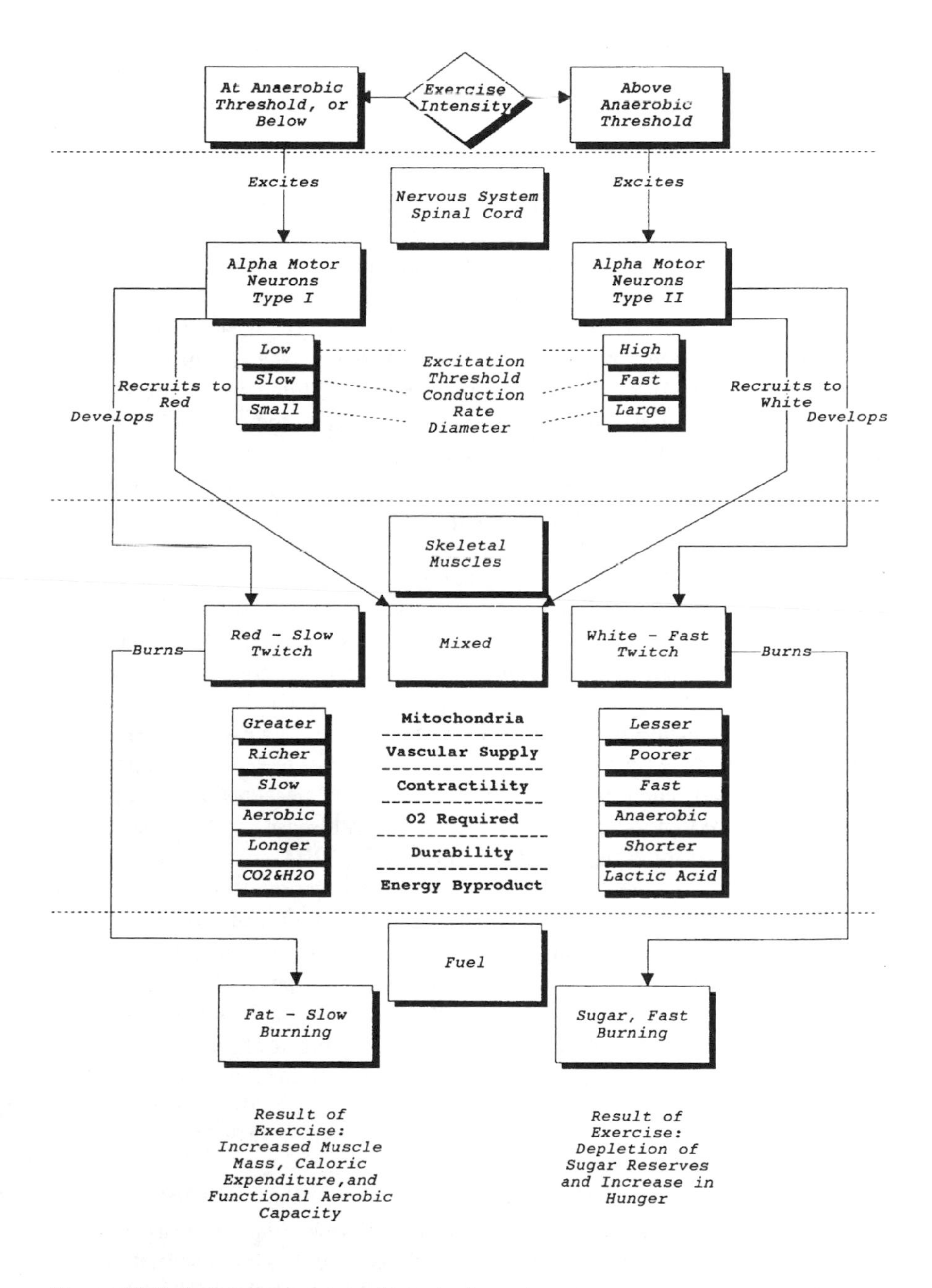

Figure 18–1 Body's Responses to Exercise Intensity

as a reserve system to be recruited in times of high metabolic demand or inadequate oxygen delivery.

Anaerobic Threshold and $\dot{V}O_2$ Max

The relative contribution of red muscle fibers (aerobic metabolism) and white muscle fibers (anaerobic metabolism) varies with the intensity of exercise. Figure 18–2 illustrates the relationship between work rate and $\dot{V}O_2$ below anaerobic threshold. Red fiber recruitment is progressive and linear as the exercise intensity increases, and oxygen consumption increases linearly as well. A further progressive increase in oxygen consumption is eventually halted when the mechanisms of oxygen delivery and/or oxygen uptake are saturated. White muscle fibers are not recruited in significant numbers at low levels of exercise, and no significant lactic acid is generated. However, at moderate exercise levels or when the exercise has a high component of isometric work, white fibers begin to be recruited in large numbers, lactic acid is generated, and lactate quickly accumulates in the blood. When this occurs it is known as the anaerobic threshold (AT). Recent arguments have been made as to whether lactate during exercise in fact increases in a pattern that is mathematically "continuous" rather than as a threshold.[3] An alternative concept has therefore been postulated that the anaerobic threshold reflects an imbalance between lactate appearance and disappearance, and the term *ventilatory threshold* (VT) has been suggested as an alternative term. Irrespective of the actual mechanisms involved, "lactate does accumulate in the blood during exercise, ventilation must respond to maintain physiological pH, a breakpoint in ventilation appears to occur reproducibly, and this point is related to various measures of cardiopulmonary performance in normals and patients with heart disease."[4] Anaerobic threshold can be defined in terms of oxygen consumption ($\dot{V}O_2$ in ml/kg/min), or the work rate or heart rate at which the breakpoint in ventilation occurs. The "saturation" point is known as $\dot{V}O_2$ max, the upper limit for O_2 utilization at a particular state of fitness or training. Technically defined, $\dot{V}O_2$ max requires that a plateau occurs on a plot of $\dot{V}O_2$ versus work rate. Maximum, or peak, $\dot{V}O_2$ is simply the highest $\dot{V}O_2$ observed during a stress test. $\dot{V}O_2$ max and peak $\dot{V}O_2$ may or may not be equal, depending on the objectives of the test, the effort of the patient, and/or the effort on the part of the technician performing the test to elicit a maximal effort by the patient.

Circulation and the Lungs

Oxygen consumption, carbon dioxide production, ventilation, heart rate, and the work rates at which these physiologic variables are measured are determined by close integration of the skeletal muscle metabolism with the cardiopulmonary

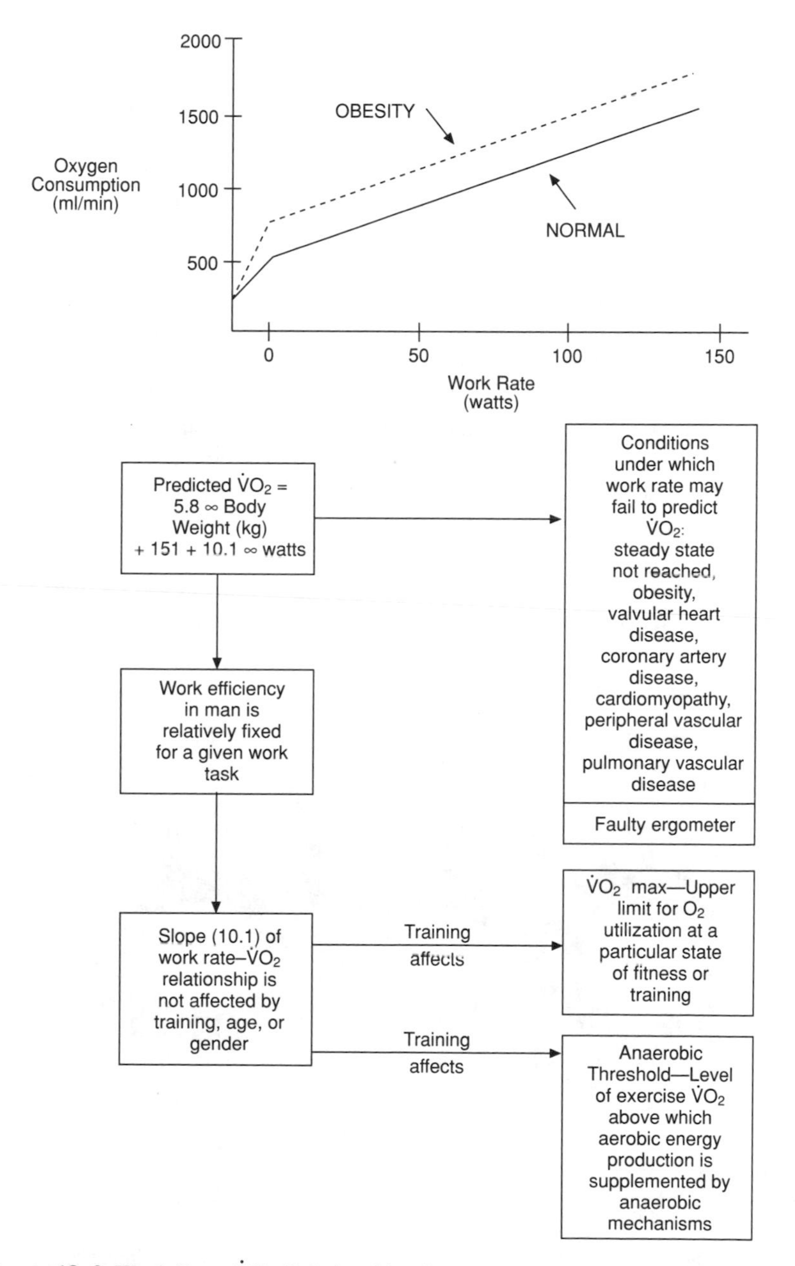

Figure 18–2 Work Rate–$\dot{V}O_2$ Relationship. *Source:* Reprinted from *Principles of Exercise Testing and Interpretation* by K. Wasserman et al., p. 15, with permission of Lea & Febiger, © 1987.

system. Wasserman has represented the gas transport mechanisms for coupling cellular (internal) to pulmonary (external) respiration as interconnected gears (Figure 18–3). At submaximal exercise intensities, the level of muscle activity determines the rate of oxygen consumption and carbon dioxide production. Through a complex signaling and response mechanism, the circulation provides an increase in blood flow that precisely matches metabolic demand. Cardiac output increases through an increase in stroke volume and heart rate, and the additional flow is directed to exercising muscle through precise adjustments in the peripheral circulation, mediated by local skeletal muscle vasodilation and constriction in other areas. Ultimately the lungs are responsible for delivery of oxygen to the blood and removal of carbon dioxide so that the link between skeletal muscle and the environment is complete. The values indicated on the right side of the gears, $\dot{V}O_2$ and $\dot{V}CO_2$, have been referred to as measures of "global function" since the function of all of the mechanisms to the left is to deliver energy-sustaining oxygen from the atmosphere and to remove the waste product of energy combustion, carbon dioxide. The local functions must all perform in concert to contribute to the global end result.

A limitation in any one of the local functions can limit the anaerobic threshold or the $\dot{V}O_2$ max. Such limitations include (1) muscle metabolism, (2) peripheral

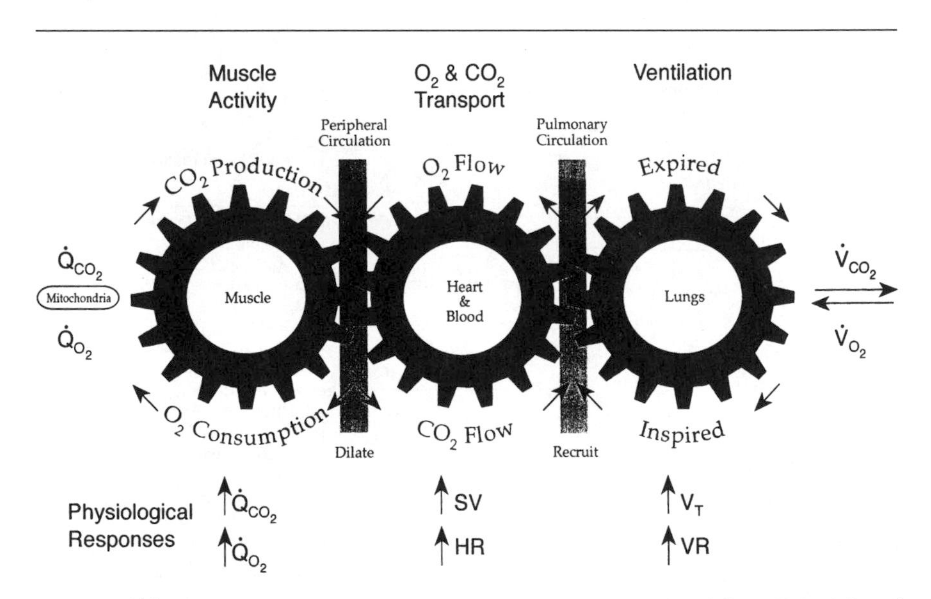

Figure 18–3 Exercise Physiology: Global/Local Function. Reprinted from *Principles of Exercise Testing and Interpretation* by K. Wasserman et al., p. 2, with permission of Lea & Febiger, © 1987.

circulation, (3) O_2 carrying capacity, (4) heart, (5) lungs, and (6) pulmonary circulation. Extensive research dedicated to resolving the question of the source of the limitation has indicated that the heart is the limiting factor. Therefore maximal oxygen consumption is determined by, and is a marker of, maximal cardiac reserve.

The relationship between oxygen uptake (what is used) and delivery (what is available) is represented by the Fick equation: Oxygen Uptake = Cardiac Output × Arteriovenous O_2 Difference. Since the maximal systemic $A\text{–}\bar{v}O_2$ difference is relatively constant across a wide range of subjects, differences in $\dot{V}O_2$ max among individuals reflect differences in cardiac output rather than differences in systemic $A\text{–}\bar{v}O_2$ difference.

$\dot{V}O_2$ max is often referred to as the level of cardiorespiratory or aerobic fitness. $\dot{V}O_2$ max varies slightly between men and women, is affected considerably by age, and is dependent on the level of training of an individual.[5]

Other Physiologic Measurements for Assessing "Local" Function

ECG—An exercise ECG is often taken to assess myocardial ischemia. A blockage of a coronary artery impedes the delivery of oxygenated blood to the heart muscle, which alters the repolarization of the heart in the ischemic areas. This can be detected by observing the changes in the T wave and ST segments of the electrocardiogram. Presence of these changes does not in all cases confirm coronary blockages, and absence of them does not necessarily rule out coronary blockages. But when these changes are accompanied by chest pain, a decline in arterial pressure, or a flattening of the $\dot{V}O_2$–work rate relationship, the diagnosis of ischemic heart disease is more certain.[6] An increase in the frequency of ectopic beats as the work rate increases also is suggestive of myocardial ischemia.

O_2 Pulse—The local function in Figure 18–3 can be described in terms of heart rate (HR) and stroke volume (SV), the volume of blood pumped in one beat of the heart. The product of these two values is cardiac output. By substituting *HR × SV* for cardiac output in the Fick equation and rearranging the formula, O_2 Pulse ($\dot{V}O_2$ /HR) = SV × Arteriovenous Difference. O_2 pulse is the amount of oxygen removed from each heartbeat. At maximal exercise, O_2 pulse is directly related to stroke volume.

Heart Rate Reserve (HRR)—HRR = Predicted Maximum HR – Maximum Exercise HR. This can also be calculated as a percentage. This value is useful for evaluating the relative stress during exercise—a small percentage reserve indicates that the patient has exercised close to his or her maximum.

Spirometry—Spirometry can be incorporated into a cardiopulmonary exercise stress-testing system by adding a software package. This can then be used to assess the "local function" of the lungs. Performed in a short period of time prior to the stress test, the lung functions of vital capacity (VC), forced expiratory volume after 1 second (FEV1), and maximum voluntary ventilation (MVV) can be measured.

Minute Ventilation ($\dot{V}E$)—This is the volume of air moving in and out of the lungs expressed as liter/min.

Breathing Reserve (BR)—BR = MVV (maximal voluntary ventilation) – $\dot{V}E$ (minute ventilation) at maximum exercise. This is the theoretical additional ventilation available at the cessation of exercise.

Dyspnea Index (DI)—DI = $\dot{V}E$ max/MVV × 100. Dyspnea is sensed when the DI exceeds 50 percent. Since a normal exercise test might terminate with a DI of 60 percent, the patient will experience shortness of breath. Except in the case of some athletes, a DI of over 70 percent will produce respiratory muscle fatigue in minutes; when over 90 percent, fatigue will occur in seconds. Both BR and DI are useful to determine which "local function" is limiting exercise capacity, the lungs or the heart.

Carbon Dioxide Production ($\dot{V}CO_2$)—This is the volume of carbon dioxide produced by the body during exercise expressed in liters/min. Carbon dioxide is a byproduct of oxidative metabolism.

Ventilatory Equivalents for Oxygen and Carbon Dioxide ($\dot{V}E/O_2$ and $\dot{V}E/CO_2$)—$\dot{V}E/\dot{V}O_2$ is an index of ventilatory efficiency. A high value characterizes the response to exercise among patients with lung disease or chronic heart failure.[7] $\dot{V}E/\dot{V}CO_2$ represents the ventilatory requirement to eliminate a given amount of CO_2 produced by the metabolizing tissues. In the presence of chronic heart failure, $\dot{V}E/\dot{V}CO_2$ is shifted upward compared to normals.[8] Many laboratories define anaerobic threshold, or ventilatory threshold, as the point at which $\dot{V}E/\dot{V}CO_2$ increases systematically without an increase in $\dot{V}E/\dot{V}O_2$.

Respiratory Gas Exchange Ratio (RER)—RER = $\dot{V}CO_2/\dot{V}O_2$. The peak value of RER is used as an index of exercise effort, with a value of greater than 1.1 to 1.15 being generally accepted as indicating near-maximal exercise effort by the patient. RER, if it is displayed in real time during the exercise stress test, is a useful endpoint for terminating an exercise test.

Arterial Blood Pressure—Vascular resistance is related to $\dot{V}O_2$ because the redistribution of large quantities of blood and oxygen to working muscle must be accompanied by vasoconstriction of other vascular beds (e.g., kidney) in order to prevent an abnormal fall in arterial pressure as exercise progresses.[9] In normal subjects performing progressive, upright exercise on a cycle ergometer or treadmill, arterial systolic and mean pressure increase, while diastolic pressure remains unchanged or falls slightly.[10]

METs—A unit called a *MET* was derived from the resting $\dot{V}O_2$ of a "normal" 70-kg, 40-year-old male, and its value is 3.5 ml/min/kg of body weight. When this same "normal" individual exercises at 2 METs, this individual would be consuming 7.0 ml/min/kg of oxygen. A MET level is frequently reported by ECG exercise stress systems, and even stand-alone treadmills, using an equation based upon presumed work rate. Wasserman lists the reasons why this value may be flawed (Figure 18–2). Froelicher also reports up to a 32 percent error in estimating $\dot{V}O_2$ max from the Bruce treadmill protocol in heart failure patients and up to 30 percent error in estimating METs from a treadmill or ergometer stage due simply to random error resulting from exercise efficiency and motivation.[11]

MEASUREMENT TECHNIQUES FOR ASSESSING "GLOBAL" FUNCTIONAL CAPACITY

Now that we have renewed our basic knowledge of how the "machine" we are studying works (human being, an obligate aerobe), this section will provide a basic knowledge of the machines that accurately assess the functional aerobic capacity of the human "machine."

Work Rate

To achieve the benefits of a stress test, the test must be performed under closely controlled conditions. The outcome of any physiological test relies upon an analysis of a set of dependent variables (ST segment changes, $\dot{V}O_2$, $\dot{V}CO_2$, $\dot{V}E$, HR, BP, etc.) that are derived from changes in one independent variable. In both stress testing and exercise training, the independent variable is work rate; thus the accuracy and reproducibility of the stress test depend on the accuracy and reproducibility of work rate.

The science of physics defines work in two ways:

1. Linear Motion: Work = Force × Distance
2. Rotational Motion: Work = Torque × Angular Displacement

Work Rate = Work/Time. For a cycle ergometer, Work Rate (in watts) = Torque (in lbs) × RPM/84.55. For a typical treadmill, as one can see, work rate cannot be measured because the force supplied by the patient cannot be directly measured.

The forcing function of any stress test is called a protocol, which is simply a work rate versus time function. There has been a considerable amount of scientific research conducted to establish exercise protocols, and which protocol is used depends on the objective of the test and the fitness of the patient.

Treadmill Protocols

Most treadmill protocols have been developed to study electrophysiologic changes during exercise and were developed prior to the commercial availability of systems to directly measure oxygen consumption, or METs. Because work rate cannot be directly measured, standardized treadmill protocols have been designed to increase the work rate in increments of "calculated METs" based upon speed and elevation. For example the "Balke" protocol begins at 2.4 METs for Stage 1, increases to 4.0 METs for Stage 2, and increases in 1-MET increments for Stages 3 to 6.[12] Other considerations include the length of the test (less than 20 minutes) and the physiologic response time to achieve steady state (approximately 3–5 minutes). Despite the lack of precision in adjusting work rate, there exists a large amount of normal data that can be utilized to classify the functional aerobic capacity of individuals based upon treadmill time (stage achieved). If the purpose of the stress test is simply to increase oxygen demand while obtaining a general estimation of METs, then estimations suffice.

With increasing utilization of direct gas exchange measurements, "ramping" treadmill protocols, such as those defined by Froelicher et al.,[13] have been developed in order to eliminate the other problem associated with pre-gas-exchange treadmill protocols—the large MET increments at the start of testing that might be intolerable for functionally impaired patients.

Patient considerations, more importantly, dictate the use of either a treadmill or ergometer. In the United States it is felt that walking is a more natural and tolerable form of exercise than cycling, which is more popular in Europe. Treadmill exercise, because it is weight bearing, recruits more exercising muscle to perform and will result in a higher $\dot{V}O_2$ max than cycle ergometers. However, if the objective of the exercise test is to document improvement in functional capacity resulting from rehabilitation, it is only important to use the same testing device throughout the course of treatment—treadmill or cycle.

Cycle Ergometer

An important point is that an exercise cycle is not a cycle ergometer. An analogous example of the difference would be to compare a weight stack to an isokinetic dynamometer. Both provide a resisting torque over the range of motion, but the latter measures these variables while the former does not. An exercise cycle of the type commonly found in health clubs provides a braking force to cranking motion; the actual work rate being produced by the cyclist is usually unknown and a function of the braking system and the pedal speed of the cyclist. A cycle ergometer for stress-testing applications should provide, in addition to a braking system:

1. a means for measuring work rate by directly measuring torque and RPM
2. a means for dynamically adjusting the braking force, which allows the patient to alter pedal speed during exercise while simultaneously maintaining the exercise protocol

3. a means for providing energy to the flywheel at the start of exercise to be able to start a test at a low work rate (zero being ideal), for testing patients with limited strength in their legs[14]
4. a means for easily calibrating torque and RPM
5. a means for reducing the amount of isometric exercise performed due to the biomechanics of rotational motion and the inertia of the flywheel

An example of one such cycle ergometer, the ErgometRx CardiO_2 Cycle, is shown in Figure 18–4.

Exercise protocols for cycle ergometers are designed in terms of actual work rate rather than METs, which is actually the dependent variable of a cardiopulmonary stress test. Cycle protocols can either be discontinuous or steady state (Weber uses 6.25-watt stages of 3-minute length), or incremental ("ramp"). A ramp protocol uses an equation of the form, Watts = Ramp Rate × Time, where the unit of ramp rate is watts/min. The slope of the ramp is selected according to the estimated functional capacity of the subject—a low ramp rate for the functionally impaired (10 watts/min) and a high ramp rate for athletes (30 watts per minute).

Other considerations for cycle protocols include the length of the test. Wasserman's protocols are designed to achieve a $\dot{V}O_2$ max in about 10 minutes. He also reports that gas exchange measurements match equally well during either incremental or steady-state exercise.

Gas Exchange

Table 18–1 summarizes types of gas exchange technologies that can make up a cardiopulmonary exercise stress testing system (CPX). Each system is made up of:

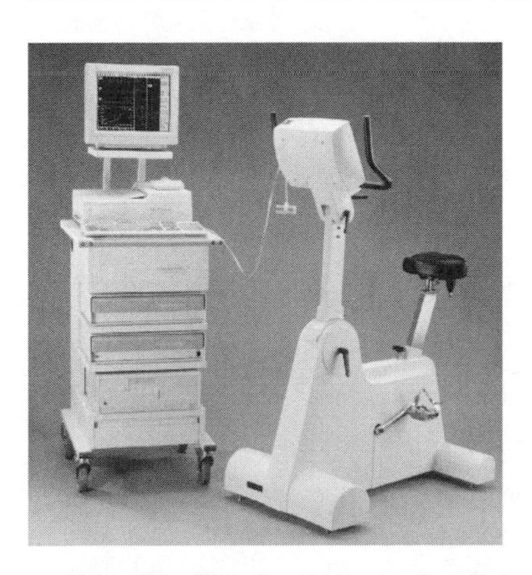

Figure 18–4 ErgometRx CardioO_2 Cycle and Medical Graphics CPX-D

1. computer and printer
2. signal interface
3. oxygen analyzer
4. carbon dioxide analyzer
5. air flow measurement
6. patient interface

The ideal CPX would measure $\dot{V}O_2$ and $\dot{V}CO_2$ at the location where gas is actually exchanged—at the alveoli. Since this is not practical, two measurement technologies—mixing chamber and breath by breath—have survived commercially. The major difference between the two is the location where the respired gases are sampled (Figure 18–5) and how frequently respired gases are measured.

The major limitation of mixing-chamber technology is due to the fact that an accurate measurement of $\dot{V}O_2$ and $\dot{V}CO_2$ requires that the patient achieve steady-state conditions. This is because the gas concentration produced by the patient is changing at a rate faster than that in the mixing chamber, the difference being a

Table 18–1 Gas Exchange Measurement Technologies

Features	*Breath by Breath*	*Mixing Chamber*
Ventilation	Expiratory	Expiratory
Gas sampling	At the mouth	In the chamber
Sample frequency	Each breath	Timed 15–60 second intervals
Information display	Real-time	Filtered and delayed
Restrictions	None	Steady-state use only
Protocol restrictions	None	Ramping is not possible
$\dot{V}O_2$ max reproducibility	Plateauing observed more readily	Patient must exert extra effort to observe plateau (due to delayed display)
AT reproducibility	Better than 5%	Limited by sample frequency and protocol choice
Submaximal testing	Practical	Not practical because endpoint values are not displayed in real time
Test time	Shortest possible	Longer, due to protocol restrictions
Cost	Same as Mixing Chamber	Same as Breath by Breath

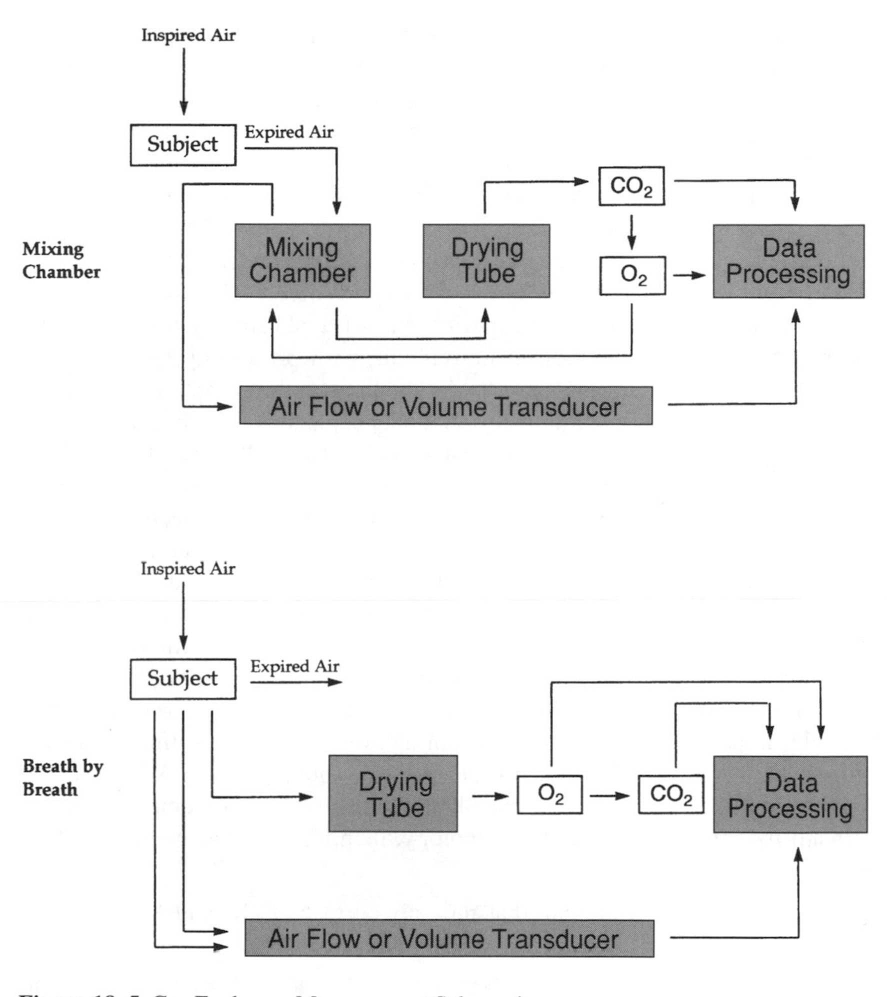

Figure 18–5 Gas Exchange Measurement Schematics

---

function of the patient's rate of ventilation and the size of the mixing chamber. Since it takes 3 to 5 minutes to achieve steady state after an increase in work rate, a mixing-chamber system can only be used in tests utilizing discontinuous protocols with 3-minute stages. This is not a problem if you have no restrictions on time, a motivated, non-symptom-limited patient, and you are only concerned with measuring $\dot{V}O_2$ max on relatively healthy subjects.

A breath-by-breath system eliminates these problems entirely by accurately measuring the oxygen consumed and the carbon dioxide produced for each breath.

The significance of this is subtle and must be stated in the context of the purpose of a stress test itself. If the purpose of the stress test is to measure "the magnitude of effort required for a particular task by an individual" and how it "is related to the individual's physical working capacity,"[15] the next question should be "how accurate do the measurements need to be?" If the answer is ± 30 percent, then treadmill time will suffice. If this is the case, there can be no economic justification for measuring gas exchange at all, since maximum capacity for sedentary or injured workers will, in all likelihood, be so low as to render the precision of the measurement meaningless (the answer is "low"—how low is moot). The concept of measuring the endpoint of functional capacity is most applicable to athletes and to determining "normal" function in individuals with presenting symptoms of dyspnea or chest pain. However, most individuals (with the exception of athletes) never exercise to the endpoint of their functional capacity—it is simply too painful. In contrast, the anaerobic threshold can be viewed as the "endpoint" for activities of daily living (ADL). The AT is the body's mechanism for saying "slow down." Since all of the algorithms developed for the detection of anaerobic threshold involve some form of curve fitting to identify an inflection point in the curve, the greater the number of data points, the more well defined the curve. A breath-by-breath system develops the greatest number of points possible for any given test—a point for each breath. The subtle difference between treadmill time, direct measurement of $\dot{V}O_2$ by mixing chamber, and direct measurement of $\dot{V}O_2$ breath by breath is the number of points needed (many) to derive only one—anaerobic threshold. In addition, the breath-by-breath measurements can be time-averaged similarly to mixing-chamber systems (i.e., 8-breath moving average), but without the inherent inaccuracy of mixing-chamber systems when using ramp protocols.

Breath-by-breath gas exchange technology, then, can be characterized as follows:

- computerization technique that measures $\dot{V}O_2$, $\dot{V}CO_2$, and VT for each breath
- rapidly responding gas analyzers (less than 150 ms total delay from time of patient expiration)
- infection control for patient interface
- comfortable patient interface
- ease of setup and use by system operator

An example of a commercially available system employing such technology is depicted in Figure 18–4.

Global Aerobic Testing and Evaluation (GATE) Technology

Higginbotham has collected gas exchange data on control subjects from multiple medical centers and has found the reproducibility to be ± 5 percent.[16] Because of the reliability of breath-by-breath gas exchange measurements, anaerobic threshold can be measured even in functionally impaired individuals. This, in turn, gives rise to the clinical practicality of performing submaximal stress tests. It should be emphasized that this is only possible with a breath-by-breath cardiopulmonary-stress-testing system in combination with a precision ergometer of the type previously discussed. GATE, then, refers to this unique combination of the appropriate forcing function (ergometer), measurement instrument (breath-by-breath CPX with real-time measurement display), and testing protocol (ramp). Table 18–2 compares GATE and treadmill testing. Once again, the major difference between the classical treadmill test and GATE is the endpoints chosen for the test itself. When utilizing GATE, measurements of RER and HR can be observed in real time, and the test can be terminated well before exhaustion is reached. Even in normals, this occurs in only a few minutes at low-ramp rates. In instituting this clinically, a medical advisory committee is helpful in establishing safe testing parameters. Initial physician consultation indicates that a licensed health care provider with CPR certification could safely acquire GATE data as a part of a functional capacity evaluation.

In addition to measuring actual functional aerobic capacity, a CPX will typically print gas exchange values, using both ml/kg/min and METs. Also, the report will provide a column of values of percentages predicted. These values allow the test reviewer to see how the actual values compare to predicted normal values for an individual of the same height, age, weight, and sex. A sample final report generated by a Medical Graphics CPX/D is included (Figure 18–6).

Rehabilitation and Training Techniques

The use of any set of measurements is only beneficial if (1) the measurements in and of themselves reliably indicate the current status of the process being measured; (2) once measured, the process can be modified in some way to achieve a desired, quantifiable objective; and (3) the measurements are reproducible enough to substantiate whether the modifications, once implemented, have achieved the sought-after objective. The process we have been discussing so far is the "gas transport mechanisms for coupling cellular (internal) to pulmonary (external) respiration," depicted in Figure 18–3, of a patient.

Using the rationale above, then, GATE can be

1. performed during the initial functional capacity evaluation (FCE)
2. used to determine what, if any, aerobic exercise therapies are needed to improve the anaerobic threshold and peak $\dot{V}O_2$ of the patient
3. used to document improvement, or lack of it, resulting from the chosen therapy (e.g., work conditioning)

Table 18–2 GATE versus Treadmill Testing

	Global Aerobic Testing and Evaluation	*Treadmill Testing*
Measurement purpose	Substrate utilization, AT, $\dot{V}O_2$ max; and HR and WR at each	ECG, BP changes, estimated maximum aerobic capacity
Goal	Determine capacity for activities of daily living and peak exercise capacity	Determine cardiovascular limitations to exercise
Endpoints of test	Submaximal RER > 1.05 HR > 70% predicted SBP > 180 mmHg DBP > 100 mmHg All other absolute and relative indications	Maximal—all published absolute and relative indications
Restrictions	Simple health history prior to testing	Medical supervision with ECG monitoring
Duration	Less than 10 minutes	Unknown
Patient tolerance	Painless	Painful
Safety	Safer than unmonitored exercise	Some risk
Crash cart needed	No	Yes
Personnel requirement	Licensed health care practitioner with current CPR certification	Licensed health care practitioner and physician
Reproducibility	Better than ±5%	Less than ±30%
Exercise prescription	HR or WR at AT; $\dot{V}O_2$ max	Some % of maximum HR
Use in evaluating therapy	In some cases required, based upon reproducibility of data and patient tolerance for repeat testing	Not recommended, based upon poor reproducibility of data and poor patient tolerance for repeat testing

The previous paragraph can be described as a "closed-loop control algorithm," which is an engineer's way of describing what the medical profession would refer to as "rehabilitation." This process should be familiar to the reader, since FCE data could be substituted for GATE data to describe current practice for diagnosing and treating musculoskeletal dysfunction.

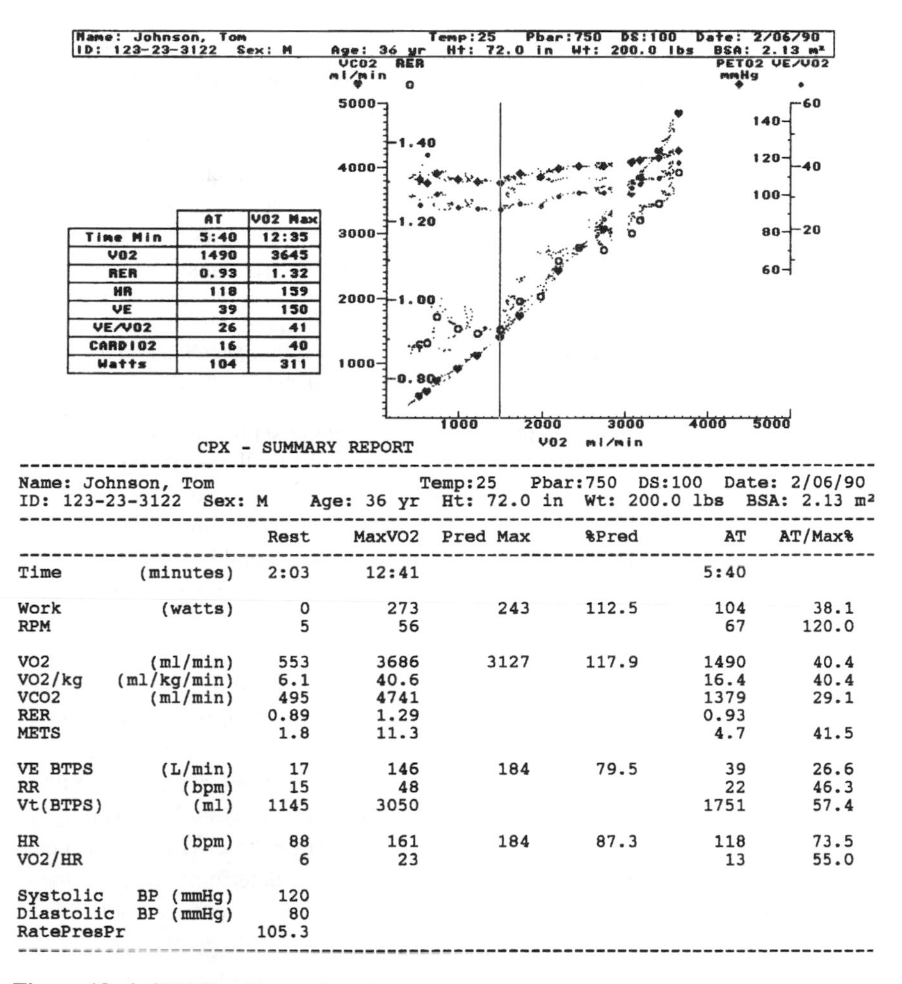

	AT	VO2 Max
Time Min	5:40	12:35
VO2	1490	3645
RER	0.93	1.32
HR	118	159
VE	39	150
VE/VO2	26	41
CARDIO2	16	40
Watts	104	311

CPX - SUMMARY REPORT

Name: Johnson, Tom Temp:25 Pbar:750 DS:100 Date: 2/06/90
ID: 123-23-3122 Sex: M Age: 36 yr Ht: 72.0 in Wt: 200.0 lbs BSA: 2.13 m²

		Rest	MaxVO2	Pred Max	%Pred	AT	AT/Max%
Time	(minutes)	2:03	12:41			5:40	
Work	(watts)	0	273	243	112.5	104	38.1
RPM		5	56			67	120.0
VO2	(ml/min)	553	3686	3127	117.9	1490	40.4
VO2/kg	(ml/kg/min)	6.1	40.6			16.4	40.4
VCO2	(ml/min)	495	4741			1379	29.1
RER		0.89	1.29			0.93	
METS		1.8	11.3			4.7	41.5
VE BTPS	(L/min)	17	146	184	79.5	39	26.6
RR	(bpm)	15	48			22	46.3
Vt(BTPS)	(ml)	1145	3050			1751	57.4
HR	(bpm)	88	161	184	87.3	118	73.5
VO2/HR		6	23			13	55.0
Systolic	BP (mmHg)	120					
Diastolic	BP (mmHg)	80					
RatePresPr		105.3					

Figure 18–6 CPX Test Report Sample

Why Perform Global Aerobic Testing and Evaluation?

Before further discussion of how to use GATE data in the rehabilitation process, a discussion of why is relevant. GATE data are now required before a patient can be placed on a transplant list (peak $\dot{V}O_2$ less than 15ml/kg/min is the requisite), and the FDA has mandated the inclusion of GATE data in documentation submitted for marketing of pharmaceuticals designed to treat heart failure and certain rate-adaptive pacemakers. A large-scale study in Australia documented a reduction in the perioperative mortality rate for patients over 60 having major surgery by measuring the AT prior to surgery.[17] The basis for this is that surgery is, in effect, a major stress test for the organs of respiration and that without sufficient functional aerobic capacity when going into surgery, such patients cannot reasonably be expected to survive the surgery. As Dr. Older has stated, "It saves lives and saves money." The point is that there is a precedent for using GATE data as the "gold standard" for defining therapeutical outcomes. The first answer to the question why, then, is:

I. GATE data are recognized as a gold standard for measuring global aerobic function objectively to document the efficacy of treatment.

Other considerations for incorporating GATE data into physical therapy FCEs include legal, ethical, and cost/outcomes analysis. While there are no clear-cut answers, any procedure that can improve patient outcome and reduce overall treatment cost will be a winner. While it is difficult to prove both, most would agree that reducing the time it takes to perform any given task would save money. One way of saving time is to perform the task more efficiently by using objective measurements. The most relevant example is the use of computerized instruments by therapists to measure musculoskeletal function. Range of motion is not estimated from the patient's age, height, weight, or sex—it is measured. But these variables frequently are used to predict exercise heart rate ranges to form an aerobic exercise prescription.

Do estimates or predicted values used to base exercise prescriptions have limitations? In a study by Coplan et al.,[18] 50 healthy subjects were studied to compare the exercise heart rate ranges, based on peak HR, age-predicted maximal heart rate, and peak oxygen consumption. The study sought to determine which of these methods produced exercise heart rate ranges that conformed to recommendations from the American Heart Association and the American College of Sports Medicine for determining exercise intensity. Those guidelines recommend the use of a percentage of maximal heart rate or oxygen consumption needed to achieve an aerobic training effect. The study concluded that "exercise prescription based on predicted maximal HR was of little value, regardless of the percentage used to determine target heart rate."[19, p.832] Fully 50 percent of the subjects fell out of the directly measured range when using any percentage of the age-predicted maximal heart rate. When using 75 percent of predicted maximum heart rate, 9 patients would have been exercising beyond their anaerobic threshold, and 18 would have been exercising at an intensity level at which no training effect could be achieved by these patients.

The second reason for performing a submaximal stress test can almost be paraphrased from the concluding statement of this study:

II. Exercise prescriptions should be based upon direct assessment of the anaerobic threshold, and GATE data provide a framework to assess the results of exercise prescription.

One successful strategy for basing exercise prescriptions on direct measurement of anaerobic threshold has been demonstrated in a study by Casaburi et al.[20] Nineteen patients with COPD in a physical-therapist-supervised, inpatient rehabilitation program performed two cycle ergometer CPX tests at a low and a high work rate for 15 minutes or to tolerance and also an incremental exercise test to tolerance. Identical tests were given before and after a five-day-per-week cycle ergometer training for eight weeks: one group exercised for 45 min/day at a high work rate, the other group for a proportionately longer period of time at a low work rate. The low work rates were established at 90 percent of the AT work rates for each member of the low-work-rate group, and the high work rates were established at 60 percent of the difference between the AT and peak $\dot{V}O_2$ work rates. Each group was allowed up to three rest periods of 10 minutes duration. Even though each group performed the same amount of actual work during training, when tested after training, the high-work-rate training group experienced an increase in lactate threshold of 24 percent versus 10 percent for the low-work-rate group. In another study by Casaburi et al.,[21] 10 normal subjects were tested, trained on a cycle ergometer using a similar strategy using exercise work rates between those at AT and peak $\dot{V}O_2$, and retested. For the 10 subjects, on the average, anaerobic threshold increased by 38 percent, and peak $\dot{V}O_2$ increased by 15 percent.

The third reason for using CPX data, then, is:

III. Exercise prescriptions that are designed using work rates that are above the anaerobic threshold work rate and below the peak $\dot{V}O_2$ work rates produce a greater physiologic training effect with less training time for the therapist and the patient.

Most important of all, because of error, variability, and limited utility of all other methodologies for determining aerobic capacity,

IV. GATE technology provides a clinically useful method to reliably quantify a symptom-limited patient's physiologic capacity to perform activities of daily living.

Using GATE To Assess a Patient's Ability To Return to Work

The metabolic and cardiac demands on an individual are related to characteristics of task, such as the work intensity, type of work, size of muscle mass employed, the work-rest cycle, and environmental conditions, as well as characteristics of the patient, including cardiovascular function, skeletal muscle training, and psychological factors.[22] GATE technology can eliminate any questions about an individual's functional aerobic capacity for performing work. Answering what the individual's requirements for work actually are is another issue.

The Haskell article offers some suggestions to determine an individual worker's job requirements. One recommendation states that "patients should demonstrate an exercise performance free of significant cardiac dysfunction that is at least twice the average energy requirement and 20% more than the expected peak energy requirement encountered on the job."[23, p.1033] Included in this article is a chart of Energy Requirements of Occupations that lists the average energy requirement for each job expressed in METs (Figure 18–7).

Anaerobic threshold, again, is a measure of the exercise intensity that is continuously sustainable, tolerable, and safe. Allowed to exercise at their own preferable rate, individuals performing occupational tasks work at an intensity of about 40 percent of their $\dot{V}O_2$ max, which is below their anaerobic threshold.[24] By expressing an individual's anaerobic threshold in terms of METs, instead of ml/kg/min, the precise MET level a particular individual is capable of is defined. The value of $\dot{V}O_2$ at the end of the test, peak $\dot{V}O_2$ for a submax test, or $\dot{V}O_2$ max for a maximal test, defines partially or totally the MET level the individual is capable of performing. If a plumber, for example, is tested and found to have an anaerobic threshold of 3 METs, he cannot meet even his average energy requirement (3.5 METs, from the table) for his job and needs further rehabilitation. If this same plumber is found to have an anaerobic threshold of 4 METs but his peak $\dot{V}O_2$ is 6.5 METs, then, according to the cited recommendation, he has not demonstrated that he can tolerate an energy requirement that is twice the average requirement (7 METs). What has been described is a borderline case at best. In practice, if the plumber had an AT of 7 METs, a conservative interpretation of exercise tolerance would conclude that the plumber could return to work safely, at least from the standpoint of his aerobic condition.

How To Develop an Exercise Prescription To Improve Aerobic Function

Strategies for improving the functional aerobic capacity of workers can be adapted from similar strategies for achieving the same by athletic trainers. The *Textbook of Science and Medicine in Sport*[25] defines four applicable principles:

- *Principle of Individuality.* Physiologic profiles may differ considerably, and any general training program is likely to be unsuitable for some group members. Wherever possible, the physiologic status of each worker should be assessed in relation to the demands of the job, and individual training programs designed accordingly.
- *Principle of Specificity.* Training should be specific to the physiological adaptations required at a particular time. The skill of a good therapist lies partly

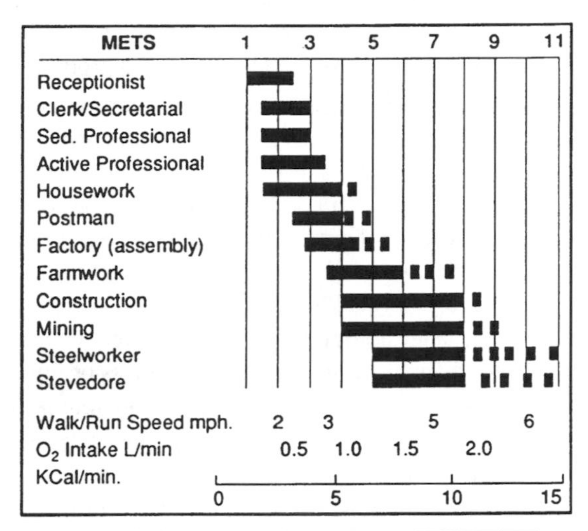

Recreational Competitive

Activity METS 1 3 5 7 9 11 13 15 17

Sitting
Standing
Driving Car
Bowling
Golf
Volley Ball
Badminton
Hunting
Fishing (Wading Upstream)
Skating
Calisthenics
Rope Skipping
Swimming
Tennis
Canoeing
Cycling
Dancing
Skiing Alpine
Skiing Cross-Country
Squash/Handball

Walk/Run mph 2 3 4 5 6 7.5 9 10 12
One mile in minutes 30 20 15 12 10 8 6.4 6 ≤
KCal/min.
0 5 10 15 20

Figure 18–7 Occupational and Recreational MET Tables. *Source:* Reprinted from *Clinical Exercise Testing* by N. Jones, pp. 113–114, with permission of W.B. Saunders, © 1988.

in identifying the physiological adaptations needed at various times and in designing training programs specifically geared to producing them.

- *Principle of Progressive Overload.* Training effects occur through adaptation to stress, and the imposition of unusual demands on a physiological system can evoke changes that make the system cope better. Consequently a stimulus that is initially stressful may soon cease to be so, and its continued application may fail to produce further change. To gain maximum benefit from training, workloads must be gradually adjusted upwards as adaptation takes place. However, the workloads should not be increased in the absence of adaptation, since a physiological system subjected to excessive stress may break down rather than adapting, resulting in illness or injury.
- *Principle of Reversibility.* Cessation of training results in a loss of adaptation, and some effects of training regress more rapidly than others. Training effects are lost more quickly than they are gained and as a consequence workers should be encouraged to continue some form of relevant activity beyond rehabilitation.

These principles and the studies by Coplan and Casaburi[26] provide an excellent framework for the process of rehabilitation, which is depicted in process flowchart form in Figure 18–8.

Referring to the numbered steps:

1. The patient is requested to complete a health questionnaire and sign a consent form. If the patient has not passed a musculoskeletal evaluation or if potential risks are identified on the health questionnaire, the participant cannot continue with the test until he or she has been cleared by his or her physician. The testing procedure is explained to the patient.
2. A resting blood pressure is taken, and spirometry is performed. The patient is further instructed to maintain a comfortable pedal speed as the computer gradually increases the work rate. At the time the subject's RER equals 1.1, the test will be terminated. Other reasons to terminate the test are if the heart rate exceeds 70 percent of patient's age-predicted maximum, if the BP reaches 180 mmHg systolic and/or 100 mmHg diastolic, or if the systolic BP falls more than 15 mmHG.
3. The patient's AT and peak $\dot{V}O_2$ are observed and compared to the patient's occupational energy requirement from an appropriate table of MET values.
4. If the patient's measured values compare favorably with the estimated values from the table, the patient needs no further study or aerobic conditioning.

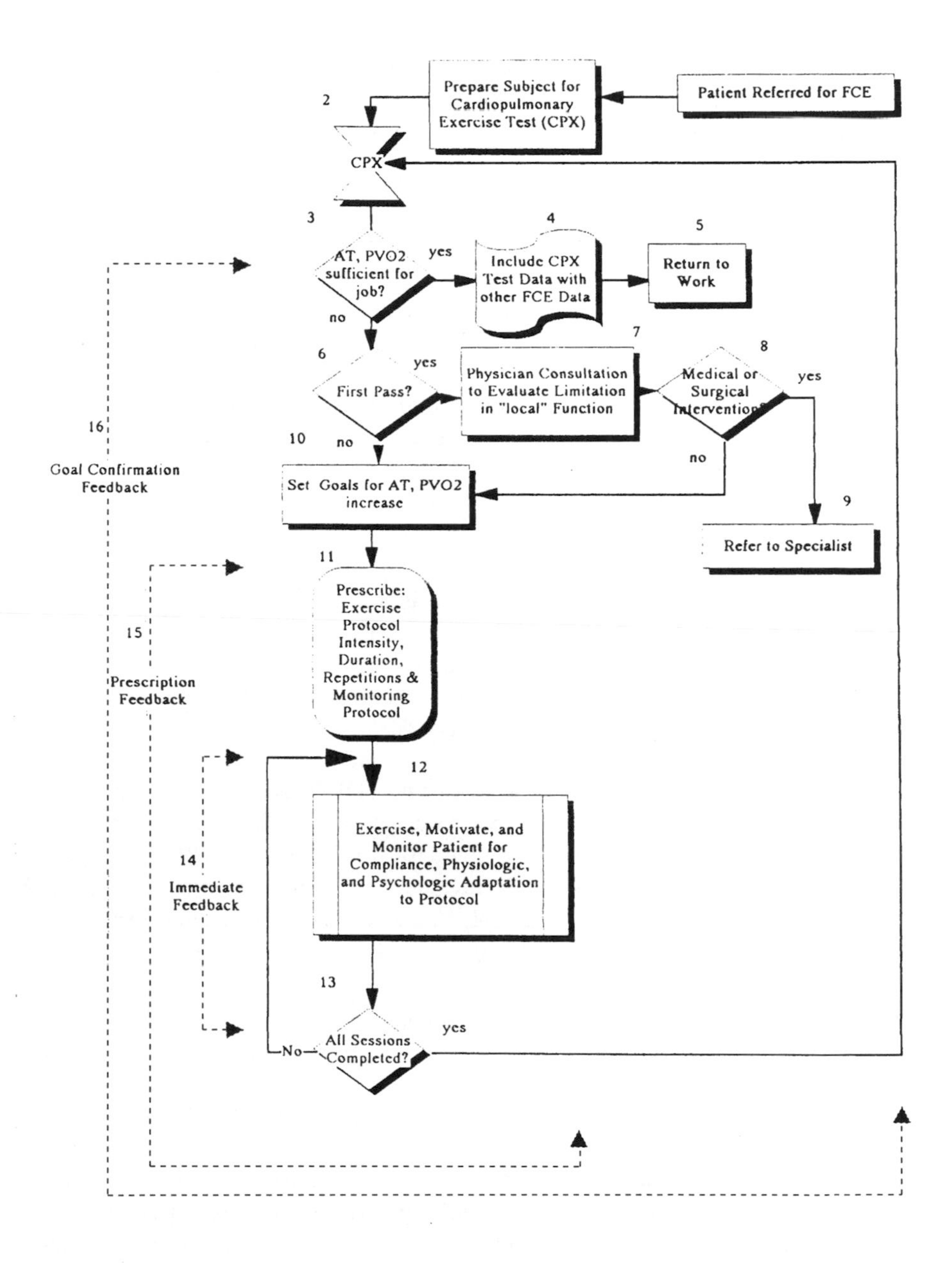

Figure 18–8 CPX-Based Functional Capacity Evaluation and Rehabilitation

5. The patient is returned to work or continues other physical rehabilitation with the knowledge that the patient is aerobically fit.

6–9. This is an "escape path" for the patient who is found, after consultation with the referring physician, to have a medical problem in addition to that of being aerobically deconditioned.

10. The CPX test data, along with the occupational MET level required by the patient's occupation, represent the goals to be achieved by rehabilitation—Steps 11–16. The desired outcome of this rehabilitation is to increase the patient's AT and peak $\dot{V}O_2$ to minimum standards for his or her occupation as efficiently as possible. The task of rehabilitation is now defined in objective, quantifiable terms and is indicated in Figure 18–8 as the "outer loop" of the control diagram, "Goal Confirmation Feedback" (16).

11. The strategy to achieve the goal is an exercise prescription. As indicated, a prescription needs to consider the exercise type, intensity, duration, and number of repetitions. The measured AT and peak $\dot{V}O_2$ are the patient's current intensity limits. It is recognized in studies[27] that increasing these values requires exercise training above the exercise intensity at AT. Deciding how much and how long is the decision of the clinician, whose skills and judgment are important in making this decision. At this time, there exist no hard and fast rules about defining the "ideal" training "window" of exercise intensity. However, if there were, it would be defined by these same parameters—AT and peak $\dot{V}O_2$. Therefore a "Prescription Feedback" loop (15) is provided as a check on the efficacy of the prescription. The prescription can be defined in terms of either an HR or work rate at AT. The advantages and disadvantages of each technique are summarized in Figure 18–9. An example of a work-rate-based exercise prescription is presented in Figure 18–10.

12–13. Depending on whether HR- or work-rate-regulated exercise is chosen, the patient exercises in a clinic or elsewhere for the chosen number of repetitions. It is important to provide as much "Immediate Feedback" (14) as possible to the patient at this point. Feedback is also important to the supervising therapist. Since the determination of exercise tolerance is part science and part art, it is important to be able to blend a sufficient level of patient motivation and observation by the therapist into the exercise prescription. Again, objective, quantifiable observations are best. Rate of perceived exertion (RPE) and HR can be used to meet this criteria, and they can be measured easily and inexpensively. With exercise training, both values should decrease (at the same level of intensity, or elapsed time within the exercise protocol). Adherence can indicate exercise tolerance.

	Heart Rate	Work Rate
Intensity Regulation		√
Equipment Cost	√	
Patient Compliance		√
Monitoring During Training		√
Protocol Reproducibility		√
Choices for Exercise	√	
Training Feedback		√

Figure 18–9 Heart-Rate- versus Work-Rate-Regulated Exercise Training

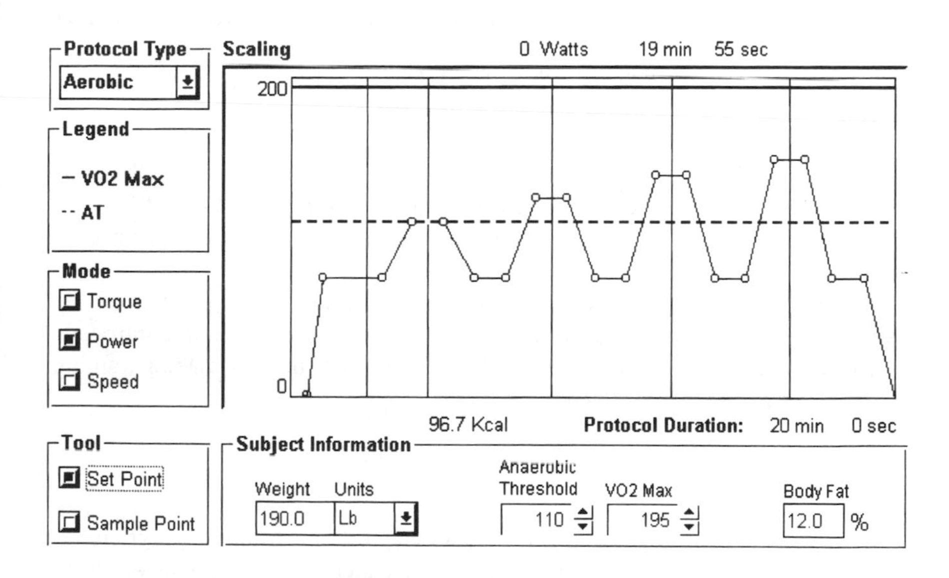

Figure 18–10 Work-Rate-Based Exercise Prescription (ExerScribe)

Poor adherence suggests that the exercise is too intense or the patient is poorly motivated. An example of one such report, from the ErgometRx ExerReview software, is included in Figure 18–11.

At some point a decision needs to be made to retest the patient. If the patient has adhered poorly to the exercise prescription or if exercise tolerance has not notice-

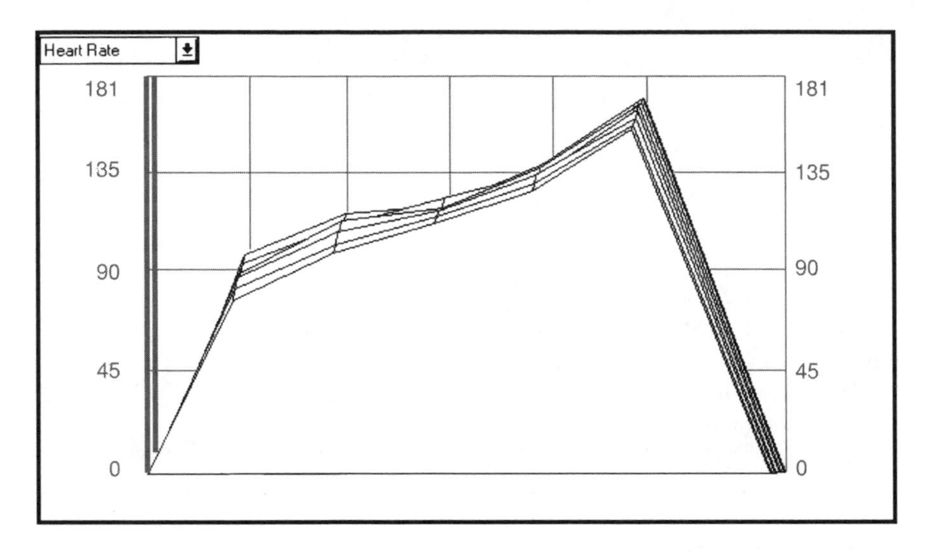

Figure 18–11 Exercise Training Progress Report (ExerReview)

ably improved, then it can be assumed that the goals of the prescription will not be achieved. If this is suspected or confirmed with monitoring of HR or RPE, then a retest would be a waste of time. If progress appears to have been made, the patient is retested at Step 2.

By using the information collected in the intermediate phase of the exercise training, the exercise prescription can be altered. This phase will tend to introduce an additional measure of objectivity into the training task, thereby making training less of an art and more of a science.

CASE STUDY*

Mr. Jones was scheduled to enter the six-week work-conditioning program. He was hoping to return to work as a heavy equipment mechanic. Diagnosis was traumatic torticollis with neck and shoulder pain. He had this problem for over a year and was restricted from most activities of daily living secondary to increased pain and associated headaches. A functional capacity evaluation was done one month earlier, and recommendations had been work conditioning as well as some one-on-one treatment for the myofascial issues. It was noted that Mr. Jones was severely deconditioned and reached 80 percent of his maximum predicted heart rate

*This case study is provided by Kristine F. Kerr, MS, PT.

early on in both the stair- and ladder-climbing tasks. His heart rate had previously increased with some of the other FCE tasks, but he had never been within 15 bpm of his training maximum.

Prior to entering the work-conditioning program, the client was preevaluated utilizing the Isernhagen FCE, a work-related functional physical assessment, and he was given a submaximal stress test. Mr. Jones' measured anaerobic threshold provided both the therapist and Mr. Jones with an objective indicator of his current functional aerobic capacity in terms of both heart rate and work rate. Mr. Jones was tested using a bicycle ergometer that was programmed for two minutes of warmup at 0 watts followed by a ramp protocol of 10 watts/minute. The test was stopped when the respiratory exchange ratio exceeded 1.0.

Mr. Jones' test revealed that he had already reached anaerobic threshold prior to the start of an increase in ergometer work rate, an indication of severe deconditioning. The work of simply moving his legs without resistance was enough to exceed his anaerobic threshold. His heart rate at anaerobic threshold was 115 bpm, and his MET level was 1.5 (see Table 18–3).

The work-conditioning program includes a total body-conditioning program, lifts and carries, education, and job simulation. Total body conditioning entails both strengthening and aerobic activities, during which heart rate is monitored. Clients are encouraged to work at the heart rate at anaerobic threshold during program activities.

The work rate at anaerobic threshold was used to establish a bicycle ergometer exercise protocol, which was stored on an easily accessible data key. Initially, the protocol required Mr. Jones to exercise for 15 minutes at 25 watts. He was able to

Table 18–3 Case Study (Mr. Jones): Effects of Six-Week Work-Conditioning Program

	Pre-W.C. Test	*W.C. Initial*	*Post-W.C. Test*	*After Return to Work*
Strengthening	weak/deconditioned		met job standards	
Anaerobic/aerobic abilities	0 watts, 1.5 METs, heart rate of 115 bpm	tolerated 8 minutes of the 15-minute, 25-watt, 1.5 MET protocol	75 watts, 4.0 METs, heart rate of 95 bpm (job standard = 6.4 to 7.5 METs)	met job standard at 7.5 METs
Job simulation			met job standards	

tolerate only 8 minutes at the start of work conditioning. As Mr. Jones exercised, his heart rate and rate of perceived exertion (RPE) were recorded on the data key as well. As the work rate became too easy, according to the RPE, the protocol intensity was increased. At the end of the work-conditioning program, Mr. Jones was able to complete an exercise protocol of 15 minutes at 70 watts.

At the end of his work-conditioning program, Mr. Jones demonstrated his functional ability to perform all FCE tasks without difficulty on a continuous basis. The anaerobic threshold was also measured at this time and showed considerable improvement. It occurred at 4.8 METs at a heart rate of 95 bpm and an exercise intensity of 75 watts. The data accumulated on his key was printed to provide objective information about his work-conditioning experience such as compliance, heart rate, and RPE. He would be able to return to work without weight restrictions, and he expressed an opinion that he felt capable of returning to work without problems.

His anaerobic threshold data could be compared to the MET levels given by the U.S. Department of Labor for classifying the general strength demands of his occupation. Mr. Jones' job as heavy equipment mechanic was in the heavy category, which was estimated at a 6.4 to 7.5 MET level.

Recommendations were made for work modification because Mr. Jones' MET level performance was below required levels. It was recommended that he be allowed periodic breaks in activity so that he would not become overly fatigued and risk reinjuring himself; and that he also continue with some type of aerobic exercise.

Mr. Jones' anaerobic threshold was measured one month later. His employer had allowed him to take breaks as needed, and he had continued exercising independently. The retest showed that he had increased his anaerobic threshold to 7.5 METs, the approximate energy requirement of his particular job as outlined by the Department of Labor.

Conclusions regarding the use of highly accurate, programmed aerobic training and evaluation are:

1. The aerobic exercise training combined with work conditioning increases work function.
2. Accurate measurement produces accurate MET work projections, allowing pertinent return-to-work recommendations.
3. With safe return to work, MET levels can be measured on an ongoing basis to facilitate removal of work restrictions.

CONCLUSION

Accurate clinical testing of a worker's physiologic capacity to perform work, either on a continuous basis or for peak requirements likely to be encountered on the job, has been a missing link in the rehabilitation of the injured worker. Anaerobic threshold is the measurement of the individual worker's functional capacity for performing continuous activity. Peak $\dot{V}O_2$ is the measurement of the individual worker's functional capacity for peak energy expenditure. Both can be obtained easily in a clinical setting with affordable, commercially available technology.

GATE data may also have other uses such as job-related prework screening: matching workers to work demands is preventative in the population of healthy workers. Testing assists the professional in closing the gap of knowledge in order to match the worker and the work demand.

NOTES

1. V. Froelicher et al., *Exercise and the Heart* (St. Louis: Mosby-Year Book, Inc., 1993), 32.
2. K. Wasserman et al., *Principles of Exercise Testing and Interpretation* (Philadelphia: Lea & Febiger, 1987), 15.
3. Froelicher et al., *Exercise*, 40.
4. Ibid.
5. P.O. Astrand, Quantification of Exercise Capability and Evaluation of Physical Capacity in Man, *Progress in Cardiovascular Diseases* 19 (1976):51–67.
6. Wasserman et al., *Principles*, 29.
7. Ibid., 38.
8. Ibid.
9. K.T. Weber and J.S. Janicki, *Cardiopulmonary Exercise (CPX) Testing in Heart and Lung Disease* (Philadelphia: W.B. Saunders, 1986), 238.
10. K.T. Weber and J.S. Janicki, Cardiopulmonary Exercise Testing for Evaluation of Chronic Cardiac Failure, *American Journal of Cardiology* 55 (1985):22A–31A.
11. Froelicher et al., *Exercise*, 33.
12. Weber, *Cardiopulmonary Exercise (CPX) Testing in Heart and Lung Disease*.
13. Froelicher et al., *Exercise*, 34.
14. Wasserman et al., *Principles*, 61.
15. W.L. Haskell et al., Task Force II: Determination of Occupational Working Capacity in Patients with Ischemic Heart Disease, *Journal of the American College of Cardiology* 14, no. 4 (1989):148.
16. M.B. Higginbotham, personal communications.
17. P. Older et al., Preoperative Evaluation of Cardiac Failure and Ischemia in Elderly Patients by Cardiopulmonary Exercise Testing, *Chest* 104 (1993):701–704.
18. N.L. Coplan et al., Using Exercise Respiratory Measurements To Compare Methods of Exercise Prescription, *American Journal of Cardiology* 58 (1986):832–836.
19. Ibid.
20. R. Casaburi et al., Reduction in Exercise Lactic Acidosis and Ventilation as a Result of Exercise Training in Patients with Obstructive Lung Disease, *American Revue of Respiratory Disease* 143 (1991):9–18.

21. R. Casaburi et al., Effect of Endurance Training on Possible Determinants of VO_2 during Heavy Exercise, *Journal of Applied Physiology* 62, no. 1 (1987):199–207.

22. Haskell et al., Task Force II; U.S. Dept. of Labor, *Handbook for Analyzing Jobs* (Washington, D.C.: Bureau of Labor Statistics, 1985).

23. Haskell et al., Task Force II.

24. Ibid., 1027.

25. J. Bloomfield et al., *The Textbook of Science and Medicine in Sport* (Champaign, Ill.: Human Kinetics Books, 1992).

26. Coplan, Using Exercise Respiratory Measurements; Casaburi, Reduction; Casaburi, Effect.

27. Coplan, Using Exercise Respiratory Measurements; Casaburi, Reduction; Casaburi, Effect; J.A. Davis et al., Anaerobic Threshold Alterations Caused by Endurance Training in Middle-Aged Men, *Journal of Applied Physiology* 46, no. 6 (1979):1039–1046.

Chapter 19

The Kinesiophysical Approach Matches Worker and Employer Needs

Laurie J. Johnson

In discussions of the injured worker, the employer, and the ideal return-to-work process, the importance of cooperation and the "team approach" is often stressed. The two members of this "team" most likely affected by a work injury, and therefore the two biggest "players," are the worker and the employer. Unfortunately, due to the numerous complexities of the workers' compensation system, the relationship between the "players" and the "team approach" often erodes during the return-to-work process. Workers often feel alienated from their employer and fellow workers. Relationships with family and friends may also become strained. Injured workers may have concerns about whether they will be able to return to work and in what capacity they will be able to work, or they may fear occurrence of reinjury when they do return to work. Conversely, employers are concerned that the worker return to work as soon as possible and in a safe capacity so as to avoid reinjury. They are also concerned about the economic impact of a work injury. In the absence of the injured worker, they may need to hire a replacement worker as well as incur other workers' compensation costs. Because the economic loss per year resulting from the direct care of work-related injuries is probably greater than 15 billion dollars,[1] employers' concern for disability cost control is justified. Also, both the worker and the employer are often confused as to what the most efficient and effective medical care should be, what appropriate modified worker duty might be, and when return to full-day work is most safe and appropriate.

One means of answering these questions that is often employed by the injury management team is the kinesiophysical functional capacity evaluation (FCE) because it can match both worker and employer needs. Because this particular approach to injured worker evaluation is nonadversarial, unbiased, and objective, both the worker and employer can benefit from the information it provides. The FCE can foster a cooperative relationship between worker and employer that helps facilitate constructive return-to-work efforts by the entire work management team.

DEFINITION OF THE KINESIOPHYSICAL APPROACH

The kinesiophysical method of evaluation focuses both on the physical body and on its functional movement to determine a person's abilities and limitations. It is an approach that looks at the *whole* worker. Inherent in this approach to evaluation are kinesiophysical principles including evaluation of muscle and joint function in relationship to strength, endurance, speed, coordination, and safety. This approach is similar to other rehabilitation evaluative approaches in that the skilled physical or occupational therapist is in control of the testing.[2] The kinesiophysical approach relies on medical objectivity and not on client subjectivity as the final determination of physical ability. It is because the kinesiophysical approach is medically based that it is preferred in FCEs. The results of a kinesiophysical evaluation are widely used by physicians, employers, workers, attorneys, and insurers in the determination of return to work. The kinesiophysical approach utilizes the kinesiological and medical knowledge of a physical therapist or occupational therapist while testing for a body's maximum, safe, and functional abilities. This is achieved by observing the use of body parts under various stresses, noting the conditions of strength, endurance, coordination, pace, efficiency of movement, and safety of the injured worker while performing physical activity.[3] Once the therapist has knowledge of the body's functional abilities and limitations, the results can be extrapolated to the essential functions of the work that the worker needs to perform. The results of the kinesiophysical FCE identify safe, functional activity levels.

The kinesiophysical approach also incorporates the principles of dynamic testing. It assumes that joint and muscle motion should be observed unrestricted by mechanical equipment. Actual functional work activities can be simulated in order to most closely mirror actual work and stresses on the body that occur during specific work activities. The kinesiophysical evaluation can also identify inconsistencies in the performance of the injured worker within a criteria-based physical/ functional evaluation. These inconsistencies may or may not be related to the work injury. A tenet of the kinesiophysical approach is that physical strengths and limitations should correlate with functional strengths and limitations. If the evaluator does determine that the client is choosing to limit his or her activity or manipulate the evaluation results, the evaluator can objectively document this. However, it is not considered within the realm of expertise of a physical therapist or occupational therapist to interpret the client's motivations or reasons for subjective inconsistency of performance or self-limitations.

COMPONENTS OF THE KINESIOPHYSICAL APPROACH

The strength of the kinesiophysical FCE in meeting the needs of the worker and employer lies in its basic components:

1. emphasis on safety during the evaluation and at the workplace
2. a medically based musculoskeletal evaluation
3. a knowledgeable evaluator who is in control of the evaluation
4. worker and employer education
5. clear written and verbal communication of evaluation results
6. evaluation that includes endurance (usually given over two days)
7. a focus on matching worker abilities with the demands of the job

Safe Procedures

Safety of the worker is stressed in the kinesiophysical evaluation approach. Ethically, professionals working with clients are responsible for ensuring to the best of their ability and knowledge that the client is not harmed while in their care. Safety procedures such as following established lifting standards must be adhered to while evaluating using the kinesiophysical approach.

Initial Physical Assessment

Preceding the performance of functional activities, the kinesiophysical approach stresses the importance of obtaining a brief history from the worker as well as a physical assessment. This physical assessment will alert the evaluator to any possible contraindications such as high blood pressure or neurological deficits that would affect the worker's safety. This musculoskeletal evaluation alerts the evaluator to the physical problems that may correlate to functional inabilities.

The Evaluator Is in Control

Because the kinesiophysical approach is based on medical and kinesiological information, the evaluator is in control of the test situation. The evaluator observes objectively the signs of effort: recruitment of accessory muscles; changes in balance, body mechanics, coordination, heart rate, and respirations; postural accommodations; and efficiency of movement. Maximum function can be verified objectively. Lack of physical signs of effort is noted. The kinesiophysical FCE looks at injured workers holistically and functionally. It is based not on workers' ability to evaluate themselves subjectively, but on the evaluator's unbiased objectivity. The evaluator's control of the evaluation does not prevent but fosters a nonadversarial relationship with the injured worker by promoting cooperation and mutual respect because it is based on objectivity.

Client Education

This component relates strongly to the safety component. If the worker does not recognize safe work positions or safe work levels, it is the goal and responsibility of the evaluator to educate in safe and appropriate work practices. If worker education is necessary during the FCE, the evaluator will document how particular education may have affected test performance. Education and safety also ensure that documented test results reflect maximum *safe* effort. If appropriate, the therapist may note in the report recommendations that education must be continued and safe procedures practiced in a controlled setting prior to return to work.

Objective Report

Upon completion of a kinesiophysical FCE, a report is generated based on the objective observations and physical/functional correlations of evaluation findings. The recipients of these reports are assured that the functional findings reported reflect the worker's full effort and safe maximum abilities, unless noted otherwise. The summary report communicates results and recommendations clearly in terminology and a format that all team members can understand. Often a conference is recommended where all parties involved can verbally discuss and ask questions regarding the evaluation results.

Two-Day Testing

Because endurance and reliability are important, the kinesiophysical FCE is performed over a two-day period. The second-day opportunity to evaluate the worker can verify the accuracy of the results of the first day of evaluation. It may also reveal that the worker cannot replicate Day 1 function due to a deconditioned status or other physical problems resulting from the activity performed the first day. This two-day approach simulates actual work conditions more closely, and worker and evaluator are better able to understand the effects of work on physical function over a period of time.

Evidence to support the use of a two-day FCE was demonstrated in a study conducted at the Isernhagen Clinic in Duluth, Minnesota. Thirty charts were randomly selected to review the results of the first day of evaluation of weighted FCE activities versus the results of the second-day performance of the same activities. It was discovered that one in every five charts reviewed documented either increased or decreased abilities from the first day of evaluation, and in all cases, workers were working to their maximum abilities on both Day 1 and Day 2. The effects of evaluation over time resulted in a decrease in the worker's ability to tolerate weighted activities in three of the charts evaluated. In two of the charts, the workers improved slightly in the amount of weight they could tolerate in lifting and carrying activities. In these five evaluations, incorrect information would have been documented as the worker's maximum safe abilities had there not been a second day of evaluation.

Matching Worker Abilities with Work

Before the FCE is administered, it is important for the evaluator to be knowledgeable regarding the worker's job demands. This information is often obtained through a written job description or by performing a functional job analysis. In order that proper work and rehabilitation recommendations be made, it is important to have an understanding of the physical demands of the work. If there is a match between the worker's abilities and the job demands, there can be an immediate, safe return to work.

NEEDS OF THE EMPLOYER AND WORKER FOLLOWING WORK INJURY

When a worker becomes injured while performing his or her job, the immediate concern of the employer and worker should be the safe return of the worker to work. The first step towards this goal should be evaluation of the worker by a medical professional. This medical professional will determine if intervention is necessary. This may possibly be the most critical point in the return-to-work process. At the time of initial medical intervention, the focus of medical persons, the worker, employer, and other "players" in the return-to-work process should be on return to function and return to work rather than on symptoms and pain. If this is the focus, successful return to work will be more likely. The employer should encourage the worker to return to work as quickly as he or she can. This early return to work can be facilitated in several ways, one of which is use of the kinesiophysical FCE approach. Following a workplace injury, employers should seek for their employees medical care from practitioners familiar with physical body functions and the physical demands of work. If it is indicated, a kinesiophysical FCE, coupled with knowledge of the critical demands of the worker's job, will assist the worker and employer in meeting their various return-to-work needs.

Worker Needs/Concerns

If workers are to return to work as quickly as possible, it is imperative that they feel comfortable reporting to their supervisors immediately any suspected work-related injury. The medical intervention should be supplied by a physician who understands the workplace. Depending on the nature of the injury, particularly if it is musculoskeletal, the worker may be referred for acute care physical therapy, may seek out chiropractic care, or may be sent home for bed rest. In a number of cases, the employee will return to work in a matter of days. But if the employee is away from his or her work for even a week, he or she may have already begun to lose strength. If there is any question about whether the employee is ready for work, what level of work he or she can undertake, or if the employee has had previous injuries on the same job, there is indication for a kinesiophysical FCE.

As mentioned previously, when the kinesiophysical FCE is performed by an injured worker, the evaluator should have knowledge of the essential functions or critical demands of the job. Many questions regarding the worker's needs can be answered with the results of a kinesiophysical FCE.

Questions Requiring Kinesiophysical Answers

1. *What Are the Worker's Safe Functional Abilities?*

Because safety is a mandatory component of the kinesiophysical FCE, the recipient of the evaluation results can be assured that what is documented reflects the client's safe, maximum effort and capacity. Many times a worker referred for an FCE will not know what he or she can do safely. There may have been restriction from the physician not to lift over a certain weight level for a period of time, or he or she may be fearful that "doing too much" will result in reinjury. Under the observation of the skilled therapist who is adhering to safe procedures, the client is encouraged to work to his or her maximum safe level of ability. Appropriate pacing and efficiency of movement are also evaluated. The therapist, using objective criteria, will stop the client when safe levels are reached. The worker is made aware of what he or she is safely able to do at the time of his or her evaluation. The evaluator gives ongoing feedback to the worker so that he or she also knows what must be improved in order to function safely. The worker may also come to understand that "pain" and "function" can be separate entities and that he or she can still perform functional activities safely even while experiencing some discomfort.

Armed with information that reflects the worker's safe levels, the evaluator can make a determination of whether these abilities match the demands of the job. Often the worker returns to work with no additional interventions. He or she has demonstrated the ability to meet the physical demands of the job. Beyond this, the worker has also been evaluated for safe techniques for performing functional activity, pace and body mechanics have been evaluated, and often specific job-simulated activities may have been observed as an adjunct to the standardized FCE protocols. The worker has been evaluated over two days so that if work and movement from the first day affects second-day activity, this can be observed and documented.

The FCE is performed to provide information to the worker as well as to the employer, physician, and insurer. Because the evaluator is communicating with the worker during the FCE, the worker should be fully aware of his or her functional abilities at the end of the evaluation. The worker will know what the safe limits are and will know that to prevent reinjury he or she must work within these limits.

2. *What If the FCE Indicates That the Worker's Functional Abilities Do Not Match the Demands of Work?*

If the results of the FCE indicate that there is not a job match, this is where the FCE can assist in effective case management. Upon completion of the FCE, recommendations are made by the evaluator for the physician, rehabilitation consultant, and insurer, based on the results and the job demands. Keeping in mind that an early safe return to work is the healthiest outcome for a worker, it is desirable if the employer can provide modified or different work that is functionally appropriate. This may be temporary or permanent depending upon the severity or permanency of the original injury. Waiting until a worker is completely recovered before returning to work can be detrimental and may turn a minor injury into a long-term disability. It is to the advantage of both the worker and employer if the worker can return as soon as possible. Often just a minor modification of the worksite can facilitate a more timely return to work.

Physicians find the FCE information extremely helpful because they are most often the professional called upon to release the worker to work following an injury. The physician can now use objective information to do this. Additionally, it is not often realistic for the physician to visit the worksite or review job descriptions. Functional information that is related to the actual demands of work will assist in making objective, safe return-to-work decisions and ensure a more successful return to work.

When functional limitations persist that would prevent *any* kind of return to the workplace, the kinesiophysical FCE is valuable in that it will provide information as to what the persistent physical limitations are that continue to interfere with or limit the worker's functional abilities. The kinesiophysical approach correlates findings from the musculoskeletal/physical evaluation with the functional abilities and limitations. This medical information will help to determine what appropriate and efficient rehabilitation is needed so that the worker can reach his or her maximum physical function. The rehabilitation should be aimed at reaching specific functional goals and ultimately return to work.

The overall impact of the kinesiophysical FCE is felt not only as an evaluation tool but as an opportunity for the workers to learn something about themselves. They can walk away from this evaluation confident that if they adhere to safe technique on the job, they can do their work safely. Or they may learn that they need to practice safe lifting techniques, alternate activities, or monitor pace, or will understand why additional strengthening, stretching, or physical conditioning is necessary before they can return to work. They become a part of the "return-to-work team" and participants in the return-to-work process, as well they should be.

Employer Needs/Concerns

Employers' concern for the safety and health of their employees should not begin when the employee is injured, but before. This concern begins with a strong and effective safety and prevention program and with making information on injury management procedures available to all workers. An attitude of trust and concern for employees will only help in dealing with work injuries. Industry and employers can show they care about the welfare of their employees by providing safe working conditions and by managing disability effectively. Unfortunately, even when an employer has implemented injury management and safety programs, has safety policies and procedures, and has genuine concern for the welfare of the worker, there is no 100 percent guarantee of avoidance of future on-the-job injuries. Employers then must be concerned with returning injured workers to work safely.

Employers are often caught "between a rock and a hard place" in the return-to-work system. On the one hand, they must be concerned about company production and efficiency as well as the financial impact a work injury will have. On the other hand, they are also concerned about the well-being of their workers both before and after a work-related injury. They need the injured worker to return as soon as possible to his or her work position, and yet they don't want to risk reinjury to the worker by returning him or her too soon. They recognize that the worker needs to be out of work long enough to fully recover from the work injury, yet they must be concerned about possible lost production, replacement wages, overburdening of other workers with additional responsibilities, and the workers' compensation costs the company may incur.

The challenge to the employer appears to be:

1. Return workers to work quickly and safely following an injury.
2. Keep company costs down.

If a worker should become injured while performing his or her work duties, there are effective and efficient ways of managing the return-to-work process. As mentioned previously, the worker should feel comfortable reporting an on-the-job injury immediately. The employer can facilitate this early injury reporting via a policy of trust and action. Action should be immediate, and may involve emergency medical care evaluation by a physician, rehabilitation, and possibly time off work to sufficiently heal and recover from the injury.

If questions persist regarding the worker's ability to return to work, or there are questions regarding how to proceed with rehabilitation, a kinesiophysical FCE would be appropriate. It is in the best interest of both worker and employer that proper case management begin immediately. This may save the company from unnecessary workers' compensation costs and keep the worker connected with the workplace and actively participating in the return-to-work process.

These are the questions that the kinesiophysical FCE can help the employer answer:

1. *Is my worker getting the best and most appropriate medical care?* Because a kinesiophysical FCE is medically based and includes a musculoskeletal evaluation, the evaluator will have knowledge of underlying pathology that may be interfering with function. If other treatment would be more appropriate before any strengthening, aerobic conditioning, or endurance training proceeds, these recommendations will be made. If the worker is not in the acute medical phase, conditioning may be an appropriate next step.
2. *Is my employee ready to return to work?* When direct return to work is the desired outcome, a job description is obtained or a functional job analysis is done. Information from these sources informs the evaluator of the employees' essential job functions. A worksite visit is especially beneficial since the evaluator can then assess the job tasks and essential work demands, the work stressors, and whether the employee can perform the job tasks in an ergonomically safe environment. Armed with this critical job information, the evaluator can then realistically compare the worker's functional abilities and limitations with the actual job demands. Employers can discuss, with an evaluator who is acquainted with the work, the worksite and what is required of the employee on the job. The evaluator will prepare for the employer an FCE report that can be easily read and understood by nonmedical persons. The report will reflect the worker's *safe* capacities and will compare these to the requirements of the job. If there is a job match, the employee may be able to return to work with no further intervention, or rehabilitation can proceed with very specific recommendations, goals, and time lines. Without the information gained from the worker performing an FCE, the employer risks returning a worker to the job before he or she is ready. This could result in reinjury or prevent a worker from recovering to maximum function. Conversely, the employer may get a worker back to work without further intervention because it has been determined that it is functionally safe for the worker to return. The physician (who typically releases the employee to work following injury) need not hesitate in returning the worker to work because there is objective, functional, work-related information available. The worker is not denied return to work or kept on light or restricted duty unnecessarily.
3. *What is the likelihood of reinjury and is the job safe for other employees?* Following an FCE, if it is determined that there is not a job match, the likelihood for reinjury will be high if the employee returns to the job without the proper intervention. Another area to consider is that perhaps the job demands that do not match the injured worker's abilities are demands that are difficult for even healthy workers to perform. If the evaluator and em-

ployer can communicate openly and honestly regarding the worksite demands, they may together design a job modification that will (1) reduce likelihood of reinjury to the injured employee, (2) make the job safer for other employees, and (3) help get the injured worker back to work sooner.

The results of the FCE may also indicate that the injured worker's abilities do match with other jobs within the company. This could facilitate earlier return to work in either a temporary position or a new permanent job. The likelihood of cost savings to the employer increases when an employee gets appropriate medical care as soon as possible and returns to the worksite as soon as possible.

4. *How does referring an injured worker for a kinesiophysical FCE evaluation reflect on the referrer?* Because the kinesiophysical FCE stresses worker safety, education, objectivity, and a nonadversarial approach, employers are demonstrating their concern for the worker's well-being by referring him or her to medical personnel who perform it. When the evaluator and employer are concerned with the safety of the worker both during the evaluation and on the job, as well as with promoting safety education and safe body mechanics on the job, this reflects positively on the employer. Most workers will assume that their employer is truly concerned about their well-being and that they are valued as workers when their employer utilizes medical evaluations that are also concerned with safety and a healthy work environment.

MATCHING THE WORKER AND EMPLOYER NEEDS

The employer and worker do have common needs—financial needs, productivity needs, health needs, and safety needs. It is essential that the employer have workers who are fit for work and safe in their jobs so that the company is productive. Workers need to be fit and safe for work so that they can earn a living and maintain an active lifestyle at work, home, and play. The following is an example case study illustrating how the kinesiophysical FCE can meet the needs of the worker and the employer.

Case Study

A 40-year-old male construction worker sustained a low back injury on March 15, 1993. An FCE was performed on May 10–11, 1993.

Because laborers often do not have transferable skills to lighter work, their earning power after an injury is often decreased unless they are retrained. However, in this particular case, the FCE results determined that this laborer could perform safely the essential job functions of his original job with only one modification to his job duties (because FCE results indicated a floor lift maximum of 75 pounds, he would need assistance for lifts from floor over 75 pounds). He was back at work and was able to return at the wage he had been earning and to a job that was satisfying to him. He did not need to take a cut in pay, nor did he need to spend time retraining for a job that might have paid lower wages and may have been less professionally satisfying. The employer regained a skilled, trained, and dependable worker while saving the costs of retraining, temporary partial or total disability payments during training, and workers' compensation costs for a worker out of work. Not only did the worker return to work, but because the suggested modification was implemented, the worker was not reinjured. This also prompted the employer to institute this modification for the entire company, and over a period of time, overall injury rates decreased.

This case study demonstrates how objective measurements of functional abilities can result in cost savings for the employer and minimize loss of income for the employee. The worker and employer were aware of the worker's safe abilities and the demands of the job. Both learned of an easier and safer way to do a job task that in turn benefited all employees. The kinesiophysical FCE, by being objective, practical, and realistic, can serve the needs of both employees and employers. By working together as a nonadversarial team, employee and employer *can* meet their common need: safe, timely, and cost-effective return to work.

NOTES

1. L.J. Johnson and M. Miller, Establishing a Functional Capacity Assessment Program, in *Occupational Health Services, Practical Strategies for Improving Quality and Controlling Costs,* ed. W.L. Newkirk (Chicago: American Hospital Publishing, 1993), 163–180.
2. S.J. Isernhagen, *Functional Evaluation and Work Hardening Perspectives: Contemporary Conservative Care for Painful Spinal Disorders* (Philadelphia: Lea & Febiger, 1991), 335.
3. S.J. Isernhagen, Functional Capacity Evaluation, in *Work Injury Management and Prevention,* ed. S.J. Isernhagen (Gaithersburg, Md.: Aspen Publishers, Inc., 1988), 139–191.

Chapter 20

Contemporary Issues in Functional Capacity Evaluation

Susan J. Isernhagen

UP TO THE PRESENT

Since the early 1980s, functional capacity evaluation (FCE) has played an important role in the return-to-work process. Medical providers and employers have responded to the need to evaluate the functional capacity of the injured worker. The need has driven the development of many types of FCE in order to fill the return-to-work gap. One functionally oriented test based on sound rehabilitation evaluation principles was discussed in *Work Injury: Management and Prevention.*[1] Other tests and methods have also been presented.[2]

Three separate paths developed for FCEs in the 1980s. They were reactions to the popularity of functional evaluations in the work injury spectrum and took diversion paths due to pressures on medical providers during those years.

The first version path was "catch the malingerer," a negative slant on functional evaluation predicated on the belief that injured workers were very likely to be "faking" their injuries in order to obtain secondary gains. This was fueled by the frustration that employers and insurance companies and some medical professionals felt at the adversarial nature of the workers' compensation system. In the zeal to find people with "nonorganic" problems, tests and measurements were designed to accentuate inconsistencies or pain behaviors.

In the mid-1990s, with a more proactive approach to solving work injury problems, many industries, physicians, and employers realized that by looking for the bad in their injured employees they would most often find it. A more benevolent, cooperative system fostered in the mid-1990s has slowly diminished the desire to negatively label injured workers. Rather, a more positive approach has been sought that can actually produce a positive case resolution. The negative influence of setting one player against another in the workers' compensation system by making injured workers' problems seem illegitimate did a great deal to accentuate the litigation and barriers to return to work. Now that this practice has been seen as not reaching the goal of FCEs, it has become less common because of poor outcomes for all parties.

The second path was use of the scientific or quantitative method to help bring objectivity to the return-to-work process. It was theorized that if an X-ray could detect bone healing and allow an injured worker with a fractured leg an opportunity to return to work at the proper time, other measurements could be used to quantify performance objectively so that injured workers could be returned appropriately.

One such method of science was developed for the purpose of evaluating workers' compensation back problems. The trunk-testing equipment that was popular in the mid- and late-1980s in the United States (discussed in detail in Chapter 27) was claimed to be linked to function. It was theorized that by measuring back strength, one could predict functional activities such as lifting and carrying and that in addition there was an opportunity to "catch the malingerer," which brought philosophy number one into the quantification picture as well.

As mentioned above, with the advent of a more positive approach toward injured worker management and a lessening desire to create an adversarial relationship, the use of high-tech equipment to find inconsistencies has lessened. Nevertheless the measurement of inconsistent patterns has been important because a positive proactive method can be found to deal with inconsistencies that may be either physical or psychological. It now is seen as a method to help identify clients' abilities rather than to negatively label them.

On the other hand, studies have found that the results of isolated physical testing, such as isometrics and isokinetic trunk testing, are not directly related to function.[3] When one reviews the complex components of any work task, including strength, endurance, coordination, balance, multiple muscle groups, multiple joint segments, aerobic reserve, and influence of the injured part, it is relatively easy to see why quantification of only one portion of the body cannot predict function. Consequently functional capacity evaluation is now recognized not to be possible through these means.

Whole-body, full-functional testing for the aim of physically matching the worker's abilities with the physical demands of the job worked well in the 1980s and into the 1990s. With the advent of the Americans with Disabilities Act requiring that testing be "job-related," a third path, the positive approach of job matching through FCEs, has seen considerable growth. The method of evaluating workers as a way of finding their true function and matching this with the critical essential demands of the job was always seen as a logical approach, and in the mid-1990s is now a strong choice of those seeking positive case resolution that can be understood by all parties and that will lead to a healthier workforce.

In the mid-1990s the advent of structuring case management forced the concept of managed care into the workers' compensation system. This integrated scientific, monetary, and employee considerations. All workers' compensation programs would now have to be effective (valid, cost-efficient, and able to return workers to appropriate work). Consequently one of the three pathways—the functional matching pathway—became the strategy of choice.

WHAT GOES AROUND COMES AROUND

Evaluation of function in any other arena, such as neurological deficit (stroke, head injury, multiple sclerosis, etc.) or sports (football, baseball, track, etc.) has always matched the individual patient, client, or athlete to his or her required functions. The stroke patient has been evaluated for return to the home. The football player has been evaluated for return to the game. In these cases, professionals acknowledge that they are dealing with a whole person who must return to a holistic activity.

Therefore the original definitions of FCE relating to work matching have received increasing acceptance.

> *Functional:* Meaningful, useful. In this context, *functional* indicates purposeful activity that is an actual work movement. It implies a definable movement with a beginning and an end, and a result that can be measured.

Comment: Because work activities are complex sequences of body movement, and body movement is a complex activity of multiple joints and muscles, test items must require complexity and intensity of work movement. Conceptually it would appear that this is why segmental measurements (strength testing, isometrics) have not been able to predict full function (see Chapter 27). The body is complex, can use adaptive movements to compensate for tightnesses and weaknesses, and may produce function abilities unrelated to what one would predict from isolated testing. Therefore FCE continues to use functional activities as its base.

> *Capacity:* Maximal ability, capability. *Capacity* indicates existing abilities for activities, including a maximal function able to be used.

Comment: There is a relationship between the percent of maximum voluntary capacity utilized and the ability to maintain work function over time.[4] The higher the percent of maximum function required, the less tolerance for endurance or repetition. Therefore an FCE that will be utilized to match a worker to the requirements of an eight-hour job must determine the person's maximum ability. The evaluator utilizes the level of effort for comparison to the frequency of work activity (See Table 20–1). Therefore actual maximums need to be determined if accu-

Table 20–1 Level of Work and Frequency Categories Used in FCE

	Frequency Category				
	Never	*Rarely*	*Occasionally*	*Frequently*	*Constantly*
Percent of workday that task may be performed	Must never or can never be performed	1–5% of 8-hour day	6–33% of 8-hour day	34–66% of 8-hour day	67–100% of 8-hour day
Number of hours per workday that task may be performed	Must never or can never be performed	Performed rarely during the workday	Restricted ability to no more than 2 hours total of this activity	Restricted ability to no more than 5.5 hours total of this activity	Generally unrestricted ability to perform
Level of effort	Unsafe	Maximum	Heavy	Moderate	Light

Note: These categories are chosen with the understanding that safety and ergonomic principles will be followed (including rest breaks, proper body mechanics, etc.). When evaluating injured workers, normative data cannot predict the level of safe work function. Therefore the "effort levels" must be obtained by individual evaluation of each test item.

rate work prediction is desired. Testing an activity only to the level of a job description or to a self-limited level will not produce accurate information on which to project repetitious work.

> *Evaluation:* Systematic approach including observation, reasoning, and conclusion. Going beyond monitoring and recording, the evaluation process implies an outcome statement that is explanatory, as well as an objective measurement of the activity.

Comment: A return-to-work process is complex and sometimes adversarial. In order for medical information to be most highly utilized, the professional involved in the evaluation should interpret the results. Tests that merely produce graphs or data without work-related interpretation are not helpful to the return-to-work process. Rather, evaluation of the client, resulting in a report that clearly describes observations, conclusions, and work recommendations, will be helpful to all parties who read the information. In addition, workers who are to return to work must feel that they have been "evaluated" and not merely given a test that does not take into consideration their own physical uniqueness or their work.

OUTCOME DATA

The issues in functional capacity evaluation are most meaningful once we study outcome statistics. Very little outcome data of functional tests have been available. Research is beginning and will continue to be critical in determining the proper role of and type of FCEs for the future.

Outcome studies of FCE objectives were measured at two clinics, one in Minnesota and one in Louisiana. This process and full data are described in Chapter 40. The positive outcome is described here to generate observations and questions.

FCE Quality Outcomes Measurement Program

Retrospective phone surveys of 119 consecutive clients who were referred for FCE were reviewed for outcome. These reviews consisted of medical record chart audits combined with telephone interviews at six months post FCE. In order to study effectiveness of the FCE alone, 89 clients who received no other form of restorative treatment following the FCE (i.e., physical therapy, work hardening or work conditioning) were studied.

For clients who received FCE but *no* other subsequent intervention, the *lost time from date of injury to FCE averaged 9.6 months*. Previous studies had shown that once clients had been off work beyond six months and were considered "chronic," their chances of returning to work were very low, and that if they did return to work it would be only with extensive and expensive rehabilitation.

Sixty-four percent of these clients returned to work based on the results and recommendations of the FCE only (with no other intervention). Despite preconceptions, these "chronic" patients *returned to work* based *only* on the functional capacity evaluation. Return to work required agreement by at least three parties:

- *the doctor* . . . who had to sign the release (which had not been signed for nearly a year)
- *the employer* . . . who had not been able to accept the employee back at work
- *the worker* . . . who had to make the decision that he or she felt capable of returning to work and trust the FCE report

In such a high percent of work return based on a functional test only, one can recognize the power of an effective test. Perhaps the skills of therapists—to do comprehensive testing to match functional capability with functional demands—have actually been overlooked in searches for more complicated (and expensive) answers.

Of those clients who returned to work, 77 percent returned to their previous jobs, either at the same or a modified level, and a total of 91 percent returned to work at the same place of employment. The combination of the FCE's clarity in defining modifications and the employer's willingness to modify the job creates a positive outcome.

Two-thirds of clients who returned to work were full time status. This is a desirable placement result, considering duration of time off work.

More than two-thirds of these clients were satisfied with their work level at six months post FCE, and several wished to work at a higher level. When returned to work at appropriate functional levels, clients are satisfied performing their tasks, which contributes to employee morale and work performance.

Only half the workers felt welcomed back to the workplace. Industry has made strides in overall attitudes regarding the return-to-work process. However, employers and peers at the job site can create an adversarial relationship. This is one aspect of work injury management that continues to need to be addressed.

Of clients who returned to work 87 percent reported discomfort at the work with which they had to cope. However, they had identified that *pain* and *function* are not the same. The psychophysical approach, which focuses on pain rather than function, can limit return-to-work capability. The kinesiophysical approach utilized here allows for differentiation between pain and function and focuses on the *ability* of the person to safely, functionally perform the job tasks required.

In the telephone interviews, none of these workers indicating discomfort said they wished to quit work. In many chronic cases, the worker may never be pain-free. An approach that helps workers differentiate between safe function (even in the presence of discomfort) and unsafe or harmful function empowers workers to be functional within safe limits in the presence of discomfort.

Over three-fourths of clients who returned to work knew the FCE helped place them back to work. It is estimated that FCE was also used in the other cases as well, as the clients did return to work.

Over 90 percent of all clients had a better understanding of their abilities. This FCE is interactive, as the therapist discusses the findings with the client during and after testing. Workers make their own decision regarding return to work. This acknowledges the goals of client education and empowerment.

Over 90 percent were encouraged by the FCE information to improve their physical condition. At the end of the FCE, clients know and understand their physical limitations and this knowledge often encourages them to improve their physical function.

Two-thirds of the workers felt the FCE clarified their work abilities at levels they already knew. One-third were surprised at the results but stated they were pleased to know their true abilities. This indicates that when professionally evaluated, the worker's true perception and the evaluator's match most of the time. However, if a psychophysical test had been used, there would have been an error in 32 percent of cases. The workers who received new functional perceptions believed the FCE results and knew their true safe function had been determined by the test, not by themselves alone.

If chronic clients who have not returned to work are problematic in the system, perhaps it is because many of them have not had the opportunity to have a good FCE. An evaluation or test is only as good as its understandability, the credibility it holds, and whether it meets its objective.

If the objective of a functional capacity evaluation is to facilitate the return-to-work process, then a return-to-work rate of 64 percent in "chronic" injured workers must be worth investigating as it facilitated the successful case resolution. The FCE studied is of a relatively short duration (five hours over two days) and relatively inexpensive (averaging $600 when these data were collected). This may indicate that returning a comprehensive, cost-effective evaluation is a better answer than extensive multidisciplinary psychosocial interventions that may also be effective but may be unnecessary.

THE TWO METHODS OF TESTING

There are two primary philosophical approaches. A comparison of these approaches is found in Exhibit 20–1. (See Chapter 19 for specific definitions and examples.)

Exhibit 20–1 Comparison of Kinesiophysical/Psychophysical Approaches to FCE

Psychophysical

Clients stop test at their discretion, thus putting them in charge.

- Test is subjective.
- Test tells clients *they* are responsible, not the evaluator, and may actually facilitate lack of effort due to fear.
- Test is medically/legally weak.
- Both overmotivation and undermotivation may deter ability to find actual maximal ability.

Safety is "assumed" because client is in charge.

- Client may get injured during test.
- If information gained is incorrect, injury may occur when client is returned to work.
- Client assumes that unsafe body mechanics are acceptable because evaluator does not correct them.
- The therapist who documents "unsafe" or "poor" body mechanics and yet lets the client continue may have written proof of negligence.

Test does not identify the physical problem.

- No history or physical is provided.
- Therapist is unable to determine undiagnosed or nonidentified contraindications before testing. Injury (negligence) may be a consequence.
- Test cannot correlate physical limitations with functional performance.
- Test is medically/legally weak.

Kinesiophysical

Therapist is an educated evaluator who is in charge of the test.

- Using criteria, the evaluator observes and identifies performance levels.
- If clients do not put forth full effort, they are confronted in a positive manner. Most often they give better effort. If not, this is documented.
- Evaluator is "professional" and communicates with client in a manner that fosters trust and cooperation.

Safety is evaluated and enforced.

- Safety during testing is linked with safety on return to work.
- Results given are based on specific criteria of safe maximum function.
- The client is empowered to perform safely and maximally, both during the test and at work.
- The therapist ultimately has the responsibility to ensure patient safety. This follows professional ethical and legal protocols.

Test follows protocols of physical assessment prior to testing.

- Initial history is taken in addition to medical records review.
- Initial physical assessment is performed prior to testing:
  a. to discover any contraindications to testing—client safety is paramount;
  b. to ascertain physical problems related to functional limitations.
- Test follows professional guidelines.
- Test gives credibility to functional limitations.

In the kinesiophysical approach, the evaluation is performed by a medical professional with a background education in kinesiology and pathology to evaluate normal and dysfunctional movement patterns. The results therefore have credibility in the medical process. If there is a psychosocial difficulty that interferes with true function, the presence of this nonphysical factor will be identified. This does not cause clients to be viewed negatively, however. Instead, they are treated as legitimately injured individuals. This deters an adversarial evaluator-evaluee relationship from forming. Clients are more likely to give full effort when they feel that they are "safe" in the evaluator's hands. In addition, while it is suspected that many injured workers wish to "underreport" their physical abilities, it is also true that many workers are willing to work beyond their safe maximum without realizing they have exceeded safe limits. FCE recommendations for safe activity levels decrease the likelihood that these workers will be reinjured upon return to work.

If clients feel unsafe, not believed, or the subject of suspicion, they may react negatively during testing. If, on the other hand, they are received with statements that their condition is not in doubt, and that they will be evaluated objectively and safely, then they will more likely give full effort with the perception that they will be honestly and objectively evaluated. When the client goes through a full objective evaluation with feedback and explanation of the results, self-efficacy is the result. This is possibly one reason that such a high percentage of the previously mentioned chronic patients have returned to work.

Therefore the kinesiophysical approach has the following parameters:

- It is based on medical physiological principles that are understood by the physician who must release the worker to go back to work.
- It provides the worker with a safe framework in which to do maximum effort.
- It allows the employer to understand that if the worker is limited, it is due to a physical cause, not "pain" or "self-limitation."
- It gives physical reasons for limitations. This is helpful in identifying what modifications could be made to the work or whether the worker should or could be rehabilitated.
- If rehabilitation is chosen, the physical limitations will already be pinpointed by the functional evaluation, thus saving time and ensuring accuracy in the work rehabilitation process.
- Open discussion of each functional ability as the testing proceeds ultimately facilitates a match between the worker's own perception and the FCE. It becomes a powerful learning experience for the client. No treatment is given, but the information gained by going through functional evaluation is very empowering to the worker in making the conscious decision about returning to work.

FUNCTIONAL CAPACITY EVALUATION, RECOMMENDATIONS: RETURN TO WORK AND COST SAVINGS

To evaluate cost-effectiveness of an effective FCE, actual cases were evaluated for potential losses to the employer, the worker, and the system. The determinations were derived by Jack Casper, MS, QRC, CIRS, of Casper Vocational/Rehab, Duluth, Minnesota.

Case Study 1

The injured worker is a 42-year-old (female) circuit board electronics tester referred for functional capacity evaluation. The employee had a medical history of cervical discectomy and fusion, and myofascial pain in the neck and shoulder. The initial date of injury was indicated to be May 1989, with surgery in April 1991. Previous treatment had included chiropractic care, physical therapy, work hardening, Nautilus, and whirlpool. The employee had been off work approximately eight months. An FCE was performed and her job description was also explored to match critical demands of the job with the client's ability to perform work tasks. FCE recommendations indicated that if the employer was able to make worksite modifications, that employee could tolerate her job. The FCE was performed September 1992, at a cost of $600. The on-site worksite analysis for job exploration cost $375.

At six months follow-up, the employee had returned to the job with modifications. She was working full time, five days a week. Arm slings suspended from the ceiling were used to modify the worksite at a cost of $300.

This would have been a case of permanent and total disability if functional restrictions had not been accurately measured and modifications not made. Experience indicates that cervical problems many times do affect manual manipulation capacities and that the results of these problems can be devastating. Reaching/handling activities are primary physical components of about 90 percent of all jobs—more so in lower-skilled occupations. Any functional restrictions on upper-extremity use, even light use (such as in the case of the identified employee), can drastically alter access to most jobs. In these cases, accurate measurement of total functional capacities is an absolute necessity in implementing a return-to-work plan. It is extremely important to concentrate efforts on placement with the original employer; attempting placement in general labor where 90 percent of jobs are immediately ruled out is a near-impossibility.

Modifications will obviously have to be made in nearly any work setting, and can only be made by initial in-depth evaluation of functional capacities. The implemented modifications allowing return to work without wage loss may seem rather simple, but are an ingenious response to measured functional restrictions.

Without functional measurements and modifications, the employee might well have been designated as a permanent and total disability case, with compensation paid indefinitely. At the very least, the employee would need retraining. An appropriate program would be difficult to identify in working around the extremity restrictions and might likely constitute another 2.75 years of benefits (considering program identification, school attendance, and job search), school costs, and appropriate qualified rehabilitation consultant costs. This total expense would be estimated at $43,500 if not for the outcome derived through accurate functional capacity measurement and subsequent modifications. Even $43,500 is a minimal expense. If the situation were designated as permanent and total, the expense could well add to several hundred thousand dollars.

Cost Savings: $43,500–$100,000+

Case Study 2

The injured worker is a 45-year-old male welder with primary diagnosis of low back pain due to a bulging disc who was referred for functional capacity evaluation. The initial date of injury was indicated to be October 1991. Primary diagnosis was L-4,5/S1 bulging discs and lumbosacral sprain. Previous treatment included physical therapy, traction, therapeutic exercise, and stabilizing belt. The employee presented for the FCE in August 1992; he had not worked since date of injury (10 months). FCE was performed in August 1992 at a cost of $600. Job descriptions were explored to evaluate the match between critical demands of the job and physical work strengths. (Job descriptions were supplied by employer.) FCE determined there was a match between work and worker.

At six months follow-up, the employee was still at his previous job with modifications that required no cost. He was working full time, five days a week.

The use of FCE for skilled industrial workers can be critical in effectively returning them to work. Typically, these types of workers are highly paid and do not have skills that are transferable to light or sedentary job categories. Cost savings in these cases can only be accomplished through a well-defined functional evaluation allowing the employer to make exacting and safe modifications to the job and allowing for the employee to continue working safely and comfortably.

Welding occupations are especially difficult to work with since they involve not only lifting of heavier weights but also stooping, bending, and awkward positioning in the normal course of a workday. In the general field of vocational rehabilitation, there is not a high rate of success in placing injured welders in an open industrial market without generating substantial wage losses. New employers are unwilling to hire welders with limitations when there is a glutted market of healthy welders who can tolerate all exertional demands. Alternative placement (outside of welding) would concentrate on entry-level employment.

To avoid wage loss and compensation expense, it is critical to work with the employer and change job tasks for the injured welder. This is not as easy as it may sound since the employer also expects productivity for the wages being paid. A "make-work" response will not succeed in the long run.

In this particular case, the patient had postinjury capacities well defined (through functional capacity measurement) so that the employer could more easily identify safe job tasks that the employee could perform. Without a well-defined functional capacity measurement, the employee would probably be forced into an open labor market at entry-level wages. Wage loss could well amount to over $16,000 per year for the employee, with temporary partial compensation costs exceeding $11,000 per year on an ongoing basis.

Retraining may not be a viable option in this situation, considering the employee's age plus the fact that his industrial background would not serve to sharpen aptitudes necessary for a successful training venture. Temporary partial compensation payments are projected over five years for a total of $55,000 in costs without the intervention of an accurate FCE. Accuracy of the FCE is critical in balancing a productive return to work for a welder with safe job tasks.

Cost Savings: $55,000 to employer; $80,000 to employee

MEDICAL-LEGAL ISSUES

With the emphasis upon return to work, work injury liability, and discrimination issues, information regarding a worker's functional abilities is desired by attorneys in workers' compensation cases and also in personal injury cases. For this reason, FCEs have been utilized in hearings, depositions, and courtrooms to delineate a person's physical work abilities. Credibility of an FCE is like credibility of any other medical testimony. The evaluation must be based upon objective medical evidence and must be performed by a qualified medical practitioner with credentials, experience, and training (in functional evaluation).

The following aspects of FCE lead to credible testimony. (Also see Chapter 30).

1. *Objectivity.* If a kinesiophysical approach is used and the therapist has criteria, measurements, and observations made during the test, then the outcome of the test is based on the professional's evaluation and criteria.[5] If, on the other hand, the psychophysical approach is used, in which clients are asked when they want to stop or the test is related to "pain," then the information is seen as more subjective and the therapist evaluator may not be seen as the expert. For example, if the therapist testifies, "He stopped at 20 pounds because he said his back hurt," this does not produce the level of credibility of "I ended the test at 30 pounds when subluxation was noted to the right shoulder. Safety is a criterion; therefore, the 25 pounds is the maximum safe lift."

In this example, the criteria used were safe, functional lifts. The client did not stop the test; the evaluator did, for the physical reasons noted. In this way, the judge and jury will understand that the therapist is a professional, is trained to observe, uses safety criteria in testing, and has identified a physical reason as the cause of the limiting factor.

2. *Ability to be the definitive functional test.* At times when physicians' "opinions" differ from the actual functional capacity results, the depth and comprehensiveness of the test, with the background and expertise of the examiner, will be used by the judge and jury to determine that actual evaluation has more weight than "opinion." For example, the statement "It is 4°F outside as measured by a thermometer" is better than the statement "It is really cold," even if the latter statement is made by a weather forecaster. The practitioner with the measurement and evaluation has more information than the practitioner with a "guesstimate." Thus, the FCE, when properly and comprehensively done, will have information that is more dependable than an opinion derived from an office visit.

The physician often will use an FCE as a foundation for return-to-work estimates in legal cases. The physician will also want to know that the FCE is medically/legally credible and not based on "subjective" or unreliable information. The physician will count on the medical-legal credibility of the test.

3. *Interrelationship of functional capacity items.* Many times the evaluation is based upon a holistic look at the person's "employability." Therefore the evaluator must be able not only to look at each individual functional test but to integrate them into overall work capacities. For example, evaluation of reaching, lifting, and overhead work positions as individual tasks scored separately would provide the reader with only scattered information. If the job description requires the worker to reach into a machine, turn controls, bring an item out of the machine, and then place it overhead on a shelf, the functional evaluator who synthesizes the activities into recommendations of entire work movement patterns will be more valuable than someone merely reporting isolated tests. The FCE must be designed to facilitate this integration.

When deciding a return-to-work case, it is very important that specific questions be answered about work ability. The FCE, therefore, must focus on whole-body, entire-functional-movement patterns so that work can be compared to the physical ability. This is one reason that segmental evaluation or isolated functional testing has not been helpful in predicting function.

4. *Credentials of the evaluator.* If the FCE is based on the skills of a physical therapist or occupational therapist, then the credentials of that therapist will be brought into court to back up his or her ability to do functional evaluation. A therapist who uses a standardized format and can discuss policies, procedures, and criteria will have strength behind his or her testimony. A therapist who has also undergone "training" or additional education on functional testing will be seen as having more advanced information rather than the therapist who has developed a

simplistic test. While years of experience in therapy in general are important as back-up, the more specific training and education a therapist receives on functional evaluation, the stronger his or her credentials will be in court.

5. *Videotaped evaluations*. Another use of functional capacity evaluation utilized for judge/jury case resolution is that of the videotaped functional evaluation. When specified segments of a videotape FCE are used, they must be brief, concise, and to the point, showing the functional abilities and limitations of the client. A method developed at Isernhagen Clinics, Inc., has been used in case resolution to facilitate case settlement without needing to go to court and in court itself. The videotaped functional evaluation process follows this format:

a. An evaluation of the patient/worker is performed to establish physical abilities, limitations, and the aspects that are most relevant to be videotaped.
b. A professional video company is chosen to videotape the proper segments.
c. Specific segments are videotaped to demonstrate most clearly the client's functional performance and how it is linked to physical abilities and limitations. Worklike activities are stressed. The videotape may be taken at work for a client who is currently working at home, if he or she is homebound, or in the therapy clinic.
d. In the editing studio the therapist edits the videotape (12–18 minutes optimum) and voices over the pertinent segment so that the attorneys or case resolution adjudicators will be able to determine the extent of physical abilities and limitations. If a picture is worth a thousand words, then a videotape is often worth many more when it comes to explaining a person's functional ability.
e. Copies of the videotape are generally sent by the referral source to the appropriate case resolution parties. Case resolution may result when it is clear that a trial is not actually necessary to demonstrate loss or maintenance of abilities.

Because of the litigious nature of the workers' compensation or personal injury system, functional evaluators should always anticipate that a functional evaluation may need to be utilized in a hearing or courtroom situation. Therefore careful attention to detail in the writing of the report, clarity of comments, and the ability to objectively back up all observations and conclusions should be present in a report. In actuality, writing "for legal case resolution" is an excellent way for functional evaluators to improve not only their testing procedures but their reports.

Ensuring Safety in Functional Testing

Injuries in functional evaluation can happen and have happened, requiring us to address an important issue. How can therapists ensure safety for the workers they are testing and also protect themselves from negligence/malpractice suits if an injury is claimed during the clinical testing?

In order to ensure spinal safety in functional testing, the first thing that must be done is to define what spinal safety is. Whatever is defined as safe may need to be defended in a court of law. Therefore definitions should be backed up by the scientific literature. Questions therapists might ask themselves are:

1. Do I have a standard of spinal testing that I teach in back schools?
2. Do I go out to industry to instruct workers in safety or proper body mechanics, and if so, what do I teach?
3. What is the lifting and handling information that I give out in my handouts to patients?
4. What do my colleagues teach and report as spinal safety in their back schools and industrial education programs? I understand I will be judged by what a "prudent therapist" would do in similar functional testing circumstances.
5. What does the scientific literature tell us regarding spinal safety at work?[6]

There must be a conscious decision as to whether safety precautions will be utilized in functional testing. If a person gets injured in an FCE, not only will policy and procedures be reviewed, but it is likely that the scientific literature regarding what a "safe lift" is will be brought into a court case.

The question "Can you ensure spinal safety in functional testing?" can be answered by examining how spinal safety may be compromised. Factors linked with higher incidences of back injury are

1. lifting with a bent or twisted spine
2. high acceleration with a lift
3. horizontal distance of the weight from the worker's center of gravity
4. weight capacities beyond a person's strength
5. poor coupling of the hands with the load
6. an asymmetric lift of a heavy magnitude

The following methods are necessary to improve safety in functional testing:

1. The client must wear clothing that will allow the spine curves and extremities to be viewed during functional testing.
2. Controlled movement patterns should be identified by observational criteria. Momentum, control, acceleration, and velocity can be evaluated to indicate whether there is a controlled safe movement pattern.

3. Heart rate monitors are important since there are potential cardiac risk factors in heavy work or in deconditioned clients.
4. Determining perceived exertion of the client is helpful. The client should always be part of the testing procedure and be able to give the therapist feedback over and above the observational criteria. The therapist, however, remains in charge of the testing, identifying safety and test endpoints. This make a good match in ensuring both client and therapist participation in safety.

It is the choice of individual therapists to decide whether to monitor safety in functional testing. Policies and procedures should be clearly written to establish whether safety is addressed and if so, how it is addressed. It should also include the type of therapist intervention needed if safety is compromised. Documenting "poor body mechanics" or "unsafe lifting" is not a defense in the case of a negligence suit, but may be used, instead, as documentation that will provide information for the plaintiff in a negligence or malpractice case to state that a therapist did not follow the "safety" that is a standard of care.

JOB MATCHING METHODS

One very important aspect of an FCE is to give employers the information they wish to know—"Can Joe go back to work?" If a job description is available, this must be included in the report so that there is no question about physical functional matching. Table 20–2 is an example of a job match grid, which is extremely helpful as part of the report in identifying work abilities and limitations.

If a job description hasn't been provided, the functional evaluator who is dedicated to return to work will try to obtain one, either by doing on-site job analysis or by identifying a credible job analysis available in writing. Employers and vocational counselors may also have to obtain a job analysis. It is not objective if the client gives the job analysis report since this is not gained through objective evaluation but rather perception.

If the purpose of the FCE is to identify areas of work that the client can do, there should be specific reporting of physical abilities and physical limitations relating to work so that the job placement specialist can accomplish the job matching.[7] Giving graphs, charts, and so forth is not as helpful as making definitive statements about abilities and limitations.

The Americans with Disabilities Act (ADA; 1990) recognizes that the individual (disabled or not) has a right to be tested only on job-related activities. The Americans with Disabilities Act intention interpretation may lead to a conclusion that the use of normative data is not appropriate. Functional normative data have

Table 20–2 Job Match Grid for a Maintenance Person in a School District

Critical Job Demands	Physical Work Strengths	Job Match
1. Use of long-handled tools—occasionally	R hand grip = 40 lbs occasionally L hand grip = 90 lbs occasionally	Yes
2. Painting and cleaning surfaces with whole-body reach—occasionally	Forward bend standing = continuously	Yes
3. General inspection of school areas	Walking = continuously	Yes
4. Floor to waist list of boxes and pails: 1–35 lbs—occasionally	Floor to waist lift = 40 lbs occasionally	Yes
5. Front carry of boxes, tools, etc. 1–40 lbs—occasionally	Front carry = 60 lbs occasionally	Yes
6. One-handed carry of pails (R or L) 1–40 lbs—rarely	R = 20 L = 55	Yes
7. Stand—frequently	Standing to tolerance: continuously	Yes
8. Walk—frequently	Walking tolerance: continuously	Yes
9. Squat—occasionally	Repetitive squat: continuously	Yes
10. Forward bend—occasionally standing	Forward bend in standing: continuously	Yes
11. Waist to over shoulder lift—1–40 lbs—occasionally	Waist to overhead lift: 5 lbs occasionally	No
12. Overhead painting and cleaning—occasionally	Elevated work: occasionally	Yes
13. Use of hand tools—frequently	R and L upper-extremity coordination: continuously	Yes

Source: Copyright © 1993, Isernhagen and Associates, Inc., Duluth, Minnesota.

high standards of deviations which make it inaccurate for any one individual (see Table 20–3).[8] They also may lead to inaccurate assumptions. For example, a capable worker who falls in the "lower percentiles" may be seen as disadvantaged, and an employer may be less likely to hire this person even though he or she can perform essential functions of the job. Conversely, the worker who scores in the "higher percentiles" may be viewed as capable even though he or she is not able to do essential functions of the specific job. Therefore the use of percentiles or normative ranges may give the wrong impression to the employer, the physician, and perhaps to the worker himself/herself.

The ADA facilitates the use of FCE job-matching recommendations,[9] as it encourages workers to be evaluated for essential functions of specific jobs. It also gives strength to job modification recommendations that can be seen as "reasonable accommodation." Proactive FCEs and ADA match is their intent.

Table 20–3 Functional Normative Data on Physical Tasks for Injured and Uninjured Subjects

		Averages of Performance					
		Injured Female Patients			*Injured Male Patients*		
Tests		*Y*	*M*	*A*	*Y*	*M*	*A*
Stand-up Lift (lbs)	MN	20.73	17.48	13.15	37.20	32.60	26.77
	SD	11.37	13.54	11.58	19.80	19.05	16.90
Level Lift (lbs)	MN	29.56	24.69	19.63	46.26	40.09	32.19
	SD	17.43	17.10	13.65	22.07	22.27	16.41
Weight Carry (lbs)	MN	30.20	24.46	20.41	48.63	42.13	32.75
	SD	15.72	13.67	14.12	23.54	23.04	17.71
		Uninjured Female Subjects			*Uninjured Male Subjects*		
		Y	M	A	Y	M	A
Stand-up Lift (lbs)	MN	41.92	28.37	14.32	67.41	64.44	51.85
	SD	9.08	13.35	19.94	19.89	19.05	13.59
Level Lift (lbs)	MN	59.16	49.51	48.72	82.70	82.24	73.66
	SD	12.66	14.01	12.65	17.10	18.05	16.98
Weight Carry (lbs)	MN	61.28	51.92	52.84	84.66	84.24	77.55
	SD	14.82	13.11	13.13	16.49	16.70	16.98

Note: A, advanced age category (aged 56 to 75); M, middle age category (aged 35 to 55); Y, young age category (aged 18 to 34).

Source: Courtesy of FCA Network, Polinsky Medical Center, 530 East Second Street, Duluth, Minnesota.

CHALLENGES FOR THE FUTURE

Redefining the role of the FCE as a facilitator of return to work by matching work capabilities with job demands has been a turning point. One future challenge is to encourage all FCE processes to be an objective link between worker and the work.

Keeping the end users in mind, all FCEs should be objective and understandable for employer, physician, vocational counselor, and most importantly, the worker.

Outcome studies should continue until only the most effective types of FCEs are utilized.

FCEs should continue to be used as an avenue of job modification when necessary. This is ADA and OSHA compliant and diminishes human and dollar costs of lost time.

Teamwork among therapist, physician, employer, insurance company, and vocational counselor should put the worker's abilities at the center of the case so that matching with essential functions is timely and accurate. When done accurately, this promotes not only a less costly injury process but also the reinstatement of a "healthy" worker placed appropriately to stay healthy.

NOTES

1. S. Isernhagen, Functional Capacity Evaluation, in *Work Injury: Management and Prevention,* ed. S. Isernhagen (Gaithersburg, Md.: Aspen Publishers, Inc., 1988), 139–192.
2. M. Miller, Functional Assessments, *Work* 1, no. 3 (1991):6–11; G. Key, Work Capacity Analysis, in *Physical Therapy,* ed. M. Scully and M. Barnes (Philadelphia: J.B. Lippincott Co., 1989), 652–667; L. Matheson, *Work Capacity Evaluation* (Santa Ana, Calif.: Employment Rehabilitation Institute of California, 1982), 64–78.
3. J.C. Rosencrance et al., A Comparison of Isometric Strength and Dynamic Lifting Capacity in Men with Work Related Low Back Injuries, *Journal of Occupational Rehabilitation* 1, no. 3 (1991):197–205.
4. S. Rodgers, Job Evaluation in Worker Fitness Determination, in *Worker Fitness and Risk Evaluation,* ed. J. Himmelstein and B. Pransky (Philadelphia: Hanley & Belfus, 1988), 219–240; S. Rodgers, Matching Worker and Worksite–Ergonomic Considerations, in Isernhagen, *Work Injury,* 65–79.
5. D. Wise, The Functional Capacity Evaluation Summary Report, *Orthopedic Physical Therapy Clinics of North America* 1 (1992):99–104.
6. U.S. Dept. of Labor, *Back Injuries Associated with Lifting,* Bulletin no. 2144 (Washington, D.C.: Government Printing Office, August 1982); G.B. Andersson, Epidemiologic Aspects of Low Back Pain in Industry, *Spine* 6 (1981):53–60; W.S. Marras and G.A. Mirka, Trunk Responses to Asymmetric Acceleration, *Journal of Orthopedic Research* 8 (1990):824–832; S.M. McGill and R.W. Norman, Dynamically and Statically Determined Low Back Moments during Lifting, *Journal of Biomechanics* 18 (1985):877–885; W.S. Marras et al., The Role of Dynamic Three-Dimensional Trunk Motion in Occupationally Related Low Back Disorders: The Effects of Workplace Factors, Trunk Position, and Trunk Motion Characteristics on Risk of Injury, *Spine* 18, no. 5 (1993):617–628; S. Isernhagen, Back Schools, in Isernhagen, *Work Injury,* 19–28; S. Kumar, Lifting and Ergonomics, in *Ergonomics: The Physiotherapist in the Workplace,* ed. M. Bullock (London: Churchill Livingstone, 1990), 183–212; J. Kelsey and A. Golden, Occupational and Workplace Factors Associated with Low Back Pain, in *Back Pain in Workers,* ed. R. Deyo (Philadelphia: Hanley & Belfus, 1987), 7–16.
7. S. Isernhagen, Past and Present of Industrial Physical Therapy in the Work Injury Management Spectrum, *Orthopedic Physical Therapy Clinics of North America* 1 (1992):1–6; S. Isernhagen, Return to Work Testing: Functional Capacity and Work Capacity Evaluation, *Orthopedic Physical Therapy Clinics of North America* 1 (1992):83–98; S. Isernhagen, The Role of Functional Capacity Assessment after Rehabilitation, in Bullock, *Ergonomics;* S. Isernhagen, Functional Capacity Assessment and Work Hardening Perspectives, in *Contemporary Care for Painful Spinal Disorders,* ed. T. Mayer et al. (Philadelphia: Lea & Febiger, 1991).
8. C.H. Nygard et al., Musculoskeletal Capacity of Middle Aged Women and Men in Physical, Mental and Mixed Occupations: A 3.5 Year Follow Up, *European Journal of Applied Physiology* (1988); S. Isernhagen, Functional Capacities Assessment—Return to Work Research (Paper pre-

sented at the Proceedings of World Confederation of Physical Therapy, May 1987); K. Timm, Isokinetic Lifting Simulation: A Normative Data Study, *Journal of Orthopaedic and Sports Physical Therapy* 5 (1988):156–166.

9. S. Isernhagen, Pertinent Laws for the Physical Therapist in Work Injury Management, *Orthopedic Physical Therapy Clinics of North America* 1 (1992):125–132; M. Rothstein, Legal Considerations in Worker Fitness Evaluations, in Himmelstein and Pransky, *Worker Fitness,* 209–218; S. Rhomberg and N. Bernstein, Employment Screening: Legal and Clinical Considerations, *Work* 1 (1990).

Chapter 21

Functional Capacity Evaluation: Case Studies

Ann M. Walker

The functional capacity evaluation (FCE) is a strong component in the safe and successful return-to-work process. The FCE report is the critical piece of information needed to determine the specific functional abilities and/or limitations of the client. This client-specific information is extremely important to the following people for the these reasons:

1. *Referring Physician:* Often physicians are unable to take the additional time needed to evaluate the client's functional abilities/limitations. They are many times restricted to medical information that *describes* the client's *medical condition* (e.g., MRI report, X-rays, CT scans) and have no specific information describing the client's *function*. This dilemma becomes additionally frustrating to physicians when they are asked to make a decision to return their client to work with specific job restrictions outlined (e.g., weight lifted or sitting/standing tolerance). Add to that the individual client's attitude about returning to work, which may or may not be realistic, and physicians can find themselves in a very frustrating quandary.

The FCE report provides the physician with standardized, measurable information regarding the client's functional ability—for example:

- specific work postures that the client is able or unable to assume
- safe workloads and ability to tolerate repetitions
- the integrity of the client's musculoskeletal condition and recommendations for additional total body conditioning or other measures that will enable the client to safely tolerate a return-to-work status
- specific job modifications to allow for a safe return to work

The physician can pair this information describing the client's functional condition with the client's medical condition and make recommendations for a safe return to work.

2. *Referring Rehabilitation Counselor/Nurse:* Responsible for the case management of the client, the referring counselor is faced with identifying the demands of the job (if there is a specific job available) and matching it with the abilities of the client to allow for a safe and timely return-to-work process. In addition, the counselor must juggle the often heated and complex issues that surround the case:

- communication between client, employer, physician, therapists, chiropractors, attorneys, and other interested parties
- attitudes of the above
- safe, cost-effective, and timely return to work
- appropriate job match

The FCE report provides the counselor with the most valuable or critical piece of information—*can the client do the job?* The FCE will answer the following questions for the rehabilitation counselor:

- *What specific job modifications, if any, need to be made by the employer before the client can return to work?* The rehabilitation counselor and/or employer can see that modifications are made to allow for a safe and successful return to work.
- *What specific movements/postures cause an increase in symptoms and may need to be modified by position changes, stretch exercises, and so forth?* The rehabilitation counselor, supervisor, or safety director can review appropriate work behaviors that will allow the client to physically tolerate the present job demands and prevent further injury.
- *Does the job match the functional capabilities of the client, or is a different job level going to be needed?* The rehabilitation counselor, with the employer, can now begin the return-to-work process or job placement process as indicated.
- *Will the client need any further reconditioning before returning to work?* The physician and the rehabilitation counselor can now set up the time line for the return-to-work process, therapy recommendations, and so forth, and coordinate with all related parties.

3. *The Employer:* Many times left in the dark, yet financially responsible for the injured employee, the employer must be a *strong, informed, integral player* in the return-to-work process. The FCE report can answer the following questions for the employer.

- *What are the specific abilities and limitations of the injured employee?* Armed with the above information, the employer can now safely return the employee to a job that clearly meets the employee's abilities. The employer may have to make some job modifications to allow for a safe return to work or may be unable to find a job within the employee's restrictions and further outside job placement may have to be investigated.

- *Do I have to make some ergonomic changes in my workplace to make it a safer place for all my employees to work?* Incorporating the ergonomic recommendations from the FCE report, the employer can now look back to the injury log book to determine if there is a pattern to past injuries or if injuries revolve around specific workstations. The employer can then begin to take steps to redesign out the potential for injury in the work areas.
- *Should we have a specific preventative education program for the use of good body mechanics available to all of our employees?* Armed with the medical/functional information from the FCE report, the employer can now design a preventative education program for his or her employees that further reduces the incidence of cumulative trauma or potential for further reinjury in the company.

4. *The Client:* The most important person in the return-to-work process, the client, needs to clearly understand what he or she can and cannot do in the *return-to-work* endeavor. The FCE will clearly be able to answer his or her questions of:

- *Is it safe for me to return back to my job and not get reinjured?* Having gone through the FCE and conference, the client should now have a clear understanding of what motions or postures aggravate or increase his or her symptoms. He or she can now begin to take the personal responsibility to prevent further reinjury on the job by following the FCE recommendations concerning proper body mechanics, weight and repetition limitations, and exercise breaks at work and at home.
- *Will my employer understand and support my abilities and limitations at work?* Working together as a team from the FCE conference back to the worksite, all parties are now aware of the client's abilities and limitations. The FCE report clearly outlines the client's abilities and limitations to aid in the smooth transition back to the job site.

5. *The Insurer:* Responsible for paying the expenses while the injured worker is out of work, the insurer will now be able to answer the following questions:

- *Will we need to put this employee into job placement, or can he or she return to the present job?* From the clear recommendations of the FCE, the claim representative now can estimate the cost and time line needed to fund this client's case.
- *Does the client possess the abilities to safely return to his or her present job even if the client chooses not to return to it?* The claim representative can now effectively have the employer offer the client his or her previous job,

and if the client refuses to return to work the case can be settled and closure attained.

6. *The Attorney (if involved):* In an effort to protect the client's rights in a safe return-to-work process, the attorney will need to know:

- *Are the demands of the job in line with the client's abilities, and will the client be safe in his or her return-to-work process?* The attorney can now support the client in a safe return-to-work process, knowing that specific job modifications, proper body mechanics, and total body conditioning will take place for the client within a certain set time line to allow for a safe and smooth transition back to work.
- *If the client can't return to his or her present job, how limiting are his or her deficits in regard to work ability and daily life skills?* The attorney will have a clear report or picture of this client's functional abilities/limitation and how these will further affect his or her abilities to ever return to work (light, sedentary, medium, heavy), as well as the client's daily functional skills. This information will aid in the settlement procedures for the injury.

THE MANY FACES OF THE FUNCTIONAL CAPACITY EVALUATION: INDIVIDUAL CASE STUDIES

The FCE, then, serves many different functions for each interdisciplinary team member in the return-to-work process. The following FCE case studies illustrate how the information from the FCE report was used to alleviate specific problem situations for each client.[1]

Case Study 1

Background

This 50-year-old man with a diagnosis of lumbar strain was referred to determine if he could return to his present job of shovel operator for a small mining company. Contributing factors surrounding the case included:

1. The mine was publicly scheduled to go on layoff one week after the client's accident.
2. No one saw the employee get injured at work.
3. The client's physician had a reputation of not returning clients back to work in a timely manner in this small town.
4. Client was previously tested (one month prior) on a psychophysical FCE with results of a 10-lb lifting restriction and yet was seen carrying a 50-lb beer keg to his car without any noticeable stress.

The client was admitted for the FCE using the kinesiophysical approach. It was clearly explained that the evaluator would keep him safe during the lifting components of the test, but that the evaluator would make the determination as to when the test would stop. If the client chose to stop the test before this time, he most certainly could do so, but it would go down in his report that he chose to self-limit and did not lift to his maximum ability. Client was in agreement, and the evaluation was completed (see Exhibits 21–1 and 21–2).

Exhibit 21–1 Functional Capacities Evaluation Summary Report, Case Study 1

FUNCTIONAL CAPACITIES EVALUATION SUMMARY REPORT
Isernhagen Clinics, Inc.

NAME: John Jones

ADDRESS: 2530 Apple Street
Duluth, MN 55812

FCE DATES: April 16–17, 1992

DATE OF BIRTH: May 13, 1942

REFERRAL SOURCE: Mary Mon, Qualified Rehabilitation Consultant

DIAGNOSIS: Lumbar strain

DESCRIPTION OF TESTS:

The client participated in a two-day (5-hour) Functional Capacity Evaluation. It is noted that client drove 2½ hours to reach FCE site each day.

COOPERATION:

Client arrived early for the testing on both days and cooperated fully in all activities requested.

CONSISTENCY OF PERFORMANCE:

Client demonstrated consistent performance and reproducible results comparing his physical assessment evaluation to his functional performance, and between Day 1 and Day 2 of the test.

PAIN BEHAVIOR:

During the evaluation the client verbalized that he was experiencing pain symptoms in his low back. However, the client was safe and functionally able to tolerate the tests without substitution patterns or protective pain behaviors.

SAFETY:

Client demonstrated a safe performance including using proper body mechanics, appropriate pacing, and a strong awareness of safe techniques throughout the entire testing procedure.

QUALITY OF MOVEMENT:

Client demonstrated smooth coordinated movements throughout the testing procedure.

continues

Exhibit 21–1 (continued)

SIGNIFICANT ABILITIES:

1. Client demonstrated good overall strength, lifting, and carrying abilities.
2. Client demonstrated good tolerance to climbing, balancing, elevated work, trunk rotation, crawling, kneeling, crouching, squatting, sitting, standing, and walking.
3. Client demonstrated solid awareness of good body mechanics and safe pacing.

SIGNIFICANT DEFICITS:

1. Client demonstrated a decreased tolerance to sitting and standing in a *forward bent* position.

JOB DESCRIPTIONS EXPLORED: Shovel Operator

CRITICAL JOB DEMANDS	PHYSICAL WORK STRENGTHS	JOB MATCH
1. Able to tolerate 7–8 hours of sitting (not in forward bent position)	Sitting tolerance within normal limits	Yes, allow for frequent stretch breaks
2. Climb 14–20 (1 ft) steps	Tolerates repetitive stair climbing	Yes
3. Push levers to control shovel (approx. 5 lb push force)	Push force 80 lb maximum	Yes
4. Must be able to lift 20 lbs horizontally occasionally	Able to lift 40 lbs horizontally occasionally	Yes

SUMMARY:

1. The client tolerated five hours of demanding physical tasks during the Functional Capacity Evaluation in addition to driving 2½ hours to and from the evaluation site on both days of the test.
2. Client was able to reproduce the same results on Day 2 as Day 1, indicating test results to be reliable.
3. The client demonstrated very solid functional capabilities throughout both testing days.
4. His physical abilities do match the critical job demands of Shovel Operator.

SPECIFIC RECOMMENDATIONS:

______________________________ Evaluator

4/20/92
______________________________ Date

Courtesy of Isernhagen Clinics, Inc., Duluth, Minnesota.

Exhibit 21–2 Functional Capacity Evaluation Form, Case Study 1

FUNCTIONAL CAPACITY EVALUATION FORM
Isernhagen Clinics, Inc.

John Jones
Client Name
4/20/92
Date

	Percent of 8-hour Day						
Item:	0	1–5	6–33	34–66	67–100	Restrictions	Recommendations
WEIGHT CAPACITY IN LBS							
Floor to waist lift		60	40	30	20		
Waist to overhead lift		50	40	25	15		
Horizontal lift		50	40	25	15		
Push		80	60	40	20		
Pull		90	70	45	25		
Right carry		50	40	30	20		
Left carry		50	40	30	20		
Front carry		50	40	30	20		
Right hand grip		110	85	55	30		
Left hand grip (*dominant*)		116	90	60	30		
FLEXIBILITY/POSITIONAL							
Elevated work					✓		
Forward bending/sitting				✓			Allow for frequent position changes
Forward bending/standing				✓			

continues

Exhibit 21–2 (continued)

	Percent of 8-hour Day						
Item:	0	1–5	6–33	34–66	67–100	Restrictions	Recommendations
Rotation sitting					✓		
Rotation standing					✓		
Crawl					✓		
Kneel					✓		
Crouch—deep static					✓		
Repetitive squat					✓		
STATIC WORK							
Sitting tolerance					✓		
Standing tolerance					✓		
AMBULATION							
Walking					✓		
Stair climbing					✓		
Stepladder climbing					✓		
Balance					✓		
COORDINATION							
Right upper extremity					✓		
Left upper extremity					✓		

Evaluator ____________ Date ____________

Source: Copyright © 1989, Isernhagen Work Systems, Duluth, Minnesota.

Follow-up

Upon completion of the FCE, the referring counselor requested the evaluator to attend the next physician appointment with the client and counselor. The counselor related that she had not previously been successful in gaining a release for the client to return to work from this physician and that decisions had been made based on the client's report of symptoms.

At the next physician appointment, the evaluator presented the FCE results and recommendations. After reviewing the above, the physician inquired if the client needed to take additional pain medications to tolerate the test. The client stated that he did not need to take any additional pain medications. The physician then agreed to release the client back to work safely based on the finding of the FCE. He stated to the client that his hands were tied as far as allowing the client to stay out of work any longer.

The referring counselor was now able to get a release to safely return this client back to his previous job *without upsetting the patient/physician relationship* that this particular physician was comfortable with in this small town.

Case Study 2

Background

This 25-year-old female with a diagnosis of bilateral carpal tunnel surgeries was referred for an upper-extremity FCE to determine if she could increase her work hours at her present job from four hours per day to eight hours per day. Her major complaint was that her hands swelled after four hours of work and she couldn't tolerate working any longer. The evaluator asked the referring counselor to have her work her four hours prior to coming to the FCE. Upon arriving for the FCE the evaluator documented that client reported her hands weren't swollen. On both days the client worked four hours prior to the test. The client was given an additional hour of job simulation *after* the FCE test on Day 1 and Day 2 without any increase in symptoms. A volumetric test was done pre- and post-FCE on Day 1 and Day 2 with no increase in swelling from the FCE tasks noted.

Client History

Functional Level. Client reported that she had difficulty with vacuuming and with doing any heavy lifting—for example, carrying grocery bags or picking up a coffee pot or a large skillet. She stated that both of her hands were weaker than before she had her surgeries, the right being worse than the left.

Pain Level. Client reported that her pain progression began with swelling, progressed to an ache, then became pain. On a scale of 1 to 10, pain progressed from a Level 3 to a Level 7. She stated that it took 1 to 2 hours of resting her hands before the pain subsided. She said she usually didn't lift over 10 pounds because then she experienced pain.

Work Status. Client was at the time working four hours a day assembling garments. She stated that there was some repetition in this job, but she was able to take frequent stretch breaks as needed, and she believed it to be a good fit for her as far as job task matching. She stated that she did not wear her splints at work because they were too restrictive. She said that after 2½ to 3 hours of work her hands began to swell and then they started to ache and become tingly and cold. She then began to lose the strength and had trouble driving home after work. Client reported that she was able to do the job until her hands began to swell, at which point she needed to stop.

Physical Assessment

Coordination, bilateral grip strength, bilateral pinch strength, and sensation were all within normal limits. Client presented with normal range of motion and muscle strength for the shoulder, elbow, forearm, wrist, and finger motions. Phalen's test produced tingling sensation in fingers bilaterally after three seconds. Girth measurements were taken pre- and post-test with results reported in Table 21–1. See Exhibit 21–3 for an FCE Summary Report.

Follow-up

With the results of the upper-extremity FCE, the referring counselor was able to meet with the client and employer to discuss a plan of gradually increasing the client's hours at work. It was at this time that the client was able to identify that the actual reason for her not being able to tolerate more than four hours of work was not related to her wrist problem after all, but was related to a substance abuse problem. The employer was able to provide his employee with support in this area

Table 21–1 Girth Measurements for Case Study 2

Location	*Pre-FCE*	*Post-FCE*
	11:00 AM	1:05 PM
Right wrist		
over the styloid process	6 in	6 in
over the palm	7¾ in	7¾ in
Left wrist		
over the styloid process	6 in	6 in
over the palm	8 in	8 in
Volumeter measurements		
Right arm	254 ml	254 ml
Left arm	256 ml	256 ml

Exhibit 21–3 Functional Capacities Evaluation Summary Report, Case Study 2

FUNCTIONAL CAPACITIES EVALUATION SUMMARY REPORT
Isernhagen Clinics, Inc.

NAME: Mary Shlat

ADDRESS: 2123 North Ash
Duluth, MN 58301

FCE DATES: May 11–12, 1992

DATE OF BIRTH: November 29, 1950

PHYSICIAN: Paula Tob, Qualified Rehabilitation Counselor

REFERRAL SOURCE:

DIAGNOSIS: Bilateral Carpal Tunnel

DESCRIPTION OF TESTS:
Client participated in an Upper Extremity Functional Capacity Evaluation with additional job task analysis totaling four hours done over a two-consecutive-day period.

COOPERATION:
Client demonstrated cooperative behavior in that she was willing to work to her maximum abilities throughout testing on Day 1 and Day 2 of a two-day modified test.

CONSISTENCY OF PERFORMANCE:
Client demonstrated consistent performance throughout her testing procedure as compared to the findings from the physical assessment on the musculoskeletal evaluation as well as between Days 1 and 2 of testing.

PAIN BEHAVIOR:
Client demonstrated appropriate symptom response during the testing procedure. She stated when she had mild discomfort but did not focus on her pain inappropriately.

SAFETY:
Client demonstrated a safe performance, including using proper body mechanics, appropriate pacing, and general awareness of safe techniques.

QUALITY OF MOVEMENT:
Client demonstrated a smooth and coordinated movement throughout the testing procedure.

SIGNIFICANT ABILITIES:
1. Above average bilateral coordination skills.
2. Excellent awareness of safe body mechanics and proper wrist positioning.

SIGNIFICANT DEFICITS:
1. General deconditioning of upper-body strength, which leads to moderately low weight capacities.

continues

Exhibit 21–3 (continued)

JOB DESCRIPTIONS EXPLORED: Garment Assembler

CRITICAL JOB DEMANDS	PHYSICAL WORK STRENGTHS	JOB MATCH
1. Assemble .5 lb garments on hangers frequently throughout the day	Able to grasp up to 10 lbs bilaterally frequently	Yes
2. Lift .5 lb garments from waist to 45 in from the floor frequently throughout the day	Able to lift 5 lbs waist to shoulder height frequently	Yes
3. Fold .5 lb garments and stack in a pile from 30 to 45 in height frequently throughout the day	Able to lift 5 lbs waist to shoulder height frequently	Yes
4. Sit up to 6 hours per day	Able to sit continuously throughout the day	Yes
5. Stand up to 6 hours per day	Able to stand continuously throughout the day	Yes

SUMMARY:

Client presented to the Modified Upper Extremity Functional Capacity Evaluation after working four hours at her present job, which is assembling garments. Upon arriving for the test, client stated that her hands were not swollen compared to what they usually are when she has to quit working due to swelling problems. Client participated in 2 days, 2 hours per day of vigorous upper extremity functional capacity evaluation consisting of twisting, fine finger manipulation, stress to the upper arm and wrist areas with lifting subtest after four hours of work at her present job. Client showed no increase in swelling as noted in the pre- and post-test for the volumeter and girth measurement test. Fine motor coordination which was done after lifting subtest scores were in the high average range, indicating that her coordination was not functionally affected.

RECOMMENDATIONS:

1. Client participate in gradual increase in hours at her present job from 4 to 6 to 8 hours.
2. Client continue to take frequent stretch breaks and position changes throughout the day.

Evaluator ____________________

Date 5/13/92

Courtesy of Isernhagen Clinics, Inc., Duluth, Minnesota.

through the company's substance abuse program, and the client has now returned to an eight-hour workday.

CONCLUSION

It is clear that the FCE can be a valuable tool in the return-to-work process. It serves a distinct role for each team member. The FCE provides the vital piece of information that each team member needs to make an informed and safe decision regarding the abilities and limitations of the injured employee, and to answer the proverbial question, "Can the worker do the job or not?"

NOTE

1. Case studies are taken from a Functional Capacity Evaluation conducted by Isernhagen and Associates, Inc., Duluth, Minnesota, 1988.

Chapter 22

Work-Hardening and Work-Conditioning Perspectives

Linda E. Darphin

Work-hardening and work-conditioning programs have developed in response to a nationwide need for better management of worker compensationable injuries. Until very recently, medical practitioners often treated injured workers in a very passive and unproductive manner. A clear analogy of our deficits in the management of the injured worker is drawn when we compare the treatment of the injured athlete to the treatment of the injured worker.

The injured athlete is typically treated very aggressively. If a college football player is injured in the middle of a football game, a team of medical professionals often including physicians, athletic trainers, physical therapists, and others, is rushed into the game to assess the athlete's injury. The athlete is assisted off the field, and immediately ice and compression or the appropriate modalities are applied to the injured part and the patient is stabilized. Typically, the athlete is rushed to an emergency room where X-rays or other diagnostic testing can be performed. If the athlete undergoes a surgical procedure the next morning, he is typically exercising a few hours after surgery. Immediately, the athlete continues on an aggressive exercise program for the uninvolved joints, while the involved joint is rested in the appropriate manner. The athlete is returned to the sidelines during practice and continues to participate during practice and games until he is able to begin practicing and playing with the team. While he is healing on the sidelines, the athlete continues to work out with the team in any exercises or activities he is cleared to perform. Mentally and emotionally he stays as a part of the team, and physically he continues to work out as aggressively as allowed by his treating practitioners. Early mobility, strengthening, and return to activity are encouraged by all involved.

Note: A special thank you to Judy Needham Fontenot for her help with typing and editing.

Compare that with the past treatment of the injured worker. Often, an injured worker would be seen by a general practitioner (who very frequently was not specialized in occupational medicine). Past theory in the treatment of low back pain was rest. Often the injured worker would be told to go home and rest and return to the physician in three to seven days, depending on the philosophy of the treating physician. Instead of being actively involved physically and/or psychologically in the work "team's" activity, the injured worker often went home and rested in positions that were most likely not helpful to his or her recovery. For example, someone with an acute bulging disc might rest in a prolonged flexed posture in his or her recliner for sustained periods of time while he or she was off work. Since the injured worker did not have any productive activity to perform, he or she often smoked more cigarettes, ate more junk food, or participated in activities that did not encourage a healthy recovery. The injured worker thus emotionally adjusted to the sedentary lifestyle as opposed to being encouraged to be active as soon as possible.

This obvious discrepancy in the treatment of injured individuals resulted in the development of more aggressive attitudes toward treating the injured worker. Many practitioners perceived that injured athletes get well quickly and injured workers have slow recoveries. Although some of the fault may lie in our workers' compensation system, our legal system, and monetary motivation for return to work, we as medical practitioners need to also assume responsibility for poor management of work-related injuries in the past. Work-conditioning and work-hardening programs have evolved to assist injured workers in returning to work as quickly and safely as possible.

DEVELOPMENT AND ACCREDITATION

The term *work hardening* became identified as an intervention service for the industrially injured client in the 1970s and 1980s. Leonard Matheson developed a work capacity evaluation, and work-hardening programs naturally followed to bridge the gap between an identification of the client's deficits and return to employment.[1] Polinsky Medical Rehabilitation Center in Duluth, Minnesota, developed and marketed a program to train physical therapists in performing functional capacity evaluations in the early 1980s.[2] This launched physical therapists into the field of work rehabilitation. Many different disciplines were working in this area, including vocational evaluators, psychologists, vocational specialists, occupational therapists, and others. This diversity of providers and philosophies created inconsistency in program definitions, and reimbursers were concerned about the value of treatment being received. While some programs labeled a single-disciplined, 1- to 2-hour exercise program as work hardening, others were identifying a full-day interdisciplinary program as work hardening. In an attempt to develop consistency in a wide variety of programs, the Commission on Accreditation of Rehabilitation Facilities (CARF) defined and developed standards for work-hardening practice. The CARF definition for work hardening as defined in the 1992 *Standards Manual* is as follows:

> Work hardening is a highly structured, goal-oriented, individualized treatment program designed to maximize the person's ability to return to work. Work hardening programs are interdisciplinary in nature with a capability of addressing the functional, physical, behavioral, and vocational needs of the person served. Work hardening provides a transition between the initial injury management and return to work, while addressing the issues of productivity, safety, physical tolerances, and work behaviors. Work hardening programs use real or simulated work activities in a relevant work environment in conjunction with physical conditioning tasks. These activities are used to progressively improve the biomechanics, neuromuscular, cardiovascular/metabolic, behavioral, attitudinal, and vocational function of the person served.[3]

In order to become accredited by CARF, an organization must comply with the work-hardening standards in addition to program and administrative standards.

Many practitioners providing work-hardening and work-conditioning services found these non-work-hardening standards cumbersome, expensive, and not necessarily related to the quality of care provided. Physical therapists frequently provided successful return-to-work services as a single-discipline program and did not comply with the multidisciplinary definition of CARF. To further inflame the issue, three state workers' compensation boards and departments began mandating CARF accreditation in order for a provider to be reimbursed for work-hardening services. For example, the three states (Florida, Ohio, and Oregon) pay for work hardening only if the provider is CARF accredited. Since CARF's accreditation standards use an institutional multidisciplinary model of care, this excludes most private physical therapy practices from becoming accredited.[4]

In 1991, the American Physical Therapy Association (APTA) established the Industrial Rehabilitation Advisory Committee (IRAC) to classify the levels of work rehabilitation to accurately reflect contemporary practice, to standardize terminology, and to address the needs of clients, providers, regulators, and payors. In response, IRAC developed guidelines for work-conditioning programs for injured workers with only physical problems and subsequently developed guidelines for work-hardening programs as a viable alternative to the CARF standards. These guidelines went out for field review and then were adopted by the APTA Board of Directors.

These guidelines began by identifying two distinct services. Work conditioning follows acute care and addresses the physical needs of those unable to return to work. Work hardening is for a limited number of clients who also have behavioral or vocational needs, and is interdisciplinary in nature. Program comparisons between the two services are made in Table 22–1.

IRAC defined *work conditioning* as a work-related, intensive, goal-oriented treatment program specifically designed to restore an individual's systemic, neuromusculoskeletal (strength, endurance, movement, flexibility, and motor control), and cardiopulmonary functions. The objective of the work-conditioning program is to restore the client's *physical* capacity and function so the client can return to work.[5] Work conditioning is most appropriate for individuals who are beyond the initial healing time following injury, which is typically three to six weeks from date of injury, unless necessity is justified, and continue with physical impairment and functional abilities.[6] *Work hardening* was defined by IRAC as "a highly structured, goal-oriented, individualized treatment program designed to return the person to work. Work-hardening programs, which are interdisciplinary in nature, use real or simulated work activities designed to restore physical, behavioral, and vocational functions. Work hardening addresses the issues of productivity, safety, physical tolerance, and worker behaviors."[7] Typically, entry into work hardening is not sooner than 12 weeks post date of injury unless necessity is justified. Entry into work hardening requires a functional capacity evaluation or work-hardening entrance examination.[8]

Specific guidelines were outlined by IRAC to determine clients' eligibility, provider responsibility, program content, and program termination for both work-hardening and work-conditioning programs. In order to facilitate understanding of how these services integrated with traditional therapies, a continuum of care was developed (Figure 22–1).

Table 22–1 Guidelines for Programs in Industrial Rehabilitation

Work-Conditioning Program	*Work-Hardening Program*
Addresses physical and functional needs that may be provided by one discipline (single-discipline model)	Addresses physical, functional, behavioral, and vocational needs within an interdisciplinary model
Requires work-conditioning assessment	Requires work-hardening assessment
Utilizes physical conditioning and functional activities related to work	Utilizes real or simulated work activities
Provided in multihour sessions up to: 4 hours/day 5 days/week 8 weeks	Provided in multihour sessions up to: 8 hours/day 5 days/week 8 weeks

Source: Reprinted from *Guidelines for Programs in Industrial Rehabilitation* by the American Physical Therapy Association Industrial Rehabilitation Advisory Committee with permission of the American Physical Therapy Association, © 1992.

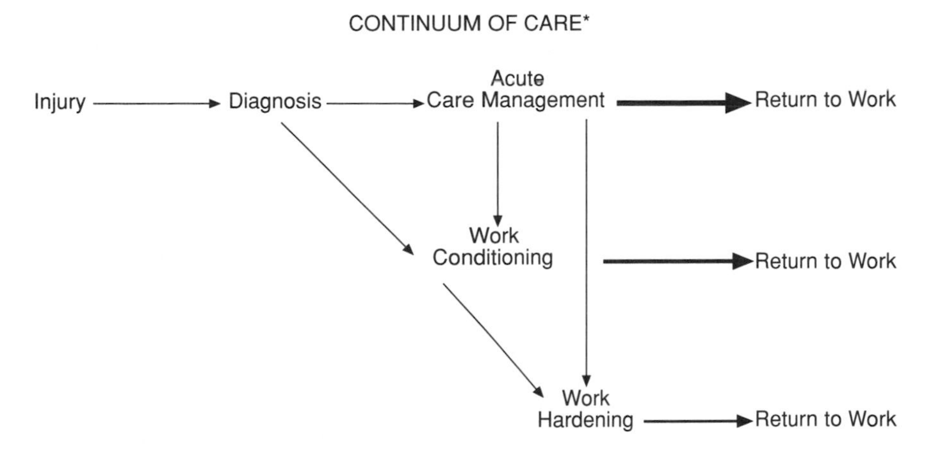

*It is recognized that most injured workers will not require progression through all phases of care in order to return to employment.

Figure 22–1 Continuum of Care in Industrial Rehabilitation Programs. *Source:* Reprinted from *Guidelines for Programs in Industrial Rehabilitation* by the American Physical Therapy Association Industrial Rehabilitation Advisory Committee with permission of the American Physical Therapy Association, © 1992.

DISCIPLINES INVOLVED

The disciplines commonly associated with work-hardening programs include physical therapy, occupational therapy, psychological/social services, and vocational services. Professional services, such as nutrition, alcohol and drug dependency, remedial education, and medication reduction may be integrated into a program or referred to as deemed necessary by the treating team. To achieve a successful return-to-work goal, it is imperative to work with a variety of "players in the game." These players may include, but not be limited to, insurance adjusters, physicians, attorneys, employers, and the client's family. IRAC states, "The client should be the center of the team. All activities should be driven by the needs of the client as perceived by the team and should be communicated frequently and thoroughly to the client."[9] The program manager must understand the needs of each consumer and communicate to the consumer in a way that meets his or her needs.

The insurance adjuster must be assured that the program is cost-effective and geared toward timely return to work. Employers will want to be assured that the employee can return to work quickly and safely at the highest level of function possible. They are interested in high productivity and the prevention of reinjury that results in lost time. Often by communicating the benefits to an employer, the therapist can encourage modified work or "light duty" return to work. The employer will feel more secure returning an injured worker to work if the worker's abilities and limitations are clearly defined.

Physicians also are concerned about their responsibility to return the injured worker to work safely. Objective information the therapist can provide in terms of the client's physical abilities, motivation, participation, job requirements, and functional limitations is extremely helpful in making a final determination on return to work. It is important to establish oneself as an unbiased evaluator in this often-litigious arena. One should clearly communicate to both plaintiff and defense attorneys that the findings are objective and not influenced by the fact that the attorney may be referring and paying for the client's care. Attorneys need a credible witness if their cases go to trial and appreciate integrity in practice. It is often very helpful for the therapist to help educate an attorney regarding how to interpret and effectively use the information.

All team members need not be present at all times. Typically, in work-hardening programs, occupational and/or physical therapists are the core team members that see all clients, and a psychologist and/or vocational counselor evaluate the client and then serve the client's needs as determined by the team. If a client does not have a job to return to, the vocational counselor may perform vocational testing to develop a rehabilitation plan for the client. As suggested by Williams, the rehabilitation counselor can make the difference between a cost-effective program and an inefficient one. An experienced rehabilitation counselor is needed to promote return to work while attempting to satisfy the somewhat divergent and potentially competing needs of all parties in the recovery process.[10] A ratio of one professional (physical or occupational therapist) for every six to eight workers receiving work-hardening services is recommended.[11]

The professionals who work in work-hardening and work-conditioning programs need to have the appropriate schooling, degrees, licensure, and certifications to be able to evaluate and develop treatment programs for injured workers as deemed appropriate by their state laws and their professional associations, with the addition of specialized training in the area of work rehabilitation.

A work-hardening program needs therapists who are specialized and designated to the work-hardening program and who are not seeing a general load of clients simultaneously. In the work-conditioning program, therapists can often see clients in the acute stage and progress them into the work-conditioning stage, but they need to be available at all times for the clients in the work-conditioning program. Most facilities find it most beneficial to have therapists committed and designated to the work-conditioning and work-hardening program who are not also seeing a variety of acute clients. It often takes therapists time to acclimate to work-hardening and work-conditioning programs in terms of developing good communication and management skills for dealing with the work-injured individual. The therapist needs to develop techniques that address behavioral as well as physical issues, and often needs to work successfully with a team of practitioners. For this reason, it is not recommended that therapists rotate through a work injury program on a short-term time schedule.

WORK REHABILITATION PROGRAM MODEL

The program model for most work rehabilitation programs begins with the client being evaluated for two to three days, with each day being one to three hours long. During this time, the client goes through an orientation process to learn the principles of work rehabilitation and to develop common goals with the work rehabilitation team.

After the initial evaluation, the client begins the work rehabilitation program. Some programs begin with three days a week and progress to five. Other programs begin daily. A program may begin with two to four hours a day and progress to four to eight, depending on the client's needs and the program consumer's needs.

The client's daily activities are often structured in a manner to include the following: group exercises or warmups, individual exercises, individual work-simulation tasks or projects, and rest breaks (see Figure 22–2). Often clients repeat cycles of tasks similar to the requirements of a job, which involve doing the same physical activities repeatedly.

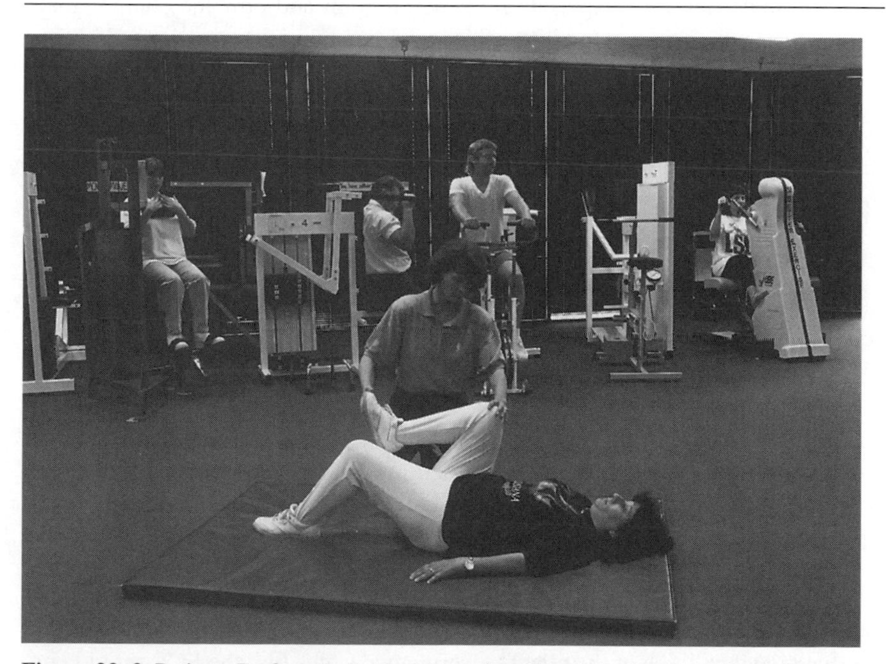

Figure 22–2 Patients Perform Individual Exercises Taught by Physical Therapists. Courtesy of Occupational Performance Center, a division of Baton Rouge Physical Therapy, Baton Rouge, Louisiana.

One method of balancing cost-effectiveness with quality outcomes is to have the client attend a four-hour-per-day program for the majority of his or her treatment stay and progress to eight hours in the last week of care. An eight-hour-day approach is a more valid indicator of a client's ability to perform work over an eight-hour day.

WORK REHABILITATION RULES AND GUIDELINES

It is important for a center to establish a rehabilitation environment that simulates a real work environment. That can be established by developing appropriate space and physical facilities, but equally as important, by creating a worker's milieu. A common method that facilities use to create this environment is work rehabilitation guidelines. Work rehabilitation guidelines are reviewed with the client, who then signs the guidelines indicating his or her agreement. By discussing expectations initially, the therapist can often head off future conflict or clarify current misunderstandings of how the client's expectations for rehabilitation may differ from the work rehabilitation program goals. The goal of work rehabilitation programs is to increase a client's functional ability so he or she can return to work. A client's goal may be to eliminate all of his or her pain. The therapist must help the client understand the expectation of the program and how the client's goals and the program goals can mesh to achieve an agreeable outcome. If a client remains pain focused and does not quickly develop function-based goals, the assistance of a social worker or psychologist may be necessary to evaluate issues that may impede a client's progress. Exhibit 22–1 shows typical work-hardening guidelines used at the Occupational Performance Center in Baton Rouge, Louisiana.

EVALUATION

Before work-hardening activities can begin, a thorough evaluation must be performed. Terminology for this return-to-work evaluation includes *functional capacity evaluation (FCE)*, *functional capacity assessment*, and *work capacity assessment*. A thorough standardized FCE should include the components of a *subjective history* and *objective musculoskeletal evaluation* followed by *functional testing*. This test should be performed in a safe manner by qualified therapists. The state of Maryland's Joint Committee on Industrial Services states that a typical FCE requires four to eight hours of client testing over one or more days and should take place within the context of competitive employment. The FCE incorporates a comprehensive neuromusculoskeletal examination and an assessment of worker behaviors and function, and culminates with the measurement of the current functional abilities of the client.[12] Isernhagen states that the FCE should emphasize the following:

Exhibit 22–1 Guidelines for a Work-Hardening Program

Your daily program will last 4–8 hours. We will increase the time you will be here on a daily basis.

You may choose to come at either 8:00 AM or 1:00 PM. It is imperative that you be here on time.

Once a week, you will have a conference with your therapist. New goals will be set every week. If these goals are met, you may continue in the program. If they are not met, you will be discharged.

Please wear comfortable clothes. If you are required to wear any specific type of clothing on the job (i.e., steel-toed boots), please advise your therapist and wear them.

Treat this as you would your job:

- Be on time.
- Do not miss appointments (if you have 3 unexcused absences, our policy is to inform the physician and discontinue your program until otherwise notified).
- Adhere to scheduled breaks. We have one 15-minute break mid-morning and one mid-afternoon.
- Clean up your work/exercise area when you finish and return all equipment to its designated area.
- Consider this program as your return to work. You will be working with us, your "coaches," who will slowly progress you to increase your activities and who will monitor your safety and response to these activities.

You may be discharged from the program for the following reasons:

- Failure to progress.
- Lack of cooperation demonstrated by tardiness, missed appointments, and/or poor motivation.
- Unsafe or careless actions during exercise and/or work-simulated activities.

You will get out of this program only what you put into it, so give your maximum effort. We are here to help you. We look forward to working with you and believe through work hardening you can achieve your best potential in regard to return to work.

I, _______________, fully understand the rules and regulations of the work-hardening program and will abide by them as I participate in the program.

_______________________________   _______________________________

Client's Signature   Date   Witness' Signature   Date

Source: Reprinted from *Work Hardening Manual* by L.E. Darphin with permission of Isernhagen Work Systems, © 1989.

1. enhancing the skilled observation powers of a physical therapist and an occupational therapist
2. providing criteria on which to evaluate, interpret, and score functional abilities
3. providing a thought process and scientific background for interpreting functional information into an eight-hour workday

4. providing a logical comprehensive report that clearly individualizes the person[13]

The FCE is the baseline data from which deficits are noted for the therapist to outline a specific program. The client's program should identify the specific deficits outlined in the FCE and work to alleviate these deficits. The FCE should also assess worker behaviors through consistency checks on specific tests of symptom magnification on nonorganic pain behavior. The identification of these behaviors is important in developing a treatment plan and expectations for return to work. The presence of symptom magnification may affect the validity of the FCE findings. The purpose of identifying these behaviors should not be to label an individual but to develop a treatment plan to address and improve upon inappropriate behaviors. Often when clients are allowed to experience the ability to function for extended periods of time in a safe environment, their fears can be dampened and their behaviors improved.

ADMISSION CRITERIA

Admission criteria should determine which clients are appropriate or inappropriate for work-conditioning/work-hardening programs. The referral sources need to be educated in the criteria. Criteria may include the following:

1. Client must not have any medical condition that prohibits a client's participation.
2. Client must be medically screened (this may be determined by state law or facility guidelines).
3. Client must agree to participate in the program.
4. Client must have physical condition that prohibits client from working.
5. Client must have potential to benefit from the program.

The need for a physician referral prior to a physical or occupational therapist's performing an FCE is determined by individual state laws. Some states allow evaluation and/or evaluation and treatment without a physician's referral. However, since many of the clients undergoing an FCE have prolonged and often unresolved medical complications, it is recommended that the therapist receive clear communication from a physician and communicate clearly to a physician to determine any possible contraindications and to determine the appropriateness of the client's receiving an FCE or participating in a work-hardening program.

PROGRAM COMPONENTS

Four basic components included in work-hardening and/or work-conditioning programs consist of work simulation, strengthening exercises, aerobic conditioning, and education. Each of these goals is appropriate for clients who have been out of work for a significant period of time.

The work-simulation component of work-hardening programs can be intricately involved and often includes actual equipment obtained from a job site. Work-simulation equipment seen at the Occupational Performance Center, a division of Baton Rouge Physical Therapy in Louisiana, includes a carpentry work-simulation activity called a work cube (see Figure 22–3), plumbing work-simulation activities using a pipe tree, railroad equipment donated by Union Pacific Railroad, scaffolding, bricklaying equipment, a semitruck cab and vibration simulator (see Figure 22–4), and a variety of other equipment. A work-hardening program may begin with an emphasis on exercise, but by the end of it two-thirds of the program should be work simulation. The other one-third may be exercise, education, or psychological or vocational intervention. In work-conditioning programs, the client's work activity is typically simulated through exercises and/or similar positions and activities. For example, if a client is returning to wallpaper hanging, which requires prolonged overhead work, the work-simulation tasks may involve performing upper-extremity exercises overhead for time frames similar to the activities required on the job. Work conditioning is primarily exercise oriented throughout the client's stay in the program.

Figure 22–3 Work-Simulation Task at a Work-Conditioning/Work-Hardening Program Center. Courtesy of Occupational Performance Center, a division of Baton Rouge Physical Therapy, Baton, Rouge, Louisiana.

Figure 22–4 Work-Simulation Task at a Work-Conditioning/Work-Hardening Program Center. Courtesy of Occupational Performance Center, a division of Baton Rouge Physical Therapy, Baton Rouge, Louisiana.

The majority of clients in work-conditioning and work-hardening programs demonstrate aerobic deconditioning. An aerobic conditioning program should be developed for each client to improve cardiovascular status. Many programs use a basic cardiovascular screen to rule out inappropriate clients for submax cardiovascular exercise and communicate with the client's medical doctor. Each facility should develop guidelines for acceptable blood pressure and heart rate ranges for program participation.

Strengthening exercises are developed based on the client's needs as determined by a musculoskeletal assessment. The client is instructed in strengthening exercises for specific muscular-group weaknesses, which are identified in the evaluation. General muscular strengthening is enhanced using exercise classes or exercise equipment (See Figure 22–5).

Clients often participate in group as well as individual educational programs. Back school classes are taught in work-hardening programs. Work-hardening programs that deal with clients who suffer from cumulative trauma disorders educate the client about prevention of these disorders. Formal classes can be held or clients

Figure 22–5 Aerobic Conditioning at a Work-Conditioning/Work-Hardening Program Center. Courtesy of Occupational Performance Center, a division of Baton Rouge Physical Therapy, Baton Rouge, Louisiana.

can be given daily tips on nutrition, back care, fitness, and return-to-work issues (see Figure 22–6).

Often psychosocial classes are taught to deal with nonphysical issues that may prohibit or slow a client's return to work. These include symptom magnification, employee relationships, feelings of entitlement, stress relief, and relaxation techniques.

Vocational specialists may also hold educational classes for clients to discuss ways of goal setting and achieving return to work, including developing a resume and capitalizing on transferrable skills.

SPACE

In order to be accredited by CARF, the facility must have a designated work area and designated equipment for the work-hardening program.[14] Matheson states that typically work-hardening programs require 1,200 square feet or more.[15] The Occupational Performance Center, a division of Baton Rouge Physical Therapy, has found that 5,500 square feet can accommodate up to 15 work-hardening clients, including evaluation areas and offices for professional and support staff. The physical space most suited for work-hardening programs typically in-

Figure 22–6 Back School Class at a Work-Hardening Program Center. Courtesy of Occupational Performance Center, a division of Baton Rouge Physical Therapy, Baton Rouge, Louisiana.

cludes a climate-controlled area with exercise equipment that may look similar to a health club's, as well as an area that is used primarily for work-simulation activities (see Figures 22–7 and 22–8). The work-simulation area may not be climate controlled and may be easily used for "dirty" activities. The area needs to have a wide variety of job-simulation activities that reflect the needs of the local community and local workers. It is beneficial for workers to have access to indoor and outdoor work-simulation activities. The space needs to have quiet and private areas for testing and evaluation, as well as appropriate office space for the case management activities, conferencing, and documentation requirements.

Regarding work conditioning, less square footage per client is necessary since work simulation is less emphasized and standard conditioning (aerobic and muscular) is more emphasized. Also, work conditioning is generally a four-hour-per-day program, so twice the client population is possible since two groups can be seen on the same day, doubling the capacity of the space. While work hardening often has permanent work-simulation stations, work conditioning often uses multipurpose adaptable stations and customizes the simulation with actual work equipment brought in for a client and then removed once the client is discharged.

Figure 22–7 View of the Occupational Performance Center's Facility for Work-Hardening Programs. Courtesy of Occupational Performance Center, a division of Baton Rouge Physical Therapy, Baton Rouge, Louisiana.

EQUIPMENT

As industrial rehabilitation has grown as a specialty, new products have developed to capitalize on the needs of this industry. Many products are helpful and functional. However, products can be expensive and unnecessary. A facility needs to determine the needs of its clients and its program goals to obtain equipment that is appropriate and cost-effective. It is not necessary to purchase extremely expensive equipment in order to be effective in returning injured workers to work. Types of equipment that are beneficial for the program include

- Traditional strengthening equipment such as a leg press, hamstring curl, abdominal crunch, back extension, biceps/triceps, and overhead press. Also multipurpose pulley and weight sets allow individual exercises to be set up. It is beneficial for the equipment to work the muscle groups in functional positions instead of isolated muscles. For example, a leg press machine is a more functional method of increasing quadriceps strength than sitting leg extension. Circuit training is a method of increasing strength and conditioning that fits with the work-hardening therapy of functional activities. The low-aerobic-level workout simulates a real work environment for many jobs.
- Aerobic equipment, including bike, treadmill, or upper-extremity ergometer.

Figure 22–8 View of the Occupational Performance Center's Facility for Work-Hardening Programs. Courtesy of Occupational Performance Center, a division of Baton Rouge Physical Therapy, Baton Rouge, Louisiana.

- Actual work-simulation equipment such as ladders, crates, bricks, carpentry supplies, tool boxes, hand tools, work tables, scaffolding, etc. For departments that can justify cost and volume, therapists may choose computerized work-simulation equipment such as the BTE or Work Set.

COMMUNICATION

The client's program manager or primary treating therapist will communicate orally and in writing to the referral source and other appropriate parties. A signed permission form should be obtained from the client prior to releasing medical records to any party.

A written initial work-hardening evaluation or FCE should be completed at the onset of the program. Treatment goals and a time-based plan should be included. On a daily basis, a record of attendance and activities performed by the client should be kept. At least weekly, the therapist should write a detailed note outlining the subjective information, objective information, assessment, and plan.

The client's progress is communicated in writing on a one- to three-week basis depending on the clinic's procedures. A final discharge report should outline the changes from the initial evaluation, how the client met or did not meet the program's stated goals and improvements in the client's musculoskeletal and functional abilities. The discharge evaluation states if the client is released to return to work and at what level of work. One can compare the client's discharge abilities to a functionally based job description or a job site analysis and make recommendations about returning to the client's previous job. If the client does not have a job to return to, a recommendation for client's release to return to work at a specific level is made. The Department of Labor uses the sedentary, light, medium, heavy, or very heavy categories as listed in Exhibit 22–2. Clients who do not have a job to return to can benefit from the assistance of a vocational specialist

Exhibit 22–2 U.S. Department of Labor Classification of Work Levels

Sedentary Work—Exerting up to 10 pounds of force occasionally (Occasionally: activity or condition exists up to 1/3 of the time) and/or a negligible amount of force frequently (Frequently: activity or condition exists from 1/3 to 2/3 of the time) to lift, carry, push, pull, or otherwise move objects, including the human body. Sedentary work involves sitting most of the time, but may involve walking or standing for brief periods of time. Jobs are sedentary if walking and standing are required only occasionally and all other sedentary criteria are met.

Light Work—Exerting up to 20 pounds of force occasionally, and/or up to 10 pounds of force frequently, and/or a negligible amount of force constantly (Constantly: activity or condition exists 2/3 or more of the time) to move objects. Physical Demand requirements are in excess of those for Sedentary Work. Even though the weight lifted may be only a negligible amount, a job should be rated Light Work: (1) when it requires walking or standing to a significant degree; or (2) when it requires sitting most of the time but entails pushing and/or pulling of arm or leg controls; and/or (3) when the job requires working at a production rate pace entailing the constant pushing and/or pulling of materials even though the weight of those materials is negligible. *Note:* The constant stress and strain of maintaining a production rate pace, especially in an industrial setting, can be and is physically demanding of a worker even though the amount of force exerted is negligible.

Medium Work—Exerting 20 to 50 pounds of force occasionally, and/or 10 to 25 pounds of force frequently, and/or greater than negligible up to 10 pounds of force constantly to move objects. Physical Demand requirements are in excess of those for Light Work.

Heavy Work—Exerting 50 to 100 pounds of force occasionally, and/or 25 to 50 pounds of force frequently, and/or 10 to 20 pounds of force constantly to move objects. Physical Demand requirements are in excess of those for Medium Work.

Very Heavy Work—Exerting in excess of 100 pounds of force occasionally, and/or in excess of 50 pounds of force frequently, and/or in excess of 20 pounds of force constantly to move objects. Physical Demand requirements are in excess of those for Heavy Work.

Source: Reprinted from Dictionary of Occupational Titles, U.S. Department of Labor, Fourth edition, pp. 1012–1013, 1991.

throughout the work-hardening program to determine transferrable skills and assist with reemployment.

PROGRAM EVALUATION/OUTCOME MEASUREMENTS

The goal of the program evaluation program is to determine if the work rehabilitation program is meeting the needs of the clients and their referral sources. To make this determination, the program evaluation process should assist in finding out which portions of the work rehabilitation program need modification to improve outcomes and enhance the quality of care. Each program needs to develop a system of analyzing outcome data that will identify strengths and weaknesses of the program. Specific key areas need to be identified for information to be gathered. The data need to be gathered on a regular basis and analyzed to determine what changes will be recommended. For example, if it is noted that a large number of clients are discharged from the program within the first week, then perhaps admission criteria need to be revised or the referral sources need to be educated as to the appropriate clients for admission. IRAC identifies elements that they recommend to be included in a systematic data collection process. These include the common types of data that are collected in work-hardening programs (Exhibit 22–3). An outcome-oriented system is imperative to prove that a program merits reimbursement and provides services in a cost-effective and quality manner.

CONCLUSION

In conclusion, work rehabilitation programs have developed over the last 20 years as a need for improved treatment and management of the workers' compensation client has been identified. These programs have successfully served the community in helping to return injured workers to work as quickly and safely as possible. These programs involve not only a multidisciplinary but an interdisciplinary approach, in which a variety of specialists work toward a common goal of return to work. Work rehabilitation programs should be selected through an initial evaluation process. For those clients needing comprehensive multidisciplinary services, work hardening is the appropriate choice. For physical-oriented work problems, work conditioning is an appropriate choice.

In order for a program to be successful, it must have administrative support at all levels throughout the organization. Numerous programs have failed because the administration has perhaps offered support in terms of finances and buying equipment but has not understood the need for a dedicated and specialized staff who understand the needs of these populations and are committed to seeing through the program's success.

Market research to determine the needs of one's community will provide information about the possible viability of a program. Outcome studies and program

Exhibit 22–3 APTA Guidelines for What Data To Collect on Work-Conditioning/Work-Hardening Programs

ELEMENTS:

1. Identify services provided as either Work Conditioning or Work Hardening.
2. Demographic data:
   a) Age
   b) Gender
   c) Race and ethnicity
3. Occupational and injury data
   a) Primary and secondary diagnoses
   b) Work status prior to injury
   c) Has the client received other treatment for this injury prior to entering the Work Conditioning or Work Hardening program? If so, identify all disciplines involved.
   d) Date of injury
      —Same injury
      —New injury
   e) Date Work Conditioning or Work Hardening program was initiated
   f) Is there a target job awaiting the client?
   g) Time off work
   h) Length of the program
      —Hours/day
      —Days/week
      —Total days
4. Discharge data
   a) Total charges billed for the program
   b) Program status (terminated or discharged) regarding return to work
      —Same employer or different employer
      —Previous job or different job
      —Full time or part time
   c) Client status at time of program termination/discharge
   d) Referrals for additional services not available in the program
   e) Payment source
      —Workers' compensation board
      —Private insurance

Source: Reprinted from *Guidelines for Programs in Industrial Rehabilitation* by the American Physical Therapy Association Industrial Rehabilitation Advisory Committee with permission of the American Physical Therapy Association, © 1992.

evaluation processes are critical to the growth and development of a program in that they identify the strengths and weaknesses, and help develop short- and long-term strategic goals and plans for change. With all the appropriate components and processes in place, work hardening and work conditioning achieve their goal and fill a niche in the profession of successfully returning injured workers back to the workforce cost-effectively and safely.

NOTES

1. L.N. Matheson, *Work Capacity Evaluation: A Training Manual for Occupational Therapists* (Trabucco Canyon, Calif.: Rehabilitation Institute of Southern California, 1982).
2. Polinsky Medical Rehabilitation Center, *Functional Capacities Assessment Training Manual* (Duluth, Minn.: Polinsky Medical Rehabilitation Center, 1983).
3. Commission on Accreditation of Rehabilitation Facilities (CARF), *Standards Manual for Organizations Serving People with Disabilities* (Tucson, Ariz.: Commission on Accreditation of Rehabilitation Facilities, 1992).
4. American Physical Therapy Association (APTA) Department of Reimbursement, Workers' Compensation Focus Group (Private Practice and Orthopedic Sections of the American Physical Therapy Association), *You and Workers' Compensation Reform*, 29 March 1993.
5. American Physical Therapy Association (APTA) Industrial Rehabilitation Advisory Committee, *Guidelines for Programs in Industrial Rehabilitation* (Alexandria, Va.: 1992).
6. American Physical Therapy Association (APTA) Joint Committee on Industrial Services, *Standards for Performing Functional Capacity Evaluations Work Conditioning and Work Hardening Programs* (Alexandria, Va.: American Physical Therapy Association, June 1993).
7. APTA Industrial Rehabilitation Advisory Committee, *Guidelines*.
8. APTA Joint Committee, *Standards*.
9. L.E. Darphin et al., Work Conditioning and Work Hardening, *Orthopedic Physical Therapy Clinics* 1, no. 1 (1992):105–124.
10. J.M. Williams, The Rehabilitation Counselor: A Critical Member of the Work Hardening Program Staff, *Journal of Private Sector Rehabilitation* 4, no. 1 (1989):39–44.
11. M.T. Ellexson, *Work Hardening Programming* (Paper presented at the Annual Conference of the American Occupational Therapy Association, Indianapolis, May 1987).
12. APTA Joint Committee, *Standards*.
13. S.J. Isernhagen, Functional Assessment: Computerized or Not? in *Garvey Gram* (St. Paul, Minn.: Garvey Company, February 1993).
14. CARF, *Standards Manual*.
15. L.N. Matheson, *Work Capacity Evaluation* (Anaheim, Calif.: Employment and Rehabilitation Institute of California, 1987).

Chapter 23

Behavior Modification with the Injured Worker

Elizabeth J. Green

Work behavior does not automatically begin at society's definition of work age, nor does it spontaneously exist once internal conflicts are resolved or an injury is stabilized. Traditionally, therapists and other clinicians have approached dysfunction from an occupational behavioral frame of reference that has the therapist involve clients as active partners in the identification of their problems, in acknowledging their strengths, in setting achievable goals, and in planning a course of action.

Behavior modification work programs should have two main components: (1) identification of the areas of poor functioning that will prevent the client from achieving or sustaining employment—that is, a functional capacity evaluation; and (2) action taken to help the client overcome these deficits, namely, work conditioning, work hardening, or work/worksite modification.

An active approach means individually centered rehabilitation programs. Different programs for clients with different problems are necessary. Improving productivity is the best developed area of work rehabilitation. Two main reasons for poor productivity are (1) working slowly, and (2) not working continuously. The use of incentives to increase productivity has been studied and found to be successful. The incentive needs to be found that is powerful enough to affect the client's behavior and that the therapist can regulate. Staff attention is an incentive that the staff can control and that is usually valued by the client. An example might be for staff to give comments to the client, providing positive social reinforcement when they note that his or her program is being performed on schedule and meeting set goals. If productivity falls, this feedback is withheld by all staff, with only one staff member encouraging the client to improve.[1]

The therapist using behavior modification treats disorders not by concentrating on inner conflicts and their diagnoses, but rather by seeking to alter maladapted behaviors resulting from faulty learning. The disordered individual was reinforced or conditioned to act in ways that did not meet the individual's needs or were not socially acceptable. For example, an employee who has observed unsafe lifting and material handling practices and whose supervisor has placed more emphasis on speed than safety on the job may develop habits of incorrect lifting. Therefore the treatment focus is on changing behaviors and modifying the observable, measurable manifestations. The general goal of this approach is to provide environments that promote learning or relearning for a client in whom the learning of effective behavior never took place or was lost. The therapist must target behavior changes and revise the treatment program as needed.

Treatment methods include shaping by successive approximation, chaining, prompting, fading, modeling, role playing, and behavioral rehearsal.

- *Shaping by successive approximation* uses reinforcement in steps, building from the absence of the desired behavior to its successful use.
- *Chaining* is a series of related simple behaviors reinforced to establish a larger, more complex behavior. When using chaining, the client is taught the last step first and continues learning in this backward sequence until the entire chain of behaviors (i.e., performing the complete job) is mastered.
- *Prompting* involves directing the client's attention to the task, such as guiding the client from one workstation to the next or verbally requesting the client to go to the next workstation or activity scheduled.
- *Fading* is the gradual cessation of these prompts so that the client will use the environment to evoke appropriate behavior, such as observing a task needing to be performed and approaching it with correct, safe techniques.
- *Modeling* is demonstrating the desired behavior for the client, such as correct pacing or body mechanics.
- In *role playing*, the client practices a role he or she normally does not perform without the risk of being in "real life," or at the job site—for example, performing an essential job function using correct techniques and possibly adapted equipment.
- *Behavioral rehearsal* combines modeling with role playing. Both positive and negative reinforcement and stimulus control are used to increase the frequency of desirable behavior. An example is having the client demonstrate to the group of clients his or her performance of a critical demand of the work, using newly taught skills of correct postures and safety. This could be videotaped and replayed for positive and negative feedback from the client and his or her peers.[2]

Roles are positions in and necessary functions of social groups, including places of employment. When people fill roles, their behavior is influenced by the expectations of the role. They develop an identity that the role provides, and they contribute to the maintenance and development of the social group, or workplace. Occupational therapy and other specialties have focused on including a socialization process in which expectations for performance are conveyed, role behavior is facilitated by participation in groups where roles can be differentiated, and role transition may be precipitated by allowing people opportunities to try out the skills and tasks associated with new roles and to develop confidence. Overall, they prepare people for the choices that must be made to facilitate role changes, such as the change from patient back to employee. They use behavior modification, a type of open system of restoring order or function to the injured worker in which the following principles apply:

1. Seek to increase productive action.
2. Elicit action commensurate with the patient's state or organization.
3. Use occupations that embody characteristics of normal occupation.
4. Provide adequate feedback on output and its effects.
5. Facilitate increasing levels of performance.
6. Use occupations suited to the person's reorganizational requirements—that is, appropriate job simulations.

Especially in industrial rehabilitation, the therapist must ensure that these performance components are included in the total program:

1. Provide skilled training relevant to the contexts of the client's daily life and work.
2. Follow a developmental sequence in skill training—that is, fundamental teaching of body awareness such as trunk stabilization techniques before applying new postures during a work activity.
3. Acknowledge roles, interests, and values in choosing skills to be learned.
4. Consider the neurological and musculoskeletal constituents in regenerating skills, which can be initially identified from the functional capacity evaluation performed as an entry evaluation prior to work conditioning/work hardening.[3]

Empowerment refers to the degree a person is permitted to take charge of and be a major determining factor in his or her own life in contrast to being controlled by society or by a specific situation, such as at the worksite. Feelings of being "less than" or "not sufficient" internalized by some clients are caused by the general attitude of society. Through behavior modification, therapists can assist individuals to provide their own sense of empowerment to overcome such feelings and return to a productive life.[4]

Medicine and the allied health specialties have undergone radical changes in the last two decades that have dramatically altered the way health care is conducted, the image of health specialists, the function of health care in communities, and the relationship between health care professional and client. These changes reflect changes in society of increasing technologic advances in addition to increasing social distance, separateness, anomie, and a commercial mentality. More health care specialties address segments of the whole client. Very importantly, this has changed the amount of time a health professional has to spend with a client.

Reimbursement systems and procedures put health professionals under constant pressure to reduce time and money spent on each client. Studies have shown that what might be termed the social effect of specialist-client interaction, or the specialist's ability to listen and to act interested in what the client has to say, has direct benefits on recovery and well-being. Therapists also are less often from the community they are working in, thereby lacking firsthand knowledge about the people and having no appreciation for the whole life of the client. This can lead to viewing the client after a trauma as a segment of the medical process without appreciation of who that person was or is.[5]

According to Carl Rogers, the following conditions are necessary for positive regard for the other:

1. unconditional positive regard for the other
2. empathetic understanding of how the other feels
3. genuineness or congruence[6]

Empathy has many definitions. It can be used to mean sympathy and compassion. It has been understood to include the projection of one person's consciousness onto another person. Empathy usually is viewed as a connection between people that is comforting and caring. However, there has been much discussion about whether we can truly know the suffering of others and therefore whether we can help them without sharing their suffering. We must be aware that we may have difficulty being empathetic with people we dislike or disrespect, or with those whose values, customs, beliefs, or ideals are different from ours.[7] Patients often say that health care providers involved in their care fail to see the personal consequences of patients' illnesses or disabilities. They ignore patients and dismiss their concerns. They fail to show, even in small ways, that they are persons who feel, who participate in their patient's pain. Instead they engage in distancing behaviors and harmful withholdings: they are silent and aloof. Patients say that these behaviors discourage them when they are in much need of encouragement.[8]

A therapist approaching the client using behavior modification techniques assumes that what has to be treated is what the patient actually does, without consideration given to what might be the patient's unconscious impulses or underlying conflicts. What a person does is primarily what the person has learned to do. Many workers learn bad habits, just as a cat in the laboratory can be taught to be neurotic. The therapist therefore has the attitude that the client is a joint product of his or her physical endowment and of the molding influence of the series of environments through which he or she has passed. Each environment, each exposure to stimulation, has modified, through learning, the character of the client to a greater or lesser extent. The goal of the therapist is to modify the behavior of the client and to change what the client does. With these methods, the therapist helps the client unlearn old habits that are maladaptive and learn new habits that work better and are safer. The simple use of terms implies various things to the person one is working with: for example, *patient* implies sickness while *client* does not. Therefore, when industrial rehabilitation is focused on health and function, the term *client* better portrays positive growth as opposed to the curing of sickness. This approach requires the therapist to create conditions for growth; the clients, however, are responsible for changing themselves. The therapist tries to create an atmosphere of a warm, accepting, and positive guidance that encourages positive changes in the client.[9]

Throughout the rehabilitation process, a client is encouraged to report any fear or pain so it can be documented and discussed. This communication is essential and helps develop trust in the health provider. To maximize function, clarification of the activity is needed for the client, with assurances that only safe performance will be allowed and that the therapist will stop the client and document the physical reasons if there is a limitation and that maximum function is reached. If the client stops before reaching maximum function, the client is clearly informed that maximum function has not been reached, told that he or she is still safe, and encouraged to continue. If he or she chooses not to continue, the test item is labeled as "self-limited," and the client is aware of that report.[10]

The behavior modification approach and the theory of operant conditioning have been written about in detail. When using this approach with the injured worker, staff members primarily withdraw attention when talk of pain or pain behaviors occur. This can be done simply by breaking eye contact and then, if necessary, attempting to change the subject. If this is unsuccessful, the staff member may briefly leave the area, returning as soon as the pain behavior stops. When "well" behavior is observed, it can be reinforced with increased staff attention. This approach can also be followed by others who are in contact with clients, such as the rehab nurse, company nurse, and spouse, to further promote clients' positive behavior rather than reward "sick" behavior, such as complaining about how painful it is to get up from a chair and therefore having someone bring them their meal. Clients are encouraged to support each other, and no negative talk is allowed in the rehabilitation area. The team itself must serve as a model for the desired behaviors.[11]

Human behavior is not fixed but flexible. Change can occur in the personality if the individual desires it to happen. As new goals are reached, clients increase their sense of self-worth, and this in turn alters the way they deal with the world, their work, themselves, and others. To obtain these goals, clients must deal with their unpleasant feelings of doubt, feelings of worthlessness, and depression, the opposite of involvement and productivity. A growth-producing experience includes learning, self-understanding, knowledge of others, and active involvement in the world. Through behavioral modification, a client can take control of his or her well-being and enjoy a productive life.

NOTES

1. A.K. Briggs and A.R. Agrin, *Crossroads: A Reader for Psychosocial Occupational Therapy* (Rockville, Md.: American Occupational Therapy Association, Inc., 1983), 212–215.
2. R. Barris et al., *Psychosocial Occupational Therapy* (Laurel, Md.: Ramsco Publishing Co., 1984), 75–77, 241–247.
3. Ibid., 241–247.
4. J. Rothman and R. Levine, *Prevention Practice* (Philadelphia: W.B. Saunders Co., 1992), 419–426.
5. Ibid.
6. C.R. Rogers, The Necessary and Sufficient Conditions for Personality Change, *Journal of Consulting Psychology* 21 (1957):95–103.
7. C.C. Nedelson, Ethics, Empathy, and Gender in Health Care, *American Journal of Psychiatry* 150 (September 1993):9.
8. S.M. Peloquin, The Depersonalization of Patients: A Profile Gleaned from Narratives, *American Journal of Occupational Therapy* 47 (September 1993):9.
9. M.K. Holland, *Psychology: An Introduction to Human Behavior* (Lexington, Mass., D.C. Heath, Co., 1994), 227–228.
10. S. Isernhagen, *Work Injury Management and Prevention* (Gaithersburg, Md.: Aspen Publishers, Inc., 1988), 161.
11. D.W. Krueger, *Rehabilitation Psychology* (Gaithersburg, Md.: Aspen Publishers, Inc., 1984), 339–341.

Chapter 24

Integrated Work Therapy in the Small Practice

Richard L. Smith

Rehabilitation of injured workers ranges from simple and straightforward to difficult and complex. Injured workers can be referred for work rehabilitation with acute pain, postsurgical pain, or chronic pain, having a multitude of diagnoses ranging from overexertional strains to multiple trauma and neurological deficits. Cumulative trauma disorders affecting the spine and extremities also challenge the therapist. The specialist must assess the problem, determine a treatment plan, and manage the case in a cost-effective work therapy regimen.

THE SMALL TEAM APPROACH

Comprehensive multidisciplinary treatment is sometimes appropriate for severely involved patients and chronic pain sufferers. Chapter 22 in this book focuses on the large institution-based rehabilitation programs that include in-house physicians, physical therapists, occupational therapists, vocational specialists, and mental health professionals. Not all injured workers need this level of intensive treatment, however.

When injured workers have a relatively uncomplicated psychologic or vocational status, but physical limitations preclude returning to work, referral to the small clinic is efficacious. Small clinic therapy is an appropriate and cost-effective therapeutic approach for common orthopedic diagnoses.

The industrial therapist working with work injures in a small clinic must be a specialist. The therapist should spend the majority of his or her time working with industrial clients in order to understand the myriad complexities of the system. The provider must know when and how to network with other professionals in order to achieve the optimum outcome.

Work injury management can be accomplished in the private sector, using a "small team approach" (Exhibit 24–1). A small team has one major advantage:

efficiency. Communication is quick, consistent, and concise. Goal-directed return-to-work efforts can be streamlined so as not to waste precious time and money.

The essence of work rehabilitation in a small clinic is to develop effective injury evaluation, rehabilitation, and prevention programs that span the work injury management spectrum.[1] A knowledge base and practice model are needed by therapists to serve as foundations for the spectrum. The concept of the model presented here has evolved from work done by expert physical therapy practitioners and presented at national conferences.[2]

MOVEMENT SCIENCE

The systematic approach to work injury management requires an understanding of movement science, a common body of knowledge made up of six scientific core disciplines (Figure 24–1). Understanding these six interrelated disciplines is necessary to appreciate both health, when all disciplines are functioning normally, and pathology, when one or more disciplines break down.

Most work injuries and cumulative trauma disorders are movement disorders. Integrity of the anatomical or structural aspects of the musculoskeletal system is compromised when there is too much movement, resulting in overuse, or too little movement, resulting in disuse and atrophy of tissues.

Skeletal muscle physiology is involved with all functional work movement. Functional strength and endurance depend on muscle efficiency, muscle coordination, and motor learning.

Exhibit 24–1 Composition of Small Teams

1. Health Care Team (Caregivers)
    - Primary Care Physicians
    - Specialist Physicians
    - Physical Therapist or Occupational Therapist with Industrial Rehabilitation Skills
2. Health, Safety, and Ergonomics Team (Prevention and Accommodation Experts)
    - Employer Representative
    - Employee Representative
    - Industrial Engineer or Nurse
    - Therapist with Ergonomic Skills
3. Vocational Team (Evaluators and Return-to-Work Facilitators)
    - Rehabilitation Counselor
    - Human Resources or Disability Manager
    - Case Manager or Rehabilitation Nurse
    - Therapist with Functional Testing Skills

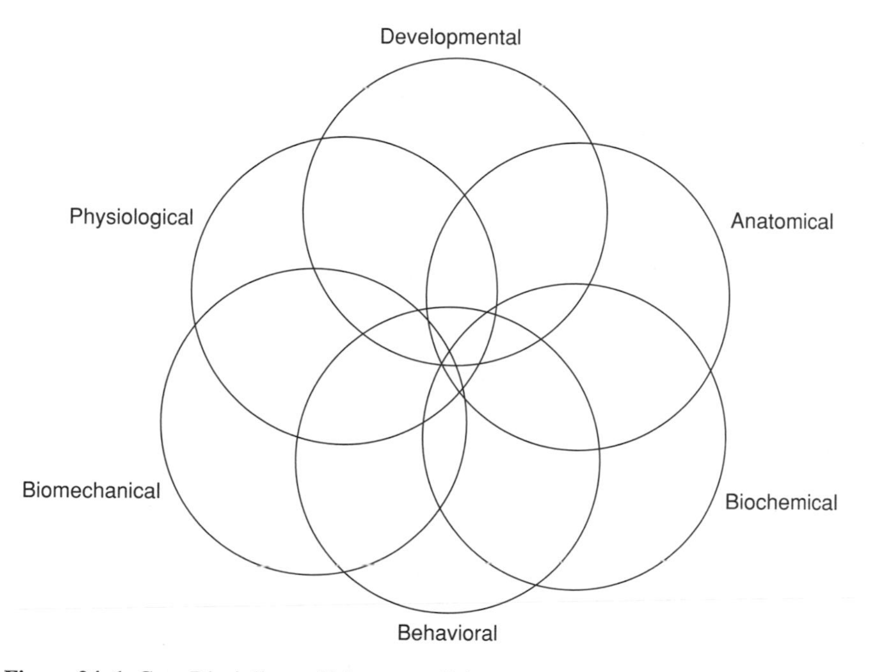

Figure 24–1 Core Disciplines of Movement Science

Cardiopulmonary physiology is involved with the worker's aerobic capacity, metabolism, and energy cost of heavy exertion during strenuous work. Ability to perform work depends on the capacity of the system to deliver fuel and oxygen to the working muscles—that is, the efficiency of oxygen uptake, including pulmonary ventilation, cardiac output, and oxygen extraction, and the physiological mechanisms that regulate these functions.[3]

Muscle performance factors combined with biomechanical aspects of movement affect the work capabilities of workers. Safe, proper body mechanics are important for healthy work movement. Unsafe or abnormal body mechanics cause physical stress, which can lead to movement dysfunction, injury, or disease.

Biochemical homeostasis is necessary for optimum health. Wellness and fitness for duty depend on a healthy immune system, the proper balance of hormones, and a bloodstream free of drugs, toxins, and chemicals.

The behavioral discipline of movement science relates to how the brain organizes, plans, and carries out functional work movement. The resulting intentional action can be smooth, safe, and controlled, or dangerous and potentially injurious.

Three of these six disciplines, the physiological, anatomical, and biomechanical disciplines (Figure 24–2), make up the foundation for the kinesiophysical method

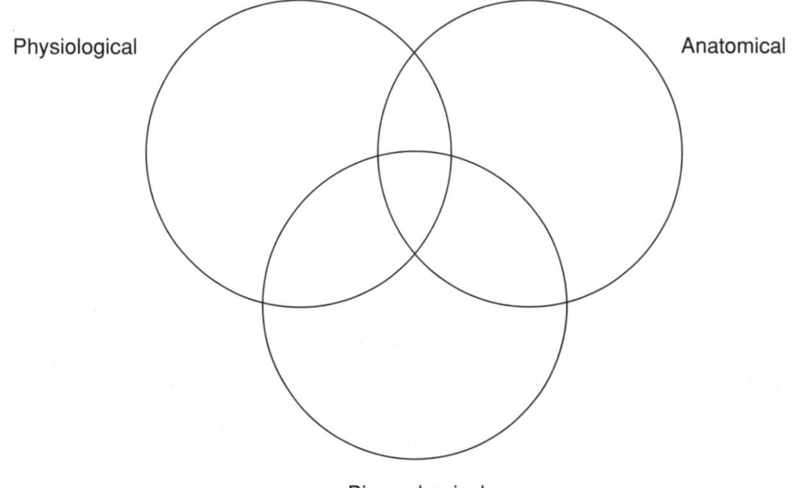

Figure 24–2 The Three Disciplines That Make Up the Kinesiophysical Foundation for Functional Capacity Evaluation

of functional capacity evaluation.[4] All six disciplines work together in a state of well-being, and all six break down in a state of work injury or disease. Together, all six disciplines make up the movement scientific basis for the work therapy model.

INTEGRATED WORK THERAPY MODEL

The integrated work therapy model (Figure 24–3) focuses on goal-directed work function to guide the therapist's intervention and decision making. The goal-directed work function can be a specific work task or many work tasks.

To achieve optimum work function, the industrial therapist must first understand the various influences on functional work movement. Understanding how age, injury, disease, and environmental and behavioral characteristics affect the worker's resultant movement gives the therapist an appreciation for how the worker moves and determines whether the resultant movement is safe or unsafe.

The effect of normal aging on biologic body changes has been studied and reviewed.[5] Many physical changes in older workers are due to decreased activity levels, but individuals age at markedly different rates. Remaining active is extremely beneficial for older workers and reduces injury-associated risk factors. Older workers are safer workers and, in heavy industry, have a lower incidence of injury.[6]

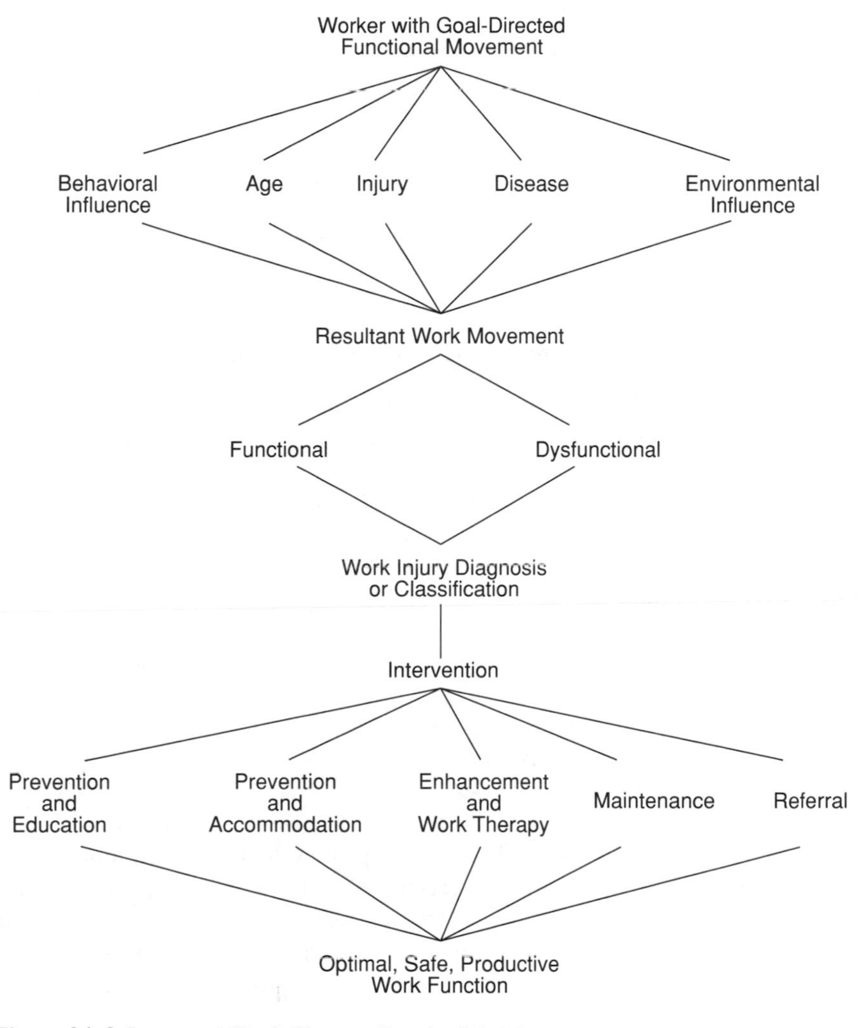

Figure 24–3 Integrated Work Therapy Practice Model

Injury and disease influence the worker's movement. Rarely does an individual go through life without an injury, and the worker's history of personal and occupational injuries needs to be considered in evaluating his or her movement. Overexertions and cumulative trauma disorders are obvious causes of movement disorders. Lung and heart disease can result from occupational exposure and can affect movement.

Physical and environmental factors such as temperature, humidity, and tools influence work movement and productivity. Attitudes of coworkers, unions, and supervisors affect the work ethic of each worker and may have a positive or negative effect on a worker's movement, pace, and safety and work practices.

Another very important factor influencing a worker's movement is psychological status. A happy, motivated worker moves differently than a depressed or stressed worker. There is no doubt that attitude affects posture and that constant stress leads to disease. Having normal or abnormal illness behavior influences the worker's financial, career, and life goals.

Evaluating Functionality and Arriving at the Work Injury Diagnosis

The resultant work movement, influenced by these five moderating factors, can be functional or dysfunctional (Figure 24–3). The result can be a change in the movement of the whole person or an isolated body part. Functional movement may or may not be perfect, but in the work environment it must be safe. Dysfunctional movement is abnormal and can be unsafe.

The purpose of functionality assessment is to arrive at the work injury diagnosis or classification. In evaluating functionality, the therapist examines the worker's functional strength, endurance, range of motion, speed, and coordination. Functional strength is different from muscular strength. For example, quadriceps strength as tested by manual muscle testing or by dynamometry may show "normal" strength. Examples of functional tests of quadriceps strength would be squatting or climbing ability. A worker can have "normal" quadriceps strength and not be able to squat or climb, or, vice versa, he or she may have a strength deficit as measured by dynamometry and have good ability to squat or climb. Tests of functional quadriceps endurance would include repetitive squatting or climbing.

Functional range of motion also affects movement patterns. Full pain-free range of motion following an injury is not always possible, yet restricted mobility can still be efficient for work demands. For example, when segmental spinal motion is lost following a spinal injury or surgery, the resultant movement may be functional or perhaps dysfunctional. If the worker moves through spinal range of motion smoothly, with good whole body mobility, he or she will be able to function productively in spite of a residual joint restriction. On the other hand, if the resultant movement is consistently rigid, spasmodic, or lacks coordination, the worker will not be productive and may even be prone to reinjury.

Standardized whole body tests of functional speed and coordination can be general or task-specific.[7] Standardized manual dexterity and functional ability tests are also available.[8] While speed of movement is important for many jobs, such as typing, data entry, or production work, the qualitative assessments of smoothness or productivity are equally important.

Work Injury Diagnosis and Classification

Assessment of the status of resultant work movement leads to the work injury diagnosis or classification and subsequent intervention (Figure 24–3). If the worker's whole body strength, endurance, and mobility are safe for a given work demand, no treatment is indicated, other than education and maintenance, for optimum work function. If the whole body strength, endurance, or mobility is dysfunctional, intervention will be necessary.

Work injury syndromes can be classified on the basis of dysfunction or loss of function, and these classifications can guide the therapy (Table 24–1).

INTERVENTION

The intervention is specific to the classification of dysfunction, depending on the functional goal. Enhancement of dysfunctional strength, endurance, and mobility through work therapy is indicated if improvement in functionality is necessary for safe productive work. Once functionality is accomplished, maintenance of the worker's health, attitude, and motivation level is paramount to keeping the worker safe, active, and performing. Prevention of deconditioning complications and chronic pain is achieved by education. Referral may also be indicated.

Prevention and Education

Client education is the most important component of intervention and includes education of the client regarding proper body mechanics, conservation of energy, and pain management. Instruction in specific anatomy and injury prevention methods can be offered to the worker as he or she progresses through the program. Back school and cumulative trauma disorder school are examples of programs that will reduce chances of reinjury. This information should be practical and simple to understand rather than too scientific or academic. The client should receive instruction in how to avoid injury and fatigue using proper posture in movement patterns, pacing, and preventive exercise procedures.

Table 24–1 Examples of Work Injury Classifications

Work Injury Classification	*Traditional Diagnosis*	*Dysfunction*
Lumbar flexion syndrome	Discogenic low back pain	Pain with lumbar flexion
Cervical extension syndrome	Facet joint arthritis	Pain with cervical extension
Wrist flexion syndrome	Carpal tunnel syndrome	Pain with wrist flexion
Shoulder elevation syndrome	Subacromial impingement	Pain and restricted ROM
Knee flexion syndrome	Patellar tendinitis	Pain with loaded squatting

Perhaps the most essential component of education during work rehabilitation is the differentiation of pain, function, and safety. Although pain is important, it must be separated from function during work therapy. Differentiation must be made between "hurt" (the response of healthy tissues to therapeutic procedures) versus "harm" (a process that is destructive and should be heeded as a warning sign). The individual is often fearful of increased symptoms, and needs reassurance, support, and encouragement that symptoms often decrease as strength and endurance improve. Active participation will be necessary for the worker to feel "good" and realize the physiologic benefits of conditioning.

Prevention and Accommodation

When functional limitations or reduced capacities result in a mismatch to the job description, the therapist working in a small clinic has two options for case resolution. One option is to visit the worksite and to use ergonomic expertise and knowledge of how the body works to modify the work or worksite. For example, special equipment, assistive devices, and redesign of the work space can prevent reinjury. This job accommodation requires communication with a health, safety, and ergonomics team (Exhibit 24–1) including the worker's employer, as well as safety and engineering personnel.

If functional limitations have the potential for improvement, the second option for a job match is to raise the work capacity of the client through work conditioning.

Enhancement and Work Therapy

The health care team may simply be the treating physicians and the physical therapist or occupational therapist (Exhibit 24–1). After the initial acute treatment of the injury, the therapist informs the physician and employer by a progress report or phone call when the worker is safe to return to either modified or full duty, and the desired outcome is accomplished.

Additional work therapy may be appropriate. While the acute medical care was directed at treating the client's pathology, signs (ROM, strength), and symptoms (pain, muscle spasm), work conditioning is directed at enhancing the client's work potential (functional strength, endurance, and productivity).

Two goals of work therapy are to increase the worker's safe work capacity and, concurrently, to objectively measure the client's abilities and limitations as improvement occurs.

Improving the functional status of the worker is accomplished by a systematic therapeutic regimen. An individualized program is designed to advance the client progressively from lower levels to higher levels of work to improve physical capacities related to the performance of work tasks. Work-conditioning programs are provided in multihour sessions available up to five days a week for a duration of up to eight weeks.[9] Systematic work therapy requires target performance goals to be set on a daily and weekly basis in order to provide the client with meaningful, practical, and sustainable levels of function.[10]

Both low-tech and high-tech equipment can be used. The program is goal directed and monitored at each session so that at any given point in the rehabilitation process the therapist knows the client's status. For example, if increased manual-material-handling capacity is the objective, the therapist sets incremental increases in weight capacities in the program. If increased bending tolerance is the objective, incremental bending time goals are set. Gains can be measured. When a plateau in physical progress occurs, the therapist will have a clear assessment of the client's current functional ability.

Maintenance

Once a maximum level of function occurs, it will be very important to keep the worker physically and mentally fit. If the worker is not working, and because speedy resolution of the case is not always possible, a maintenance conditioning program will be necessary until reemployment or return to work occurs. If the worker is working, health maintenance is vital. Both situations require instruction in using home exercise equipment and exercise facilities to maintain physical strength and fit-for-duty status. Also, maintenance of a positive attitude will require timely follow-up by the vocational team (Exhibit 24–1).

Referral

The small clinic therapist must know his or her limitations, and realize when it is appropriate to include other professionals on the case. Vocational team networking (Exhibit 24–1) with vocational counselors, rehabilitation nurses, and psychologists can be as simple as a phone conference, or problem solving at a staffing in the office with or without the client. Psychosocial barriers that interfere with recovery[11] must be identified early on in the program. Some clinical problems, such as depression, can be effectively dealt with only by well-trained psychologists. The physician and insurer must be informed of the nature of the problem so that a decision can be made whether treatment should be authorized to resolve the issue.

In summary, the integrated work therapy practice model takes the work movement dysfunction out of the traditional treatment approach and into a progressive intervention mode. For a given category of work injury, specific functional limitations in ability to perform specific critical job demands guide the intervention.

Once the functional limitations are eliminated and/or accommodated, the worker can return to productive, safe employment.

Safe and productive goal-directed work function is the objective. If the appropriate intervention has occurred, the dysfunctional work movement will be diagnosed, treated, and restored for the optimal work capacity of the worker.

MULTIPLE ROLES OF THE INDUSTRIAL THERAPIST: COACH, ARTIST, AND DIPLOMAT

The small practice therapist contributes to work injury management by preparing the client physically and mentally to return to work. To achieve success, the therapist must have a creative, energetic attitude combined with patience, tenacity, and diplomacy.

Coach

A sports medicine philosophy to work rehabilitation allows the therapist to assume the leadership role of coach. Physical reconditioning and psychological readiness are the cornerstones of all successful rehabilitation programs. Many industries are like sports and often demand high performance skills and near-maximum cardiovascular and muscular efforts. Firefighters, like decathletes, must have high levels of strength and speed. Loggers, like soccer players, need endurance, agility, and quickness. Steel workers, like gymnasts, must be able to balance, climb, and work in contorted postures.

One of the most successful coaches of all time was the Green Bay Packer football coach Vince Lombardi. Lombardi was known for his ability to instill motivation and a winning attitude in his players' lives, both on and off the football field. His five rules of self-motivation can be applied to work therapy:

1. Develop mental toughness.
2. Take the initiative.
3. Stay in shape.
4. Use time wisely.
5. Make that second effort.

Mental toughness was Lombardi's keystone to all success. When players were facing adversity, whether they were behind and time was running out, or their bodies were aching from bumps and bruises, Lombardi would profess, "Never give up." He asserted, "You carry on no matter what the obstacles. You simply refuse to give up. When the going gets tough, you get tougher. You refuse to let failures get you down."[12]

The industrial therapist can use his or her attitude to encourage his clients. When faced with aches, pains, and frustrations, the worker must become more determined, more persistent, and more resolved. Mental toughness requires dedicated mental concentration and takes practice to develop. Through repetitive use in training, determination can be learned and mastered to overcome mental and physical hurdles.

Lombardi defined mental toughness as self-discipline or disciplined will, a state of mind. It can almost be called character in action. Barriers thrown in the way of an injured client endowed with mental toughness are a stimulant rather than a depressant. They are challenging. The worker refuses to admit defeat and develops personal motivation to overcome any problem. His or her mental effort is focused on the desired goal: return to work.

In football, the team that controls the ball carries out its game plan, and in rehabilitation, the worker who takes the initiative demonstrates the power and ability to follow through with a rehabilitation plan. A worker needs to be encouraged to immediately initiate action, start projects and tasks, and begin the next activity. Once back to work, the individual will be responsible for demonstrating appropriate initiative and work behavior.

A high level of physical fitness is essential for optimal performance in both the sports arena and the industrial setting. Work-conditioning programs use weight training and aerobic exercise to develop the injured worker's physical capacities. Good physical condition also breeds self-confidence. Athletes know when they have the endurance to play the whole game. With conditioning, workers increase their work confidence and prove to themselves that they can tolerate a full day's work.

To Lombardi, making good use of time meant utilizing every second to get maximum results. There is regular time and Lombardi time. With Lombardi time, if you're ten minutes early for work therapy, you're still late. This may seem a bit extreme, but the philosophy goes beyond being punctual for appointments. Lombardi time utilizes every minute of the rehabilitation session to maximize effectiveness. Time is of the essence. The clock is ticking, and the precious moments cannot be wasted. The sooner the worker returns to work, the sooner he or she can realize a productive and meaningful existence again.

Lombardi's favorite rule was, don't give in, go that extra step, make that extra effort. The coach said, "I have to sell my players on themselves, on forgetting their small hurts because they're a part of life. I have to sell them on making this team, this season, this game, and each individual play the most important thing in their lives."[13] Encouraging injured workers to live up to their potential requires that the therapist take on the roles of both coach and counselor. Facilitating the individual's drive and commitment is extremely challenging. But if clients can reach this highest level of motivation, they will "make the team" or, in other words, stop at nothing short of complete successful return to work.

Lombardian enthusiasm carries two caveats in the work therapy setting. Injured clients should be taught that muscle soreness is to be expected and is normal. They sometimes need help in differentiating healthy soreness from injury aggravation. Inappropriate overachieving or overly enthusiastic patients also need monitoring by therapists to prevent carelessness and possible reinjury.

A primary responsibility of the industrial therapist is to motivate injured workers to accept responsibility for pushing themselves. Lombardi's principles of self-motivation can be used by therapists to motivate injured workers to push themselves in therapy, ensuring optimum outcome and potential for return to work.

Diplomat

To be effective in work injury management, a therapist must often use the highest diplomacy skills. Juggling the wishes and needs of the worker and at the same time handling disputes and conflicts, and pressures from employers, insurers, and attorneys requires a delicate balance and belief in what is right for the individual. Earning the worker's trust and establishing a positive rapport with the worker takes honest communication and openness. Once the worker understands that the therapist has good intentions to provide quality work rehabilitation tailored to the worker's needs, cooperation and compliance will occur. By demonstrating willingness to improve, the worker gains the therapist's objective support.

Sometimes resolution of a client's status can be more important than rehabilitation. And in rare instances, a financial settlement is all the individual desires. This situation and other psychosocial barriers to return to work require specific methods of arbitration for successful outcomes.[14]

If the client is confused about his or her diagnosis, all questions must be answered before proceeding. The client must be assured of medical stability by the physician and given assurance by the treating therapist that he or she will not be allowed to perform activities that would be unsafe or cause reinjury. The resulting outcome will be the client's confidence that he or she is "in good hands."

Artist

According to the dictionary, an artist is a person who creates works of art, or who performs his or her work as if it were an art.[15] Artists are people who take ordinary materials such as paint or clay and create masterpieces for the rest of us to enjoy. Like an artist who takes mundane materials and builds a work of beauty, work therapists must be creative with their everyday equipment and supplies to restore optimum function in a worker.

The equipment used in small clinic work rehabilitation can possess high-tech or low-tech qualities. Basic strengthening does not require expensive, sophisticated machinery. Much of the high-cost spinal strengthening equipment that therapists purchased in the 1980s became obsolete in just a few short years. In fact, some of the least costly and most effective strengthening exercise regimens have centered on spinal muscular stabilization regimens, requiring minimal special equipment.[16] More advanced spinal muscular stabilization techniques use free weights, elastic tubing, and pulley systems.[17]

For work-related activities, inexpensive equipment and supplies can be purchased from the local hardware store. The worker can also bring in actual equipment from his or her job, allowing for the most realistic work simulation.

Cybex (Cybex, Inc., Lumex, N.Y.), LIDO (Loredan Biomedical, Inc., Davis, Calif.), Biodex (Biodex Corp., Shirley, N.Y.), and KINCOM (Chattecx Corp., Chattanooga, Tenn.) all provide excellent high-tech isokinetic testing and exercise equipment. The Isostation B-200 (Isotechnologies, Inc., Hillsboro, N.C.) measures isoinertial strength. Each of these machines has unique features and has been found to be reliable, but most testing and exercise regimens are not job-specific. The usefulness of trunk-testing machines for work assessment has recently been questioned.[18] Two pieces of high-tech equipment appropriate for work therapy are the Lido WorkSET (Loredan Biomedical, Inc.) and the BTE Work Simulator (BTE Company, Hanover, Md.). These two robotic work simulators can be used to reproduce actual work task demands in many industries and can quantify the amount of work that is done during work therapy.

SMALL CLINIC SUCCESS

The spectrum of work injury management requires professional leadership from small clinic therapists. These professionals must offer efficient, cost-effective intervention as caregivers, prevention and accommodation experts, evaluators, and vocational facilitators (Exhibit 24–1). The integrated practice model requires meaningful intervention in order to influence results. When this is done at high levels, small practice consultants with expertise can accomplish the ultimate functional outcome: optimal, safe, productive work.

In summary, integrated work therapy in the small clinic requires a movement scientific basis as a foundation, a functional classification of the work injury, and an artistic, diplomatic approach to intervention. Working as a small team member, with dedicated Lombardian mental toughness, the small clinic therapist can make an excellent contribution to successful rehabilitation of injured workers.

NOTES

1. S.J. Isernhagen, The Spectrum of Work Injury and Its Management, in *Work Injury: Management and Prevention,* ed. S.J. Isernhagen (Gaithersburg, Md.: Aspen Publishers, Inc., 1988), 1–6.
2. S.L. Burkart et al., *The Clinical Practice of Physical Therapy in the 21st Century* (Philadelphia: Executive Communications, Inc., January 21, 1992); S.L. Burkart et al., Integrated Physical Therapy Practice Model (Paper presented at the meeting of the American Physical Therapy Association, Denver, June 1992).
3. P.O. Astrand, *Textbook of Work Physiology,* 2nd ed. (New York: McGraw-Hill, 1977), 451.
4. S.J. Isernhagen, Functional Capacity Evaluation: Rationale, Procedure, Utility of the Kinesiophysical Approach, *Journal of Occupational Rehabilitation* 2, no. 3 (1992):157.
5. J.A. Coy and M. Davenport, Age Changes in the Older Adult Worker, *Work* 2, no. 1 (1991):38; S.J. Isernhagen, An Aging Challenge for the Nineties: Balancing the Aging Process against Experience, *Work* 2, no. 1 (1991):10.
6. L.E. Franits and G.W. Meader, The Older Worker and Incidence of Injury: Trends in Heavy Industry, *Work* 2, no. 1 (1991):29.
7. S.J. Isernhagen, *Functional Capacity Evaluation* (Duluth, Minn.: Isernhagen Work Systems, 1989).
8. J. Bear-Lehman and B.C. Abreu, Evaluating the Hand: Issues in Reliability and Validity, *Physical Therapy* 69, no. 12 (1989):1025.
9. American Physical Therapy Association, *Resource Guide: Industrial Physical Therapy* (Alexandria, Va., 1993), 1–15.
10. L.E. Darphin et al., Work Conditioning and Work Hardening, *Orthopedic Physical Therapy Clinics of North America* 1, no. 1 (1992):105; G. Swanson, Functional Outcome Report: The Next Generation in Physical Therapy Reporting, in *Documenting Functional Outcome in Physical Therapy,* ed. D.L. Stewart and S.H. Abeln (St. Louis, Mo.: C.V. Mosby, 1993), 101–135.
11. Darphin et al., Work Conditioning.
12. V. Lombardi, *Second Effort* (Videotape, Dartnell Corporation, Chicago, 1977).
13. Ibid.
14. Darphin et al., Work Conditioning.
15. *American Heritage Dictionary of the English Language* (New York: American Heritage Publishing, 1969).
16. J.M. Irion, The Use of the Gym Ball in Rehabilitation of Spinal Dysfunction, *Orthopedic Physical Therapy Clinics of North America* 1, no. 2 (1992):375.
17. T. Sweeney et al., Cervicothoracic Muscular Stabilization Techniques, *Physical Medicine and Rehabilitation: State of the Art Reviews* 4, no. 2 (1990):335.
18. M. Newton and G. Waddell, Trunk Strength Testing with Iso-Machines Part I: Review of a Decade of Scientific Evidence, *Spine* 18, no. 7 (1993):801.

Chapter 25

Work Rehabilitation: The Importance of Networking with the Employer for Achieving Successful Outcomes

Barbara A. Larson

With skyrocketing costs dictating major reform in U.S. health care policy, and the increased role of industry in determining the health benefits of its employees, work rehabilitation service providers must continually reevaluate their position in the marketplace. The future of work rehabilitation will be determined by the providers' ability to demonstrate positive outcomes. A major key to their success will be their ability to develop strong networks with employers and assist them in returning employees to work.

Mark Zitter, president of the Center for Outcomes Information, a San Francisco–based research group, describes the profound changes in the way health care will be administered, "all the major health-care constituencies—payers, providers, employers, regulators—are increasingly focusing on the outcomes that health products and services generate."[1] Individual state workers' compensation regulatory bodies are mandating cost reduction. The Medical Services Review Board of the Minnesota Department of Labor and Industry issued a preliminary Statement on Outcomes and identified three components found to be measurable and appropriate for determining effective medical care in the workers' compensation system: return to work , cost, and worker function.[2]

Job-related injuries and sickness in 1991 cost public and private employers an estimated $62 billion in workers' compensation expenditures, nearly triple the amount spent in 1980. At the present growth rate, costs will nearly triple again by the year 2000. The average medical cost per claim paid by workers' compensation insurers has increased by an average of 14 percent a year. In 1990, it was $6,511. The average wage-replacement cost per claim to cover an injured employee's salary during that time missed from work was 11 percent a year; it was $12,833 in 1990.[3]

LITERATURE REVIEW

A review of the literature shows the value of return-to-work programs. These programs have established philosophies and methods of treatment that serve as models to the current practitioner. King studied program performance data on 22 work programs in Wisconsin over an 11-month period.[4] She specifically obtained data related to demographics of gender, age, occupation, insurance coverage, diagnosis, services received, patterns of attendance, and psychological services received. Data were collected on 928 clients following discharge from the 22 programs. King reports that more than half of the clients (59 percent) upon discharge from work-hardening programs returned to their same job with or without modifications, 8 percent obtained an alternative job in the same company, 3 percent obtained an alternative job in the same company with modifications, 2 percent changed occupations, 9 percent were referred for vocational services, 18 percent had no return to work, and 1 percent were retrained.

Mayer et al. showed significant return-to-work statistics in studies of a functional restoration program with behavioral support.[5] They compared 98 patients with low back injuries who entered a treatment program with 56 not treated. After two years 87 percent of the treatment group were working and only 41 percent of the nontreated group were working. Hazard reports the results of a treatment program of functional restoration with behavioral support in a one-year prospective study where 90 patients were studied.[6] At year's end, 81 percent of program graduates, 40 percent of the dropouts, and 29 percent of those denied the program had returned to work. In both programs patients assumed an active role in their rehabilitation.

A program at the San Francisco Spine Institute studied a group of patients treated nonoperatively for lumbar herniated disc.[7] Patients participated in an aggressive treatment program that included back school, exercise training to teach spinal stabilization, flexibility, weight training, and aerobic conditioning. A 92 percent return-to-work rate was reported, with 90 percent of the patients returning to their previous occupations. Lichter reported on 120 patients chronically disabled with back pain who entered a comprehensive return-to-work program: 58.3 percent either returned or improved their work status.[8] Kuhn and Kneidel, in a study looking at early entry into work hardening, found that out of 52 patients with work-related back injury who had entered work hardening, 50 percent returned to work at the same job or a different job with a different employer.[9] Seven individuals enrolled in school or vocational programs after work hardening. Nineteen patients remained unemployed after work hardening.

The importance of documenting the efficacy of return-to-work outcomes not only will be the basis for continued reimbursement but will serve as a guideline to evaluate treatment methods. Darphin states, "A systematic method of program evaluation is imperative to enable the organization to identify the results of its services and effects of its programs. Basic outcome measures, such as return to work rates, can be researched and provided to referral sources and consumers."[10]

An example of outcome data is presented in Table 25–1. Therapists must demonstrate that work-rehabilitation services will result in a cost benefit to the employer.

NETWORKING WITH THE EMPLOYER

The importance of a mutual understanding between the therapist, employer, and employee of the critical demands of the job and the circumstances under which the employee is allowed to return to work is critical. The goals of work rehabilitation are clearly established in the literature.[11] How one goes about achieving those goals depends on the provider's ability to instill ownership in the employee for his or her job and to involve the employer in the return-to-work process. Performing this role successfully requires a partnership between medical practitioners and industry.[12] Therapists are involved not only in return to work, but in safe work practices, reasonable accommodations, and assisting the employer to make the determination of continued employment or an acceptable alternative.

With the passage of the Americans with Disabilities Act (ADA), therapists will be involved in assisting employers to meet their legal responsibility in hiring or returning the disabled worker to work.[13] Of the approximately 43 million disabled people in the United States, 70 percent are unemployed.[14] The law mandates that "an employer must provide a reasonable accommodation to the known physical or mental limitations of a qualified applicant or employee with a disability unless it can show that the accommodation would impose an undue hardship on the business." The injured worker can fall under the ADA if he or she has an impairment

Table 25–1 Outcome Data from a Work Rehabilitation Program

Number of patients considered	68
Total work-hardening visits	1595
Average program cost	$2,130
Average number of visits	22
% of clients returned to work following program	74%
Same job, same employer	42%
Different job, same employer	25%
Different job, different employer	6%
Job search	23%
Unable to return to work	3%

Courtesy of the Work Center, a service of Orthopaedic Sports Physical Therapy, Stillwater, Minnesota.

that "substantially limits a major life activity," or has a "record of" or is "regarded as" having such an impairment. He or she must also be able to perform the essential functions of a job currently held or desired with or without an accommodation.[15]

Stockdell and Crawford suggest an Industrial Model of Rehabilitation Management to facilitate earlier employer involvement in the qualified disabled employees return to work under the ADA (Figure 25–1).[16] This model has broad applicability. It includes issues regarding safety and productivity that apply to the able-bodied worker as well as the worker with a disability.[17] In applying this Industrial Model to case management of the injured worker, it is important to note that the appropriate course of action following acute medical care is determined by a functional or work capacity evaluation or task analysis. Proper placement to vocational or work rehabilitation can then be made, or the employer can be contacted for return to work.

JOB SITE ANALYSIS

Job site analysis increases our ability to design work-simulation tasks that closely align with the actual job components as well as determining modifications or accommodations. It also affords therapists the opportunity to establish communication with the employer. Without close communication with the employer, employee ownership and responsibility will not be transferable to the workplace. Case resolution and successful continued employment will be greatly diminished.

COMBINING FUNCTIONAL CAPACITY AND PHYSICAL WORK DEMANDS

The following case examples demonstrate that provider, employer, and employee networking is essential in directing a work rehabilitation program to successful completion. Different methods were used to analyze jobs based on the individual employee and specific job requirements of the company.

Case 1—Carpal Tunnel Syndrome

Job Title: Preparing and assembling medical diagnostic kits.

Injury: Employee was experiencing symptoms related to bilateral carpal tunnel syndrome. She not only was having difficulty sleeping due to numbness and tingling in her fingers but was experiencing exacerbation of symptoms at work. The symptoms were felt to be related to the repetitive nature of her job.

Overall Process

Initial meetings were held with the company occupational health nurse and the employee's immediate supervisor. Following the plant walk-through, an on-site

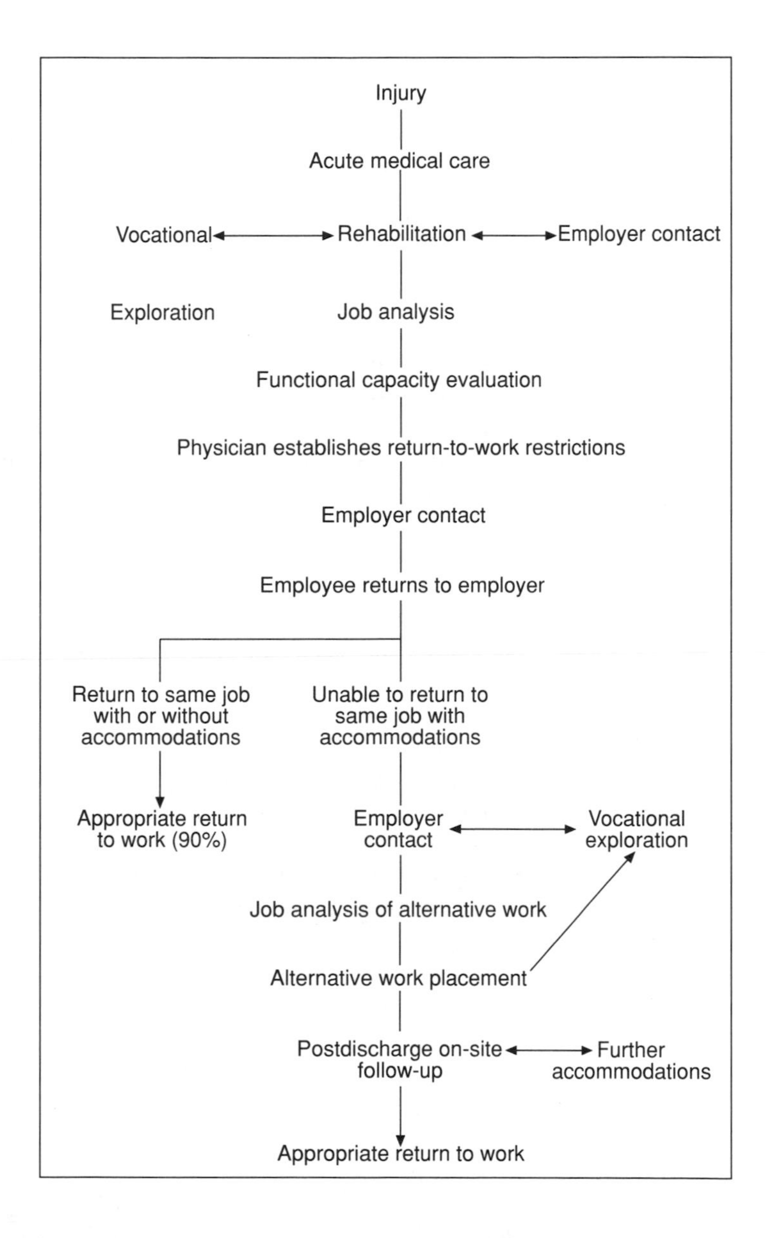

Figure 25–1 The Industrial Model of Rehabilitation Management. *Source:* From S.M. Stockdell and M.S. Crawford, An Industrial Model for Assisting Employers To Comply with the Americans with Disabilities Act of 1990, *American Journal of Occupational Therapy*, Vol. 46, p. 429. Copyright © 1992 by the American Occupational Therapy Association, Inc. Reprinted with permission.

job analysis was performed. Four jobs were identified as problem jobs. These jobs were observed, videotaped, and analyzed, and ergonomic risk factors were identified. The employee participated in a program on cumulative trauma injury prevention with application to her specific work tasks and work area.

Job Performance Component Descriptions

Tipping and Capping: This job is done sitting, handling small parts consisting of a cap and a bottle (about the size of an insulin bottle). There are 808 bottles in one tray. The first section involves securing a cap to a bottle. The employee holds handfuls of caps with her left hand and places caps on the bottles with her right (dominant) hand. To secure a cap to a bottle, she alternates pushing her middle and index fingers and her thumb against the cap using a palmer pinch. The second section involves securing caps to the bottles by screwing them on. Employee holds several caps with her left hand and places a cap on the bottle just tightly enough to hold it. She then lifts the bottle out of the tray and holds it with the left hand while tightening the cap with the right hand.

Placing Bottles in Packing Trays. Taking bottles off the line and placing them in the packing trays. These are Styrofoam trays and bottles resembling insulin bottles. The task is done standing, using both hands working at waist to mid-chest height.

Weighing the Packets. This task is done sitting. A small scale is placed on the table, and small packets are held with the right hand, placed on the scale with the left hand, and then flipped over and dumped into a bucket next to the scale.

Placing the Slides in the Slide Machine. Work position is seated, facing the machine. A box of small parts is held in the employee's lap. The slides are on the tray held in the left hand. The tray weighs about one pound. The employee feeds the slides into the machine about one per second. The reach is to chest level to insert the slides into the machine pocket.

Identified Stress Factors

1. static positioning of left arm when holding caps or slides
2. repetitive finger manipulation and finger pressure required when pushing in or screwing the caps onto the bottles
3. the pressure on the palm when holding the caps or box of slides
4. the number of wrist, hand, and finger manipulations per day

Comments and Corrective Ideas

1. Provide forearm support to reduce the static positioning required during tipping and capping and slide placement.
2. Reduce the pressure on the fingers. Instead of squeezing each cap to secure it on the bottle, place several caps loosely on the bottles one at a time. Place

a flat piece over them and push down with one or both hands in a fisted (thumbs-up) position to secure all the caps at one time.

3. When securing the bottle caps, eliminate some of the motions: have the caps on the table rather than holding them; take out the bottle and tip it sideways, allowing the wrist to remain in a neutral position.
4. Set the box of slides on a cart to eliminate the static wrist and forearm position required to continuously hold the box of slides.
5. Incorporate pause/stretch breaks while working on different lines.

Outcome

A meeting was held with the employer following the job analysis. The company ergonomist addressed the worksite issues. The employee participated in three sessions for education in cumulative trauma prevention and returned to her job. The suggestions in the job analysis had been implemented.

A few weeks later the employer contacted the therapist. The employees, after the therapist's suggestion on an alternative way to do the tipping and capping, had found a way to perform the job using equipment at the workstation designed for another purpose. The solution belonged to the employees, but the idea for the process was introduced by the therapist. This alternative method for tipping and capping not only eliminated all repetitive movements on that particular job but speeded up the process and saved the company a considerable amount of money through increased productivity.

Case 2—Degenerative Disc Disease

Job Title: School Janitor.

Injury: This 52-year-old male had been working as a school janitor in an elementary school system for the past 11 years. He had had back problems in the past, and it was becoming increasingly difficult for him to perform his job due to back pain. He was responsible for cleaning all classrooms and hallways, including dusting, mopping, and emptying trash. Tasks such as washing walls and windows and cleaning desks were done twice a year. This individual was seen for 25 work-hardening sessions over five weeks. At the time of his program he had been off work for four months.

Overall Process

After an initial work-hardening evaluation the problems that surfaced were in the area of forward bending (this individual was 6 feet, 8 inches tall, which increased his difficulty with most of the work tasks) and the twisting and turning involved in mopping, sweeping, transporting, lifting, and emptying the 33-gallon trash bins into the dumpster. This employee was extremely thorough and meticulous in his work.

The work-hardening program included conditioning, strengthening, stabilization, education, and work simulation. A job analysis was performed. The specific job components were outlined and problems were identified. Work-simulation tasks were then developed, and the therapist worked with the employer on possible accommodations.

Accommodations for this employee were as follows:

- Lighter mop heads
- Extension handles for mops, brooms, and dusters
- A handle was fabricated to insert in the 33-gallon trash bins, allowing him to push and pull two trash bins at the same time without forward bending (Figure 25–2). This had been a major problem due to the employee's height. The handle has been made from steel pipe and wrapped in duct tape. Pipes were purchased at a local hardware store. The cost was insignificant, and if the handle was lost it could be remade by the client without difficulty (Figure 25–3).
- Assistance would be available for emptying trash bins into dumpsters.
- Employee would be allowed to incorporate position/stretch breaks into his daily routine.

Outcome

At the conclusion of his program the employee had incorporated the safe work techniques he had learned during his program. He understood the concept of using only the amount of muscle effort a task required and was trying to "let go" of his need to be a perfectionist all the time. Following the program the employee returned to work full time.

Case 3—Overuse Syndrome

Job title: Second Finisher, Sewing Machine Operator.

Injury: Overuse Syndrome; Carpal Tunnel Release. The employee had been a seamstress in the same company for 18 years. She was paid a minimal base salary and made her money on the number of "pieces" she produced. This employee was a high producer. Her job duties included fabrication of coveralls, jackets, and vests for hunters. She was originally diagnosed with tendonitis in 1988, was briefly off work, and was treated with anti-inflammatories and a wrist brace. She returned to work full time but continued to experience problems. After approximately six months she underwent carpal tunnel surgery. She was off work six weeks and

Figure 25–2 Handle Extensions Made To Eliminate Forward Bending When Pushing Trash Barrel

Figure 25–3 Employee Demonstrates Pushing Trash Barrel with Modified Handle

---

returned half time light duty. She attempted full time, continued to have difficulty, and was removed from work and referred to work hardening. She was seen for 28 work-hardening sessions over six weeks.

Overall Process

Initial evaluation and job site analysis were conducted to determine capabilities and areas of concern related to job requirements.

The critical demands of the job included

- cutting several layers of cloth, pinning fabric
- sewing fabric pieces together on an industrial sewing machine
- sewing in zippers
- lifting fabric bundles weighing up to 45 pounds

Employee's problems and goals were as follows:

Problems	*Goals*
1. Decreased wrist flexibility, which limited ability to handle material as it ran through the machine	1. Increase wrist flexibility to run material through machine
2. Decreased tolerance for sustained grip and pinch, which affected ability to grasp material to feed through machine	2. Increase endurance for feeding material into machine
3. Decreased tolerance for sustained firm gripping, which limited ability to use seam ripper or cut fabric	3. Increase ability to use seam ripper and cut fabric
4. Decreased ability to lift 45-pound clothing bundles	4. Increase lifting capacity to 45 pounds
5. Increased edema in response to right-hand resistive activities	5. Teach edema reduction techniques
6. Decreased understanding of cumulative trauma	6. Educate in appropriate pacing, relaxation techniques, body mechanics, and exercises for cumulative trauma reduction

The employee brought her sewing machine to The Work Center. Part of her program included a "sewing circuit" that involved operating the sewing machine, pulling zippers, using the seam rippers, cutting material, using clippers, and feeding material. Initially she was limited to 10 minutes of sewing work. By the end of her program she had increased her tolerance to 4 hours of sewing.

The following accommodations were made regarding return to work:

- Employee would return to work 4 hours per day, 3 days per week for 2 weeks, and would gradually increase.
- She would be allowed to take position/stretch breaks throughout her shift.
- She would rotate jobs to a task using alternate muscle groups from the sewing machine job.
- No overtime would be required.
- A pair of special scissors were purchased for employee's use at work (Figure 25–4). These scissors have stainless steel blades and are designed to cut several layers of fabric with little effort required from the user. Prior to the request to authorize the scissors, the client used them during her work-hardening program to determine their effectiveness.

Outcome

The employer agreed to the above recommendations. Employee returned to work.

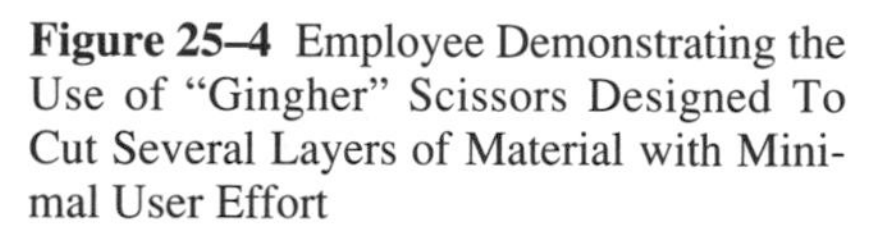

Figure 25–4 Employee Demonstrating the Use of "Gingher" Scissors Designed To Cut Several Layers of Material with Minimal User Effort

Cost Analysis

The above semiskilled labor jobs paid between $300 and $400 per week. The three individuals involved in the case studies received both medical benefits and wage replacement. Two of the three positions required worker replacement. Through the work rehabilitation program, the employees returned to work, thus eliminating wage replacement benefits, replacement costs for additional employees, rehabilitation costs, and reliance on the medical system. As a result, workers' compensation costs were reduced.

The individual in the assembler position was off work for three weeks receiving conservative treatment for carpal tunnel symptoms. The work rehabilitation program included an ergonomic job analysis and three sessions of work rehabilitation for cumulative trauma education and injury prevention. The work rehabilitation program also included implementation of the ergonomic suggestions resulting from the job analysis. Wage replacement for the time off work plus costs from the work rehabilitation program were approximately $2,025. Without the job site analysis and work rehabilitation, intervention costs would have increased considerably. The employee might have returned to work, but the ergonomic recommendations would not have been made, thus increasing the potential for additional problems and more lost time from work. Besides the wage replacement and medical benefits, an employee replacement cost would have been added.

The janitor was off work seven months. Approximate costs for wage replacement, eight weeks of work rehabilitation, and employee replacement were $20,300. Without work rehabilitation, return to work would have been greatly affected or possibly eliminated. If time off work were doubled due to lack of work rehabilitation intervention, wage replacement and worker replacement alone would have been $22,000. Estimated medical costs would have increased this amount to approximately $38,000.

The sewing machine operator had been off work for four months. Wage replacement, work rehabilitation, and the additional costs of adding another employee to fill the position came to approximately $9,000. With an additional four months off work, employee replacement, wage replacements, and increased need for medical services, estimated costs would be in the $17,500 range.

None of the above projections takes into consideration increased medical costs, other than a work rehabilitation program, or increased case management costs. With sedentary lifestyle and loss of worker traits because of absence from work, the rehabilitation process often takes longer and involves a greater number of professionals to achieve a successful return to gainful employment. The accommodations and modifications discussed in the previous case studies are further illustrated in Table 25–2.

CONCLUSION

The provider relationship with the employer is basically a structured communication process. The elements of this process include a timely interaction, ongoing communication, and methods to overcome problems as they surface. Contact with the employer should begin once it has been determined that a work rehabilitation program is indicated. The purpose of the contact is to gain support and commitment of the employer, promote a "team" structure, and establish a working relationship.

Once the initial contact has been established, it is imperative to maintain regular communication with the employer. A problem identification process should be discussed. If these things are in place, miscommunication and lapses of time between the employee's program discontinuation and return to work can be minimized. If during the work rehabilitation process you determine that an employee will be unable to return to his or her previous job, the employer should be notified as soon as possible. This is important in directing the employee to appropriate case resolution. It is important to the employer from a cost standpoint in paying medical and indemnity benefits, determining whether reasonable accommodations are indicated, and making future staffing decisions.

Regular reporting of the employee's status regarding potential for work will optimize maximum benefit from the program. Every case does not result in return

Table 25–2 Summary of Cases 1–3 and Their Interventions

Job/Medical Diagnosis	*Problem*	*Employer/Therapist Interaction*	*Accommodation*
Case 1			
Assembly Worker/ Assembling Medical Diagnostic Kits Dx: Carpal Tunnel Syndrome	1. Repetitive pinching and gripping 2. Static hand positions holding trays 3. Static wrist position holding trays 4. Repetitive assembly work	Employer initiated contact as soon as employee was off work. Agreed with therapist to have employee participate in a cumulative trauma education program. Allowed therapist to videotape work stations. Incorporated ergonomic recommendations into workplace.	1. Alternative method using machine to place caps 2. Forearm support to reduce static holding 3. Set boxes of slides on tray/cart to reduce static wrist positions 4. Incorporate pause/stretch breaks and job rotation
Case 2			
School Janitor Dx: Degenerative Disc Disease	1. Forward bending, twisting when mopping, sweeping, pushing, pulling, or emptying trash bin into dumpster 2. Twisting and pulling when sweeping	Employer contacted after referral to work rehabilitation. Job site visited prior to setting up work simulation. Employer agreed to recommendations suggested by therapist.	1. Extension handles on mops, brooms, trash can; assistance lifting trash bin into dumpster 2. Lighter mop head; alternate light and heavy tasks
Case 3			
Sewing Machine Operator Dx: Carpal Tunnel Release	1. Cutting through several layers of fabric 2. Static positioning 3. Frequent overtime	Employer initially concerned about slowed productivity and loss of income with suggested modification and changes/ recommendations for the job. Employer was kept informed of employee's progress. Gradual return to work was initiated.	1. Special scissors to reduce effort needed to cut fabric 2. Job rotation, pause/stretch breaks 3. No overtime required

to work at the same job. However, the outcome of the work rehabilitation program should not be a surprise to the employer if this is not the case. Work hardening and work conditioning are not segmented treatments. They are a rehabilitative process.[18] The whole is more important than the parts. How each part fits together is more important than the perception of each part. The quality of the work rehabilitation program and its processes determines the quality of the final outcome.[19]

While we may be successful in returning the injured worker to work, it is the therapist who views the employer as the final consumer with whom continued success can be achieved. Involving the employer in the work rehabilitation process maximizes the time, effort, and dollars that each case absorbs. It facilitates intelligent decisions in providing accommodations, and placing and scheduling employees in jobs within their physical capabilities. It promotes success by helping the return-to-work program be more job-specific, addressing the employer's concerns about future employment, and helping the employer plan staffing changes when the employee returns to work.[20] When set up and implemented properly, return-to-work programs can provide numerous benefits to both employees and the company.

NOTES

1. M. Zitter, Outcomes Assessment: True Customer Focus Comes to Health Care, *Medical Interface,* May 1992:32.
2. Minnesota Dept. of Labor and Industry, *Statement on Outcomes Draft* (St. Paul, Minn.: State of Minnesota, August 1992).
3. J. Cicalis and P. Going, Industrial Therapy: Breaking Down the Barrier, *Work* 3 (1993):73.
4. P.M. King, Outcome Analysis of Work-Hardening Programs, *American Journal of Occupational Therapy* 47 (1993):7.
5. T.G. Mayer et al., A Prospective Two-Year Study of Functional Restoration in Industrial Low Back Injury: An Objective Assessment Procedure, *JAMA* 258 (1987):1763.
6. R.G. Hazard et al., Functional Restoration with Behavioral Support: A One-Year Prospective Study of Patients with Chronic Low Back Pain, *Spine* 14 (1989):157.
7. J.A. Saal and J.S. Saal, Nonoperative Treatment of Herniated Disc with Radiculopathy—An Outcome Study, *Spine* 14 (1989):431–437.
8. R.L. Licter et al., Treatment of Chronic Low Back Pain, *Clinical Orthopedics* 190 (November 1984):115–123.
9. M.P. Kuhn and T.W. Kneidel, Is Work Hardening Effective, *Industrial Rehabilitation Quarterly* 3 (1990):45.
10. L.E. Darphin, Success: How To Achieve It with Work Injury Programs, *Work* 1, no. 4 (1991):46.
11. King, Outcome Analysis; Mayer et al., Prospective Two-Year Study; Hazard et al., Functional Restoration; Saal and Saal, Nonoperative Treatment; Licter et al., Treatment; Kuhn and Kneidel, Is Work Hardening Effective.

12. J.W. King, An Integration of Medicine and Industry, *Journal of Hand Therapy* 3 (1990):45–49.
13. Equal Employment Opportunity Commission (EEOC), *A Technical Assistance Manual of the Employment Provisions (Title 1) of the Americans with Disabilities Act* (Philadelphia: Hanley and Belfus, January 1992).
14. S.J. Isernhagen and H. Fearon, *Functional Job Analysis Manual* (Duluth, Minn.: Isernhagen Work Systems, 1992).
15. EEOC, *Technical Assistance Manual.*
16. S.M. Stockdell and M.S. Crawford, An Industrial Model for Assisting Employers To Comply with the ADA, *American Journal of Occupational Therapy* 46 (1992):428–433.
17. EEOC, *Technical Assistance Manual.*
18. S.J. Isernhagen and L.E. Darphin, *Work Hardening Manual* (Duluth, Minn.: Isernhagen Work Systems, 1989).
19. D. Day, How Is Business and Industry Addressing Ergonomics? (Paper Presented at Applied Ergonomics: Controlling Overexertion Injuries at Work, Philadelphia: Meeting Planners, 1991).
20. S.J. Hunter, The PT and Workers' Compensation, *Clinical Management* 10 (1990):23.

Chapter 26

Clinical Case Study: Successful Management of an Injured Worker

Sara B. Chmielewski and Daniel J. Vaught

Steven Dwayne Swenson (name has been changed), a 40-year-old millwright technician, sustained a work-related injury on April 25, 1993. At that time, he was assisting fellow employees in the transfer of 50-pound dust collector tubes onto a slurrifier deck. He was standing in a wooden crate that was suspended in the air by a fork truck adjacent to the deck. Mr. Swenson states he fell out of the crate after it unexpectedly slid forward on the forks of the fork-lift. He fell backward approximately nine feet and landed on a concrete floor, striking his head on a pipe. The blow resulted in a five-inch star-like pattern laceration over the left parietal cortex. He experienced a brief loss of consciousness and was subsequently transferred by ambulance to a local hospital.

Mr. Swenson's head wound was sutured, and his report of cranial, chest, back, and right-sided rib discomfort was assessed. X-ray study of the ribs revealed no evidence of fracture, and the chest X-ray revealed no evidence of active cardiopulmonary disease. Cervical and lumbar spine X-rays revealed no fracture. Early degenerative osteophytes were observed anteriorly at level C5–6. Multiple contusions were diagnosed in addition to the complex scalp laceration. He was medicated for pain and kept overnight for observation. He was discharged from the hospital on April 26, followed by visits to his general practitioner. On the third day post injury, Mr. Swenson reported a persistent headache experience with dizziness, some memory loss, and difficulty with fine motor skills. He also reported mid-thoracic pain. A CT scan of the brain was obtained and revealed no intracranial bleed or subdural hematoma. A thoracic spine X-ray revealed a 40 percent T-6 compression fracture. Examination by a neurologist revealed the presence of a cerebral concussion without significant intellectual phenomena.

Mr. Swenson's head wound healed without complication. His headache experience and constant dizziness subsided over the course of approximately five weeks. He experienced occasional residual right-sided thoracic back pain, which he described as "sharp and knife-like." He felt his back was "catching" at times. Physical therapy was then initiated.

Mr. Swenson began physical therapy treatment at a local hospital on June 6, 1993, at a rate of two to three sessions per week. Initial treatments included hot packs, electric stimulation, massage, and ultrasound. This progressed to an exercise program, and within one month he was able to ride an exercise bicycle and perform thoracic stretching and trunk strengthening with pulleys. Treatment with hot packs, ultrasound, and massage continued because of ongoing thoracic back pain. The program later included myofascial release techniques in addition to ultrasound and massage, which continued through August 26, 1993.

As a result of continuing physical improvement and the nature of his required work tasks, it was determined that Mr. Swenson would benefit from industrial rehabilitation to facilitate his return to work. As a maintenance millwright technician, Mr. Swenson is responsible for the inspection, dismantling, cleaning/lubricating, repairing, and replacement/installation of a variety of mechanical equipment. In addition, he performs layout work and cuts all types of pipelines, fittings, and fixtures in equipment maintenance and construction. Tools and equipment include drill press; power saw; chain saw; burning equipment; portable drill and grinder; miscellaneous hand tools and measuring devices; welding equipment and fixtures; bending, threading, and cut-off machines; soldering and testing equipment; dies; blueprints; and scaffolding.

During the course of an average eight-hour workday, he sits for one to one and one-half hours, walks for one and one-half hours, and stands for five hours. The amount of standing and walking varies depending upon the assigned job task. Bending/stooping activities are required frequently to continuously. He is frequently required to reach above shoulder level, kneel, balance, push/pull, and twist. Climbing stepladders and straight metal ladders attached to machinery is required frequently. Crawling is required occasionally. Frequent lifting/carrying up to 34 pounds is required, with occasional lifting/carrying from 35 to 100 pounds. On the job, he uses his feet for repetitive movements operating foot controls. Repetitive simple/firm grasping is also required. The working environment is continually dirty and usually wet, and the floor is occasionally covered with taconite pellets. Potential hazards include falling from equipment while making repairs, and exposure to hot materials.

Mr. Swenson was referred to an industrial specialist physical therapy clinic on September 4, 1993, by a neurologist for evaluation and treatment of left scapular and thoracic pain, and assessment for a work-conditioning program. At his initial evaluation, Mr. Swenson reported back pain, memory loss, and loss of fine hand coordination. A thorough history of his injury and medical care was reviewed, and his present complaints explored. He described pain along his entire left spine, particularly from the vertebrae to scapula, with complaints of "locking" and sudden loss of breath. He described sharp pain across his scapulae with sneezing. He also complained of a "stretching pain" from his left buttock into the posterior mid-calf. Functionally, Mr. Swenson stated that he was unable to lie in a prone position with his spine extended, or on his left side propped up on his left elbow. He experienced occasional low back pain after sitting or standing for longer than one hour. He stated that he was capable of walking two miles in 30 minutes, although prior to his injury he had participated in several 25-mile fund-raiser walks. He reported that he was unable to swim secondary to his pain. He slept a maximum of three hours uninterrupted, and then would wake up and have to change positions. His usual night's sleep consisted of moving frequently between his bed and recliner.

Objectively, Mr. Swenson presented with forward head posture, and holding his left shoulder and scapula higher than the right. His cervical active range of motion (AROM) was within normal limits (WNL), with the exception of extension, which produced complaints of dizziness. End-range cervical flexion produced "stretching" discomfort. Trunk AROM was WNL in extension and bilateral side bending. It was limited in flexion with a perception of "hamstring stretching," and limited in bilateral rotation, with pain greater to the right than left. Neurologically, sensation was found to be slightly diminished along his entire left side, upper and lower extremities, when compared to the right. Deep tendon reflexes (DTRs) were intact and symmetrical, and strength was intact and equivalent in his upper and lower extremities. AROM of bilateral upper extremities was WNL with the exception of right internal rotation, only slightly limited when compared to the left. Palpation and mobilizations revealed significant tenderness and decreased posterior-anterior (P-A) mobility of the T4–11 and L4–5 vertebrae, especially T4–7, with significant muscle hardening through bilateral scapular regions. The initial assessment and resultant recommendations: Mr. Swenson was an excellent candidate for the work-conditioning program, with occasional but regular physical therapy treatments for myofascial release and vertebral mobilizations.

Mr. Swenson then completed a pre-work-conditioning evaluation on September 11, 1993, with the results compared to his post-functional capacity evaluation (FCE) in Table 26–1.

Mr. Swenson began the work-conditioning program on September 14, 1993, and participated three days per week from two and one-half to three and one-half hours per day for the first two weeks. His program was then increased to five days per week from Week 3 through 6, and increased gradually to four hours per day. During his six-week program, Mr. Swenson attended 25 of 26 sessions offered, with one excused absence. His work conditioning program, with gradual progression throughout the six-week period, consisted of the following:

1. Education classes for the prevention of back injury and cumulative trauma, including instruction in safe postures, techniques, and body mechanics, and anatomy and physiology of the human musculoskeletal system, with focus

Table 26–1 Pre/Post-Work-Conditioning Evaluation

Item	Pre-Work-Conditioning Evaluation					Post-Evaluation				
	Percentage of 8-Hour Day					Percentage of 8-Hour Day				
WEIGHT CAPACITY IN POUNDS	0	1–5	6–33	34–66	67–100	0	1–5	6–33	34–66	67–100
Floor to waist lift		50	30	10	0		70	45	25	10
Waist to overhead lift		10	10	5	0		50	30	20	10
Horizontal lift			NOT TESTED				85	50	30	10
Push		100	75	50	25		113	85	57	28
Pull		77	58	39	20		80	60	40	20
Right carry			NOT TESTED				70	50	30	10
Left carry			NOT TESTED				70	50	30	10
Front carry		50	30	20	10		85	50	30	10
Right hand grip			NOT TESTED				96	72	48	24
Left hand grip			NOT TESTED				85	64	43	21
FLEXIBILITY/POSITIONAL										
Elevated work			X							X
Forward bending/sitting				X						X
Forward bending/standing				X						X
Rotation sitting		left			right					X
Rotation standing		left			right					X
Crawl				X						X
Kneel				X						X
Crouch—deep static					X					X
Repetitive squat					X					X
STATIC WORK										
Sitting tolerance										X
Standing tolerance										X
AMBULATION										
Walking					X					X
Stair climbing					X					X
Step ladder climbing					X					
Balance				X						X
COORDINATION										
Right upper extremity					X					X
Left upper extremity					X					X

on understanding and prevention of injury. Classes were 30-minute sessions three days per week.

2. Stretching exercises for the entire body, with particular focus on his thoracic and cervical regions.
3. Aerobic conditioning.
4. Total body conditioning, three sets of 30 repetitions at moderate weights for all primary muscle groups.

5. Functional tasks, including
   a. Thirty repetitions of lifting 35 pounds from waist to eye level
   b. Thirty repetitions of lifting 45 pounds from waist to floor level
   c. Ten repetitions of carrying 35 pounds a distance of 50 feet
6. Job simulation tasks for a millwright technician:
   a. Kneeling and working in a forward bent position at the bolt box for 15 minutes, five minutes of this with five-pound cuff weights on each hand to simulate working with tools
   b. Climbing a straight ladder and crawling around in a limited crawl space
   c. Elevated work from a six-foot stepladder
   d. Hoisting a 30-pound simulated chain fall up to a crawl space
   e. Balancing on an unstable platform (Chattexc Balance System) to simulate walking and working on an uneven surface (taconite pellets).
   f. Simulated foot control clutch work.
   g. Using a sledgehammer in various positions with focus on returning the sledgehammer to a pick-up position following the blow. Mr. Swenson contacted his employer with regard to obtaining a tool necessary for job simulation tasks, and the employer willingly provided a sledgehammer.
   h. Supine lying, working overhead on a bolt box to simulate mechanic work.

During his six-week work-conditioning program, Mr. Swenson participated in eight physical therapy sessions that included myofascial stretching for the cervical, thoracic, and scapular regions, P-A Grade II-IV vertebral mobilizations, middle and lower trapezius strengthening, and cervical extension ROM.

Mr. Swenson was discharged from the work-conditioning program on October 23, 1993, and he underwent an FCE on October 27, 1993 (see results in Table 26–1).

On November 5, 1993, six months after Mr. Swenson's injury occurred, he was released to return to full-time employment as a millwright technician within the guidelines of the FCE. His physician recommended no climbing over 20 feet, and this was easily accommodated by the employer. The use of an industrial back brace was recommended by the physical therapist and implemented by the employee to provide increased low back stability when performing heavy lifting/carrying activities. An independent exercise program was set up at a local health club to allow Mr. Swenson to maintain general flexibility, strength, and aerobic conditioning. Two follow-up on-site job visits were completed by the nurse rehabilitation consultant during the first month of employment. Mr. Swenson demonstrated consistent use of proper body mechanics and reported no significant physical concerns with regard to the performance of his job tasks. He did report some degree of ongoing right thoracic soreness.

Follow-up employer contact was made, and Mr. Swenson's supervisor stated that he was pleased to have him back on the job. He reported no concerns regarding work attendance or job performance. Following the completion of a 30-day return-to-work monitoring period, Mr. Swenson's rehabilitation file was closed.

Many factors affected the successful resolution of Mr. Swenson's complex work injury and eventual return to work placement. Due to the severity of Mr. Swenson's injuries (40 percent vertebral compression fracture and head injury), the acute phase of physical therapy at a local hospital was delayed for approximately six weeks. The utilization of modalities, myofascial release, and stretching exercises successfully prepared Mr. Swenson for the start of his work-conditioning program. The physician responded by accurately timing his order for a work-conditioning program. The rehabilitation consultant coordinated a meeting with the work-conditioning physical therapist prior to the initiation of the program in order to explain Mr. Swenson's initial injury and treatment. Detailed medical/rehabilitation records and job description information were also provided. Weekly telephone and/or in-person contact with the client and physical therapist was maintained to monitor Mr. Swenson's progress, and the employer was regularly updated.

Mr. Swenson's positive attitude and strong desire to reach his maximum physical potential was an important factor in the resolution of his injury. A positive employer/employee relationship was revealed early on, and Mr. Swenson demonstrated a strong work ethic and sincere desire to return to his millwright position. The employee was supported throughout his entire program regarding adjustment to disability and preparation for return to work. Specifically, he was encouraged to focus on function versus pain, which allowed him to successfully achieve his physical goals. A trusting physical therapist/client relationship was established at the outset and provided a safe environment to foster functional outcomes.

Chapter 27

The Role of Equipment in Evaluation and Treatment of Occupational-Related Injuries

Dennis D. Isernhagen

The age-old dilemma of what role equipment does or should play in an occupational rehabilitation program continues to be debated. The debate is heating up and is being fueled by the changes underway in health care. These changes include state and federal laws governing reimbursement for rehabilitation, the Americans with Disabilities Act, changes in workers' compensation regulations, increasing utilization of rehabilitation professionals as expert witnesses, and the ever-present pressure to remain competitive in an increasingly competitive market.

The constant changes and advancements being made in technology add more to the dilemma. The claims that manufacturers make about the reliability and validity of their equipment to determine physical impairment, plus their statements that their equipment can identify the "malingerer," add to the overall confusion as to what, if any, equipment should be purchased.

Adding further to the dilemma is the need to operate an efficient practice. Equipment can enhance the therapeutic experience and improve the efficiency of the clinician. Conversely, however, the equipment may provide inaccurate information or become a nonfunctional focal point of a clinician's practice methods. Equipment should only be used as an adjunct to a practice and should not be the sole focus. The thirst to be "scientific" should not overshadow the need to develop and practice the "art" of health care.

The intent of this chapter is to provide an overview of some of the factors a clinician needs to take into consideration when faced with the decision to buy equipment for an occupational rehabilitation program. The utilization of equipment for both evaluation and treatment will be explored. A series of preliminary questions needs to be asked before the decision to purchase is made. It should be kept in mind that the occupational rehabilitation program is focused on evaluating and treating a person's function as it relates to the performance of work.

PRELIMINARY QUESTIONS IN SELECTING EQUIPMENT

1. *What is the intended purpose of the equipment to be considered?*

- Is it because your competitor has one so you need to get one? Too often clinicians make the decision to purchase a piece of equipment solely on an analysis of their competition rather than on their practice and client needs.
- Is it based on a perceived or real need to be more scientific in clinical decisions (have "data" to back decisions)? As stated earlier, many clinicians want their profession to be seen as more scientific—thus the perceived need to present data. Too frequently the data presented are just that, "data." They are often presented in a manner that the reader has difficulty understanding. What is critical is the need to have data that are useful in answering referral and return-to-work question(s).
- Is the decision to purchase based on a need to add more variety to the treatment program? This is often a stated need by clinicians involved in occupational rehabilitation. When making a selection based on this need, one should make sure the equipment provides for functional activity and not isolated movement. Equipment that can be utilized for multiple functions will be more advantageous.

2. *Is the test or treatment procedure(s) to be administered safe for client(s)?*

Are there enough safeguards to ensure that client(s) will not get injured? The potential of injury to a client should always be a major concern. This is a concern not only for the client but for the therapist administering the test or treatment procedure. Lawsuits can be a disaster to all parties concerned.

3. *Is the equipment simple and practical in its use?*

Some high-tech equipment can be very difficult and time consuming to set up for use. This needs to be taken into consideration when personnel costs are considered. In addition, if the equipment is going to be utilized by clients for therapeutic purposes, can they operate it safely? This will free the therapist and promote independence of the client.

It is also noted that complicated equipment is often avoided by rehabilitation personnel who do not feel educated in its operation or do not wish to take treatment time to set it up.

Concerning practicality, one thing that should be kept in mind when selecting equipment is, does the movement transfer into the client's functional needs to accomplish his or her job? Also, can the activity be performed in some manner that the client can duplicate on his or her own away from the clinic? This is important to foster independence and encourage compliance with therapeutic recommendations.

4. *Is the equipment cost-effective?*

Too often rehabilitation professionals find themselves in a real quandary when they have purchased equipment that performs limited functions. They find they are not getting their money back on their initial investment and need to find more patients to use the equipment on. This can lead to inappropriate testing and/or overutilization of therapeutic procedures. Equipment that can perform multiple functions is more valuable than a single-function apparatus. Each purchase needs to be evaluated as to its role in a clinic's overall purpose and goals.

Current trends are in reduced reimbursement for rehabilitation services, thus making the cost-effective portion of the buying decision even more important.

UTILIZATION OF EQUIPMENT FOR EVALUATION

As in most areas of health care, technology has made advancements in the evaluation of work-related injuries. The majority of this equipment has focused on the most common work injury, back injuries. The cost of high-tech equipment is substantial, and its use in a practice needs to be carefully evaluated.

In order to make an appropriate decision about high-tech equipment that claims to measure function, one needs to understand some of the concepts behind the equipment found in today's market. The first thing that needs to be taken into consideration is the intended purpose of the equipment. The purpose usually falls into one of two categories, measurement or functional evaluation.

Equipment utilized for *measurement* purposes needs to be able to limit the number of variables that may interfere with the intended measurement. The limitation of variables requires limitation of joints and planes to be measured at one time. This limitation of variables increases the "science" of the findings. However, this may limit the use of the data generated to their application to "functional movement" as related to a specific job function.

Apparatus to be used for *functional evaluation* purposes will have a high degree of variability. Evaluation takes into consideration a broad scope of movements to accomplish a specific activity. This makes evaluation an art as well as a science. An example is in the evaluation of a person's ability to lift. The following characteristics need to be assessed as they pertain to the function of lifting:

- pace
- speed
- speed change (acceleration/deceleration)
- use of momentum
- synchronized muscle use
- different muscle use
- different muscle length at each degree of motion

- different joint stress at each degree of motion
- balance
- coordination
- heart, lung capacity
- six major joints
- all back and neck joints
- multiple muscle involvement
- changing base of support
- psychological impact

Another issue affecting equipment choices is the impact that the Americans with Disabilities Act will have on the utilization of high-tech equipment in evaluating an individual's ability to perform specific jobs or job functions. A major question arises when high-tech equipment is used to make a determination if an individual can perform the essential physical functions of a job. If the individual is able to perform exactly the same movements with the exact resistance and patterns that are required on the job, then the equipment may be used for determination of return to work or for prework screening. If, however, the equipment is unable to duplicate this activity, the legal community may imply that this testing does not comply with the ADA.

OVERVIEW OF COMMONLY USED HIGH-TECH EQUIPMENT

Types of equipment and methods commonly used in evaluation of work-related injuries are isometric, isotonic, isodynamic, and isokinetic. The following analysis focuses on the use of the stated equipment for back injuries. The findings can also be extended to uses of the equipment for other parts of the body that are under evaluation.

Isometric Testing

Isometric testing for the determination of an individual's ability to perform manual work has been used to obtain a baseline. This type of testing involves the static exertion of a force against a fixed resistance. It is performed at zero speed with no observable motion or functional movement. The muscle length remains constant. It has been one of the primary types of preemployment screening used for jobs that require manual material handling. It can be measured by low-tech dynamometers. Also, the formulas that have been devised using isometric testing are presently used in the software of much of the high-tech equipment presently on the market.

One advantage of using a low-tech version of isometric testing is that it can be performed almost anywhere. There is minimal equipment needed to conduct the test, and it only requires a small amount of space. The amount of time to perform the testing is relatively brief. Reproducibility of the findings can be easily determined because of the ability to reproduce each testing position.

The high-tech testing devices using isometric force produce reports that include graphs and charts comparing the individual being tested to normative data and other measurements. This presents a "generic" (not individualized) comparison.

There are several disadvantages to the use of isometric testing. One of the primary disadvantages is that the results do not simulate "functional motion." Strength can only be tested in one position and not throughout the entire functional movement. The findings provide no correlation to the individual's endurance. The testing procedures can be biased by psychological interferences.[1] There is not a perfect relationship between the force produced and the dynamic functional strength performance that is required for a specific job or task. There have been reported injuries to clients while undergoing isometric testing. The normative data used for comparison are gathered from subjects without pathology. This raises the question of the applicability of the findings to subject(s) who have specific pathological conditions (e.g., injured workers).

Isotonic Testing

This form of testing involves the measurement of a person's ability to maintain voluntary muscle contraction while the resistance remains constant throughout a range of motion. When a muscle contracts against a fixed load, the tension produced in the muscle varies as it shortens or lengthens through the available range of motion. The contracting muscle is subjected to varying amounts of resistance through the range of motion. Dynamic strength, muscular endurance, and power can be measured with isotonic testing.

There is a wide variety of equipment, procedures, and protocols available for performing isotonic testing and exercise. The testing procedures can be performed with equipment or manually by a therapist. The work performed is through a range of motion that can be done at a functional speed. The work performed can be objectively documented as to the strength, endurance, and power that a specific muscle group can perform. The testing can mimic, to a certain extent, a segment of the essential physical functions of a job.

Isodynamic Testing

The uniqueness of isodynamic testing is that it can measure triaxial (three-dimensional) movement patterns. The term *isodynamic testing* was developed by Isotechnologies, Inc., a manufacturer of triaxial testing equipment. The equipment that has been developed can test the motions of flexion, extension, side bending, and rotation either separately or combined in a number of different patterns. Research using this equipment has demonstrated that muscle fatigue can compromise coordination and control. This has demonstrated that as an individual's primary muscles fatigue there is deviation in movement patterns and an increased use of accessory muscles. One of the major advantages of isodynamic testing is its ability to measure multiplane movements or isolated motion. It can identify and measure substitution patterns that occur as a result of muscle fatigue.

As with the other types of high-tech equipment, the use of isodynamic testing to determine "functional" motion is questionable. The anatomic stabilization that is required restricts the normal movement that occurs when an individual performs a task. This is especially noted in setting an individual up for testing: the lower extremities are strapped in an extended position, allowing movement only in the upper part of the body. This raises a major question concerning the validity of the claim that this equipment can measure functional capacity.

The reliability of the equipment is also in question. Most of the studies that have been used with this and other high-tech equipment use normal subjects. The tests are not conducted on the individuals for whom the equipment was designed to test, those with pathology.

Isokinetic Testing

Isokinetic testing is one of the more common types of testing used by rehabilitation providers. The technology behind this method of testing has advanced significantly over the past two decades. Isokinetic testing is a form of dynamic exercise in which the speed of muscle shortening or lengthening is controlled throughout a range of motion. Since the velocity of the movement is always constant, the resistance to the movement will vary throughout the range of motion. Isokinetic exercise is often referred to as *accommodating resistance exercise.*

There is a wide variety of isokinetic equipment on the market today. Units have been manufactured that can test or treat one specific extremity or can be used on multiple extremities. Isokinetic equipment can provide maximum resistance at all points in the range of motion as a muscle contracts. The equipment will accommodate for a painful arc of motion. High-speed and low-speed testing can be performed. It provides a good record of the data generated. Some of the high-tech equipment provides an entire report, thus limiting the amount of time that a therapist needs to spend analyzing and interpreting the test results.

There is a previously stated concern regarding the use of high-tech equipment, namely, its cost. This type of equipment is the most expensive of the high-tech apparatuses. In addition, dedicated space is required to house these units. Set-up time can be lengthy, especially when multiple muscle groups are being tested or exercised. Some of the lifting tasks do not reflect real-world situations. The equipment cannot be used for a home exercise program, thus requiring the client to be dependent on the health care provider for his or her total rehabilitation program. It has been demonstrated that psychological variables can influence the performance of individuals using isokinetic equipment.[2] The motion used with this equipment is isolated movement and is not functional when applied to a real-world situation.

Outcome Research

There is continual research being conducted on the reliability and validity of high-tech equipment as it relates to "functional movement" and in particular as it relates to an individual's ability to perform a specific physical task or job. The research is being challenged by its application to the "real world," in particular the Americans with Disabilities Act.

In a recent scientific literature review on "iso-testing" of dynamic trunk strength related to low back pain, Newton and Waddell pointed out some interesting conclusions to keep in mind when considering the purchase of high-tech equipment for evaluation purposes:[3]

- Too many of the manufacturers of "iso-testing" equipment refer to unpublished abstracts or internal publications that do not meet the standards of peer-reviewed scientific journals. This is directly tied to the marketing and sale of the equipment in question.
- The testing protocols used in the research are often not standardized or clearly described. Different protocols are often used for "normal" subjects and "patients."
- There is lack of information on the instructions, encouragement, and feedback that are given to the subjects being tested. This has been demonstrated as a significant variable in the research performed by Matheson et al.[4]
- There is a lack of reliability studies in "patients." Since the application is to be administered to this group of individuals, it would only stand to reason that a comparison should be drawn between this group and the "normal" group.
- There is a lack of evidence on the relationship between pain and the performance demonstrated on high-tech equipment.
- There is a lack of evidence on the relationship between high-tech measurements and the psychological variables that may interfere with the testing procedure.

The conclusion of the Newton and Waddell study was that "after a decade of investigation, there is inadequate scientific evidence to support the use of either isokinetic or isoinertial testing of trunk muscle strength in pre-employment screening (pre-work screening), routine clinical assessment, or medical-legal evaluation."[5]

UTILIZATION OF EQUIPMENT FOR TREATMENT

All of the equipment and procedures discussed above may also be used for treatment purposes. The advantages and disadvantages discussed also apply to using the equipment for treatment purposes. The major drawback to utilizing high-tech equipment for treatment is its cost. The equipment itself is expensive, and its operation becomes expensive when it requires personnel to set the client up and make various adjustments. One person can be treated at a time, and this can make the overall operating cost high. In addition, much of this equipment does not simulate real-world situations. It often provides nonfunctional exercise that is directed to one body segment. It does not allow the body to move in synchronized motion, as is required when a person performs a work task.

Some manufacturers have gone to great lengths to develop high-tech equipment that will simulate various work tasks such as lifting. They have actually attached boxes, crates, and other items that an individual may use to their equipment so that "data" can be collected while the individual performs a given task. The main question that one can ask, however, is, does the cost justify the outcome? Does a therapist get better results when he or she utilizes high-tech equipment, or have we developed a false sense of security that better therapy is provided because of technology?

A therapist can provide a more effective treatment program with less sophisticated equipment. The purchase of equipment should be based on its functional application. The movement that the equipment produces should simulate a functional activity that will assist the individual in accomplishing a task. Whole-body activities provide real-world situations. Exercising a single body part that is weak is necessary to restore it to a normal level. In addition, the therapist needs to incorporate the extremities' use into a functional activity that simulates a task that the client must perform upon return to his or her job. This does not require high-tech equipment.

In planning the rehabilitation program, the therapist needs to know all he or she can about the job the client will be returning to and what the physical requirements are. Only until this is known can the appropriate selection of equipment be made. One of the most effective methods of treatment is to have the actual equipment available for the individual to use while under the supervision of the treating therapist. This will enable the therapist to observe the individual using the equipment and make recommendations to improve function. It cuts the guesswork out of predicting the individual's ability to perform the task required. It also allows the therapist the opportunity to recommend changes in the equipment or how it is used that will assist not only the injured worker but other workers who may be utilizing the same equipment.

The functional, realistic equipment that the individual must use is usually inexpensive to obtain. Requesting the employer to provide this equipment is usually met with much support. This not only assists the injured employee but establishes good communication with the employer that is critical in the return-to-work process. Once the therapeutic session is completed, the equipment can be returned to the employer. This requires no purchase by the therapist and no ongoing space requirements.

CONCLUSION

The utilization of equipment for evaluation and treatment in an occupational rehabilitation program needs to be considered carefully. High-tech equipment can be very expensive and may not provide the best outcome from either a therapeutic standpoint or a financial standpoint. In the rapidly changing health care environment, one must seriously consider equipment purchases. In addition, the use of high-tech equipment for evaluation purposes may have some major legal implications, especially in light of the Americans with Disabilities Act. The use of computerized equipment for functional testing for prework or return-to-work determination is questionable. The ADA requires that functional testing must be "work related." If these tests do not simulate the actual work tasks, they may be inappropriate or illegal under the ADA. Validation of an apparatus' ability to test the essential functions of a wide variety of jobs is possible and desirable if the cost is not prohibitive.

The future of testing and the decision to use high-tech or low-tech equipment are being influenced by the shrinking reimbursement dollar and the need to demonstrate cost-effective outcome. In addition, serious thought should be given to the use of high-tech equipment in making employment decisions and in other medical-legal situations where an individual's "functional capacity" is in question. There is a lack of evidence that such equipment can make a "real-world" determination.

NOTES

1. M. Battie and S. Bigos, Industrial Back Pain Complaints, a Broader Perspective, *Orthopedic Clinics of North America* 22 (1991):273–282; A. Papciak and M. Feuerstein, Psychological Factors Affecting Isokinetic Trunk Strength Testing in Patients with Work Related Chronic Low Back Pain, *Journal of Occupational Rehabilitation* 1 (1991):95–104; M. Robinson et al., Physical and

Psychosocial Correlates of Test-Retest Isometric Torque Variability in Patients with Chronic Low Back Pain, *Journal of Occupational Rehabilitation* 2 (1992):11–18.

2. M. Parnianpour et al., The Relationship of Torque, Velocity, and Power with Constant Resistive Load during Sagittal Trunk Movement, *Spine* 15 (1990):639–643.
3. M. Newton and G. Waddell, Trunk Strength Testing with Iso-Machines, Part 1: Review of a Decade of Scientific Evidence, *Spine* 18 (1993):801–811.
4. L. Matheson et al., Effect of Instruction on Isokinetic Trunk Strength Testing Variability, Reliability, Absolute Value, and Predictive Validity, *Spine* 17 (1992):914–921.
5. Newton and Waddell, Trunk Strength Testing, p. 809.

SUGGESTED READING

Gervais, S., et al. 1991. Predictive Model To Determine Cost/Benefit of Early Detection and Intervention in Occupational Low Back Pain. *Journal of Occupational Rehabilitation* 1:113–131.

Hart, D. 1992. Practical Decisions under the ADA. *Clinical Management* 12:105–110.

Isernhagen, S. 1990. Exercise Technologies for Work Rehabilitation Programs. *Orthopaedic Physical Therapy Clinics of North America* 1:361–374.

Isernhagen, S. 1990. Computerized Wounders. *Physical Therapy Today* 13, no.1:56–59.

Lancourt, J. and M. Kettelhut. 1992. Predicting Return to Work for Lower Back Pain Patients Receiving Workers' Compensation. *Spine* 17:629–640.

Marras, W., et al. 1993. The Role of Dynamic Three-Dimensional Trunk Motion in Occupationally-Related Low Back Disorders. *Spine* 18:617–628.

Parnianpour, M., et al. 1988. The Triaxial Coupling of Torque Generation of Trunk Muscles during Isometric Exertions and the Effect of Fatiguing Isoinertial Movements on the Motor Output and Movement Patterns. *Spine* 13:982–991.

Perry, L. 1991. Preemployment Strength Testing, Implications in Sex Discrimination. *Work* 1, no. 3:32–36.

Slane, S. 1992. Computerized Back Testing: Making Informed Choices about Equipment Needs. *Advance Rehabilitation* July/August:25–30.

Ulin, S., and T. Armstrong. 1992. A Strategy for Evaluating Occupational Risk Factors of Musculoskeletal Disorders. *Journal of Occupational Rehabilitation* 2:35–50.

Chapter 28

Getting a Handle on Motivation: Self-Efficacy in Rehabilitation

Leonard N. Matheson

Motivation is widely recognized by front-line clinicians as an important determinant of success in rehabilitation. Clients who are "motivated" seem to be more successful in rehabilitation programs than clients who are "not motivated." It is enjoyable to work with clients who are motivated, but it is frustrating to work with clients who are not motivated. Similarly, the most successful rehabilitators often are described as "great motivators," while those who only seem to tolerate their work are often heard to complain that their clients are poorly motivated and won't cooperate.

It is clear that motivation is an important issue. But how much do we know about motivation? How do people become motivated? How do people develop motivational problems? How do great rehabilitators identify, measure, develop, or otherwise interact with motivation as a clinical issue?

Motivation has been an important focus of the field of psychology for decades. As a consequence, it has become a formidable topic. Although science can provide us with a tremendous amount of information about it, developing an understanding of motivation that can be useful to the clinician requires that we narrow our scope of study and concentrate on its most important and useful part, the "why" of motivation: Why do people do what they do? If the clinician understands the "why" that underlies the client's behavior, the clinician will be better able to understand the client and to provide the assistance that will be most likely to bring about optimum benefit. This is the value of the concept of *self-efficacy*. It provides a window on motivation into which we can peer that will provide us with information that we can use to become great rehabilitators.

DEFINITION AND DESCRIPTION OF SELF-EFFICACY

Like any important idea, self-efficacy is easy to understand and difficult to master. It is simple but it has depth. Self-efficacy is the belief that one is *competent* to handle a situation that one considers to be important.

A key aspect of being human is the need to feel competent. White defines competence as "efficacy in meeting environmental demands."[1, p.297] Competence is the ability to interact effectively with the environment while maintaining individuality and growth.[2] Competence is defined in terms of the person's ability to control his or her environment. White believes that there is an intrinsic drive in humans to influence the environment that provides motivation for exploring, manipulating, and acting on the environment. He uses the term *urge towards competence* to emphasize the basic nature of this drive. He places the need for competence on a level with the need for satiation of hunger, establishment of security, and sexual satisfaction. If we think about it, these needs are ancient parts of the human's sense of competence. Based on Darwinian theory, the human's ability to successfully compete (to achieve competence) underlies the species' current level of development.

Self-efficacy is an important cornerstone of healthy adulthood. Albert Bandura, who has done much of the early important work on this topic, points out that self-efficacy is based on the person's *perceptions* of competence and that these self-perceptions affect psychosocial function. Individuals' perceptions of their abilities affect how they behave, their level of motivation, their thought processes, and their emotional reactions to challenging circumstances.[3]

Christiansen notes that the development of competence is based on "the experience of occupation or doing . . . by learning skills and strategies necessary for coping with problems and adapting to limitations." Christiansen reports that "there is considerable agreement that the single characteristic of the individual which has the greatest influence on performance is one's sense of competence."[4, p.9,19]

Bandura argues that self-efficacy beliefs form an important component of motivation in that they encourage the person to attempt new activities. Self-efficacy beliefs also keep the person from attempting tasks that may be dangerous or risky. Self-efficacy beliefs usually affect cognitive functioning through the joint influence of motivational and information processing operations. People's self-efficacy beliefs determine their level of motivation as reflected in how much effort they will exert in an endeavor and how long they will persevere in the face of obstacles.[5]

The urge to experience self-efficacy is also the basis of the person's socialization. Aided by reinforcers such as money, social recognition, and improved security, society is able to channel individuals' urges towards competence. Society takes advantage of the urge towards competence to shape behavior (such as working for a living) that meets its needs. However, these reinforcing mechanisms can break down when a person experiences chronic illness and disability. Such a breakdown can have disastrous results for both the person and society.

EMOTIONAL RESPONSE TO PAINFUL IMPAIRMENT: THE "FEEL GOOD TRAP"

A painful and incapacitating injury is frightening. It produces fear for the immediate consequences of the injury. It produces anxiety about the future if the injury results in impairment that may be sufficient to bring about disability. "Disability anxiety" is especially troublesome for the person with an activity-based symptom disorder, someone for whom activity brings about pain.

In order to cope with this anxiety, the individual initially denies the disabling consequences of the impairment. Denial is an important defense mechanism. It is useful to the person in that it helps to handle the emotional discomfort of the injury's probable consequences. However, it also traps the client and the caregiver in that the caregiver is inclined to avoid confronting the denial with a rational challenge by focusing on palliative measures—helping the client to feel better. This is accepted by the client not only because he or she wants to feel better physically but also because he or she wants to avoid confronting the future if the future contains tangible losses.

A focus on palliative measures leads to two problems. The first is that the client gradually becomes deconditioned as he or she postpones the physical challenge that is entailed with learning about functional limitations. The second is that the client with an activity-based pain disorder will gradually develop a set of self-perceptions that are more functionally limited than is usually appropriate for prophylaxis. Pain will be taken as an indicator of harm by the client rather than merely as an index of challenge. The client often has the expectation (and certainly has the desire) that pain is unnecessary. The client may believe that avoidance of pain is a function of the skill of the caregiver. The caregiver who asks the client to experience pain as part of therapy is in danger of being replaced by the client who cannot trust the caregiver's guidance and foresight. Often it is easier for the caregiver to continue with palliation and not risk the client's ire. However, the wise caregiver knows that in the long run the client must be challenged and the pain of the challenge must be borne.

It is difficult for both the client and the caregiver to move away from palliation as the primary focus of treatment. This can be accomplished by the great rehabilitators, but it requires an extraordinary effort by both the client and the caregiver to move the focus of treatment toward challenging the client's functional limits. As the functional limitations are challenged, usually in a functional capacity evaluation (FCE), the client often will become emotionally distraught. This is because the client's denial is being confronted. During the FCE, the skillful caregiver will couple this challenge with accomplishment to reestablish self-efficacy.

FUNCTIONAL COMPETENCE AND DISABILITY

Perceived functional competence is an important component of the disabled person's self-concept. In general, people know that they are able to perform tasks that do not exceed their abilities. They are also aware of their strong disinclination to perform tasks that entail aversive outcomes, such as painful symptom experiences.

The disabled person's perception of functional self-efficacy arises from experience. Christiansen reports that "the extent to which individuals are able to develop a positive sense of self and belief in their autonomy is largely based on their successes in dealing with environmental challenges."[6, p.19–20]

It appears that once impairment has stabilized, the disabled person begins explorational activity to reestablish functional self-efficacy. In this process, various activities that are important to the person are attempted. If the explorational activity is successful, the behavior is reinforced and additional exploration at a higher level of challenge is undertaken.

Success shapes subsequent behavior through the development of expectancies about one's ability to perform. However, if the exploration is unsuccessful, the behavior is negatively reinforced. The immediate effect of this negative reinforcement is that exploration is less frequent. A more enduring effect of unsuccessful exploration is the self-referent belief that this experience has helped to develop, a belief about functional inefficacy. This will negatively affect the impaired person's self-perception and self-referent reports to others and will become an important part of his or her adjustment to the disability.

If a self-referent belief has a rational basis, the person's psychological adjustment will be more likely to have a rational basis and will reflect a realistic fund of personal resources. If the belief is irrational, the psychological adjustment will be more likely to have an irrational basis. In situations in which the person with an irrational self-referent belief is struggling with pain, a gradual erosion of self-confidence takes place. Given enough time without rational input, he or she will develop a self-referent belief system that will be quite negative and will reflect an impoverished fund of personal resources. This will occur regardless of whether or to what degree the person is actually functionally limited. If the client is receiving palliative care, there is no opportunity to gather information about actual functional capacity because the focus is on avoiding pain. On the other hand, when the client is being challenged to perform activities in spite of the pain, he or she will be able to gather experiential information and develop a set of self-referent beliefs that is rational. The healthy self-referent belief system that emerges contains complete sets of messages such as

- Some activities that I do cause pain and others do not cause pain. I know which are which.

- Some activities that cause pain may be doing harm, and I shouldn't do them.
- Some activities that cause pain may be beneficial, and I should do them.
- Some activities that cause pain don't cause harm, and I can do them if I am willing to put up with the pain.

This set of beliefs provides the client with a host of benefits, including improved self-confidence, a greater sense of self-control, and improved ability to plan for the future. It provides the basis of a healthy emotional adjustment to the impairment and any residual disability that may be present.

The relationship between pain, challenge versus palliation, and self-efficacy has been studied in various disability groups. Shoor and Holman, in work with individuals suffering from osteoarthritis and rheumatoid arthritis, measured perceived self-efficacy for control of arthritis. These researchers found that self-efficacy was inversely related to future pain and disability. That is, higher levels of self-efficacy were related to lower levels of both future pain and future disability.[7]

In the area of cardiac rehabilitation, Ewart et al. demonstrated the relationship between self-efficacy and functional competence. These researchers provided subjects who were cardiac patients with a pain-limited treadmill test. They found that pretest self-efficacy predicted subsequent performance on the treadmill test and also that the performance on the treadmill test subsequently affected post-test self-efficacy. In fact, a measure of post-test self-efficacy was significantly related to subsequent level of activity in the nonclinical environment, whereas the functional capacity demonstrated in the treadmill test itself did not predict such levels of activity.[8]

The more pain people expect, the lower their self-efficacy for pain-related movement. Council et al. studied perceived self-efficacy and patients' expectations for pain response to activity as potential predictors of impaired function. In this study involving patients suffering from chronic low back pain, the Movement and Pain Prediction Scale (MAPPS) was used to develop predictor variables of subsequent movement ratings based on videotape review. The MAPPS uses stick figure drawings to describe five successive stages of 10 simple movements involving the trunk and lower extremities. Predicted level of performance was presumed to reflect self-efficacy, while the patient's prediction of pain was presumed to indicate the person's pain response expectancy. Both of the expectancy measures were correlated with global measures of pain and physical impairment. A multiple regression procedure on the MAPPS scores indicates that predicted level of performance accounted for so much variance that the pain expectancy variable improved the prediction negligibly. These authors report that "it is possible that efficacy ratings were better predictors of performance than pain response expectancies because they accounted not only for variance due to expected pain, but also to physical harm expectancies."[9, p.329]

Dolce et al. examined the relationship between self-efficacy expectancies concerning pain tolerance and a cold pressor test. Self-efficacy was consistently related to increased pain tolerance, while there was no significant consistent relationship between pain ratings and pain tolerance. That is, in spite of the pain rating, tolerance times were predicted by self-efficacy ratings.[10]

Is tolerance of pain an ability unto itself? Ajzen and Fishbein found that these predictions were statements about the length of time that the subject was *willing* to tolerate a painful stimulus. These predictions were statements of the person's intention to tolerate pain rather than actual predictions of ability. In this sense, pain tolerance is not a capability as much as it is an intention. The person who anticipates greater benefit from tolerating pain predicts that he or she will tolerate a painful stimulus longer than the person who expects less benefit. Because intention is an immediate determinant of voluntary behavior, predictions of pain tolerance are usually accurate. Thus prediction of pain tolerance may not assess perceived capability as much as it reflects the person's perception of the benefit to be gained from "putting up with" the pain.[11]

Baker and Kirsch present a model in which self-efficacy mediates the effects of expected pain and other outcome expectancies. These researchers found that training and coping skills significantly affected pain tolerance but did not have a corresponding effect on self-efficacy. Consistent with the previously mentioned findings of Council et al., they also found that pain expectancy and self-efficacy were very highly correlated. Pain expectancy was correlated with both reported pain and tolerance, but self-efficacy was correlated only with tolerance. These authors found that pain tolerance was enhanced by cognitive coping training and by motivational manipulation and that these two effects were additive. Self-efficacy was a moderately strong predictor of tolerance, while pain expectancy was a very strong predictor of reported pain. Self-efficacy did not predict reported pain, but the effect of pain expectancy on tolerance was entirely mediated by self-efficacy.[12]

Self-Efficacy, Pain, and Disability

Over decades of work with disabled people, the author has come to the realization that *beliefs supplant knowledge, but are not recognized by the person as other than knowledge*. The disabled person's belief concerning his or her abilities extends far beyond the narrow and temporarily rational basis upon which it was first established. The person who has activity-related symptoms usually experiences substantially more reinforcement for a belief in functional inefficacy than for a belief in functional efficacy. As this person experiences life after an injury, many behaviors result in pain and discomfort. Depending on the reinforcement of his or her environment, the disabled person can develop a belief that pain is a primary limiting factor and therefore will establish a level of activity that provides relief from pain. This may so limit the person that he or she is unable to return to work and resume other social roles. Thus this belief system becomes a major impediment to the success of the person's rehabilitation program. Kirsch points out that in the face of a painful consequence, if a task is to be performed, it is common for the person to say that he or she "cannot" perform the task, even though the person will subsequently acknowledge that he or she can perform the task if it is absolutely necessary.[13]

The strength of a person's self-efficacy beliefs may be a consequence of the degree to which the early explorational behavior resulted in an aversive consequence. Take, for example, the person who has activity-related low back pain: an attempt to perform a lifting task that results in excruciating pain and a brief inability to stand or walk may be so frightening that it leads to a belief that the person is unable to perform such activity or will be reinjured. If the aversive experience was particularly intense, this belief will be strongly held and resistant to change. When this client enters a rehabilitation program, he or she will report (and demonstrate) very low levels of lift capacity. These are the clients who frequently demonstrate gains in tasks such as lifting on the order of 100 percent after only one week of treatment. Does this reflect an actual increase in the client's underlying biomechanical capacity? Probably not. It is more a reflection of the client's increased self-efficacy.

Why do some clients have such low levels of performance that such impressive improvement is possible in so short a time? Principally because many people resist utilizing their recently restored biomechanical capacity. This occurs through two interacting mechanisms. The first is diminished exploration as the client refrains from attempting tasks that are similar to the tasks that resulted in the aversive experience. The second is a resistance to information that contradicts the client's self-efficacy beliefs. Information that is supplied by caregivers concerning the client's probable level of ability is discounted in the face of this previously established belief. The client appears to be recalcitrant or uncooperative.

Self-efficacy beliefs both color and reflect disabled clients' abilities and limitations, their interests and preferences, and their perception of aversive stimuli, discomfort, and pain. To a great degree, self-efficacy beliefs are intertwined with all of the components of disabled clients' life circumstances so that the beliefs become both cause and effect. Disabled clients' perception of themselves as being disabled controls what they will attempt by controlling the range of what they consider to be viable behavioral options and by limiting the extent to which any options may be pursued.

The Pain-Perception Hierarchy

The relationship between the person's perception of his or her pain and subsequent behavior can be construed in terms of a hierarchy. The "Pain-Perception Hierarchy" is presented below:

	Self-Perception	*Behavior*
Level 5	This activity causes painful reinjury . . .	. . . so I have to avoid it.
Level 4	The pain won't let me do this activity . . .	. . . so it's no use to try it.
Level 3	I can't handle the pain if I do this activity . . .	. . . so I won't try it.
Level 2	The pain if I do this activity is not worth it . . .	. . . so I don't want to do it.
Level 1	When I do this activity I hurt . . .	. . . so I will do it only if necessary.

The self-referent statements made by the person as he or she progresses through a rehabilitation program allow placement at one or another level of this hierarchy. Note that the self-perceptions are each activity related and based on the person's knowledge of the relationship between pain and activity. This is the key to successful rehabilitation. Some activities will be at Level 5, other activities will be at Level 3, and so on. The client needs to know which activities are which. The client is assisted to develop a rational self-perception through successful experiences with activities lower on the hierarchy. By developing more rational self-perception in relation to these activities, the client develops the ability to negotiate with his or her symptoms and is also encouraged to attempt activities that are higher on the hierarchy.

Christiansen argues that competence is based on environmental mastery that is achieved through occupational activity, occupation being defined broadly as "purposeful involvement."[14, p.15] Kielhofner describes a model of human occupation that is a continuum of occupational function. Function ranges across the following spectrum:

1. Achievement—Role performance is above the expected standards.
2. Competence—Performance is adequate to the demands of the situation.
3. Exploration—Behavior that promotes discovery of one's individual characteristics and potentials.
4. Inefficacy—Performance does not meet standards that lead to personal satisfaction.
5. Incompetence—One's performance is consistently inadequate.
6. Helplessness—One is totally unable to act on one's environment.[15]

The disabled person's ability to respond to a traumatic injury that is severe enough to cause disability is dependent on several factors. These include:

- *Preinjury psychological adjustment*—Research has shown that individuals who become chronically disabled due to pain are more likely to have experienced problems with emotional adjustment and/or personality disorders earlier in life than the normal population. Polatin et al. studied men and women who suffer from chronic low back pain. On the basis of formal diagnostic criteria, 57 percent of the patients had an Axis I clinical disorder *before* the onset of their back injury. Depression was present in 39 percent of the patients, substance use disorder in 34 percent, and anxiety disorder in 14 percent on a preinjury basis.[16]

- *Perception of severity of the injury*—The person's perception of the severity of the injury is, of course, somewhat dependent on the actual severity in terms of tissue damage, pain, disfigurement, and the "drama" of the precipitating incident itself. However, perception of the severity of the injury is also dependent on the manner in which the injury is handled by emergency personnel and acute caregivers. Additionally, the person's perception of severity of the injury is dependent on his or her level of sophistication, knowledge, and expectations about the future impact of the injury on aspects of the person's life that are valued.
- *Perception of the severity of functional limitations*—The person's perception of functional limitations will affect his or her adjustment to the injury by the expectation of how those functional limitations will affect maintenance of social, familial, and vocational roles. The importance of early intervention to set these role expectations has been reported by Matheson et al.[17]
- *Acceptance of guidance from the treatment staff*—The disabled person's ability to effectively utilize information and guidance from the treatment staff affects his or her ability to respond healthfully to the injury on both a primary and a secondary basis. On a primary basis, staff input that is acceptable can be followed more closely and is thus more likely to be of actual benefit to the person. On a secondary basis, staff input that is acceptable is useful in developing expectations and goals for the future. Emotional adjustment to the injury on an intermediate-term basis will be intimately related to the degree to which the person's expectations for resumption of usual and customary roles are able to be approximated.
- *Messages provided by treatment staff*—The treatment staff's values concerning the balance between symptom relief and functional restoration will be communicated to the client with the first contact. If the message is that "pain is the primary limiting factor to which we will pay attention" or that "you can resume activity as soon as this pain is under control," the client who has chronic activity-related pain will often become trapped in an inactive pain-avoidant role. In spite of later treatment efforts of the caregivers to encourage the client to "push through the pain" to reestablish activity levels, the earlier-developed values concerning pain avoidance will restrict the client's efforts.
- *Family and social support*—The degree to which the person's family and close social network support gradual resumption of usual and customary roles affects his or her healthy emotional response to the injury on both a

primary and secondary basis. On a primary basis, the person is able to participate in more activities sooner as a consequence of this support. Thus the resumption of usual and customary roles is more rapid. On a secondary basis, the person's morale is more easily maintained, resulting in a positive effect on caregivers and family members that redounds to the client's benefit. O'Leary describes research by Taylor et al. in which the spouse's perception of the patient's efficacy was also considered.[18] She concludes:

> This study . . . points to the importance for health care workers of examining relevant factors in the patient's environment when planning rehabilitation programs. The finding of enhanced recovery in those patients who together with their husbands were able to experience . . . their physical efficacy may be extended to other patients who suffer from acute or chronic disability or trauma.[19]

- *Duration of medical care*—The length of care following a serious injury appears to be inversely related to the success of the person's healthy adaptation to the injury. That is, the longer the duration of care, the less likely it is that the person will be successful in adapting. In addition, Matheson et al. reported a strong inverse relationship between self-efficacy and length of time since injury among individuals entering a rehabilitation program. The greater the length of time since injury, the lower the ratings of self-efficacy. Furthermore, the greater the length of time since injury, the less likely it would be that significant improvement would be experienced by the person as a consequence of involvement in a functional capacity evaluation.[20]

Problems with self-efficacy among disabled populations have been recognized for many years. Parsons first defined the "sick role" as one that is conferred on the person who is ill and is actively involved in treatment. While the patient is in the sick role, he or she is allowed to temporarily escape from other role responsibilities.[21] Mechanic introduced the concept of "illness behavior" as an idiosyncratic response to symptoms given an individual's unique make-up and personality. This is a category of behaviors that are found by people in the sick role.[22] Mechanic reports that the purpose of illness behavior is "to make an unstable and challenging situation more manageable."[23, p.189]

Pilowsky defines abnormal illness behavior as that which continues to dominate other social roles in spite of information provided to the patient concerning his or her disability indicating that active treatment has concluded and that a return to modified or alternative social roles is now appropriate.[24] The behaviors that comprise the patient role appear to be learned within the context of a social support system.[25] The familial context for such a role has been supported by research by Moss and by Rickarby et al.[26] That such behavior is effective has been reported by Waddell et al.[27]

Matheson defines "symptom magnification syndrome" as a self-destructive pattern of behavior that is learned and maintained through social reinforcement. This pattern of behavior is composed of reports and/or displays of symptoms, the effect of which is to control the life circumstances of the sufferer. A substantial minority of individuals who are disabled by activity-related pain suffer from the symptom magnification syndrome.[28] In an earlier study of patients sponsored by workers' compensation who had been off work for an average of 2.1 years at entry into a rehabilitation program, Matheson reported that 24 percent were found to meet formal classification criteria for the symptom magnification syndrome.[29]

Problems with self-efficacy as a consequence of chronic pain also lead to psychopathology. Among individuals who suffer from chronic pain, the incidence of major depression ranges from approximately one-third to more than one-half.[30] Polatin et al. found that 97 percent of men and women who suffer from chronic low back pain also received a diagnosis of somatoform pain disorder. In addition, 64 percent fulfilled criteria for major depression, 36 percent for psychoactive substance use disorders, and 19 percent for anxiety disorders. All of the somatoform pain disorders were diagnosed after the injury; none were present prior to the injury. Twenty-nine percent of the patients developed depression after the injury, with a modest preponderance of these being females.[31] Turk and Salovey report that depression in patients with chronic pain is a consequence of the chronic pain and the inabilities and lack of confidence in dealing with the pain. In this situation, the absence of competence can be thought of as leading to a loss of self-efficacy in even the role of patient that may naturally lead to the onset of depression.[32] Feuerstein et al. report that "many patients with chronic lumbar pain display a set of clinical features including affective pain experience, increased subjective distress, preoccupation with somatic concerns, low self-esteem, and disease conviction.[33, p.267] It appears that psychological disturbance is an important part of a chronic painful disability. Only when this is recognized by caregivers can adequate treatment be provided.

MEASUREMENT OF SELF-EFFICACY

Several excellent reviews of instruments that are used to evaluate the psychological and behavioral consequences of chronic pain and disability have been performed.[34] Feuerstein et al. recommend that psychometric devices be developed to address the cognitive-perceptual factors that may be related to chronic low back pain.[35] Millard reviews 14 questionnaires that have been used in the field of industrial rehabilitation to assess pain-related disability. These questionnaires are grouped in terms of measures of disability of back pain, disability of pain without reference to site, and disability of illness without reference to pain. The psychometric properties of each instrument are reviewed with an emphasis on reliability and validity studies for each instrument. Reliability is generally quite good and has been well studied. In contrast, validity has not been well studied and, when studied, yields inconsistent results. Most notable is the absence of concurrent validity studies in which questionnaires have been studied at the same time and results compared. A uniform method to assess disability is recommended as one means to compare findings from the different devices. Millard concludes that many excellent questionnaires are available, obviating the need to derive additional questionnaires in that "the existing instruments sample a wide universe of content and can be adapted to virtually all conceivable settings."[36, p.300]

Of all the instruments reviewed by these authors, none measured self-efficacy. This may reflect the recency of development of this concept, but it probably also reflects the difficulty of its measurement. O'Leary reports that self-efficacy should be considered in terms of level, strength, and generality. She defines level of self-efficacy in terms of the person's expected performance attainment. Strength is defined in terms of the expressed confidence that a person has that he or she can attain an expected level. Generality is defined in terms of the number of domains of function in which people judge themselves to be efficacious.[37]

Measurement of self-efficacy is also complicated by the fact that self-efficacy affects the person's perception of himself or herself and thereby affects his or her self-report. As a consequence, optimal measures of self-efficacy are always relative. That is, self-efficacy is best measured when the person is used as his or her own yardstick. Normative measures of self-efficacy can be somewhat useful as long as the person is a member of the reference group. Measurement of self-efficacy can be accomplished by a structured interview, structured questionnaire, or card sort that is designed to tap those items that may be pertinent to the person. Such methods are useful as long as it is possible to exclude items that are not relevant and to be reasonably assured that all possibly relevant items have been included. Because information concerning self-efficacy is based upon elucidation of patterns of performance, a substantial amount of information must be collected. To the degree that responses can be made simply, they can be easily gathered and will be reliable.

Lorig et al. report on a 20-item scale that they developed to measure self-efficacy in people with arthritis. Each item is rated by the evaluee in terms of current level of efficacy for the particular area of competence along a 90-point scale from "very uncertain" to "very certain." The instrument comprises three subscales that concern self-efficacy for physical function, for controlling arthritis symptoms, and for pain management. Of the three scales, physical function was most highly related to disabled health status. Additionally, physical function items demonstrated significant concurrent validity in a blind study. These researchers also reported a high association between increases in perceived physical self-efficacy and diminution of reported pain.[38]

Gallon performed a retrospective telephone study on patients four to six years after each had been evaluated by a multidisciplinary pain rehabilitation program. Perception of disability was measured by an 18-item scale in which the subject was asked to indicate whether his or her status had "increased, decreased, or remained the same" for each item. Nine of the 18 items involved various levels of physical activity. An additional three items involved time in resting, napping, and general activities outside the home. The validity of the scale was demonstrated through strong statistical relationships between perceived disability as measured by the instrument and variables such as not working, continuing to receive disability payments, utilizing narcotic or sedative medication, and having received back surgery.[39]

Osuji reported on an "acceptance of loss scale" with a general sample of disabled adults in an industrial rehabilitation center. This research demonstrated that individuals who prematurely terminated rehabilitation programs were significantly less accepting of loss and had worse punctuality and attendance without regard to severity of disability.[40]

Armentrout studied the impact of chronic pain and hospitalization on self-concept. This study utilized the Tennessee Self-Concept Scale. A physical health scale and an "Identity Self" scale appeared to be closely related and were the most significantly affected by the experience of hospitalization.[41]

The Sickness Impact Profile has been carefully studied.[42] Deyo et al. found that this instrument was not sensitive to changes in the status of rheumatoid arthritis within individual patients.[43] Deyo and Diehl found that the Sickness Impact Profile was not able to distinguish between low back pain patients who had no change in their status over a course of treatment from those who had improved, but that it was able to differentiate those who had deteriorated.[44]

Another approach to the measurement of functional self-efficacy bases measurement on its cognitive aspects, reflected by a set of self-perceptions that is cohesive. Perception of function as it may affect ability to work has been assessed through the RISC Tool Sort, a deck of 212 cards, each of which shows a picture of a different tool.[45] The evaluee sorts the cards into different categories, depending on perceived ability to use the tool. A version of this was developed and made commercially available as the WEST Tool Sort.[46] More recently, the Loma Linda University Activity Sort[47] and the Spinal Function Sort[48] have become available. The latter is a gender-balanced test that assesses the evaluee's perception of functional capacity for tasks in generic manual-material-handling activities across a broad range of physical demands that progress from light to heavy in terms of the U.S. Dept. of Labor Physical Demand Characteristics of Work system, as in Table 28–1.

Table 28–1 Physical Demand Characteristics of Work Chart

Physical Demand Level	*Occasional (0–33% of the workday)*	*Frequent (34–66% of the workday)*	*Constant (67–100% of the workday)*	*Typical Energy Required*
Sedentary	10 lbs	Negligible	Negligible	1.5–2.1 METS
Light	20 lbs	10 lbs and/or Walk/Stand/ Push/Pull of Arm/Leg controls	Negligible and/or Push/Pull of Arm/Leg controls while seated	2.2–3.5 METS
Medium	20 to 50 lbs	10 to 25 lbs	10 lbs	3.6–6.3 METS
Heavy	50 to 100 lbs	25 to 50 lbs	10 to 20 lbs	6.4–7.5 METS
Very Heavy	Over 100 lbs	Over 5 lbs	Over 20 lbs	Over 7.5 METS

Source: Copyright © 1993 Leonard N. Matheson, PhD.

The Spinal Function Sort (SFS) is presented in the form of a booklet containing 50 items. Each item is composed of a drawing of a person performing a task and a simple written task description. A sample drawing is presented in Figure 28–1, and a list of selected task descriptions is presented in Table 28–2.

The evaluee is instructed to "look at each drawing and read the description. On the separate answer sheet, indicate your current level of ability to perform the task." The answer sheet provides a five-point rating from "Able" to "Restricted" to "Unable." There is also a sixth rating that is depicted as "?" and indicates "I don't know." The purpose of the SFS is to quantify the disabled person's perception of his or her ability to perform work tasks.

A large multisite study of disabled males and females demonstrated a relationship between time since injury and both rating of perceived capacity and the degree to which this rating was able to be improved with short-term intervention. This study found that the longer the time since injury, the lower the score, which reflected the degree to which chronicity was likely to be coupled with lower levels of self-efficacy. Perhaps more importantly, this study ($N = 172$) also found that the earlier clients receive service, the more likely that the SFS score could be improved by functional capacity evaluation services.[49] This relationship is depicted in Figure 28–2.

A similar tool, the Activity Card Sort (ACS), has been developed by Baum.[50] The ACS is a Q-sort of 61 activities typically performed by older adults. With the ACS it is possible to calculate the person's previous level of engagement in activities and to compare it with the current level. This method does not place value on the difficulty or frequency of the activity, nor does it discriminate whether the

Sweep with a Kitchen Broom.

Figure 28–1 Sample Spinal Function Sort Drawing. *Source:* Copyright © 1989 Performance Assessment and Capacity Testing, Mission Viejo, California.

Table 28–2 Spinal Function Sort Sample Task Descriptions

Item	*Description*
2.	Retrieve a small tool from the floor.
4.	Push and pull a shopping cart.
7.	Lower a 10-pound milk crate from eye level to the floor.
20.	Trim shrubbery with a hedge trimmer.
21.	Change a light bulb overhead.
32.	Get out of an automobile driver's seat.
39.	Sweep with a push broom.
44.	Lift a 50-pound tool box from the floor to a bench.
47.	Lift a 100-pound milk crate from the floor to eye level.

Source: Reprinted from *Spinal Function Sort* by L.N. Matheson and M. Matheson with permission of Performance Assessment and Capacity Testing, Mission Viejo, California, © 1989.

person performs the activity independently or with another person's help. The ACS score reflects the proportion of activities that have been given up after the injury.

As self-efficacy matures as a research topic, additional measurement devices will be developed. Measurement of self-efficacy is the foundation of its clinical development.

TREATMENT TO DEVELOP SELF-EFFICACY IN REHABILITATION

Biopsychosocial Model

Waddell describes a "biopsychosocial model" for treatment of disability that differentiates true impairment from the patient's perception of functional limitation. Waddell presents disability as a construct that is based on the person's attitudes and beliefs. In Waddell's model, disability is learned rather than a natural and inevitable consequence of serious impairment. The incidence of impairment becoming disability is directly related to how issues such as pain and self-efficacy are handled by the treatment staff. The treatment staff reinforces certain attitudes, beliefs, and behaviors about the relationship between impairment and pain and disability. The degree to which this is done carefully is related to the degree to which the client will develop a rational self-perception and optimal self-efficacy. The degree to which self-efficacy is managed in a work injury management program will directly relate to the program's success.[51]

Self-efficacy is an especially important determinant of the disabled person's ability to benefit from rehabilitation. Christiansen reports that "the extent to which individuals are able to develop a positive sense of self and belief in their autonomy is largely based on their successes in dealing with environmental challenges."[52]

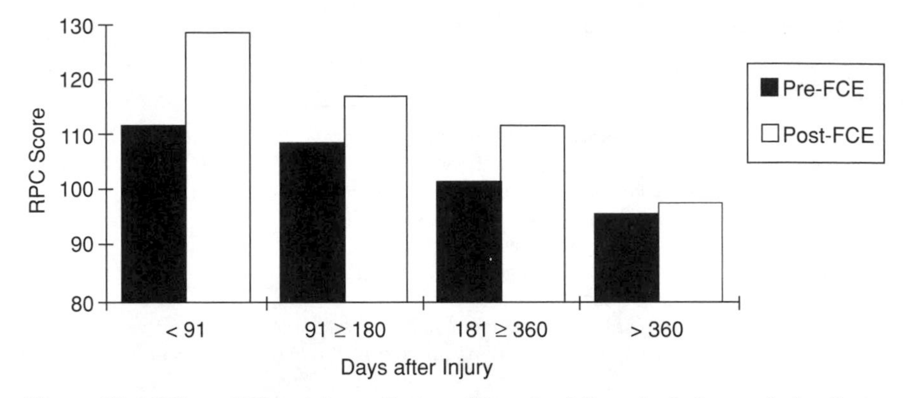

Figure 28–2 Effect of Chronicity on Rating of Perceived Capacity before and after Functional Capacity Evaluation Services

Smith argues that a sense of self-competence and hopefulness about one's abilities provides the basis for an orientation to life that involves the person in a "benign spiral of increasing competence and fulfillment" and that helplessness and hopelessness are "at once the root problem requiring therapeutic assistance and the chief obstacle to its solution."[53, p.13]

The biopsychosocial model is an important viewpoint that is shared by increasing numbers of rehabilitation professionals. It presents a serious challenge to additional biomedical approaches to rehabilitation. The biopsychosocial model is certainly not without its detractors, but Ogden-Niemeyer and Jacobs predict that it will be adopted across a broad spectrum in the future. They anticipate that this will occur in response to continuing shortfalls in the current biomedical approach to treatment that have led to increases in the cost of workers' compensation without attendant improvements in outcome.[54] Among Isernhagen's "global concepts" to guide therapists in strategic planning is the statement that "each worker's perception of functional capacity for work will interface with actual physical capacity."[55]

Self-Efficacy Training

Bandura's theory of self-efficacy also postulates that self-perception is a mechanism through which the person can exercise influence over his or her own behavior.[56] Bandura reports that "among the mechanisms of personal agency, none is more central or pervasive than people's belief about their capabilities to exercise control over events that affect their lives. Self-efficacy beliefs function as an important set of proximal determinants of human motivation, affect, and action."[57] The stronger people's perceived self-efficacy, the higher the goals set for themselves, and the firmer their commitment to them. Self-efficacy training recognizes that self-perception is the basis of self-efficacy and that self-perception can be shaped by appropriate experience. Self-efficacy training takes advantage of the person's urge towards competence and increases the efficiency with which the person moves towards competence. This results in improved motivation in terms of both level of effort expended by the client and the client's focus.

Self-efficacy can be effectively developed in industrial rehabilitation programs that combine behavioral and cognitive psychological treatment with work conditioning, work hardening, or functional restoration. The level of treatment depends on a combination of the degree to which the individual has psychologically decompensated subsequent to the injury, the degree to which the deconditioning syndrome has developed, and the degree to which the medical aspects of the case are well understood.

A sports medicine work-conditioning approach is useful for individuals who require education, information, and physical restoration. A work-hardening approach can be undertaken on a secondary basis for those individuals who require physical reactivation, vocational planning, and behavior modification. Functional restoration can be undertaken on a tertiary basis for those individuals who require physical reactivation, psychological adaptation, vocational planning, and behavior modification.

Saunders and Anderson, Andersson and Frymoyer, and Saal and Saal describe early treatment intervention for painful soft tissue injuries that focuses on education of the patient and the development of the patient's responsibility for symptom control.[58]

Darphin et al. report on the use of a sports medicine model to enhance motivation in an industrial rehabilitation program. They draw parallels between the rehabilitation process for an injured worker and the sports competition experience for the athlete. They describe a social reinforcement system to develop "goal-oriented work behavior."[59]

Mayer et al. report on the success of a functional restoration program that enhances the self-efficacy of individuals who suffer from chronic low back pain. Results from this program indicate that 87 percent of the graduates were working two years post discharge compared with 41 percent of a nontreatment comparison group.[60] Similar results are reported by Hazard et al.[61] In these programs, patients receive aggressive physical restoration paced by frequent functional testing to provide both the therapist and patient with information concerning current level of function. As part of the physical restoration program, patients perform work-simulation activities that are graded and progressive and that provide patients with specific information concerning ability to handle their target jobs. Concerns about return to work are addressed through psychotherapeutic intervention, which is short-term and involves stress management, family counseling, and a cognitive-behavioral approach to individual counseling.

Self-Efficacy Training Strategies

The basis for development of self-efficacy in a treatment program is that the treatment program holds as its central value respect for the client and the development of his or her potential. With this value at its core, the program will naturally place expectations on the client that will be both realistic and growth enhancing. Within the context of this value system, several self-efficacy training techniques can be applied.

Progressive Functional Challenge

Self-efficacy training requires that the client be willing to face a meaningful challenge within a controlled clinical environment and have a means to measure his or her performance in response to this challenge. Subsequently, the performance must be interpreted accurately by the evaluee. In the self-efficacy training environment, all consequences are successful as long as they are clear and well understood. Failure can only be a consequence of tasks that are left unfinished.

Evaluation is an important part of the self-efficacy training process. The development of early and accurate expectations about the extent of the functional limitations and the impact of these functional limitations on the person's ability to work is of paramount importance. Adjustment to the injury begins within hours after the injury. Functional testing that can be performed soon after injury provides much-needed information to the patient that the patient, the patient's family, and the patient's employer will utilize to set appropriate expectations. If this information is not available or is incorrect, expectations will nevertheless be set given the perceptions of the patient, family, and employer. Experience has shown that misperception of expectations is common. Hart describes the importance of objective measurement of function in industrial rehabilitation. He cautions that "the exact physical impairment or maximal functional level cannot be determined accurately without knowledge of the level of patient motivation. . . . Consistency checks should be used to confirm the reliability and validity of all measurements.[62]

Frequent functional evaluation helps the caregiver to properly pace the level of progressive challenge. In addition, evaluation provides feedback that allows the person's urge towards competence to generate correction. Correction can be suggested or shaped by the therapist, who acts as a communicator and interpreter of feedback and as a guide in the exploration of new alternative forms of responses.

Functional Skills Training

Kirsch found that self-efficacy depends on whether the task under consideration is one that requires only skill or one that requires the subject to overcome an aversive stimulus such as fear.[63] When the rating of efficacy refers to a skill issue, such a rating cannot be altered by the offer of incentives. In contrast, when the efficacy rating refers to the subject's willingness to perform a task, such a rating is readily affected by incentives. Self-efficacy for nonaversive achievement tasks is not greatly affected by incentives. Improvements in self-efficacy require that the person's skills improve if the task is perceived by the person to be physically challenging or aversive. That is, the person will not develop a sense of self-efficacy for a task that he or she is able to perform unless the person is assisted to develop skills to perform this task in an improved manner. To the degree that skill development takes place, performance of the task will result in improved self-efficacy.

Development of task-related skills requires that the person participate in tasks under supervision. Instruction on task performance may be effective. Biofeedback to diminish the level of pain that is experienced as the task is undertaken may be useful. Modification of the task to decrease its ergonomic demands and thus the symptom response to the task may be useful. The degree to which the client is able to "own" the manner in which task performance is improved will be directly related to the degree to which he or she subsequently develops a sense of self-efficacy concerning the task.

Work Function Theme Development

The person's perception of function can be thought of in terms of "work function themes," the rules that guide participation in work activities. These rules are learned throughout life and are constantly modified, based on experience. Information that is gained in unique circumstances is generalized to other circumstances. As a new situation is approached, work function themes guide the client's participation. For example, the new warehouse person who approaches a refrigerator carton with the responsibility to move the carton from the loading area and place it on a truck may not have had previous experience with such a task. However, this person's experience with other similar tasks is brought to bear. If the person believes that this cannot be handled independently, alternate action is taken. *The individual's response to limit involvement in tasks that cannot be undertaken safely is based on his or her own commitment to avoiding reinjury within the context of his or her personal work function themes that are based on prior experience.*

The problem encountered by the person who has chronic activity-related pain can be interpreted in terms of the work function themes that have been developed by this person. Take the example of the warehouse person who has many years of experience in lifting and carrying heavy loads. If this person experiences a low back soft tissue injury that is traumatic and quite painful, the experience itself will tend to modify the person's work function themes in a conservative direction. To the degree that the experience was painful, frightening, or generally aversive, the limitation of the work function theme will be irrational given the actual functional demands of the task. The ecological purpose of this limitation is to prevent the person from experiencing a reinjury. Unfortunately, because the chronically pain-disabled person has only the aversive experience of the event as a guide, the self-limitation is often more pronounced than is appropriate. In order to develop rational work function themes, the chronically pain-disabled person must perform progressively demanding tasks successfully, experiencing pain that can be maintained within limits that he or she understands do not connote reinjury.

Goaling Process

The Goaling Process is an important part of self-efficacy training. Goaling helps the client to establish a future orientation, develop a rational basis for planning, and receive positive feedback from his or her community. A goal is a distinct, complete, and clear communication about one issue that makes life more satisfying. The goaling process is important to the client for several reasons. Perhaps the most significant value to the client is that it assists the client to communicate clearly with the program staff those aspects of life that are important to him or her. Given this information, the client and program staff can work with greater coherence to move towards the client's goals.

As the client participates in the goaling process, it will become clear that he or she has numerous goals, some of which are big and some of which are small. It is important to recognize that all of these goals have meaning to the client and that progress towards each will be valued by the client. This will increase the value of the program for the client and will increase the client's self-efficacy. Self-efficacy will generalize from one small goal achievement to have a positive effect on the client's ongoing attempts to achieve other goals.

The goaling process has four steps that can be accomplished quite easily by any rehabilitation team member. The rehabilitation professional may find that the process will feel unnatural to clients. As a consequence, the client may require a considerable degree of encouragement to complete the process. Each of the steps must be completed in this order:

1. *Structured interview*—The caregiver and client meet in a quiet room. The caregiver asks the client, "What do you want most out of a job?" The caregiver records the client's responses and works with the client to develop individual goal statements that meet the following criteria:
   a. Each goal is presented as a complete but simple sentence.
   b. Each goal is listed in the present or future tense.
   c. Each goal is easily understood and unambiguous.
   d. Each goal is stated in a positive manner. Negative statements are not allowed. Goals such as "I don't want to make less than $8.00 an hour," or "I don't want to live in the city" are not allowed. They should be restated

in the affirmative. In this example, acceptable goals might be, "I want to earn at least $8.00 an hour," or "I want to live in a rural setting."

2. *Developing priorities*—Each of the individual goals is listed as it is developed by the caregiver and the client. After 12 to 15 goals have been identified, the list is presented to the client and the client is asked to identify which of the goals is least important. In fact, all of the goals have importance, but there is an inherent priority that will not be readily apparent to the client. After the client selects the goal that is least important, the caregiver continues with the client to identify which of the remaining goals is least important. This "negative prioritization" is quite useful and will result in some surprises. At some point during this process, the client may have difficulty in deciding and may state, "All of these are most important." The caregiver should persevere and encourage the client to identify which of the remaining goals is least important, confirming that all of the goals have importance, but that one of the remaining goals is less important than the others.
3. *Significant other review*—After the goal list has been organized by priority, a copy is made. The original is provided to the client to take home and review with at least one significant other. The client is encouraged to make any changes that may be appropriate in the goals, including deleting or adding goals. The order of the goals may be changed. Wording may be changed. The client is to return with the goal list after this review.
4. *Development and publication of the goaling document*—After the client returns with the goal list that has been reviewed, a formal list is prepared. An example of a formal goal list appears in Exhibit 28–1. Copies of the goal list are made. The client is provided with the copies and is instructed to distribute the copies to as many people as he can, but no less than a certain number selected by the caregiver.

This is a very challenging experience for the client. It will involve some level of risk and concern about embarrassment. It is also potentially very rewarding. The client is encouraged to provide a goal list to other treating professionals, important family members, friends, and former colleagues or coworkers. One copy should be placed prominently in the client's home, on the refrigerator door or medicine cabinet. The caregiver should structure this step so that it is actually completed. The tendency of many clients will be to brush this aspect of the Goaling Process aside and not follow through. Experience with several thousand clients has shown that this step is crucial. It also sets the tone for the relationship between the caregiver and the client. The Goaling Process has taken place early in the therapeutic relationship. The client has revealed much of what is important about himself to the caregiver. The degree to which the caregiver takes the client's goals

Exhibit 28–1 Sample Goal List

Goal List
John Smith
May 12, 1993

1. To have health insurance for my family.
2. To have a safe job.
3. To have stable employment over the years.
4. To earn $1,700 per month.
5. To let my wife work only one job.
6. To get my car fixed.
7. To have good school clothes for my son.
8. To be respected by my wife and family.
9. To save something each month for retirement.
10. To visit family in Texas at least one time each year.
11. To be respected by my coworkers.
12. To be a coach, assistant coach, or umpire in Little League again.

seriously and respects the client's goals models the behavior that the caregiver seeks from the client. That is, to the degree that the client can appreciate and respect his or her own goals, they will be an important part of the rehabilitation process. If the client does not respect his or her own goals, the stage will be set for an absence or a minimization of respect of the goals or the program and the caregiver.

Clients respond to the Goaling Process in a manner that reflects their life situations. Teenagers respond differently from adults, who respond differently from older adults, who respond differently from people who are experiencing terminal illnesses. For the person who has a work injury and is not of retirement age, the responses to the goaling process will center on work because work is the primary vehicle by which the person's goals will be achieved.

Symptom Negotiation Training

Experience has shown that clients who are chronically disabled by pain typically do not effectively negotiate with their symptoms. They perceive their symptoms to be outside of their control or minimally controlled by them. This is a major threat to the client's self-efficacy. Failure in predicting symptoms leads to failure in controlling symptoms, which leads to a sense that the person is unable to control himself or herself and the environment. This leads to a general absence of self-efficacy that has important negative consequences for rehabilitation. Symptom negotiation training is a central part of the work injury management program for these individuals.

Symptom negotiation training is based on a simple idea: If symptoms can be predicted, they can be managed. In order to achieve this, the caregiver must set up situations in which the symptoms occur on a predictable basis and for which prediction and control of the symptoms can be accomplished by the client. An excellent example is the situation encountered frequently in cardiac rehabilitation programs for people who have stable disabling angina pain. The pain has a close relationship to activity because it is a signal that a segment of the heart is not getting sufficient oxygen in the activity level undertaken by the client. Rather than learning to avoid the activity in order to avoid the symptom, the client is taught to grade the level of pain and relate the pain level to the heart rate response at that point in time. Linking of the symptoms to the heart rate provides the client with the ability to predict the symptoms and thus to measure his or her participation in activity. In the normal course of events, the client will use this information to structure involvement in the activities to maintain a level of pain that is acceptable given the payoff from participating in the activity itself.

The situation with an individual who suffers from musculoskeletal pain is more complicated in that the symptom response is often delayed and not as clearly related to a particular activity or level of activity. However, the relationship between symptom and activity exists and can be identified in most cases. There are several strategies to help the client negotiate with symptoms. Some of the most useful are

1. *Progressive loading*—A common method of symptom negotiation training involves a task that presents progressive demands to the client on a graded basis. The starting point and gradient of increase is controlled by the caregiver in order to develop a clear pattern of the relationship between activity and symptom. An excellent example of such a task is a progressive lifting test. The evaluee may begin at 10 pounds of load to perform a lift over a restricted vertical range and gradually increase the load and/or the vertical range and/or the frequency under the caregiver's direction. The purpose of this gradual increase is to not only identify the client's lifting capacity but to develop a better understanding of the relationship between lifting tasks and the client's symptom response. Experience shows that the symptom response will not be idiosyncratic but will have a specific and dependable pattern. There are many other types of progressive demand tasks that can be used. The selection of the task depends on several factors. Early in the program, the task should relate to a symptom response that is relatively easily identified and controlled. This may not be the most important of the symptoms in a particular situation. However, success in developing negotiation strategies with a simple symptom-activity combination will generalize positively to more complex and difficult symptoms and activities.

2. *Work pacing*—A large majority of clients who are disabled by activity-related pain approach activity as an "on or off" experience. That is, the person works at one pace until he or she is no longer able to work and then stops. The caregiver can work with the client to slow the pace of work or, perhaps more effectively, interrupt the normal pace of work with "micro-breaks." Micro-breaks last 30 to 90 seconds and can be scheduled every 5 to 15 minutes. They are intended to interrupt the flow of the activity in order to allow the person to stretch, change posture, lower the heart rate or cardiovascular workload, or in other ways decrease the immediate demand that the task is placing on the client. A kitchen timer with an alarm can be used to signal the micro-break. A token reinforcement behavior modification system may be useful to encourage the client to utilize the signal and take a micro-break.
3. *Tool or job modification*—Many clients who are disabled by activity-related pain have not learned the value of working smart rather than working hard. They continue to perform the task in the same manner that they used prior to the injury or illness. While in some cases, this may have been the most efficient means of performing the task, other means are available that may be nearly as effective or, in some cases, more effective. The focus of these is to allow the client to return to a reasonable level of productivity through the completion of the task in a modified manner. Modification of hand tools, power work tools and stationary equipment, the workstation, or the job tasks themselves can be undertaken by the client with consultation from the caregiver. A complete listing of possible tool and job modifications is beyond the scope of this chapter. However, the general structure should follow in this hierarchy:

   a. Biomechanical alignment
      - Workstation design
      - Tool selection
      - Posture
      - Body mechanics

   b. Job task control
      - Self pace
      - Task load—pace
      - Task load—duration
      - Task load—force

   c. Work duration control

- Minimum task set duration
- Minimum work period duration
- Workday productivity preparation

This hierarchy is based on the employer's hierarchy in terms of the degree to which each of these modifications can be accomplished. Biomechanical alignment is usually most easily accomplished while work duration control is least easily accomplished. Within the context of biomechanical alignment, posture and body mechanics are more easily achieved than tool selection, and tool selection is more easily achieved than workstation design.

Self-efficacy training is an effective strategy for improving the likelihood that rehabilitation will be successful with individuals who are chronically disabled due to motivational problems. However, even with individuals who are only temporarily disabled and do not evidence motivational problems, self-efficacy training can be useful and will become a valued part of the work injury management professional's repertoire. The value of this approach is that it recognizes the importance of self-efficacy in the development of motivation and recognizes the importance of motivation in the success of rehabilitation.

CONCLUSION

Motivation is an important component of every rehabilitation program. The degree to which the health care professional is able to manage and develop the client's motivation will be directly related to the success of the program. It will also be related to the satisfaction with the program experienced by both the client and by the professional.

One approach to developing motivation is provided through self-efficacy training. Self-efficacy training is based on the idea that the individual develops a sense of competence by effectively interacting with the environment to successfully perform tasks that are meaningful to him or her. This can be accomplished in spite of problems such as activity-related pain because humans possess an "urge towards competence" that can be developed and focused. The professional's appreciation of this important need sets the stage for harnessing it for rehabilitative purposes.

Work injury management programs are uniquely positioned to assist disabled individuals to develop self-efficacy and, as a consequence, both develop and focus the disabled person's motivation. Through appropriate challenges, with support and guidance from the caregiver, the disabled person can develop a perspective of himself or herself as a person who is competent and who, in spite of significant functional limitations, is able to interact effectively with the world and return to an optimum level of productivity.

NOTES

1. R.W. White, Motivation Reconsidered: The Concept of Competence, *Psychological Review* 66 (1959):197–333.
2. R.W. White, The Urge toward Competence; *American Journal of Occupational Therapy* 25 (1971):271–274.
3. A. Bandura, Self-Efficacy: Toward a Unifying Theory of Behavioral Change, *Psychological Review* 84 (1977):191–215.
4. C. Christiansen, Occupational Therapy: Intervention for Life Performance, in *Occupational Therapy: Overcoming Human Performance Deficits,* ed. C. Christiansen and C. Baum (Thorofare, N.J.: SLACK Inc., 1991); chap. 1.
5. A. Bandura, Self-Efficacy Mechanism in Human Agency, *American Psychologist* 7, no. 2 (1982):122–147.
6. Christiansen, Occupational Therapy.
7. S.M. Shoor and H.R. Holman, Development of an Instrument To Explore Psychological Mediators of Outcome in Chronic Arthritis Patients, *Transactions of the Association of American Physicians* 97 (1985):325–331.
8. C.K. Ewart et al., Effects of Early Postmyocardial Infarction Exercise Testing on Self-Perception and Subsequent Physical Activity, *American Journal of Cardiology* 51 (1984):1076–1080.
9. J.R. Council et al., Expectancies and Functional Impairment in Chronic Low Back Pain, *Pain* 33 (1988):323–331.
10. J.J. Dolce et al., The Role of Self-Efficacy Expectancies in the Prediction of Pain Tolerance, *Pain* 27 (1986):261–272.
11. I. Ajzen and M. Fishbein, *Understanding Attitudes and Predicting Social Behavior* (Englewood Cliffs, N.J.: Prentice Hall, Inc., 1980).
12. S.L. Baker and I. Kirsch, Cognitive Mediators of Pain Perception and Tolerance, *Journal of Personality and Social Psychology* 61, no. 3 (1991):504–510.
13. L. Kirsch, Efficacy Expectations or Response Predictions: The Meaning of Efficacy Ratings As a Function of Task Characteristics, *Journal of Personality and Social Psychology* 42, no. 1 (1982):132–136.
14. Christiansen, Occupational Therapy.
15. G. Kielhofner, *A Model of Human Occupation: Theory and Application* (Baltimore: Williams & Wilkins, 1985).
16. P.B. Polatin et al., Psychiatric Illness and Chronic Low-Back Pain. The Mind and the Spine—Which Goes First? *Spine* 18, no. 1 (1993):66–71.
17. L.N. Matheson et al., The Interdisciplinary Team in Cardiac Rehabilitation, *Rehabilitation Literature* 36, no. 12(1975):366–376.
18. A. O'Leary, Self-Efficacy and Health, *Behavior Research and Therapy* 23 (1985):437–451.
19. Ibid., p. 447.
20. L.N. Matheson et al., Development of a Measure of Perceived Functional Ability, *Journal of Occupational Rehabilitation* 3, no. 1 (1993).
21. T. Parsons, *The Social System* (New York: Free Press of Glencoe, 1951), chap. 10.
22. D. Mechanic, Response Factors in Illness: The Study of Illness Behavior, *Social Psychiatry* 1 (1975):11–20.

23. D. Mechanic, The Concept of Illness Behaviour, *Journal of Chronic Diseases* 15 (1982):189–194.
24. I. Pilowsky, Abnormal Illness Behavior, *British Journal of Medical Psychology* 2 (1969):347–351; I. Pilowsky, A General Classification of Abnormal Illness Behavior, *British Journal of Medical Psychology* 51 (1978):131–137; I. Pilowsky, Abnormal Illness Behaviour, *Psychiatric Medicine* 5, no. 2 (1987):85–91.
25. K.M. Gil et al., Social Support and Pain Behavior, *Pain* 29, no. 2 (1987):209–217.
26. R.A. Moss, The Role of Learning History in Current Sick-Role Behavior and Assertion, *Behavior Research and Therapy* 24, no. 6 (1986):681–683; G. Rickarby et al., Abnormal Illness Behavior As a Required Family Role, *Psychiatric Medicine* 5, no. 2 (1987):115–122.
27. G. Waddell et al., Chronic Low-Back Pain, Psychologic Distress, and Illness Behavior, *Spine* 9, no. 2 (1984):209–213; G. Waddell et al., A Concept of Illness Tested As an Improved Basis for Surgical Decisions in Low-Back Disorders, *Spine* 11, no. 7 (1986):712–719.
28. L.N. Matheson, Symptom Magnification Syndrome Structured Interview: Rationale and Procedure, *Journal of Occupational Rehabilitation* 1, no. 1 (1991):43–56.
29. L.N. Matheson, Symptom Magnification Syndrome: A Modern Tragedy and Its Treatment. Part One: Description and Definition, *Industrial Rehabilitation Quarterly* 3, no. 3 (1990):1, 5, 8, 12, 23.
30. D. Fishbain et al., Male and Female Chronic Pain Patients Categorized by DSM-III Psychiatric Diagnostic Criteria, *Pain* 26 (1986):181–197; W. Katon et al., Lifetime Psychiatric Diagnoses and Family History, *American Journal of Psychiatry* 42 (1985):1156–1160; K. Krishnan et al., Chronic Pain and Depression. I: Classification of Depression in Chronic Low Back Pain Patients, *Pain* 22 (1985):279–287.
31. Polatin et al., Psychiatric Illness.
32. D.C. Turk and P. Salovey, Chronic Pain As a Variant of Depressive Disease: A Critical Reappraisal, *The Journal of Nervous and Mental Disease* 172 (1984):398–404.
33. M. Feuerstein et al., Biobehavioral Mechanisms of Chronic Low Back Pain, *Clinical Psychology Review* 7 (1987):243–273.
34. Ibid., R.W. Millard, A Critical Review of Questionnaires for Assessing Pain-Related Disability, *Journal of Occupational Rehabilitation* 1, no. 4 (1991):289–300.
35. M. Feuerstein et al., Biobehavioral Mechanisms.
36. Millard, Critical Review.
37. O'Leary, Self-Efficacy.
38. K. Lorig et al., Development and Evaluation of a Scale To Measure Perceived Self-Efficacy in People with Arthritis, *Arthritis and Rheumatism* 32, no. 1 (1989):37–44.
39. R.L. Gallon, Perception of Disability in Chronic Back Pain Patients: A Long-Term Follow-Up [see comments], *Pain* 37, no. 1 (1989):67–75.
40. O.N. Osuji, "Acceptance of Loss" and Industrial Rehabilitation: An Empirical Study, *International Journal of Rehabilitation Research* 10, no. 1 (1987):21–27.
41. D.P. Armentrout, The Impact of Chronic Pain on the Self-Concept, *Journal of Clinical Psychology* 35, no. 3 (1979):517–521.
42. Bergner et al., The Sickness Impact Profile: Development and Final Version of a Health Status Measure, *Medical Care* 19 (1981):787–805.
43. R.A. Deyo et al., Measuring Functional Outcomes in Chronic Disease: A Comparison of Traditional Scales and a Self-Administered Health Status Questionnaire in Patients with Rheumatoid Arthritis, *Medical Care* 21 (1983):180–192.

44. R.A. Deyo and A.K. Diehl, Measuring Physical and Psychosocial Function in Patients with Low-Back Pain, *Spine* 8, no. 6 (1983):635–642.
45. L.N. Matheson, *RISC Tool Sort* (Woodland Hills, Calif.: Rehabilitation Institute of Southern California, 1979).
46. L.N. Matheson, *WEST Tool Sort* (Signal Hill, Calif.: Work Evaluation Systems Technology, 1981).
47. D. Anzai, *Loma Linda University Activity Sort* (Long Beach, Calif.: Work Evaluation Systems Technology, 1984).
48. L.N. Matheson and M. Matheson, *Spinal Function Sort* (Mission Viejo, Calif.: Performance Assessment and Capacity Testing, 1989).
49. Matheson et al., Development of a Measure.
50. C.M. Baum, The Effect of Occupation on Senile Dementia of the Alzheimer's Type and Their Careers (Ph.D. diss., Washington University, 1993).
51. G. Waddell, A New Clinical Model for the Treatment of Low-Back Pain, *Spine* 12, no. 7 (1987):632–644.
52. Christiansen, Occupational Therapy, p. 19.
53. B. Smith, Competence and Adaptation: A Perspective on Therapeutic Ends and Means, *American Journal of Occupational Therapy* 28, no. 1 (1974):11–15.
54. L. Ogden-Niemeyer and K. Jacobs, Trends in Work Practice for the Twenty-First Century, *Orthopedic Physical Therapy Clinics of North America* 1, no. 1 (1992):149–154.
55. S.J. Isernhagen, Challenge for the Future, *Orthopedic Physical Therapy Clinics of North America* 1, no. 1 (1992):177.
56. A. Bandura, Human Agency in Social Cognitive Theory, *American Psychologist* 44 (1989):1175–1184.
57. Ibid., p. 1175.
58. R.L. Saunders and M.A. Anderson, Early Treatment Intervention, *Orthopedic Physical Therapy Clinics of North America* 1, no. 1 (1992):67–74; G.B.J. Andersson and J.M. Frymoyer, Treatment of the Acutely Injured Worker, in *Occupational Low Back Pain: Assessment, Treatment and Prevention,* ed. M.H. Pope (St. Louis: Mosby Year Book, 1991), 183–193; J.A. Saal and J.S. Saal, Nonoperative Treatment of Herniated Lumbar Intervertebral Disc with Radiculopathy: An Outcome Study, *Spine* 14 (1989):431.
59. L.E. Darphin et al., Work Conditioning and Work Hardening, *Orthopedic Physical Therapy Clinics of North America* 1, no. 1 (1992):107.
60. T.G. Mayer et al., Objective Assessment of Spine Function Following Industrial Injury: A Prospective Study with Comparison Group and One-Year Follow-up, *Spine* 10, no. 6 (1985):482–493; T.G. Mayer et al., A Prospective Two-Year Study of Functional Restoration in Industrial Low Back Injury: An Objective Assessment Procedure, *JAMA* 258 (1987):1763–1767.
61. R.G. Hazard et al., Functional Restoration with Behavioral Support: A One-Year Prospective Study of Patients with Chronic Low-Back Pain, *Spine* 14, no. 2 (1989):157–161.
62. D.L. Hart, Standards in Tests and Measurements, *Orthopedic Physical Therapy Clinics of North America* 1, no. 1 (1992):81.
63. Kirsch, Efficacy Expectations.

Chapter 29

Pain and Return to Work: Turning the Corner

Matthew Monsein and Robert B. Clift

Returning an injured worker back to the workplace either to his or her original position or to an alternative position within appropriate restrictions is the ultimate goal of rehabilitation. This is particularly challenging when dealing with chronic pain patients (arbitrarily defined as individuals out of work with pain lasting more than six months). While representing a small percentage between 3 and 10 percent of all work injuries, these individuals account for an estimated 80 to 90 percent of the total dollar amount spent in the workers' compensation system. This dollar amount includes medical, administrative, legal, and particularly lost wage reimbursement.

Pain by definition is a subjective personal experience of discomfort known only to the individual. Because of its subjective nature, quantifying this experience remains highly problematic. For example, while imaging studies will demonstrate anatomic changes and EMGs will show electrophysiologic alterations, it is well known that there are a rather high percentage of both false positives and to a lesser degree false negatives with these studies. None of the current diagnostic tests actually measure pain.

Therefore in evaluating pain, clinicians are dependent on both the subjective reports of the patient—descriptions of location, quality, intensity, and duration of pain—and the outward manifestations of pain—pain behaviors such as guarding, weight shifting, limping, abnormal gait patterns, limited range of motion, and reliance on braces, canes, or other devices.

It is important to recognize that these types of behaviors are often normal physiological responses to a painful condition—for example, the guarding of an extremity associated with reflex sympathetic distrophy (RSD), or limping associated with osteoarthritis of the hip.

On the other hand, "excessive" pain behaviors such as continuous grimacing, inability to sit for more than several minutes in the absence of structural disease, or excessive doctor shopping suggest poor pain tolerance and the presence of significant emotional distress. Thus an individual's expression of pain is ultimately a manifestation of a complex process integrating primitive nociceptive stimuli with higher cortical functioning. Particularly in the case of chronic pain, psychosocial factors significantly affect the interpretation of these signals and the ultimate manifestation of pain, or more properly pain behaviors. Some of the most important variables are listed in Table 29–1.

It is important to draw a distinction between *chronic pain* and the *chronic pain syndrome*. Chronic pain is simply the brain's interpretation of nociceptive input from the body's periphery. The chronic pain syndrome, however, is the individual's behavioral response to pain—a response that is determined not only by the experience of pain but also by the patient's emotional state, personality factors, and social contingencies.

The chronic pain syndrome includes all the changes that take place when chronic pain enters an individual's life. These changes include, but are not limited to, a loss of activity tolerance, physical deconditioning, disturbances of mood and behavior, changes in interpersonal relationships, and the loss of preferred social and recreational pursuits. Finally, there are changes in the vocational sphere, such as loss of job, decreased productivity, and impaired work motivation. It is this combination of physiological, psychological, and social factors that determines an individual's response to an injury or painful condition. *Failure to consider the patient's social and psychological uniqueness often leads to incorrect, or even detrimental, treatment initiatives.*

To provide a better understanding of psychological issues, the remainder of this chapter will give an overview of types of psychological dysfunction. In addition, a model for working with pain patients will be presented to assist the therapist in dealing with these challenging individuals.

PSYCHOLOGICAL DISTURBANCE

Disturbances of Mood and Attitude

Depression is probably the most commonly encountered emotional complication of chronic pain. Usually it is part of the patient's emotional response to the losses that have accrued in his or her personal life. When it persists, there is an insidious alteration in brain chemistry that may require antidepressant medication. This same alteration in brain chemistry apparently generates still further decreases in pain and activity tolerances. If the depressive component is untreated, the patient often fails to move forward in rehabilitation, whether it be physical reconditioning or the search for new employment.

Table 29–1 Important Psychosocial Factors Involved with the Chronic Pain Syndrome

Mood Disorders (Adjustment Disorders)	*Neurotic Behavior Patterns*	*Personality Disorders*	*Other Factors*
Depression	Compulsive pain behaviors	Passive-aggressive	Age
Anger	Obsessive pain thoughts	Perfectionistic	Education
Pain-related irritability	Somatization of stress	Melancholic	Secondary gain
Fear	Conversion hysteria	Histrionic	Employee anger
		Hyperactive	
		Inadequate	
		Chemically dependent	
		Codependent	
		Antisocial	
		Other	

Irritability is another frequently encountered complication of chronic pain. It appears to be a nearly inevitable consequence of any persistent, noxious stimulation. In animal behavior research, irritability (e.g., attacking a cagemate) can be induced by exposing the experimental subject to a noxious stimulus from which it cannot escape. In humans, chronic pain can function as this type of unavoidable stimulus. Pain-related irritability also induces the impatience and loss of frustration tolerance that family members often describe as a "personality change" in the sufferer.

Accumulated feelings of *anger and resentment* are a nearly ubiquitous feature of the chronic pain syndrome. These socially undesirable emotions arise for the same reasons as depression: namely, they are a response to accumulated losses. Often the anger and depression are so intertwined that the tears seen are as much a reflection of anger and frustration as depression. Note that not all health care professionals are temperamentally suited to working with angry patients. In a pain clinic setting it is important to hire staff and therapists who can tolerate the emotional intensity of the angry pain patient.

Anger and depression often develop from the *fear and anxiety* that become part of patients' lives. Fear of the unknown, or a lost sense of control over one's personal outcome, is a recurrent theme. Fear of future physical functioning is often manifest in patients' perception that their physical condition is deteriorating. While the orthopedist may be interpreting a patient's degenerative disc disease as part of the normal aging process, the patient hears the word *degenerative* and visualizes a wheelchair in his or her future.

Fear of future economic well-being often impels pain patients toward symptom magnification—not because they are dishonest, but because they are insecure about their future capabilities, and want to accumulate an economic cushion in anticipation of impending hard times.

Behavior Disturbances

Excessive *pain behaviors* are often seen in individuals with psychologic distress. Examples include moaning, groaning, grimacing, and unnecessary guarding. *Doctor shopping* is also a form of pain behavior. While excessive pain behavior is often motivated by fear (discussed above) and secondary gain (discussed below), it is important to recognize that it can also exist simply as a "bad habit" that persists in accordance with the laws of classical conditioning.

Pain talk is a verbal form of pain behavior. Psychologically, it has two maladaptive drawbacks. First, it interferes with the essential process of distracting oneself from pain. The brain has some limits in the number of perceptions it can process simultaneously. If attention can be directed to matters other than pain, many sufferers will, subjectively, experience less discomfort. However, sufferers cannot be distracted from that which they are constantly reminding themselves of through their own compulsive pain talk. Thus most comprehensive pain clinics try to extinguish pain talk through distracting socialization and nonreinforcement.

A certain amount of pain talk may be necessary. For example, the patient may need to communicate a new symptom to the physician, and the injured worker may need to apprise the employer of physical restrictions. However, excessive pain talk can drive away the patient's social support system. Many marriages break up because the well spouse "got tired of hearing about it." In addition to resulting in social isolation, pain talk can cause a potential employer to stereotype the patient as an unreliable complainer.

Somatization of stress is another behavioral disturbance frequently encountered in persons with chronic pain. A vicious cycle can develop when the stress associated with chronic pain causes increased tension and guarding, which in turn produces increased pain. It is not unusual for the stress of a marital problem, job search, or training program to produce a subjective increase in perceived pain or precipitate an objectively determinable "flare-up." Pain and stress management counseling as well as relaxation training are important tools for dealing with the somatization phenomenon.

The concept of *secondary gain* is important to understanding neurotic aspects of pain behavior. Secondary gain refers to the reward value of a particular class of behaviors. Thus the compulsive talker maintains a steady stream of verbal behavior to avoid hearing any negative feedback. Similarly, compulsive pain behavior can have reward (functional) value. Sometimes, the rewards are obvious; other times, subtle. For example, pain behavior can be a way of communicating to the family, "Don't abandon me." It may be a means of telling the physical therapist, "I'm scared. Don't push me too fast." To the Social Security judge, the message may be, "Let me retire. I can't provide for myself." To the prospective employer, the pain behavior may be communicating, "Don't hire me. I really can't work."

Occasionally, behavior disorders develop into frank neuroses. Neuroses are compulsive, maladaptive, anxiety-reduction behaviors. In pain patients, it is not uncommon to encounter *panic attacks* and *agoraphobia* as severe forms of social withdrawal. *Conversion hysteria* involves the unconscious inhibition of a voluntary behavior, such as using one's arm.

One of the authors recently evaluated a woman who displayed hysterical paralysis of her dominant, right upper extremity. An anesthesiologist diagnosed reflex sympathetic dystrophy and proposed a dorsal column implant. Physical examination, however, revealed no clinical evidence of RSD, whereas psychological assessment identified numerous secondary gain factors consistent with a hysterical reaction. Specifically, the patient's complaints enabled her to stay at home with a new baby while maintaining an income through wage loss benefits. Also, she obtained increased attention and concern from a previously philandering spouse.

Disturbances of Reality Contact

Psychosis is an emotional disorder in which the hallmark feature is the loss of contact with reality. While psychotherapy may provide the psychotic individual with some goal-directed support, these disorders, for the most part, respond best to therapy with the major tranquilizers. One must always remember that a mentally ill individual can develop a genuine chronic pain problem. However, it is not unusual for psychotic individuals to develop *somatic delusions* or *delusional pain*.

For example, a patient presented to one of the authors complaining of intermittent buttock pain that she described as "electrical shocks." Pain relief was obtained by unplugging her television set. It was eventually learned that the patient held the paranoid delusion that the police were transmitting "electrical rays" through the power lines into her TV. She believed this was covert punishment for having a child out of wedlock. Most psychotic pain syndromes are more subtle and can be understood only through a thorough psychological evaluation.

Personality Disturbances

Individuals with personality disturbances have a highly enduring, persistent style or pattern of behavior that is either personally or socially maladaptive. Unfortunately, many of the patients treated in a chronic pain clinic suffer from a significant personality disorder.

The *passive-aggressive personality* tends to express feelings of anger by behaving in a manner that is opposite or contrary to what is expected by those with whom he or she is angry (usually authority figures). Frequently, the health care provider can sense the passive-aggressive individual's resentfulness. However, when the provider tries to address the issues, the patient smilingly responds, "Everything's fine."

The passive-aggressive personality has its origin in childhood when a parent is so controlling that the only power the child can exert is to behave opposite to expectations. It is not unusual for passive-aggressive individuals to behave in opposition to their own interests just to frustrate an authority figure. Exaggerated pain complaints often become a passive-aggressive means of avoiding undesired vocational responsibilities or retaliating against an employer or compensation provider. These individuals are often adept at making other people appear to be at fault.

The *perfectionistic personality* is compelled to put more effort into solving a problem than most people would find appropriate. Perfectionists often have a hard time completing tasks because the outcome is never quite good enough. Developmentally, this disorder originates when a child overcompensates to please a shaming parent. Perfectionism also develops through role modeling. That is, many perfectionists have parents who were themselves compulsive, rigid, and inflexible.

Perfectionists are sometimes identified by their fastidious grooming, or by their speech patterns, which are overly precise, pedantic, and obsessional. They are particularly prone to develop *psychophysiological*, or *psychosomatic disorders*, such as peptic ulcers or chronic headaches. Often their headaches and facial pain syndromes relate to repressed feelings of anger—an emotion that perfectionists often find unacceptable.

In rehabilitation, perfectionists are often "all or none" in their attitude and effort. They ascribe to the saying, "It's not worth doing unless it's done right"—which can become an excuse for not even attempting activities that would be of rehabilitative value. Another problem that perfectionists have is that of "overdoing it" when feeling relatively well, thereby instigating a "flare-up." Perfectionists can be extremely compulsive in their "doctor shopping," ever seeking a practitioner who will cure their problems.

The *melancholic personality* struggles with chronic low self-esteem, pessimism, and unhappiness. In general, melancholics are quickly discouraged by unpleasant life circumstances while showing a muted or cynical response to situations that, for others, would be pleasurable. It is often difficult to differentially diagnose the melancholic from a person who is suffering from situational depression. In general, the melancholic is more unrealistic and persistent in his or her verbalizations of helplessness and hopelessness.

While there is probably a hereditary component to chronic melancholia, it can also be induced by growing up in a *dysfunctional family* in which the child was exposed to constant hypercritical parenting, thereby engendering deep-seated feelings of shame, self-doubt, and pessimism.

Melancholics tend to have very low pain tolerance. The experience of chronic pain often confirms their hopeless outlook on life. Not surprisingly, these individuals can be difficult to motivate in rehabilitation. Thus the rehabilitation process usually requires a great deal of externally imposed structure. These individuals are often no stranger to psychotherapy. Often they benefit from additional psychotherapy that can relate skills they have already acquired to pain management. Exercise and antidepressant medication can also have an energizing effect. Where possible, passive modalities of physical rehabilitation should be avoided, as these seem to reinforce an already strong passive-dependent orientation.

The *histrionic personality* may be the most commonly encountered personality disturbance in the pain treatment community. Histrionics are noted for overusing repression and denial concerning their "true" feelings (into which they have little insight) while dramatizing the superficial. Physical complaints are characteristically out of proportion to underlying medical findings.

These individuals often show a conspicuous failure to improve in the face of all appropriate medical care. Despite dramatic complaints of discomfort, their outward demeanor is often that of good-natured cheerfulness (la belle indifference). Often they narcissistically thrive on the attention they derive from family and health care providers. There are usually other secondary gain factors, involving the fulfillment of underlying, immature dependency needs.

The *hyperactive personality* typically displays chronic impulsiveness and distractibility, along with hyperactive speech and thought processes. In other words, hyperactive people talk fast, think fast, and do things fast. They have a very low tolerance for frustration and, not surprisingly, low pain tolerance as well. Because their pain experience is more intense, they express their complaints more dramatically. As a result, they are more likely to become addicted to pain medications. Generally speaking, addictive medications are contraindicated with this group.

Of all the personality disorders, this may be the most genetically determined. Any parent of more than one child has probably observed an energy difference between any two of their children. Hyperactive children often have so much interaction with their environment that they inevitably elicit an aversive punishing response from adults. Many children with attention deficit disorders (ADD) become hyperactive adults with a low tolerance for frustration and negative attitudes toward authority.

Because of impulsiveness and poor self-pacing, these individuals are highly *accident prone*. Not only are they more likely to be injured in the first place, but they are more likely to become reinjured during a return-to-work effort. These individuals often respond well to insight-oriented therapy and development of self-pacing strategies. Relaxation training can also be of benefit.

The *inadequate personality* is usually very poor at solving the problems common to adult life. People with this pattern of disturbance often feel inadequate in social situations and show subtle defects in social judgment. Often they are seen as "multiple problem" people. Like many personality disorders, the inadequate personality evolves from a mixture of hereditary and environmental forces. On one hand, parental role models were usually themselves inept at the challenges of adult life. However, cognitive deficits, such as dyslexia, are often evident.

Inadequate personalities tend to look at problems with pain in a concrete, "black and white" manner. They may have difficulty comprehending scientific explanations of their discomfort and may gravitate toward the mechanistic model offered by chiropractors. They usually mistrust those who would have them "accept" a problem while embracing those who promise to "fix" their difficulties.

These individuals tend to live under constant fear of failure and, as a result, easily gravitate toward a disabled lifestyle as a face-saving means of accounting for their feelings of rejection by coworkers and society. There is often a premorbid history of poor adjustment to the world of work. When asked to set vocational goals, they can quickly enumerate a long list of "can't do's." Frequently, there is avoidance and a lack of initiative in the job search.

The *chemically dependent personality* is characterized by a current or past history of compulsive and potentially self-destructive abuse of addictive substances. There are two subtypes—the *practicing* and the *recovering* abuser. Just because individuals are abstinent from their drug of choice does not mean they are free from the character defects that are part of the chemically dependent personality. This is particularly the case with so-called *dry drunk* behavior.

Both former and current drug abusers often display a low tolerance for frustration and, as a result, a low tolerance for pain. Pain behavior quickly becomes an excuse to use or procure the drug of choice, whether it be obtained from a bartender, liquor store, or physician. Pain behavior also becomes a means of avoiding activities (e.g., exercise, work) that interfere with "using."

For the practicing chemical abuser, meaningful rehabilitation cannot take place until the patient first establishes a stable period of abstinence. Inevitably, the first step is drug withdrawal (detoxification) or, in the case of addictive analgesics, a medication reduction program. Virtually no pain treatment or pain management can be effective when the patient is "using." Vocational rehabilitation efforts become similarly futile. Even with primary chemical dependency treatment, the rate of recidivism is high.

Codependent personalities tend to lose their sense of self-identity by allowing another person to take control of their life. The classical example is that of the vocationally unskilled woman who stays with a physically abusive male breadwinner in order to provide for the children. There is also a great deal of codependency in the workplace. Economic need forces many people to stay in an unsatisfying job in which they feel unappreciated and poorly rewarded. Codependent adults have often modeled their behavior on that of the parent with whom they most strongly identified. Early victimization experiences also promote codependency. When children are abused, they learn that passivity and nonassertiveness are the best way of surviving.

Chronic pain sufferers often develop a codependent relationship with their own pain. That is, they hate the pain but hold on to their suffering because it brings them attention or compensation that they fear they would not otherwise receive. Being dependent on that to which you are a victim is known as a *hostile-dependent* relationship. Hostile-dependent relationships often exist between the injured worker and compensation providers. Frequently, codependent individuals start out as victims of dysfunctional parenting and end up feeling like victims with any authority figure they encounter, including the physician, rehabilitation professionals, and insurers.

Counseling can help codependent personalities develop some insight into how they replicate earlier victim experiences in their rehabilitation relationships. Just as the 12-step principles of AA can help the alcoholic recover from chemical dependency, Al-Anon can help the codependent break the cycle of enablement, victimization, and resentment.

Individuals with an *antisocial personality* disorder display behavior that is markedly characterized by authority-resistant attitudes and values. This behavior pattern originates in the preadolescent years when children grow up with insufficient or unhealthy supervision. In effect, these children raise themselves and inevitably adopt impulsive, ego-directed rules of conduct.

There is a high risk of *malingering* in this group. Even when objective medical findings are present, symptom magnification, elaboration, and fabrication are characteristic. Pain behavior quickly becomes a vehicle for achieving personal goals. Usually these are monetary goals, but avoidance of undesired responsibilities is also common. As with most of the personality disorders, rehabilitation must be highly structured, with explicit consequences for failure to comply with a formal rehabilitation plan. Administrative claim resolution is often more effective in restoring function than anything offered through the medical rehabilitation community.

It is beyond the scope of this chapter to cover all of the personality disorders that might be encountered when treating chronic pain. For example, we have not discussed the paranoid, self-defeating, obsessive-compulsive, borderline, or schizoid personality. Note also that few individual patients fit neatly into one psychodiagnostic category.

WORKING WITH THE PAIN PATIENT

Working successfully with chronic pain patients is extremely challenging and requires skills not typically taught in the classroom setting. In addition to an understanding of physiology, awareness of the interplay of psychologic, social, and economic factors is critical to successful rehabilitation. Skill in the arts of negotiation and communication is at least as important as an understanding of body mechanics. Patients are often manipulative, resistant, and noncompliant. Dealing with these types of behaviors may prove incredibly frustrating to a concerned therapist.

Many professionals involved in the treatment of pain patients seem to assume that the psychological component of the chronic pain syndrome is an *adjustment reaction*. In other words, one tends to think of the chronic pain sufferer as a premorbidly "normal" individual who is responding to pain-related losses with depression, anger, worry, fear, and so on. When the chronic pain syndrome truly is an adjustment (or loss) reaction, the patient is often highly responsive to pain clinic treatment and to rehabilitation in general.

However, many of the patients treated *do not present simple loss reactions* and in fact have significant preexisting disturbances of adjustment and character. They often have hidden agendas and display manipulative behaviors that can readily compromise rehabilitation. These individuals will not do well in a pain clinic (or in rehabilitation) unless the psychological factors are well understood and the rehabilitation plan is adjusted accordingly. Moreover, these individuals will not necessarily be appreciative of the efforts of their health care providers.

Because health care professionals may value achievement highly, they may assume that patients are as motivated to get better as providers are to help them. The reality, however, is that patients often have personalities and motivations very different from the provider. It is especially important in dealing with the character-disordered pain patient to avoid projecting self-values onto the patient. Patients come from all walks of life and present a wide variety of personality styles. Unless there is accommodation of the treatment plan to these styles, professionals may be naively setting the stage for a poor treatment outcome.

While there are no fail-safe formulas for dealing with these varied individuals, there are several simple principles that, if followed, can help "turn the corner" and move the client toward the goal of successful rehabilitation and return to work.

Avoid Countertransference

Countertransference may be thought of as the feelings evoked in the therapist as a result of the interaction with the client. For example, an argumentative or passive-aggressive patient often engenders in the therapist feelings of anger and frustration. If that individual patient thrives on confrontation, then the provider becoming irritated and angry only reinforces and gives ammunition to the patient, reinforcing the patient's perception that the provider is indeed incompetent. Instead of the relationship being therapeutic, channels of communication only break down further, with the patient assuming that the therapist is "a jerk" and the therapist likewise convinced that the patient is "a crock."

Likewise, working with an hysteric who is severely limited by pain but appears to be sincerely trying may evoke excessive feelings of sympathy towards that individual. While feeling sympathetic may allow the therapist to feel virtuous and caring, it does not help the patient improve and often negatively reinforces the patient's perception of disability.

Thus working with these individuals requires that the therapist be very clear about his or her own feelings of frustration, anger, mistrust, sympathy, cynicism, sexual attraction or repulsion, superiority, and so forth. Once these feelings are acknowledged, it is important to try to relate to the patient in a supportive but not enabling or overly sympathetic manner. One should avoid punishing behaviors as well. In order to do so, it is important to try to stay objective and to set clear and appropriate boundaries and expectations. It is important to remember that the so-called chronic pain syndrome is actually a number of syndromes that are determined not only by physical disease but by the person's personality. William Osler once wrote, "It is more important to know what kind of man has the disease than to know what kind of disease a man has."

Set up Clear Structure and Boundaries

Pain is a symptom, *not a progressive disease*. One of the important aspects of any successful pain management or work-hardening program is to provide the appropriate cognitive tools to the patient so that he or she can understand that in the chronic stage experiencing pain (hurt) is not the same as causing further tissue damage (harm).

Conventional wisdom is that pain is to be avoided. Experiencing pain evokes conscious and unconscious fear. Often chronic pain patients have seen numerous physicians, undergone multiple diagnostic tests, been given numerous diagnoses, and in spite of extensive treatment not responded with either elimination or amelioration of their symptoms.

Frequently, they are given conflicting and at times contradictory advice from clinicians. For example, with low back pain, one therapist may recommend flexion exercises, another extension exercises, and still another aerobic conditioning.

Therefore, it is important, when outlining a return-to-work plan, to make it crystal clear to patients that they will continue to experience discomfort and that this does not mean they are ultimately doing themselves harm.

With this understanding, it then becomes central to establish specific rehabilitation goals that are not in and of themselves limited by the patient's experience of pain. While it is critical to involve patients in this decision-making process, individuals must not be allowed to control the return-to-work plan and to limit themselves simply by their experience of discomfort.

For example, a typical work-hardening plan should consist of specific hours or a specific number of days a week and for a specific time with specific incremental increases (i.e., four hours a day, five days a week × two weeks, then increasing one hour a day per week). This would be opposed to telling the patient to "try four hours but if it hurts too much just work up to your tolerance." Not every work-hardening plan will prove successful and a plan may require modification, but again it is important that any modification be outlined clearly and precisely.

A strong attempt should be made not to go backwards. Be extremely cautious in recommending decreasing hours once a pattern of work has been established, even if the patient is having difficulties. This type of behavior can easily undermine and appear contrary to previous messages given, as well as reinforce the patient's own perception of disability.

Unfortunately, this approach does not always prove effective, and at times one is forced to say simply that the patient is limited by his or her self-report of pain rather than stating that the patient can or cannot perform an activity.

Involvement of Other Important Players in the Rehabilitation Process

An injured worker's expression of pain or pain behavior is often that individual's only leverage in dealing with a very complex system. Workers cannot state that they do not want to return to work because they do not trust their employers, that they are afraid they will be laid off once they return to work, that they have been doing the same job for 35 years and really want to retire, or that they have young children at home who require special attention.

For example, a 62-year-old factory worker with a back injury may find it very demeaning to take a light duty job. He may feel that he does not have the financial resources to retire, and his only leverage in avoiding what he perceives as an unpleasant experience is through his expression of pain. He says his back pain is too severe to do this job even though based on his objective findings this should not be the case.

Likewise, many patients who initially presented with severe incapacitating pain but now show a marked diminution in pain behaviors continue to complain of persistent pain following a settlement (in which they received a small financial reward) because they were able to extricate themselves from environments they perceived as hostile.

In order to address these critical psychosocial issues, involvement of all concerned parties—that is, the insurance representative, the rehabilitation counselor, and particularly the patient's attorney and employer—is mandatory. Negotiating an acceptable win/win plan that will allow the patient to return to work is a much more desirable outcome than the administrative conference with its adversarial atmosphere. Helping the injured worker and the employer understand each other's concerns and work together often represents a critical factor leading to successful rehabilitation.

If, on the other hand, the employer and the employee do not communicate well, then there is really little likelihood of successful rehabilitation. Often in these cases, the employer sees the employee as someone who is sandbagging or malingering, and the employee feels that the employer is dishonest and really is only providing a job with the hope that either the employee will quit or the employer will find an opportunity to fire him or her. They tend to blame each other, and, as in a bad marriage, divorce or a formal settlement that will extricate the individual from the compensation system may represent the best approach to resolving differences. In these circumstances, which unfortunately are not so unusual, no amount of physical therapy, work hardening, or counseling will prove successful in addressing the core issue, which is the dysfunction and adversarial relationship between the worker and his employer.

CONCLUSION

In performing and interpreting functional capacity evaluations on individuals with chronic pain, an understanding of psychosocial factors is essential. These various factors may be much more critical than mechanical ones in predicting outcome. Factors predictive of poor outcomes include an individual who sees himself or herself as never returning to the workplace, evidence of significant psychopathology, an extremely mistrustful relationship between the patient and the employer, and evidence of chemical dependency.

These variables do affect an individual's performance and his or her level of motivation. Certainly, addressing correctable factors such as drug dependency or depression prior to a functional capacity assessment is preferable. Other issues such as underlying personality disorders or resentment toward the employer are much more difficult to address and may prove impossible to change. Assessment by a skilled psychologist or pain specialist prior to the functional capacity evaluation may prove helpful for identifying and modifying these issues. Understanding these common personality disorders, recognizing one's own emotions and feelings toward these individuals, setting clear and appropriate rehabilitation goals, and involving all concerned parties in the rehabilitation process are important in maximizing success. However, in spite of our best efforts, these individuals will continue to remain a rehabilitation challenge.

SUGGESTED READING

Fishbain, D., et al. 1986. Male and female chronic pain patients categorized by DSM-III-R. *Psychiatric and Diagnostic Criteria of Pain* 26:181–197.

Hanner, L., et al. 1991. *When you're sick and don't know why*. Minneapolis: DCA Publishing Co.

Politan, P.B., et al. 1993. Psychiatric illness and chronic low back pain: The mind and the spine, which goes first. *Spine* 18:66–71.

Sternbach, R.A. 1974. *Pain patients: Trades and treatments*. New York: Academic Press.

Chapter 30

Testifying in Court: You and Your Records

William D. Sommerness

Evaluating and treating clinicians are important in areas of the work injury beyond the generally accepted treatment techniques learned in school and practiced in the clinic on a daily basis. The fundamental concept is that the injured person needs to return to work as soon as possible. Issues of safety and ability to return are a given; what is not a "given" is whether the clinician understands, accepts, and is proficient at working within the legal framework to accomplish that goal. If the individual professional is willing to learn these skills, the patient will be fortunate indeed.

The clinic must foster an environment that recognizes the legal ramifications of the return-to-work situation as an important part of the medical-legal system and must provide the evaluators with the training, education, experience, and guidance needed to work in the field. If the process is not important to the clinic or hospital administration, it is not likely to be an important area for the individual clinician. When that happens, the patient is the loser in the short run and the clinic-hospital is the loser in the long run.

There is a corollary here that must be understood: the lawyer and his or her paralegal staff have an equal responsibility. They must recognize the importance of the medical observations and treatment options as identified by the therapist, the doctor, and the balance of the medical team. In a very real sense, the injured worker needs the help of both professional groups. Just as the patient can suffer a poor "medical" result, he or she is in just as much trouble when the legal team decides to not take the time to learn what the medical team is requiring in the return-to-work efforts. Negative outcomes may also be caused if the legal team chooses to view the medical team as "the enemy" who cannot be trusted, should not be consulted, and is to be avoided whenever possible. It is important to recognize the truth of the matter: the lawyer and his or her staff can really frustrate and make the return to work impossible when they choose to, or do so by ignorance.

An uncooperative and/or poorly trained legal staff will undo the best efforts and work of the medical team, if given the chance, much too often. Therefore legal staff need their own training sessions and guidance.

The opportunity is for the medical and legal systems to work together. If each set of professionals understands the other's system, ethics, and role, there will be enhancement of the client's chances of fair medical treatment and resolution.

YOUR CREDIBILITY

This section highlights the most fundamental legal asset, credibility. If it were about your honesty, it would be entitled, "Your Honesty." It is not because it is not about whether one is honest or dishonest; most clinicians' integrity and honesty are not an issue. Credibility is much more than that and needs to be discussed and understood. The other parts of the system will be watching to see if clinicians are "credible" or not.

Witnesses who sell their testimony are not credible. One does not sell testimony or written reports: what one sells is timc. Credibility means that a professional comes to the same conclusions and opinions about the injured worker regardless of who is paying for the time.

Credibility means, in the simplest terms, believability. If, for instance, all you do is examinations for attorneys who represent injured employees, sooner or later people are going to question your credibility. Bias is to prejudice what credibility is to honesty. Prejudice is a conscious act of intent; bias is impropriety. If all you do is independent medical evaluations (IMEs) for insurance companies or their lawyers, you shouldn't be surprised if someone questions the motives as being protective of that financial relationship rather than concerned with ascertaining the true condition of the employee that has been examined. The key is to find a balance: do work for both the insurance industry and the attorneys who represent injured people. An astute lawyer has credible medical people targeted to work with the injured person before the other side can use those same people for an IME. Lawyers know whether the person/clinic is credible, so they want to have the person do the evaluation. If medical professionals choose to sell their testimony, they will be busy for awhile because, unfortunately, there are always people looking to buy that kind of assistance. But they will not last. Sooner or later it will catch up with them and they will be finished as experts. By selling out as "professional witnesses" clinicians will have a predictably short shelf life. They will become used goods rapidly. Credible medical professionals, on the other hand, are sought for their competence, honesty, and objectivity by both sides.

Credibility means how clinicians view themselves within the medical team. If nonphysicians or nonspecialists feel they don't have anything important to say because they do not have the right letters after their names, they are correct: they don't have anything important to say—how sad! Credible people believe in themselves and in their own professional skills. Evaluating and treating therapists, for example, have specific and extensive training in the area of function. An orthopedist or neurosurgeon does not have these credentials. The secret is to learn when to stand up and be counted. The therapist's task is not to read X-rays or do surgery; that is the province of the surgeon. The therapist does have a singularly important role to play, however, on the issues of return to work, function, and work restrictions that are particularly the heart of that profession.

Credibility means that clinicians recognize their strengths as professional practitioners in the health care field instead of focusing on what they are not. For example, if all lawyers wanted in the medical-legal system was the opinions of doctors, only doctors would be included in the process. That is not the case, and therapists must believe in their abilities, training, and expertise to participate in the process to an extent that no other health care professionals can. This expertise is needed by the patient and the system trying to get the patient back to work. Therefore both doctors and functional evaluators, such as therapists, are required.

A common theme in health care reform seems to be that all health care providers will have one thing in common: the need to prove results. Outcome studies/data will be an important factor in determining where patients will be allowed to go for therapy. Does it make any sense to do anything else? The clinicians who validate the return of patients to work at a reasonable cost will be the providers of choice. In a very real sense, this is just another way of saying they will have professional credibility. Obviously, to do any of this, clinics will have to have the data to substantiate their successes in economically returning patients to productive lives. This need for data will be fundamental in establishing credibility in the medical system, and the data can and will be used by the clinician in the medical-legal system, for the clinician and clinic should be able to verify the success they have had in returning people to work. The data supporting the actual return of people to work will go a long way in establishing credibility when clinicians are asked to substantiate their reasons for concluding that a person will or will not be able to return to work. Put another way, the number of people returned to work settings serves as part of clinicians' credibility in cases where they testify that the patient cannot return to work. Success in the marketplace makes clinicians more believable in their answer to the hypothetical expert opinion question.

Credibility is absolutely essential to the clinician in the medical-legal system. Credibility has many facets, all of which are important. It is critical to the impression people have of the clinician as a messenger in the medical-legal system; it is just as critical to the message. Without credibility the clinician is nothing; with it, the clinician's professional conclusions and opinions are given the weight they deserve. If there is no credibility in the clinic's operation, there can be none in the legal aspect of its practice: its professional reputation is on view for all to see.

Clinicians work hard to obtain credibility in their practice; they shouldn't give it away in dealing with attorneys and insurance companies.

THE EVALUATION

In its truest sense, the clinician's chart and credibility are what the practice is all about. That is to say, how the individual clinician charts is not likely to change from patient to patient, but is a daily task that becomes second nature. If one clinician's charts are pulled at random and compared with charts of other clinicians, they will usually be identifiable because charts represent one clinician as easily as individual fingerprints: the attention to detail (or lack thereof), the use of common phrases and professional jargon, the method of writing clearly or ambiguously, and the clarity of opinions are all unique in some degree to each clinician. The reputation of the clinic is well known long before the particular case in the medical-legal system. Similarly, clinicians' credibility is established within their clinic, among their former patients, among their peers, and in their "community" long before they become significant players in the medical-legal arena. What clinicians do, how they do it, when they do it, and their overall approach to the whole process will, in time, identify clinicians within the system just as clearly as they are now recognized within the walls of their clinic.

The evaluation has many facets, all of which are important. Clinicians need to make sure of exactly what they are being asked to do. The possibilities include cases in which they are asked to comment on a patient they have seen for an extended period of time and whom they recognized from the beginning as a case on which they would have to testify or, at a minimum, write a narrative report with specific conclusions and recommendations. The patient who is hurt in an industrial accident is going to need all the clinician's skills—the recognized evaluation skills and the important charting skills. The clinician should recognize from the beginning that sooner or later he or she is going to be asked when this individual can go back to work and, if not, why not. Why a chart doesn't address that concern from the very beginning is hard to understand. A second possibility is the patient who is evaluated for what amounts to an IME. This is usually a one-time evaluation; the clinician's preparation is usually good, the charting accurately reflects observations, and thus the conclusions are better understood.

The functional capacity evaluation (FCE) is the single most important test in the medical-legal system. It is consistent, it is documented, it is what has given the clinician the criteria needed for professional opinions needed by the other partners in the medical-legal system.

A bit of history is important. The issue of whether a person can return to work is not of recent import only; this has been a concern as long as there have been workers' compensation systems in place. It was not that many years ago that the only accepted testimony on the subject came from a physician. The system was serviced by many caring physicians who did the best job they could with the training and time they had; by and large, they brought a good deal of common sense to the system. To be sure, there were no shortages of company doctors who would regularly testify that everyone was able to go back to work regardless of the facts; and it was not hard to find doctors who seemed to always find reasons why nobody could return to work. The system seemed "used" because it was being used by people more concerned with their own financial interests than those of the individual employee, the company paying the bills, or the system's integrity.

The next step was the utilization of specialist physicians, in particular the orthopedist; the general practitioner was deemed not as qualified to testify about the ability to return to work as this group of specialists. Clearly, since certain injuries allow or don't allow for the return to work, that was a reasonable choice; however, not every person who has surgery is forever removed from the workplace; they don't even all need retraining for another occupation! The great majority of people can and do return to work and their homes for productive and enjoyable lives. The medical-legal system is where many of those decisions are made. Who is going to influence the decision makers?

Therapists are in the best position to determine function. After all, isn't that what it is all about—evaluation of the functional abilities and capabilities of the person? Several items acting in concert have provided this opening for therapists. First, physicians have by and large avoided medical-legal involvement; it takes time, and they are not in control of the system. Before the widespread use of videotaped depositions, their testimony required a personal appearance at a hearing or a court proceeding, and this scheduling was difficult at best. Doctors haven't historically lined up with fervor for cross-examination. Attorneys quickly learn to ask the doctors about their criteria for conclusions on weight restrictions, return-to-work date, function, and job capabilities. Physicians don't enjoy being asked to explain the basis for their opinions. Of course, there are exceptions to the rule, and in every system there is a group of doctors who enjoy the testifying experience and who do well at it. Imagine, however, the compensation referee or judge who sees the same doctors appearing time and again for injured people, and a similar group of doctors and clinics always appearing for the employer and insurance companies. It gives their credibility a bad name, to say the least.

At the same time, the costs of the systems are going nowhere if they aren't going up; is there a state legislature anywhere not being asked to "do something" about the high cost of premiums paid by industry and their insurance carriers? Similarly, in the auto accident systems, many states went to a "no fault" system of compensation; the paying of basic wage losses and medical expenses without need to incur the additional expense of litigating the issue of who was at "fault." In the work injury system, with the costs exorbitant, the doctors in the main not wanting to testify, and the professional witnesses for both "sides" playing a major part of the problem instead of the solution, the time was right for a breath of fresh air: the therapist-based FCE. The FCE has brought criteria where before there was cynicism, brought consistency in testing where before there was consistency in certain doctors always testifying for one side or the other, and, finally, has brought some hope for the person being tested. With the FCE, the employee, the employer, and the insurance company all have a basis to believe in performance evaluation. That testing has replaced the short conversation in the doctor's office with a scribbled note regarding the order to return to work and the weight restrictions, if any. The critical people here were the patients: they knew the system was not listening to them and, in many cases, not caring. The therapist spends significant time with the patient, and observes the successes and failures of the patient over a matter of hours and days, so when the patient is given the "bottom line" he or she has a basis in fact to believe it.

This is all to say that the evaluation process needs an awareness of the importance of what the therapist does in comparison with what has historically been done with the injured patient. The secret is to build on that advantage—thus, the importance of learning how to prepare to testify, write reports, and conduct oneself at the deposition or before the workers' compensation referee.

THE REPORT

Given the choice, the attorney will take a clinician who can write a good report over one who thinks he or she will be a star on the witness stand, but who can't write a report. It is easier to teach how to defend a well-written report than it is to try to teach how to handle a cross-examination of a poorly written report by a well-prepared attorney. The written report is what will be used to defend the clinician's under-oath opinions.

The chart needs to contain what is needed to write a clear and concise report; if there is not a good foundation, there is no way the house will stand any pressure. Similarly, a report that has more observations in it than the chart supports is bound to be suspect and will be the subject of strong cross-examination. Reports that contain observations different than those in the chart are also suspect. If the report contains conclusions not substantiated in the history and observations in the chart, the cross-examination is not hard to predict. The best way to learn how to write reports is by writing reports, but it is important to get the basics down pat. A bad report is a lot like a good report in the sense that it is hard to describe, but lawyers know it when they see it!

This will all be theory until after the witness's first time at supporting a bad report when under oath. All the reading and talk will not impress a witness as being under oath will impress. Then the clinician will have a new interest in clarity, conciseness, and completeness.

The written report is, in its simplest terms, nothing more than the narrative form of the clinician's chart notes and entries. The report is expansive where the chart notations are short and to the point. That is not to say the report should be verbose. The key is to focus on the audience. In the chart, the audience is internal, the clinicians and colleagues who next pick up the file and determine the patient's treatment modalities and progress. The chart is how medical personnel communicate with each other.

The report speaks to a different audience: people who may have some familiarity with the medical picture but who aren't classically trained in the specialty medical professions. The lawyer, insurance adjuster, employer, and workers' compensation judge all have more than a passing familiarity with the subject matter, but the report must educate the reader. Conclusions without an adequate history and explanation of the clinician's evaluation will not be well received. If, after sending out the report, one receives calls asking for an explanation of "what does it mean," "why do you conclude what you do," or "what kind of testing did you do," then the clinician is not writing the kind of report the medical-legal system needs. Clarity, conciseness, and completeness are the big three; more than that is inappropriate, less is incomplete.

CONCLUSION

The goal of the medical-legal system is to return people to work. A secondary goal, particularly in the field of clinician evaluation and treating the patient, is to teach employees proper body function so they do not reinjure themselves and return to the system. The system of insurance and risk payors have a right to expect a system in which the injured person is promptly and fairly evaluated and treated; they have the right to expect a system that fosters return to work as soon as the person can safely do so. The injured person expects no less.

Professionals in the legal and health care systems need to step back occasionally and realize their efforts are part of the system, not the system itself. If lawyers, insurance company representatives, employers, employees, and physicians minimize the role of the clinician/evaluator, clinicians need to ask themselves if they are enabling that event.

Particularly since the advent of the functional capacity evaluation, clinicians have criteria to back up their functional opinions. Credibility is the most important concept to understand and utilize. Credibility of data now exists supporting the functional testing concepts; the credibility of clinicians who have worked with hundreds of workers, assisting their return to their work, cannot and should not be ignored; there exist many credible clinicians who have not only withstood the rigors of proper report writing and testifying in depositions and trials but have been stellar leaders in their profession.

The return to work of the injured person can only be accomplished when the parts of the system understand and respect the role the other professions play in that process. Cooperation can only come after understanding. The leaders in the field of clinical evaluation and treatment have come a long way in establishing their rightful place in that process. Now it is up to the profession as a whole to expand on that and not lose the objectivity that training, observation, and years of "hands-on" experience bring to the system. The decision is that of every clinician and clinic; the outcome should be clear, but it will not come without setbacks and questions. The profession is up to the task.

Chapter 31

The Tertiary Care Level: The Problem Patient

Rebecca Cox, Pam Garcy, Gordon Leeman, Amy Santana, and Tom G. Mayer

Concepts of nonoperative care for work injuries are changing rapidly. In the past, the term *physical therapy* was loosely applied to any nonoperative management not specifically performed by physicians. A multitude of developments over the past two decades, however, have led to segmentation that permits definition of several levels of care tied to time since injury.

Primary care refers to that provided during the acute phases of work injury, and usually intended for symptom control. This includes, but is not restricted to, the so-called "passive modalities," such as electrical stimulation, temperature modulation, and manipulation. A variety of early assisted-mobilization and educational programs are also included in care on this level. Primary care is generally provided by a single therapist, with a limited number of treatments applied to a large number of patients entering the medical system.

Secondary care refers to that provided to a smaller number of patients not responding to initial primary treatment, whose more longstanding symptoms pass into the postacute or immediate postoperative period. In these cases, it is now recognized that *reactivation* is the common need, with programs generally adapted to provide appropriate exercise and education as the primary modalities, but often assisted by additional passive modalities still utilized for symptomatic control. In many cases, the secondary level of care lends itself to programmatic consolidation, particularly toward the end of the postacute period, in which interdisciplinary consultation may be available for specialized situations, but is not required on site or in the majority of cases. The primary treaters in this phase are usually physical therapists and occupational therapists, with the available consultants including physicians, psychologists, social workers, disability managers, and/ or chiropractors (depending on the venue or community standards).

In the small number of cases not responding to secondary care and becoming chronic, or undergoing complex surgical procedures, *tertiary care* is the final option. Tertiary care involves *physician-directed, interdisciplinary team care, with all disciplines on site and available to every patient.* The CARF (Commission for Accreditation of Rehabilitation Facilities) pain management guidelines most closely identify common standards of tertiary care programs, although such treatment may be diverse and eclectic. The picture is further confused by the fact that not all "pain clinics" provide tertiary care, just as not all "work-hardening" programs provide interdisciplinary consultative services needed to fulfill CARF definitions for this particular service. Thus, it is necessary to remain true to program definitions concerning service provision in order to have hopes of achieving the goal of *quality of care*. It is with this in mind that a specific program, *functional restoration* (available at the Productive Rehabilitation Institute of Dallas for Ergonomics [PRIDE]), will be discussed as an example of tertiary care in work-related injury. The intent is to give the reader a direct understanding of the roles of each member of the interdisciplinary team within this particular type of tertiary care.

THE ROLE OF PHYSICAL THERAPY IN TERTIARY CARE

The physical therapist (PT) plays a vital role in tertiary care that differs from that in primary or secondary care. In the acute phase of injury, the PT's main areas of concern include pain reduction, inflammation control, and protection of the injured tissues. Once the healing process is well underway and symptoms have lessened, restoration of function of the injured area, while keeping pain in check, becomes the primary goal. If total ability to function does not return once the healing phase is completed, secondary treatment is initiated to address residual dysfunction. The goal of treatment in this subacute phase is directed at return to preinjury ability (work, athletics, etc.) and includes education in self-maintenance.

Tertiary care is designed for those patients unable to reach full recovery with customary primary or secondary intervention and within a reasonable amount of time. These are the chronic patients who continue to demonstrate physical dysfunction and complain of ongoing severe pain without confirmation of diagnostic tests. We know now, on the basis of much research, that this group can demonstrate a variety of barriers to recovery, including psychological and vocational as well as physical. For complete return to productivity, a treatment approach that addresses and disarms each barrier is essential and requires an interdisciplinary team.

Physical therapists must have certain qualifications to function well as part of the team in tertiary care. Obviously, they must have excellent orthopedic skills as well as good, basic knowledge of neurology and pathology. While always considering the anatomy of injury and the physiology of healing, they must also understand nonphysiological signs[1] and chronic pain symptomatology. They must become well acquainted with not only the pathology of injury but also the sequelae of injury. Additionally, they must have a clear understanding of, and the ability to sympathize with, the chronic pain personality and condition. They have to be able to believe that, almost without exception, these patients are not malingerers and that the psychological and vocational dysfunctions that interfere with physical function can be managed. Finally, general knowledge of the workers' compensation system and the key players who affect the patient's case is important. Physical therapists must be able to alert the appropriate professional to the patient with specific issues and/or be quick to intervene themselves when they recognize certain "red flags." They must define the issues of rehabilitation in the context of the vocational environment and must understand that, here, medical decisions often have legal implications.

Therefore the responsibilities of the physical therapist in tertiary care have psychological and vocational components as well as physical. These responsibilities can be outlined in three broad areas: (1) physical issues, (2) educational issues, and (3) motivational issues.

Physical Issues

For a patient to become chronic usually means that his or her symptoms and/or responses to treatment have been confusing to health care professionals, resulting in lack of clear direction in *managing* his or her rehabilitation. Thus knowledgeable, confident, and decisive physical management is imperative for successful recovery. To begin with, the physical therapist must be sure he or she has a clear understanding of the patient's injury—the structural defect, surgical correction, and any residual deficit. Next, as with patients in acute and subacute phases, it is important for the PT to understand pain and inflammation as they relate to rehabilitation. With progressive resistive exercise, tissue irritation is common, resulting in mild swelling, muscle spasm, and so forth. This happens in rehabilitating the chronic patient, and the therapist should be on the lookout for these objective signs. This becomes especially important when evaluating pain: the PT must differentiate chronic pain from acute pain and inflammation to make appropriate treatment decisions. Without the presence of these more objective signs, the evaluation of pain can be more difficult with these patients, who tend to be symptom magnifiers.[2] Since swelling or other objective signs do not always accompany an acute occurrence, the PT must also rely on the evaluation of pain patterns and effects of mechanical stimuli.[3] An acute episode can be managed through temporary exercise modifications and anti-inflammatory treatment. Once the pain has been identified as nonphysiological or chronic, pain reduction techniques should be employed without interruption of the progressive exercise program.

The next important physical issue is that of physical progression. One problem in the chronic patient is that the normal feedback system is lost: pain does not necessarily indicate harm or excessive exercise. Therefore physical therapists must depend on objective measures of physical capacity to establish baseline exercise parameters.[4] Once these have been defined, standard sports medicine principles can be applied to achieve the rehabilitation goals of improved muscle strength and endurance, increased joint range of motion, and improved aerobic capacity.[5] When therapists are convinced they are not hurting patients, they can confidently push patients to progress in exercise even in the presence of pain. Progressive exercise accomplishes two things in the chronic patient: true physiological changes and changes in pain perception. Both of these things must happen for the patient to overcome disability.

The deconditioning syndrome has been described as both cause and result of disability.[6] Not surprisingly, most chronic patients present with severe deconditioning due to postinjury inactivity as well as predisposing risk factors such as obesity. Therefore the usual physical profile of the chronic patient is of an out-of-shape, overweight, injured person who has little preinjury experience with exercise. The physical therapist must know the patient's complete physical condition to anticipate rehabilitation hazards and outline appropriate exercise expectations. This also applies to patients who may present with other physical or medical problems requiring special monitoring. Next, there is the problem of inhibition: prolonged disuse causing slowed neuromuscular responsiveness and faulty physiological feedback, resulting in active loss of function out of proportion to the physical dysfunction. Inhibition should be addressed quickly to ensure that exercise becomes effective, but it often requires a process of patient education and trust building.

Educational Issues

Like other members of the treatment team, the physical therapist in tertiary care is responsible for dispelling misconceptions and giving information to the injured worker. Often these patients feel distrust towards the medical community and have developed their own ideas regarding the origin of and "cure" for their pain. Basic information in anatomy and physiology along with specific information about present physical deficits can help explain to patients what they feel. This education in conjunction with monitored exercise can help reduce the fear of reinjury and teach patients that pain does not necessarily mean harm.

Pain should be discussed openly and honestly, but always with instruction in pain reduction techniques. Thus patients learn to expect certain patterns of pain and begin to take responsibility for their own pain. This technique of patients managing their own pain has been called "symptom negotiation."[7] While not pain-free, patients can gain control over their pain. The physical therapist must also teach basic exercise principles that patients should apply for a lifetime. Rehabilitation should be the beginning of lifelong exercise habits. The exercise program becomes a classroom of learning proper exercise techniques and planning for a future home exercise program.

Motivational Issues

The population of chronically disabled patients is typically a "nonmotivated" group. The physical therapist must become both coach and counselor to help them succeed in rehabilitation. PTs in tertiary care rely heavily on the psychologist for suggestions in ways to approach the patient about exercise concerns. All patients need encouragement to accomplish their rehab tasks. With clues about the subtleties of each personality, the therapist can understand behavior and be supportive while ever challenging patients to extend themselves further physically. While usually at an unconscious level, motivational issues such as ambivalence about return to work will often be manifested physically; that is, one will observe poor effort that has no physiological basis. An understanding of behavior along with repetitive physical evaluation assists the therapist in boldly and confidently confronting ongoing motivational issues. In the most serious cases, confronting the patient with his or her lack of effort is done with the whole treatment team. When done carefully but honestly with support and encouragement, this technique becomes a way for patients to face the barriers to recovery, and frees them to perform physically in a way that becomes consistently effective.

In conclusion, the role of physical therapists in tertiary care is to ensure effective physical improvement in the midst of psychological and vocational challenges. They must be analytical enough to distinguish physical limitation from poor motivation and patient enough to encourage the poorly motivated to change. While being responsible for specific rehabilitation of the injured anatomic area, they must take the holistic view of improving the way each patient functions in the social environment.

THE ROLE OF OCCUPATIONAL THERAPY IN TERTIARY CARE

In the two phases of treatment that precede tertiary care, 80 to 90 percent of the injured workers recover and return to work. However, for the chronic pain population a third phase of treatment is necessary. This is known as tertiary care, which mainly treats the chronic pain population (those people that do not recover with traditional therapy). The occupational therapist (OT) is challenged to work with these difficult patients, who display a number of barriers to return to a functional lifestyle. Some examples of these chronic traits are psychological barriers, uncertainty about future job plans, and lack of self confidence with regard to physical functioning.

These traits create a patient who may be difficult to motivate towards rehabilitation and requires a cohesive team approach in order to be successful. The occupational therapist plays an important role on the rehabilitation team. The main goals of occupational therapy are to (1) gain the patients' trust, (2) motivate the patient to increase physical progression, and (3) modify behavior (pain, work, social).

Gaining the patients' trust is the first and most important goal for the OT to accomplish because without trust the rehabilitation process will not be successful. Many times chronic pain patients feel they have been misled by the medical and/or workers' compensation system, and often this is not their first experience with rehabilitation. So they ask, why will this therapy work when others have not? To gain trust, occupational therapists take on the roles of supporter, educator, nurturer, and encourager. They support the legitimacy of the patient's injury by acknowledging and understanding pain complaints. They educate the patient about the functions of the body and how it responds to rehabilitation and also teach the positive effects of using ice and stretching to decrease and control pain. They let the patient know the goals of rehabilitation (to increase function) so that there are no secrets or surprises. They also nurture the patient and encourage physical progress.

In the beginning phase of rehabilitation, physical progress can be slow. The patient must gently be encouraged to move. This takes patience and finesse on the part of the occupational therapist. The therapist must carefully "push" the patient to increase physical function and compliment any signs of progress made. Progress is most important functionally, but can be noted as improved physical appearance or affect. The next step to be taken, once trust has been established, is to motivate the patient to increase ability to perform whole-body task performance.

The OT works with the treatment team to find out how to individually motivate each patient. Many times there are a number of nonphysiologic barriers to rehabilitation that make this chronic patient population difficult to motivate. Examples of nonphysical barriers might be fear of returning to work, fear of reinjury, or family and financial problems. Understanding these barriers makes it possible to motivate patients based on their individual needs. Examples of different techniques are gentle encouragement and support, firm limit setting and structure, and repeated education and assurance.

Motivating the patient goes hand in hand with increasing physical progression. Physical progression in occupational therapy is based on functional whole-body activities such as lifting, bending, twisting, reaching, pulling, pushing, climbing, walking, and carrying. A baseline test is performed to determine functional capacity. The test consists of quantifiable whole-body lifting using the PILE technique and a lift task as well as an obstacle course measuring other functional activities like bending, reaching, crawling, pushing, and pulling. Once the tests have been interpreted, a training program can be established in the above-mentioned functional areas.[8] Daily progression with increased weight and/or repetition of activity is encouraged. The patients are guided in a positive atmosphere to push themselves through the pain threshold to the "other side" of functioning.

The OT should focus on what patients can do today despite the pain they are feeling. Pain symptoms localized in the injured areas are usually magnified and can be dealt with safely as progression increases. Pain is addressed through education and information related to the body's healing process. It is also evaluated for physical signs so that program modifications can be temporarily made. At various stages in the program the patient is tested to determine if physical progress is being made (outside the training area). Testing can also assist in showing how much to push a patient in training. Testing, consistent with training, can show good effort on the patients' part that confirms they are trying to succeed. The goal of OT functional progression is to meet or exceed previous job demands, permitting return to work and leisure activities.

Modifying pain behavior in the chronic patient is one of the most challenging goals to achieve. Pain is an escape from facing nonphysical barriers because it cannot easily be observed. All pain is legitimate; however, the reasons for increased pain may or may not be organic. Regardless of OT perceptions, the patient is evaluated for physical signs, but then is encouraged to move forward functionally and cope with the pain in a constructive way. The patient is taught to take responsibility for controlling pain episodes by using ice and stretching techniques. Pain episodes generally arise during OT training sessions with work simulations because the patient relates this activity to returning to work (which can be terrifying). The goal for modifying behavior is to enable patients to work independently and take responsibility for themselves with regard to pain and injury. The occupational therapist is trained to observe work feasibility behavior such as whether the patient can work within a time frame, focus on tasks, sequence in proper order, organize time, and so forth. These behaviors are subtly addressed during work-simulation activities. A work-simulation plan is created, based on past and/or future job plans, in which the patient is an active participant. Work simulations are timed or repetitious activities tied to specific job requirements and position tolerances individualized to the needs of each patient. During this time, suggestions are made on the use of proper body mechanics and energy conservation techniques. Behavioral observations should also be reported to the treatment team as possible "red flags" that might be obstructions to returning to normal activities of daily living. Suggestions on job modifications can be approached with the medical case manager, whose role is to be involved directly with the employer and the patient. In this final phase of tertiary treatment the patient should be able to independently and confidently achieve his or her job simulations and return to work.

In conclusion, the occupational therapist initiates the rehabilitation process by nurturing and supporting the patient in order to gain trust. Following this, the patient is educated and encouraged to grow and move forward functionally, coordinating with PT attention to the injured "weak link" anatomic area. Finally, the therapist must modify behavior to create an independent, confident, and productive patient to return to society as productive as his or her residual impairment permits.

THE ROLE OF THE PSYCHOLOGIST IN TERTIARY CARE

Health care professionals are becoming increasingly aware of the need to address psychosocial variables in the rehabilitation of patients with chronic low back pain.[9] Many of these patients suffer from diagnosable clinical syndromes, such as major depressive disorders, somatoform disorders, substance abuse disorders, and anxiety disorders, and recent data suggest that the onset of certain psychological syndromes are present prior to the onset of chronic back pain.[10] In addition, chronic pain has repeatedly demonstrated the considerable toll it can exact on a person's mental health, and these factors can interfere with the patient's return to a productive lifestyle.[11] It therefore becomes imperative that care providers allow the patient to have the opportunity to overcome psychosocial difficulties that could slow or prevent their recovery. This section will attempt to describe how psychological assessment and treatment are incorporated into the functional restoration program at PRIDE.

The Initial Assessment

Prior to entering treatment, the patient is interviewed by a professional in the Psychology Department. The intake interview serves to identify and address issues that may be critical to the patient's treatment (see Exhibit 31–1). In order to thoroughly assess the presence of such issues, preliminary psychological assessment data are obtained and later integrated with interview findings. Tests include the quantified Pain Drawing, the Millon Visual Analog Scale, the Beck Depression Inventory, the Symptom Check List-Revised (SCL-90-R), and the Hamilton Rating Scale. The use of these instruments with this population is discussed in detail elsewhere.[12] The clinician then communicates with the patient and other professionals regarding the patient's specific needs.

It is not infrequent that patients will present with a combination of stressors they are negotiating, psychological symptoms, and pragmatic difficulties that can interfere with program participation. For example, a patient may present with depressive symptomatology, family stressors, financial strain, transportation problems, and concerns about his or her ability to reenter the work force. In such a case, the psychology professional can serve both as a consultant to the patient (e.g., by beginning to educate the patient regarding the symptoms and treatment of depression) and as a liaison between the patient and other professionals (e.g., communi-

Exhibit 31–1 Critical Issues Investigated during the Intake Interview

COUNSELING CONSULT
Critical Issues

NAME: ______________________ DATE: __________

COUNSELOR: ______________________

CRITICAL ISSUES:

1. Distance from clinic.
2. _____ level of clinical/subjective depression with/without vegetative signs. (B = _________ H = _________)
3. _______________ level of anxiety/tension/stress.
4. Absence of psychological discomfort with current situation due to [A] Denial, [B] Habituation to disability, [C] Possible secondary gain, [D] Probably invalid paper and pencil testing.
5. Financial disincentives/comfortable financial situation.
6. Financial stressors.
7. Cognitive impairment/cannot rule out neuropsychological difficulties.
8. Post history of dependency issues/suspect dependency issues (drug or alcohol abuse).
9. Surgery not ruled out in patient's judgment.
10. Child care problems.
11. Marital difficulties.
12. Extraneous litigation causing anxiety (pending criminal charges for theft, etc.)
13. Significant level of pain sensitivity and/or fear of reinjury.
14. Family overprotective/family does not support functional restoration.
15. Alienation from [A] Medical profession [B] Workplace [C] Legal system [D] Insurance company.
16. Use of pain medications and/or muscle relaxants.
17. Language barrier/limited understanding.
18. Prior athletic background.
19. Extreme inactivity/extended period of disability.
20. Patient appears overweight and/or deconditioned.
21. Multiple medical difficulties, including __________, __________, and __________.
22. Has auditory and/or visual problems, with possible thought disorder.
23. Former job is not available/no future work plans.
24. Patient continues to work despite injury.
25. Rule out Axis II.
26. Other:
27. Other:
28. Other:

Source: Copyright © 1992 PRIDRIVE, Inc.

cating with the doctor regarding the patient's mental status, communicating with the disability counselor regarding the patient's financial status and transportation problems, etc.). In this way, the therapist can assist in focusing both the patient and staff on coping with those issues that might undermine the patient's recovery.

Preliminary Program

In the preliminary program or *preprogram*, the focus is upon educating patients, preparing them to engage in the intensive phase (discussed below), and stabilizing them psychologically. During this two- to six-week period, the Psychology Department provides patients with brief individual psychotherapy, an introduction to group therapy, and classes in stress management. Counseling serves to address the critical issues identified in the initial interview, including psychopharmacologic monitoring, while considering general psychological issues (such as the patient's mental status) and aligning patients' goals with the treatment program's goals. The therapist attempts to ascertain whether treatment recommendations are being followed by inquiring and monitoring patients (e.g., are patients tapering narcotic or tranquilizing medications? Are they taking any prescribed antidepressant regularly? Do they understand the rationale of stretching/icing?). He or she will also frequently attempt to address pragmatic issues with patients (such as child care, living arrangements, or transportation). In addition, the therapist helps patients to develop pain coping strategies, to distinguish pain due to injury from pain due to improving function, to direct patients' energies toward improving function, and to help patients improve compliance. During this period, patients may reveal financial, legal, and/or interpersonal disincentives to recovery, to which the clinician must be vigilant and responsive. The clinician must also be astute regarding the presence of substance abuse and malingering, which may be more appropriately addressed in other settings. Treatment is coordinated with other staff members during staffings and consultations.

Intensive Phase

Following the preprogram phase of treatment, patients enter a three-week intensive phase. They are assigned to an individual counselor with whom they meet at least three times per week based on individual needs. Initially, the patients undergo a complete psychological evaluation. This assessment includes the administration of (1) the Wechsler Adult Intelligence Scale-Revised (WAIS-R), a measure of intellectual functioning; (2) Trails A & B, a general neurological screen, which is integrated with WAIS-R and history to determine the need for further neuropsychological testing; (3) the Minnesota Multiphasic Personality Inventory, a well-researched self-report true/false questionnaire providing information relevant to treatment; (4) the Structured Clinical Interview for the DSM-III-R, a clinician-administered interview providing diagnostic information; (5) a clinical interview, an unstructured interview focused on history and return-to-work issues; and (6) readministration of tests previously discussed for monitoring. Cross-cultural issues are taken into account during interpretation. Patients are provided with feedback, and salient information is communicated with staff.

Treatment includes individual therapy, biofeedback for stress management, support groups, and family classes. Treatment is short term, goal oriented, and directive. Emphasis is placed upon helping patients return to a productive lifestyle through overcoming skill deficits (e.g., assertiveness training, social skills development, stress inoculation, relaxation training, and building coping repertoires). In addition, patients are educated regarding common emotional responses to injury and pain. These include depression, anger, anxiety, fear, entitlement, and helplessness.[13] They may be taught to distinguish between productive behavioral responses and self-defeating responses, such as denial, blaming, acting out, or complaining. Progress is monitored through weekly readministration of various test instruments and interviews.

Supplementing individual treatment are support groups, which help to decrease interpersonal isolation, reinstill a sense of self-efficacy, and enhance coping. Family classes help to educate the patient with respect to family dynamics, while encouraging autonomy and discouraging the patient's adoption of a "sick role." Care is continually coordinated with staff during regular meetings.

Discharge and Follow-up

Prior to discharge, the therapist and patient work to develop a support network outside PRIDE. In addition, by the time of discharge, most patients have a concrete work plan and living arrangements coordinated. Some patients will need additional assistance and are referred to other health care providers. This determination can be made on the basis of clinical information and discharge test results. When patients return for follow-up care, a brief follow-up session ensues to reinstill important concepts learned during the intensive phase.

THE ROLE OF DISABILITY MANAGEMENT IN TERTIARY CARE

Disability management (also termed medical case management) in tertiary care differs markedly from that in primary or secondary care. Tertiary case management has, as its overriding priority, the goal of vocational and social reactivation.[14] The medical case manager will therefore generally be a master's-level vocational rehabilitation professional.

The process of vocational reactivation—that is, movement from disability to productivity (specifically important in work injuries)—is, as with all tertiary care services, carried out with physician direction and support. The "pathology" in the chronically disabled patient frequently has, as its primary component, vocational discord,[15] which makes it crucial to treat this aspect of the patient's condition. The disability manager therefore provides repeated emphasis on increase of social function rather than pain relief,[16] while the physician is always available for support to address the myriad of physical symptoms that invariably present themselves as vocational issues unfold (and conversely, frequently become less pronounced as vocational issues become resolved). Resolving these issues is by no means an easy task with an individual who is habituated to disability. The tertiary medical case manager will therefore need to possess counseling skills and be familiar with the psychological mechanisms at work. Various forms of resistance will appear, and negotiating vocational movement with a patient becomes very much akin to a psychotherapeutic relationship.[17]

The primary responsibilities of the disability manager fall into three broad categories: (1) program compliance and performance; (2) communication, including postprogram support and tracking; and (3) occupational planning and sequencing. There is much overlap among these areas of responsibility, and each is a process that begins to be addressed from the initial contact and develops through to the end of treatment. As with all aspects of tertiary care, these duties are carried out in the context of the interdisciplinary team.

Program Compliance and Performance

Generally, chronically disabled patients will find themselves in rather dismal financial circumstances and/or heavily dependent on (or depended upon by) others. Therefore the case manager will initially problem solve with the patient on issues of transportation, child care, finances, and so on. There may be referral to an external agency, and the case manager will have a list of all the support service agencies in the area. Although flexibility is frequently called for, two guiding principles serve the case manager here:

1. Don't solve the problems for the patient. Generating a sense of personal control (and responsibility) starts here.
2. Attendance and timeliness in treatment are mandatory and essential to work reintegration.

The case manager is the tertiary care "voice of social responsibility." He immediately contacts a "no-show" and addresses tardiness. This is typically done through a hierarchy of verbal to written contracts, culminating in a "no-tolerance contract" after which any further violation results in termination of treatment. These contracting principles also apply to performance in training, testing, and other medically directed activities. All contracting is done with the participation of the other team members. Although successfully turning the chronically disabled around requires this "tight structure," total rigidity in enforcing rules would be fruitless. Fear, pain, and resistance typically respond to a supportive, nurturing, but "tough love" approach.

Communication

The medical case manager functions as the primary arbiter between the doctor and the patient, the patient and the external "players," and the treatment team and the external community. Aligning goals and providing clear communication regarding direction and duration of treatment are vital from the beginning. The external community may consist of other medical care providers, attorneys, employers, state commissions, third-party payors, and rehabilitation professionals, any of whom could conceivably "pull the plug" on treatment, leaving the patient to become even more entrenched in chronicity. Chronically disabled patients themselves may go to great lengths to maintain disability by splitting the various players in the disability system (employers, insurance, attorneys, etc.) off against each other. They can often be very effective in sabotaging their own treatment. Goals for each "player," however, may be quite divergent. For this reason, the case manager must be an effective negotiator and astute clinician on the lookout for hidden agendas. Where goals cannot be aligned, clear explanation and justification for tertiary care goals must be communicated. Tertiary care goals are stated as follows:

1. Return the patient to productivity.
2. Maximize function, thus minimizing pain.
3. Reduce or eliminate future medical utilization.
4. Avoid reoccurrence of injury.

The case manager communicates to the patient his or her rights and benefits as mandated by applicable laws. This does not constitute "legal" advice but merely a statement of the facts as laid out in pertinent state, federal, or other applicable statutes. On the negative side, the disability manager makes clear the consequences of certain actions such as medical noncompliance and, *gradually*, the concept of treatment termination. Employer communications are typically important to the degree that the patient will be returning to work there. The disability manager conducts a job analysis and negotiates the possibilities of short-term light duty, an alternative position, or transition to full-time/full-duty employment. Contradictions, such as employer hostility accompanied by having a position available for the injured worker to return to, are a prescription for failure. Alteration in either the hostility or the work plan is advisable.

Attorney communication can be invaluable where compliance/trust issues are involved (where attorney goals are mitigating the effects of injury and returning to productivity). It is important for the disability manager to communicate tertiary care goals, including the concept of maximum medical recovery at the end of tertiary care so the attorney can counsel his or her client appropriately.

Certainly the third-party payors involved are interested in outcomes, and as a neutral medical care provider, the disability manager presents goals independently (with documentation of effort and compliance) to representatives of both employer and employee.

When other medical care providers are involved, splitting their ranks becomes a very common tactic for the patient reluctant to disengage from disability. The disability manager will communicate often with these other providers so that the patient hears the same message from all involved. This communication is also vital to ensure case resolution, which is implicit in tertiary care, as the highest level of restorative care available.

Long-term tracking of, and ongoing communication with, functional restoration patients serves two functions:

1. *Outcome monitoring:* Such tracking provides continuing support for functional restoration as valid and effective in accomplishing return-to-work and medical utilization goals.
2. *Clinical utility:* Postprogram quantitative evaluations (of physical and functional capacities) provide feedback and motivation for patients to maintain their physical gains and functional status.

For those patients who do not return for postprogram testing, support and tracking are done by phone contact. In person or by phone, this contact is done at three months, six months, one year, and two years following completion of functional restoration, and has emerged as a significant clinical and research component of functional restoration.

Occupational Planning

As stated earlier, vocational discord is often at the heart of chronic disability with an industrial injury population. The process of vocational planning starts gradually and culminates with a sound occupational plan and sequence (and a backup plan) mandated as part of the medical protocol during the mid- to latter phases of treatment. Ambivalence regarding the vocational plan can usually be seen as an increase in symptom complaints and/or lack of physical progress in OT/PT areas. Two principles here should guide the disability manager:

1. *"The patient owns the plan."* The occupational plan is not a spoon-fed reiteration of what the counselor thinks the patient should do. It is driven by the

patient's work history, vocational aspirations, skills, real life considerations, motivation, and reality-testing abilities.[18]

2. *The decision to return to a previous job (or any job) is a personal one, not a medical mandate.* Functional restoration is about *giving options*, not taking them away.

The vocational plan may take on any of several faces, the point being to capitalize on momentum to return to productivity immediately after functional restoration is complete. Interest testing, retraining options, skills assessment, transferable skills analysis, job exploration, labor market survey, and job history are just a few of the vocational reactivation tools that may be utilized by the case manager. If a previous job is available to the injured worker and he or she can do it, then returning to that job may be preferable for economic, work record, and momentum considerations. State agencies and other support services are solicited for retraining, on-the-job training, or job search assistance, but are most effective when such assistance is contingent on attaining medical goals necessary for maximizing function.

Self-employment is not usually advisable after functional restoration, although it may not be an unrealistic long-term goal. The chronically disabled patient must start to reform the habits of enduring full-time employment, adapting to a regular work schedule, and responding to authority as part of his or her recovery. These things are not easily accomplished in a self-employment situation, which can quickly destroy the dramatic momentum frequently seen immediately following functional restoration.

The disability manager in a tertiary care setting assists in treating an aspect of the chronic pathology that appears to be related to chronic disability: that of occupational discord. Left unattended, chronic patients will be found expending great energy and resources to hang on to their only perceived security—the disability system, be it workers' compensation, social security, long-term disability, or welfare. This will serve neither patients nor the society in which they live.

THE PHYSICIAN'S ROLE IN TERTIARY CARE

The physician is responsible for medical direction, providing a "captain-of-the-ship" leadership role for the interdisciplinary team. Together with nurses and/or physician's assistants as physician extenders, the physician plays a very specific role in the different phases of treatment. The medical direction is usually provided by an individual who specializes in rehabilitation care, but may also emerge from training in physiatry, orthopedic or neurological surgery, psychiatry, occupational medicine, or internal medicine. At one time or another, the skills of all these specialists are required.

The Preprogram Phase

During this phase and its antecedent evaluation, the primary goal is making a diagnosis and instituting appropriate treatment to address any remaining pathological issues. In tertiary care, however, structural issues have usually already stabilized or been adequately addressed with surgery, normal healing, and/or the physiological sequelae of permanent impairment (e.g., scarring, degenerative arthritis). As such, the history and physical examination, appropriate structural testing, and additional invasive treatment occupy a limited period of time. On the other hand, identification of physical (the injured "weak link") and functional (whole-body) capacity deficits consistent with the deconditioning syndrome, and identification of psychiatric/psychosocial barriers to recovery are more vital to the diagnosis. Most chronically disabled workers have all three: a pathophysiological, functional, and psychosocial diagnosis. The interpretation of the quantitative evaluation for physical/functional deficits and the counseling consult for psychosocial barriers to recovery assist the physician in preparing the preprogram treatment plan.

Based on the degree of physical deficit and psychosocial involvement identified, a preprogram period is determined, identifying duration and frequency of visits within specific utilization limits. Certain goals are targeted by the physician prior to commencement of the intensive phase:

1. elimination of habituating medications (narcotics, tranquilizers, etc.)
2. stabilization on appropriate non-habit-forming neuroleptic and/or anti-inflammatory medication
3. stretching and physical disinhibition to permit participation in the aggressive "sports medicine" progressive resistance training of the intensive phase
4. providing any remaining structural diagnostic or invasive treatments, acceptable to the patient, that might alter the natural history of the disability process

Intensive Phase

During the three-week intensive phase, with 8 to 10 hours of daily treatment, a great deal of coordination is required through the daily visits of the physician and/or physician extenders. While most training and education is provided by the interdisciplinary team, physician availability for handling specific medical issues is essential. Patient perception of "injury" is commonplace and must be addressed. Medication issues requiring adjustment and changes are also vital. Classes on pathophysiology, physical testing procedures, and patient access for questions and answers are vital aspects of the physician relationship to the patient cohort.

Twice weekly, patient staffings are held, providing the entire treatment team an opportunity to meet with the physicians for in-depth analysis of treatment progress, and planning for intermediate steps. The formal staffings follow previous team meetings of a smaller, nuclear team involved with treating component patient cohorts. In the formal staffings, presentation by each member of the treatment team, consistent with the patient's program level, takes place. At the conclusion of the patient presentation, discussion takes place concerning the remaining short- and long-term problems, and the priorities for treating them. In this way, the treatment plan initially embarked upon is constantly being compared to the results of that treatment, with modifications taking place as required, based on actual patient performance and test reevaluation.

Discharge and Follow-up Phases

When the patient is discharged from the intensive phase, a period of time for preparation for work reintegration is termed the *follow-up phase*. Medical utilization during this phase is generally low, with the intensity and duration of services being determined by the patient's physical/functional capacity, the match between job demands and these capacities, job availability, patient attitude, and transferrable skills. Within six weeks at the latest, a medical endpoint can be established, appropriate limitations identified, a rational occupational plan implemented, and a fitness maintenance program taught. The interdisciplinary team works on continued training of the injured area, work simulation, and psychosocial preparation for reintegration into a more normal set of social responsibilities consistent with premorbid behaviors. During this period, medication maintenance continues to be a specific problem, often tied to the patient's tendency to relapse.

The physician plays an important role in helping the patient overcome the anxieties of this period by remaining "steady in the boat" while avoiding the various rabbit trails leading to resumption of disability behaviors that characterize common patient efforts to "jump ship." The patient continuing to weather the course is usually a grateful recipient of functional restoration. The one who is, for whatever reason, committed to disability as a solution to ambivalence about being a productive citizen is always resistant, usually resentful, and, most unfortunately, often litigious. The physician's documentation stands as evidence of the interdisciplinary team's well-intentioned efforts to motivate the patient to the highest functional levels and the opportunity to resume relatively normal social functioning. Significant financial issues relative to the compensation laws for *permanent impairment* and/or *permanent disability* are an important part of the physician documentation and conceptual framework. Providing appropriate documentation to the nonmedical disability system (insurer and attorney representatives of the employer and employee respectively) to assist in resolving financial issues, future medical benefits, vocational rehabilitation opportunities, and future employer-provided benefits is a vital task of the physician. This frequently necessitates involvement in medico-legal procedures, including subpoenas, letters of testimony, depositions, and even court appearances. This involvement "goes with the territory" of dealing with an essentially adversarial workers' compensation system in which the interests of employer and employee may be divergent, and sometimes irreconcilable.

CONCLUSION

This chapter has attempted to provide a microscopic dissection of the functioning of the interdisciplinary physician-directed team involved in functional restoration, a model tertiary care program. Such specialized care is intensive, requiring the involvement of expensive professionals providing specialized treatment within a setting of work injury not commonly understood as "catastrophic illness." As such, care must be taken to keep the focus of tertiary care narrow enough to provide this type of care only to those patients who need it, while still preserving the continuity of care from secondary treatment so that long periods of ineffective and time-consuming therapy do not need to take place before tertiary treatment for appropriate patients is invoked. If the hierarchy of primary, secondary, and tertiary nonoperative treatment can be understood by physicians and allied health care providers and can be integrated into the preoperative planning by the surgical community, great strides in patient outcomes can be anticipated for the 1990s and beyond.

NOTES

1. G. Waddell et al., Non-Organic Physical Signs in Low Back Pain, *Spine* 5 (1989): 117–125.
2. L. Matheson, Integrated Industrial Rehabilitation (Paper presented at a one-day symptom magnification syndrome workshop, Dallas, November 1992).
3. J. Porterfield and C. DeRosa, *Mechanical Low Back Pain: Perspectives in Functional Anatomy* (Philadelphia: W.B. Saunders Co., 1991).
4. T. Mayer et al., Use of Noninvasive Techniques for Quantification of Spinal Range-of-Motion in Normal Subjects and Chronic Low-Back Dysfunction Patients, *Spine* 9 (1984): 588–594; S. Smith et al., Quantification of Lumbar Function—Part 1: Isometric and Multi-Speed Isokinetic Trunk Strength Measures in Sagittal and Axial Planes in Normal Subjects, *Spine* 10 (1985): 757–764.
5. M. Pollock et al., eds., *Exercise in Health and Disease: Evaluation and Prescription for Prevention and Rehabilitation* (Philadelphia: W.B. Saunders Co., 1984).

6. T. Mayer and R. Gatchel, *Functional Restoration for Spinal Disorders: The Sports Medicine Approach* (Philadelphia: Lea & Febiger, 1988).
7. Matheson, Integrated Industrial Rehabilitation.
8. L. Curtis et al., Physical Progress and Residual Impairment after Functional Restoration, Part III: Isokinetic and Isoinertial Lifting Capacity (in press); N. Kishino et al., Quantification of Lumbar Function—Part 4: Isometric and Isokinetic Lifting Simulation in Normal Subjects and Low-Back Dysfunction Patients, *Spine* 10 (1985): 921–927; T. Mayer et al., Progressive Isoinertial Lifting Evaluation, Part I: A Standardized Protocol and Normative Database, *Spine* 13 (1988): 993–997; T. Mayer et al., Progressive Isoinertial Lifting Evaluation, Part II: A Comparison with Isokinetic Lifting in a Disabled Chronic Low Back Pain Industrial Population, *Spine* 13 (1988): 998–1002; T. Mayer et al., Progressive Isoinertial Lifting Evaluation: Erratum Notice, *Spine* 15 (1990): 5.
9. L. Conte and T. Banerjee, The Rehabilitation of Persons with Low Back Pain, *Journal of Rehabilitation* April/May/June (1993): 18–26.
10. P. Polatin et al., Psychiatric Illness and Chronic Low-Back Pain: The Mind and the Spine—Which Comes First? *Spine* 18 (1993): 66–71.
11. D. Fishbain et al., The Prediction of Return to the Workplace after Multidisciplinary Pain Center Treatment, *Clinical Journal of Pain* 9 (1993): 3–15.
12. Mayer and Gatchel, *Functional Restoration.*
13. Ibid.
14. W. Stolov and M. Clower, eds., *Handbook of Severe Disability* (Washington, D.C.: U.S. Dept. of Education, Rehabilitation Services Administration, 1981).
15. A. Cobb, ed., *Medical and Psychological Aspects of Disability* (Springfield, Ill.: Charles C. Thomas, 1978).
16. R. Goldenson, ed., *Disability and Rehabilitation Handbook* (New York: McGraw-Hill, 1978).
17. B. Caplan, ed., *Rehabilitation Psychology Desk Reference* (Gaithersburg, Md.: Aspen Publishers, Inc., 1987).
18. H. Rusalem and D. Malikin, eds., *Contemporary Vocational Rehabilitation* (New York: University Press, 1976).

Chapter 32

Rehabilitation of Chronic Musculoskeletal Disorders at the Clinic of Rheumatology and Institute of Physical Therapy, University Hospital, Zurich

Michael Oliveri, Marie-Louise Hallmark, and Heinz O. Hofer

The concept of *deconditioning syndrome*[1] has led to a new understanding of rehabilitation for chronic musculoskeletal disorders. The sports medicine approach, functional capacity evaluation, and work hardening[2] have had an important influence on recent rehabilitation concepts in Switzerland.

In Europe, training as a therapeutic tool has strong roots. The use of mechanical means for the application of exercise in therapeutics was first introduced and employed in a comprehensive manner by Dr. Gustav Zander in Stockholm in about 1857.[3] A modern concept of therapeutic training based on biomechanical principles of manual therapy was developed by the Norwegian physiotherapist Holten.[4] Two other Norwegian physiotherapists, Evjenth and Gunnari, transferred their experience with training athletes to therapeutic training with deconditioned patients. They designed the sequence training system, well known in the United States and Canada.[5]

At our clinic, carefully controlled therapeutic training is regarded as a basic element in the rehabilitation of patients with chronic musculoskeletal disorders. Additional treatment modalities (manual therapy, deep massage, electrostimulation) can help to achieve better results in rehabilitation. Psychological support or social and vocational consultation are offered to patients when psychosocial barriers seem to interfere with training progress.

Education of proper body mechanics and correct handling of load (Low Back School) are also part of our rehabilitation program. If load factors at the workplace are considered too high, work area analysis and job modification are carried out. Functional capacity evaluation and work-related training have been introduced in

---

We thank Professor B. Michel for the review of the manuscript and Agnes Verbay for her assistance in preparing the photographs.

our clinic. For further establishment of these work-related procedures, we will have to address problems created by the social and economical environment, particularly the Swiss insurance system.

THERAPEUTIC TRAINING

The purpose of therapeutic training as a separate treatment unit is to enhance patients' muscle and joint functions and tissue tolerance to stress as well as the patients' psychophysical attitudes toward the demands of a work-hardening program or work itself.

In our clinic, therapeutic training starts using the Medical Training Therapy (MTT) concept introduced by Holten.[6] The training devices used are mainly pulleys and small weights in the form of dumbells and barbells. The program involves 8 to 10 exercises. Due to the high degree of possible variation of each exercise,[7] the program can be adapted very specifically to chronic or acute situations and tailored individually. The main variables for the exercises are listed in Table 32–1.

According to the purpose of the respective exercise, the appropriate number of repetitions is chosen (Table 32–2).[8] The required weight should be up to 80 to 85 percent of the maximum weight for a defined number of repetitions. Usually the training includes two to three sets of the chosen 8 to 10 exercises.

Local stabilization is an essential precondition for movement control. Therefore training very often starts with locally directed resistance (short leverage arm). Lo-

Table 32–1 Variations for Exercise Performance

Variables of Exercise	*Possibility for Progression in Training*
static or dynamic relative to the body region that should be trained	static → dynamic
number of repetitions/weight	high number of repetitions/small weight → low number of repetitions/high weight
leverage arm	short → long
range of movement	small → entire range of motion
speed of movement	slow → fast
intermittent or continuous resistance (lowering the weight)	intermittent → continuous
body positioning (e.g., lying, sitting, standing)	supported → unsupported

Table 32–2 Goal of Exercise and Corresponding Number of Repetitions for Each Set

Goal of Exercise	*Number of Repetitions*
flexibility; local metabolism and exchange of chemicals; pain control	>40
muscular endurance	25–40
muscular strength and endurance	15–25
muscular strength; local stabilization; tissue resistance for high load	<15

cal strength and coordination will be trained before beginning exercising movements with long leverage arm and more global functions (Figures 32–1 to 32–8). In some conditions in which movement causes irritation (e.g., narrow spinal canal), a training with only 5 to 10 repetitions, high resistance and limited range of movement may be adequate.[9] In addition to MTT, therapeutic exercises from Functional Kinetics, introduced by the Swiss therapist Klein-Vogelbach, are used.[10]

When patients get used to training and have made some progress, they are introduced to the sequence training exercises.[11] More load is applied.

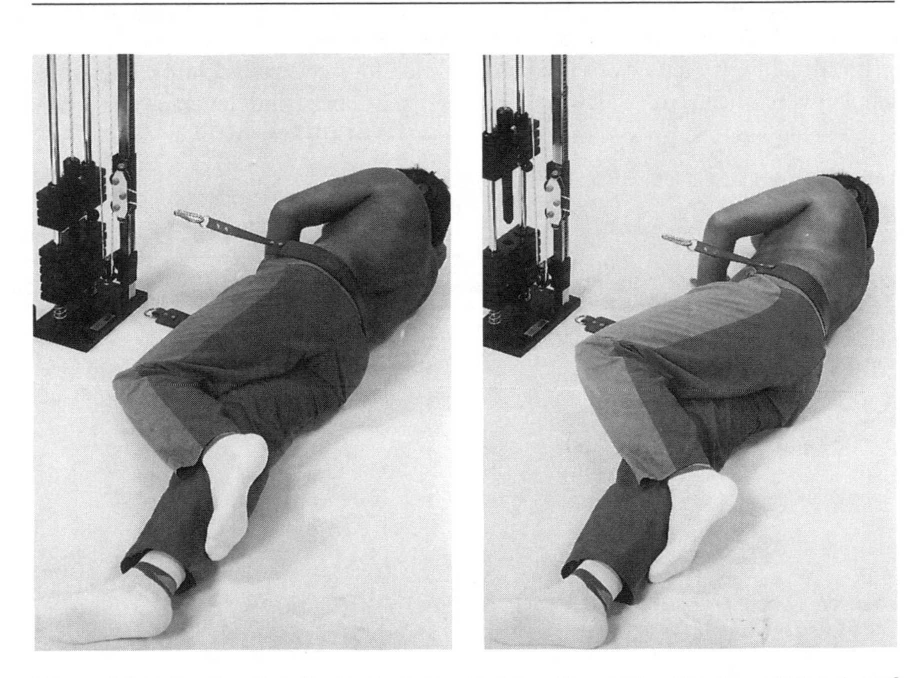

Figure 32–1 Lumbar Rotation in the Lying Position: Forward and Backward Rotation of the Pelvis

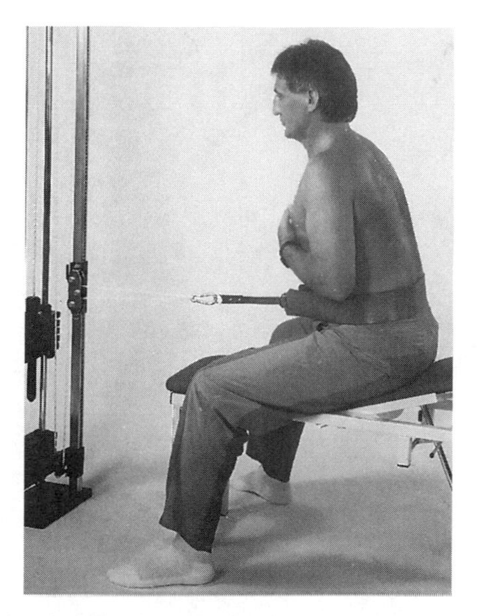

Figure 32–2 Lumbar Stabilization Training in the Sitting Position with Short Leverage Arm: Forward and Backward Bending with Straight Back

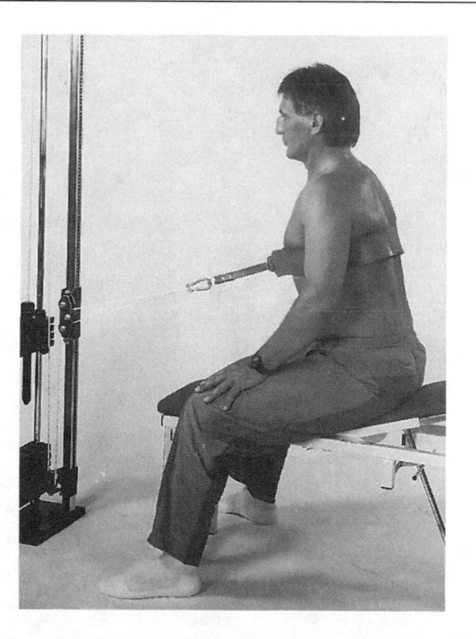

Figure 32–3 Lumbar Stabilization Training in the Sitting Position with Long Leverage Arm: Forward and Backward Bending with Straight Back

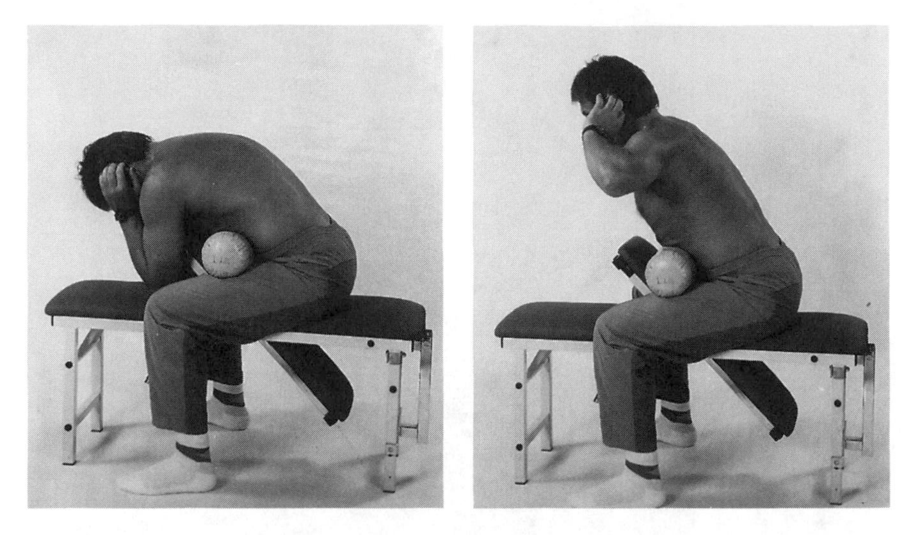

Figure 32–4 Upper Trunk Extension: Lumbar Spine Stabilized

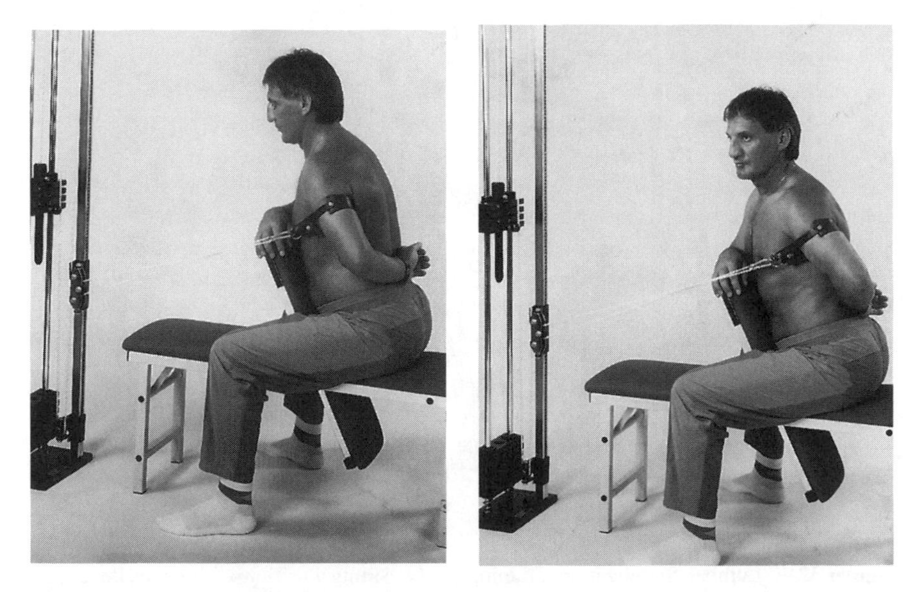

Figure 32–5 Trunk Rotation in the Sitting Position

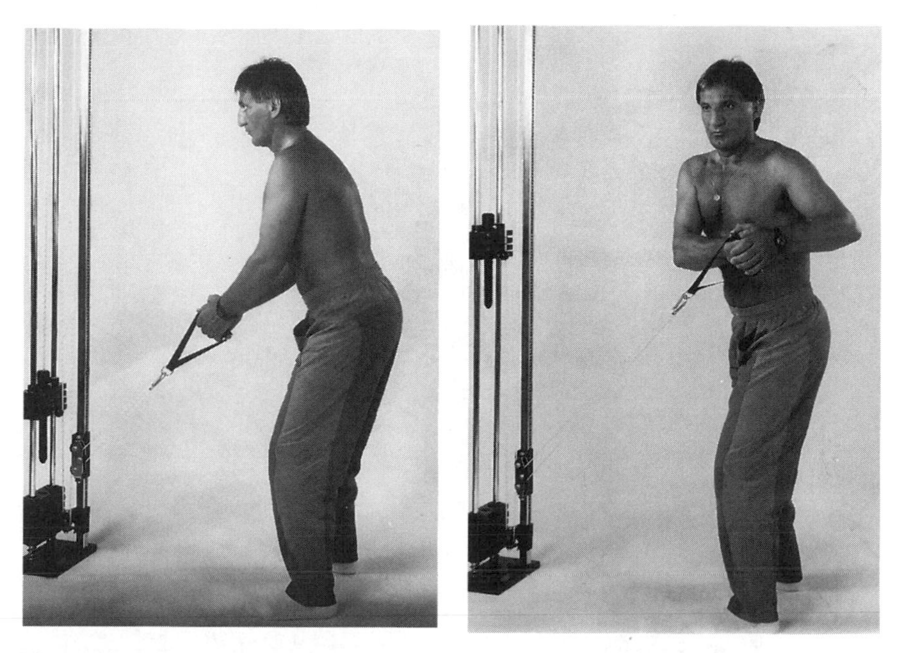

Figure 32–6 Trunk Rotation in the Standing Position with Long Leverage Arm

In addition to muscular strength and endurance training, patients perform programs for cardiovascular endurance (bicycle or upper body ergometer, low-impact group gymnastics, flipper swimming). Exercise mobility (including self-mobilization such as the McKenzie exercises[12] and muscle stretching[13]) are also part of the program.

We believe that therapeutic training itself is a useful tool to prepare the patient for later handling of workloads. For instance, training of lumbar stabilization against resistance helps to adapt to a proper biomechanic movement pattern that is needed for a correct lifting technique and for other handling of workloads. Furthermore, the patient learns about his or her actual physical performance level and limitations, learns to work independently, and realizes that gain in functional capacity is attainable.

ADDITIONAL THERAPEUTIC MODALITIES

Part of the problem of chronic musculoskeletal disorders may be not only deconditioning, but also persistent functional disorders such as internal joint dysfunction, imbalance in or nonphysiological pattern of muscular activity, pathological changes, or persistent irritation of soft tissue (including neurological

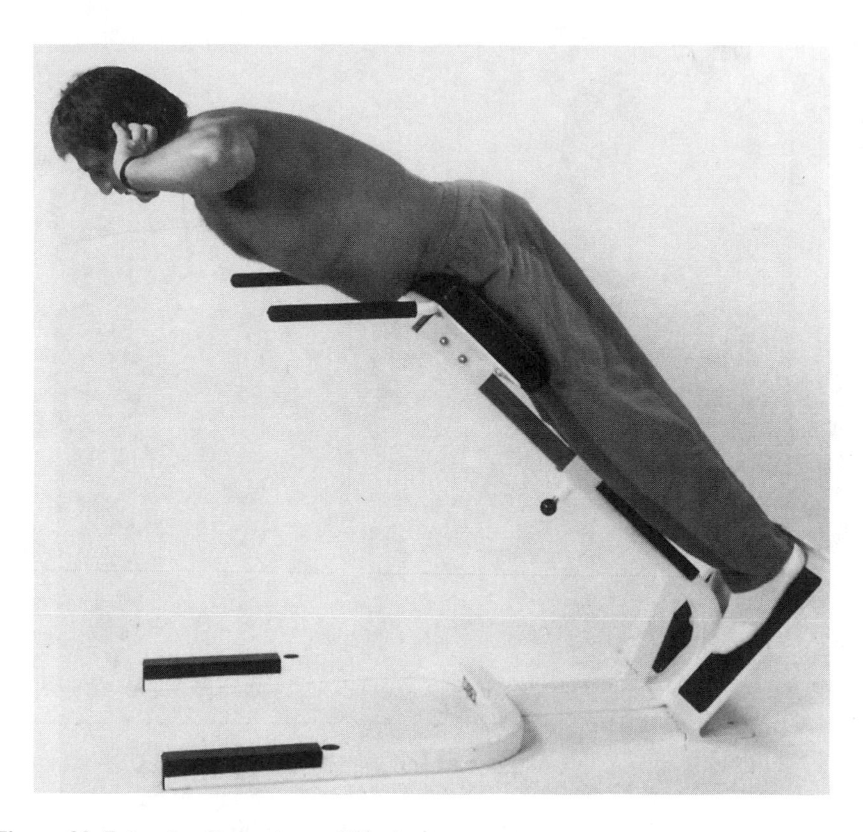

Figure 32–7 Lumbar Extension and Flexion

tissue). Our experience shows that manual therapy (as outlined by Maitland,[14] Kaltenborn[15] and Mulligan[16]) or deep massage[17] can help to achieve better results with therapeutic and work-related training. Sometimes mobilization techniques for the nervous system are also applied.[18] Electrostimulation may help to reduce the patients' medication. Passive modalities alone do not solve the problems of chronic musculoskeletal disorders and are generally used far too often. However, they may represent a useful adjunct to active therapy.

WORKPLACE ANALYSIS AND JOB MODIFICATION

Work-related factors often play an important role in inducing or maintaining musculoskeletal disorders—not only work-inherent factors such as manual lifting of material, posture, and piecework but also factors of work organization, skill to adapt the workplace to one's needs, and body capabilities. In order to recognize

Figure 32–8 Lumbar Stabilization Training: Lifting and Lowering the Weight

such factors, a qualitative and quantitative analysis of the characteristics of a worker's job demands (tools, loads, repetitions, posture, work environment) is a prerequisite.

Our assessment includes a qualitative evaluation of job characteristics such as those developed by Granjean and coworkers at the Swiss Federal Institute of Technology[19] and measurements of the anthropometric data and of the dimensions of the workplace. Based on this information, the workplace is adapted, and advice is given regarding performance if necessary.

This approach has been shown to be very useful for occupations where static posture and low loads dominate.[20] However, when occupations involve high external loads, such as construction or manual material handling, those loads have to be quantified. This means that peak load as well as total load handled per workday have to be assessed because both contribute to the exertion of the worker and the risk of injury.[21] This part of evaluation will be next introduced in our system.

Knowing the gap between actual functional capacity and the job demands will allow for better planning: we have a better idea of when and to what extent a worker can be cleared for work, and we have better guidelines for therapeutic or work-related training.

FUNCTIONAL CAPACITY EVALUATION AND WORK HARDENING

For the evaluation of our actual rehabilitation program for chronic low back pain, we have been using the following tests: mobility measurements with use of the inclinometer,[22] isometric[23] and isokinetic[24] strength and endurance tests, Progressive Isoinertial Lifting Evaluation,[25] and cardiovascular endurance tests.[26] These tests have a good reliability and are useful for comparing physical performance capacity at the beginning and at the end of the program. However, because they do not correspond well to normal function and demands, they do not appear to be appropriate for the evaluation of work ability. For this reason, we introduced the functional capacity evaluation.

In Switzerland, evaluation of work ability is usually performed as follows: either evaluation is based on clinical observations by the clinician and therapists (mobility measurements, radiography, observation of movement patterns and behavior), or the patient is referred to a center where different professions are evaluated in praxis over three to five weeks.

In contrast, the uniqueness of the functional capacity evaluation system lies in the following distinct features:

- comprehensive, systematic, and standardized evaluation
- realistic stress tests and quantification of tolerance to physical stress
- easy applicability to an outpatient setting
- work-related tests that can be adapted to specific work demands
- feedback for the patient about his or her ability and limitations
- report promoting communication and consensus about the client's problems and the further goals and procedures

Further experience and research will show if items of the functional capacity evaluation system are also suitable and reliable enough for scientific evaluation of rehabilitation programs.

To build up a work-hardening program at our clinic, a first attempt with work-related training in small groups of patients within our rehabilitation program was made. With this pilot program we wanted to find out if our Swiss or immigrant patients would accept work-simulation training similar to what is successfully used in the United States. Our experience so far is that patients cooperate in such a program under the following preconditions:

- Much information and education is necessary to convince them that work simulation is not a dull game just to keep them busy but an effective tool to gain more work ability in a partly realistic but still therapeutic setting. In this setting, the physical stress level and the duration and speed of performance can be much better adapted to actual ability than in real working conditions.
- Work simulation should be as specific as possible for the work that the patient usually performs.
- The patient must believe that he or she has a chance of return to work and that to an important extent gain is related to his or her own effort and self-responsibility.

Further introduction of functional capacity evaluation and work-hardening programs in Switzerland depends in the first place on support from the insurance companies. The problem with the Swiss insurance system is that it includes many different types of insurance companies, and the company that pays the costs for treatment and intense rehabilitation usually is not the same as the one that gets the profit from successful rehabilitation by saving costs of workers' compensation.

Another problem of the insurance system interferes with comprehensive rehabilitation programs: the insurance companies don't have any way of exerting economic pressure on patients participating in a functional restoration program. Patients can be absent from the program, see another doctor or therapist, or refuse further participation in the program without financial consequences. Therefore a lot of energy and time has to be spent to ensure their compliance.

DEFINITION OF OUTCOME CRITERIA FOR REHABILITATION

The return-to-work rate has been the main outcome criterion in many recent studies. It is evident that this criterion is very important considering the high costs of treatment and workers' compensation. But it is also influenced by a number of factors not related to patients' functional capacity, such as the unemployment rate and the degree of flexibility on the part of the employer. Therefore the return-to-work rate is not the only objective tool for measuring the success of a rehabilitation program. Besides, being off work is not the only relevant parameter for chronicity and severity of disorders: some patients manage to stay at work in spite of moderate or even severe chronic disorders, either because their workload is not too heavy or because flexible working conditions were offered to them. For this group of patients, as well as for patients with not yet severe but beginning chronic disorders, there is a need for outcome criteria to evaluate the success of rehabilitation and conditioning programs aiming at reduction of complaints and prevention of further chronic deterioration.

Which outcome criteria have to be taken into consideration? Should judgment be based primarily on tests of physical performance capacity? Or should it be based on patients' self-estimation of their health condition and satisfaction with the treatment? If one agrees that physical tests and patients' self-ratings are both important, outcome assessment of rehabilitation should be based on a composite score of several such factors, and further research should be directed towards evaluation of valid and reliable multifactorial scores.

NOTES

1. T.G. Mayer and R.J. Gatchel, *Functional Restoration for Spinal Disorders: The Sports Medicine Approach* (Philadelphia: Lea & Febiger, 1988).
2. S.J. Isernhagen, *Work Injury: Management and Prevention* (Gaithersburg, Md.: Aspen Publishers, Inc., 1988); S.J. Isernhagen, Functional Capacity Evaluation and Work Hardening Perspectives, in *Contemporary Conservative Care for Painful Spinal Disorders*, ed.T.G. Mayer et al. (Philadelphia: Lea & Febiger, 1991), pp. 328–345; Mayer and Gatchel, *Functional Restoration.*
3. R.T. McKenzie, *Exercise in Education and Medicine*, 3rd ed. (Philadelphia: W.B. Saunders, 1924).
4. O. Holten, Medical Exercise Therapy—The Basic Principles, *Fysioterapeuten* 58 (1991):27–32.
5. H. Gunnari et al., *Sequence Exercise: The Sensible Approach to All-Round Fitness* (Oslo: Dreyers Forlag, 1984).
6. Holten, Medical Exercise Therapy.
7. R. Gustavsen and R. Streeck, *Training Therapie* (Stuttgart: Thieme, 1985); M. Oliveri, Die Konservative Behandlung der Instabilität/Hypermobilität in der Lendenwirbelsäule, in *Aktuelle Probleme in Chirurgie und Orthopädie*, ed. A. Benini and F. Magerl (Bern: Huber, 1991), pp. 128–137; M. Oliveri et al., Bewegungstherapie, *Schweizerische Rundschau für Medizin (PRAXIS)* 12 (1992):359–374.
8. H.P. Faugli, Medizinische Trainingstherapie, in *Muskuläre Rehabilitation: Beurteilung Motorischer Funktionen*, ed. D. von Ow and G. Hüni (Erlangen: Perimed, 1987); J. Weineck, Optimales Training, *Leistungsphysiologische Trainingslehre* (Erlangen: Perimed, 1983).
9. Oliveri, Die Konservative Behandlung.
10. S. Klein-Vogelbach, *Therapeutic Exercise in Functional Kinetics* (Heidelberg: Springer, 1991).
11. Gunnari et al., *Sequence Exercise.*
12. R.A. McKenzie, *The Lumbar Spine: Mechanical Diagnosis and Therapy*, 4th ed. (Waikanae: Spinal, 1981); M. Oliveri, Die Behandlung des Lumbovertebralen Syndroms nach McKenzie, *Schweizerische Rundschau für Medizin (PRAXIS)* 73 (1984):735–740.
13. O. Evjenth and J. Hamberg, *Muscle Stretching in Manual Therapy. Vol. 2. The Spinal Column and the TM-Joint*, 2nd ed. (Uppsala: Alfta Rehab, 1988); O. Evjenth and J. Hamberg, *Muscle Stretching in Manual Therapy. Vol. 1. Extremities*, 2nd ed. (Uppsala: Alfta Rehab, 1988); O. Evjenth and J. Hamberg, *Auto Stretching: The Complete Manual of Specific Stretching* (Uppsala: Alfta Rehab, 1990).
14. G.D. Maitland, *Vertebral Manipulation*, 5th ed. (London: Butterworths, 1986); G.D. Maitland, *Peripheral Manipulation*, 3rd ed. (London: Butterworths, 1991).
15. F. Kaltenborn, *Manual Mobilisation of the Extremity Joint*, 4th ed. (Oslo, Norway: Olaf Norlis Bokhandel, 1986).

16. G.P. Grieve, *Mobilisation of the Spine: A Primary Handbook of Clinical Method*, 5th ed. (London: Churchill Livingstone, 1991); B.R. Mulligan, *Manual Therapy—"NAGS," "SNAGS," "PRP's," etc.* (Wellington: Hutcheson, 1989).
17. J.G. Travell and D.G. Simons, *Myofascial Pain and Dysfunction: The Trigger Point Manual. The Upper Extremities*, Vol. 1 (Baltimore: Williams & Wilkins, 1983); J.G. Travell and D.G. Simons, *Myofascial Pain and Dysfunction: The Trigger Point Manual. The Lower Extremities*, Vol. 2, 2nd ed. (Baltimore: Williams & Wilkins, 1992).
18. D.S. Butler, *Mobilisation of the Nervous System* (Melbourne: Churchill Livingstone, 1991).
19. E. Grandjean, *Physiologische Arbeitsgestaltung. Leitfaden der Ergonomie. Inhaltsverzeichnis*, 3rd ed. (Thun: Ott Verlag, 1979).
20. H.O. Hofer et al., Ergonomic Intervention in the Workplace: Experiences from 360 Patient Assessments, in *Occupational Musculoskeletal Disorders: Occurrence, Prevention and Therapy*, ed. K. Fehr and H. Krüger (Basel: Eular, 1992), 91–106.
21. G.D. Herrin et al., Prediction of Overexertion Injuries Using Biomechanical and Psychophysical Models, *American Industrial Hygiene Association* 31 (1986):579–586.
22. J. Keeley et al., Quantification of Lumbar Function, Part 5: Reliability of Range-of-Motion Measures in the Sagittal Plane and an In Vivo Torso Rotation Measurement Technique, *Spine* 11 (1986):31–35; Mayer and Gatchel, *Functional Restoration.*
23. M. Oliveri and M.L. Hallmark, Ergonomics and Rehabilitation, in *Occupational Musculoskeletal Disorders: Occurrence, Prevention and Therapy*, ed. K. Fehr and H. Krüger (Basel: Eular, 1992) pp. 123–133.
24. T.G. Mayer et al., Qualification of Lumbar Function. Part 2: Sagittal Plane Trunk Strength in Chronic Low-Back Pain Patients, *Spine* 10 (1985):765–772.
25. T.G. Mayer et al., Progressive Isoinertial Lifting Evaluation I. A Standardized Protocol and Normative Database, *Spine* 13 (1988):993–997; T.G. Mayer et al., Progressive Isoinertial Lifting Evaluation II. A Comparison with Isokinetic Lifting in a Disabled Chronic Low-Back Pain Industrial Population, *Spine* 13 (1988):998–1002; T.G. Mayer et al., Progressive Isoinertial Lifting Evaluation: Erratum Notice, *Spine* 15 (1990):5.
26. L.A. Golding et al., *Y's Way to Physical Fitness*, 3rd ed. (Champaign, Ill.: Human Kinetics, 1989); M.L. Pollock et al., *Exercise in Health and Disease: Evaluation and Prescription for Prevention and Rehabilitation* (Philadelphia: W.B. Saunders, 1984); B. Villiger et al., *Ausdauer* (Stuttgart: Thieme, 1991).

Chapter 33

Occupational Rehabilitation and Functional Capacity Evaluation in Australia

Sharyn McGuire

THE AUSTRALIAN HEALTH CARE SYSTEM

Since 1983 Australia has had a partially socialized system of health care called Medicare. In theory this is funded by a 1.25 percent levy on all personal income, but in fact escalating costs have meant that the federal government has needed to supplement this with large sums from consolidated revenue.

Most major hospitals in Australia are state owned and provide totally free treatment. Up to 1993 this has resulted in a high and relatively uniform standard of emergency and essential care. However, there are increasing waiting periods for free elective treatment, for example, orthopedic surgery. Office visits to medical practitioners are also funded by Medicare on a fee-for-service basis although doctors have the option of charging the patient a contribution. About 40 percent of Australians choose, in addition to paying the universal Medicare Levy, to pay for private health insurance. This provides for treatment in private hospitals—the major advantage being faster access to elective treatment and choice of treating specialist.

The present Labour government would prefer a more socialized health care system in the interest of tighter budgetary control, and there are some feelings of concern being expressed by the medical profession in this regard. The Liberal (conservative) opposition party favors a bigger role for private health insurance and free market forces. In general terms, the Australian system lies somewhere between highly socialized systems, such as those in place in Scandinavia, and the current system in place in the United States, with many parallels with the Canadian health care system.

Occupational therapists and physiotherapists in Australia are employed in either government hospitals or clinics or private practices. In private practices their fees are usually underwritten by private health insurance but not by Medicare, which only covers doctors' fees.

In all of the states and territories of Australia, treatment for workplace injury is covered by various systems of compulsory workers' compensation insurance that is the responsibility of the employer. The fee structure and method of remuneration are totally separate from Medicare and private health insurance. Occupational rehabilitation services are generally covered by workers' compensation insurance schemes.

LEGISLATIVE BACKGROUND AND WORKERS' COMPENSATION SYSTEMS

In Australia prior to 1985 little incentive was available for injured employees to return to work since the adversarial system that was in place emphasized lump-sum compensation for a disability incurred rather than return-to-work benefits. Since rehabilitation at that stage was not part of the corporate philosophy, many companies believed they were not in a position to offer part-time or selected duties as part of a return-to-work program, and did not want workers back on the job until they were fully fit. This emphasis, together with increasing numbers of lump-sum settlements in cases where common law liability for negligence on behalf of the employer could be established, led to rapidly escalating costs of workers' compensation insurance. This was generally perceived as a threat to the viability of industry.

Australia's first rehabilitation-based workers' compensation scheme appeared in 1985 with the introduction of the Work Care scheme in the state of Victoria. Six other jurisdictions have since introduced new workers' compensation legislation, and all place emphasis on rehabilitation. The new systems are attempting to cut down costs involved in provision of workers' compensation and include limitation of the right to sue at common law in favor of improved weekly benefits. Emphasis on rehabilitation is seen as part of cost-effectiveness measures.

Rehabilitation of injured workers enshrined in legislation is therefore a relatively new concept in that each state has its own legislation, with the first laws appearing in 1985. Provisions were included by legislation in Victoria in 1985, South Australia and Northern Territory in 1986, New South Wales in 1987, Tasmania in 1988, Queensland in 1990, and Western Australia in 1992.

Major provisions for rehabilitation in various Australian jurisdictions fall into the following main categories:

- Establishment of rehabilitation programs with administration by statutory bodies. In some cases sanctions are in place against employees who refuse to participate in rehabilitation programs.
- Employer duty to reemploy or provide suitable work for an injured employee.
- Employer liability for provision of rehabilitation services.

Different approaches are being taken between states in legislation concerning rehabilitation provision. For example, in New South Wales, each employer is required to establish and maintain a rehabilitation program suitable for that particular workplace and to be responsible for the management of each rehabilitation case. This can be compared to approaches in place in other states where the administrating body (i.e., state authority) takes greater responsibility for individual cases.

In most other jurisdictions the administering authority requires that the injured worker undertake a rehabilitation program, and the worker's failure to undertake such a program can lead to compensation payments being reduced or stopped.

Also in some states the employer has a legal obligation to provide suitable employment for an injured worker. In New South Wales and Victoria, dismissal of an incapacitated employee within six months of injury is an offense. These are some examples of state differences in legislation that affect successful return to work.

The rehabilitation of Commonwealth (federal) government employees comes under a scheme known as Comcare Australia. This was introduced in 1988 and made the Commonwealth government liable for compensation and rehabilitation of all its employees. Each federal department or authority makes contributions to finance the scheme, and this results in the injured worker being provided with a weekly payment plus payment of medical expenses. Lump-sum payments are made available only for permanent impairment, and under this system onus is placed on the employer to take all reasonable steps to provide suitable employment in cases where a rehabilitation program has been deemed appropriate. Therefore there are two main systems used throughout Australia:

1. The state system, which is specific to legislative requirements in each state
2. The Commonwealth or federal system, administered by Comcare Australia and covering all employees of the Commonwealth government—that is, federal departments such as the Departments of Defense, Education, Employment and Training, and Taxation

With the reform of workers' compensation in the 1980s, there has been a parallel introduction of more modern and comprehensive occupational health and safety legislation. This incorporates the concept of a general duty of care and emphasizes a more cooperative approach between parties in the workplace.

INCENTIVES FOR RETURN TO WORK

Currently in Australia the major external incentive for an injured worker to return to normal full-time duties is financial. There are several factors here, one of the major ones being the worker's eligibility to receive "penalty" payments, or extra payments for overtime and shift work. When an injured employee is returned to the workplace on alternative or selected duties, he or she is not eligible to receive these extra payments, and does not become eligible until undertaking normal full-time duties that have been approved by a medical practitioner. In some situations this can mean that workers who have relied on their penalty rate payments to double their earnings, as is common in jobs such as mining, are subject to immediate halving of wages. This obviously acts as a major incentive for a successful and fast return to work.

It is generally accepted that an early return to work in any capacity leads to more satisfactory maintenance of work performance and a higher rate of success in rehabilitation programs. However, when a worker is placed back on duties outside his or her interests or general skill levels, feelings of incompetence and dissatisfaction are common and may eventually result in the worker's resignation or an adversarial situation that leads to a lengthy workers' compensation case.

The term *light duties* is now losing favor in Australia, although its use is still quite common and can still cause some confusion. For a worker to be returned to suitable duties, the preference is that such duties be chosen from within the usual job scenario and be known as *selected duties*. Otherwise, completely alternative duties or modified normal duties are provided.

It is unlikely that the Australian Medicare system is a factor in hindering people from returning to work. This is because an occupationally compensable injury is treated completely separately from an illness or injury otherwise incurred. On the other hand, the relatively free availability of benefits under the national social security system could be seen as a factor affecting return-to-work incentives. Under this system, when weekly workers' compensation payments have been reduced or ended, or when a lump-sum payment has been made, the worker can still receive adequate financial support and medical services.

An adversarial legal system is still in place in some states, and this also can act as a deterrent in the return-to-work and rehabilitation process. Workers can become convinced that to demonstrate their true physical capacity throughout the process would be detrimental in terms of claim settlement because the case could be lost or a reduced sum be awarded if a higher level of capacity was demonstrated at any time. Some workers have fallen into a "sick role" or an "us and them" mentality under an adversarial system in which it is important that reduced functional capacity be demonstrated. This leads to increased time off work and lowered success rate in rehabilitation.

Under the Commonwealth workers' compensation system (Comcare Australia), a generous and supportive approach is used with injured workers. This is different from the state system in that Comcare has a policy whereby no lump-sum payout is available and income support is made in regular wage installments, until retirement if necessary.

REHABILITATION PROVIDERS

Under the relevant legislation, each state has an accreditation and monitoring system for organizations providing rehabilitation services. There are several types of rehabilitation providers:

- Commonwealth rehabilitation services operated by the federal government with regional units throughout each state.
- Providers in private practice. Staff requirements in private practices vary between states and can consist of medical practitioners, occupational therapists, physiotherapists, rehabilitation counselors, and trained rehabilitation coordinators.
- Employer-based rehabilitation agencies—for example, hospitals, universities, or state government departments providing in-house rehabilitation services.

In some state systems, all employers are obliged to nominate a designated rehabilitation coordinator on site. This is usually a company employee who holds a permanent position within the organization such as paymaster, occupational health nurse, or personnel officer. In the Commonwealth system each federal government department nominates a rehabilitation case manager who acts as liaison with external rehabilitation providers in establishing return-to-work programs, and is trained in rehabilitation case management by Comcare Australia.

METHODS OF REFERRAL TO REHABILITATION SERVICES

Because there is generally a legislated requirement in Australia for injured workers to be offered rehabilitation and return-to-work programs, there is a wide range of approved referral sources. These include employer companies; insurance companies who are responsible for management of workers' compensation and rehabilitation funds; treating medical practitioners; lawyers acting on behalf of the injured worker; employers; government or insurance agencies; and workers themselves. When referrals are received by a rehabilitation service provider approval must usually be gained from the relevant body—for example, the workers' compensation insurance company—for initial and ensuing assessments and rehabilitation programs to be funded. Advice of each referral is then provided to the appropriate workers' compensation body in each state for statistical purposes.

THE ROLE OF FUNCTIONAL CAPACITY EVALUATION (FCE) IN REHABILITATION AND RETURN-TO-WORK PROGRAMS

Throughout Australia, FCEs are conducted by occupational therapists and physiotherapists. There is currently a wide variety of systems in use, and selection of the system to be used is made by the individual provider organization. Traditionally it has been the role of occupational therapists in Australia to conduct these assessments in occupational rehabilitation, with many physiotherapists now becoming involved. In some states there are still very few physiotherapists conducting FCEs.

As a return-to-work tool, the FCE is often used as the first step to define current physical status of the worker, identifying physical abilities and restrictions and thereby providing information to match the worker's current capacity to job requirements. The necessity for the provision of selected or alternative duties or modifications is also identified at this stage. The FCE can be used as the initial contact point with the injured worker, but currently it is more usual that an initial assessment interview is undertaken with the worker, initial contact made with the treating medical practitioner and employer, and then a rehabilitation program established that includes an FCE for return-to-work purposes.

The results of a comprehensive and objective FCE can be used to modify work tasks so that the injured employee will be returning safely to his or her job within well-defined physical limits. These results also identify necessary modification of the worksite to include ergonomic workstations or application of inexpensive changes that will enable an employee to return safely to a job. This can include such simple methods as raising the height of a work bench or stacking shelf. Specific measurements of worker abilities and restrictions also allow for an informed upgrading of a return-to-work program, with prescriptions including the nature and duration of duties required at the job. Rest breaks and postural change periods can also be prescribed to avoid aggravation of injuries and to ensure a more successful outcome in return to work. Once measures of current limitations are obtained through the FCE, the ensuing monitored and graded rehabilitation program can be assessed regularly throughout until normal and/or full-time duties are achieved. Often an injured employee will have been returned to duties without specific measurements of abilities and restrictions being obtained. This can lead to difficulties in the worker's ability to maintain performance: for example, the worker may be placed on duties that are regarded as "light" but that may require prolonged positioning, which is an aggravating factor for his or her particular injury. Where such difficulties are experienced and the worker is not able to achieve success in upgrading duties, an FCE is a valuable tool to obtain baseline measurements of functional capacity and therefore to avoid a sense of failure. This places the worker in the more confident position of being able to achieve a gradual increment in duties and hours.

In Australia, it is the treating medical practitioners, including family doctors and medical specialists, who have the authority to approve the worker's reentry to the workforce. Often where medical practitioners do not have specific information concerning the worker's abilities and limitations, they will be placed in the position of signing a release certificate nominating general guidelines such as "light duties only" or "not to lift more than five kilograms." It is possible for this information to be misconstrued by or not helpful to employing companies since there is no specification as to how often that particular weight can be lifted and no observation of the consequence of lifting that weight in a repetitive manner. It is also then up to the rehabilitation coordinator or supervisor on site to decide what constitutes "light duties." This can become problematic when light duties are seen as desktop work but the employee has suffered an injury that results in the inability to sit for extended periods of time without aggravation of the condition occurring. In such cases, an objective FCE can be used to assist medical practitioners in nominating specific duties, duration of postures, and weight limits. Aggravation caused by repetitive movement is also an important factor to be considered in a successful return to work, and a well-measured FCE will provide relevant information to the treating practitioner for use in the workplace recommendations. It is often the case that the rehabilitation coordinator on site does not have a health sciences background, and the standardized measures available from an objective FCE can also be used by the coordinator to select a group of duties to be offered to the returning employee.

Workers who have been assessed in an FCE have the opportunity of working over two days with a highly skilled assessor who is able to identify when a safe maximum capacity has been reached on each test item. Since this maximum capacity is identified by observation of body mechanics during assessment and not by self-reporting, the worker is aware that the result of the FCE is a true measure of restrictions and abilities. This has the effect of allowing the worker to feel more confident in the use of the measurements obtained and to achieve some sense of control over his or her situation. This is extremely important information for the injured worker as well as the monitoring rehabilitation coordinator in that safe limits are now set, and the risk of overstepping limits and aggravating the condition is removed. Likewise, the worker discovers that there is no need to keep him- or herself safe by underachieving and working below safe limits.

Other important information in this area relates to the identification of physical restrictions aside from the presenting injury. These restrictions may be the result of an old injury or degenerative process and may still be having an effect on overall worker performance. In the case of a middle-aged employee with a low back injury and arthritic changes in the knee joints, the task of lifting from the floor is compromised not only by the compensable back injury but also by the instability caused by decreased weight-bearing ability in the knee joints. In this case, to keep the worker safe, job modifications would be required relevant to both the back injury and the knee condition. Identification of overall physical restrictions allows for prescribed reconditioning programs, workplace modifications, and selected duties programs. This information also allows employers to have a greater understanding of difficulties being experienced on site and of the benefits of a total approach to a safe return to work.

Another important consideration in occupational rehabilitation is the individual worker's use of safe work practices. Throughout the two-day test, preferred for reliability, the worker's use of safe body mechanics and safe work practices can be thoroughly assessed. This FCE can then be used as a tool for education in safe work practices and further assist in ensuring a safe return to work and avoidance of aggravation or reinjury.

Other valuable information to be gained from an objective approach to functional capacity evaluation lies in observation of the use of overt pain behaviors or self-limiting behaviors. These behaviors are often adopted by people who have become confused by their involvement in the workers' compensation system, who are afraid to return to work or to participate in physical activity due to the risk of reinjury, or who are apprehensive about a recurrence of pain. Once the use of these behaviors has been identified, it is possible for the assessor to discuss their use with the worker throughout the FCE and therefore to provide that worker with an opportunity to eliminate, alleviate, or change the use of self-limiting behaviors. This is particularly relevant in an objective FCE, which is conducted over a two-day period and allows for thorough observation of the worker's approach to physical activity and use of body mechanics. Where it has been possible to identify the use of pain behaviors or self-limiting behaviors, it is also then possible to prescribe relevant treatment. This can involve referral to a pain management program, to assertiveness training, to supportive rehabilitation counseling, or to monitored exercise programs, all aimed at increasing confidence levels and reducing the need for the use of inappropriate behaviors. If the worker is able to overcome the use of inappropriate behaviors through such treatment, it is sometimes beneficial for that person to be referred at a later date for more meaningful evaluation of functional capacity.

It is possible to achieve an immediate result during the FCE with workers in this category if the assessor maintains a supportive approach to confronting the worker and is able to provide information about the measured effect of behaviors on physical performance and therefore on achieving a return to work. There are many external factors influencing those who have been injured at work. These include psychosocial and financial factors, and as it is only the worker who can give the reason for the use of the behaviors, it is not always possible for any positive outcome to be achieved until workers' compensation claims are settled. However, provision of information to these workers from objective observation of physical performance can be very powerful in achieving behavior change. In Australia there are still adversarial workers' compensation systems in place in some states, and these can have a major effect on the cooperation levels that an injured worker is willing or able to demonstrate.

The extent to which FCEs are used in occupational rehabilitation in Australia varies from state to state. Factors affecting this are the ready availability of funding for FCEs, the dominant model of rehabilitation in place, and the differing backgrounds of case managers. For example, in New South Wales, an FCE, a medical assessment, an initial assessment, and a worksite assessment can be used as an automatic entry into establishing a rehabilitation program, and can be performed without prior approval from the workers' compensation insurer. When the program is formulated, it is then submitted to the insurance company/fund manager for approval, and when that approval is received the formal program commences. In some other states an FCE cannot be undertaken except as a preapproved component of a total program.

Rehabilitation providers are closely monitored for costs of services, and costs are compared between provider agencies. This can discourage some organizations from including an apparently costly FCE in a program, even though its exclusion may lead to worse outcomes for the injured worker and to decreased cost effectiveness.

FUNCTIONAL CAPACITY EVALUATION IN CLAIMS SETTLEMENT

FCEs are being used increasingly in Australia as a tool for assessment of worker incapacities and negotiation of settlement of workers' compensation claims. This is particularly so with the two-day (five-hour) objective testing procedure, which provides recommendations concerning worker or worksite modification required for eventual return to work.

Information matching worker capacity to job requirements is important in the area of negotiating claim settlement in that a true picture of the feasibility of return to work for each injured employee can be made clear to all parties. This system also clearly defines the level of cooperation demonstrated by the worker and the use of any inconsistent physical presentation of self-limiting behaviors. Likewise the objective measures used allow all parties to feel confident in the results where there is a high level of cooperation and a consistent physical presentation. Such an FCE assists specifically in determination of the percentage disability that has been incurred by the worker and the effect of percentage disability on employability. Reemployment issues can be fully considered by those responsible for making judgments on workers' compensation cases.

Again the differences in workers' compensation legislation between each state have resulted in different usage of FCEs in settlement. In the states with a dual system in place involving (1) a no-fault statutory system that is dependent on the injury occurring at work and (2) the common law system, which is dependent on negligence on the behalf of the employer being proved, the FCE has been seen as providing the court with information that has been demonstrated on cross-examination to be objective. This means that the court no longer needs to depend on subjective information and can be better placed to make a more accurate assessment.

At present, the judiciary often receive widely differing opinions from different medical practitioners. These opinions are usually based on the 30- to 60-minute review of the worker's history and the results of a short physical examination. Some medical practitioners are now finding that results from the five hours spent with the worker during an FCE provide objective baseline measures of current abilities and restrictions that enable them to make a clearer comparison of the worker's physical status to job requirements or to suitable alternative employment. However, the most common approach in adversarial systems is still for several medical reports to be presented, and FCEs are only recently beginning to be used. This is important since the claim settlement is based not on the possibility of return to preaccident duties but rather on the capacity of the worker to undertake alternative employment. In some states, the market availability of nominated or alternative jobs is taken into consideration, whereas in others it is not. Also, in some states, settlement is decided purely on the basis of percentage impairment, for example, "50 percent impairment of function of the arm below the elbow." Such an approach does not take a holistic view of whole-body or whole-person function, let alone relate it to other factors such as vocational aptitudes, availability of retraining, or labor market realities.

Where the system allows its use, objective information and results obtained from a standardized FCE applied over five hours and two days of testing and incorporating assessment of 29 different work items, together with results of two physical examinations, are of great assistance in clarifying the current physical status of an injured worker and the matching of that physical status to preaccident or possible alternative employment. Where the results of an FCE are to become an integral part of the judgment concerning percentage disability and capacity for return to preaccident or alternate duties, it is essential that these results be objective and based on observation of changing body mechanics over time, weight limits, and limits on frequency in activities, and that they are not in any way seen to be based on subjective client perception.

One recent case study taken from a court settlement involved the failed return to work of a sewing machine operator who had sustained a significant low back injury in a fall on site. Since the injury, new computerized machines had been installed in her job, and medical opinion was that she would be able to return to her normal duties since the new machines required operation only by light lateral knee pressure while the operator was in the sitting position. Because little hip or leg movement was required to operate the on/off mechanism on the new machinery, general opinion was that this worker would be able to return to her full-time normal duties. However, the FCE results indicated that this worker would be precluded from returning to the normal duties of any sewing machine operator due to her physical restrictions in sustaining a position of prolonged sitting. Forward flexion and rotation of the trunk were also found to be significantly restricted, leading to increased lumbar symptomatology throughout the FCE. These postures were physical requirements of the job with the new machine in place and were shown on video film in court, so that comprehensive discussion about the physical requirements of the job was made available to the sitting judiciary at the time. This was important information to non-medically-trained adjudicators who might not have previously considered postural factors in the worker's ability to maintain work performance.

BENEFITS OF A COMPREHENSIVE, OBJECTIVE FCE TO THE INJURED WORKER

An FCE has specific benefits that are made available to the worker throughout the assessment procedure and that can be used throughout rehabilitation.

Educational Benefits

Because this evaluation takes place over two days and a five-hour period, there is an excellent opportunity for the assessing therapist, having identified the use of unsafe body mechanics or inappropriate pacing to demonstrate and explain to the worker the need for modification of unsafe work practices and the effect of continued use of those practices on aggravation of any injury. This is also important in preventive measures in reducing risk of reinjury on return to work.

Behavior Change

This assessment also provides the worker with an opportunity to change the use of inappropriate behaviors. The nature of the FCE allows for the objective identification of self-limiting physical behaviors and pain behaviors. Once these have been identified, they can be explained to the worker, together with explanation of their likely effect on physical and psychosocial rehabilitation. The worker can, at the time of assessment, be provided with the opportunity to change behaviors, and this may result in a shorter program. However, if the worker is unable to respond at the time, prescription of appropriate treatment can be recommended, for example referral to pain management programs prior to return to work.

Active Role

The achievement of specific measures of capabilities and restriction also allows for the worker to take more responsibility and undertake an active role in his or her own rehabilitation. This is because the worker's self-perception of limitations is no longer the basis for participation in physical activity.

Further Medical Investigation

In some cases, where the worker's function has been found at FCE to be more impaired than would be expected according to medical diagnosis, further investigation has been recommended by the assessing therapist and undertaken by the treating medical practitioner, resulting in the diagnosis of previously unsuspected pathology that has then laid the basis for more appropriate management of the case. This obviously has an effect on a successful return-to-work outcome.

Other Benefits

There are also benefits to the employer and the insurer organizations involved in workers' compensation cases. Following an FCE, steps can be taken to prevent reinjury or aggravation of the condition, identification of over- or underachieving behaviors can be made and addressed, and a clearer understanding of requirements of a prescribed rehabilitation program is possible. A higher success rate in return-to-work programs and a shorter period of time away from the workplace result. For insurer agencies, the need for re-referrals for medical opinion may be reduced once physical restrictions and abilities are clearly identified.

Prescription of work-hardening or physical reconditioning programs can be the most critical factor in achievement of a successful return to work. Identification of all current physical restrictions, those due to the compensable injury and those due to other causes (e.g., sports injury, arthritis) means that a complete picture of the worker's functional ability is available. Often postural changes have been adopted by the worker in order to compensate for the effect or function of a previous injury. This will now in turn be affected by the occurrence of the work injury and its physical consequences. In this case, a worker can be placed at greater risk of reinjury or of failure to achieve return to work, if all limitations and their effect on overall function have not been identified and treated accordingly. The specific measurements and identification of all physical causes for restriction obtained in an FCE provide the basis for specific prescription of work-hardening programs. The report also indicates whether return to work should initially commence with part-time or selected duties, or full-time modified duties. Work-hardening activities can be matched in the return-to-work requirements and upgraded under supervision. This ensures success in that the worker is monitored regularly and the program upgraded as physical status allows.

Exercise programs can also be prescribed for individual workers and measured restrictions. These can be run in conjunction with work hardening and part-time or full-time duties. Again, maximum results will be achieved by the injured worker if exercise is matched to specific physical restriction.

Measurement, monitoring, and prescription of work-hardening and reconditioning programs remove the guesswork from rehabilitation. It is often the case that a work-hardening or reconditioning program is funded under workers' compensation as a treatment modality. However, where standard programs are provided, such as a list of exercises aimed at a lumbar injury, other physical restrictions are not taken into account as limiting whole-body function and remain untreated. This results in lower success rate of the program due to the worker's inability to maintain exercise or to achieve an improvement in physical status. This can be detrimental to return to work, since concerns are raised about worker motivation or availability of suitable work. Rehabilitation can become lengthy, and the worker can be referred to several different agencies in an attempt to resolve what appears to be a difficult situation.

The prescriptive information made available through a thorough FCE can have a major impact on worker and employer confidence levels. The worker's performance in work hardening can be recorded and matched to ability in particular work tasks. Expense of lengthy and inappropriate treatment interventions is reduced and the success of return to work enhanced. Currently in Australia there is no major emphasis placed on supervised work-hardening programs. Although several organizations offer these programs and/or reconditioning programs, it is rare that these are prescribed to match FCE results. Where this has been the case, worker improvement (physical and psychosocial) has been seen to be immediate, with steady progress towards rehabilitation.

CASE STUDY

One major benefit of the use of a comprehensive kinesiophysical FCE is the identification of all physical restrictions, since there are often restrictions apart from those due to the presenting injury. For example, in a recent case involving a nurse, severe restrictions were found on evaluation aside from those due to the compensable low back injury. This worker had sustained four lumbar injuries, requiring time off work on each occasion, before being referred for FCE and a rehabilitation program. The FCE indicated marked lumbar restrictions and increases in lumbar symptomatology over the two days of testing. This evidence of the significant effect of work had not before been made available to parties involved in this woman's rehabilitation. The functional capacity evaluator was able to recommend further medical investigation to the treating doctor, and a disc prolapse was diagnosed as a result. Previous diagnoses referred to soft tissue and facet joint injury, and there had been many concerns raised at the worker's inability to resume normal duties over a long period of time.

However, the FCE also indicated restrictions in the neck and shoulder areas. Physical examination revealed generally reduced levels of muscle strength in these areas with restrictions in all movements, and as the FCE progressed there were physical signs of increasing symptoms in the neck and shoulder. This indicated that the worker was compromised not only by the low back injury but also by previous and unresolved injury in the upper back, neck, and shoulders.

As a result of the FCE and specific identification of these multiple restrictions, a fully supervised exercise program was prescribed for this woman. She was assigned a personal physiology instructor, and her program was aimed at increasing confidence levels, muscle strength and endurance, flexibility, and range of motion in all affected areas. Appropriate prescription of this program would not have been possible if all physical restrictions had not been identified at a comprehensive FCE. The worker was enthusiastic in her approach to her job and the employer very supportive, but until her true functional capacity was revealed, she had been unable to maintain her work performance despite provision of "light duties," and it had not been possible for anyone to provide her with prescribed rehabilitation.

CONCLUSION

The concept of occupational rehabilitation requirements being legislated is relatively new in Australia, so that different legislative and administrative systems exist throughout the country. Models of rehabilitation also differ as yet, with a limited uniformity of approach.

Various FCE systems are in use, many being generic in their approach, and there is no standard agreement about when or how they should be used in return-to-work programs. Their use and status is very much at a developmental stage. FCEs are used in return to work and claims settlement, and only to a small extent in prescribing work hardening, since work hardening is not yet an established component of occupational rehabilitation. There is a strong future, however, for continued progress in FCE use in the return-to-work process.

Part III

Administrative Concerns

Administrative control and practices are critical for clinical and industrial success. Who runs the programs, who measures the programs, and who decides the future directions?

Frank Leone begins this section by involving the reader in concepts of occupational medicine programs developed based on thorough marketing and evaluation. Dennis Isernhagen shares examples of how the correct programs need to be applied to the industrial needs of the community. Susan Abeln provides the reader with a most interesting look at the management point of view by discussing the compelling issues that employers face and that medical professionals often miss.

From industry, Deena Pease supports and expands the concepts of benevolence, strong work injury management, and subsequent indications of reduction in workers' compensation costs through good work injury management programs.

Regarding new issues important for the twenty-first century, Dave Clifton discusses issues related to reimbursement and outcome. His point of view brings us many futuristic possibilities. Gloria Gebhard evaluates the reactions of state departments of labor and industry to the work injury management crisis.

Michelle Wiklund brings her expertise on outcome measurement in work rehabilitation to an administrative level when she discusses the legal and reimbursement responsibilities for strong outcome measurements. Melanie Ellexson and Barbara Kornblau round out the administrative concerns with a cohesive look at the Americans with Disabilities Act. Interestingly, the ADA espouses all of the prevention, work injury management, and administrative principles that have been discussed by the experts in both this book and *Work Injury: Management and Prevention.*

The work injury management of the future will be leaner but not meaner. The leanness will come from studying and implementing management techniques that provide the best outcome. The meanness or negativity should be removed as we find that encouraging workers, employers, and medical professionals to work together for self-actualization of the worker will ultimately provide the opportunity to reduce both the human and financial costs of work injury.

Chapter 34

Developing Occupational Medicine Programs and Systems

Frank H. Leone

Relationships between health care providers and employers are becoming more complex as we enter a new era in health care. Discrete occupational health services are being replaced by comprehensive health and safety management services; understanding the art of developing an occupational health program that is genuinely responsive to employers' needs has never been more important. This chapter will review emerging trends and strategies in the development and refinement of occupational programs and delivery systems.

THE EMPLOYER-PROVIDER RELATIONSHIP COMES OF AGE

The development of a well-conceived provider-based occupational health program is contingent upon an understanding of the evolving health care provider-employer link.

Freestanding and hospital-based occupational health programs are growing dramatically and are likely to be a driving force—perhaps *the* driving force—in our nation's health care system during the next 10 years. This growth is occurring at a time when both providers and employers are undergoing fundamental transitions. Employers are finding it more difficult to compete in the global market and often feel overwhelmed by uncontrollable health care costs. Recent attempts by employers at dealing with these problems include downsizing the workforce, tightening benefits packages, or turning toward various types of managed care efforts. In many instances, these approaches have met with limited success.

Health care providers also must deal with a competitive and challenging market. They are subject to restrictive reimbursement policies, and are aware that new approaches to service delivery are necessary in order to prosper. As providers seek new business venues they are inevitably attracted to the employer market because of its almost limitless size, increasing influence, and comparatively strong reimbursement.

As the balance of power in health care inexorably shifts from the government to the private sector, close associations with employers become a paramount positioning strategy for both individual and institutional health care providers. A well-conceived occupational health program or delivery system represents an extraordinary opportunity for physicians, clinics, and hospitals to position themselves in accordance with the currents of change that are affecting the delivery and finance of health care services.

The emerging provider-employer partnership concept is not limited to contracting and government mandate. Logic dictates that the workplace will ultimately become the optimal setting for the provision of safety training, health monitoring, and the management of acute and chronic conditions as our nation becomes oriented toward the prevention of deleterious health and safety habits that result in injuries and illnesses.

Several vibrant societal and health care delivery trends strongly support the notion of more active provider-employer relationships:

- *Environmental Awareness*—The pendulum is swinging from environmental neglect to environmental action. This shift inevitably places a greater emphasis on the workplace. We should expect more grass roots efforts to address exposures to workplace hazards and greater utilization of the workplace as a forum for health care delivery and education.
- *Regulatory Activity*—Recent and forthcoming occupational health and safety regulations on both the state and federal levels provide vast new service and consulting opportunities for provider-based occupational health programs. A recent example is the 1989 Department of Transportation drug-screening initiative; many exempt employers have followed the spirit of this initiative, and drug screens are now required by many companies.
- *"Office-Based" Occupational Health*—We are witnessing a greater emphasis on occupational health concerns that are based in settings other than the classic factory. Cumulative trauma and other repetitive motion injuries, indoor air pollution, and stress-related conditions are emerging as leading occupational health concerns that affect workers of all types.
- *Broader Delivery Systems*—Genuinely comprehensive provider-based occupational health programs are likely to be the wave of the future. A model hospital-based or clinic-based program would treat and rehabilitate injured workers, conduct screenings, manage broad-based worksite wellness programs, address cost containment issues, and be the primary source of health, safety, and regulatory information. Indeed, the actual occupational health product should not be a single service or bundle of services, but a provider-employer relationship that provides the full spectrum of interventions necessary to achieve optimal workplace health status.

- *New OSHA Directions*—It appears that the Occupational Safety and Health Administration (OSHA) is poised to move forward with considerably more vigor than at any time in its history. Likely areas of increased emphasis include smoking in the workplace, hazard communication, motor vehicle safety, ergonomics with an emphasis on cumulative trauma, work-induced stress, and drug/substance abuse. In addition, many states, and eventually OSHA itself, are adapting standards that require the formulation of joint labor-management safety committees at all companies with a minimum workforce size (e.g., 11 employees). Such committees are traditionally empowered to review safety programs, conduct inspections, and make health- and safety-oriented recommendations.
- *Managed Care*—The era of fee for service in workers' compensation is rapidly diminishing in many states. Reimbursement for workers'compensation-related medical care is becoming more tightly controlled and will ultimately force providers and employers to become more prevention oriented. There is clear movement toward managed care in workers' compensation.
- *Greater Employer Sophistication*—Individual employers and business coalitions are becoming more astute purchasers of health care for their employees. Employers are beginning to view continuity of care, breadth of services, and the case management of injured workers as more important than fees in their selection of a provider. Employers also seem to be looking for providers with personnel who are well trained in occupational medicine.

Health Care Reform and the Provider-Employer Relationship

Change breeds opportunity. The likely direction of change in the health care delivery system seems to augur well for provider-based occupational health programs:

- As our system moves from a competitive mode to a more integrated, collaborative model, occupational health programs may serve as microcosms of the larger health care system. An occupational health clinic can serve as a classic primary care gatekeeper that refers to specialist provider panels, interacts with carriers and employers, and otherwise provides leadership in overall health management.
- The shift from a classic medical model to a proactive, prevention-oriented model puts provider-based occupational health programs in a relatively strong position. With its emphasis on injury and illness prevention, occupational health is inherently a preventive discipline.
- The burden of supporting 35 to 37 million uninsured Americans is almost certain to fall on employers—through either some type of mandate or a pay-

roll tax. In either case, this responsibility will undoubtedly provide employers with greater influence in directing the health care system.

A Brief History of Occupational Medicine Programs

Occupational health continues to grow and change dramatically. Hospital-based and clinic-based programs, as well as some individual practitioners, are rapidly establishing a dramatic presence in the marketplace. Screening programs are becoming commonplace, and consumers are better able to differentiate the quality of competing programs. Consulting services are beginning to overshadow injury management as the centerpiece of effective and financially viable provider-based programs.

Although the work of Bernadino Ramazzini (*Diseases and Occupations,* 1713) is generally considered the birth of occupational medicine, little real association between health and the workplace occurred until the industrial revolution in the eighteenth century and later the establishment of workers' compensation laws in the United States during the early twentieth century.

The impetus to occupational health in the modern era, however, is traced to the passage of the Occupational Safety and Health Act in 1970. Although the rate of progress within OSHA has disappointed many, its very existence has more readily defined occupational health in the eyes of employees and set the stage for greater advances in workplace health and safety.

Of particular interest to provider-based programs is the movement toward the externalization of occupational health that began in the early 1980s. By 1983, health care providers began to realize that change was in the wind with the implementation of the prospective payment system (PPS). With the advent of diagnostic-related groups and the need to assume a greater outpatient posture, many providers began to view the employer market as a viable diversification strategy. Indeed, between 1984 and 1993, provider-based programs have experienced frenetic growth.

Alternate Service Delivery Settings

External providers come in many shapes and forms. Many, perhaps most, hospitals maintain some type of occupational health program, either through their emergency department, elsewhere within the hospital, or in one or more freestanding facilities affiliated with the hospital. In most communities, hospital-affiliated programs compete with one or more freestanding, independent occupational health clinics and/or urgent care centers that offer occupational health services. Many individual practitioners also reputedly serve as externally based company physicians by treating and managing injuries and providing a variety of screening examinations.

Hospital-Based or Affiliated Programs

In the late 1980s, with the advent of the prospective payment system, many hospitals recognized occupational health as a prudent diversification strategy and entered the market in large numbers. In the past, injury care and other occupational health-related services were loosely provided out of hospital emergency departments. Today, injury care is commonly bundled with other services to create a well-defined product line.

Furthermore, hospital-affiliated programs have matured and are expanding services to include screenings, education, and rehabilitation services, along with injury treatment management and other core services. A hospital base can provide several advantages in the competitive occupational health market:

- Hospitals invariably possess a breadth of services to offer a comprehensive approach in one setting.
- Hospitals typically have the financial resources to withstand a development period that may last 12 to 18 months.
- Hospitals can utilize input from existing personnel in management, finance, and marketing to offer immediate support for a program.

Freestanding Occupational Health Clinics

Freestanding occupational health clinics—with no formal ties to a hospital—are also growing rapidly. Such clinics can include nonoccupational ambulatory care in addition to occupational medicine. Many successful clinics are large and multispecialty, while others may involve only a few physicians or a single practitioner.

The freestanding occupational health clinic is likely to provide greater autonomy and tends to give employers an impression of greater efficiency, easier access, and lower fees than services provided by hospitals. In addition, many primary care physicians seek to increase their patient base by offering occupational medicine services as an adjunct to an existing primary care practice.

As employers become more prudent users of occupational health services, clinics will need to broaden their services and become even better acquainted with the workplace. The era of the freestanding "injury care mill" that lacks an appreciation of the gamut of occupational health responsibilities is thankfully beginning to slip away.

Consulting

The dramatic shortage of physicians with training in occupational medicine provides yet a third practice option—that of consulting. The occupational health consultant can provide consultative services to provider groups as well as employers. Physicians may be part-time employees under contract with a specified daily rate or may receive fee-for-service payments covering a defined series of services.

An occupational health consultant is well advised to develop a relatively narrow niche of expertise and provide replicable services to large numbers of employers. Examples include medical and workplace surveillance, educational training, health and safety policy and program development, and regulatory compliance counseling. Cost containment activities such as workers' compensation loss control and managed care are frequently discussed with occupational health consultants.

A MODEL OCCUPATIONAL HEALTH PROGRAM OF THE FUTURE

Despite the rapid evolution of provider-based programs, there may well be light at the end of the model program development tunnel. A responsive occupational health program of the future is likely to emphasize the following elements:

- *The Community Care Concept*—Health care delivery seems poised to move out of the hospital and physician's office and into the community. Community care is likely to include home health care, health education and screenings with the school system, and the workplace as a delivery locale for a wide range of preventive, treatment, and rehabilitation services. One might envision many more workplaces housing formal medical clinics, conducting a broad array of general and targeted worker screenings, providing a coherent and ongoing series of health education training programs, and achieving a better linkage between health status and work performance.
- *Delivery Networks*—Leading provider-based occupational health programs are likely to have a centrally located "Occupational Health Center" surrounded by a series of satellite facilities or clinics that feed into that center. Such a center might contain, for example, a full array of work-hardening and occupational rehabilitation services, high-tech equipment for worker screenings, and administrative offices. The concept of satellite facilities is also consistent with the general trend in health care toward easier access and more ambulatory care options. The advisability of establishing multisite clinics in a given market is closely related to worker population size and the industrial mix; scarcely populated or service-oriented areas are understandably less likely to support this concept.
- *Market Dominance*—As employers become better equipped to discriminate among multiple occupational health provider choices, poorly run or narrowly focused programs are likely to disappear. Most markets will be limited to one or two predominant providers or provider networks. Such a narrowing is consistent with, or may even be a forerunner to, the emergence of successful provider groups that are selected primarily on their ability to generate cost-effective, prevention-oriented outcomes.

- *On-Site Mini-Clinics*—The prosperous program of the future will most likely have a series of small clinics located within targeted companies. In many instances these clinics will replace company-run clinics as employers recognize the value associated with contracting with an external program (e.g., the elimination of fringe benefits and overhead, and greater flexibility in utilizing part-time personnel). In addition, employers are likely to be attracted by the convenience of having many services provided within their company.
- *Broad-Based Services*—The leading programs of the future will most likely offer services spanning the patient care continuum. There will almost certainly be an emphasis on prevention, including comprehensive workplace assessments and employee screening, injury care management, and state-of-the-art rehabilitation. The occupational health product will be the employer-provider relationship.
- *Capitated Services*—Provider-employer relationships of the future are likely to be based on contractual arrangements. Providers will assume complete management of workplace health and safety needs for medium and small employers, including regulatory compliance, health care cost containment, and health and safety management.

Initially the trend is likely to emphasize management contracts based on set fees; in due time it is more likely that compensation will be capitated and that the provider will assume some degree of risk sharing. For example, the effective provider would probably benefit from an arrangement in which it received a percentage of cost savings accrued as a result of a reduction in injuries and illnesses or lost work time associated with injuries or illnesses. If savings were in fact not incurred, the provider would be at risk and not compensated (save for the costs of direct medical treatment). A sound comprehensive program that intervenes to prevent injuries, manage them effectively when they do occur, and provide astute rehabilitation for more chronic injuries should reduce costs *for most employers* and thereby benefit financially from this arrangement.

ASSESSING AND UNDERSTANDING COMMUNITY OCCUPATIONAL HEALTH NEEDS

Understanding the Employer Perspective

Like any business, provider-based occupational health programs should strive to understand the needs of the constituencies they serve. Arguably, the constituents in question are employers and, by extension, their workforce. A proper understanding of employers, then, involves both their global perspective and their specific needs and expectations.

In general, employers are becoming more discriminating in their selection of external providers of health care services to their workforce. Furthermore, many large employers are downsizing their staff by eliminating in-house medical personnel and electing to contract out for services that were formerly provided in house.

A provider-based program should endeavor to understand the employer perspective and adjust its scope of services and promotional techniques accordingly. There are several recurring themes in the process of selecting a provider:

- *Centralized Services*—Employers want programs that offer a broad range of services but are coordinated by a single individual. In theory, a large hospital, with its vast array of personnel and services, should be able to offer a menu of services more effectively than other, smaller providers. However, chronic staff and resource inefficiencies at hospitals frequently short-circuit quality service delivery in the eyes of employers and make independent, freestanding programs appear more desirable.
- *Closer Relationship*—When looking toward the future, employers tend to envision closer relationships with health care providers. They foresee ongoing scrutiny of the workplace and more collaborative health care cost containment efforts. Given that expectation, programs should promote close working relationships and the partnership concept rather than a selection of discrete services.
- *Staff Expertise*—Employers are drawn to programs staffed by personnel with experience and/or specialized training in occupational health. In some instances, board certification in occupational medicine is desirable or even necessary, but more frequently it is not essential. Certification probably has added significance in service settings in which employers frequently deal with complex exposure issues, or in cases where there is direct competition with other programs that utilize board-certified personnel. However, it is not uncommon for a provider-based program to insist on board-certified personnel, only to find that they lack the people skills that are required to effectively manage a program.
- *Direct Sales Works Best*—When asked what marketing techniques would be most likely to draw a positive response, employers invariably cite one-on-one direct sales. Programs often make the mistake of placing too much emphasis on advertising, direct mail pieces, and educational events, to the detriment of direct face-to-face contact with prospective clients.
- *Complimentary On-Site Education*—Employers often emphasize the need for a provider to visit their workplace in order to train their workforce in such fundamental areas as emergency preparedness, first aid, and basic lifting techniques. Development of a generic one-hour presentation replicable at different companies provides an excellent entree to developing relationships

with management. Such sessions familiarize individual workers with your program and ultimately provide an essential community service.

The Art of Market Assessment

Successful business ventures are invariably built upon a formal written marketing plan. The plan defines the nature and scope of the venture's mission, outlines specific steps and a timetable to achieve objectives, and provides a financial framework for the program. In health care, for the most part, such planning tends to be comparatively poor or infrequent.

A systematic market assessment is the core element of a marketing plan. The assessment traditionally encompasses an internal review and an external survey of employers to measure the competitive market and obtain specific intelligence on employer needs, practices, and volumes.

The competitive analysis should provide a program with profiles of key institutional and individual competitors, including their product/service lines, fees, and reputed strengths and weaknesses. Often, too much emphasis is placed simply on who competitors are and what they do. It is common for a program to overestimate the market activity of other providers and become preoccupied with real or perceived competitors to the detriment of its own product development.

One of the best sources of competitive intelligence is consumer research that provides up-to-date information on the services, market share, and characteristics of each competitor. Discussions with internal staff and current and prospective clients and continuous feedback from sales personnel are useful supplements to formal consumer research.

The Employer Analysis

High-quality, professional consumer research is an excellent way to profile employer wants, needs, and practices, and to obtain a priority list of the most highly targeted employers.

Consumer research can be both qualitative (e.g., focus groups) and quantitative (e.g., employer surveys). A combination of both techniques provides a program with optimal insight regarding its employer market.

Professional focus group research in occupational health has several requisites:

- a manageable size of five to seven participants per session
- at least two sessions to control for respondent bias
- an experienced moderator with a health care orientation
- adequate notice and incentives for prospective participants
- a neutral site (e.g., a restaurant or hotel)
- participant anonymity

The qualitative information obtained during focus group sessions is of considerable value in understanding the attitudes and perceptions underlying consumer behavior. Furthermore, it provides an excellent means for identifying and documenting program or institutional weaknesses or limitations. Transcripts from focus group sessions are an excellent vehicle for sensitizing senior management and physicians to community misconceptions about their services or the existence of service voids. For example, employers frequently elaborate on why no local provider has been addressing their needs adequately. Such findings often tend to provide strong justification for a new or revitalized program.

Unlike qualitative research, which provides in-depth information, quantitative research provides statistical summaries on market potential. The most productive way to conduct quantitative research for provider-based occupational health programs is invariably through employer telephone surveys. Mail surveys generally produce inadequate response rates (20 to 25 percent at best), and personal interviews are invariably too expensive.

A well-conceived, professionally executed telephone survey results in a response rate of about 70 percent and permits probing and in-depth questions. A prototype questionnaire should elicit information concerning

- currently received occupational health services
- specific needs and desired services
- current occupational health providers, on a service-by-service basis
- desired program attributes
- perceived institutional or programmatic strengths
- perceived institutional or programmatic limitations
- a business and demographic profile
- volume data concerning injury incidence, lost work time, screening examinations, and other services

The compilation of market research data can be of value to a program in two main ways:

1. Service volume data provide programs with the ability to target employers with high injury rates or significant needs for other services.
2. Comments on desired attributes provide a program with a company-by-company and aggregate summary of what is important to employers in the market. Such insight is valuable in positioning a program as well as subsequently dealing with individual employers. Examples of provider selection factors include
   - proximity to the worksite
   - immediate worker status feedback

- comprehensive services
- fees
- staff training and expertise
- 24-hour-a-day services
- worker preference
- written patient status reports
- the availability of on-site services
- strong rehabilitation services
- streamlined billing procedures
- the provider's reputation

Employer Outreach

Direct Sales

Provider-based occupational health sales are driven by simple but compelling premises:

- Employers are universally beset by work injuries and illnesses that create an enormous economic burden.
- A high percentage of these injuries and illnesses are preventable or can be managed more effectively.
- External health care providers are well positioned to address these concerns and achieve significant health care cost savings for employers while achieving a superior health status for the workforce.

The real occupational health product is therefore a *single* broad-based approach to health and safety management rather than a disparate menu of occupational health services. The essence of effective sales is a program's ability to create a consumer-driven product superior to other options and to effectively communicate this superiority to prospective consumers.

Successful sales is contingent upon a program's ability to differentiate its product. Several steps are involved in the classic art of product differentiation:

1. *Scrutinize the Market*—Product differentiation begins with an understanding of the market, including employer wants and needs, and competitor services and penetration. Market research is essential.
2. *Define the Product*—Once a provider understands the practices, needs, and voids of a market, it can more easily package existing services, create new ones, and develop a product that is responsive to consumer needs.
3. *Identify Actual Competitive Advantages*—A provider should identify or create *genuine* competitive advantages for each service. For example, com-

petitive advantages associated with work injury management might include a sophisticated patient tracking system, 24-hour service, or the training and background of the injury management team.

4. *Develop a Compelling Benefit-Oriented Justification*—Simply enumerating competitive advantages is generally not enough. A program should carefully "script" a compelling argument on behalf of each competitive advantage. For example,

   > Because of our sophisticated computerized patient tracking system, we can stay fully abreast of where each of your employees are in the system, move them through the system more efficiently, and provide you (the employer) with useful summary reports. Since wage replacement costs associated with lost work time represent 65 percent of your workers' compensation costs, a 20 percent reduction in lost work time should reduce these costs by at least 13 percent (65 percent × 20 percent).

5. *Apply These Competitive Advantages*—The benefit-oriented justification associated with each advantage should then be assimilated into sales presentations and promotional materials.

Once the occupational health product is defined and differentiated, the actual sales effort may begin. The following principles apply to effective occupational health sales:

- *Recruit Carefully*—Occupational health programs frequently place inadequate emphasis on recruiting sales personnel. The program should insist on previous sales experience for this full-time position, ideally within health care or occupational health services.
- *Incorporate Incentives*—Programs should not underestimate the value of a strong, well-compensated salesperson. Monetary incentives (in the form of either a bonus or a commission) are important. Incentives should be quota based, easy to calculate, based on total revenues (rather than total calls or encounters, or new companies "acquired"), and provided frequently (e.g., quarterly).
- *Maximize Direct Sales Time*—A salesperson should meet with employers at least 20 hours each week, with the remaining time devoted to preparation and follow-up.
- *Stratify Target Employers*—Sales strategy should reach out to three different employer groups: high-profile larger employers, mid-sized employers (50–500 employees), and the broad remaining universe of employers. The first group may require highly customized services, lengthy negotiation periods, and team selling; the latter group involve ongoing direct mail efforts.

- *Target Employers*—The sales plan should assess the market in order to determine which employers can be expected to generate high volumes and target those employers for priority contact. The more systematic the targeting, the greater the likelihood of increased patient volumes.
- *Understand the Product*—The occupational health salesperson should understand the occupational health macro-environment—that is, how employers should address the entire patient care continuum and what changing delivery and finance mechanisms are likely to mean to the employer-provider relationship.
- *Perfect the Art of Information Gathering*—The art of sales involves the ability to ask the right questions, listen carefully, probe as appropriate, and develop solutions that match employer/employee needs with clinic capabilities.
- *Identify and Solve Problems*—Invariably sales involves problem solving. The real problems may not be injuries, illnesses, or even an unsafe workplace, but the *actual costs* associated with these problems.
- *Sell Benefits, Not Features*—Board-certified personnel, a computerized injury worker tracking system, and 24-hour service availability are features. The fact that these features should save an employer money is the benefit.
- *Prepare for Objections*—Employer objections to prospective provider-based occupational health services are generally finite (satisfaction with current provider, location, minimal need) and can be anticipated. The salesperson should develop written responses to each anticipated objection and be prepared to respond as confidently as possible.
- *Provide Written Proposals*—Sales calls conducted by an occupational health provider representative should be followed up with written benefit-oriented proposals. Proposals tend to be more effective when they portray the provider's track record and are specific in terms of fees and other commitments.

Supplemental Marketing Techniques

The actual sales effort should be supplemented by several concepts or activities:

- *Effective Promotional Materials*—Occupational health programs should develop collateral materials on quality stock, featuring photographs, staff biographies, and third-party testimonials from satisfied employer clients and/or previous clinic patients.
- *Employee Outreach*—Occupational health programs should market to the employee as well as the employer. Providing employees with program membership/identification cards can help streamline paperwork and frequently gives the individual a sense of affiliation with a person. Educational programs such as emergency preparedness or first aid for employees at the worksite are also useful in this regard.

- *Educational Seminars*—Periodic breakfast, luncheon, half-day, or full-day seminars on a variety of key issues (e.g., "Getting the Injured Worker Back to Work," "Legal and Ethical Issues Associated with Worker Screening") are a useful way to publicize a program and generate revenue at the same time.

ATTAINING OPTIMAL SERVICE DELIVERY

Developing an Array of Services

Although this chapter has defined the real occupational health product as broad health and safety management that consists of numerous components, an examination of the more basic, common components is useful.

Work Injury Management

Work injury management is the bread and butter of most provider-based occupational health programs. The provider of work injury management services must recognize that such services are more than a variant of primary acute care. To be effective, work injury management requires an understanding of the workplace, an emphasis on the origin of the injury or illness as well as the cure, and a commitment to both short- and long-term patient case management.

The goal of injured worker tracking and case management is the reduction of lost work days and the costs associated with them. This goal should be measured in some fashion: that is, the employer should receive quantitative feedback on savings associated with the provider's interventions. Most employers have readily available baseline data on lost work time injuries and lost workdays. These data should be tapped by the provider at the outset of their relationship with an employer and used in subsequent years.

If a program is hospital affiliated, it should also tap into emergency department activity. Work-related injuries seen in the emergency department should be handled in the same manner as those seen during regular clinic hours. Once systems are in place for managing work-related injuries, half of the work is done.

An effective program also needs to provide services that prevent injuries and illnesses. This is where screening examinations, workplace surveillance, and consultative services enter the picture.

Screening Services

Despite concerns by some occupational health providers that the Americans with Disabilities Act of 1992 would make preplacement screening examinations obsolete, the use of such examinations does not appear to have appreciably diminished. Many employers realize that such examinations are useful injury prevention tools and an excellent vehicle for acquiring health status information on a new hire.

The classic array of screening examinations includes medical surveillance and periodic physicals, annual (formally executive) examinations, health risk screening programs, and drug screens.

Consulting and Workplace Surveillance

Despite generally noble intentions, many occupational health programs provide the basic array of work injury treatment and screening services with little or no attention to the origin of the maladies that they are treating. A program should supplement its basic product line with an array of consulting services that address the broad array of physical, chemical, biological and psychological conditions that may be present. Providers should move away from the service menu concept toward a management contract mentality. To make a difference for most employers in the long term, a provider should develop and implement a strategy involving ergonomic analysis, medical surveillance, health education, and worker-management planning and collaboration.

Selecting a Service Location

One of the first hurdles for a nascent occupational health program is the selection of a service location. Usually the choice involves a hospital-based location and/or a choice among a series of freestanding clinic locations.

If cost were not a consideration, the freestanding option would be preferable in most situations. However, the cost of leasing or buying a freestanding clinic, plus potentially high configuration costs, can serve to undermine the service advantages of a clinic being located off site. It is useful to survey potential clinic users in order to determine what degree of resistance, if any, they may have toward each option under consideration. In general, several rules apply:

1. A hospital-affiliated program is less likely to prosper in a busy emergency department, while an emergency department with smaller volumes is more likely to emulate an ambulatory care clinic.
2. If a freestanding facility is preferred, it is better to lease than to buy existing space.
3. A clinic site should be near the epicenter of the market's industrial base, accessible and visible from major freeways or thoroughfares, and in a location with easy and ample parking.
4. An occupational health clinic can be comfortably integrated with a nonoccupational health ambulatory care clinic. In most cases the two clinic types should be separated into distinct areas, with, if possible, separate entrances. If the ambulatory care population is from a distinctly different socioeconomic class than the occupational health population, integration of the two clinic types is reputably more difficult.

Attaining Effective Injured Worker Case Management

A provider-based program requires two essential ingredients to effectively manage the treatment of injured workers: a strong information system and highly skilled personnel to conduct record review, coordinate each aspect of the work injury management system, and input and manage all related information. Three individuals are customarily required to provide case management services: the case manager(s), the physician(s), and the data clerk.

The case manager is normally responsible for monitoring and managing, but not for rendering a medical opinion. Strong interpersonal skills and knowledge of the workers' compensation system are recommended. The physician(s) should review all open cases for appropriate diagnosis, treatment prescription, and fair prognosis. The data entry clerk, whether patients are tracked manually or by computer, needs to be responsible for the paperwork, including data generation and manipulation.

Effective injured worker case or work injury management depends on a number of factors. Approximately 50 to 60 percent of workers' compensation costs are directly associated with lost work time. Therefore effective management is likely to get the injured worker back to work more expeditiously and consequently reduce the wage replacement costs associated with his or her care. In addition, a greater continuity of care takes place when an individual or team is responsible for monitoring the care. Careful monitoring of a patient should result in a superior overall quality of health care.

With regard to information gathering and storage, when information is organized in a single database, it tends to be of higher quality. For example, employers find it appealing to have one individual responsible for monitoring the progress of an injured worker from the day of injury to safe return to the workplace. Central coordination of the injured worker's treatment allows a program to maintain a database by which to measure results.

Finally, an effective injured worker tracking system intervenes in cases before they become eligible for workers' compensation and consequently reduces the overall number of clients. Furthermore, by expediting the return-to-work process for injured workers, patient tracking prevents chronic cases from falling into long-term disability.

Unless the volume of injured workers being tracked is minimal, some type of computerized system is desirable. Although the system can be customized or integrated with an existing system, it may be advisable to consider one of several excellent occupational health software systems available on the market, since they are constantly being improved and are designed to assist programs with clinic management, finances, and numerous other program needs.

Service Quality

Given the breadth of the product line, and the complexity and novelty of the provider-employer relationship, service quality takes on special importance. Several principles should be followed in order to ensure optimal service quality.

Weekly Staff Meetings. Internal communication within programs is frequently inadequate and can be addressed through systematic, well-organized weekly staff meetings. For example, personnel might meet every Monday from 7:00 to 8:00 A.M. A typical meeting should review:

- operational procedures and special problems
- concerns identified through patient satisfaction surveys and other client intelligence-gathering mechanisms
- quality assurance issues
- sales and marketing strategies and activities during the previous week
- staffing issues

Monthly Summary Reports. A one-page program summary report should be developed and distributed to senior management, relevant department heads, and program administration. Monthly volumes, years-to-date volumes, and historical comparisons should be highlighted; a half-page narrative on primary program accomplishments might also be included.

Weekly Sales Report. Sales representatives should be responsible for developing a sales summary report for each week of activity. The report should include the following:

- number of sales calls (week and year to date)
- number of individual clinic tours (week and year to date if applicable)
- a summary of each sales call, including contact person, results, and action steps

Operational Protocols. Protocols are program-specific, dynamic, and evolutionary. They should be developed and updated on a continuous basis, and maintained both on a computer and in formal protocol books. The following represents a series of operational, communication, and physician referral protocols that are adaptable to a variety of programs:

1. *Patient Satisfaction Instrument*—An occupational health clinic should provide every patient with an opportunity to complete a short (e.g., index-card-size) questionnaire assessing his or her satisfaction with the encounter and the care received. Two basic questions are recommended: a 1-to-5 scaled response reflecting patients' satisfaction with their level of care and a 1-to-

5 scaled response reflecting patients' satisfaction with their overall experience. Clinic patients should be invited to provide comments on the back of the card.

2. *Annual Employer Client Questionnaire*—Programs should send a mail questionnaire to the primary liaison at all employers who have used program services. The questions should solicit the following information:
   - an assessment of the quality of each service (a scale of 1 to 5 is useful)
   - additional health and safety needs of the employer/workforce
   - suggestions on how your program might improve services
   - names of other contacts within the company that should be on the provider mailing list
3. *Quarterly Calls to Your 40 to 50 Highest-Volume Employers*—Program management should call each of the program's 40 to 50 highest-volume employer clients every three months to ask them at least two basic, open-ended questions: "Are we serving your needs effectively?" and "What can we do to improve our services?" Feedback from these calls should be categorized and shared at weekly staff meetings.
4. *Provide Immediate Injury Care Treatment Feedback*—A central feature of your external communication protocols should be a guarantee that employers will receive verbal feedback regarding their injured workers within hours of their care encounter:
   - *In the Clinic*—A liaison at each company should be identified through a "sign-up" mechanism. This individual should be called by clinic staff, as a matter of routine, shortly after the encounter, with a simple status report (e.g., diagnosis, treatment plan, expected return to work). Calls should be made to employers of patients seen during the last clinic hour of the day. Written summaries should be sent via mail within 24 hours of the time of service.
   - *Through an Emergency Department (If Applicable)*—All patients in an emergency department associated with a freestanding occupational health clinic during nonclinic hours should also receive follow-up. The program coordinator should routinely obtain contact sheets each morning and call employers with a status report. A follow-up letter should include information about the clinic network and the sign-up form.
5. *Offer Faxed Summaries of Screening Examinations*—Employers should receive a faxed summary of all preplacement, drug, or other physical examinations performed in their behalf. Confidentiality constraints need to be implemented; code numbers or social security numbers are often employed in lieu of names.

6. *Establish Specialist and Primary Care Physician Referral Policies*—A program's medical director and network physicians should take the lead in developing referral panels of primary care and specialist physicians. Involvement should be open-ended provided that referral panel physicians are willing to
   - accept referrals within 48 hours in most circumstances
   - send or fax patient care reports back to your case manager(s)
   - pass scrutiny through quality assurance reviews

   All clinic patients requiring referrals should be asked if they have a preference, and if so, the referral should be made to the preferred physician. If a preference is not stated, the referral will be rotated among panel physicians unless preempted by geographical considerations.

Attention to service quality often separates the mediocre providers from those that are most likely to meet the needs of employers over the long haul.

Internal Staffing

Staffing of a provider-based occupational health program usually varies by service mix, program location, and volume. A core staff of a medical director, program director, and one or more sales representatives is generally applicable.

Ideally, a program should have a full-time medical director, although 0.20 to 0.50 FTE is common. If the medical director is full time, about 60 percent of his or her time should involve direct patient care and the remainder involve worksite visits, consulting, and administrative matters.

A physician who is board certified in occupational medicine can be exceptionally valuable to a program. Since there are fewer than 1,000 board-certified physicians in practice, it is frequently not practical and usually not necessary to retain a board-certified physician. Proper and continuous training are essential, and completion of a short course program is recommended.

The program director (or coordinator) needs to be a full-time position. The program director should have sufficient stature and authority to coordinate what is inherently a multidepartmental service. Because of the case management/patient tracking responsibilities usually associated with this role, a nursing or other paraprofessional background is useful. However, successful program coordinators frequently tend to exhibit an energetic, motivational, and people-oriented personality rather than a specific training track.

At least one full-time sales or marketing representative is essential. The old adage "If you don't make the call you won't make the sale" is applicable. A full-time sales representative, dedicated entirely to occupational health sales, is recommended.

All things being equal, it is preferable to recruit/appoint a sales representative with a sales background rather than a background in health care or occupational health. Pharmaceutical sales or a sales background in other service industries (e.g., hotels) is often useful. Programs are well advised to cross-train the program director and sales representative.

Recruitment and Compensation Issues

Whenever possible, compensation should be leveraged: that is, the compensation package should include some incentives.

Several principles apply to effective compensation packages:

- They should be as *simple* as possible. For example, a sales representative might receive a base salary of $24,000 per year plus a percentage of any increase in revenues over the previous year, which would likely result in total compensation between $35,000 and $50,000 or more.
- The incentive package *should not be capped.*
- Incentives should be *paid frequently* and the incentive plan reviewed frequently, such as once a quarter.
- Incentives should be based on something that is *easily measurable* and reflects the *full product line* of the program. For example, incentives based on total revenues are more meaningful than those based on the number of client companies enrolled in the program.

CONCLUSION—THE OCCUPATIONAL HEALTH OPPORTUNITY

Employers are emerging as key figures in the changing health care delivery system. Their role as gatekeepers and financial supporters of the health care system presents an extraordinary opportunity for the providers who work with them.

There is much to motivate a health care provider to join the ranks of a growing and highly evolutionary specialty that places a premium on prevention by recognizing the workplace as a perfect venue to address the nation's health through the reduction of workplace health and safety hazards and the practice of genuine preventive medicine.

There are many potential advantages available to the provider that makes a commitment to the occupational health market:

- *A Delivery Network*—Collaboration and integrated systems seem to be the health care delivery direction of the 1990s. A broad occupational health program permits a provider to begin the formulation of such systems.
- *Positioning*—A strong program positions a provider now for subsequent formal relationships with employers and serves as a defensive maneuver against other provider-based programs.

- *Expansion of Primary Care Revenue*—Occupational health programs attract new primary care revenue through direct injury care, associated revisits, family referrals, and a more lucrative payor mix (direct employer billed and workers' compensation).
- *Referrals, Ancillaries, and Admissions*—National research of provider-based occupational health programs indicates that 19 percent of initial work injury encounters result in specialist referrals, and approximately 9 to 12 percent of all screening examinations result in primary care or specialist referrals. The same research indicates that 3.2 percent of initial injury treatment encounters result in a hospital admission.
- *Reimbursement Issues*—Occupational health programs offer services that frequently generate contractual revenues that do not rely on an increasingly constrictive reimbursement system. Examples include physician and nurse placement programs, wellness contracts, industrial hygiene services, and other consulting services.
- *Cost/Risk Ratio*—The typical cost to a provider of starting and maintaining an occupational health program is extremely low compared to that of other common provider initiatives.

Our present health care system is prohibitively expensive and irresponsibly tolerant of unnecessary morbidity and mortality. An ideal way to address these conditions is through a genuine commitment to preventive medicine using the workplace.

Health care providers possess the personnel, community image, and expertise to effect positive change in mutual benefit of the employer, the workforce, and the provider. Given astute planning and proper dedication, provider-based occupational health programs are likely to play a central role in the new era of health care delivery.

Chapter 35

Building Occupational Health and Rehabilitation Programs on the Consumer's Actual Needs

Dennis D. Isernhagen

Health care providers spend significant amounts of money and time developing various occupational medical programs for specific populations. They may purchase expensive equipment, add space to house a program, attend workshops and seminars to learn the newest techniques, and develop beautiful brochures to promote their new program. They are especially proud when they are the "first" in the area to offer a new program or service. When they are not the first, they will attempt to implement programs similar to their competitors' so that they can remain competitive.

So why do so many of these new and exciting programs fail or not meet their founders' original expectations? Why do so many of these programs, when first started, grow and then plateau or decline?

Health care providers often take the position that they know what the consumer needs and wants. This attitude is best illustrated by an analogy to the movie *Field of Dreams*. In this movie an Iowa farmer who is a diehard baseball fan hears a spirit that tells him that if he converts his cornfield into a baseball field, all of the past great baseball players will come to play. The spirit voice says "Build it and they will come." The farmer does what the voice tells him to do, and they come. Many health care providers demonstrate this same "Build it and they will come" attitude.

Today's health care business is entirely different from what it was in the past, when such an attitude would work. The changes that are taking place in health care are rapidly altering the business environment. Underutilized health care and reimbursement for services are causing hardships to the health care providers who built programs that were not tied into the needs, wants, and desires of the consumer. Consumers are becoming much more knowledgeable about health care issues and are demanding more from health care professionals than they did in the past. They want proof that the health care services that are provided will be beneficial and will be provided in the most economical way.

This chapter is intended to provide assistance to health care providers who are in the process of developing or expanding their occupational health and rehabilitation program. It will address some of the steps that need to be taken to gather information to make decisions and to guide the development of a successful program. It will also provide information on the national trends that are occurring in occupational health and rehabilitation.

ANALYZING THE MARKET

Today's health care provider, more than ever before, needs to analyze the marketplace before investing time and money in a new or expanded product or service. This is something to which most health care providers are not accustomed. An analysis of the health care provider's market will provide significant information from which an informed decision can be made. The analysis itself should be viewed as an investment. The time and money spent on this process will minimize the risk of failure, avoid expensive mistakes, and maximize the potential for success.

GOALS OF THE MARKET ANALYSIS

Before the process of analyzing the market begins, the health care provider needs to identify the goals it would like achieved from the process. These may include but will not be limited to the following:

1. to identify the present and future market needs and trends in occupational health and rehabilitation
2. to provide an analysis of the provider's current rehabilitation programs—their strengths and weaknesses
3. to identify what should be added or modified to accommodate present and future occupational health and rehabilitation needs
4. to identify how the health care provider can increase its market presence and increase utilization of its occupational related services
5. to identify reimbursement issues that will affect the program, such as health maintenance organizations, preferred provider organizations, managed care organizations, workers' compensation laws and regulations, and third-party payors
6. to identify who the competitors are and determine what programs they offer and what their strengths and weaknesses are

7. to determine if a specialized market niche can be developed. This is particularly important when there are multiple providers in the market
8. to identify resource needs such as staff, space, and equipment to support the program
9. to identify possible networking opportunities with other health care providers who are engaged in similar programs or who can offer a service that will enhance the overall program
10. to identify the most effective and efficient method of promoting and marketing the program
11. to determine the fiscal picture of the program in order to determine start-up costs and the projected return on the investment

Additional goals should be developed that are specific to the health care provider and its overall company philosophy and goals.

Identifying the Needs of the Consumer and Referral Source

It is important for the health care provider to identify the consumers (payors) and referral sources. These will include employers, insurance companies, rehabilitation or vocational specialists, physicians, chiropractors, and attorneys. Each of the consumer/referral groups will have different needs for an occupational health and rehabilitation program.

The methods that can be used to identify these needs can include telemarketing, questionnaires/surveys, focus groups, or one-on-one discussion. The personal, face-to-face contact that a focus group or one-on-one discussion offers is the most effective method of obtaining accurate and necessary information. These two methods not only gather information but also send a very strong message to the consumer and referral source that the health care provider considers them valuable to the program and wants their input. This fosters support for the program through active participation in the development process. When the program is implemented, the consumers and referral sources will be ready and willing to participate.

Employers

Prior to communicating with employers, health care providers need background knowledge about the business and industry in their community. The information that will be the most helpful relates to the types of business and industry that are in the area (manufacturing, construction, retail/wholesale, government, etc.). This should include the number of different companies, where they are located, how many employees they have, how long they have been in business, and so on. Such information is available through the chamber of commerce, area universities/colleges, and other business and labor organizations. These organizations can also provide projected economic changes for a period of time into the future. Most of these organizations put together information such as population projections, economic projections, and other demographic information generic to the area.

Knowing the make-up of the business community, the health care provider can project which companies potentially have the highest risk for work-related injuries. This projection can be made based on the average lost days of work due to injury that a specific industry experiences. The statistics in Table 35–1 are representative of the average for specific business groups.

The health care provider should also be familiar with some of the needs and problems that other employers have voiced regarding medical and rehabilitation services. This will be of assistance in preparing questionnaires and interviewing the employer. Their concerns and needs usually include the following:

- They often do not receive adequate or useful information from medical professionals regarding the injured employee's ability to return to work.
- They are frustrated with health care providers who make return-to-work decisions without accurate knowledge of the worker's job and its physical requirements.
- They are frustrated that access to health care is often delayed because of the lack of priority given to work-related injuries by the health care providers.
- They would welcome input from a rehabilitation specialist on how to modify a job if it means the employee could return to work earlier and without the risk of reinjury.
- They welcome rehabilitation providers to visit the worksite to gain a better understanding of the work and the environment in which the job is per-

Table 35–1 Annual Lost Days Due to Injury (per 100 Employees)

Type of Company	*Average Lost Days*
Agriculture, forestry, fishing	112.0
Mining	119.5
Construction	147.9
Transportation & public utilities	134.1
Manufacturing	120.7
Wholesale/retail trade	65.6
Services, finance & insurance	56.4
Real estate	27.3

Source: Statistics are reprinted from *Accident Facts, 1992 Edition*, by the National Safety Council, Vol. 43, p. 49, 1992.

formed. They would also welcome recommendations as to how the job could be performed more safely.

- They are frustrated with the high cost of health care and believe that a provider who could objectively demonstrate its effectiveness and justify its costs would best meet their needs.

Insurance Companies

Insurance carriers are critical to an occupational health and rehabilitation program. They are one of the key parties who will direct where health care will be provided. They are interested in identifying providers who deliver high-quality and cost-effective care. Their concerns and needs usually include the following:

- They are most interested in the rehabilitation providers who can demonstrate positive outcomes. These outcomes are primarily based on ability of the provider to get the worker back to work with a demonstrated financial savings to the insurer.
- They are very supportive of a provider who can return the injured worker back to work in a rapid and safe manner without a risk of reinjury.
- They, like the employer, are frustrated with the lack of communication with the health care provider. They would appreciate timely reports from the health care provider.
- They want objective information from the provider that will support and justify their compensation claims.

Rehabilitation or Vocational Specialists

Rehabilitation or vocational specialists are professionals who interface with the employer, the employee, and the medical specialist in the return-to-work process. They manage the return-to-work process in cases that have gone on for an extended period of time. Their concerns and needs usually include the following:

- They, like the insurance carrier, are looking for occupational health and rehabilitation providers who can provide services that demonstrate positive outcomes. These outcomes are based on the ability of the rehabilitation provider to return the individual back to work quickly and safely. They are charged with spending insurance company money and need to be assured that they are getting the best for the insurer's money.
- They want a provider who will communicate with them in a timely and accurate manner.
- They want accurate information that will assist them in making recommendations to the insurer as to additional health care services required for the individual to return to work or in closing the case.

- Their own reputations depend upon safe return to work, and they welcome conferences and other indications of coordinated effort.

Physicians

In most jurisdictions, physicians have the legal responsibility of making the return-to-work determination. Many physicians do not like this responsibility because of the lack of objective information to support their decisions. They will often be conservative in their decision because of the lack of accurate information about the worker's job. Their concerns and thoughts are:

- They are supportive of a therapist providing information that will help them in making a determination for an injured worker's ability to return to work. They are particularly interested in having a therapist visit the worksite and determine the physical requirements of the job.
- They want the therapist's evaluation of the functional ability of their patient.
- They want the therapist's physical findings, such as range of motion, muscle strength, and functional movements, that may be used in disability determinations.
- They want recommendations from the therapist as to the appropriate therapy required for their patient to return to work safely.

Attorneys

Attorneys are becoming a larger consumer of the services offered by occupational health and rehabilitation programs. They recognize the credibility that the provider can supply to justify their cases that involve work or personal injury. The plaintiff attorney will be seeking objective information from the rehabilitation provider documenting functional levels. This will influence a decision in making a financial settlement for his or her client. The defense attorney will be seeking information that will demonstrate an injured client's abilities to perform specific jobs or tasks so that he or she can return to work as soon as possible with the least limitations and disability possible.

The attorneys will be looking for the following:

- Functional capacity evaluations that will provide information that can be used in a medical/legal situation. Such evaluations should involve objective and clear documentation that substantiates specific findings and should offer recommendations that are based on sound objective information.
- Information that will document an individual's functional ability to return to work or not return to work.
- Information that will provide specific measurement of ability or disability in terms of muscle strength, range of motion, endurance, or other objective measures.

- Recommendations regarding what can be done to resolve a disability. Usually a case will not be closed until the rehabilitation question is satisfactorily answered and a client is determined to be at maximum medical improvement (MMI).
- A professional opinion by the rehabilitation professional on certain aspects of the case. This is over and above the objective information that is found in an evaluation. The rehabilitation professional needs to provide this only when asked in a medical/legal situation.
- Information that is objective and will stand up in court.

Rehabilitation professionals who choose to work in the area of occupational health and rehabilitation can expect to interface with attorneys and the medical/legal system on a more frequent basis than in any other area of rehabilitation. How aggressive the provider wants to be in marketing medical/legal services needs to be carefully evaluated. Not all rehabilitation providers enjoy or are qualified to provide the required medical-legal services. If a decision is made to develop this area, caution should be taken to be as neutral as possible. The rehabilitation provider should make an effort to provide services for both the defense side and the plaintiff side. It would not be beneficial for the provider to gain a reputation of working for only one side.

NATIONAL TRENDS IN OCCUPATIONAL HEALTH

Major trends that are occurring in occupational health need to be taken into consideration when developing a program that will be successful. Knowledge of the trends will be of assistance when gathering information about the needs and concerns of the payor and referral sources. It will help in analyzing this information and in developing a program that will satisfy consumer needs, resolve their problems, and be reflective of present and future trends.

Present Trends

1. Successful programs in occupational health and rehabilitation offer a broad scope of services to the work injury management spectrum. The relationship between insurance companies, employers, and health care providers is closer than ever before. Prevention of work-related injuries is now seen as important an issue as treatment.
2. There is increasing pressure for the health care provider to interface with employers and employees at the worksite.
3. Prevention programs, ergonomics, early intervention, and rehabilitation after the injury require a "team" approach that includes the employer, the employee, and the health care provider.

4. Successful work injury management programs are neutral. They are not viewed as a vehicle of the employer or the employee. They are not advocates of one side or the other, but rather advocates of proper functioning of the entire system. Since the system should be directed toward safe work for the employee and early "safe" rehabilitation and return to work, everyone benefits. If the health care provider is ever seen as "an employer arm" or a "union arm," its ability to function effectively is lost.
5. The successful occupational health and rehabilitation provider will work diligently to prevent "disability" of employees. Disability is a very negative outcome for all parties concerned. Case management is a part of the approach necessary to prevent disability following injury. There must be attention to psychosocial disability as well as medical disability.
6. The health care provider needs to focus on the ability of the injured worker more than the disability. What can the individual "safely" do at work that will keep him or her connected to the work environment? How can his or her function be improved by either rehabilitation or job modification? The labeling of an individual as "disabled" contributes to negative outcomes.
7. The occupational health and rehabilitation provider needs to be legally credible in evaluation and the management of care. Clients of rehabilitation programs will be involved in legal issues relating to their injury. Rehabilitation providers must be able not only to answer referral questions but also to be objective and clear. The process and the outcome of rehabilitation need to stand up to the strong scrutiny of the legal system.
8. Cost-effectiveness is a very important issue. In general, the cost is not as important as the effectiveness of the rehabilitation program. As a result, effective programs do not have to be the lowest cost provider. Effective programs will be cost-effective because of their outcomes. Quality will always be seen as worth the cost.
9. The consumers of occupational health and rehabilitation services are the insurance companies, the workers' compensation systems, attorneys, and the employers. Physicians and rehabilitation providers must work together since their bills are being paid by the referral sources. There must be an understanding of this so that the rehabilitation provider does not think that the physician is the consumer. The physician is at as much risk of being selected or rejected by the payor source as is the rehabilitation provider.
10. Communication between the rehabilitation provider, employee, employer, physician, and insurance carrier is vital to successful programs. Information should be exchanged among all parties on a regular basis when rehabilitation services are being provided. Open dialogue will foster trust and cooperation, which are key elements in the return-to-work process.

Program Content

When assessing what the consumer of occupational health services wants, needs, and is willing to pay for, the health care provider must keep in mind what services can be developed and provided. Occupational health and rehabilitation programs should consider offering the following continuum of programs and services. It is important for the provider of occupational health and rehabilitation programs to ensure that there is easy access to these services by business and industry. This list of programs and services is not all-inclusive but includes the foundation elements. Additional services and programs will only expand upon this foundation and provide a more attractive offering. Some of the program elements are discussed in greater detail in other parts of this book.

In the process of analyzing the present market, one needs to explore if the consumer is currently receiving the following services. Providers need to know the consumer's level of satisfaction with these services, whether these services are considered important to the consumer, and, if consumers are unaware of the services, whether they are interested in learning how the services may benefit their company.

Prevention Programs

Going to the worksite to prevent the injury is a critical part of the work injury management spectrum. The following are prevention programs and national trends.

Ergonomic Job and Job Site Evaluation. One of the first steps in providing effective work injury prevention programs is for the health care provider to have a good understanding of the job, its physical requirements, and the environment in which it is performed. An analysis of the job and job site will identify potential hazards that may cause musculoskeletal injuries. Ergonomic modification of the job site or method by which the job is performed can be done to reduce the potential of injury, or reinjury in the case of the injured employee who is returning to work.

A job analysis is valuable not only for prevention purposes but for an effective rehabilitation program when an injury does occur. The analysis will provide the attending therapist with information that will allow for the development of a more appropriate rehabilitation program. It will also provide information from which a return-to-work decision can be made.

The following areas should be explored with consumers when assessing their needs and desires for ergonomic job and job site analysis.

1. What are the common injuries that employees have in the various businesses and industries in the area? This will provide an idea of the types of

injuries and those that are related to cumulative trauma, for which ergonomic modifications would be beneficial.

2. Do employers have ergonomic programs in place? If so, who is responsible for them? If they do not have programs, is this something they feel would be beneficial to their company? This will provide information about the employers' emphasis on ergonomics and their knowledge of how programs of this type can be of benefit to them and their employees.
3. Do physicians have a good understanding of the physical requirements of the job(s) that their work injury patients perform? Would this information be helpful in making return-to-work decisions? Would physicians return an individual back to work following an injury sooner if they knew the job was safe or could be modified to safely accommodate the individual's present abilities?

   Many physicians do not have a good understanding of the physical requirements of a particular job. They usually rely on the patient's description, which may be over- or understated. They would appreciate having a more accurate description of the job and whether it could be modified to accommodate their patient. In addition, physicians are more willing to make the return-to-work decision when the analysis of the job is performed by a health care provider who knows the patient's functional abilities and the physical requirements of the job.
4. What emphasis does the third-party payor place on prevention programs? Would an ergonomic analysis of a job in which an injury has occurred be reimbursed if it were part of the return-to-work plan? Would it be reimbursed if it were performed as a prevention program?

   Third-party payors are becoming more supportive of prevention programs especially when they demonstrate a decrease in injuries and provide a way to return injured workers back to work more quickly and safely. This will ultimately save them and their client money.

Functional Job Description. Job descriptions are often overlooked as an injury prevention tool by employers. Job descriptions in the past have been primarily designed to identify the training, tasks, and skills required to perform the job. Job descriptions today have taken on an expanded role in business and industry. With the advent of the Americans with Disabilities Act (ADA), this new role is very critical. The job description as a prevention tool is known as a *functional job description.* This job description identifies the essential physical requirements of the job, as required by the ADA to discourage discrimination in employment practices with regard to qualified disabled individuals. It also assists the employer in selecting individuals who have the physical ability to perform the tasks required and in matching the worker to the work.

In order to develop a functional job description, one must be able to perform a job analysis. Reasonable accommodations (ergonomic modifications) should also be identified to prepare for a qualified disabled individual who applies for the job.

The following areas should be explored with consumers to determine their needs and desires to have these services made available to them.

1. Do employers have job descriptions for all positions within their company? Are they written in a way that identifies the physical requirements of the job? Do they satisfy the intent of the ADA? How are the job descriptions used in the employment process? Are job descriptions ever provided to health care providers in order to make a return-to-work decision?

   Often the job descriptions that are used by employers are primarily task oriented. They describe the duties that the employee must perform but do not address the physical requirements of the job. Most employers are aware of the ADA and know that a functional job description would be beneficial to them in complying with this law.

   The process of developing a functional job description can also be of benefit to an employer. It is probably the first time that the employer has looked at a job(s) in this manner. When assessing a job for its physical requirements, one should ask the question, can this job be performed with less physical stress? If the answer is yes, the job should be modified for all employees regardless of their ability. This will decrease the potential of injury and can improve productivity, which will eventually contribute to the profitability of the company.

2. Do physicians ever receive a job description from an employer that identifies the physical needs of the employee to perform the job? Would a functional job description be helpful in making a return-to-work decision?

   One of the criticisms from physicians is that they do not have a good understanding about the job that their patient performs and often make their decision on subjective information. They would welcome an accurate job description that addressed the physical requirements of the job.

Prework Screening. Evaluating an individual's ability to physically perform a job or task is the first step an employer can take in preventing injuries. This process will match the worker appropriately with the work.

The Americans with Disabilities Act has made significant changes in how screening can and must be done. Guidelines have been developed to ensure that the screening is done properly and is not discriminatory toward qualified disabled individuals. These guidelines have improved the method of screening and have enhanced the validity of the screening process when it is performed.

When assessing the consumer needs for prework screening, the health care provider should explore the following areas.

1. Do employers use prework/preplacement screening in their employment process? If employers do not, is this something that would be of interest to them?

   Prework screening would be of particular benefit to those employers whose jobs include manual material handling or other moderate to heavy physical labor. Employers who have a high work injury experience rate would be interested in better matching of the employee with the work.

   Employers should also keep in mind that prework screening not only is beneficial for new hires but can be used for job transfers and employee reinstatement after injury.
2. When prework screening is utilized, is it functionally relevant to the job? Does the screen accurately simulate the essential functions of the job, or is the same screen used on all employees regardless of the job? Prework screening must be functional, job-specific screening rather than health risk screening. Many employers are performing or contracting for screening that does not accurately screen the essential functions of the job. This is particularly true when the screening is performed by using computerized strength-testing equipment. The validity of this type of screening is questionable and could easily be challenged.
3. Are all prospective employees screened for a job in the same manner? It is essential that the screening be conducted the same for all applicants unless the individual is a qualified disabled individual for whom an accommodation needs to be made if reasonable. Employers cannot screen only those prospective employees who they suspect may have difficulty.
4. Are the records for screening considered confidential and kept in separate files from the personnel files? ADA requires that medical-related information on employees be kept in a separate and locked file.

Prevention Education Programs. A must in prevention education is the involvement of the workers in problem solving and in taking responsibility for their own actions on the job. This must be combined with safe work practices and a safe workplace. Education programs need to be conducted at the worksite and presented in a way that is applicable to the work the employees perform. The education program should be geared to increase awareness of the causes of injuries, especially cumulative trauma injuries. The programs should be designed not only to inform employees but to teach them how to correct or prevent a problem.

The following areas should be explored with the potential consumers of work injury prevention education programs.

1. Do employers provide prevention education programs to their employees on a regular basis? If so, how are these provided and by whom? Most employers provide prevention education programs because they are required to under Occupational Safety and Health Administration (OSHA) regula-

tions. They utilize many different resources and approaches. A number of outside resources can provide companies with educational materials or on-site presentations. How satisfied are the employers with their present arrangements?

2. How does the employer determine the effectiveness of the education program? Are there criteria to determine the success of the prevention program offered? It will be important for the health care provider to know how employers determine the success of an injury prevention program. Is this an area where the health care provider can assist them by developing questionnaires, pre- and post-training assessments, or noted reduction in accidents and/or injuries?
3. How often are prevention education programs provided? What topics have been covered in the past? It is helpful to find out what topics are important to the employer. It is also appropriate to explore other topics that employers may want to add to their present offerings.

Return-to-Work Programs

Early return to work through early intervention is critical. While there is now awareness of the importance of early intervention with return to modified duty, opposite problems are now beginning to surface. In some cases people are not fully healed and are returning so early that they have not clinically been restored to full function. This leads to either reinjury or overutilization of modified work. Perennial light duty is one of the problems seen with return to work before full rehabilitation is done. Reinjury is also one way to measure when intervention has returned people to work too early. Therefore, even though early intervention is the goal, there must be quality assurance measures to guard against unsafe early return.

The following areas should be explored with the potential consumers of return-to-work programs to help determine their needs and desires of obtaining services from the health care provider.

1. Do the employers presently use modified or light duty jobs for employees who have had a work-related injury? Are there specific jobs that have been identified as "light duty," or are these selected on the abilities of the individual? Is the employer willing to modify a job on a temporary or full-time basis for an employee with limited ability?

   Many employers are resistant to having injured employees return to work until they can perform the job at "100 percent." They are concerned about reinjury and about individuals being put on "light duty for life." They are also concerned about the impact on other workers if the injured employee cannot do the job.

This information will be of assistance to the health care provider in identifying those employers who encourage return to work with modified jobs and those employers who would benefit from more information and education on the economics of modified jobs. It will also provide information about those employers whose employees may require more work conditioning or work hardening to return to work.

2. Do the employers need assistance in identifying jobs appropriate for employees with limited ability, or do they prefer just to know what the limitations (restrictions) are and to identify the jobs themselves?

   Employers often would like assistance in identifying jobs that can be performed by an injured worker if this would get the individual back to work quicker without the risk of reinjury. They are also interested in any modifications of the present job that can be made to accommodate the individual and allow him or her to return more quickly to the original job. This information will be beneficial for the health care provider to identify the best method to provide return-to-work programs. It is also an opportunity to let the consumer know that the provider is willing to come to the worksite as part of the return-to-work process.

3. Do employers receive adequate information from the health care provider to make an appropriate determination for a temporary or permanent job assignment? What information would be helpful to the employer in getting an individual back to work more quickly and safely?

   One of the most consistent complaints from employers is that they do not receive adequate information from the health care provider. There is very little communication between the health care provider and the employer except for work restriction. This is one area where the health care provider can be of significant benefit to the employer. Developing an effective communication process with all parties concerned will provide a most appreciated service.

4. How does the employer and/or the health care provider determine when a person should be progressed from light duty back to full duty? Does the health care provider offer assistance in monitoring the injured employee after the return to modified duty is made, is the employer left with that decision, or is it a "light duty for life" situation?

   One problem that arises with employers who have light duty jobs is that often employees who are put into these jobs never get out of them. Employers would be very receptive to the health care provider who could provide them with assistance in closer monitoring of these individuals so that they do not become stuck in these jobs for an overextended period of time.

5. Do health care providers visit the worksite prior to and/or at the time of return to work to gain an understanding of the job and to see if the individual can actually perform the job safely? Does the health care provider educate the injured employee and the employer on precautions? Is the employer open to making permanent or temporary modifications to the job to facilitate a rapid and safe return to work?

   A factor in the return-to-work process is the potential of reinjury of an employee because of an unsafe or poorly designed work area. A standard procedure for return to work should be an evaluation of the injured worker's workstation, especially when cumulative trauma has caused injury or is suspect in the cause. In cumulative trauma cases, it is critical that the worker report symptoms early. This will require a combination of educating the worker in early symptoms and getting the cooperation of employers to facilitate changes that can prevent cumulative trauma from being severe.

Return-to-Work Evaluation

Functional capacity evaluation should be an integral part of the return-to-work process. The evaluation should have a relationship to return-to-work for specific work activities such as lifting, carrying, and bending. It is no longer acceptable for functional evaluations just to give numbers, graphs, repetitions, or comparison to data banks. There must be some outcome of what this means in the return-to-work process. It needs to be specific to the individual and the job in question. The findings should be projected for an eight-hour workday. The functional evaluation must be able to stand up in court. Credability, accuracy, and objectivity are critical. The written report must be easily read and interpreted by the individual(s) receiving it.

Functional capacity outcome statements should involve one of the following:

1. The individual's physical abilities match the job description; therefore return to work is indicated on the physical level.
2. The individual's physical function has deficiencies; therefore the individual needs the intervention of work hardening/work conditioning.
3. The individual's physical deficiencies cannot be changed with work hardening/work conditioning; therefore work modification or worksite modification is indicated.
4. Functional capacity evaluation indicates significant problems; therefore case resolution can be accomplished (this may include an impairment rating).

The following areas should be explored with potential consumers regarding functional capacity evaluations.

1. Do physicians make referrals to health care providers for a functional capacity evaluation? If so, what is their level of satisfaction with the testing performed? Does it provide adequate information from which they can make a decision about return to work? How difficult is it to schedule a patient to receive a functional capacity evaluation?

   More physicians are becoming aware of the value of functional capacity evaluations that compare the individual's functional ability with the job in question. They are interested in evaluations that provide easily understood reports, not reports with many graphs and tables that are difficult to understand. They also want reports that are brief and specific.

2. Are employers aware of functional capacity evaluations? Do they understand the value of the evaluation in returning an individual to work? Do they receive copies of the report, and/or are they included in a conference with the individual administering the evaluation and the employee?

   An increasing number of employers are aware of functional capacity evaluations. Their level of satisfaction is usually mixed, depending on their experiences. There are often questions about the validity of the evaluation when compared to the actual physical requirements of the job in question. This is especially true when the evaluation does not compare its results to the actual job. Employers often do not receive the results of the evaluation other than a statement on restrictions. They usually appreciate being able to meet with the evaluator to ask specific questions pertinent to the job.

3. Do insurance carriers, rehabilitation consultants, or vocational counselors make referrals for or receive functional capacity evaluation reports? If they do, what is their level of satisfaction? Do the reports provide assistance in the return-to-work determination and/or in the settlement of a particular case?

   Most insurance companies, specifically those dealing with workers' compensation, are aware of functional capacity evaluations. Many, however, do not understand that there are variations of functional testing and do not have a good grasp on these differences (psychophysical versus kinesiophysical). They are concerned about the accuracy of the evaluation, especially those evaluations that are conducted in one to two hours or even one day when the job is performed eight hours, five or more days in a week. They express a dislike for lengthy reports that are difficult to understand and that are not specific to the job(s) in question. They are concerned about the availability of evaluations and how rapidly a test can be conducted. Time is very important to these consumers.

4. Do attorneys use functional capacity evaluations? What is their impression of the usefulness of these evaluations in preparing their case(s)?

An increasing number of attorneys, both plaintiff and defense, are seeing the effectiveness of using functional capacity evaluations. This is especially true in the testing of a person's ability or inability to perform a specific job or task. They want evaluations that are backed up with objective findings that relate to the specific job or task. They would prefer a standardized evaluation that has a proven record and stands up under the rigors of the legal system.

Work Rehabilitation Programs

Both work hardening and work conditioning are seen as viable alternatives in work rehabilitation. Definitions and guidelines for work hardening and work conditioning have been developed by the American Physical Therapy Association and are available through that organization. Work hardening/work conditioning should be five days a week for at least four hours per day. When a worker is able to perform safely for four hours at a level that will enable him or her to return to work safely, he or she should do so on a part-time basis. This will be dependent on the employer's willingness to cooperate with this plan. This approach will promote the early return-to-work concept. The worker, if returned to work for four hours, can finish off his or her day (an additional four hours) at either the rehabilitation facility, an on-site fitness program, or a job within the company that requires less physical ability. The critical point is rapidly returning the individual physically back to work, an outcome that facilitates keeping the individual mentally and physically fit through connection with the work environment.

The design of the work-hardening/work-conditioning program is most critical. Four aspects to be included are strengthening, work simulation, education, and aerobic conditioning. To be truly successful, measured outcomes as indicators of quality and cost-effectiveness will be the best assurance of reimbursement.

When analyzing what the consumers of work-hardening and work-conditioning programs want, the health care provider should explore the following areas.

1. Do the consumers have an awareness of work-hardening and work-conditioning programs? If so, what is their impression of their effectiveness in the return-to-work process? Are the programs that they are aware of cost-effective? Do the providers of this service communicate with all parties concerned?

   There are often mixed impressions among consumers of the value of work-hardening programs. This is especially true of insurance carriers. They question the cost-effectiveness of these types of programs. They can be very supportive of a facility that can demonstrate that its programs are successful at returning individuals back to work in a timely manner. Employers question the effectiveness of these programs for some of the same reasons. A growing number of employers see the value of work hardening/work conditioning and want this type of program to be put into place inside the company. This is an area that the health care provider should explore, especially with large employers (500 employees or more).

Quality Enhancement (Outcome Monitoring)

Effectiveness and outcome studies regarding the services delivered by an occupational health and rehabilitation program are critical. Insurance companies are asking for return-to-work statistics. Data need to be collected to validate outcome results. In general 70 to 80 percent return to work is considered a good outcome.

Return-to-work discharges become increasingly important. Discharge at the worksite should be a goal. This will enable the therapist to evaluate the client actually performing the job in the work environment and thus make recommendations on changes and to communicate with the employee and employer about safeguards and other such matters.

In assessing the needs of the consumers for occupational health and rehabilitation programs, the following questions should be explored with regard to program quality and outcome:

1. Do the rehabilitation programs that are presently utilized provide documentation on the quality of their services? Is this important in the employer's decision to use a specific health care provider?

   Most consumers, especially insurance carriers and employers, are becoming more interested in how effective a health care provider's programs are. They want documentation that will demonstrate effectiveness. They prefer "outcome"-based information. The consumer of the future will require that this information be available and maintained by the health care provider on an ongoing basis.

2. How do consumers determine the quality of the health care services provided to them and their employees?

   The criteria that consumers use will vary depending on the consumer needs. The health care provider needs to have the consumer's input to design methods that will demonstrate its ability to meet the consumer's expectations.

CONTINUUM OF OCCUPATIONAL HEALTH AND REHABILITATION CARE

In the 1980s it was enough for providers of occupational health and rehabilitation services to offer only one of two services such as work hardening, functional capacity evaluation, or rehabilitation programs. In the late 1980s, however, rehabilitation providers began to expand programs to focus attention on total work injury management and rehabilitation services. This expanded contact with the employer and the worksite. There was a move toward multifaceted programs such as ergonomics, prework screening, injury prevention, and early intervention services.

The complexity of services, the increase in the number of providers and the nature of workers' compensation systems made the management of the continuum of care difficult. As a result, progressive occupational health programs began to monitor claims for workers' compensation for individual employers. This improved case management is the cornerstone of a good occupational health rehabilitation program.

In the 1990s, competition for occupational health and rehabilitation has become stronger. The dollars available to treat injured workers will be directed toward programs that have more of a "specialty" in occupational health and rehabilitation.

Programs will continue to need to provide more comprehensive individual services and better communication with all parties concerned.

CONCLUSION

Health care providers who want to develop an occupational health and rehabilitation program or who want to expand/improve their present program should thoroughly analyze their market. Knowing what consumers want and identifying the problems they want resolved are critical. This information will help form the basis from which to build a successful program. The program should attempt to resolve the problems that are identified in the analysis. The analysis will also provide the criteria that will eventually be used in measuring outcome/success of the program's ability to meet the market needs.

Chapter 36

Risk Management Consulting: Blending the Analysis of Problems with the Development of Solutions

Susan H. Abeln

Workers' compensation costs have become a major concern for most employers. The Second Biennial Towers Perrin Survey Report on Workers' Compensation, entitled *Regaining Control of Workers' Compensation Costs*[1] shows almost 70 percent of survey respondents reporting that workers' compensation claims are "somewhat" or "significantly" threatening their financial results. Recent risk management trade magazines have had numerous headlines such as "Workers' Compensation Problems" and "Health Care Costs—a Risk Manager's Worst Nightmare!" They often go on to detail a number of critical concerns that should be of interest to health care providers as well as employers. These routinely include topics such as spiraling health care costs, the increasing need for *effective* workplace safety efforts, the growing misuse of the workers' compensation system for nonwork injuries, and the need for every employer to develop and adopt workers' compensation cost containment strategies.

All these concerns are not too surprising when one considers that

1. The direct costs of industrial workers' compensation programs currently exceed $68 billion per year, are doubling every five years,[2] and are estimated to exceed $140 billion by the year 2000.
2. Industry and government sources estimate that direct costs from workers' compensation across all industries averaged 2.57 percent of payroll in 1992.
3. The "true costs" of a company's workers' compensation program are conservatively estimated at four to six times the dollars actually paid for direct costs.
4. Medical costs currently make up over 40 percent of industry's total workers' compensation costs, and these medical costs are increasing at a rate well above general medical inflation.[3]

5. Over 40 billion lost workdays are claimed against industries and businesses each year.[4]
6. Twenty percent of workers' compensation claims account for more than 85 percent of the costs.[5]

Again, it is not surprising that many employers are shifting to PPOs and HMOs when laws permit or to other managed care options to cover their workers' compensation costs. Precertification for medical care and utilization review programs have likewise increased dramatically.

While most employers realize they have a workers' compensation problem, they are not exactly sure of its real dimensions and are therefore uncertain of the best strategy to use to fight its increasingly negative effect on their businesses. This is where health care providers and/or loss control specialists can and should come in. Yet attempts to discuss, sell, or implement interventions that have been documented to be effective frequently seem to fall on deaf ears. At times, it seems that no employer or payor is listening to or understanding what is said. Somehow, health professionals have failed to build a successful business relationship.

Let's take a minute to review an essential ingredient in most successful relationships—business or otherwise. Most people, I am sure, would agree that trust must exist between friends or the relationship withers. Lewis and Pucelik suggest that this is also true in business relationships.[6] They state that "a business deal can not be consummated unless there is a mutual trust based on the belief that each party is being understood by the other."[7] They go on to say that the development of trust can only begin when both parties get the sense that they are being fully understood and that they and the other party are truly "speaking the same language."[8] How can one get employers and payors to understand the effective intervention strategies being proposed? How can one understand what these people believe to be the major problems within their facilities? Further words from Lewis and Pucelik provide a clue: "Understanding implies that you can 'join' a person at his own model of the world. This is important because people tend to operate as if their model of the world is the real world. Understanding is the crucial bridge between their model of the world and ours."[9] There are several ways one can "join" people at their own model of the world, but Lewis and Pucelik suggest that the simplest is to use their model's language to describe their perception of their world. This, they say, "will pave the way for highly effective and influential communication to occur."[10] This does *not* mandate that one accepts the other's model as one's own, only that one demonstrate understanding of the other's world by one's willingness and the ability to join the other in speaking the other's language. Consultants must accept the responsibility of being willing and able to "join" each of their clients in the clients' model of the world and then to help. To do this effectively, consultants must always encourage industrial clients to recognize and give voice to their perceptions of the unique characteristics of their particular situation. Only then—

when there is an understanding of their world and their perceptions of the problems associated with that world—will a consultant be able to successfully interest them in the benefits that can be provided.

Working with a business client as a potential problem solver can, however, create some very real problems for a risk management consultant. As stated above, the consultant and client probably don't currently see the world the same way or speak the same language. Range of motion figures, measurements of torque or power, maximum lifting limits, or even research studies aren't very persuasive to an individual interested in bottom line costs or lost time indices. And not every individual in the client's organization is going to be interested in the same things. The deeper one gets into the infrastructure, the more detailed the analysis of any problem will probably have to be. The task is further complicated by the fact that there is a need to gain understanding of the situation by gathering information from varied sources: the chief financial officer (CFO) of the client, the insurance company/claims administrator who handles the workers' compensation coverage for the facility, the human resources department that coordinates the facility's workers' compensation, the risk manager who traditionally trends injuries within the facility, the loss control consultant, and appropriate others. Each of these sources has its own view of the world. Each perceives the situation and the workers' compensation problems with his or her own bias, and each speaks primarily the language of his or her specialty. The consultant's task is to communicate with all parties who can add information to the assessment (in their language, not the consultant's), gather all the pertinent information, decide on the priorities for "treatment," and then develop a plan of action. At the same time, an eye must be kept trained on the fact that the "fix" must address the problem as each in the client company sees it.

To help one successfully complete this formidable task in the real world, a not-too-unreal case study is presented step by step using Abeln's Workers' Compensation Investigative (WCI) model.

CASE EXAMPLE: THE UNINFORMED INFIRMARY

The Uninformed Infirmary is a 350-bed hospital that is part of a for-profit chain. The chain—Sickly and Saintly—has recently realized that its expenditures for workers' compensation throughout its 10 hospitals now exceed its expenditure for medical malpractice. The chief financial officer (CFO) estimates that the corporation's annual workers' compensation costs exceeded $3.375 million in 1992. This is a full 4.5 percent of its $75 million annual payroll. Yet industry and government sources estimate that direct costs for workers' compensation across all industries averaged only 2.57 percent of payroll in 1992.[11] Even you or I can see that the corporation appears to have a problem. Or, looking at the situation positively, we could say they seem to have a *very* large opportunity for improvement. Recognition of this problem leads the corporation's president to a no-holds-barred confrontation with the CFO and to her own extensive review of the financials. She then calls an emergency management meeting and "asks" each of the hospitals' executive directors (EDs) to thoroughly review the workers' compensation experience of their facility and, if need be, to design and present to the president and the remainder of her senior management team, within four weeks' time, a plan of action to more aggressively manage workers' compensation costs within the hospital.

Now the ED at each of the facilities has a problem. And it is likely that by day's end, the departmental managers and others within each facility will also have a problem.

For the purpose of this discussion, we will focus on only one facility—one that is a potential client of yours—the Uninformed Infirmary. As will soon become apparent, this facility and its executive director Joseph J. St. Clair are definitely facing a big challenge. And that's where you come in. You are not only the rehabilitation provider who contracts with the hospital, and the owner and key therapist of the premier rehabilitation clinic in town for the treatment and prevention of work-related injuries; you are also a de facto department head of the Uninformed Infirmary. Because of this relationship, you have been asked by the executive director to lead a team of managers within the organization in a review of past workers' compensation experience and, if necessary, to devise a way to fix any problems. The situational review as well as any needed "fix" is to be reported to Mr. St. Clair within three weeks' time so he can pass it up the line to the corporate president.

Now you have a problem. Or maybe it is an opportunity. But whatever you choose to call it, it is critical for you and your team to fully understand both the corporate president's and Mr. St. Clair's perception of the problem so that you can offer a situational review and a plan of action that each of them will understand, relate to, and accept. In other words, you must "join" the corporation and the key management of the facility at their own models of the world and then build the bridge between their world and yours by active listening and effective communication. This undertaking is critical to create an ideal climate for positive growth and change and to become recognized and utilized for the many strengths and resources that you can lend to the fight.

At this moment, some professionals might be more than a bit frustrated by what is being proposed. After all, the proposition is that you find out what the client—in this case, most of "the brass" at the facility—thinks is wrong when they just got through giving you the job of finding out what really is going on and how to fix it—if, in fact, you can even find it. It sounds like a waste of time. But it's not. Most people just are not persuaded by facts that are contrary to their beliefs. It always helps to know where someone is coming from before you try and tell them where you want them to go.

In clinical practice, successful treatment always demands a clear and concise identification of the problem. Just as it makes little sense to treat injured patients without a full understanding of their symptomology, their reactions to standard tests, and their impairments and functional limitations, it makes even less sense to treat this hospital's workers' compensation problems without a full understanding of the situation from the key players' perspective.

Enough of theory. Let us continue with the case study and take a look at the client's perceptions of the world. Their thoughts about the workers' compensation world are as follows:

- *Corporate President, Penelope Peregrine:* As managed care has advanced and made inroads into the communities where Sickly and Saintly facilities are located, pretax profits for the corporation have been reduced to a 20-year low—3.7 percent. My contract expires next year, and I'm afraid that with these record low profits I will be replaced by the board of directors at the next board meeting. The CFO's discovery of our rapidly escalating workers' compensation costs offers me a glimmer of hope in reducing the cost of doing business and ultimately of becoming more profitable in the ever-tightening health care marketplace. My review of the financials shows that all of the facilities have suffered from increases in losses. The Uninformed Infirmary, one of the largest of the 10 hospitals, has experienced far worse results than any of the other facilities. I'll bet it is because of their high turnover. No matter. Unless St. Clair proposes a solution that I believe can have an immediate positive effect, I am going to terminate him and hope that a new executive director can rescue our profits and my job.
- *The Uninformed Infirmary's Executive Director, Joseph St. Clair:* This facility has been making record revenues every quarter since I took the rein. Yet I know there have been costs for this higher-than-expected level of income. Turnover of full-time and part-time employees (excluding those who work per diem or at a type of registry) has exceeded 40 percent in the last year. I have heard from corporate of the high cost of extreme levels of turnover (greater than 25 percent is considered extreme in this organization). When I think about the management of the facility at the department level, it shocks me to realize that I have lost and replaced more than half of the "seasoned" managers with new graduates in the past two years. In fact, in my four-year tenure, I have lost and replaced my senior management team at least three complete times. Each of these senior managers got a very generous severance package (paid for from the corporate coffers). Some stated publicly that they had jumped ship with golden parachutes. Also, I seem to

recall that according to the most recent employee satisfaction poll, 76 percent of employees are frightened/dissatisfied and no longer feel good about recommending the hospital to a family member or friend. Furthermore, Mrs. Peregrine, the corporate president, looked right at me during most of her discussion of workers' compensation costs. I wonder if our experience is worse than that at the other facilities? I'd better find out what is going on and get my act together.

WORKERS' COMPENSATION INVESTIGATIVE (WCI) MODEL

As with clinical decision making, the process of discovering the true nature of an organization's workers' compensation problem is complex. It is often very difficult to understand someone else's world, and consequently it is even harder to join them there and lead them across the bridge to a shared understanding. Several individuals within rehabilitation have proposed decision-making models using standard methodology to reduce the complexity and to optimize the results of clinical decision making. A similar decision-making model has been developed for clarifying the features of an organization's workers' compensation problem and optimizing the results of any decisions made—the Workers' Compensation Investigative (WCI) model. In the WCI model, the characteristics from various tests of the situation help identify the true scope and breadth of the situation. This model is shown in Exhibit 36–1. For the purposes of this case study, this model will be used to do the situational analysis needed and to develop the facts for the presentation to the senior management of the hospital and possibly to the senior management of the corporation—Sickly and Saintly. Throughout the entire case study, once a pivotal fact about the client's model of the world (a fact that may shape client perception of their workers' compensation situation) is turned up, it will be emphasized and reference will be made of the need to hold onto the information in preparation for the report to management.

IN SEARCH OF THE BOTTOM LINE

Review of the model suggests that one's first step to understanding and joining most clients' models of the workers' compensation world is to determine how workers' compensation costs hit the bottom line. In this case study, we will explore how this effect is felt at Uninformed Infirmary and how that in turn affects the corporation. Let us return to your assignment: to lead a team of managers within the organization in a review of past workers' compensation experience and, if necessary, to devise a way to fix any problems.

Prior to calling the first team meeting, you decide to spend part of the day in conversation with the hospital's CFO, Mike Meticulous. The purpose of the meet-

Exhibit 36–1 Workers' Compensation Investigative (WCI) Model

1. Determine how workers' compensation costs affect the financials of the corporation and/or client organization (this may include where they can be found on the financial reports).
2. Review available cost data for appropriate (3–5 years) time frame to determine direct workers' compensation costs on a calendar- or fiscal-year basis. (This review should include a review of allocations if workers' compensation costs are allocated to the local facility from a corporate entity.)
3. Calculate workers' compensation costs as a percentage of total payroll and compare with national average.
4. Calculate employers' frequency index using an appropriate time frame. (Review OSHA 200 log and number of hours worked by all employees of the employer.)
5. Calculate the lost case index for the same time period. (Review as above.)
6. Calculate the lost day index for the same time period. (Review as above.)
7. Compare the above indices to national and industry averages as published by the Bureau of Labor Statistics and the National Safety Council.
8. Access or create and review facility-specific trending of losses (these can be trended from loss runs or from incident reports).
9. Identify five areas of major concern.
10. Interview managers and staff to expand understanding of these critical areas and probe for hidden areas of concern.
11. Working with key personnel and focusing on the employer's objective, prioritize critical areas.

Source: Copyright © 1993, Susan H. Abeln, Strategic Healthcare Alternatives, San Clemente, California.

ing is for you to get a handle on how workers' compensation costs have affected the hospital in the past few years. It is also important to understand how that effect is seen and felt by the facility and corporation. And if the CFO knows how dollars spent on injured employees are translated into a number that is reflected in the bottom line, all the better. Prior experience in dealings with operations, accounting, and risk management personnel of the outpatient office's industrial clients should provide clues as to where any potential problems might be found.

After a short meeting, you discover that the hospital's workers' compensation expenditure is allocated to it from the corporate office as a separate line item under the general category of benefit expenses. Mike says that the allocations are based on a rolling average of three years of the facility's claims using the total incurred value for these claims (what has been paid on each claim and what is in reserves to pay for the remaining activity related to that claim). He also explains that he and the Director of Human Resources recently discovered that the dollar amount that has been paid on claims as well as the dollars held in reserve can be most easily found in a monthly report submitted to them by their claims administrator. This report is called the detail loss run. Both the Human Resource Department and Mike receive and file these reports monthly. Mike further states that the corporation also performs some additional comparisons prior to finalizing the allocations. It is his belief that corporate first compared each facility's annual total incurred cost to his payroll for each of the three years; if a facility's latest experience differed greatly from its norm (positively or negatively), corporate made corresponding adjustments to its preliminary numbers. Then the corporate CFO compared each facility's experience to the total corporation's experience. Again, if the site's experience was outside of the corporate norm, the corporation made further adjustment prior to finalizing the allocations. Mike's offhand mention of these adjustments suggests that these calculations are of secondary importance to the objective of understanding corporate management's model of the world—a world in which the effect of workers' compensation costs on profit is of prime interest.

UNINFORMED INFIRMARY'S RECENT EXPERIENCE

Now that the means by which workers' compensation costs hit the bottom line and the relationship between the dollars spent on injured employees and the bottom line are understood, it is time to look at the facility's experience. To be consistent with corporate practice, this look must include at least the last three years of experience at the facility. Therefore one needs to know how many injuries occurred in the last three years at Uninformed Infirmary, what the costs were for these injuries (Mike calls this claim frequency and severity), how this affected this hospital's allocation in the same time period, and finally how these allocations compare to the rest of the corporation. Mike provides the numbers as detailed in Table 36–1 in answer to your questions and gives you the number of the CFO of a sister hospital of about the same size (Healthy Haven Hospital) to provide meaningful comparison data. The other hospital's data are shown in Table 36–2. As one can see, Uninformed Infirmary's costs as well as their allocation have gone up approximately 100 percent a year for each of the last three years. It is also interesting to note that their percentage of the corporation's costs has also risen dramatically. This is critical for the report and presentation.

Table 36–1 Workers' Compensation Costs: The Uninformed Infirmary

	Number of Claims	*Total Incurred Cost*	*Allocation*	*% of Corporate Costs*
1990	64	$287,000	$0.3 million	23.7%
1991	93	$594,750	$0.6 million	27.1%
1992	134	$1,383,490	$1.2 million	35.6%
Total	291	$2,265,240	$2.1 million	
Average	97	$755,080	$0.7 million	28.8%

Source: Copyright © 1993, Susan H. Abeln, Strategic Healthcare Alternatives, San Clemente, California.

Table 36–2 Workers' Compensation Costs: Healthy Haven Hospital

	Number of Claims	*Total Incurred Cost*	*Allocation*	*% of Corporate Costs*
1990	73	$145,151	$0.20 million	15.8%
1991	55	$128,307	$0.16 million	7.2%
1992	69	$198,921	$0.25 million	7.4%
Total	197	$472,379	$0.61 million	
Average	66	$157,460	$0.20 million	10.1%

After the meeting, you need to consider one additional fact about the workers' compensation allocation process at the Uninformed Infirmary. Workers' compensation expense is not allocated to departments as part of the financials that they are to manage. It is processed only as part of the overall facility's expense. While this is common knowledge, somehow it just hasn't been considered important before now. Reflecting on this fact, you realize that no manager within the facility except a senior manager has access to specific information related to workers' compensation costs. Therefore it is safe to say that no department manager understands, appreciates, or can attempt to manage the costs of his or her injured workers. This information may prove key in clinical decision making about this facility's workers' compensation problem and the action plan recommended to the ED. This also is critical for the presentation.

COSTS AS A PERCENTAGE OF PAYROLL

From the CFO, you have information regarding the allocations as well as a copy of the information stating the total incurred costs for each of the last three claim years. With these figures in hand and the task clearly defined, the second meeting of the day takes place—this time with the Director of Human Resources. The immediate task is to obtain the facility's payroll for the last three years. A review of the WCI model suggests that more extensive time will need to be spent with the director or the individual she has designated as the workers' compensation coordinator to complete a statistical review of the facility's performance. But first things first. After only five minutes with Erma Erving, the director, she has found all the required in-house information in a standard report she calls the payroll register. A call to her counterpart at Healthy Haven for their payroll information provides the rest of the data needed to complete the necessary calculations. The results of these calculations are found in Table 36–3. Looking at these calculations and recalling that according to Towers Perrin, the national average for all industries for 1992

Table 36–3 Workers' Compensation Costs as a Percentage of Payroll

The Uninformed Infirmary

	Number of Claims	*Total Incurred Cost*	*Payroll*	*Costs as % of Payroll*
1990	64	$287,000	$11.4 million	2.5%
1991	93	$594,750	$18.6 million	3.2%
1992	134	$1,383,490	$20 million	6.91%
Total	291	$2,265,240	$50 million	

Healthy Haven Hospital

	Number of Claims	*Total Incurred Cost*	*Payroll*	*Costs as % of Payroll*
1990	73	$145,151	$12 million	1.2%
1991	55	$128,307	$16 million	0.8%
1992	69	$198,921	$17 million	1.1%
Total	197	$472,379	$45 million	

Source: Copyright © 1993, Susan H. Abeln, Strategic Healthcare Alternatives, San Clemente, California.

was 2.57 percent (it is unfortunate that this author is not aware of industry-specific averages), it is obvious that some information on costs as a percentage of payroll must be included in the report.

INDEXES, INDEXES, INDEXES

Steps 4 through 7 of the WCI model (Exhibit 36–1) again explore a facility's performance, but this time using standard measures of performance as published by the Bureau of Labor Statistics and the National Safety Council. These standard measures use information readily available to the employer such as the recordkeeping information that OSHA (Occupational Safety and Health Administration) mandates be kept and be available for a period of at least five years by each employer. The measures are called the frequency index, the lost time case index, and the lost day index. Quite simply, once calculated, they provide a tool to measure each employer's frequency of recordable injuries, the number of cases that involve lost time, and total days lost on lost time cases as compared to other employers. Because this is done with a simple calculation that standardizes all of these measures to the rate per 100 FTEs (full-time equivalents), it is possible to compare employers' performance, or lack thereof, across an industry nationally (in this case, health care) and across all industries. The calculation necessary to develop these indexes can be seen in Figure 36–1 and is really quite basic.

For the three indexes, you simply divide the number of OSHA-recordable events in the time period (injuries and illnesses as indicated on your OSHA 200

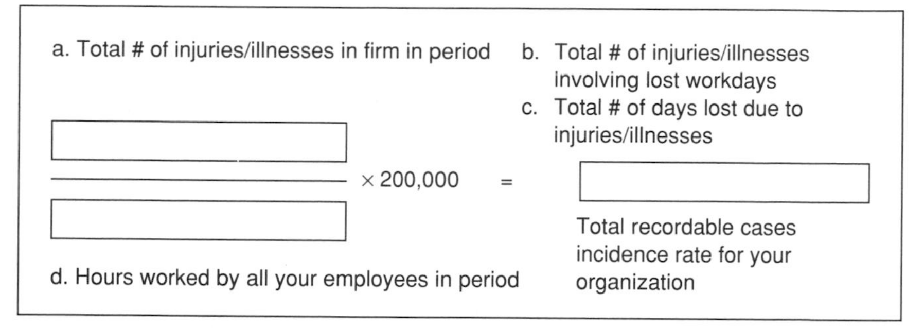

Figure 36–1 Frequency, Lost Day, and Lost Workday (Severity) Incidence Rate Formula. To determine the frequency (total recordable) incidence rate of injuries and illnesses, provide the information asked for in the boxed formula above and calculate the result. The same formula is used to compute the lost workday case incidence rate and the severity rate. Simply replace item (a) in the formula with item (b) for the lost workday case incidence rate and item (c) for the lost workday (severity) rate. Item (d) is the base number of hours and remains the same. This incidence rate is the total number of injuries and illnesses per 100 full-time employees. (The figure of 200,000 in the formula represents the equivalent hours of 100 full-time employees working 40 hours per week, 50 weeks per year, and provides the standard base for incidence rates.) *Source:* Copyright © 1993, Susan H. Abeln, Strategic Healthcare Alternatives, San Clemente, California.

log for the frequency index; the number of cases that involve lost time per time period as indicated on your OSHA 200 log of the lost case index; and the number of actual days lost for each injury that occurred or will occur until the injured employee is permanent and stationary or has achieved maximum medical improvement for the lost day index) by the number of productive hours worked for that same time period (this does not include benefit hours) and multiply by 200,000. Exhibit 36–2 will explain to you how to do this calculation for the Uninformed Infirmary. Table 36–4 contrasts the three indexes for the Uninformed Infirmary, Healthy Haven, and Sickly and Saintly. Specific industry (health care) national indexes, as published by both the Bureau of Labor Statistics as well as the National Safety Council, are found in Table 36–5.

As can be seen in Table 36–6 and Figure 36–2, Uninformed Infirmary, as expected, has a greater frequency of injury than the national average for hospitals. However, in looking at their lost case index as shown in Table 36–6 and Figure 36–3, you will see that the number of cases that involve lost time at Uninformed Infirmary is lower than the national average. However, the interesting factor, and the factor that drives the allocation of costs to the Uninformed Infirmary is the calculation that is designated as the number of lost days per 100 FTEs. As illustrated in Table 36–6 and Figure 36–4, the Uninformed Infirmary loses an average

Exhibit 36–2 Calculation of Incidence Rate for Uninformed Infirmary, 1992

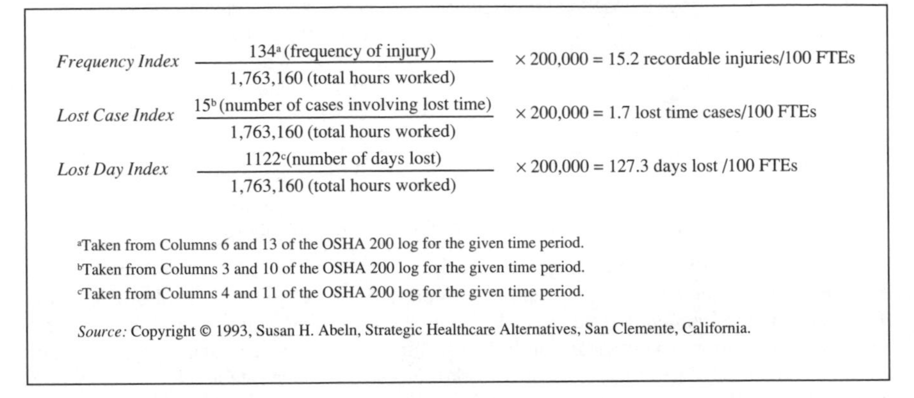

Frequency Index $\dfrac{134^{a}\text{ (frequency of injury)}}{1{,}763{,}160\text{ (total hours worked)}} \times 200{,}000 = 15.2$ recordable injuries/100 FTEs

Lost Case Index $\dfrac{15^{b}\text{ (number of cases involving lost time)}}{1{,}763{,}160\text{ (total hours worked)}} \times 200{,}000 = 1.7$ lost time cases/100 FTEs

Lost Day Index $\dfrac{1122^{c}\text{ (number of days lost)}}{1{,}763{,}160\text{ (total hours worked)}} \times 200{,}000 = 127.3$ days lost /100 FTEs

[a]Taken from Columns 6 and 13 of the OSHA 200 log for the given time period.
[b]Taken from Columns 3 and 10 of the OSHA 200 log for the given time period.
[c]Taken from Columns 4 and 11 of the OSHA 200 log for the given time period.

Source: Copyright © 1993, Susan H. Abeln, Strategic Healthcare Alternatives, San Clemente, California.

Table 36–4 Incidence Rates for Uninformed Infirmary, Healthy Haven, and Sickly and Saintly

	Uninformed Infirmary	*Healthy Haven*	*Sickly & Saintly*
Frequency of injury/100 FTEs	15.2	8.3	10.9
Cases with lost time/100 FTEs	1.7	3.6	4.4
Number of days lost/100 FTEs	127.3	79.0	74.3

Source: Copyright © 1993, Susan H. Abeln, Strategic Healthcare Alternatives, San Clemente, California.

Table 36–5 National Health Care Norms

	Frequency Index	*Lost Time Case Index*	*Lost Workday Index*
Bureau of Labor Statistics[a]	11.5	4.3	81.5
National Safety Council[b]	7.5	3.9	72.0

[a]These national rates are taken from the Bureau of Labor Statistics' 1991 rates specific for the hospital industry.
[b]These national rates are taken from the National Safety Council's 1992 rates specific for the hospital industry.

Table 36–6 Comparison of Uninformed Infirmary to National Levels of Performance

	Uninformed Infirmary	*National Average[a]*	*Comparison*
Frequency of Injury/100 FTEs	15.2	11.2	35.7% above avg.
Cases with Lost Time/100 FTEs	1.7	4.3	60.5% below avg.
Number of Days Lost/100 FTEs	127.3	81.5	56.1% above avg.

[a]These national rates are taken from the Bureau of Labor Statistics 1991 rates of the hospital industry.

Source: Copyright © 1993, Susan H. Abeln, Strategic Healthcare Alternatives, San Clemente, California.

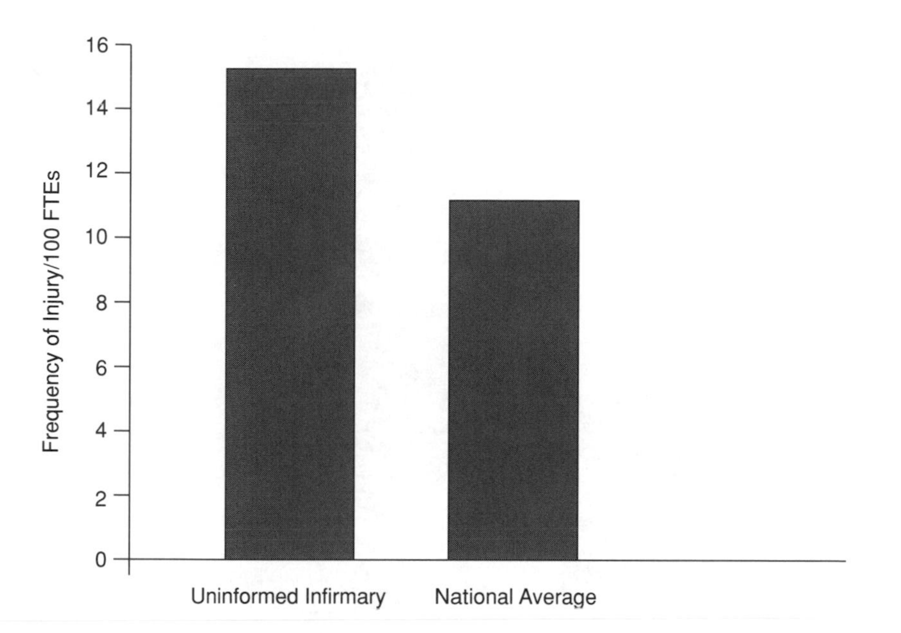

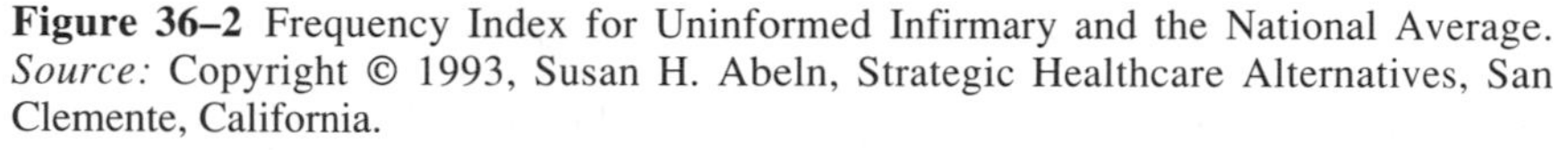

Figure 36–2 Frequency Index for Uninformed Infirmary and the National Average. *Source:* Copyright © 1993, Susan H. Abeln, Strategic Healthcare Alternatives, San Clemente, California.

---

of 127 days per every 100 FTEs per calendar year. This is more than the national average.

A major piece of the problem identification is now complete because you know that this facility has a dramatic problem as it relates to the frequency of injury and to lost days (number of days lost per lost time case). It is also evident that the action plan recommended to the ED must effectively address and propose a management of these two issues. Continuing to build, in your mind's eye, a customer-focused explanation to the ED for a strategic proposal, you again mentally note this information for the final report.

PREPARING A PRESENTATION FOR THE TEAM

Steps 1 through 7 of the WCI model are now complete. The fact-finding process has provided a disturbing picture of the Uninformed Infirmary's workers' compensation patterns. It is now almost certain that when you share this facility information with Mr. St. Clair and Mrs. Peregrine, they will both understand that the facility has a problem (an opportunity for improvement) and will also see the im-

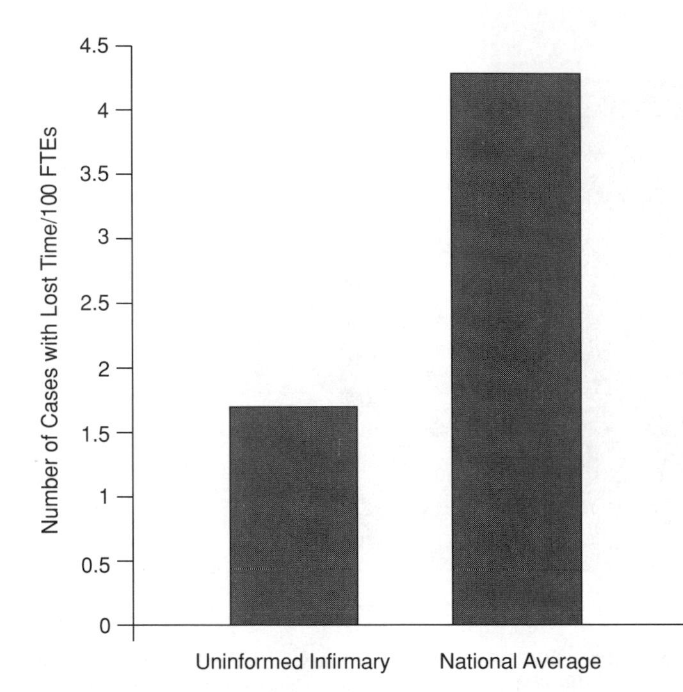

Figure 36–3 Lost Time Case Index for Uninformed Infirmary and the National Average. *Source:* Copyright © 1993, Susan H. Abeln, Strategic Healthcare Alternatives, San Clemente, California.

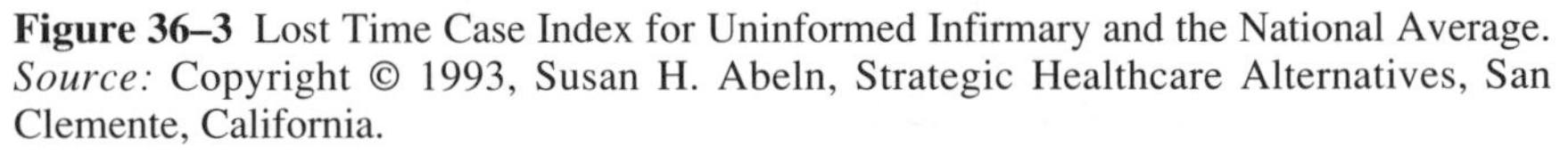

mediate need for a facility-specific action plan and most probably for a schedule for implementation. You must present this information in terms that can be understood and related to—that is, you must build a bridge of communication. But for the presentation to succeed, you must not only build the bridge but cross it in order to help the clients across. As a leader, you are just about ready to meet with the other members of the team. It was worth waiting until now in order to (1) determine if there was a problem that they as a group needed to spend their time and energy on and (2) define the problem, the framework, and the complicating issues of the problem more specifically. Now is the time to pull in the team's expertise and intimate knowledge of the facility in order to create a plan of action that addresses the Uninformed Infirmary's specific problems and is consistent with the hospital's culture. Before calling the meeting to share the preliminary information with the team of people assigned to this task, it would be helpful to summarize the major facts gathered thus far and then to attempt a draft of the presentation.

The key facts are:

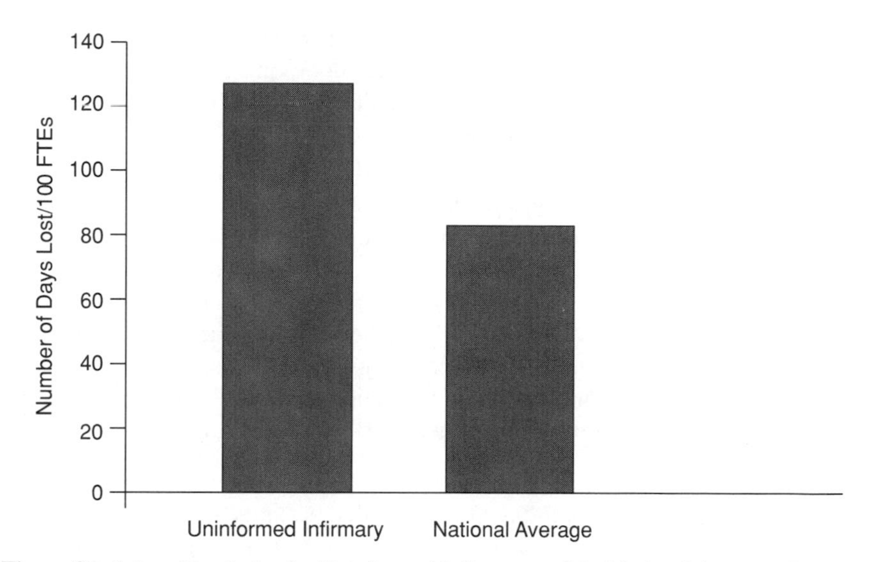

Figure 36–4 Lost Day Index for Uninformed Infirmary and the National Average. *Source:* Copyright © 1993, Susan H. Abeln, Strategic Healthcare Alternatives, San Clemente, California.

- Costs for workers' compensation at Uninformed Infirmary have risen dramatically within the last three years. Direct costs have risen from $287,000 to over $1.3 million. The infirmary's costs as a percentage of the corporation's costs increased from 23 percent to over 35 percent. Total direct costs for the last three years total $2.265 million, and over 2.1 million dollars has hit the bottom line in allocations.
- Most experts consider the direct costs of workers' compensation to be only one-quarter to one-sixth of the true cost of workers' compensation to an employer.
- As a comparison, Healthy Haven's costs have increased only 37 percent in these last three years—with their direct costs rising from $145,151 to $198,921. Healthy Haven's costs as a percentage of corporate costs have actually gone down 54 percent. Their total direct costs for all three years equate to less than $500,000 and their allocations have only totaled $0.61 million for this time period.
- The Uninformed Infirmary also has seen an increase in the percent of payroll that workers' compensation costs represent—from a slightly higher than average 2.5 percent to 6.9 percent of payroll in calendar year 1992. National

average for all industry is 2.57 percent. The national average for health care facilities was not available. Healthy Haven—living up to its name—had workers' compensation costs totaling a paltry 1.06 percent over the last three years. That is half the national average.

- OSHA, the Bureau of Labor Statistics, and the National Safety Council calculate the national averages for industries for frequency of injury per 100 FTEs, frequency of cases that involve lost time per 100 FTEs, and days lost per 100 FTEs. The Uninformed Infirmary does not compare well in two of the three instances. The frequency of OSHA-recordable injuries per 100 FTEs is 35.7 percent higher at this facility than at other similar facilities across the country, and the number of days lost because of a work-related injury is 56.1 percent higher than the national average, while actual cases incurring lost time are considerably below national average.

Now that the facts are laid out, the draft of the presentation logically follows. In creating this first attempt, care should be taken to present the facts in the language of the client and thereby to bridge the gap between the client and the consultant. A draft presentation might begin like this:

> A bit over four weeks ago, the senior management of this corporation realized that the corporation's expenditure for work-related injuries was greater, by far, than their expenditure for medical malpractice. My team's charge was to determine in what way the Uninformed Infirmary was contributing to the corporation's workers' compensation costs and if necessary to design a strategy for immediately reducing those costs and preventing the present situation from escalating further. Table 36–1 shows that the costs for workers' compensation at Uninformed Infirmary have risen dramatically within the last three years. Our total expenditures for this time period exceeded $2.2 million dollars, but Table 36–2 shows that over that same time period our sister hospital—Healthy Haven Hospital, approximately the same size as we are—spent less than $500,000. Our share of the corporation's total workers' compensation expenditure has also risen to the point where we are responsible for more than one-third of the corporation's total expenditure. Since most experts consider the direct costs of workers' compensation to be only one-quarter to one-sixth of the true cost of workers' compensation to an employer, our true cost of workers' compensation at Uninformed Infirmary over the last three years has been over $9 million.
>
> We know that we have added a lot of employees in the last three years. Table 36–3 compares our expenditures and Healthy Haven's to each of our payrolls for the same time period. Not only did our increase in em-

ployees not explain our increased costs, but as this comparison clearly shows, our costs are also rising compared to our payroll and now almost triple the national average for workers' compensation costs as a percentage of payroll. We calculated our frequency of injury per 100 FTEs, our frequency of lost time cases per 100 FTEs, and the critical number of days lost per 100 FTEs. You can see from the charts shown that while we have a higher frequency of injuries than the national averages for hospitals, we have a much lower than average number of cases that include lost time. This is important because lost time costs are generally equal to 60 percent of the cost of workers' compensation. However, the news is not all good. In the few cases we have with injuries that involve lost time, the length of time that each of our injured employees is losing is significantly greater (56 percent) than the national average. Each week that one of our employees loses in time due to a work-related injury, we are paying him or her $448/week to remain at home.[12] Many experts report that injured employees learn helplessness from this experience.

TRENDING

The first meeting with the group occurs the next day. The attendance is better than expected and includes the Director of Human Resources; the Safety Officer; the Risk Manager/Director of Quality Improvement; her guest, the hospital's loss control consultant; the Workers' Compensation Coordinator; two nursing supervisors; and the Employee Health Nurse. The leader presents a now well-rehearsed initial problem analysis and a statement of the team's purpose—to create a plan of action that addresses the Uninformed Infirmary's specific problems and is consistent with the hospital's culture and to do so in the two and a half weeks that remain before the presentation to Mr. St. Clair. The group is amazed that the hospital's record is so poor and wants more specific information on the types of injuries occurring so they can decide upon the most beneficial strategies to suggest for the action plan. Most of the individuals included in the meeting feel they already know what injuries are occurring, but the Risk Manager/Director of Quality Improvement, Mary Marble, cautions them to look only at real data and not to assume anything. Mary volunteers to abstract the data they want from the detail loss runs provided by the claims administrator and a manual review of the past year's employee injury reports in cooperation with the facility's loss control consultant, Suzanne. Upon review of Abeln's Workers' Compensation Investigative (WCI) model, the group decides that they want to know the frequency and severity (as measured by the total incurred value) of the hospital's accidents and/or illnesses. They feel that a bridge is needed to help others in the organization understand what really is going on, as well as to help the team to develop and then get approved an effective combination of intervention strategies. The elements they want to review include:

- Frequency and severity of injury by result: What was the end result of the injury? Did the employee receive a bruise, a strain/sprain, a laceration, a burn?
- Frequency and severity of injury by cause: What was the employee doing at the time of injury—lifting, pushing, reaching, bumping into something, spilling a chemical, slipping?
- Frequency and severity of injury by result and cause: Of the strains/sprains that occurred, what percentage happened when an employee bumped into the wall, slipped, etc; of the lacerations, what percentage happened when the employee was struck by a patient, etc.?
- Frequency and severity of injury by body part: What part of the body was getting injured—the shoulder, knees, fingers, thoracic spine, cervical spine, lumbar spine?
- Frequency and severity of injury by length of employment: Were only new employees getting hurt, or was it the employees who had worked at the hospital for a long time and perhaps were getting old?
- Frequency and severity of injury by position: Who was getting hurt—was it the housekeepers, the RNs, the nurses' aides?
- Frequency and severity of injury by department by position: If the nurses were getting hurt, was it the ICU nurses, or the surgery nurses?
- Frequency and severity of injuries by amount of money that had been spent and/or reserved on each case. This is frequently called a stratification of losses and simply answers the question of how many of your claims cost a small amount of money and how much of your cost of claims was due to low-cost claims, versus how many claims cost a significant amount (greater than $50,000 per claim) of money and how much of the total cost of claims was due to high-cost claims.

Two days later, Mary and Suzanne return to the team with the abstracted information. Their presentation of the information to the group uses strong visual images in order to clearly present the information and attempts to answer each question that the team has posed. The first graphics they share, Figures 36–5 and 36–6, demonstrate that over each of the last three years the most frequent injuries at Uninformed Infirmary have resulted in lacerations, with strains/sprains taking a close second. However, the most costly work-related injuries at the facility have been those that resulted in strains and sprains (see also Table 36–7). If both Mr. St.

Clair and Mrs. Peregrine of the corporate office are primarily interested in reducing the costs associated with work-related injuries, the team *must* propose a long-term and a short-term solution to the costs of strains and sprains within the facility.

Upon seeing these numbers, the team begins to murmur quietly about their failed back belt and back school program. However, the murmuring stops suddenly when Figures 36–7 and 36–8 are placed on the overhead. These two graphs and the attached spreadsheet (Table 36–8) examine the cause of each injury. As can be seen, the most frequent cause was "struck by" (a classification most often used for needlesticks), with the second most frequent injury cause being "slips/falls." The most costly causes of injuries showed a variation of this theme with slips/falls being the most expensive for two of three years and push/pulls being second in cost.

It has become apparent to the group that to truly address the client's model of the world, they will have to address the sprains/strains and slips/falls. How these

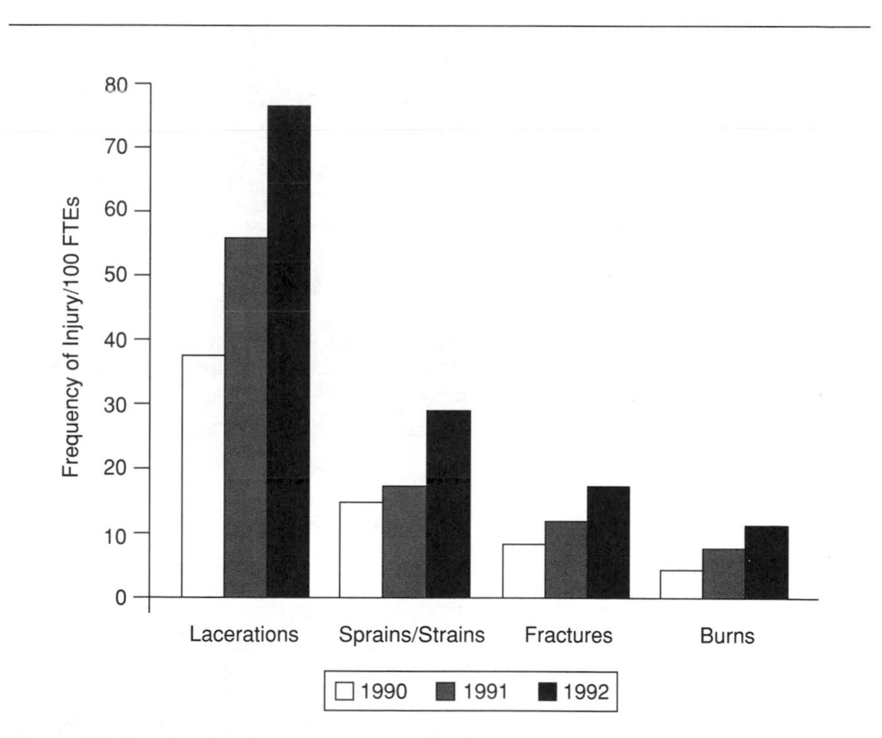

Figure 36–5 Frequency of Injury for Uninformed Infirmary by Result of Loss, 1990–1992. *Source:* Copyright © 1993, Susan H. Abeln, Strategic Healthcare Alternatives, San Clemente, California.

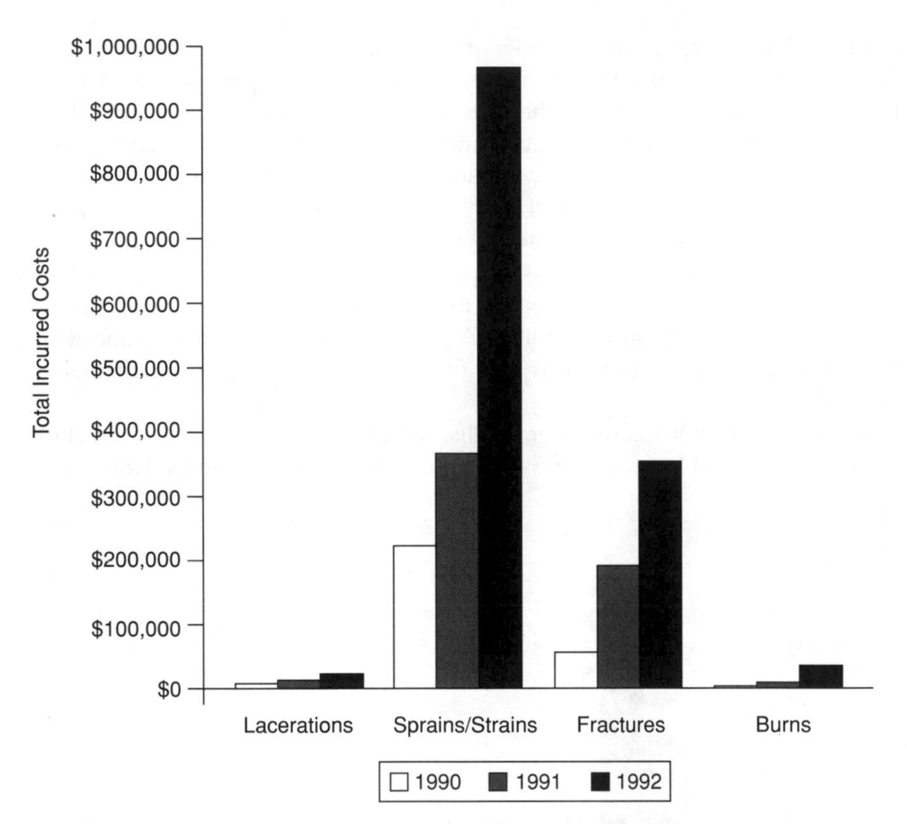

Figure 36–6 Severity of Injury (As Measured by Total Incurred Cost) for Uninformed Infirmary by Result of Loss, 1990–1992. *Source:* Copyright © 1993, Susan H. Abeln, Strategic Healthcare Alternatives, San Clemente, California.

Table 36–7 Frequency and Severity of Injury at Uninformed Infirmary by Result of Loss

	Frequency			*Total Incurred Costs*		
	1990	*1991*	*1992*	*1990*	*1991*	*1992*
Lacerations	37	56	77	$6,012	$11,812	$23,435
Sprains/Strains	15	18	29	$226,765	$374,130	$967,711
Fractures	8	12	17	$52,610	$199,422	$356,485
Burns	4	7	11	$1,613	$9,386	$35,859

Source: Copyright © 1993, Susan H. Abeln, Strategic Healthcare Alternatives, San Clemente, California.

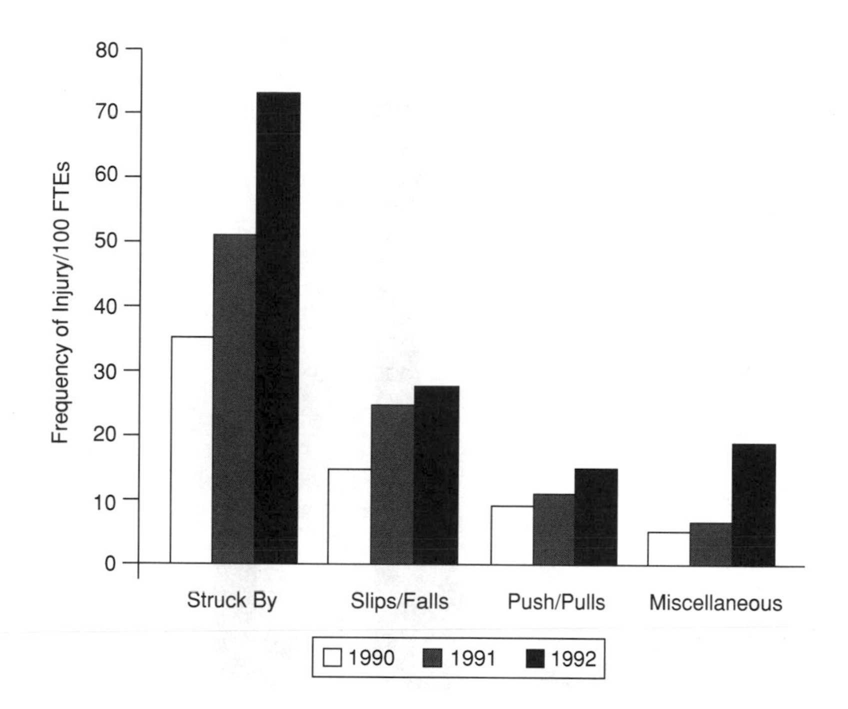

Figure 36–7 Frequency of Injury for Uninformed Infirmary by Cause of Loss, 1990–1992. *Source:* Copyright © 1993, Susan H. Abeln, Strategic Healthcare Alternatives, San Clemente, California.

---

two are related is the next question the group asks. Again, Mary and Suzanne have anticipated their question. As can be seen from Figures 36–9 and 36–10, and from the spreadsheet in Table 36–9, the facility's frequency of strains/sprains is caused first by slips/falls, secondarily, by overhead reaching, third, by pushing and pulling, and fourth by lifting and bending. Yet in severity, slip/falls is the first, pushing/pulling the second, lifting/bending the third, and overhead reaching the fourth most severe cause of strains/sprains.

What do all these data have to do with our attempts to discuss, sell, or implement interventions? What do they have to do with building a successful business relationship by understanding the clients and "joining" them at their model of the world? Experience dictates that this type of trending information helps both the consultant and the employer understand the factors that shape their world. Once one has a shared understanding of the situation, the most effective treatment intervention becomes apparent to both the consultant and the employer. Just suppose

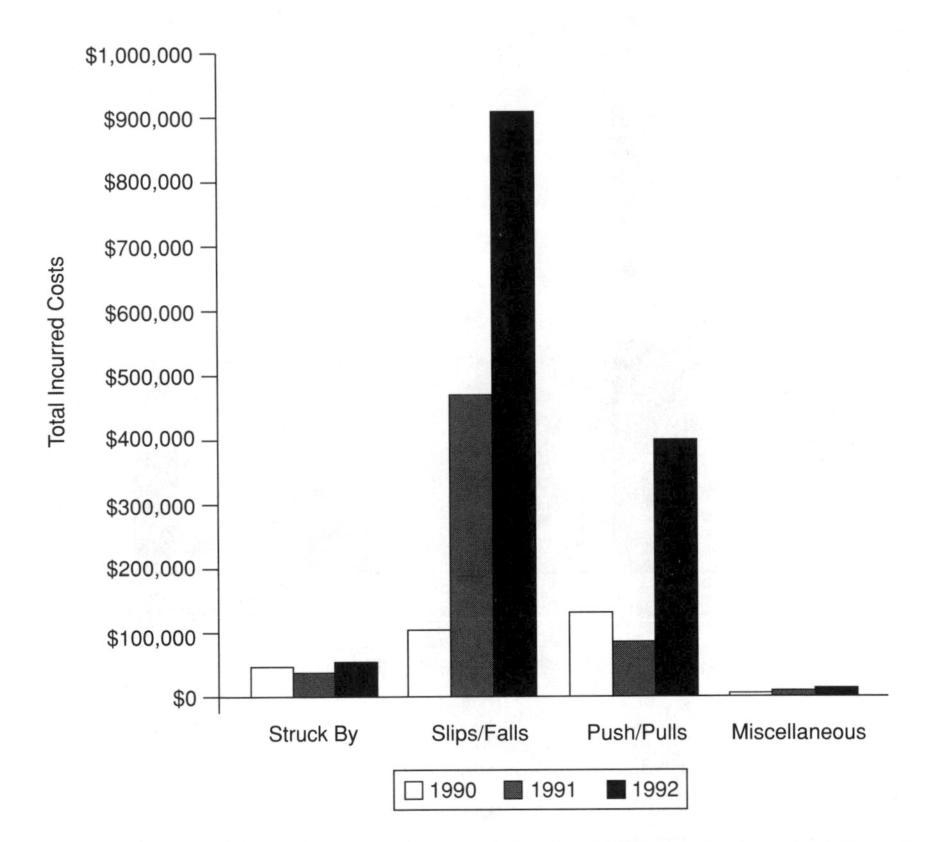

Figure 36–8 Severity of Injury (As Measured by Total Incurred Cost) for Uninformed Infirmary by Result of Loss, 1990–1992. *Source:* Copyright © 1993, Susan H. Abeln, Strategic Healthcare Alternatives, San Clemente, California.

Table 36–8 Frequency and Severity of Injury at Uninformed Infirmary by Cause of Loss 1990–1992

	Frequency			*Total Incurred Costs*		
	1990	*1991*	*1992*	*1990*	*1991*	*1992*
Struck By	35	51	73	$43,432	$31,351	$49,837
Slips/Falls	15	25	27	$101,587	$473,552	$913,445
Push/Pulls	9	11	15	$134,892	$80,765	$404,420
Miscellaneous	5	6	19	$7,089	$9,082	$15,788

Source: Copyright © 1993, Susan H. Abeln, Strategic Healthcare Alternatives, San Clemente, California.

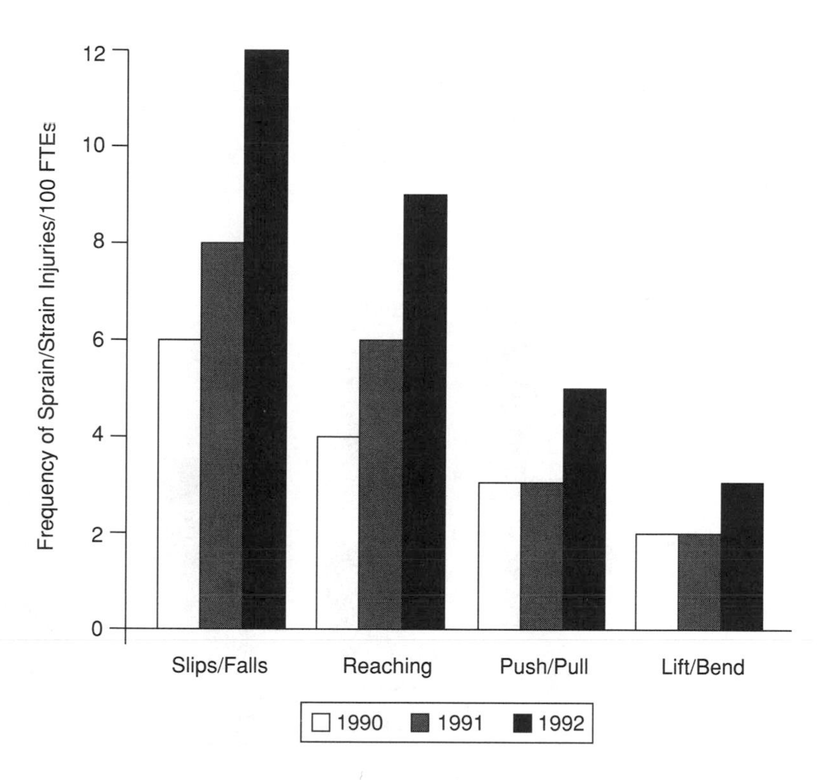

Figure 36–9 Frequency of Strain/Sprain Injuries for Uninformed Infirmary by Cause of Loss, 1990–1992. *Source:* Copyright © 1993, Susan H. Abeln, Strategic Healthcare Alternatives, San Clemente, California.

one had not done this type of research and, as a rehab provider, had come to the Uninformed Infirmary proposing to assist them in reducing their workers' compensation costs. After completing the preliminary research, the provider had successfully persuaded the client that immediate action was needed because the most frequent type of injury was strains/sprains. The provider then proposed to them a second back school program as the cost-effective "fix" for their workers' compensation costs. Would this program have had any real effect on their costs? Would the costly injuries due to slips/falls decrease? How would that sale have positioned the provider for maintaining and/or strengthening his or her relationship with the client in the future? One would venture to say not very well.

Let us return to our team. The results of this type of trending have provided the group with tremendous insight. Such is also the case with the trending that exam-

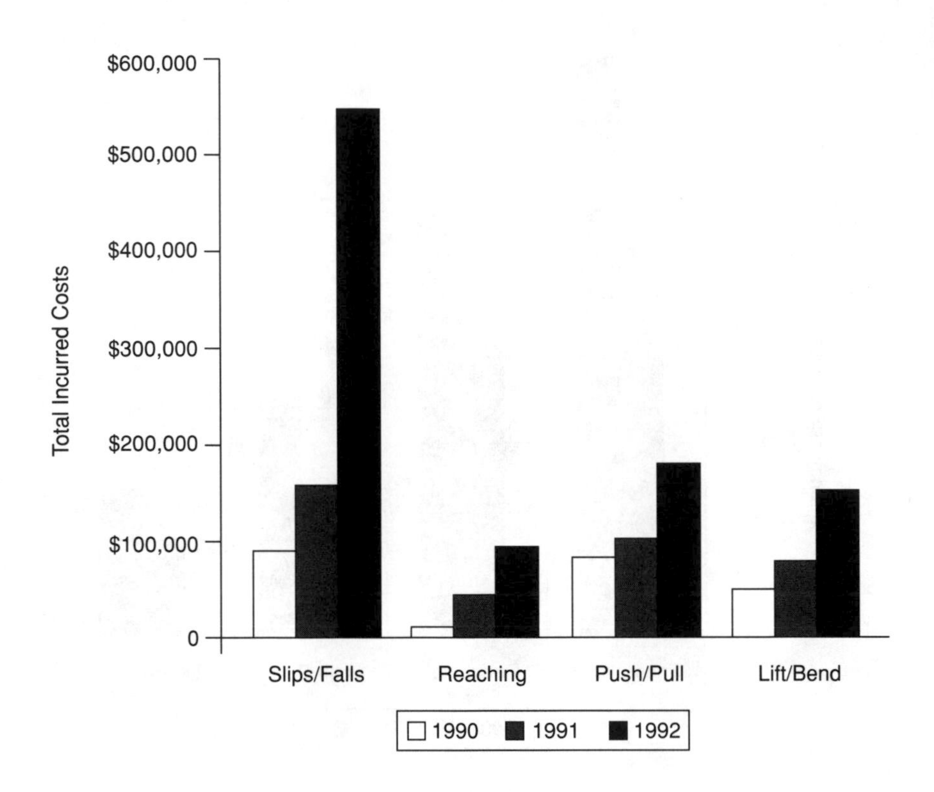

Figure 36–10 Severity of Strain/Sprain Injuries (As Measured by Total Incurred Costs) for Uninformed Infirmary by Cause of Loss, 1990–1992. *Source:* Copyright © 1993, Susan H. Abeln, Strategic Healthcare Alternatives, San Clemente, California.

Table 36–9 Frequency and Severity of Strain/Sprain Injuries at Uninformed Infirmary by Cause of Loss, 1990–1992

	Frequency			*Total Incurred Costs*		
	1990	*1991*	*1992*	*1990*	*1991*	*1992*
Slips/Falls	6	8	12	$85,465	$158,427	$547,339
Reaching	4	6	9	$17,865	$38,432	$93,236
Push/Pull	3	3	5	$78,342	$101,332	$175,577
Lift/Bend	2	2	3	$45,093	$75,939	$151,559

Source: Copyright © 1993, Susan H. Abeln, Strategic Healthcare Alternatives, San Clemente, California.

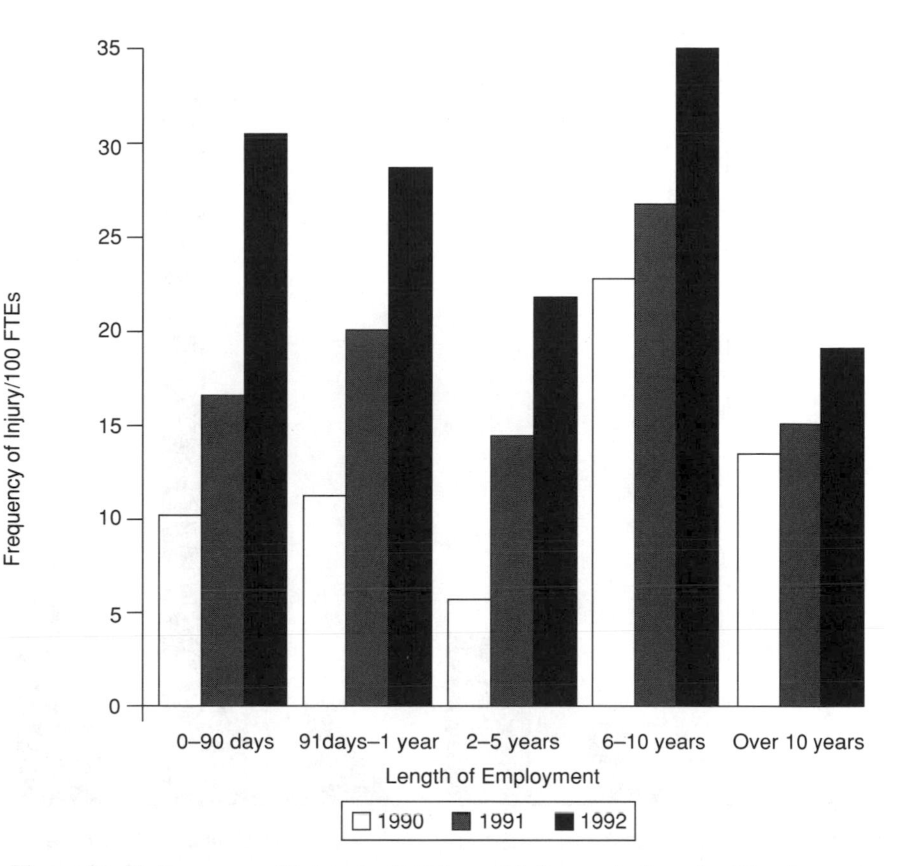

Figure 36–11 Frequency of Injury for Uninformed Infirmary by Length of Employment, 1990–1992. *Source:* Copyright © 1993, Susan H. Abeln, Strategic Healthcare Alternatives, San Clemente, California.

---

ined the body part injured. In spite of earlier advice concerning assumptions, the group, assuming they knew the results of this trending before it had even been discussed, try to push Mary and Suzanne to skip over the information. But, wonder of wonders, their assumptions about which body part is injured most frequently and most severely are only partly correct. The most frequent body part injured is indeed the back. The most severe and thus the most costly injuries for this facility, however, are those involving the shoulders and knees. Hmmm. The last two surprises for the group come from the analysis of the people who are getting injured and the departments they are from. Prior to understanding that the most costly injuries are slip/fall injuries that result in strains/sprains, the team had assumed

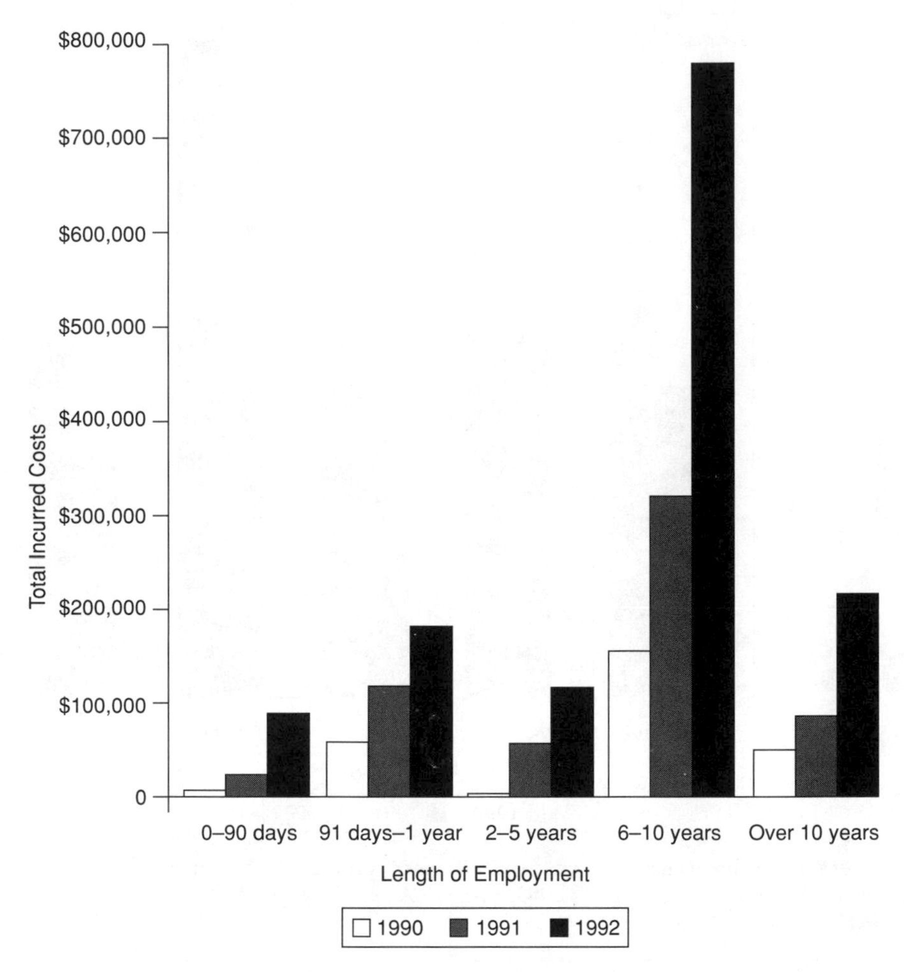

Figure 36–12 Severity of Injury (As Measured by Total Incurred Cost) for Uninformed Infirmary by Length of Employment, 1990–1992. *Source:* Copyright © 1993, Susan H. Abeln, Strategic Healthcare Alternatives, San Clemente, California.

that the nurses in the rehab unit would be the people who received the most injuries and cost the facility the most money. They are surprised to see that the most costly injuries are those to the Operating Room (OR) and Recovery Room (RR) techs.

The group originally requested two more trends—one stratified the frequency and severity of losses based upon how long employees had been employed at the time of injury, and one stratified the frequency and severity of losses based upon the total incurred value of the claim. The results are again illuminating. One will

Table 36–10 Frequency and Severity of Injury at Uninformed Infirmary by Length of Employment 1990–1992

	Frequency			*Total Incurred Costs*		
Employment	*1990*	*1991*	*1992*	*1990*	*1991*	*1992*
0–90 days	10	17	31	$8,234	$27,420	$84,321
91 days–1 year	12	20	27	$50,018	$113,332	$180,458
2–5 years	6	14	22	$1,105	$46,057	$110,453
6–10 years	23	27	35	$160,548	$326,460	$788,238
Over 10 years	13	15	19	$67,034	$81,481	$220,020

Source: Copyright © 1993, Susan H. Abeln, Strategic Healthcare Alternatives, San Clemente, California.

notice from Figures 36–11 and 36–12 and Table 36–10 that most injuries (and most expended dollars) concern employees who have been employed for 6 to 10 years or more. This is unusual.

Generally, 40 percent of most employers' losses are from employees who have worked within the facility for less than one year. Again, this understanding can lead a rehab provider to select a different intervention strategy for this employer than for others. It may not be as effective, efficient, or important here at the Uninformed Infirmary to conduct a prework functional assessment on all new employee programs as it might be at another facility. But it may make more sense to design and implement a wellness fitness training program for older employees trying to keep flexibility and strength or to develop and implement a comprehensive ergonomic program to eliminate many of the workstation stressors that are cumulative in nature.

The last set of graphs is the one set that is not aberrant. As one can see from Figures 36–13 and 36–14 and Table 36–11, 20 percent of the claims are responsible for over 85 percent of the cost of all claims. This illustrates the value of early intervention and rehabilitation strategies and the importance of a well-designed return-to-work program. It is almost always thus.

Before the meeting closed, Suzanne mentioned an additional trending exercise she performed but had not included in her earlier presentation. This work showed that the majority of injuries to longer-term employees occurred during their first hour at work. No one knew immediately quite how to use this last bit of wisdom, but you would be 100 percent correct to assume that some reference to it would show up in their final report.

Through this process of reviewing the facility's specific losses, your team has ferreted out a number of additional facts that will need to be kept in mind for the upcoming report to Mr. St. Clair to help explain the realities of the world of workers' compensation at the Uninformed Infirmary. They are summarized below:

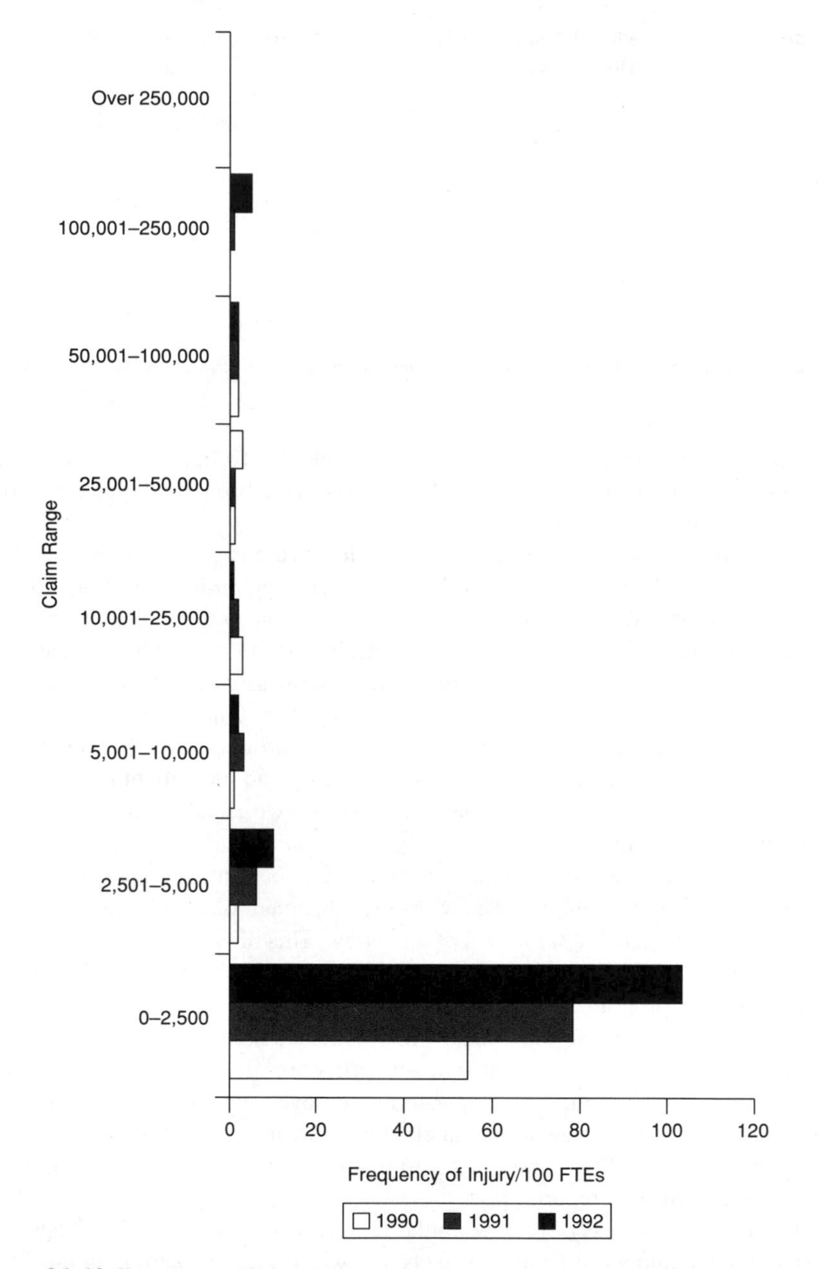

Figure 36–13 Frequency of Injury by Stratification of Losses for Uninformed Infirmary, 1990–1992. *Source:* Copyright © 1993, Susan H. Abeln, Strategic Healthcare Alternatives, San Clemente, California.

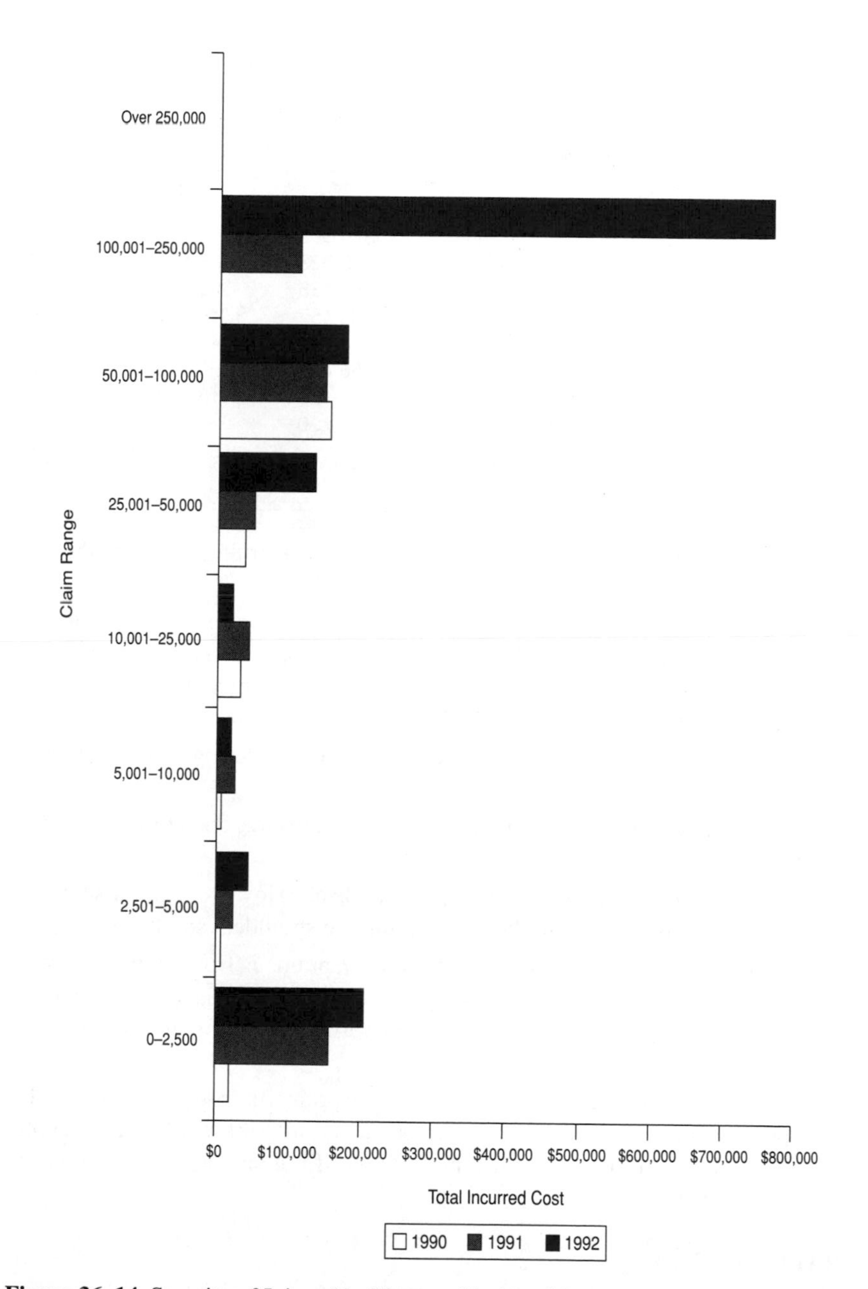

Figure 36–14 Severity of Injury (As Measured by Total Incurred Costs) for Uninformed Infirmary by Stratification of Losses, 1990–1992. *Source:* Copyright © 1993, Susan H. Abeln, Strategic Healthcare Alternatives, San Clemente, California.

Table 36–11 Frequency and Severity of Injury at Uninformed Infirmary by Stratification of Losses, 1990–1992

	Frequency			*Total Incurred Costs*		
Claim Range	*1990*	*1991*	*1992*	*1990*	*1991*	*1992*
0–2,500	55	78	111	$24,563	$169,798	$202,548
2,501–5,000	2	6	10	$9,323	$28,435	$48,708
5,001–10,000	1	3	2	$9,438	$28,834	$18,778
10,001–25,000	3	2	1	$38,457	$46,575	$22,900
25,001–50,000	1	1	3	$48,787	$49,776	$136,458
50,001–100,000	2	2	2	$156,432	$152,896	$178,456
100,001–250,000	0	1	5	$0	$118,436	$775,642
Over 250,000	0	0	0	$0	$0	$0

Source: Copyright © 1993, Susan H. Abeln, Strategic Healthcare Alternatives, San Clemente, California.

- Twenty percent of the claims are responsible for over 85 percent of the cost of all claims.
- Most of the injuries (and most of the expended dollars) occur to employees who have been employed for 6 to 10 years or more. This is uncharacteristic. (Strategic Healthcare Alternatives has found that 40 percent of most injuries are of employees who have worked within a facility for less than one year.)
- Most of the injuries to 6- to 10-year employees occur during their first hour at work.
- The most costly injuries within the last three years have been those to the OR and Recovery Room techs.
- The most frequent body part injured is the back. However, the most costly injuries for this facility are those involving the shoulders and knees.
- The facility's most frequent cause of strains/sprains is from employees who slipped or fell. The other causes that lead to strains/sprains (listed in hierarchical order) are injuries from reaching overhead, pushing and pulling, and lifting and bending.
- The facility's most costly cause of strains/sprains is from employees who slipped or fell. The other causes that lead to strains/sprains (listed hierarchical order) are injuries from pushing/pulling, lifting and bending, and reaching overhead.

WRAPPING IT UP

During the next three weeks you and the team focus on three remaining activities. First, you work on completing the last two steps in the WCI model—interviewing managers and staff to expand understanding of critical areas while probing for hidden areas of concern, and working with key personnel to prioritize the critical areas identified. Second, you create a plan of action that addresses the Uninformed Infirmary's specific problems as seen by Mr. St. Clair and Mrs. Peregrine. The third activity is technically the easiest, yet the team finds this aspect of their charge the most challenging. This is the task of building a bridge with their presentation to help Mr. St. Clair and the others in senior management within the organization come to a shared understanding of the world of workers' compensation at the Uninformed Infirmary, as well as to see the benefit of the team's proposed combination of intervention strategies. As part of this process, the team reviews the draft you have presented to them. They make no changes in the original draft, but they add several paragraphs to explain the results of the facility-specific trending and to incorporate the proposal of the "treatment" or intervention strategies. Let us return to the conclusion of the first draft and move forward with the inclusion of the new paragraphs.

> . . . they are being paid $448/week to remain at home. Many experts report that injured employees learn helplessness from this experience. This extremely large average number of days lost per injury may be one of the main reasons that we find that 20 percent of our claims are responsible for 85 percent of the costs, as seen in the graphs presented.
>
> We explored current practices for each employee whose injury resulted in lost time by asking some of the following questions: Are there any attempts to direct injured employees to a health care provider immediately upon injury? To which physician do they go? To which rehab provider do they go? What are the criteria in selecting their physicians and rehab providers? Have practice profiles been used to get at this information? Has a critical/clinical pathway been developed and used to help determine the standard progression of injured employees through the system? What is our normal approach for returning the employee to work? Do we have a return-to-work program? Do we have specifications for the amount of time we allow an employee to remain off work? Do we have a transitional duty program that will allow an employee to move slowly back to strength? Do we routinely communicate with that injured employee after an injury in a positive manner? Are most of these cases litigated?
>
> What we discovered was—injured employees are basically left to find their own way through the system and to react to all the barriers and influences that are thrown in their path. For that reason, we would like to propose the following as our post-loss interventions:

- Immediate direction of the injured worker to a physician provider who has been identified through outcome-based practice profiles and patient satisfaction surveys to be an outstanding provider with a return-to-work mentality.
- Immediate and routine thereafter "sentinel," nonadversarial communication from the facility/employer to the injured employee explaining the system and acting as a system facilitator and as the employee's advocate throughout the course of the claim. (This strategy alone has been shown to reduce workers' compensation costs by over 22 percent.[13])
- From initial referral to the physician, initiation of a critical/clinical pathway that includes immediate referral to rehabilitation. The pathway creates a framework for the injured worker's progression from acute injury management through work conditioning to transitional return to work.
- Aggressive use of modified/alternative/transitional work for any injured employee.

But we as a team cannot advocate *only* post-loss strategies. All research and experience indicates that the facility must also focus on pre-loss or prevention strategies. The other facility-specific trending shown in our graphs make it apparent that these strategies must focus on slip/fall prevention, especially in the surgical suites; ergonomic evaluation and job redesign of the tasks of the OR/RR techs to prevent the plethora of strains/sprains they are experiencing while pushing and pulling gurneys; and a mandatory prework stretching program for all employees to prepare them for the physical demands of their work.

For this proposed intervention strategy we have developed a cost-benefit analysis (Exhibit 36–3) to demonstrate to you and the senior management of Sickly and Saintly the return on investment that we project from this approach. The initial costs for implementation of the post-loss section of the strategy are primarily associated with increased staff time, with a small fee included for a consultant to assist in the development of the critical/clinical paths for injured employees within diagnostic categories. You will note that we anticipate that despite these time expenditures, the facility will save $800,285 in workers' compensation costs within the first year. If other facilities within the corporation have similar experience, the savings could be at least 10 times this figure because of the shared resources available. We do not anticipate as great a direct dollar savings from the pre-loss interventions, but the indirect savings from fewer injuries should be equally dramatic.

Exhibit 36–3 Cost-Benefit Analysis

It is important for a cost-benefit analysis to use pertinent details of a client's workers' compensation experience. Uninformed Infirmary's 1992 frequency incidence was 15.2/100 FTEs. Of these injuries, over 65 percent were from strains and sprains. The average cost for these strains over the last three years varied between $15,117 and $33,369. The cost/claim for a strain injury at that facility in 1992 was $33,369. From past experience, it is projected that the interventions selected would produce a 35 percent reduction in the frequency of strain/sprain injuries in the first year. This must have a significant effect on the bottom line if we are to make a positive case for adopting these interventions.

The Present Situation

If the frequency index is 15.2 injuries/100 FTEs and the percentage of strains/sprains as a part of total injuries is 65 percent (.65), then multiplication of these two factors provides the number of strains/sprains/100 FTEs at present.

$$\frac{15.2 \text{ injuries}}{100 \text{ FTEs}} \times 0.65 = 9.88 \text{ strain/sprain injuries/100 FTEs}$$

Given that the cost/claim for strain/sprain injuries for 1992 was $33,369, multiplication of this amount by the frequency of strains/sprains will provide the cost of strains/sprains/100 FTEs.

$$\frac{9.88 \text{ strain/sprain injuries}}{100 \text{ FTEs}} \times \$33{,}369 = \$329{,}686/100 \text{ FTEs}$$

The Projected Situation

It is projected that the interventions selected will reduce the frequency of strains/sprains/100 FTEs by 35 percent (.35). To determine the projected number of strain/sprain injuries/100 FTEs, one must multiply the projected reduction in frequency times the current frequency to obtain the anticipated reduction.

$$0.35 \times 9.88 = 3.46$$

Thus the anticipated number of strain/sprain injuries per 100 FTEs will be 9.88 – 3.46, or 6.42.

The new cost of these injuries, assuming the same cost per case, will be the new frequency of strain/sprain injuries times the cost/claim:

$$\frac{6.42 \text{ strains/sprains}}{100 \text{ FTEs}} \times \$33{,}369 = \frac{\$214{,}229}{100 \text{ FTEs}}$$

Savings

$$\frac{\$329{,}686}{100 \text{ FTEs}} - \frac{\$214{,}229}{100 \text{ FTEs}} = \frac{\$115{,}457}{100 \text{ FTEs}}$$

Because Uninformed Infirmary has over 850 employees, the true savings could be estimated by multiplying the savings per 100 FTEs by 8.5. As seen below, this means the savings will equal $981,385.

continues

Exhibit 36–3 (continued)

$$\frac{\$115{,}457}{100\ \text{FTEs}} \times 8.50 = \frac{\$981{,}385}{100\ \text{FTEs}}$$

To truly demonstrate the bottom line benefit of these interventions, it is important to include the costs associated with the same interventions. The following is an analysis of both the costs and benefits.

Projected

Differential Cash Revenues (Savings)		$981,385
Less: Differential Cost Expenses (cost of intervention)		
• Consultant's fees	$8,000	
• Material purchase for surgical suite	$5,000	
• Ergonomic evaluation and job redesign by internal PT	$7,500	
• Cost of prework stretch for all employees with greater than medium lifting		
$\frac{0.08\ \text{hour}^{a}}{\text{day}} \times \frac{\$10^{b}}{\text{hour}} \times 365\ \text{days} \times 550^{c}\ \text{employees} =$	$160,600	
TOTAL COSTS ASSOCIATED WITH INTERVENTIONS	$181,100	$181,100
Before Tax Net Cash Flow		$800,285

$$\text{RETURN ON INVESTMENT} = \frac{\text{Before Tax Net Cash Flow}}{\text{Initial Investment}} = \frac{\$800{,}285}{\$160{,}600} = 498\%$$

[a].08 hour = 5 minutes.

[b]$10.00 = average hourly wage.

[c]Number of employees who have greater than a medium lifting requirement in their essential job functions.

Source: Copyright © 1993, Susan H. Abeln, Strategic Healthcare Alternatives, San Clemente, California.

This presentation was given to Mr. St. Clair two days later and went well—so well, in fact, that you now stand in front of the senior management of Sickly and Saintly as well as Mrs. Peregrine, clearing your throat and then beginning your second explanation of the team's analysis of the situation and her proposals for future action. After a tense 20-minute presentation, you sit down to answer the questions that these senior executives raise. They are few and minor. You are surprised how easily you have created a shared understanding of the workers' compensation world at the Uninformed Infirmary, and even more surprised when the team's plan of action is fully approved and in fact, the same problem-solving approach is recommended to all of the other hospitals within the corporation as good future planning. You have done it. With the team's help, you have joined Mr.

St. Clair and Mrs. Peregrine at their model of their world by using their model's language to describe their perception of that world. By doing this you have built a bridge to understanding and ensured that the interventions that are so critical will finally occur.

THE REAL WORLD

The WCI model can be used to build bridges in a slightly different way. It still is important to strive to develop the trust that can only begin when a person gets the sense that he or she is being fully understood, and that he or she and the other person are truly "speaking the same language."

External consultants must not only "join" the world of the corporate Director of Insurance and present ideas and strategies but also "join" the world of a number of people at the local facility, from people like St. Clair—the Executive Director—to the clerk who serves as the workers' compensation coordinator for the Director of Human Resources. It is only by creating an understanding that can be shared by all the parties involved in workers' compensation that the consultant can succeed in helping them address their problems. Therefore it is preferred practice, for all interactions with clients, to go to the local site armed with the majority of the data that were developed internally in the case study. Access to the data is gained through review of the client's loss data as provided by the insurer or third-party payor, through analysis of the client's OSHA 200 logs, and through discussion with any appropriate individual within the risk management or insurance department. As a consequence of this preparation, each conversation and consultation with the local client is entered with a hint as to where the problems might lie and the knowledge that the bridge to understanding the client's world has been partially crossed before the client and the consultant have met face to face.

NOTES

1. Towers Perrin, *Regaining Control of Workers Compensation Costs,* © June 1993 Towers Perrin. Available by calling 1-800-525-6751.
2. Ibid.
3. C.A. Telles, *Medical Cost Containment in Workers' Compensation: A National Inventory,* WC-93-2 (Cambridge, Mass.: Workers' Compensation Research Institute, June 1993).
4. C. Woolsey, There's More to Managing Workers' Comp Care Than Using Network, *Business Insurance,* 18 October 1993.
5. Towers Perrin, *Regaining Control.*
6. A. Lewis & F. Pucelik, *Magic Demystified: A Pragmatic Guide to Communication and Change* (Portland, Ore.: Metamorphosis Press, 1982), 15.
7. Ibid.

8. Ibid., 14.
9. Ibid.
10. Ibid., 1.
11. Towers Perrin, *Regaining Control.*
12. This is the maximum rate for temporary total disability (TTD) benefits in California as of 1993. This number will vary according to each state's statutorily defined benefits.
13. G.L. Dent, *Return to Work . . . by Design. Managing the Human and Financial Costs of Disability.* (Stockton, Calif.: Martin-Dennison Press, 1990).

Chapter 37

An Industrial Perspective on Reducing Workers' Compensation Claims

Deena E. Pease

WEYERHAEUSER COMPANY'S WORKERS' COMPENSATION HISTORY

Weyerhaeuser Company is involved in all aspects of the forest industries, thereby establishing its position as high risk relative to safety issues, with significant costs associated with workers' compensation. The number of employees has remained comparatively stable for the past two decades at approximately 40,000, located in 250 facilities in 44 states, with the majority of employees located in the Northwest and the South.

The decision to self-insure the workers' compensation risk wherever possible in the United States was made in the 1950s, with the primary focus on the cash flow advantages of self-insurance. In the early years, Weyerhaeuser determined that self-administration of claims at the local level for larger facilities would be advantageous with the responsibility given to local human resource personnel. Various third-party administrators were contracted to handle claims from smaller locations located across the country. The overall results were not centrally monitored except from a financial perspective. Consequently, with the passage of years, claims administration varied substantially from state to state and facility to facility.

During the 1970s, the threat of losing self-insurance certifications in some states became the motivator in the decision to move the responsibility from these multiple diverse claims processors to one national third-party administrator. This proved to be a wise decision from issues surrounding self-insurance certification and administrative expense. However, the costs associated with claims were not favorably affected. During the years from 1976 to 1983, the cost of workers' compensation for the company increased 300 percent. Even though this was fairly consistent with the national trend, the decline of the forest industry during this same period and the heavy hit caused by workers' compensation created the need to take a good look at the program. Thorough evaluation was given a high priority by senior management.

SELF-ADMINISTRATION

In 1982, a one-year cost containment study was conducted in an effort to gain insight into past practices and learn from mistakes. In 1983, the results of the study revealed that self-administration had many advantages that could not be duplicated by an outside resource. However, if the company was again to undertake self-administration, controls were necessary to ensure consistency and to ensure that the company's values and ethical standards were met. It was estimated that from 3.5 to 6.8 million dollars annually could be saved with self-administration—a figure that was ultimately determined to be very conservative (Figures 37–1 through 37–4).

With senior management support, the following controls were put into place:

- The cost of workers' compensation was to be charged back to the operating unit.
- More emphasis was to be placed on safety and health issues.
- A centralized claims office was to be established, staffed with professional claims representatives, that would be given final accountability for claims management.

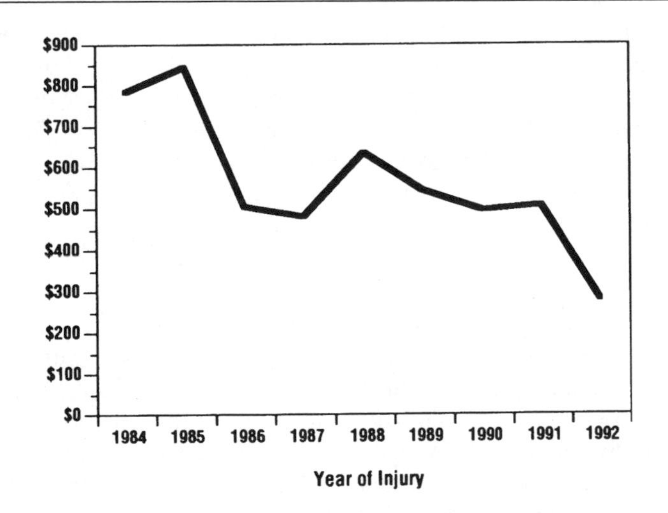

Figure 37–1 Incurred Workers' Compensation Cost per Employee at Weyerhaeuser Company, 1984–1992. *Source:* Weyerhaeuser Worker's Compensation Annual Report—1992, Copyright © 1992, Weyerhaeuser Company, Federal Way, Washington.

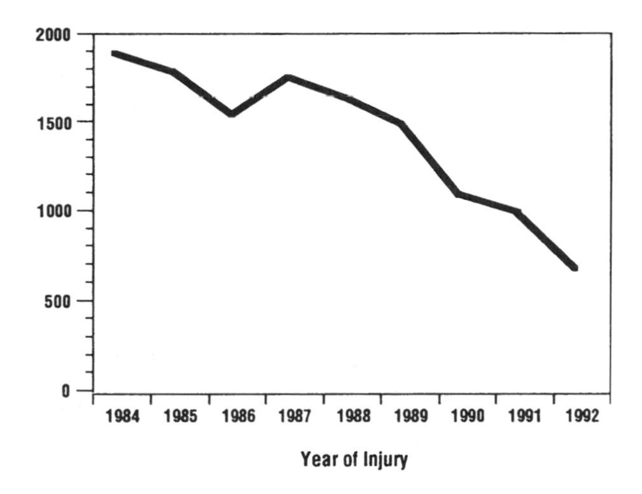

Figure 37–2 Lost Workday Injuries at Weyerhaeuser Company, 1984–1992. *Source:* Weyerhaeuser Worker's Compensation Annual Report—1992, Copyright © 1992, Weyerhaeuser Company, Federal Way, Washington.

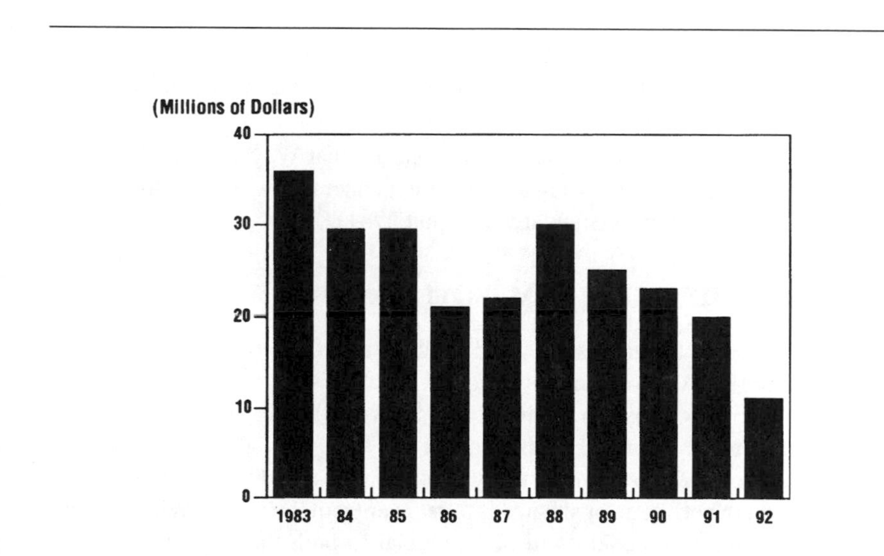

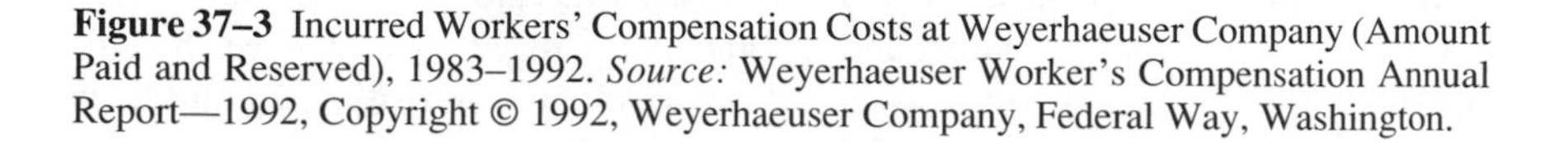
Figure 37–3 Incurred Workers' Compensation Costs at Weyerhaeuser Company (Amount Paid and Reserved), 1983–1992. *Source:* Weyerhaeuser Worker's Compensation Annual Report—1992, Copyright © 1992, Weyerhaeuser Company, Federal Way, Washington.

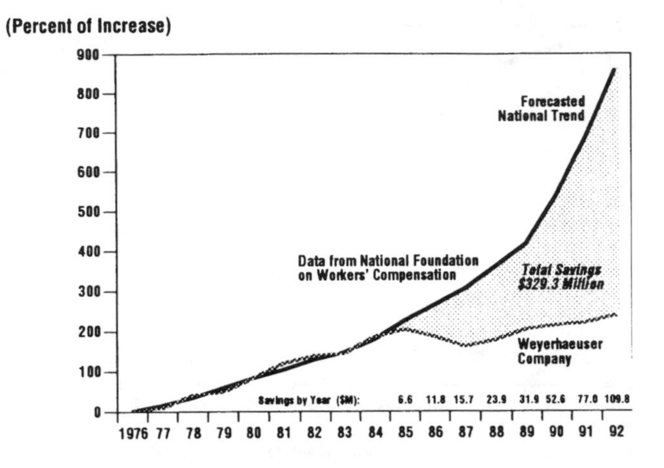

Figure 37–4 Workers' Compensation Costs, Weyerhaeuser Trend vs. National, 1976–1992. *Source:* Weyerhaeuser Worker's Compensation Annual Report—1992, Copyright © 1992, Weyerhaeuser Company, Federal Way, Washington.

- A monthly computerized statistical report was to be provided to line management.
- A workers' compensation liaison position was created in the local facility.
- Roles and responsibilities of the unit managers, supervisors, liaisons, and claims representatives were clearly defined.
- A philosophy statement was prepared to ensure that Weyerhaeuser's injured employees would be treated in a consistent manner that would conform to the company's high ethical standards (Exhibit 37–1).

SERVICE TO THE INJURED EMPLOYEE

The success of the program since 1984 is due to a shift in attitude. High respect and concern are shown to an employee following an injury.

The time following a lost-time injury causes extreme anxiety beyond the physical agony the employee may be suffering. If the employee is worried about finances, return-to-work options, or family relationships, this distress will naturally take away from the effort he or she needs to recover from the injury. Whatever can be done to reduce this anxiety will be beneficial to both the employee and the employer.

Immediately following an injury, the line supervisor will put the injured employee's well-being foremost and, if practical, will personally drive the employee in for medical treatment, with follow-up inquiry as routine. Emphasis is given that the employee is wanted back on the job as soon as it is medically fea-

Exhibit 37–1 Weyerhaeuser's Workers' Compensation Philosophy

Workers' Compensation Philosophy

Weyerhaeuser Company's philosophy is to treat injured employees with dignity and respect. The highest standard of care will be extended to our injured employees to assure rapid recovery and minimize residual disability.

Claims service will provide equitable evaluation and prompt payment of claims. To assure that Weyerhaeuser employees receive uniform administration of their claims, standardized processes and methods consistent with Total Quality will be followed.

Weyerhaeuser's savings in workers' compensation costs will come from accident prevention and early return-to-work programs. It is not the company's practice to attempt to create savings by denying or delaying an injured employee's benefits.

If a permanent disability prevents an employee from returning to work with the company, professional services will be provided to assist the employee in gaining re-employment.

Decisions made during the administration of claims will be based on careful evaluation of facts in terms of legal compliance. In cases where injury circumstances cannot be supported by facts, the benefit of the doubt will be in favor of the employee. Employees may appeal a decision made by the company and will not be treated differently because of their appeal. The claim decision will not be changed, however, unless new factual evidence is introduced.

Due to our financial strength and ability to care for injured employees, Weyerhaeuser has earned the right to be self-insured for workers' compensation. The company's managers are responsible for protecting this privilege by maintaining adherence to all workers' compensation laws and ensuring the philosophy is upheld.

John W. Creighton, Jr.
President

Source: Courtesy of Weyerhaeuser Company, Federal Way, Washington.

sible, whether returning to his or her regular job or returning to work in a modified capacity. The claims representative will contact the employee within 24 hours following a lost-time injury to describe the benefits the employee will receive, to give an explanation of the claims processing procedures, and to answer any questions. The claims rep will ask about the medical care the employee is receiving and if the employee is satisfied. A follow-up letter is sent that includes a toll-free number to call for questions the employee may have as the claim develops. The claims representative will contact the employee regularly and is often able to establish a close rapport.

Following direct contact with the employee, the claims representative is expected to make the first payment of compensation benefits within 14 days following knowledge of a claim. A late benefit payment is compared to a late paycheck. It is simply not acceptable. Timely compensation payments are consistently applied throughout the company and are given the highest priority. The claims staff's record for timeliness exceeds 95 percent overall.

The word *claimant* has been eliminated from the claims representative vocabulary. This small detail has truly set the stage for the change in attitude that claims reps now show toward Weyerhaeuser injured employees, their coemployees. Accident investigation immediately following an injury is an integral part of the responsibility of the line supervisor or designated safety committee, with primary focus on prevention. This early evaluation has also contributed toward the change in attitude toward accepting that the worker is truly injured, rather than doubting or challenging the worker at the beginning of the claim. If a claim is found noncompensable, this decision is based solely on the facts of the claim, with the employee given the benefit of any doubt.

The benefits of a caring attitude and early return-to-work practices has been beneficial to both the injured employee and the company: the employees' morale is kept high, and the supervisor has an experienced worker back on the job sooner. Medical treatment is often reduced by early return to work, with less need for physical therapy, work hardening, and vocational rehabilitation service, plus a lessened risk of drug dependency, doctor shopping, and other functional overlay aspects of the workers' compensation syndrome. The majority of the reasons for litigation will also be eliminated. This is the key to reducing workers' compensation costs and results in a win-win situation for everyone.

CLAIMS ADMINISTRATION

There are four claims offices throughout the United States: two on the West Coast, one in the South, and one on the East Coast. Each office has a high performing team in place, consisting of a regional claims manager, claims representatives, and claims assistants. Each claims representative is assigned to work directly within an assigned geographical area. Larger locations are "on line" with a workers' compensation computer network, and through this mechanism, knowledge of a workers' compensation claim is often received by the centralized claims staff the same day as the liaison. With the exception of medical-only claims, which are managed by claims assistants, the claims representative has full responsibility from the day the claim is received to the day it closes, no matter what aspect of complexity may result.

The claims representatives are more than just "adjudicators." Necessary education and training is provided for them to be full-service case managers. That is, they not only investigate the compensability of the claims assigned to them, but they represent the company in prehearing conferences wherever a layperson is allowed to do so. They make regular visits to the locations to work with plant managers, line supervisors, human resources personnel, nurses, and safety professionals to review files and discuss case management strategy. They have even been known to have sit-down sessions with union representatives and conduct educational seminars on workers' compensation for them. The claims staff is now included as part of an exchange team that conducts annual safety audits of Weyerhaeuser facilities in the United States and Canada. They are also included as members of local safety committees, providing the injury and incident data that are used for analysis and assisting in safety process development.

When expertise is required that goes beyond the capabilities of claims representatives, they are authorized to make referrals to outside resources. Nurses are contracted with for medical case management assistance for employees who may have suffered major catastrophic injury. Vocational rehabilitation counselors are needed for providing assistance to employees who are physically unable to return to their job within Weyerhaeuser. Prior to contracting with any outside resource, including investigators and attorneys, the regional claims managers conduct an interview process to ensure that the professional contractor can philosophically accept the Weyerhaeuser concept of claims management. They are also told that the claims representative will not delegate any of his or her responsibility or authority to them; but rather, they become a member of the Weyerhaeuser team as decisions are made. It is only on a rare occasion that an outside professional does not embrace the prerequisite and become an avid supporter of the program.

Medical bill and utilization review is an outside service that is also used nationally, with preferred provider organizations established in some metropolitan areas. As the PPO concept expands to the rural areas, where most of the Weyerhaeuser facilities are located, and as the National Health Care plan is developed, more involvement with medical networks is expected.

The attention the claims staff are expected to give to their locations, to injured employees, and to taking advantage of ongoing educational opportunities to remain proficient takes them away from their desk more often than is the case for most claims representatives with other companies. This results in the need to monitor their claims performance, both statistically and technically, to ensure that efficiency standards are met. It is important that the caseload of each claims representative is kept at a level that allows him or her to work outside of the office and to avoid the trap of backlog and adjuster burnout. This is accomplished by having each regional claims manager monitor cases by means of an override diary. The managers are also provided with monthly statistical claims production reports and computer query capability from the workers' compensation database.

Total quality concepts and employee empowerment are now commonly accepted practices within the department. Very recently a new program was created, called "The Best Practices" team. Two members of each claims office will visit a different claims office annually to evaluate the work product of their peers and file a report. Even though the function is similar to an audit, the focus is on the positive aspects of each office, to replicate best practices and to ensure consistency. Special recognition awards can be given to staff members based on peer recommendation.

PHILOSOPHY IN PRACTICE

The success or failure of any workers' compensation program ultimately rests on the caliber of person that is responsible for claims administration—not from a general program perspective, but rather according to how each individual manages claims on a consistent, claim-by-claim basis. Once processes are in place to control the overall program, the emphasis needs to be shifted to monitoring claims-processing performance. It was discovered that the transition to the Weyerhaeuser program, for persons with previous workers' compensation claims experience, is sometimes difficult. In contrast, those persons who have been promoted from within have developed into expert claims professionals with none of the built-in cynicism that often exists among seasoned professionals. Why? Because the Weyerhaeuser philosophy is not generally accepted by insurance companies, state funds, or other self-insured employers or TPAs. The philosophy statement is widely circulated throughout the company and is a constant reference document for the claims representatives:

> Weyerhaeuser's savings in workers' compensation costs will come from accident prevention and early return-to-work programs. It is not the company's practice to attempt to create savings by denying or delaying an injured employee's benefits.
>
> Decisions made during the administration of claims will be based on careful evaluation of facts in terms of legal compliance. In cases where injury circumstances cannot be supported by facts, the benefit of the doubt will be in favor of the employee. Employees may appeal a decision made by the company and will not be treated differently because of their appeal. The claim decision will not be changed, however, unless new factual evidence is introduced. (See Exhibit 37–1.)

This simply means that claims will not be settled based solely on the cost advantages associated with an individual claim. In the final analysis, to do so would send a clear message to our employees and their union and attorney representatives that Weyerhaeuser was an easy mark for increased monetary award based on the practice of appealing or litigating a decision. When the decision is made to adopt the concept of "we will not settle," it must be followed consistently, even if litigating will ultimately cost more than settling an individual claim. It is this aspect of the philosophy that the insurance industry has difficulty accepting; but we believe that if it were more widely accepted, the cost of workers' compensation litigation would be greatly reduced for all employers. Weyerhaeuser's experience has shown that fewer and fewer employees seek the services of attorneys, and many plaintiff attorneys will not take on a capricious case from a Weyerhaeuser employee. Further proof of this success is the win ratio of the in-house workers' compensation defense attorneys in Washington and Oregon. In the past five years, the attorneys have successfully litigated over 90 percent of the cases that have been referred to them. They have not completed this feat single-handedly, but rather by working cases that have been extremely well prepared for litigation by the claims representative.

In conclusion, the feedback received from both within and outside the company is that Weyerhaeuser's workers' compensation program is "Best in Class." This recognition is valued highly by the staff, and there exists a desire and commitment by everyone to maintain this reputation. No matter how successful the program has become, staff are actively looking toward the future and are ready to put into place any enhancements necessary to further improve the service they are able to give to Weyerhaeuser injured employees and their company.

Chapter 38

Managed Care and Workers' Compensation

David W. Clifton, Jr.

The American health care system is currently experiencing a series of convulsions that promise to reshape the delivery and receipt of services. Nowhere will these changes be more dramatic than in workers' compensation, a traditional oasis from aggressive cost controls. Managed care is clearly evolving as the dominant form of health care delivery, and as a result, workers' compensation is increasingly adopting managed care strategies. These reforms have the potential to drive many clinicians out of practice while repositioning others to take fullest advantage. An excerpt from *The Art of War* illustrates the future under managed care:

> The contour of the land is an aid to an army; sizing up the opponents to determine victory, assessing dangers and distances, is the proper course of action for military leaders. Those who do battle knowing these will win, those who do battle without knowing these will lose.[1]

This chapter will examine the historical development of managed care and workers' compensation along divergent pathways and explore the recent convergence of these seemingly contradictory systems. Health care trends and payor initiatives affecting provider reimbursement will also be examined. Lastly, a series of survival strategies will be provided to assist providers in positioning for the 1990s and beyond.

HEALTH CARE TRENDS

Paul Ellwood, whom many have described as the "grandfather" of managed care, describes four prevailing forces in the reshaping of the American health care landscape:

1. greater selectivity on the part of employers for doctors and hospitals
2. aggregation of doctors and hospitals into accountable units
3. a move toward universal coverage

4. a new approach to measure and compare the results of medical care based on outcomes[2]

The United States is the only country in the world where health care providers have relative autonomy in setting their own rates. Medicine in America has defied traditional economic principles because the suppliers, *not* the consumers, establish the cost of services. However, we are in the throes of change, particularly as it relates to power and who wields it. Payors are wresting control of power away from the provider of services and demanding proof of treatment efficacy. In today's world of limited resources and deep recession, this should come as no surprise. A new era has begun, an era in which payors everywhere are exercising the "Golden Rule"—that they who have the gold are now making the rules. Providers, if they are to survive under managed care, must subscribe to a new dictum: "In God we trust, all others bring data." As medicine continues to be managed like a business, the customers will increasingly demand value. Many providers have traditionally considered the patient as the "customer." The customer, as long as insurance was a covered medical expense, did not measure satisfaction in monetary terms. Quality care was often defined by the "feel-good" elements and not by treatment efficacy or cost-effectiveness. "Feel-good" elements included timeliness of visit, friendliness of the receptionist and staff, and free coffee. This too has changed under managed care.

Medicine has long been the only economic sector in which the consumer of its product is not necessarily the buyer or payor. For years, medicine perceived itself as immune from normal rules of commerce. As a result, the following principles did not too significantly influence medical providers' behavior:

- *Supply and demand:* In medicine more providers led to higher, not lower, costs as would be expected in a normal economy.
- *Product quality dictates market share:* In medicine this concept did not apply because both consumer (patient) and customer (payor) had little data on which to base purchasing decisions.
- *Demand creation:* Medicine was in a unique position because it told consumers what they needed and when they needed it; the patient often had little choice.
- *Access to service:* The physician was the only entry into the health care system; alternatives were limited.
- *High technology:* In most industries high technology reduced costs, but in medicine it increased costs.
- *Warranties:* The purchase of most services or products in our economic system involved some form of warranty, but in medicine there were no warranties or guarantees of results.

- *Autonomous rate setting:* the United States is the only country where health care providers have had relative autonomy to set their own rates.[3]

These are just some of the cost drivers resulting in payor attempts to curb spiraling health care cost expenditures. It is critical to gain a greater understanding of the payor's perspective if one is to survive and perhaps thrive within managed care as it continues to evolve. Dr. Reece, a pioneer of managed care, shares the following advice with providers: "The time has come for us to think like purchasers, to place ourselves in their shoes, and to persuade them that what we are doing is humane, efficient, effective, productive, and in their best interest as well as our patient's and society's."[4, p.229]

The American health care system is enormously complex, and this in and of itself has contributed to rising costs. Whenever one operates within a complex structure, there is a tendency to view the world from a self-centered perspective. Instead of seeing the world as it is, we often see the world as we are. Franke pinpoints the source of this phenomenon as it relates to health care: "Fundamental to the understanding of the problem is that each of these discrete elements of the problem (the research technology interests, the medical establishment, the legal rights system, the local hospitals) are all acting in their own self-interest which is consistent with America's political philosophy of individual freedoms."[5]

Patients themselves have contributed to this new shift in power through two mechanisms: the reduction in private payment for medical services (patients are no longer the customers), and unrealistic expectations. Patients have come to view our medical system as being able to cure all diseases, particularly their disease. The line between what is possible medically and what is reasonable care has become considerably blurred.

Although payors are aggressively regaining control over the system, it is incomplete control. As long as medical providers continue to produce "widgets," they will retain a certain degree of power.

Those who prosper in managed care will be the ones who balance benefit and cost, resulting in value. This shared power will be driven by a series of incentives for each of the distinct elements of the system.

- *Employers:* incentives that encourage the offering of employee health benefits that will provide the most cost-effective use of health care
- *Payors:* incentives that reward the ability to manage risk, negotiate purchase of health care services, and reward the use of injury prevention programs
- *Consumers:* incentives that will encourage greater participation in the process of choosing between options, foster more realistic expectations of the health care system and treatment outcomes, and encourage healthier lifestyles

- *Providers:* incentives that will reward cost-effective decisions and require greater responsibility[6]

HEALTH CARE COSTS

A 1991 *Time* magazine article entitled "Condition Critical" aptly describes the state of health care in America.[7] The economic viability of the United States is conditioned on reducing health care costs. In 1990, General Motors Corporation spent over $3.2 billion on employee health care, which is more than it spent on all of the steel put into its products. The United States spends more than 14 percent of its gross domestic product on health care.

Nowhere are rising costs a more acute problem than in the workers' compensation system. Workers' compensation loss ratios now exceed 122 percent (22 cents more paid out than collected via premium dollars). National Safety Council statistics from 1990 illustrate the scope of work-related problems (see Exhibit 38–1).

Workers' compensation costs have risen at 14 percent per year, or at nearly twice the rate for general medical costs.[8] A Minnesota Blue Cross study of workers' compensation costs versus general liability claims revealed that for similar conditions claims treated under workers' compensation were twice as costly. Although low back injury accounts for 25 percent of all workers' comp indemnity dollars, "cumulative trauma injuries" are the fastest-growing claims, accounting for 56 percent of costs in 1990 compared to 21 percent in 1982.[9] The average

Exhibit 38–1 Work Injury Statistics, 1990

- Total of 9 million work-related injuries
- 1,800,000 disabling injuries
- Total work injury costs: $63.8 billion
- $8.7 billion in medical costs
- $10.2 billion in indemnity costs
- Cumulative trauma most frequent category—146,900 cases
- Average payment for sprains/strains—
  $1,641.00 indemnity
  $3,568.00 medical
- Average payment for upper-extremity cumulative trauma—
  $1,382.00 medical
  $2,234.00 indemnity

Note: Medical costs include surgical, medical and allied medical expenses; indemnity costs include lost wages and disability payouts.

Source: Data from the National Safety Council, Chicago, Illinois, 1991.

length of disability for these claims was 8 to 10 weeks at an average cost of $190,000 in 1990 versus $29,000 in 1982.

Many variables attributed to rising workers' compensation costs will be covered in this chapter, but Berman cites four primary causes:

1. injuries that could have been prevented
2. inappropriate treatment
3. undirected care
4. poor return-to-work planning[10]

These four causes should have tremendous relevance to those who practice work injury management services. Data from Managed Comp, Inc., assert that 25 percent of every workers' compensation dollar is wasted *because the system is unmanaged.*[11]

WORKERS' COMPENSATION

> Wherever workers' compensation systems are in trouble and even in areas where they are not, medical costs are the drivers.[12, p.1]

The workers' compensation system in the United States was originally created to be a "no-fault" system and to provide "exclusive remedy" for work-related injury and disease. By design it was to protect injured workers from indiscriminate dismissal, provide medical coverage and wage replacement, and ward off liability lawsuits drawn against employers. Workers' compensation has not been a system that provided incentives for return to work for injured employees, employers, providers, or attorneys. Instead, by its very design it has encouraged protracted medical care and discouraged attenuated treatment and early return to work.

Workers' compensation governance falls within state jurisdiction. Each state has a different workers' compensation act, but they share the following attributes:

1. first dollar coverage with no employee deductibles or copayments
2. provisions for replacement of lost wages (indemnity payments)
3. provisions for permanent and temporary disability payments (partial or total)
4. rights to rehabilitation (mentioned in over 44 workers' compensation acts and mandated in some, i.e., Minnesota, California, Florida)[13]
5. access to doctor of choice (with varying restrictions)
6. unlimited medical coverage
7. burden of proof is on employers to disprove causality, medical necessity, appropriateness of care, and disability level

Workers' compensation insurance is provided through a number of mechanisms: private insurance, state funds, self-insured plans, or the federal government. According to the National Council on Compensation Insurance, nationally private insurers pay 57 percent of the workers' compensation bill, state/federal programs 24 percent, and self-insurance 19 percent. In many jurisdictions, medical costs are responsible for more than 50 percent of overall costs. For years medical providers could charge whatever fee they deemed appropriate, but this practice is gradually succumbing to the establishment of medical fee schedules. At least 30 states have adopted fee schedules as a cost containment measure.

Because of its minimal restrictions when compared to other lines of insurance (e.g., automobile liability personal injury protection caps), workers' compensation has been the victim of cost shifting. The average cost of a medical claim for a lost-time injury climbed from $1,748.00 in 1980 to $6,611.00 in 1990.[14] Hager cites the following reasons for increasing workers' compensation costs:

1. provider fraud
2. increases in medical expenses
3. increases in legal expenses
4. physician-owned auxiliary services
5. expanding definition of job-related injuries[15]

The National Council on Compensation Insurance (NCCI), a rate-setting organization, requested a 12 percent increase in premiums in 1991 followed by another 11 percent increase in 1992 to help offset persistent losses. An increasing number of workers' compensation carriers have elected to stop underwriting in high-cost states such as Texas and California. Additionally, a larger number of employers have been forced to join state funds, and an increasing number are opting for self-insurance. Self-insurance in workers' comp rose to 29 percent of underwriting in 1991 from 18 percent in 1980.[16] Under self-insurance an employer pays no premium but must be financially sound enough to pay out-of-pocket for incurred expenses. When a worker is injured, the employer must reserve funds for direct medical costs and indemnity payments. Self-insured employers may either self-administer their claims or retain a third-party administrator (TPA) for this purpose. These employers have a greater incentive for aggressive cost containment because it is now "their" money that is being taken out of everyday cash flow; they do not pay an insurance premium. Out of necessity, self-insured employers were among the first to integrate managed care strategies within workers' compensation.

Expanded Definition of Compensability

Cost escalation has followed recent expansions in the definition of "work-related" or compensable injury and illness. Isokowe et al. assert that "the number of nonspecific complaints continues to grow as the average age of the labor force increases, bringing with it a host of chronic conditions aggravated by work."[17, p.52]

Workers' compensation, despite its no-fault mission, has been the victim of increased litigation. The Workers' Compensation Research Institute (WCRI) in its 1993 Annual Report described litigation as "the modus operandi of the system." WCRI data indicate that litigation added $1.5 billion to the workers' compensation bill in 1990, and this figure has been growing at an annual rate of 22 percent since 1981.[18]

In Minnesota, similar injuries treated via workers' compensation have been shown to be twice the costs when treated under Blue Cross/Blue Shield.[19]

The fastest-growing types of claims overall involve cumulative trauma injuries/diseases (CTDs).[20] Stress claims have also skyrocketed, especially under the California system, where 99 percent of these cases are litigated.[21] In California the defense attorney bar has reported a 700 percent increase in stress cases between 1987 and 1992. Isokowe et al. illustrate the domino effect that commonly occurs in workers' compensation cases:

> The inescapable conclusion borne out of these trends [insurance losses] is that as long as prolonged treatment by providers continues to extend disability benefit periods, more injury claims will be represented by litigators, the denial of adequate premium rates will continue, medical costs will rise and large policy holders will move towards self-insurance.[22, p.52]

According to the National Council on Compensation Insurance, workers' compensation produces more than 45 percent of all property and casualty net losses but collects only 13 percent of the premiums.[23] The following is an overview of workers' compensation reforms that have occurred in pacesetting states. By workers' compensation standards, these states have undertaken dramatic steps to curb rising medical costs. Managed care applications that had their genesis in group health/employee benefit schemes lead the list of reform activities and are expected to infiltrate workers' compensation statutes throughout the country.

OREGON

- established objective criteria for determining compensability
- tightened standards for classifying injuries
- submitted impairment disputes to a panel of arbitrators
- stepped up employer safety requirements
- increased provider monitoring and accountability[24]

MICHIGAN

- maximum fee for each procedure
- practice standards contained in a 55-page rulebook

- increased substantiation through medical documentation
- cases that exceed $5,000 are automatically sent for peer review
- establishment of consortiums of state regulators, insurance companies, self-insured employers, large medical providers, and labor organizations[25]

COLORADO

- strict definition of permanent total disabilities
- impairment ratings based on AMA guidelines
- mandatory mediation for certain disputes[26]

TEXAS

- restricted legal involvement in workers' compensation
- treatment precertification following eight weeks of care

WISCONSIN

- "delawyerized" the system[27]
- enforcement of "exclusive remedy" doctrine
- clearly established criteria for settlement of cases

MASSACHUSETTS

- establishment of treatment protocols that will be the only treatments reimbursed[28]
- employees required to use a preferred provider on first visit
- reduced lost-wage payments

Controversial Issues within Workers' Compensation

> In the arena of disability management and rehabilitation of chronically ill, injured and disabled workers, one steps into a maelstrom of public and private sector policies as well as policy preferences of organized groups . . . i.e., the business community, labor organizations, professional associations and disability consumer groups.[29]

Workers' compensation system architecture breeds a plethora of controversial issues that affect the industrial health specialist either positively or adversely. Contestability of claims is usually grounded in one of the following issues:

- causality
- medical necessity for treatment
- reasonableness and appropriateness of care
- cost
- length of disability (temporary, permanent)

- extent of disability (partial, total)
- provider credentials

Every organization involved within the workers' compensation system has a perception of its problems and a proposed set of solutions. A selection of these solutions can be found in Exhibit 38–2.

Problems within the system can be placed into one of three categories: system design, perception of system elements, and execution within the system. System

Exhibit 38–2 Proposed Solutions to Workers' Compensation Problems

American Insurance Association

1. Eliminate litigation.
2. Eliminate unnecessary medical care.
3. Have reasonable benefit levels.
4. Pay for what you get.

AFL-CIO

1. Increase use of state funds.
2. Lower costs.
3. Increase efficiency.
4. Assure availability.
5. Focus on social service and responsibility.
6. Commit to loss prevention.

Alliance of American Insurers

1. Establish guidelines for reasonable and appropriate levels of treatment.
2. Limit an employee's choice of providers.
3. Allow employers/insurers to selectively contract with provider networks.
4. Implement fee schedules and dispute resolution procedures.
5. Prohibit providers from billing injured workers for any portion of the bill.
6. Require that state workers' comp administrators take action against providers who overbill.
7. Impose penalties for late payments of benefits.
8. Maintain a medical database.

National Council on Compensation Insurance

1. Focus on maximum medical improvement and earliest return to work (RTW).
2. Increase workplace safety activities.
3. Retain some employer control of medical treatment.
4. Use uniform electronic forms for reporting.
5. Set malpractice controls.
6. Assert authority to use medical cost management in all states.
7. Set fee schedules.[30]

Note: This is not an all-inclusive listing of organizations interested in workers' comp reform, nor is it a complete list of policy preferences.

design elements have already been discussed; therefore this section will focus on perception and execution.

Perceptions vary depending on whose perspective is being considered. For instance, injured employees often perceive workers' compensation as a disability policy that will reward them in lump-sum payments for disability. Some actually view it as an opportunity to "hit the lottery." Others perceive workers' comp as infinite medical care beyond the restorative phase and into the maintenance phase of care. Perceptions of the intent of the workers' comp are as varied as the players—attorneys, workers, insurers, employers, rehabilitation specialists, vocational counselors, and providers. Exhibit 38–3 lists a number of perceptions about the workers' compensation system. Like many perceptions, some have become realities of the system.

It is safe to conclude that "fundamental to the understanding of the problem is that each of these discrete elements of the problem (the research technology interests, the legal rights system, the local hospitals) are acting in their own self-interest which is consistent with Americans' political philosophy of individual freedoms."[31]

This section will explore only the perceptions of payors concerning providers of care and ultimate reimbursement for services. These perceptions run the gamut from complete satisfaction with providers to complete disdain for providers. Reimbursement or, better yet, nonreimbursement is on the minds of all providers in the midst of health care reform. If one is overcome by objections, the basis for them must first be understood. The following statement was made by the president of the National Council on Compensation Insurance during the 1992 Issue Symposium and later reprinted in an NCCI article entitled "Harnessing the Forces of Change":

> Today, with certain "fringe area" operations, irresponsible professionals have found the workers' compensation system in a big way. And

Exhibit 38–3 Common Perceptions about Workers' Compensation

- The trend has been toward liberalization of coverage to include such injuries as psychological disorders, cumulative trauma, and heart disease.
- The system is heavily adversarial.
- Tort system and workers' comp are undergoing cross-over, allowing dual compensation in some cases.
- Employers feel that they are paying for "normal aging."
- Workers' comp costs are greater than for the same injuries treated under other lines of insurance.
- Providers may have a monetary goal and not the worker's welfare in mind.

> these illicit professionals today are feeding at the trough of workers' compensation. And many don't give a tinker's damn about this system nor about the workers it is designed to serve. And these professionals have names. They are attorneys, forensic physicians, chiropractors, vocational counselors, rehabilitation specialists, physical therapists and physicians.[32, p.16]

Certainly, these comments are not intended to impugn all attorneys, physicians and the like, but are directed at the minority whose behavior has led to cost containment strategies that affect all parties.

The following quotes illustrate the intensified scrutiny of rehabilitation services. Rehabilitation misuse, overuse, and abuse have collectively spawned negative attitudes directed at its providers. Work therapists are surely targets as payors continue to seek workers' compensation cost controls.

> We're investigating fraud schemes that require cooperation on all levels—from the guys who file the phony personal injury claims to the attorneys who handle the cases to the doctors and chiropractors who make bogus diagnoses to the physical therapists who overbill to the guy who worked on the car.[33]

> Carve-outs have had the greatest impact on controlling the costs of mental health care and substance abuse benefits programs. But they are quickly popping up to target chiropractic and physical therapy.[34]

> You may want to focus your initial review on the following outpatient services that are often flagged as "high utilization" services:
>
> - scoping procedures
> - physical therapy
> - mental health services; and
> - prescription drugs[35]

These statements are not intended to imply that all rehabilitation is abusive but merely to illustrate the negative publicity that leads to intensified scrutiny. Certainly there are many examples of positive effects from rehabilitation. However, perception often becomes reality, particularly in the eyes of payors. One such perception is that "physical therapy" is synonymous with physical therapists' services. Data from many sources, including the Florida insurance department and California workers' compensation, indicate that much of what is labeled as "physical therapy" is *not* performed by physical therapists. Yet many payors assume that all physical therapy is the same irrespective of who renders the services. Likewise, payors often assume that all work therapies are equal despite the obvious differences in program philosophies and provider credentials.

Dramatic and unrelenting increases in workers' compensation costs have resulted in numerous examples of adverse impact, including employer loss of insurance underwriting, bankruptcy or staff reductions in some companies, withdrawal of carriers from high-cost states, and increased burden on state funds. These and other pressures have wedded workers' compensation to managed care principles.

MANAGED CARE

Managed care has multiple personalities. Yet for many, it has become synonymous with HMOs. Goldstein points out that the two terms are not interchangeable.[36] Managed care has never been well defined—hence the alphabet soup of terms: PPOs (preferred provider organizations), HMOs (health maintenance organizations), IPAs (independent provider organizations), EPOs (exclusive provider organizations), and POS (point-of-service) plans.

Managed care has been defined by the structure of the organization and not necessarily by the strategies employed. Goldstein provides the following definition of managed health care:

> Managed health care is health care provided in a delivery system in which a mix of organizations—provider organizations, payers, financial intermediaries, and management companies—use, in varying forms and degrees, methodologies aimed at managing the outcomes of one or more of the financing, cost, accessibility, or quality of the health delivered; and in which the decisions of providers are subject to a spectrum of treatment guidelines and parameters, potential peer review, and financial incentives and penalties.[37, p.453]

We are now seeing HMOs that are becoming PPO-like and PPOs that are becoming HMO-like. The lines between these predominant forms of managed care have blurred considerably, thereby making estimates of their numbers difficult.

The birth of HMOs was predicated on three key elements:

1. an agreement by or on behalf of a group of providers to accept responsibility for the delivery of medical services
2. a defined population of enrolled consumers
3. a specific, per-person sum of money agreed upon in advance and paid on a periodic basis

There have been few indications that HMOs have abated costs, much less improved quality of care. A recent Government Accounting Office study failed to find significant savings resulting from managed care.

IN SEARCH OF QUALITY

Measure is established as the source of wisdom.

—Jonas Salk, *The Survival of the Wisest*

Gone are the halcyon days when a provider's care went unquestioned. Gone are the days when one defined quality through "process" or "structure." We are now in an era when quality is increasingly being considered synonymous with "results" or "outcomes." There is a quality imperative running through all strata of society. Today's American business leaders have rediscovered the quality principles set forth by individuals like Juran and Deming. The Malcolm Baldridge Award has become a much-coveted prize and a stamp of approval for any company that receives it.

However, medicine, contrary to other industrial sectors, has not clearly defined what constitutes a "quality" product. Like most service sectors, medicine grapples with the problem of defining quality elements. Unique elements of medicine further compound the problem. Perhaps the greatest confounding factor is the "community context" in which medicine operates. As described earlier in this chapter, there are a number of self-interest groups, each possessing a different agenda. The "community context" refers to the variety of players commonly involved in a work injury case: the patient, physician, attorney, employer (occupational medicine and personnel), insurer, rehabilitation nurse, vocational specialist, and others. In addition to having differing agendas, these parties frequently speak a different language and use different methods of measuring "quality." For instance, the insurance company may measure quality by cost alone—the lower the cost of treatment, the higher the quality of services. The employer may measure "quality" by how quickly an injured worker returns to work, with little regard for work restrictions and functional limitations. The work therapist may measure "quality" by how much time he or she spent with the client or by provider credentials. The client may measure "quality" by some of the "feel good" elements such as timeliness of appointment or friendliness of the receptionist.

Conventional assessment of quality in health care has involved three types of measures: structure, process, and outcomes. *Structure* and *process* are terms used to define specific characteristics of a treatment plan such as staffing, square footage, and equipment. *Outcomes* is a term used to describe either the positive or the negative change in a patient's health that can be attributed to a specific course of care. Outcome is considered a more direct measure of quality than process or structure.[38] A valid outcome measure must take into account severity indices and differences commonly found in diagnoses.

It is critical to view "quality" in medicine in terms of both quality of medical *treatment* and quality of medical *management*. *Treatment* measures involve direct medical intervention, whereas *management* refers to the entire matrix within which treatment is rendered. Examples of each include:

1. treatment
   - skill of clinician
   - technology
   - provider credentials
   - appropriateness of intervention
2. management
   - documentation requirements
   - timeliness of treatment
   - rapport building
   - communication

In order to achieve a satisfactory outcome, treatment and management elements must mesh together. These are not mutually exclusive concepts, and they involve both subjective and objective elements. Frequently "good care is in the eyes of the beholder."[39]

Definitions of quality are extremely diverse. The following is a sampling of both serious and whimsical definitions:

> Quality . . . you know what it is, yet you don't know what it is. But, that's self-contradictory. But, some things are better than others, that is, they have more quality. But, when you try to say what the quality is, apart from the things that have it . . . there's nothing to talk about. But, if you can't say what quality is, how do you know what it is or how do you know that it even exists?[40]

> Quality is not that the catheter is put in well, but is it put in friendly.[41]

> Quality health care consistently contributes to the improvement or maintenance of the quality of life and/or the duration of life.

> Quality care is defined as the provision of appropriate service to the appropriate patient at the appropriate time by the appropriate person with appropriate results.[42]

Christoffersson and Moynihan contend that "determining the quality of care provided to a patient cannot be accomplished without first defining appropriate standards of care for that condition."[43, p.24] The entire medical community is weak in this regard, particularly rehabilitative medicine. Due to the paucity of practice standards and outcome measures, providers of care frequently exercise a hit or miss approach to treatment. As a result, many treatments are unnecessary and wasteful. According to Berman, "25 percent of every workers' compensation dollar is wasted because the system is unmanaged. The waste can be pinpointed to the following four areas:

- injuries that could have been prevented
- inappropriate treatment
- undirected care; and
- poor return-to-work planning"[44]

Payor Outcome Initiatives

There is a crying need for outcome measures coming from the provider community and the payor sector. On January 1, 1991, Blue Cross/Blue Shield of Minnesota became the first national insurer to directly link health care reimbursement to patient outcomes. This initial project involved 34 high-volume hospitals. The state of Maine is embarking on a project that will make it the first state to offer physicians insulation from medical malpractice suits if they use approved medical practice guidelines. The U.S. Agency for Health Care Policy Research has been charged with developing clinical practice guidelines and screens that will test if physicians who treat Medicare patients are subscribing to national standards of care. This project is not restricted to Medicare patients only but has private sector implications as well. A number of states, beginning with Pennsylvania, have enacted legislation that enables the publication of outcome data such as morbidity, mortality, and cost-based data. It is only a matter of time before this trickles down to workers' compensation and the management of work injury/illness. The John Hartford Foundation is in the process of designing automation of health care transactions and collection of data on outcomes, effectiveness, mortality, and other indicators of quality care. The goal of this project is to shift volume away from less effective providers.[45] Throughout the United States various business coalitions are publishing outcome measures; they include the New Orleans Business Coalition, Cleveland Employers and Apache Medical Systems, the Minnesota Outcomes Consortium of 24 major employers, the Managed Healthcare Association, Interstudy, and Johns Hopkins.[46] Even the Juran Institute has started a demonstration project entitled "Quality Improvement in Health Care," a collaborative effort between health care providers and industry.[47] Soon the norm in medicine will be not what the Japanese describe as *atarimae hinshitsu* or "quality that is taken for granted" but *miryokuteki hinshitsu* or "quality that fascinates."

Survival beyond the 1990s and into the next century will be conditioned on the ability to conform to technical specifications (i.e., practice standards) and customer requirements (i.e., outcome measures).[48] Batchelor and Esmond cite new evidence that high-quality care costs less than poor-quality care through the elimination of unnecessary or inappropriate services while providing better clinical outcomes, fewer avoidable complications, and greater patient satisfaction.[49] According to some, our current system of health care financing and delivery operates on a "cost-unconscious" basis.[50] It is time for providers everywhere to exercise a "gut check" by discarding unproven treatment techniques while heralding those that produce excellent outcomes at reasonable cost.

Gone are the days of open-ended fee for service that rewards providers for doing more, not less, whether or not it is appropriate. Gone are the days of free choice of providers, which has traditionally deprived the payor of control over spending. We are now in the days when high-quality providers will be rewarded.

OUTCOME MEASURES DEVELOPMENT

The first step in developing practice standards and outcome measures is to define these terms. There is a profound need for clarification of terms amidst the current semantics problem. Terms found in Exhibit 38–4 have often been used inappropriately and interchangeably to describe treatment characteristics and performance measures.

Practice parameters or guidelines are typically developed through four processes:

1. literature search
2. retrospective statistical analysis of cases
3. prospective analysis of cases
4. expert consensus

A number of different entities are engaged in the development of practice guidelines. These include:

- federal government
  - National Institute of Health
  - Centers for Disease Control
  - Office of Technology Assessment
  - General Accounting Office
  - Agency for Healthcare Policy Research
- provider groups
  - American Medical Association (AMA)
  - specialty associations
  - American Physical Therapy Association (APTA)
  - American Occupational Therapy Association (AOTA)

Exhibit 38–4 Practice Standards and Outcome Measures Definitions

algorithm: a series of steps for addressing a specific problem

appropriateness of care: the degree to which correct care is given based upon current state of the art

case-based review: the review of an individual case by a professional to determine whether the care rendered was of sufficient quality as per his or her judgment, also known as *medical record review* or *utilization review*

criteria sets: see *performance standards*

indicator: a measurement tool used to monitor and evaluate the quality of important governance, management, clinical, and support functions. Indicators may be designed for different purposes: e.g., outcome indicator, process indicator, rate-based indicator.

medical criteria sets: see *performance standards*

outcome indicator: an indicator that measures what happens or does not happen to the patient after a specific treatment(s) is rendered

performance database: an organized comprehensive collection of data designed to provide information about the quality of patient care

performance standards: methods or instruments to monitor the extent to which actions of a health care provider conform to medical review criteria and standards of quality

practice guidelines: systematically developed statements to assist the practitioner and patient decisions about appropriate health care for specific clinical circumstances, also known as practice parameters or treatment protocols

practice parameter: see *practice guidelines*

process indicator: an indicator that measures an activity that contributes directly or indirectly to patient care

process standard: a statement that defines an organization's functional capacity to provide quality care (e.g., treatment planning functions, safety management)

rate-based indicator: an indicator that measures an event for which a certain proportion of the events that occur are expected when state-of-the-art care is provided (e.g., infection rates < 3 percent, return-to-work rates > 80 percent)

screens: see *sentinel event*

sentinel event: a patient care event that requires further investigation each and every time it occurs, usually an undesirable event (e.g. "physical therapy" rendered by a non-physical therapist)

standards of quality: authoritative statements of (1) minimum levels of acceptable performance or results, (2) excellent levels of performance or results, (3) the range of acceptable performance or results

statistical profiling: an evaluation approach to quality of care that employs data from large databases as markers for quality of care (e.g., 80 percent of low back pain sufferers return to work within two weeks, compared to treatment center X's return to work average of four weeks)

treatment protocol: see *practice guidelines*

Source: Adapted from *Primer on Indicator Development and Application: Measuring Quality in Health Care*, by the Joint Commission on Accreditation of Healthcare Organizations, Oakbrook Terrace, Illinois, 1990.

- academic centers
- independent research entities
  - Institute of Medicine
  - Rand Corporation
- insurers
- foundations
- hospitals
- managed care entities
- accreditation bodies
  - Joint Commission on Accreditation of Healthcare Organizations (JCAHO)
  - Commission for Accreditation of Rehabilitation Facilities (CARF)
- state/local governments
  - workers' comp reforms
  - state insurance regulation
  - professional licensure regulation
- business coalitions
  - Minnesota
  - Central Florida
  - Washington
  - Louisiana

Practice guidelines, if they are to be meaningful, must be systematically developed and possess a number of attributes described by Lohr and Field:

- credible
- disclosed to all parties
- valid
- flexible
- clear
- reliable
- clinically adaptable
- periodically reviewed
- must have research relevance
- applicable to health systems management
- explicit

- based on science
- developed with practitioner input[51]

The relationship between practice guideline development and reimbursement for services is depicted in Figure 38–1, "The Interdependency of Research, Clinical Practice, Marketing, and Reimbursement." The development of practice guidelines is a meaningless process if they are not adopted by the provider community or used by the utilization review industry as a means of determining appropriateness of care.

PURPOSE OF OUTCOME MEASURES

Outcome measures serve a multitude of functions for a variety of health care stakeholders. For the medical provider, they serve to reduce variability and establish efficacy of specific interventions. Additionally, outcome measures convert science-based knowledge to clinical practice and strengthen the link between medical care and medical management. The consumer (payor and patient) benefits because outcome measures clarify medical choices and identify centers for excellence. Practice guidelines and outcome measures also serve a risk management function by avoiding acts of omission and commission. Lastly, outcome measures

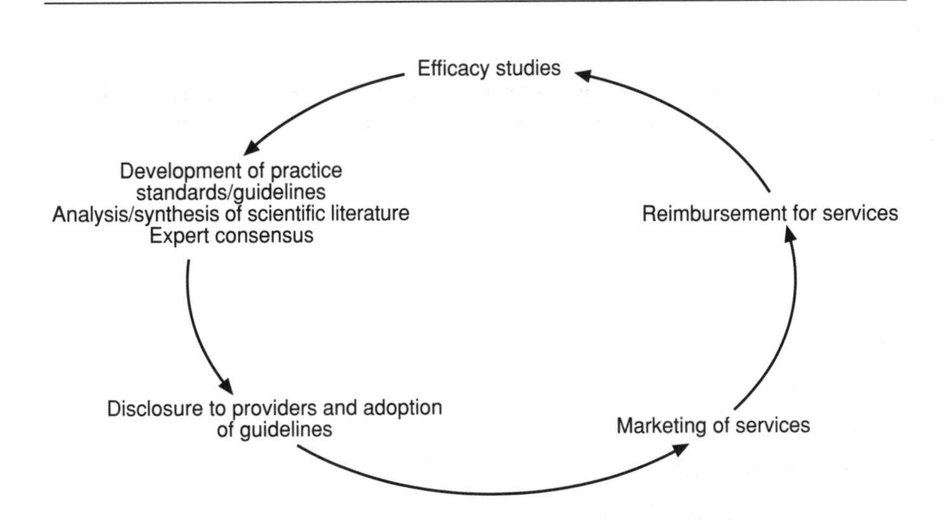

Figure 38–1 The Interdependency of Research, Clinical Practice, Marketing, and Reimbursement

will be linked to reimbursement levels. For instance, in Florida's workers' compensation system providers must prove an "acceptable placement rate" as a condition of recertification or the ability to treat injured workers.

Chassin believes that guidelines can be used effectively and impartially for education, reimbursement, quality improvement, and research purposes.[52]

POTENTIAL PROBLEMS ASSOCIATED WITH USE OF GUIDELINES

The development, disclosure, and adoption of practice guidelines are not without controversies and potential problems. Problems can be categorized as reimbursement, malpractice, and quality assessment. The following questions remain largely unanswered and will most likely be debated for years to come.

- *Reimbursement implications*
  1. Will consumers be able to selectively contract for established sets of guidelines?
  2. Will those who do not adopt published guidelines be locked out from reimbursement schemes?
  3. What constitutes a fair decision to exclude providers (either individuals or whole categories) from inclusion in health care plans?
  4. What are the reimbursement ramifications when a practice guideline is applied but an adverse outcome results?
  5. What will the payment policy be when there are conflicting guidelines covering identical conditions?
- *Malpractice implications*
  1. Will the existence of guidelines require that providers disclose treatment options as a component of informed consent?
  2. If guidelines offer malpractice protection to some, will their existence foster more, not less, litigation?
  3. What is the significance of patient preference for specific treatment options in the face of guidelines?
  4. Will regional/local or national standards prevail?
  5. What protection is afforded the provider who follows an established guideline that results in an adverse result?
  6. What are the legal ramifications when a provider's enrollment in a health plan is predicated on adoption of specified guidelines that result in a negative outcome? Who is legally responsible? Can the provider seek relief from the plan (e.g., the managed care plan)?
  7. What will be the legal significance of conflicting guidelines that address the same condition?

- *Quality assessment*
  1. What constitutes a good versus bad guideline?
  2. What constitutes a consensus?
  3. How often should guidelines be updated?
  4. How is the weight of scientific evidence determined? Which process is considered superior?
  5. When does a guideline become a standard?
  6. Which should the provider follow, a regional/local or national standard?
  7. Which should be considered, minimal levels, optimal levels, or a range?

TYPES OF OUTCOME MEASURES

Outcome measures can be divided into four primary categories: impairment-driven, disability-driven, cost-driven, and patient satisfaction measures.

Impairment Measures

Impairment measures are among the most common but least predictive measures. Impairment measures are "medically based" and address anatomical and/or physiological processes. Impairments are often the result of underlying injury or pathology. However, not all diseases lead to impairments. Examples of impairment measures include range of motion, strength, pain, swelling, sensation, and balance measures.

Disability Measures

On the other hand, disability measures are those having to do with an individual's ability or inability to return to gainful employment. *Disability* is also defined as the social consequences of an individual's impairment or functional deficits. Disability, unlike impairment, is an administrative or legal and not a medical decision. Not all impairments lead to disability. Examples of disability measures include activities of daily living, return-to-work status, reinjury or recidivism rates, and Department of Labor critical functions.

Work therapists would be well advised from both a clinical and reimbursement standpoint to adopt disability measures in favor of traditional impairment measures.

Cost Measures

Cost measures, although perhaps the greatest concern of payors, are among the measures least understood by clinicians. Clinicians are accustomed to fee-for-service reimbursement and do not routinely analyze total case costs. Total case costs in a workers' compensation situation exceed medical costs and include legal, indemnity, and administrative costs, which often outpace medical expenditures. It is clear that reimbursement will continue to be based on case costs or capitation for some time to come. Work therapists would be well advised to develop a combination disability/cost outcome measure system since under workers' compensation it is the disability status of a given client that is the primary cost driver. A client's disability can lead to medical, indemnity, administrative, and in some cases, legal costs.

Patient Satisfaction Instruments

Many patient satisfaction instruments dwell on subjective, not objective data. These measures have historically focused on "medical management," or process and structure, and less on "medical treatment" and outcomes, or results. A current trend involves the use of lifestyle inventories such as Ware's Medical Outcome Survey Short Form 36. Patient satisfaction surveys may be an important clinical element, but from a payor's perspective these tend to be less important than disability or cost measures.

The ideal outcome measures incorporate all four elements—impairment, disability, cost, and patient satisfaction. Exhibit 38–5 gives a self-assessment exercise regarding outcome measures development. This survey should enlighten work therapists about areas that need to be addressed in their current clinical practice as they relate to management of the injured worker.

MANAGED CARE AND WORKERS' COMPENSATION CONVERGE

> HMOs are all moving toward workers' compensation managed care status.
>
> —David Strand, Minnesota HMO Council[53, p.2]

In the absence of universally accepted practice standards and outcome measures, payors have resorted to a variety of cost containment strategies. Techniques have been borrowed from group health and employee benefits administration schemes and been applied to workers' compensation.[54]

The development of preferred provider networks is one managed care strategy that until recently was prohibited from many workers' compensation schemes. To date, the limited control over provider choice has restricted the growth of workers' compensation networks, but this is rapidly changing.

Simpson cautions that it is critical to have sufficient numbers of each type of provider in the managed care organization. He considers the appropriate providers

Exhibit 38–5 Outcome Measures Survey

1. Do you know the average length of stay/number of visits for your clients? ☐ Yes ☐ No
   a. Do you break this data down by: ☐ diagnosis
   ☐ procedures
   ☐ industry type
   ☐ other, please explain

2. What percentage of work hardening enrollees return to gainful employment:
   ☐ % RTW restricted duty/transitional work
   ☐ % RTW full capacity/no restrictions
   ☐ % RTW in completely different job/transfer
   ☐ % RTW with no restrictions but job ergonomically redesigned
   ☐ % don't RTW in any capacity
3. Do you keep data on:
   ☐ % program drop outs (any reason)
   ☐ % clients who expire
   ☐ % clients who retire
4. Do you track recidivism rates? ☐ Yes ☐ No
   a. What percentage return to therapy due to:
      ☐ % reinjury (alleged or demonstrated)
      ☐ % work related reinjury
      ☐ % nonwork related reinjury
5. Do you have written work hardening admission criteria? ☐ Yes ☐ No
6. Do you have written discharge criteria? ☐ Yes ☐ No
7. Do you have a written Quality Assurance and Utilization Review Program? ☐ Yes ☐ No
8. Do you compare your LOS/outcome measures to industry standards? ☐ Yes ☐ No
   a. What source of data do you use for comparison?
      ☐ state workers' compensation data
      ☐ insurer morbidity data
      ☐ DOL data
      ☐ industry specific (via trade groups, labor, coalition, professional organization data)
      ☐ other sources, please list
9. Do you provide a projection of the following to the payor and client?
   ☐ treatment frequency
   ☐ treatment duration
   ☐ daily fees
   ☐ total case costs
10. Do you ever guarantee a total case cost? If yes, under what circumstances and how do you derive your fees?

11. Do you know how your fees compare to usual, customary, and reasonable fees? If yes, which data sources do you use?
    ☐ Medicare DRGs
    ☐ state workers' compensation fee schedules
    ☐ RVS
    ☐ Health Insurance Association of America

continues

Exhibit 38–5 (continued)

☐ other insurance trade group data
☐ professional trade associations (providers)
☐ surveys
☐ other, please explain

12. Which best describes your fee calculation process?
☐ driven by what competitors charge
☐ driven by what payors will pay, i.e., managed care discounts
☐ cost based
☐ RVS
☐ determined by UCRs
☐ arbitrarily selected
☐ charge whatever the market will bear

to include "ancillary providers such as physical therapists and occupational therapists."[55] Simpson provides two reasons for combining managed care techniques with workers' compensation: (1) the cost of medical treatment is approximately 50 percent that of workers' comp; (2) workers' comp has become an adversarial system for most parties. In considering the providers for workers' compensation, he emphasizes the need to reconcile providers' experience with workers' comp. General practitioners and internists are the predominant providers in a typical group health managed care plan, whereas orthopedists, neurologists, and neurosurgeons are critical components of work injury management.

Single-service providers will be hard pressed to satisfy all of the needs of an employer or insurer. Madlem considers hospitals, physicians, urgent care centers, podiatry, chiropractic, physical therapy, and rehab services to be integral components of a workers' comp preferred provider organization.[56] However, few companies meet all of these requirements, and today's recessionary economy is prohibitive to start-up enterprises.

Strategic partnering may be the most logical and perhaps most cost-effective method of satisfying all of a customer's needs. The trend towards strategic alliances between providers, employers, and insurers is a national one and a component of many health care reform packages. Many have predicted that strategic partnering will be the predominant model of health care delivery into and beyond the 1990s.[57]

There are three predominant forms of strategic partnering. One involves a partnership between different providers in the form of mergers, acquisitions, and alliances. In this structure the lines between professions blur, and services are offered in a team or integrated fashion. Under this scenario, case rates will be paid to provider groups. This model is currently in pilot stages in Minnesota.

The second form involves a triad between employers, insurers, and providers. In this model selective contracts are negotiated based upon capitation or case rates as a means of shared risk. This reimbursement structure is displacing the per-visit discounted fee system, which is largely unproven as a cost containment mechanism. Case management decisions are jointly made between the three parties, and communication is tightly controlled. All parties are held accountable for different elements of health care delivery. For instance, the insurer is expected to reimburse only for medically necessary and appropriate care, the employer is responsible for posting preferred providers and the establishment of "transitional duty" programs, and the provider is accountable for positive outcomes following agreed-upon treatment protocols.

The third model involves like providers forming large networks in order to densify their services in key markets. These providers tend to occupy "niche" markets such as head trauma, rehabilitation, and psychiatric services. One of the more prolific examples involves the recent rage in physical therapy network development through acquisition, mergers, and franchises, and for establishment of independent networks through cooperation between regional provider networks.

Participants of all three models are attempting to secure selective provider contracts within workers' compensation, in some cases exclusive contracts. Reimbursement schemes are being borrowed from the group health/employee benefits industry as more and more workers' compensation contracts will be predicated on case rates (dollar amount per diagnosis per course of care) or capitation.

Tomorrow's rehabilitation provider will be confronted with a dizzying array of managed care strategies, some of which are identified in Exhibit 38–6.

UTILIZATION MANAGEMENT STRATEGIES IN WORKERS' COMPENSATION

It was inevitable that utilization management techniques found their way into the workers' compensation system. What is most amazing is that it took so long! Two-thirds of the growth in workers' compensation costs has occurred since 1982. During this same period the utilization review industry was in its accelerated growth phase. However, the vast majority of utilization review strategies were directed at group health, not workers' compensation.

The convergence of workers' compensation and managed care has resulted in a variety of payor initiatives. Five of the most prevalent strategies are depicted in Figure 38–2. These include case management, cost reduction, provider selection, prevention, and utilization review (UR).

Utilization Review: Industry Overview

According to Spencer, the number of utilization review entities is between 200 and 300.[58] Utilization review entities, like managed care organizations, have a

Exhibit 38–6 Payor Initiatives

UTILIZATION REVIEW: External Organizations
- fee audits
- use of disability duration tables
- rebundling of codes
- mandatory second opinions
- preauthorizations/precertifications
- reimbursement linked to outcomes
- peer review: prospective-concurrent-retrospective
- provider profiling
- case rates and capitation
- case management
- claimant surveillance

INSURANCE UNDERWRITING
- exclusion of specific coverages (e.g., TENS, thermography, surface EMG)
- mandatory second opinions for specific conditions and procedures (e.g., LBP, CTS, knee arthroscopy)
- increased deductibles
- increased copayments
- certification requirements (e.g., CARF)

CLAIMS ADMINISTRATION: Internal Staff
- fraud investigations
- internal use of medical staff for UR
- requirements of provider certifications
- case rates/capitation
- selective provider contracting in MCOs
- discount fee negotiations
- code rebundling
- fee audits
- use of disability duration tables

variety of structural and program differences. Because of the diversity in review standards, this industry is under siege itself. Regulators at the state level and trade organizations at the national level are attempting to further define the scope of this industry. The explosion in UR growth has demanded an attempt to standardize review processes and to assure that a fair, equitable, and professional approach is taken. A number of "protectionist" clauses are incorporated into a variety of statutory requirements of review entities. These include: protection of patient and provider confidentiality, the right to appeal or reconsideration, provider access to review agent, use of standardized review criteria, and the assurance that a "likes reviewing likes" process is followed. Figure 38–3 illustrates the trend towards regulation of the UR industry.

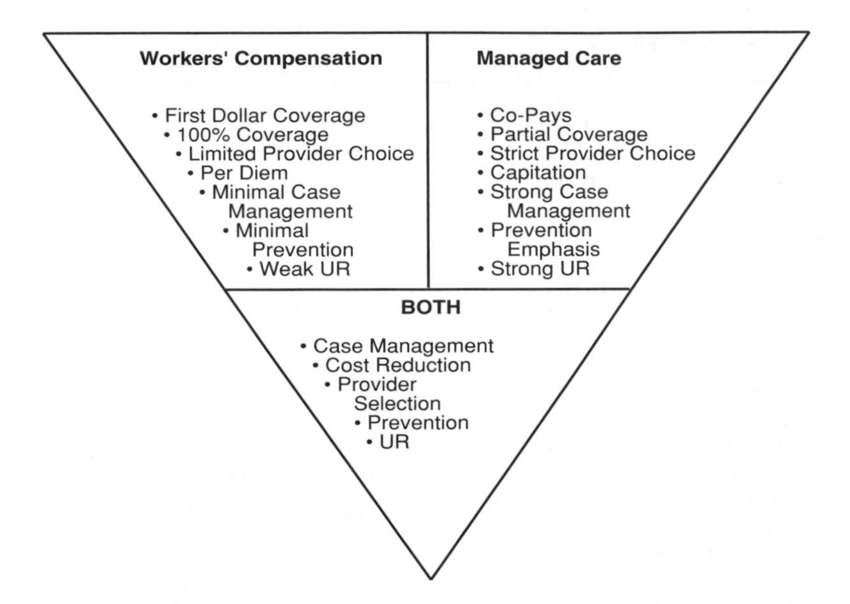

Figure 38–2 Convergence of Workers' Compensation and Managed Care

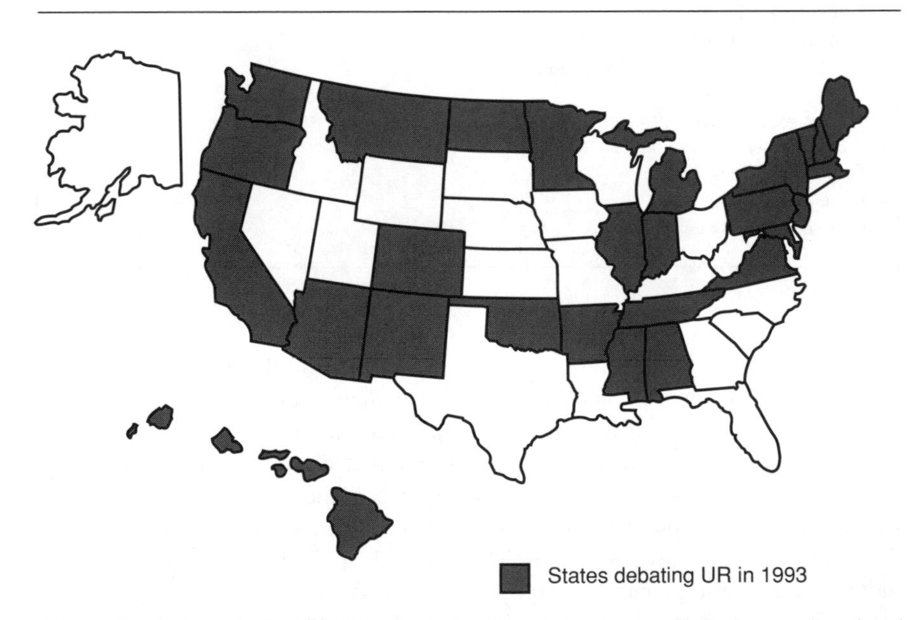

Figure 38–3 States Debating UR Legislation/Regulation during 1993. *Source:* Reprinted with permission of Faulkner & Gray, Healthcare Information Center, 11 Penn Plaza, New York, NY 10001, *Medical Utilization Review Directory*, 1994.

Twenty-eight states and the District of Columbia have passed legislation regulating the utilization review industry, particularly regarding reviewer's qualifications, review standards, and appeals processes. The American Managed Care and Review Association (AMCRA) has promulgated a certification process for utilization review organizations (UROs). These serve as national standards but are voluntary. The Utilization Review Accreditation Commission (URAC) was established in 1990 by AMCRA to develop quality assurance standards; however, these relate primarily to inpatient or hospitalization review functions. In contrast, managed care techniques for workers' compensation are specific to outpatient care.

Peak growth of utilization review organizations was seen in 1985, with half of all UROs offering nationwide services (see Figure 38–4).[59] Ownership is predominantly private, with less than 12 percent publicly owned. The largest percentage of utilization users continues to be private corporations, followed by insurance companies and third-party administrators. Faulkner and Gray's survey demonstrates that the greatest concentration of UROs is found in California, Pennsylvania, and Texas, where between 129 and 140 companies operate (see Figure 38–5).[60]

Two ineffective traditional cost containment techniques involve fee discounting and a focus on hospitalization. The by-products of these unidimensional methods have included increases in the number of outpatient visits. These two cost containment strategies have done little or nothing to curb escalating costs, particularly under workers' compensation, because they fail to address the more salient issues of

1. injury causality
2. medical necessity
3. reasonableness/appropriateness of care
4. functional outcomes of care

Since workers' compensation must prepare the injured worker for an expedient but safe return to work, *functionality* is paramount. Fee schedules and point of service fail to address this issue.

The most frequently used cost containment services for workers' compensation, according to one managed care expert, include preadmission certification, continued stay review, case management, retrospective review, provider and bill auditing, and surgical necessity review.[61] Programs that involve utilization review and not just fee schedules or a site of service focus enjoy cost savings. Seventy-six percent of employer respondents to Corporate Health Strategy's *Health Poll* reported that cost savings exceeded the cost of programs. Respondents of this survey listed the following utilization review programs by frequency of use:

- preadmission certification—94 percent
- bill auditing—68 percent

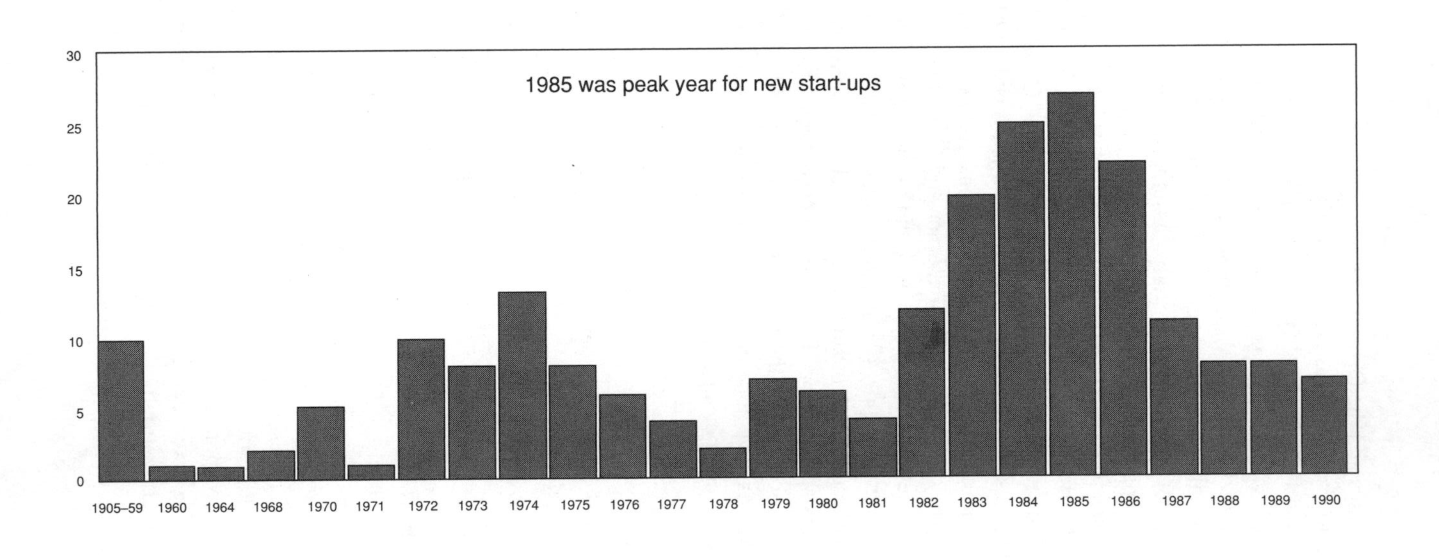

Figure 38–4 Growth of UR Industry: Number of Companies Founded, 1905–1990. *Source*: Reprinted with permission of Faulkner & Gray, Healthcare Information Center, 11 Penn Plaza, New York, NY 10001, *Medical Utilization Review Directory*, 1992.

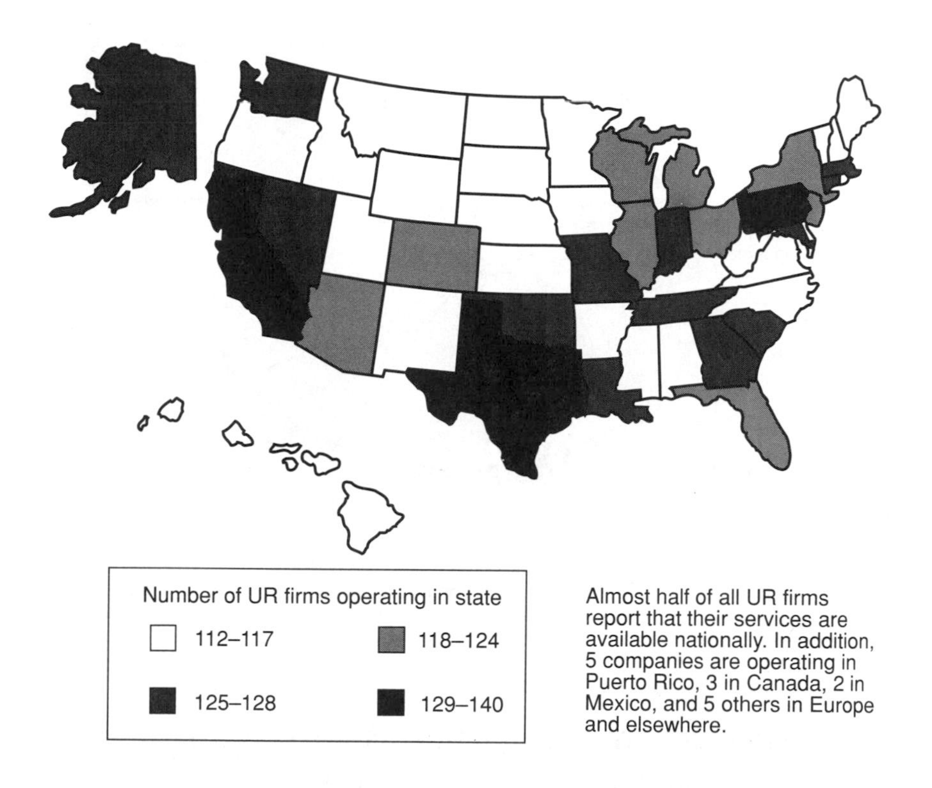

Figure 38–5 Medical Utilization Review Companies—Market Penetration. *Source:* Reprinted with permission of Faulkner & Gray, Healthcare Information Center, 11 Penn Plaza, New York, NY 10001, *Medical Utilization Review Directory*, 1992.

- discharge planning—68 percent
- telephone hotline for employees—67 percent
- retrospective review—59 percent

The inability to produce hard cost savings data was one of the biggest problems with utilization review cited by respondents.[62] Attempts to measure the "sentinel effect" of utilization review have consistently met with failure. Yet the skyrocketing costs in the absence of UR are not difficult to prove, as evidenced by workers' compensation. Many continue to extol the virtues of utilization review despite cost savings data.[63] Figure 38–6 reinforces the user confidence in utilization review strategies.

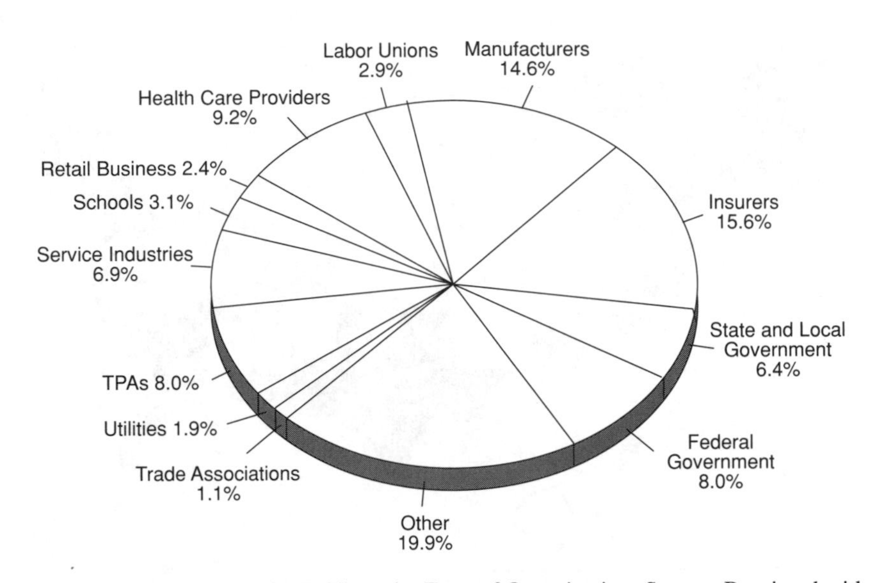

Figure 38–6 Breakdown of UR Clients by Type of Organization. *Source:* Reprinted with permission of Faulkner & Gray, Healthcare Information Center, 11 Penn Plaza, New York, NY 10001, *Medical Utilization Review Directory*, 1994.

UTILIZATION REVIEW VERSUS UTILIZATION MANAGEMENT

A continued escalation in UR growth is anticipated as more and more payors (both casualty insurers and self-funded employers) begin to manage risk in addition to financing it. Faulkner and Gray estimate the utilization review industry to be worth $11 billion. This represents a 30 percent surge over what is described as the "stagnant" 1989–1992 period. The average revenue of UR firms continues to rise and now represents $39 million. Much of this growth has occurred in the workers' compensation arena with 40 percent of UROs nationwide offering services geared towards the injured worker (see Figure 38–7).

Many employers and insurers have realized the importance of early intervention in workers' compensation claims. Once injured workers are entrenched in the "sick role," it is difficult to extract them from this mind-set. Return-to-work rates are inversely related to the length of unemployment; as unemployment increases, the likelihood of a return to work decreases. Countless authors have concluded that "to delay rehabilitation is to jeopardize rehabilitation."[64]

Delayed rehabilitation should become a thing of the past as the trend toward managed rehabilitation evolves. Providers must and will become more accountable as utilization management supplants utilization review. The plethora of individuals and corporate entities engaged in utilization management has led to an

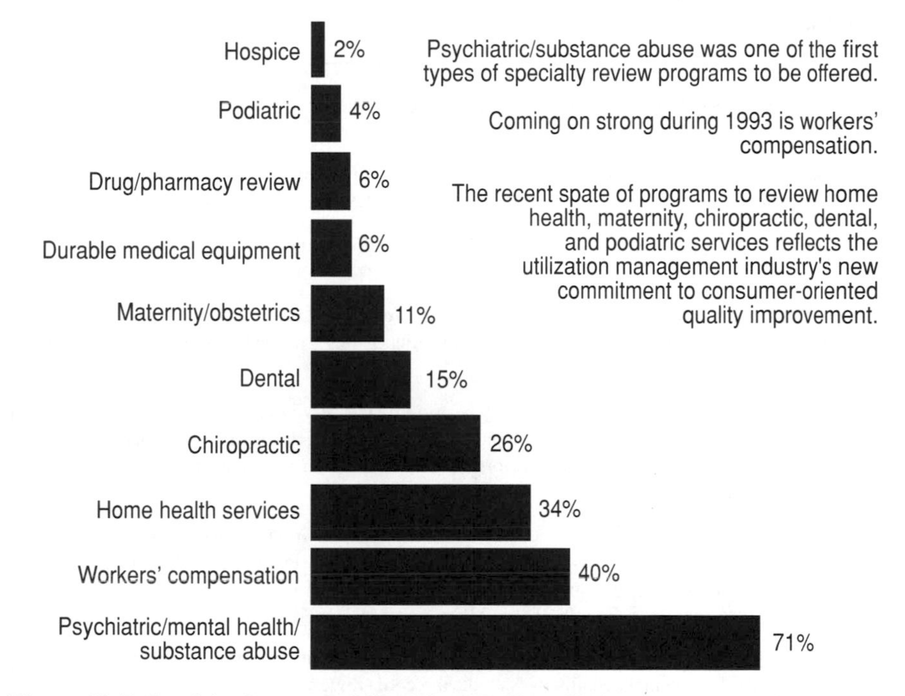

Figure 38–7 Specialty Programs Offered by UR Firms Today. *Source:* Reprinted with permission of Faulkner & Gray, Healthcare Information Center, 11 Penn Plaza, New York, NY 10001, *Medical Utilization Review Directory*, 1994.

---

alphabet soup of definitions. Exhibit 38–7 will enhance one's understanding of a growing industry.

Utilization review has yielded to utilization management. Utilization review, as the term implies, is performed forensically or after the fact, while utilization management (UM) techniques involve intervention at any point along the injury spectrum, from pre-loss (before injury) to post-loss (after injury). Additionally, UR involves little provider contact, while UM is based upon frequent provider contact through case management from admission through discharge planning. Exhibit 38–8 contrasts utilization review with utilization management while Figure 38–8 illustrates the components of a utilization management approach to cost control.

SURVIVING AND THRIVING IN MANAGED CARE

There will be winners and losers in managed care. In the face of incessant growth of utilization management, reimbursement for services has become the

Exhibit 38–7 Utilization Management Definitions

concurrent review: a system of review while care is still being provided; may involve both records review and telephonic provider contact

criteria: "current predetermined, measurable elements of health care against which an actual occurrence can be compared" (North Carolina HB 1079, Chapter 340, June 15, 1989)

critical pathways: are related to clinical guidelines by documenting the treatment approaches that lead to consistent and reliable outcomes

peer review: the review of the medical necessity, reasonableness, and appropriateness of a provider of like specialty, licensure, and experience (e.g., physical therapist reviewing a physical therapist)

private review agent: "any person or entity performing UR services for third-party payers on a contractual basis for outpatient or inpatient services. Does not include personnel or staff performing UR for their respective organizations, including insurers, HMOs, hospitals" (Florida SB 2316, Chapter 90-187, effective date October 1, 1990)

prospective utilization review: "a process whereby a third party reviews the appropriateness of and/or necessity for a health care provider's proposed admission or medical procedure for a particular problem" (Government Accounting Office, *Information on External Utilization Review Organizations*, GAO/HRD-93-22FS, Nov. 1992)

retrospective review: a system or collection of review activities performed after a treatment course has been completed; involves medical records review

utilization review: "a system for reviewing the necessary, appropriate, and efficient allocation of health care resources to be given to a patient or group of patients" (South Carolina, SB 329, Act Number 311, Laws of 1990)

utilization review plan: "a reasonable description of the standards, criteria, policies, procedures, reasonable target review periods, and reconsideration and appeal mechanisms governing UR activities performed by private review agents" (Georgia 1990HB 1813, Chapter 1257)

Exhibit 38–8 Comparison between Utilization Review and Utilization Management

Utilization Review	*Utilization Management*
"After-the-fact"	"Before-the-fact"
Second guesses care	Case management
Minimal provider contact	Frequent provider contact

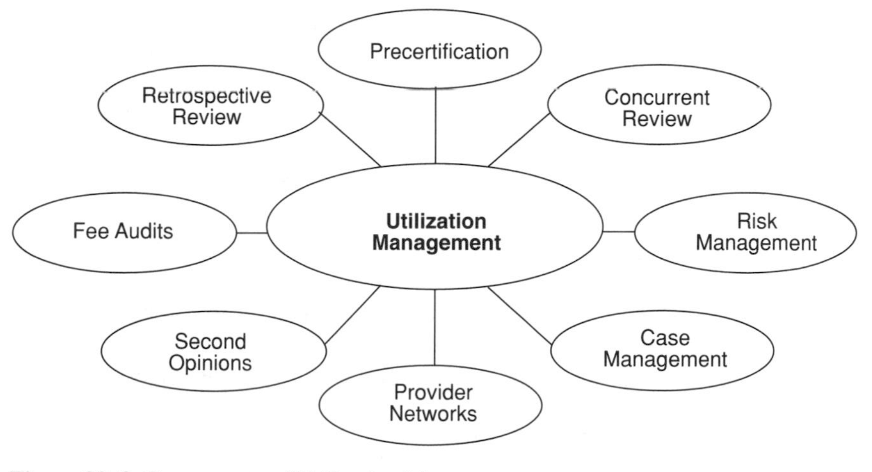

Figure 38–8 Components of Utilization Management

most seminal issue of the 1990s. At least in the foreseeable future, reimbursement will continue to shrink. Some providers will adjust their sails to direct this new wind while others will sink. Diversification of products and services will become imperative, as will the need to differentiate one's business from competitors. Nahra identifies three key market segments:

1. individuals eligible for health benefit coverage through employer plans
2. benefit claims processors
    - insurers
    - third-party administrators
    - managed care entities
3. those who purchase group health plans
    - employers
    - union trusts[65]

Coddington suggests two responses to survival in a health care market subjected to severe price competition, that is, managed care. The first is to differentiate products/services from the competition through quality definition. The second is to increase profit margins on selected products/services or to develop new services with higher profit margins.[66] Providers face two potential dilemmas within managed care: accepting fees that are too low or face being locked out of managed care organizations for not accepting low fees. Many of these new products will be outside of patient care and involve consultative roles in utilization management, education, research, and injury prevention. Most experts caution providers about competing on price because the internal costs of doing business are not necessarily consistent with the competitors'. One simply has to look at the airline industry when it is in a "price war": all lose money.

Coddington identifies one important caveat about diversification: "the experience of firms in many other industries has repeatedly shown that the farther away an organization strays from its primary business, the greater the probabilities that problems will occur."[67]

It is impossible in the scope of this chapter to provide specific advice to all providers about how to position their unique business, particularly given the alphabet soup of managed care entities and techniques currently employed by payors.

Sheldon encourages organizations to examine a number of elements critical to success:

- the number and diversity of market segments
- the number and size of competitors in these market segments
- the growth or contraction in market segments
- the number of products/services in each segment
- keys to value in each segment—financial, social
- life-cycle stage of each product in market segments, for the competition and one's organization[68]

For every trend there is always a countertrend. Positioning for the 1990s and beyond will involve the need to expand services while at the same time contracting so that core products/services truly reflect what one's organization represents. Quinn reinforces the need to balance these forces for success: "Smart strategists no longer analyze market shares and their associated cost positions ad infinitum, nor do they build integrated companies to exploit them. Instead, they concentrate on identifying those few core service activities where their company has or can develop unique capabilities."[69, p.60]

Many external forces are often beyond the control of the provider. These include legislative shifts, market preferences, growth of competitors, and access to capital. It is estimated that approximately 20 percent of Americans with health insurance are currently enrolled in managed care programs. An increasing number of these programs do not cover rehab services. These and other data may portend a bleak future for some, but for those who can balance the delivery of quality care, meaningful outcomes and cost-effective services, the market is ready with open arms.

The most important positioning mechanism for the 1990s and beyond involves the development of strategic partnerships with corporate buyers. Health care organizations must move away from the traditional referral sources and begin to partner with those who pay the bill: third-party insurers, employers, and third-party administrators. Strategic partnering goes well beyond selective provider contracting, the keystone of most managed care entities. Critical elements of strategic partnering also include the following elements:

1. long-term contracts between employers and providers
2. agreed-upon outcome measures
3. shared risks and rewards
4. agreed upon treatment protocols
5. close communication between partners to measure quality, performance, and cost-effectiveness of various programs[70]

The reader is strongly encouraged to consider the Additional Resources list in Appendix 38–A and to explore the Stakeholder Organizations and Trade Groups listing in Appendix 38–B.

NOTES

1. Sun Tzu, *The Art of War*, trans. T. Cleary (Boston: Shambhala Publications, 1991), 88.
2. P. Ellwood, *Healthweek*, 6 Nov. 1989.
3. M. Schahner, United States Has the Only Health Care System Based on Employer Plans, *Business Insurance*, 22 Oct. 1990, 58–61.
4. R.L. Reece, *And Who Shall Care for the Sick?* (Minneapolis: Media Medicus, 1988).
5. R.J. Franke, Resolving the Health Care Paradox: A Need for National Debate, 1990 Robert D. Eilers Memorial Lecture, Wharton School, University of Pennsylvania, August, 1990.
6. B.R. Tresnowski, *Healthweek*, 19 Nov. 1990, 23.
7. J. Castro, Condition Critical, *Time* Magazine, 25 Nov. 1991.
8. Medical Portion of Comp Reaches 44%, *National Council on Compensation Insurance Digest* 3, no. 4 (1992).
9. Armageddon for Workers' Comp? *Risk and Insurance* 4, no. 1 (1993):18–21.
10. H. Berman, Taking Charge of Workers' Compensation Costs, *Employee Benefit News* October (1991):35–37.
11. Ibid.
12. Medical Portion.
13. R. Dorsey, Making Rehabilitation Work, *National Council on Compensation Insurance Digest* 4, no. 4 (1988):89–99.
14. W. Hager, Harnessing the Forces of Change, *National Council on Compensation Insurance Digest* 3, no. 2 (1992):1–19.
15. Ibid.
16. Armageddon.
17. D.S. Isokowe et al., Behind the Scenes of the Workers' Comp Crisis, *Risk Management* 37 (Nov. 1990):52–60.

18. Workers' Compensation Research Institute, *1993 Annual Report/Research Review*, 101 Main Street, Cambridge, Massachusetts.
19. L. Thornquist, Health Care Costs and Cost Containment in Minnesota's Workers' Compensation Program, *John Burton's Workers' Compensation Monitor* 3, no. 3 (1990):3–26.
20. Armageddon.
21. H.J. Stevens, Stress in California, *Risk Management* 39, no. 7 (1992):38–43.
22. Isokowe et al., Behind the Scenes, 52.
23. Hager, Harnessing the Forces of Change.
24. Ibid.
25. C.A. Chapman, The Michigan Experience—Workers' Comp Cost Containment, *The Self-Insurer* 7, no. 6 (1990):20–22.
26. Hager, Harnessing the Forces of Change.
27. C.A. North, Workers' Comp Pacesetters, *Risk and Insurance*, March (1992).
28. Massachusetts Overhauls Workers' Comp Laws—OK's Use of Managed Care Strategies, *Employee Benefit News* April (1992):20–21.
29. D.E. Galvin, Employer-Based Disability Management and Rehabilitation Initiatives: A Rehabilitation Research Review (Washington, D.C.: NARC National Institute of Handicapped Research, U.S. Department of Education, 1986).
30. National Council on Compensation Insurance, Insurance Industry Sets Strategies for Success, *NCCI Update* 1, no. 2 (1991).
31. G.S. Goldstein, Defining Managed Health Care, *HMO Magazine* 33, no. 2 (1992):453–454.
32. Hager, Harnessing the Forces of Change.
33. Scam City, *Philadelphia Magazine* March (1993).
34. K. Berney, Carve-Outs: The Next Generation, *Risk and Insurance* 3, no. 9 (1992).
35. Ask the Experts, *Employee Benefit News* February (1993):42.
36. Goldstein, Defining Managed Health Care.
37. Ibid.
38. D. Acquilina et al., Using Severity Data To Measure Quality, *Business and Health* June (1988):40–42.
39. H.R. Palmer, Definitions and Data, in *Assuring Quality in Medical Care: On the State of the Art*, ed. R. Greene (Cambridge, Mass.: Ballinger Publications, 1976).
40. R.M. Pirsig, *Zen and the Art of Motorcycle Maintenance: An Inquiry into Values* (New York: Morrow, 1974).
41. R.J. Franke, Resolving the Health Care Paradox, from the 1990 Robert D. Eiler's Memorial Lecture, available from John Nuveen and Co., 333 West Waker Drive, Chicago, Illinois, 60606.
42. P. Kondela-Cebulski and M. Wightman-Sojkowski, Ensuring Quality Patient Care: Applying the JCAH Standard, *Clinical Management in Physical Therapy* 7, no. 3 (1987):26–29.
43. J. Christofferson and C. Moynihan, Can Systems Measure Quality? *Computers in Healthcare* April (1988):24.
44. Berman, Taking Charge of Workers' Compensation Costs.
45. E.M. Robertson, Managed Health Care, *Healthweek* 2 December (1991):22.
46. Coalitions To Receive Data on Hospital Quality Measures, *Employee Benefit News* June (1991):23–27.
47. The Quality Imperative, *Business Week* Special Edition, Dec. 1991.

48. Ibid.

49. G.J. Batchelor and T.H. Esmond, Maintaining High Quality Patient Care While Controlling Costs, *Healthcare Financial Management* February (1989):21–22.

50. A.C. Entoven and R. Kronick, Universal Health Insurance through Incentives Reform, *JAMA* 265, no. 19 (1991).

51. National Academy of Sciences-Institute of Medicine, Committee to Advise the Public Health Services on Clinical Practice Guidelines, in *Clinical Practice Guidelines: Directions for New Program*, ed. M.J. Field and K.N. Lohr (Washington, D.C.: National Academy Press, 1990).

52. M.R. Chassin, Practice Guidelines: Best Hope for Quality Improvement in the 1990's, *Journal of Occupational Medicine* 32 (1990):1199–1206.

53. D. Strand, *Managed Care Week* 3, no. 12 (1993):2.

54. Massachusetts Overhauls Workers' Comp Laws; R. Simpson, Managed Workers' Compensation, Cost Containment and Reform Activity Report, *National Council on Compensation Insurance* 2, no. 4 (1992):5–6; S. Warren and S. Gerst, Workers' Compensation and Managed Care, *AAPPO Journal* 2, no. 1 (1992):11–17.

55. Simpson, Managed Workers' Compensation.

56. L. Madlem, Workers' Comp PPO's, Managed Care: Making the Most of the Tools at Hand, *Employee Benefit News*, April 1992, 22–25.

57. C. Sardinha, Experts: With Greater Options, More Services, PPO's Shouldn't Be Edged Out of Reform Plan, *Managed Care Law Outlook*, April 1993, 19; J. Burns, Jackson Hole Group Offers Health Care Reform Plan, *Business and Health*, Jan. 1992; G. Mann, Move to Integrated Managed Care, *Risk Management* 40, no. 1 (1993):41–45.

58. V. Spencer, ed., *Faulkner and Gray Healthcare Information Services, 1992 Medical Utilization Review Directory* (New York: Faulkner and Gray, 1992).

59. Ibid.

60. Ibid.

61. R.J. Calhoun, Workers' Compensation Payers Turn to Cost-Containment Strategies, *Occupational Health and Safety*, April 1990, 84–87.

62. Corporate Health Strategies Inc., *The Health Poll*, 276 Post Road West, Westport, Connecticut 06880.

63. Responding to the Workers' Compensation Crisis: Can Employers Manage and Control Costs? (New York: Tillinghast, a Towers Perrin Company, January 1991).

64. Galvin, Employer-Based Disability Management; L.E. Hood and J.D. Downs, *Return to Work: A Literature Review* (Topeka, Kans.: The Menninger Foundation, 1985); M.W. Eaton, Obstacles to the Vocational Rehabilitation of Individuals Receiving Workers' Compensation, *Journal of Rehabilitation* 45, no. 2 (1979):59–63.

65. S. Nahra, *Sell Right for Success* (Plymouth, Mich.: Benefit Buyer's Guide, 1991).

66. D.C. Coddington and K.D. Moore, *Market-Driven Strategies in Health Care* (San Francisco: Jossey-Bass Publishers, 1987).

67. Ibid.

68. A. Sheldon and S. Windham, Competitive Strategy for Health Care Organizations: Techniques for Strategic Action (Homewood, Ill.: Dow Jones-Irwin, 1984).

69. J.B. Quinn et al., Beyond Products: Services-Based Strategy, *Harvard Business Review*, March/April 1990, 58–68.

70. G.L. McManus and L. Pavia, The Ultimate Alliance, *Healthcare Forum Journal*, Sept./Oct. 1991, 27.

Appendix 38–A

Additional Resources

WORKERS COMPENSATION RESEARCH INSTITUTE PUBLICATIONS (WCRI):

- Workers Compensation in California: Administrative Inventory. Professor Peter S. Barth and Carol Telles. December 1992. WC-92-8.
- Workers Compensation in Wisconsin: Administrative Inventory. Duncan S. Ballantyne and Carol A. Telles. November 1992. WC-92-7.
- Workers Compensation in New York: Administrative Inventory. Duncan S. Ballantyne and Carol Telles. October 1992. WC-92-6.
- Workers Compensation in Georgia: Administrative Inventory. Duncan S. Ballantyne and Stacey M. Eccleston. Sept. 1992. WC-92-4.
- Workers Compensation in Pennsylvania: Administrative Inventory. Duncan S. Ballantyne and Carol A. Telles. Dec. 1991. WC-92-4.
- Workers Compensation in Minnesota: Administrative Inventory. Duncan S. Ballantyne, and Carol A. Telles. June 1991. WC-91-1.
- Workers Compensation in Maine: Administrative Inventory. Dr. Richard Victor, Duncan Ballantyne, and Stacey M. Eccleston. Dec. 1990. WC-90-5.
- Workers Compensation in Michigan: Administrative Inventory. Dr. H. Allan Hunt. Feb. 1990. WC-90-1.
- Workers Compensation in Washington: Administrative Inventory. Sara R. Pease. Nov. 1989. WC-89-3.
- Workers Compensation in Texas: Administrative Inventory. Professor Peter S. Barth, Dr. Richard Victor, and Stacey M. Eccleston. March 1989. WC-89-1.

- Workers Compensation in Connecticut: Administrative Inventory. Professor Peter S. Barth. Dec. 1987. WC-87-3.
- How Choice of Provider and Recessions Affect Medical Costs in Workers' Compensation. Dr. Richard Victor and Charles A. Fleischman. June 1990. WC-90-2.
- Medical Costs in Workers' Compensation: Trends and Interstate Comparisons. Dr. Leslie I. Boden and Charles A. Fleischman. Dec. 1989. WC-89-5.
- Medical Cost Containment in Workers' Compensation: Innovative Approaches. Dr. Richard Victor, Editor. Nov. 1987. WC-87-1.
- Medical Cost Containment in Workers' Compensation: A National Inventory. 1991–1992. Dr. Leslie I. Boden, Susan M. Johnson, and Dr. Joseph C.H. Smith. Feb. 1992. WC-92-1.
- Return to Work Incentives: Lessons for Policymakers from Economic Studies. Dr. John A. Gardner. June 1989. WC-89-2.

Appendix 38–B

Stakeholder Organizations and Trade Groups

National Association of Managed Care Physicians
Innsbrook Corporate Center
5040 Sadler Road, Suite 103
Glen Allen, Virginia 23060-6124
Tel: (800) 722-0376

Workers' Compensation Research Institute (WCRI)
101 Main Street
Cambridge, Massachusetts 02142
Tel: (617) 494-1240

Institute for International Research (IIR)
437 Madison Avenue, 23rd Floor
New York, NY 10022
Tel: (212) 826-1260 or
(800) 345-8016

Group Health Association of America (GHAA)
1129 20th Street, N.W., Suite 600
Washington, D.C. 20038
Tel: (202) 778-3236

American Managed Care and Review Association (AMCRA)
1227 25th Street, N.W., Suite 610
Washington, D.C. 20037
Tel: (202) 728-0506

American Association of Preferred Provider Organizations (AAPPO)
1101 Connecticut Avenue, N.W.
Washington, D.C. 20036

National Managed Health Care Conference (NMHCC)
1000 Winter Street, Suite 4000
Waltham, Massachusetts 02154
Tel: (617) 487-6709

American College of Medical Quality
9005 Congressional Court
Potomac, Maryland 20854

Health Insurance Association of America (HIAA)
1025 Connecticut Avenue, N.W.
Washington, D.C. 20036-3998

Self Insurance Institute of America (SIIA)
17300 Redhill Avenue
Irvine, California 92714

National Council on Compensation Insurance
One Penn Plaza
New York, New York 10119

Chapter 39

Topical Medical Issues in Workers' Compensation Systems

Gloria J. Gebhard

Workers' compensation medical costs comprise only 2 to 3 percent of the actual medical dollars spent in the United States.[1] However, the annual growth rate for workers' compensation medical costs is increasing at a greater percent than non-workers'-compensation medical costs. Analysis by the Workers' Compensation Research Institute (WCRI), an independent, not-for-profit research organization, showed that between 1985 and 1990, the annual growth rate for aggregate medical expenditures for workers' compensation was 15.2 percent while the growth rate for all payers in the United States was 9.7 percent (see Table 39–1).[2] Because of this high growth rate for both general and workers' compensation medical costs, federal and state governments have developed strategies and enacted legislation in efforts to control this growth.

"Cost containment" has become the operative term at state and national levels in all areas of health care expenditures. One of the means for reducing general health care costs on the state and national level is managed competition. Managed

Table 39–1 Workers' Compensation and Total U.S. Medical Expenditures

	Aggregate Medical Expenditures, Annual Growth Rate Percent	
Time Period	*All Payors*	*Workers' Compensation*
1970–1980	13.0	14.1
1980–1985	11.2	13.6
1985–1990	9.7	15.2
All years 1970–1990	11.7	14.2

Source: Reprinted from *Medical Cost Containment in Workers' Compensation: A National Inventory, 1992–1993,* by C.A. Telles, p. 94, with permission of The Workers' Compensation Research Institute, Cambridge, Massachusetts, © 1993.

competition encourages health care providers to form large competitive organizations to supply health care at reduced rates while improving quality of care and increasing patient satisfaction. Also being utilized are managed care and other strategies such as rationing of care, guaranteed coverage, outcomes management, treatment standards, and capitation. Several workers' compensation state agencies or jurisdictions and state legislatures are adopting these programs and modifying them to fit the special needs of workers' compensation and to reduce the medical costs.

WORKERS' COMPENSATION HAS SPECIAL NEEDS

Workers' compensation is not just a medical system but also a wage replacement system. The employer is responsible not only for the medical costs incurred due to the work injury but also for the wages lost due to the work injury. The actual amount of wage replacement varies from state to state, but in all cases the employer must pay a portion of the lost wages due to a work injury. These wage loss benefits are generally referred to as indemnity benefits.

In addition to indemnity and medical benefits, the employer may also be responsible for vocational rehabilitation, retraining, permanent partial disability, total disability, and death and dependency benefits. These benefits vary from state to state but all jurisdictions require some additional responsibilities to be paid by the employer.

Because of the required payment of indemnity and other mandated benefits, the medical environment differs in the general health and the workers' compensation arenas. In the workers' compensation arena, the health care provider must assume responsibility in the decision about when and under what circumstances an employee returns to work. Often decisions must be made about the level of activity upon return to work. Sometimes there is a disagreement between the employee, the employer, and the health care provider about the readiness of an injured employee to return to work and the specific level of activity within the employee's capacity upon return to work. Because of this difference of opinion, additional health care costs are often incurred. For example, a health care provider or employer may order a functional capacity evaluation to determine an objective measure of the employee's physical ability.

Litigation and disputes also complicate and add health care costs in the workers' compensation system. Some health care providers are seen as "pro-employee" or "pro-employer." These providers are perceived as giving more or less treatment or higher or lower disability ratings depending on the employee or employer point of view. A "pro-employee" health care provider may be thought of as giving more treatment than considered necessary by some other providers to support the disability claim and substantiate the "off work" slips for the employee. A "pro-employer" health care provider may be perceived as suggesting an unusually low disability rating in order to save the employer money on disability payments. In addition, the medical-legal costs in the form of independent medical exams or adverse examinations result in additional medical costs for workers' compensation.

Activity on the national health care scene may add an additional level of complexity to workers' compensation medical costs. One option is to incorporate workers' compensation medical care into the general health care system. This may concern only reimbursement levels but also may include the utilization of services. There is a question of who would have jurisdiction in cases of dispute, the state workers' compensation agency, the health care system, or someone else. When or if this merger occurs, workers' compensation jurisdictions, employers, and health care providers are still going to have to coordinate the medical and indemnity benefits in efforts to keep medical and indemnity costs under control.

CURRENT STRATEGIES TO CONTAIN WORKERS' COMPENSATION MEDICAL COSTS

Because of the close connection between medical and indemnity benefits in the workers' compensation arena, any medical cost containment measures must be balanced by appropriate return-to-work efforts. In addition, most jurisdictions recognize that it is not enough to control overall medical costs by regulating solely the cost per service through a medical fee schedule. They must also control the utilization of medical services. Therefore most states will have more than one program in effect. For example, Minnesota has a medical fee schedule regulating the maximum reimbursement per medical service, treatment guidelines to regulate the utilization of medical services, and medical administrative rules requiring the health care provider to cooperate with return-to-work efforts.

WCRI completed a national inventory of medical cost containment strategies for the 51 workers' compensation jurisdictions in the United States (see Table 39–2).[3] This table demonstrates some of the current strategies and combinations of strategies used by the jurisdictions as of September 1992. Since there are many variations of these strategies, an explanation follows for each strategy as defined by WCRI.

Choice of Provider

In 1992, 22 of the 51 workers' compensation jurisdictions initially limited the employee's choice of treating doctor, while 41 limited the employee's ability to change providers. The initial limitation of choice is usually by requiring the employee to see a doctor of the employer's choice. Limitations on change of provider include requiring the employee to obtain permission from the employer to make any change of doctor.

Table 39–2 Common Cost Containment Strategies in Workers' Compensation as of September, 1992

Jurisdiction	*Limited Initial Provider Choice*	*Limited Provider Change*	*Medical Fee Schedule*	*Hospital Payment Regulation*	*Utilization Review*	*Bill Review*
Alabama	X	X	X	X	!!	!!
Alaska		X	X			
Arizona[a]	X	X	X			
Arkansas	X	X	X		X	X
California[a]	X	X	X			
Colorado	X	X	X	X		
Connecticut		X		X		
Delaware						
District of Columbia		X			X	
Florida	X	X	X	X	X	X
Georgia	X	X	X	X		
Hawaii		X	X	X		
Idaho	X	X				
Illinois		X				
Indiana	X	X				
Iowa	X	X				
Kansas	X	X	!!	!!		!!
Kentucky			X	!!		
Louisiana		X	!!	!!	X	!!
Maine[a]	!!	X	X	X	!!	
Maryland			X	X		
Massachusetts		X	X	X	!!	
Michigan	X	X	X	X	X	X
Minnesota		X	X	X		
Mississippi		X	!!		!!	!!
Missouri	X	X				
Montana		X	X	X		
Nebraska		X	X	X		
Nevada[b]		X	X	X	X	X
New Hampshire			!!	!!	!!	
New Jersey	X	X		X	X	
New Mexico[a]	X	X	X	X	X	
New York			X	X		
North Carolina	X	X	X	X		X
North Dakota[b]		X	X	X	X	X
Ohio[b]			X			X
Oklahoma		X	X	X		
Oregon		X	X	X	!!	X
Pennsylvania	X	X				
Rhode Island			!!	X		
South Carolina	X	X	X	X		X

continues

Table 39–2 (continued)

Jurisdiction	*Limited Initial Provider Choice*	*Limited Provider Change*	*Medical Fee Schedule*	*Hospital Payment Regulation*	*Utilization Review*	*Bill Review*
South Dakota		X				
Tennessee	X	X			!!	!!
Texas		X	X	X	X	X
Utah	X	X	X		X	
Vermont	X					
Virginia	X	X				
Washington[b]			X	X	X	X
West Virginia[b]			X	X	X	X
Wisconsin		X	X	X		
Wyoming[b]		X	X	X	X	X
TOTAL (excludes !!)	22	41	32	28	15	13
Changes from 1991 to 1992	+1	+1	+5	+6	+1	+0

[a]Arizona and California divide initial provider choice between the employer and the employee. In New Mexico, the employer or insurer can control provider choice and change during the 60 days following the injury or after that period. In Maine, effective January 1, 1993, the employer has the right to select a health care provider for the employee for the initial 10 days of medical care.

[b]Exclusive state fund.

Note: !! = Being developed. The table does not reflect strategies that the states have authorized, but rather strategies that the states have implemented.

Source: Reprinted from *Medical Cost Containment in Workers' Compensation: A National Inventory, 1992–1993*, by C.A. Telles, pp. xi–xii, with permission of The Workers' Compensation Research Institute, Cambridge, Massachusetts, © 1993.

Medical Fee Schedules

In 1992, 32 of the 51 jurisdictions used some type of medical fee schedule to set a maximum reimbursement level. The schedules vary from jurisdiction to jurisdiction according to statutory mandates and administrative rules. The major differences in the fee schedules include:

- scope of services covered
- basis of maximum reimbursement level (e.g., workers' compensation charge data, regular health care charge data, other states' data)
- fee expressed as dollar amount or relative value unit

The *scope of services* varies tremendously in the medical fee schedules. Almost all the medical fee schedules use the Current Procedural Terminology (CPT) codes developed by the American Medical Association and the Health Care Financing Administration's Common Procedure Coding System (HCPCS) as a basis to establish descriptions of medical services or supplies provided. Most schedules include the basic evaluation and management, surgery, and physical medicine codes. Some jurisdictions allow chiropractors to use the same codes as medical doctors, while others have specially designated codes for chiropractic services.

The actual *level of maximum reimbursement* varies for each jurisdiction and is based on several different factors. Some jurisdictions confer with health care providers about charges for medical services and then set a maximum reimbursement rate. Others collect data on charges either from the general health care arena or from workers' compensation medical data and set a maximum reimbursement rate based on statutory mandates (e.g., 50th percentile, median, usual and customary). Either of these methods often is accompanied by public hearings before the workers' compensation director or commissioner or an administrative law judge, at which anyone can testify for or against the published or established maximum rate.

The two ways the maximum reimbursement level appears are as *the actual dollar amount per code* or as a *relative value unit per code*. The listing of the actual dollar amount is a relatively straightforward method of stating the maximum reimbursement rate. If a relative value unit per code is used, it must be accompanied by a conversion factor expressed in a dollar amount. To get the amount of the maximum reimbursement allowed, the conversion factor is multiplied by the individual relative value unit. For example, the relative value unit for an intermediate office visit may be 1. If the conversion factor is $10.00, the maximum reimbursement allowed will be $10.00.

Each individual medical service or CPT or HCPCS code has a specific relative value unit assigned. The relative value units are established using the "base unit" of 1. This is generally equal to an intermediate office visit. All other relative value units are based on the value relative to this service. A brief office visit may be worth only one-half or 0.5 units of the intermediate office visit. In this case, the maximum reimbursement allowed will be 0.5 × $10.00 (the conversion factor) or $5.00. Conversely, if a service is more valuable than an intermediate office visit, the relative value unit will be above 1.

An advantage of the relative value fee schedule is that updates are easier to accomplish. In updating a dollar amount schedule, each code must be individually adjusted. In updating the relative value schedule, only the conversion factor is adjusted.

The current relative value fee schedules in place include the California Relative Value Schedule and the McGraw-Hill Relative Value Schedule. These relative values are based on historical charge-based data. In other words, the relative values are based on previous charges by health care providers for medical services.

Hospital Charge Regulation

In 1992, 37 of the jurisdictions commonly used fee schedules, diagnosis-related groups (DRGs), or cost-to-charge discounts to reduce hospital reimbursements.

Hospital services have limited uniformity in coding services and supplies, making it difficult to set up a fee schedule for individual services. Some jurisdictions use the CPT and HCPCS codes as much as possible and apply the schedule to hospital services as well as independent providers and clinics. The actual maximum reimbursement rate may be the same as or a percentage of reimbursement rates to independent providers and clinics. For inpatient services, some jurisdictions have developed DRGs using either Medicare or workers' compensation data. The Washington Department of Labor and Industries has developed DRGs for several types of injuries common to workers' compensation. They use data from workers' compensation medical services to set the parameters of the DRGs and maximum reimbursement rate. Another strategy for limiting reimbursement for hospital services is to apply the charge-to-cost ratio. This strategy requires extensive information on profitability of an individual hospital. A small hospital with a low level of profitability will be reimbursed at a higher ratio than a hospital that has a high profit margin.

Utilization Review

Over 10 percent of the jurisdictions use this strategy as defined by the WCRI. A jurisdiction is considered to have utilization review if:

- the workers' compensation statute requires the payor to review claims
- the workers' compensation agency reviews all or designated medical claims
- an exclusive state fund regularly reviews claims
- the jurisdiction includes a managed care program

Most jurisdictions have some type of utilization review for medical costs and services. In 1992, only 15 jurisdictions met the above criteria for inclusion in the WCRI inventory. These programs are very broad in nature and often have few guidelines to assist the jurisdictions in limiting medical costs. The most common statutory limitation on medical costs is a variation on the theme that "all medical treatment must be reasonable and necessary." Since there are few standards for what is reasonable and necessary, jurisdictions must adjudicate several disputes in this area.

Bill Review

An increasing number of jurisdictions use this strategy. Bill review means that the statute requires the payor or the state agency to review all the medical bills. As with the utilization review, most jurisdictions have some type of bill review by the payors of workers' compensation medical bills, but in 1992 only 13 jurisdictions required this activity.

The jurisdictions are not limited to the above strategies, but these are the most common ones used at this time. Also, a jurisdiction may use a variation of a particular strategy but not meet the WCRI definition of the strategy and therefore not be included on the inventory list.

SO WHAT'S NEW?

Most of the "new" cost containment activities are variations of the current strategies discussed earlier plus strategies found in the general health care arena. Several jurisdictions have developed innovative programs combining the general health care methods of cost containment and the special needs of workers' compensation. New trends in medical cost-cutting activities for workers' compensation include:

- *medical fee schedules:* includes either implementing new or revising existing medical fee schedules through improved data collection and analysis of cost of services based on resources needed, not prior charge history
- *treatment standards/guidelines:* these are often developed by health care providers in cooperation with workers' compensation agencies to limit unnecessary medical services
- *peer review:* used to control utilization of services through a health care provider's professional organizations, state licensing, or registration board
- *managed care programs:* provides for the special needs of workers' compensation and may include utilization and bill review
- *24-hour care:* designed to combine general and workers' compensation health care

Medical Fee Schedules

As of September 1992, 27 states have a medical fee schedule that sets a maximum reimbursement fee for medical services delivered under the workers' compensation system. Several other jurisdictions are in the process of implementing or revising a fee schedule for workers' compensation. The schedules vary from jurisdiction to jurisdiction according to each state's statutory mandates and administrative rules. As indicated earlier, three of the primary differences among the different schedules are (1) scope of services covered, (2) basis of reimbursement, and (3) fee expressed as dollar value or relative value unit. Developments and modifications in these areas are part of what's new in workers' compensation.

Additions to the scope of services covered comprise new territory for the workers' compensation arena. Expansion of statutory authority and improved data collection have allowed an increased number and type of services to be covered. Expansion areas include medical/surgical procedures, laboratory/pathology services, anesthesiology, and dental, pharmacy, and hospital services. Improved data collection allows for the increases in coverage in the medical/surgical, laboratory/pathology, and anesthesiology services. Jurisdictions have a better idea of what current charging practices are and can set the reimbursement rate accordingly. Jurisdictions are also getting a better idea of what other payors are reimbursing providers. In most states, Medicare and Medicaid rates are considerably lower than workers' compensation rates. The WCRI is studying the potential impact of using Medicare rates in workers' compensation.[4]

Another area of expansion is in pharmacy reimbursement. Several states are requiring that generic equivalents of prescription drugs be substituted for brand name drugs. Some jurisdictions are also using the average wholesale price as a maximum reimbursement rate plus a dispensing fee.

Reimbursement limits on hospital services appear in many proposed and current medical fee schedules. One tactic used by jurisdictions includes requiring outpatient hospital services to be subject to the same schedule as independent providers and clinics. For inpatient services, some jurisdictions are developing their own or adopting the Medicare DRGs and setting a maximum reimbursement rate. Instead of setting up elaborate DRGs or cost-to-charge systems, some jurisdictions are merely setting a percentage discount for hospital services.

The newest type of relative value medical fee schedule is the Resource Based Relative Value Schedule (RBRVS) developed by Medicare and Harvard University.[5] The RBRVS is important because several jurisdictions are in the process of adopting or are considering adoption of this schedule. In most instances the conversion factor will be unique to workers' compensation and will not be the Medicare conversion factor. As of January 1, 1993, the conversion factor assigned by Medicare is 31.001.[6]

In the RBRVS system, each service is assigned a relative value based on the resources that go into the service. The components evaluated are (1) the work involved by the professional providing the service, (2) the overhead expenses necessary to provide the service, and (3) the amount of malpractice necessary to cover the health care provider.[7] Also factored in is the geographic practice cost adjuster, which controls for variations of costs in different parts of the country.

Medicare initially published the proposed RBRVS in 1991 and implemented it for reimbursement of Medicare medical services on January 1, 1992. The values were updated in 1992 and implemented on January 1, 1993.

One of the consequences of the RBRVS is to shift payments for health care services towards the evaluation and management codes and away from the high-technology and procedural services. In practical terms, this has resulted in signifi-

Table 39–3 Relative Value Units (RVUs) for Selected Physical Medicine Codes

HCPCS[a]	*Description*	*Total RVUs*
99213	Office/outpatient visit, est	1.00
29874	Knee arthroscopy/surgery	19.08
72148	Magnetic image, lumbar spine	15.64
97010	Hot or cold packs therapy	0.47
97012	Mechanical traction therapy	0.44
97014	Electric stimulation therapy	0.44
97022	Whirlpool therapy	0.43
97024	Diathermy treatment	0.42
97110	Therapeutic exercises 30 min	0.54
97112	Neuromuscular reeducation	0.51
97114	Functional activity therapy	0.43
97116	Gait training therapy	0.45
97118	Manual electric stimulation	0.53
97124	Massage therapy	0.43
97128	Ultrasound therapy	0.44
97220	Hydrotherapy	0.80
97500	Orthotics training	0.64
97520	Prosthetic training	0.73
97530	Kinetic therapy	0.74
97540	Training for daily living	0.87
97700	Training checkout	0.81
97720	Extremity testing	0.83
97752	Muscle testing with exercise	1.08

Note: The Medicare Conversion Factor in effect on January 1, 1993 is 31.001.
[a]All numeric CPT HCPCS Copyright 1991, American Medical Association.

Source: Reprinted from the *Federal Register.* Vol. 56, No. 227, Monday, November 25, 1991, 59502-59819.

cant reductions in reimbursements to high-technology services such as magnetic resonance imaging (MRIs) and surgical procedures such as laminotomies and arthroscopies. Table 39–3 lists some examples of relative values assigned by Medicare.[8]

To determine what Medicare reimburses for services, multiply the conversion factor times the assigned relative value unit. For example, to determine the maximum reimbursement for 30 minutes of therapeutic exercise, multiply the conversion factor, $31.001, times the assigned relative value, 0.54. This equals $16.74. The actual reimbursement rate is further modified slightly by the geographic adjustment factors.

Some jurisdictions are considering or are currently requiring reporting of health care services and supplies on a uniform billing form. Most jurisdictions are adopting or modifying the HCFA 1500 (for independent clinics) and UB-93 (for hospital services) uniform billing forms. These forms are currently required by Medicare for billing of all medical services and supplies.

Treatment Standards/Guidelines

State legislatures have mandated several workers' compensation jurisdictions to develop treatment standards or guidelines for common types of work-related injuries. Many payors of workers' compensation medical bills and even some workers' compensation state agency employees are nonmedical personnel and often do not know if one, 10, or 100 treatments are reasonable for a given condition. The treatment guidelines can assist them and health care providers to determine reasonableness of care.

Several states use the guidelines merely as mileposts or benchmarks. In other words, the provider can exceed the guidelines without specific permission from the payor. Fewer states use the guidelines as a "standard": the provider cannot exceed the limits without permission of the payor and still expect to be paid for the "excessive" treatment.

Almost all jurisdictions that have guidelines or standards have developed them in conjunction with the professional medical associations (e.g., state chapters of the American Medical Association, the American Chiropractic Association, the American Physical Therapy Association, and the American Occupational Therapy Association). Some organizations such as the American Chiropractic Association and the American Orthopedic Association are developing standards for all systems of health care, not just workers' compensation.

The Minnesota Treatment Standards are a good example of encouraging the health care provider to provide active treatment early in the case and limit prolonged passive treatments.[9] The example in Table 39–4 shows an overview of the standards for Category 1 Low Back Pain: Regional Low Back Pain (i.e., pain affecting lumbosacral region with or without referral to buttocks and/or leg above knee). Any forms of passive treatment should be reevaluated at 8 to 12 weeks post initiation of treatment and terminated at the end of 12 weeks. The standards require active treatment to begin after the first week of treatment. Active treatment may include education, exercise, posture training, and/or worksite modification. These standards are not meant to be "cookbook" treatment plans but to allow flexibility for individualized responses to treatment. The provider can exceed any of these guidelines with the permission of the payor.

Peer Review

Several state chapters of the American Physical Therapy Association and the American Chiropractic Association have active peer review groups for all payors, not just workers' compensation. These peer review groups will generally follow specific procedures in evaluating a file or claim and issue a written recommenda-

Table 39–4 Treatment Parameters for Regional Low Back Pain[a] (Category 1)

Diagnosis	*0 Weeks*	*8 Weeks*	*12 Weeks*
	Treatment: 3 Phases		
	Initial Nonsurgical Management		
		Reevaluation and Surgical Management	
			Chronic Management
Initial Diagnosis	*A. Initial Nonsurgical Treatment*	*B. Reevaluation*	*C. Chronic Management*
1. History and physical examination 2. Other tests must be indicated by history and physical. • Lab tests if infection, tumor, metabolic-endocrine, systemic musculoskeletal disorder, or medication problem suspected. • X-ray if trauma, or suspect infection, tumor, inflammatory disorder; if patient is over age 50; or if provider plans high-velocity, low-amp, or grade V mobilization. 3. Tests not indicated: electrodiagnostic, thermography, surface EMG, plethysmography, electronic X-ray analysis, or diagnostic ultrasound	1. Initial nonsurgical care must be tried. 2. Includes passive, active, and injection modalities 3. Passive treatment limited to 12 weeks 4. Passive treatment may include rest, heat or cooling mods, traction, adjustment/manipulation of joints and acupuncture. 5. Active treatment must be included after first week. May include: • education • exercise • posture, work method training • worksite modification 6. Injections indicated include facet joint, facet nerve, trigger points, and SI joints 7. Medication	1. If symptoms and objective findings continue and regular ADLs not resumed 2. Reeval as early as 8 weeks, no later than 12 after beginning care 3. Purpose: Is diagnosis correct, surgery indicated? 4. Procedures may include, in accordance with rules, medical imaging, diagnostic blocks and injections, psych eval, and consultation. 5. Surgery must meet parameters: • only lumbar arthrodesis with or without instrumentation • preauthorized • after surgery passive and active treatment allowed up to 8 weeks	1. If symptoms and objective findings continue after surgery, if surgery refused, or if not a candidate for surgery 2. No further passive care or therapeutic injections indicated. 3. Psych/personality test may be performed. 4. Must include plan to discontinue or limit scheduled medications 5. Treatment may include: • home-based exercise • health club (3 months)[b] • computer exercise program (6 weeks)[b] • work hardening or conditioning (6 weeks)[b] • chronic pain mgmt (up to 20 8-hour days)[b] • psych counseling (12 sessions in 12 weeks) • durable medical equipment[b]

Note: This diagram is a summary of the rule for this condition. Refer to rule text, Minnesota Rule part 5221.6200, subp. 1 through 11, 5221.6500 (surgery); 5221.6050, subp. 9 (preauthorization). Courtesy of Sandra Keogh, Medical Policy Analyst, RMA, Minnesota Department of Labor and Industry.

[a]Pain affecting lumbosacral region with or without referral to buttocks and/or leg above knee.

[b]Provider must request preauthorization.

tion. The goal of these groups is to instruct the provider in the correct or accepted method or length of treatment. Sometimes the instruction needs to be only in the area of proper documentation of treatment and outcomes. Most of the groups do not have authority to discipline a provider, but they can refer egregious cases to the appropriate state licensing or registration board for possible discipline.

Managed Care

In the general health care arena, several managed care programs have developed in the form of health maintenance organizations (HMOs), preferred provider organizations (PPOs), independent practice associations (IPAs), and several other forms of alphabet soup organizations. These organizations generally require the patient to prepay (usually in monthly installments), have limited provider networks, require the health care providers to accept utilization standards, and have negotiated reimbursement fees for health care providers.

A handful of state legislatures have mandated managed care programs or pilot programs for workers' compensation. These programs incorporate the special needs of workers' compensation and therefore do not always resemble the managed care model found in the regular health care system.

One of the special needs of workers' compensation incorporated in the Oregon managed care plan was a requirement of the plan to see an employee within 24 hours of the managed care organization's knowledge of the need for or a request for treatment.[10] This fills the special need in workers' compensation to have the employee seen promptly and evaluated for necessity of treatment and return-to-work status. The employee is not inappropriately kept off work because he or she cannot get an appointment to see the health care provider.

The managed care program may also include provisions for utilization and bill review. These programs would be similar to the cost containment programs found in the general health care managed care plan.

Twenty-Four-Hour Care

This program would integrate workers' compensation and general health care under one administration. It would not make any difference in the health care received if the person was injured at home or on the job. An individual would be covered by one health insurance plan "around the clock" or 24 hours a day. A few states have authorized legislation for an insurer to offer this plan, but none have developed and implemented one as of yet. The potential for cost savings would be in the area of reimbursement reductions and simplified administrative procedures.

However, if an employee is injured at work, the employer is still faced with payment of indemnity benefits for lost wages and all of the problems surrounding return-to-work issues. This means that there must still be a determination if the injury occurred at work, and this may require additional decisions to be made by the health care provider. The employer still needs to coordinate the medical benefits and return-to-work activities.

Another potential problem concerns copayments and deductibles. In the general health care arena, these are used to control utilization of services. None of the workers' compensation jurisdictions allows for the employee to pay a copay or deductible. This would have to be accommodated for under a 24-hour plan.

CONCLUSION

New medical cost containment activities cannot be understood without an examination of the current practices. Most of the new medical fee schedules, treatment standards/guidelines, and peer review, managed care, and 24-hour care plans are variations of current programs in either workers' compensation or the general health care arena. Most jurisdictions use a variety of programs to combat not only the cost per medical service but also the utilization of medical services.

NOTES

1. Minnesota Department of Labor and Industry, *Health Care Costs and Cost Containment in Minnesota Workers' Compensation: A Report to the Minnesota Legislature* (St. Paul, Minn., March, 1990); L. Baker and A. Krueger, *Twenty-Four Hour Coverage and Workers' Compensation Insurance* (Chevy Chase, Md.: Health Affairs, 1993), 271.
2. C.A. Telles, *Medical Cost Containment in Workers' Compensation: A National Inventory, 1992–1993* (Cambridge, Mass.: Workers' Compensation Research Institute, 1993), 94.
3. Ibid., xi–xv.
4. Workers' Compensation Research Institute, *Annual Report/Research Review 1993*, Cambridge, Mass., 32.
5. W. Hsiao, Resource-Based Relative Values: An Overview, *JAMA* 260, no. 16 (1988):2347–2353.
6. Federal Register, Part II, Department of Health and Human Services, 25 Nov. 1991, 59512.
7. Minnesota Department of Labor and Industry, *Medical Study Implementation Action Plan: A Report to the Legislature* (St. Paul, Minn., February, 1991), 15.
8. *Federal Register*, 59635-59765.
9. Minnesota Department of Labor and Industry, *Emergency Rules Relating to Workers' Compensation; Treatment Parameters* (St. Paul, Minn., 18 May 1993), Chapter 5221.6200.
10. Oregon Department of Insurance and Finance, *Managed Care Organizations*, Oregon Administrative Rules, Chapter 436, Division 15, 436-15-030(1)(b)(E).

Chapter 40

Quality Outcomes in Work Rehabilitation

Michelle Wiklund

The demand for proof of effectiveness of return-to-work programs has strongly influenced the need for providers in work rehabilitation to participate in outcomes measurement programs. Not only does the competitive marketplace demand such information, but with the introduction of managed care into the workers' compensation system, there is the need, now more than ever, to know the value of the care and treatment provided to the disabled worker.

With this ever-increasing demand for information and shrinking resources, the industrial rehabilitation industry faces a critical challenge in devising and implementing quality management and outcomes measurement programs to enhance the quality of care and life for disabled workers. Clifton states that therapists can continue to develop their practices only by assembling evidence of the efficacy of their treatments. They must show that what they do is effective at both rehabilitating the patient and holding down overall costs.[1]

There is general consensus among leaders in the rehabilitation field that outcomes data, that is, identifying overall performance in the care and treatment of injured workers, must be a critical component of any measurement system.[2] King states that having this "performance" data implies knowledge of services provided, time and charges for these services, to whom the information should be supplied, and how to improve upon these outcomes in order to survive in the marketplace.[3]

To meet the outcomes measurement challenge, a quality management system for describing the effects of clinical services and other factors for the work-injured client is needed. Industrial physical and occupational therapy should evolve into an observational study to:

- measure a client's functional status and quality of work and life over time, using terms readily understandable by the clients and the work rehabilitation industry
- document changes over time in a client's clinical condition as a result of therapy and return-to-work programs

- ensure that data are collected in a common format, using widely accepted protocols and statistical process control
- maintain data collected in a manner that allows for comparison of client outcomes
- incorporate standardized methods of measuring industrial rehabilitation effects on the health and quality of life of the work-injured client[4]

MOVEMENT TOWARD OUTCOMES MANAGEMENT

Ellwood's Outcome Theory

Paul M. Ellwood, MD, has long been considered the leader in development of outcomes research. According to Ellwood, "Rehabilitation will be required to prove its value, as will all other types of health care services. Outcomes management will be the means used to measure the quality of health care provided by an organization. Since outcomes management will focus on the quality of care, it is likely that rehabilitation medicine will become, as a result, more focused on those therapeutic interventions that produce improvements in function and well being."[5] Providers in industrial rehabilitation are prime candidates to demonstrate that what we do is effective and that our care, treatment, and return-to-work programs restore injured clients to safe, functional work status and improved quality of life.

According to Ellwood, providers in industrial rehabilitation must begin to conduct critical evaluation of each other and determine whether the care and treatment we provide really do have a favorable effect on client outcomes. Providers of work rehabilitation services must collect outcomes information in a fashion that is comparable across facilities. In addition to clinical outcomes normally collected in rehabilitation, such as range of motion, we must also collect information about:

- the patient's ability to perform activities of daily living
- the time elapsed before his/her return to work
- evaluation of psychological well-being
- determination of the presence/absence of pain
- whether the patient can resume normal social activities

In the realm of work rehabilitation, this is to say, "Does the patient feel better as the result of treatment, can he/she return safely to the worksite, and is he/she physically and functionally able to perform required tasks?"

Focus on Outcomes Measurement

The focus on outcomes measurement has recently intensified. Outcomes measurement is a systematic data collection procedure designed to assess client care results and evaluate program effectiveness and efficiency. Effectiveness is a measure of meeting established goals, including return to work. Efficiency reflects total cost for these services and the time utilized to achieve established goals.

The American Physical Therapy Association (APTA) has developed outcome assessment measures under its *Guidelines for Programs in Industrial Rehabilitation* to assess, at a minimum, client care results and program effectiveness and efficiency. These measures were developed under the direction of the APTA Industrial Rehabilitation Advisory Committee (IRAC) to classify levels of industrial rehabilitation in three ways: to accurately reflect contemporary practice, to standardize terminology, and to address the needs of clients, providers, regulators, and payors.[6] Exhibit 40–1 identifies those data items to be collected and reported as outcome measures.

The Commission for Accreditation of Rehabilitation Facilities (CARF) is another organization that has tried to define standards for quality and outcome measurement. In September 1993 the CARF Board of Trustees approved its new "1994 Subacute Medical Rehabilitation Program" (SMRP) to be published in 1994. Regarding program evaluation, this new standard emphasizes that an organization's program evaluation system should identify the results of services and the effect of the program on persons served. Additionally, program evaluation should measure outcomes following delivery of services. CARF-accredited facilities will be subject to survey under these standards beginning in July 1994. Questions related to these standards and the outcomes to be collected may be addressed to CARF, 101 Wilmot Road, Suite 500, Tucson, Arizona 85711.

A voluntary group known as Focus on Therapeutic Outcomes (FOTO) has been established to assemble a national database.[7] FOTO is a collaborative effort between major corporate rehabilitation providers who have formed a consortium to study outcomes. It is predicted that the database will report on over 200,000 patients a year and will compile physical and occupational data on outpatient, orthopedic rehabilitation. Each participant in the project will get not only benchmark data but also individual therapist data. To collect data, measurement tools such as the Oswestry Scale for Low Back Pain, the Lysolm Knee Scale, and the Neck Disability Index scale will be used. Overall, the data are also expected to help define treatment outcomes, justify costs, and influence health care reform.

The Agency for Health Care Policy Research (AHCPR) was created by Congress in 1989 to focus on medical effectiveness and outcomes research. The agency is funding the development of guidelines in a number of areas, and they have already reported on some. In essence, its approach emphasizes what can be done to develop enhanced outcomes through comprehensive review of the literature and consensus of experts.[8]

The Uniform Data System for Medical Rehabilitation (UDS) is an organization that has developed a functional dependence measure as well as standardized demographic, financial, and diagnostic data sets to collect data.[9] The organization is

Exhibit 40–1 Work Hardening Guidelines of the American Physical Therapy Association

WORK HARDENING GUIDELINES

WORK HARDENING GUIDELINES OUTCOME ASSESSMENT:
Definition: Outcome assessment is a systematic data collection procedure to assess, at a minimum, patient care results, program effectiveness, and efficiency. Effectiveness is a measure of meeting established program goals including return to work. Efficiency reflects total cost and time utilized to achieve established goals.

ELEMENTS:

1. Identify services provided as either Work Conditioning or Work Hardening.
2. Demographic data:
   (a) Age
   (b) Gender
   (c) Race and ethnicity
3. Occupational and injury data
   (a) Primary and secondary diagnoses
   (b) Work status prior to injury
   (c) Has the client received other treatment for this injury prior to entering the Work Conditioning or Work Hardening program? If so, identify all disciplines involved.
   (d) Date of injury
      - same injury
      - new injury
   (e) Date Work Conditioning or Work Hardening program was initiated
   (f) Is there a target job awaiting the client
   (g) Time off work
   (h) Length of the program
      - hours/days
      - days/weeks
      - total days
4. Discharge data
   (a) Total charges billed for the program
   (b) Program status (terminated or discharged) regarding return to work
      - same employer or different employer
      - previous job or different job
      - full time or part time
   (c) Client status at time of program termination/discharge
   (d) Referrals for additional services not available in the program
   (e) Payment source
      - worker's compensation board
      - private insurance

Source: Guidelines for Programs in Industrial Rehabilitation, American Physical Therapy Association.

headquartered at the State University of New York in Buffalo. UDS aggregates data and reports them to subscribing facilities.

The Impact of Managed Care

Managed care also plays a major role in promoting collection of outcomes data. The current generation of managed care entities has focused on the cost of services provided.[10] The next generation can be expected to focus both on cost and quality. Managed care organizations (MCOs) are tracking and utilizing outcomes data and are using this information both to recruit practitioners and in their marketing campaigns. The data are used to promote rehabilitation services that provide high-level care and that are cost-effective.

While it may be true that outcomes are difficult to assess in rehabilitation where so much depends on the response of the patient to the treatment, that difficulty will not excuse therapists in the future from the burden. Industrial rehabilitation providers will need to supply:

- information to be analyzed to determine the techniques of patient care that can yield the best results
- information that can be used to identify opportunities to improve care
- information that identifies the value of the care provided
- information that responds to the needs of the employer and those involved in the rehabilitation of the injured worker

Put simply, data must be supplied that correlate results obtained to the cost of care: that is, did the patient get better, and at what dollar amount?

Proof of clinical and industrial rehabilitation program effectiveness can be the greatest ally in securing and retaining MCO contracts. Program outcomes can demonstrate to the MCOs the effectiveness of a profession's, a practitioner's, or a facility's treatment, and other comparative data can be used to help MCOs with the issues of cost-effectiveness.

In many states, legislation has been passed to allow business and industry to provide workers' compensation health benefits supplied through a managed care organization. The MCOs are to list health care practitioners who are able to provide industrial rehabilitation services through aggressive management programs in the return-to-work process. The ultimate goals of these programs are anticipated to be the provision of cost-efficient, coordinated care; lower total number of workdays lost; lower workers' compensation incidences, rates, and costs; promotion of overall higher productivity; and maintenance of a functional healthful workforce.

QUALITY MANAGEMENT AND OUTCOMES

The Quality Management Concept

Quality management is a concept that has been used in industry for many years to ensure production of a high-quality product. Since Berwick first suggested that quality management techniques could be applied to the health care industry,[11] there has been a strong movement in medicine to embrace this concept. The industrial rehabilitation field is now overwhelmed with trying to prove its role and its effectiveness in the continuum of care in returning injured workers to safe, functional work status.

The purpose of quality management in industrial rehabilitation is simply to show whether the "product," that is, good patient care that results in positive outcomes, is being provided. It answers the question, "How do we know we are doing a good job in returning injured workers to their jobs?"

The challenge in work rehabilitation is the definition of "quality" outcome. Rehabilitation providers often define "quality" outcome in terms of technological sophistication or therapeutic efficacy. Does the therapy or treatment or return-to-work program (work conditioning/work hardening) accomplish expected goals? Purchasers, that is, employers, third-party payors, and managed care organizations, are likely to define "quality" outcome as reduction in lost workdays, cost-effectiveness, and appropriateness. Disabled workers are likely to define "quality" outcome by such things as access to the provider, ease of scheduling visits, and treatment with courtesy and respect. In other words, the "voice" of the customer defines perceptions of quality: is the client better, and can he or she return safely to the job site?

The development of quality management programs in work rehabilitation has been impeded for these reasons:

- Industrial rehabilitative care is highly diverse and provided in many settings.
- Clinical medical records, particularly ambulatory medical records, are not normally suited to "quality" review.
- Modeling freestanding work rehabilitation programs after hospital department programs has been unsatisfactory to date.
- Postinjury care is fragmented, and the treatment rendered may be outside the employer's or therapist's control.
- Effectiveness has been difficult to measure because clinics lose control once the client is discharged from their care.

Therefore protocols must be developed and implemented to successfully measure and improve upon care provided to disabled workers. Therapists will simply have to accept that more of their time will be taken up charting patient expectations, reviewing clinical procedures, and measuring outcomes.

The Quality Management Plan

The single most powerful strategy for industrial rehabilitation providers to meet the outcomes measurement challenge will be to implement a quality management plan. Such a plan can be the foundation for the outcomes measurement program. The plan allows for systematic review and evaluation of services provided. Client demographics, treatment provided, and return-to-work program progress can be researched and provided to referral sources and consumers.

The plan should identify the stated purpose of the program, goals and objectives, the method for collecting data, manner of organization and function, and evaluation process. Attention should be paid to the manner of collecting and reporting outcomes data. Exhibit 40–2 provides a sample format for an industrial rehabilitation clinic quality improvement plan.

Work rehabilitation facilities should examine effectiveness of care not from a mode of inspection but from a continuous examination of the work rehabilitation

---

Exhibit 40–2 Isernhagen Clinics, Inc. Quality Improvement Plan

Quality Management is a concept that has been utilized in industry for many years to ensure production of a high quality product. More recently, this philosophy has extended to the health care industry. The purpose of Quality Management in health care is simply to show whether the "product," that is, good patient care, is being provided. It answers the question, "Are we doing a good job?" We at Isernhagen Clinics, Inc. need to continually evaluate ourselves by asking, "How do we know we are doing a good job?"

Isernhagen Clinics, Inc., through its administration, medical, and clinical staff, is dedicated to the provision of quality care to all its patients. In order to ensure that quality care is provided, an ongoing Quality Management and Quality Outcomes Measurement program will be established under the auspices of the Isernhagen Clinics, Inc. Quality Improvement (QI) Plan.

This written QI plan is designed to evaluate, maintain, and improve the quality of patient care provided in Isernhagen Clinics, Inc. All services will be provided in a manner consistent with the principles of professional practice, and will reflect concern for the acceptability, accessibility, availability, and cost of services. This plan will be looked upon as an evolving document, and annually will be reappraised to ensure that its purpose, goals, objectives, and overall organization adequately reflect the needs of Isernhagen Clinics, Inc. and the customers to whom it responds.

I. Purpose
II. Goals/Objectives
III. Methods
IV. Authority for the Quality Improvement Program
V. Organization
VI. Function of the QI Committee
VII. Evaluation of the Plan

process. One successful method is the FOCUS-PDCA (Plan-Do-Check-Act) approach developed by the Hospital Corporation of America.[12] The premise behind this approach is that often the breakdown in a system (here, the work injury rehabilitation system) is the result of a process breakdown across departmental or communication guidelines.

Using the FOCUS-PDCA approach, one can attempt to resolve and improve upon outcomes in an area of service delivery in the following manner:

- Find a process to improve.
- Organize a team that is involved with the process.
- Clarify current knowledge of the process.
- Understand the sources of process variation.
- Select the process improvement.

The final result should be improvement in client outcomes.

An example of the FOCUS-PDCA approach in work rehabilitation is identified in Exhibit 40–3. In this situation, in which the identified goal is delivery of industrial rehabilitation services in a timely and cost-efficient manner, the final result in looking at improvement and outcomes should be:

- increased client satisfaction
- increased return-to-work rates
- reduced number of lost workdays
- reduced average medical claim costs
- decreased workers' compensation costs

Another method used in the collection of outcomes is the gathering of subjective data. Subjective assessments, used alone or in conjunction with quantifiable outcome measures, can provide a wealth of information.[13] While some practitioners simply interview clients to assess level of satisfaction, others have designed written surveys or other methods of obtaining feedback.

Exhibit 40–4 identifies an example of a written patient satisfaction survey developed by Isernhagen Clinics, Inc., to capture basic patient satisfaction levels about all areas of the clinic—front office, exam rooms, interaction with the therapist, reporting of patient visit, and billing office interface. In some facilities, clients may be surveyed more than once. At the Center for Physical Therapy and Exercise in Nashua, New Hampshire, clients are surveyed twice: the first survey allows for continual communication with the patient on current status and opportunities to make the experience better; the second is a written survey given to the patient on his or her last visit.

Exhibit 40–3 FOCUS-PDCA Approach

Find a process to improve
- What is the process?
  - Delivery of industrial rehabilitative services in a costly and timely manner

Organize a team that is involved with the process
- Who knows the process?
  - Isernhagen Clinics, Inc. Administration
  - Director of Clinical Performance
  - Employer Representative
  - Therapist
  - Identified Pertinent Representatives

Clarify current knowledge of the process
- What do we know about the actual steps in the process?
  - Flow chart or list the procedure steps required to accomplish the job

Understand the sources of process variation
- How does the performance of the process vary over time?
  - Identify causes for variation

Select the process improvement
- How should the process be changed?
  - Select method of improvement
  - Start the PDCA process
  - Change in procedure
  - Change in policy
  - Change in staff accessibility/availability

FINAL RESULTS: IMPROVEMENT IN OUTCOMES

- Increased patient satisfaction
- Increased return-to-work rates
- Reduced number of lost workdays
- Reduced average medical claim costs
- Decreased workers' compensation costs

Source: Adapted from *FOCUS-PDCA. Hospital Quality Improvement Process: Strategy for Improvement* with permission of the Hospital Corporation of America, Quality Resource Group, 1989, Nashville, Tennessee.

Development of Work Rehabilitation Algorithms

Patients, clinicians, employers, payors, and regulatory agencies are beginning to realize that unexplained variation in clinical practice has important implications for both the cost and the quality of care. As a result, many initiatives are now

Exhibit 40–4 Isernhagen Clinics, Inc. Quality Outcomes Measurement Client Satisfaction Survey

To serve you better, Isernhagen Clinics, Inc. would like to know how you feel about the care we provide. When thinking about the therapy you receive in our clinic, please indicate your degree of satisfaction by drawing a circle around the number that most closely reflects your feelings.

	Very Dissatisfied	Dissatisfied	No Opinion	Satisfied	Very Satisfied
1. Convenient office hours to schedule visits.	1	2	3	4	5
2. Your telephone calls are handled properly and courteously.	1	2	3	4	5
3. Your telephone calls are returned by the therapist in an acceptable length of time.	1	2	3	4	5
4. The length of time spent waiting to see the therapist is acceptable to you.	1	2	3	4	5
5. The clinic is clean and comfortable.	1	2	3	4	5
6. Your dignity and privacy are respected.	1	2	3	4	5
7. Thoroughness of examination or evaluation.	1	2	3	4	5
8. You are given enough time to talk to your therapist about problems and treatment.	1	2	3	4	5
9. Your therapist/exercise coordinator explains things in words you can understand.	1	2	3	4	5
10. You are given advice about ways to improve functional activity.	1	2	3	4	5
11. Your therapist shows a caring, friendly attitude.	1	2	3	4	5
12. The clinic staff shows a caring, friendly attitude.	1	2	3	4	5
13. Your therapist or a member of the staff tells you how and when you will receive information regarding your therapy visit.	1	2	3	4	5
14. Billing questions are answered in a friendly, helpful manner by the business office.	1	2	3	4	5
15. The availability of care when the clinic is not open is acceptable to you.	1	2	3	4	5

PLEASE SUGGEST ONE IDEA TO HELP US IMPROVE SERVICE TO OUR PATIENTS.

Source: Copyright © 1992, Isernhagen and Associates, Inc., Duluth, Minnesota.

underway to develop and disseminate guidelines for clinical practice. It is anticipated that such development will lead to decreased variation in the delivery of service, improved care, better outcomes, and lower costs.

At this time, many clinical guidelines for delivery of rehabilitative care are in an algorithmic format. A clinical algorithm identifies a step-by-step procedure for making decisions about the diagnosis and treatment of clinical problems. Algorithms have many advantages, including the fact that they are concise and can provide an excellent means for communicating the process of delivering care. When implemented effectively, clinical algorithms may both improve quality and decrease costs by directing clinicians toward standardized, optimal, cost-effective strategies of care.

One of the key problems in the development of clinical algorithms to provide foundations for collection of data and measurement of outcomes in work rehabilitation is that little concrete information is available on algorithms and outcomes measurement in work rehabilitation. According to Melvin, the problem is that there is so little valid outcomes research currently available that many algorithms will have to be based upon a consensus of experts who focus on a methodology that emphasizes outcomes and efficiency.[14]

One such project is the algorithm development and outcomes collection of data being undertaken and implemented by the National Back Injury Network. The algorithm of care for on-the-road truckers injured on the job and away from their primary physicians has been developed by two panels composed of national experts. One, a panel of physicians and spine surgeons headed by Dr. Arthur White, has defined medical and surgical evaluation, care, and treatment. The other, a physical therapy panel headed by Susan Isernhagen, has defined the rehabilitation algorithm for physical therapy treatment and return-to-work program protocols. Further information about these algorithms and the outcomes data being collected can be obtained by contacting the National Back Injury Network corporate office.[15]

Algorithms of care can be developed by panels of "experts" in your own clinical facilities and treatment centers. Project teams can be identified to develop clinical guidelines and algorithms of care that reflect the standard of care and practice in your community or region. An example of an algorithm developed for treatment of low back pain by staff in an independent practice site is included in Figure 40–1.

MOVING FORWARD INTO OUTCOMES MEASUREMENT

Data Support

Data collection is extremely important and crucial to the outcomes measurement process. Data can be collected using patient surveys, clinician surveys, em-

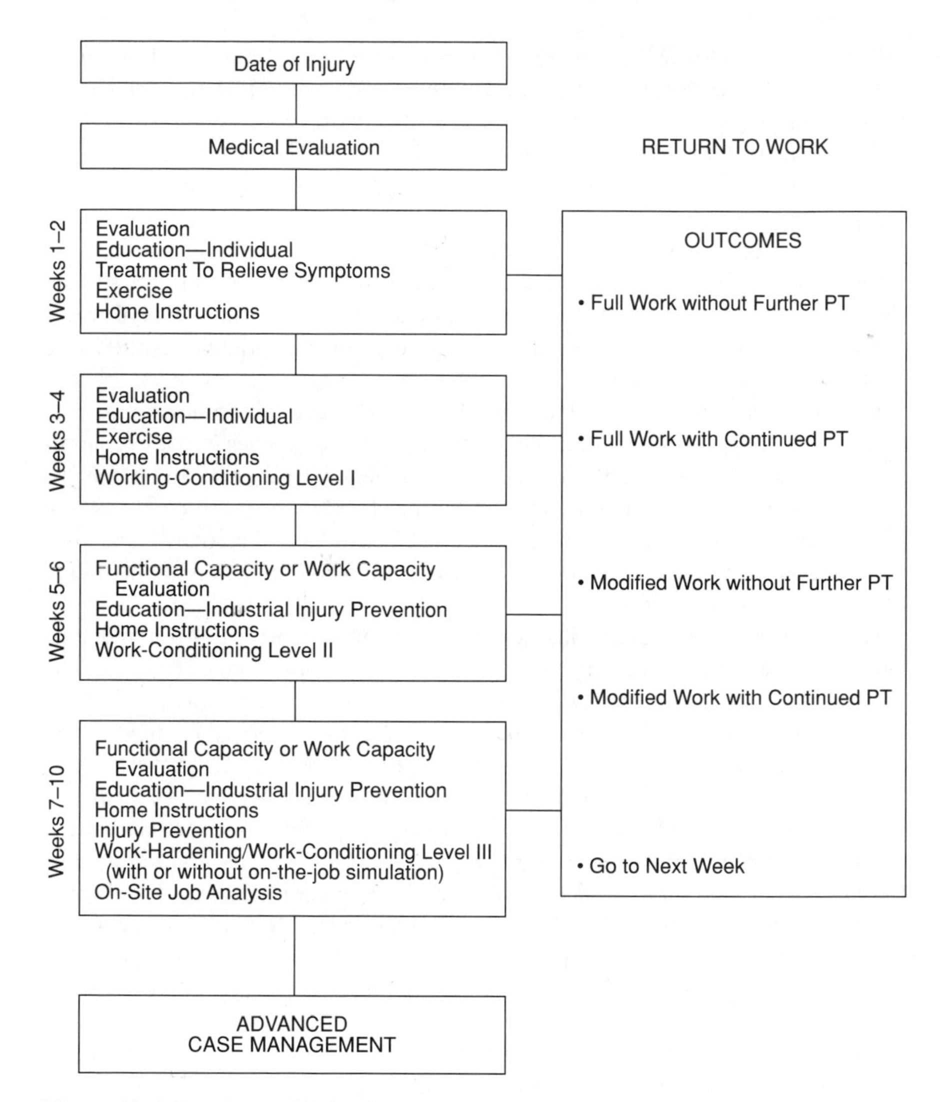

Figure 40–1 Isernhagen Clinics, Inc. Physical Therapy Algorithm (Nonsurgical) for Back Pain. *Source:* Copyright © 1992, Isernhagen and Associates, Inc., Duluth, Minnesota.

ployer surveys, chart reviews, incident reports, investigations of delays in service, and errors in service. Steffen emphasized the need to combine data on outcomes of medical care with patients' perceptions and goals of treatment.[16]

Data may be collected manually or with the assistance of computers. In general, most offices now have the capability to perform basic statistical measurement with statistical packages that can be purchased for minimal prices. There are also very sophisticated software packages available for purchase. Most hospital-based work rehabilitation programs have use of the hospital's mainframe computer to collect, store, and display outcomes data.

Capturing the Data

The purpose of outcomes monitoring and measurement is to provide a basis for effective quality improvement. In work rehabilitation, the purpose of outcomes monitoring and measurement is to gain insights for clinical and managerial decisions that will result in optimal outcomes for the work-injured client in the return-to-work process:

- improvement of the quality of care provided in industrial rehabilitation facilities
- a mechanism for integrating outcomes measurement with other management activities, such as risk management and marketing strategy
- a means by which to involve the entire work injury management team actively in the return-to-work process

The profession of physical therapy has begun to publish extensively in quality management. Ellingham of the University of Minnesota suggests beginning with a review of the APTA Standards of Practice, particularly Section VIII, on quality assurance.[17] Swanson suggests it is important to address these four outcome traits when structuring your data collection method:

- diagnosis
- therapy problem in terms of the inability to perform a function or task—that is, the disability
- client's desired outcome: the functions or tasks relevant to his or her situation
- payor's desired outcome: relationship of overall cost or value[18]

It has been said that quality is the perception of the customer. Abeln suggests questions such as these when collecting outcomes data:

- Are your clinic's treatment plans directed toward a rapid return to work for the patient?
- Are your clinic's treatment selections the most cost-effective?
- Is the client's documentation of pain written in terms that an examiner can quantify?
- Are your current industrial practices directed toward reducing the average cost per case for the customer?[19]

THE OUTCOMES MEASUREMENT REPORT

What Should We Be Looking For?

Clifton feels outcomes data will best serve those in rehabilitation if they are focused on a patient's ability to function rather on degrees of physical impairment or absence of pain. The reason for this is that function determines, for example, whether a work injured client can return to the job. And payors will respond to improvements in function. Further, payors respond more to *ability* than disability; getting workers to return to gainful employment is a key benchmark for reducing costs associated with workers' compensation.

Stockdell has coined the term *industrial model* versus *medical model* (for evaluations), and has her facility transitioning toward the focus on return to gainful employment. Her point is well taken: "If you can't justify your service by proving that the patient improved functionally, I strongly believe that more and more payors will start denying benefits."[20, p.166] She suggests revising evaluation forms to look at them more in terms of functional outcomes.

Examples of Outcomes Measurement Reports

Types of Information To Be Collected

Evaluating a program systematically will help the supplier of services to measure its clinical outcomes. Client demographics to be collected on every case might include gender, age, length of time from injury to treatment, time off work from injury to treatment, diagnosis, time in physical therapy treatment, time in work-conditioning/work-hardening programs, type of service, referral source, and overall costs associated with your program.

Additionally, various measurable criteria can identify the disposition of the work-injured client upon discharge from care:

1. Return-to-work level
   - same employer or different employer
   - previous job or different job
   - full time or part time
2. Return-to-work load
   - full time
   - part time
3. Perception of work capability
4. Job satisfaction level
5. Client use and understanding of physical abilities information
6. Employer use and understanding of physical abilities evaluation information

Sample Statewide Uniform Collection of Data

As an example, Minnesota managed care legislation was passed in 1992 allowing business and industry to provide workers' compensation benefits through a managed care organization. To respond to anticipated data requirements, a special group of therapists from Minnesota met early on to develop a functional demographic and outcomes report form. The goal was to establish criteria for measuring "effective rehabilitative care" and to establish a uniform process of functional data collection. Exhibit 40–5 identifies the form developed as a result of these task force meetings.

As of this writing, the MCOs are developing select networks of industrial medicine practitioners to provide health care services through aggressive management of the return-to-work process. The key will be the ability to provide such aggressive management *without* sacrificing quality. Therapists throughout the state are expected to respond to the credentialing requirements of these MCOs by specifically identifying the quality management and outcomes measurement programs put in place as the foundation for providing high-quality, efficient and effective work rehabilitation programs. Examples of questions that are being asked in the credentialing process include:

- Describe policies and procedures to
  a. return the client to functional work status
  b. communicate on a timely basis with injured workers, employers, and payors
  c. determine necessity of modalities used
- Do you use clinical practice parameters for services provided?
- Do you have a formal quality management program implemented in your facility?
- Do you assess patient satisfaction?
- How do you propose to resolve patient grievances?
- How do you propose to measure outcomes?
- What services set you apart from your peers in treating injured workers?
- What services do you offer to employers?
- What are the costs of services directly related to workers' compensation?
- What is the value of the services you provide?

If your industrial medicine facility has not yet been challenged to answer and respond to questions such as these, you will be shortly.

Examples of Outcomes Measurement in Private Practice

1. Isernhagen Work Systems reports preliminary research to determine the ef-

Exhibit 40–5 Functional Demographic and Outcomes Form

1. Service Provided:
2. File Number:
3. Discharge Date:
4. Demographic Data (optional): Age Gender M ( ) F ( ) Race and ethnicity
5. Employment Status at discharge:

Employment Status at discharge:	( ) FT	( ) SJSE	( ) DJSE
	( ) PT	( ) SJDE	( ) DJDE
Employment Status at 3 months?	( ) FT	( ) SJSE	( ) DJSE
	( ) PT	( ) SJDE	( ) DJDE
Employment Status at 6 months?	( ) FT	( ) SJSE	( ) DJSE
	( ) PT	( ) SJDE	( ) DJDE
Employment Status at 12 months?	( ) FT	( ) SJSE	( ) DJSE
	( ) PT	( ) SJDE	( ) DJDE

6. Prior to injury DOT Occupation Classification: ( ) FT ( ) PT
7. Labor Classification:

Labor Classification:	Sedentary	Light	Medium	Heavy	Very Heavy
at preinjury	( )	( )	( )	( )	( )
at program entry	( )	( )	( )	( )	( )
at time of discharge	( )	( )	( )	( )	( )

8. Occupational and injury data:
   Date of injury _______________.
   ( ) same injury
   ( ) new injury
   Date treatment was provided _______________.
   Date work-conditioning or work-hardening program was initiated: _______________
   Is there a target job awaiting the client? ( ) Yes ( ) No
   Lost time from job before onset/initial treatment (Days): _______________________
   Length of Program
   Number of visits _______________ Over number of weeks ______________
9. Has client hired an attorney? Yes ( ) No ( )
10. Does client have a Rehab Consultant? Yes ( ) No ( )
11. Diagnosis—ICD9 Codes Prime (   ) Secondary (   )
12. Discharge data:
    Total charges billed for the program: $_______________
    Payment source
    ( ) workers' compensation board
    ( ) private insurance
    ( ) other

fectiveness of its functional capacity evaluation. The study is used to improve the program and ensure that it is meeting the needs of its clients and referral sources regarding outcome. The review consists of a chart audit combined with a telephone interview. Program outcomes to be displayed include:

- general population demographics
- outcomes—clients who have returned to work
- outcomes—clients who have not returned to work
- client information/feedback

Exhibits 40–6 through 40–8 identify the data collection forms that are used in the outcomes measurement process. Exhibit 40–9 identifies a sample of the outcomes measurement results that have been reported to employers, rehabilitation consultants, physicians, and internal staff associated with our functional capacity evaluation programs.

Exhibit 40–6 Outcome Measurement Survey, Client Post-FCE Profile, Isernhagen Clinics, Inc.

AUDIT DATE: ________________
FCE DATE: ________________
QRC: ________________
Referring MD:________________ Co:____________ Occupation: ________________

I. INFORMATION
1. Patient name:______________________ Age: __________ Sex: M F
2. Primary diagnosis: ______________________ Code: ______________
3. Secondary diagnosis: ______________________ Code: ______________
4. Estimated time off work: ______________________ Date of injury: ___________
5. Resting pulse rate:______________________
6. BP: ________________
7. DOT CLASSIFICATION OF WORK 1 S 2 L 3 M 4 H 5 VH

II. FCE TEST SCORES

A. Weight capacity

Floor to waist lift	____	____	____	____
Waist to overhead lift	____	____	____	____
Horizontal lift	____	____	____	____
Push	____	____	____	____
Pull	____	____	____	____
Front carry	____	____	____	____
Right carry	____	____	____	____
Left carry	____	____	____	____
Right hand grip	____	____	____	____
Left hand grip	____	____	____	____

continues

Exhibit 40–6 (continued)

B. Flexibility/positional—category checked in FCE form: Use of following scale:
1 = Unable 2 = 1–5% 3 = 6–33% 4 = 34–66% 5 = 67–100%

Elevated work _____
Forward bent/sit _____
Forward bent/stand _____
Rotation/sit _____
Rotation/stand _____
Crawl _____
Kneel _____
Crouch _____
Repetitive squat _____
Sit tolerance _____
Stand tolerance _____
Walking _____
Stairs _____

C. Therapist Recommendations 1 2 3 4 5
1. Return to work—previous job, same
2. Return to work—previous job, modified
3. Return to work—same company, new job
4. Return to work—new company, new job
5. Cannot return to work—physical abilities do not match
6. Was work hardening/work conditioning recommended? Yes _____ No _____

Source: Copyright © 1994, Isernhagen and Associates, Inc., Duluth, Minnesota.

Exhibit 40–7 Isernhagen Clinics, Inc. Functional Capacity Evaluation Survey

PATIENT NAME: ________________________
Please indicate your response by checking (✔) the statement that you feel most closely describes your feelings:

IF YOU HAVE RETURNED TO WORK, PLEASE ANSWER 1–8 below:

1. Work Level:
 _____ Previous job—same
 _____ Previous job—modified
 _____ Same company, new job
 _____ New company, new job
2. Work Load:
 _____ Full time
 _____ Part time (4–8 hours/day)
 _____ Part time (1–3½ hours/day)

continues

Exhibit 40–7 (continued)

3. DOT Classification of Work
   _____ Sedentary
   _____ Light
   _____ Medium
   _____ Heavy
   _____ Very Heavy
4. Work Schedule
   _____ Five days a week or more
   _____ Less than five days a week
5. Job Satisfaction
   _____ Satisfied with current work level
   _____ Wish to work harder
   _____ Wish to not work as hard
6. Attitudes of People at Work
   _____ People at work welcomed me back
   _____ People were neutral
   _____ People made it difficult for me
   _____ Self employed, not applicable
7. Discomfort
   _____ I have no limiting discomfort
   _____ I have pain but work through it
   _____ Pain stops me from working hard
8. Capability Perception
   _____ I feel physically capable of doing all my work
   _____ I feel physically capable of doing more than half my work
   _____ I feel physically capable of doing less than half my work
   _____ I feel physically incapable of doing my workload

IF YOU HAVE NOT RETURNED TO WORK, PLEASE ANSWER 9–10 below:

9. I have not returned to work because:
   _____ I am in litigation
   _____ My old job is not available
   _____ A new job is not available
   _____ I'm not physically capable of working
   _____ Things seemed stalled—there is no progress in the return-to-work process
10. I am currently:
   _____ In the vocational counseling process
   _____ Not interested in working
   _____ In a work hardening program
   _____ In another form of treatment

WE WOULD LIKE TO ASK YOUR PERCEPTION OF HOW INFORMATION ABOUT YOUR FUNCTIONAL CAPACITY EVALUATION TEST WAS USED. PLEASE ANSWER 11–14 below:

continues

Exhibit 40–7 (continued)

11. The Functional Capacity Evaluation
    _____ Was used by others to understand how I could best work
    _____ Did not seem to be used by others
12. The Functional Capacity Evaluation information
    _____ Helped me understand my physical function
    _____ Confused me
13. Regarding education about future exercise
    _____ It encouraged me to want to get stronger
    _____ It discouraged me
    _____ It did not matter
    _____ I was not given education about future exercise
14. The Functional Capacity Evaluation information
    _____ Verified I was more capable than I thought I was
    _____ Verified I was less capable than I thought I was
    _____ Verified what I already know about my condition

Comments: Are there any comments you would like to make?

Thank you for participating in this survey.

Exhibit 40–8 Isernhagen Clinics, Inc. Survey of Validation of FCE Return-to-Work Status

If client has returned to work:

1. Are you aware of your abilities and limitations as indicated in the FCE evaluation?
   _____ Yes _____ No
2. Regarding return to work levels, when you returned to the worksite, were you working
   ____ at the level indicated by the FCE?
   ____ above the level indicated by the FCE?
   ____ below the level indicated by the FCE?
3. Regarding your current work status, are you currently working
   ____ at the level indicated by the FCE?
   ____ above the level indicated by the FCE?
   ____ below the level indicated by the FCE?
4. If your working level is different than that recommended by the FCE, why?
   ____ Supervision decision
   ____ My decision
   ____ Job change
   ____ Physical condition/fitness change
   ____ Other __________
5. Have you been off work again with a work related injury? _____ Yes _____ No
6. If the answer to # 5 is Yes, were you reinjured (same injury), or was this considered a new injury? _____ Same injury _____ New Injury

Exhibit 40–9 FCE Outcome Measurement Results, Isernhagen Clinics, Inc.

The staff of Isernhagen Work Systems reports research that quantifies the effectiveness of its Functional Capacity Evaluation (FCE) and indicates how it is used in the return-to-work system. Retrospective audits of consecutive clients who were referred for FCE were reviewed for outcome. These reviews consisted of medical record chart audits combined with telephone surveys six months post FCE. 119 clients were reviewed in this initial study; 89 clients were identified as having received no other form of treatment following the FCE, i.e., PT, Work Hardening or Work Conditioning. Time off work from date of injury to FCE was calculated to be 9.6 months. The Isernhagen FCE has a positive, objective approach that facilitates return to work through objective information, professional recommendation and dialogue with the worker. The following study relays information about these 89 clients, and demonstrates the power of the FCE as a tool in the return to work process.

FCE CLIENTS WHO RECEIVED *NO* OTHER SUBSEQUENT INTERVENTION

CLIENTS WHO RETURNED TO WORK
64 percent (57 of 89) of these clients returned to work based on the results and recommendations of the FCE, with no other intervention:
Once a client has been off work beyond six months and is considered "chronic," statistics indicate very low chances of returning to work, and then only with extensive and expensive rehabilitation.
Despite these preconceptions, these "chronic" patients *returned to work* based *only* on the functional capacity evaluation. Why?
This 5-hour, 2-part test is comprehensive and clearly describes functional abilities related to work. All parties can easily understand the specific findings. Analysis of the return-to-work process indicates that for return to work to be accomplished there must be agreement by at least 3 parties:

the doctor who must sign the release (which had not been signed for nearly a year)

the employer who had not been able to accept the employee back at work

but most importantly *the worker* who must make the decision that he or she feels capable of returning to work and who believes the FCE

Comment:
In view of such a strong work return based only on a functional test, one must first recognize the power of an effective test. Perhaps the basic skill of the therapist—to match functional capability with functional demands—actually has been overlooked.

If the client returned to work, the work level was indicated to be:

1. 32% Same company, previous job—same
2. 45% Same company, previous job—modified
3. 14% Same company, new job
4. 9% New company, new job

77 percent returned to their previous jobs, either at the same or a modified level. In total, 91 percent returned to work at the same place of employment. The combination of the FCE's clarity in defining modifications and the employer's willingness to modify the job creates a positive outcome.

continues

Exhibit 40–9 (continued)

Of those clients who returned to work, work load was indicated to be:

1. 65% Full time
2. 32% Part time (4–8 hours/day)
3. 3% Part time (1–3½ hours/day)

Two-thirds of clients who returned to work were full-time status. This is an excellent placement result considering duration of time off work.

Of those clients who returned to work, the job satisfaction level was indicated to be:

1. 68% Satisfied with work level
2. 14% Wish to work harder
3. 18% Wish to not work as hard

More than two-thirds of these clients were satisfied with their work level at six months post FCE, and several wished to work at a higher level. When returned to work at appropriate functional levels, clients can satisfactorily perform their tasks, which contributes to employee morale and work performance.

Client perceptions of attitudes at work were:

1. 55% People at work welcomed the client back
2. 15% People were neutral to the client
3. 19% People made it difficult for the client
4. 11% Self-employed, not applicable

Industry has made strides in overall attitudes regarding the return-to-work process. However, employers and peers at the job site can create an adversarial relationship. This is one aspect of work injury management that may need to be addressed.

Clients indicated discomfort level to be:

1. 14% Had no limiting discomfort
2. 58% Clients indicating having pain but worked through it
3. 28% Pain stopped the clients from working hard

Pain and *function* are not the same. The psychophysical approach, which focuses on pain rather than function, can limit return-to-work capability. The kinesiophysical approach allows for differentiation between pain and function and focuses on the *ability* of the person to functionally perform the job tasks required.

Information related to use of the FCE test results was identified for all 89 clients participating in the study. For those 57 clients who had returned to work:

Clients indicated use of the FCE by others to be:

1. 77% Used by others to understand how the client could best work
2. 23% Did not seem to be used by others

Three-fourths of clients who returned to work knew the FCE helped place them back to work. It is believed that the FCE was also used in the other cases as well, since the clients did return to work.

continues

Exhibit 40–9 (continued)

Client understanding of the test and test results was indicated to be:

1. 91% Helped the client understand physical function
2. 9% Confused the client

Regarding future exercise, the clients indicated that the FCE:

1. 91% Encouraged the client to want to get stronger
2. 9% Discouraged the client about exercise
3. 2% Didn't matter
4. 5% Offered no information regarding exercise

At the end of the FCE, clients know and understand their physical limitations and are encouraged to improve. Exercise, both musculoskeletal and aerobic, will provide a good foundation for early and successful return to work.

From the clients' perspective, the FCE

1. 14% Verified the client was more capable than the client thought
2. 18% Showed the client was less capable than the client thought
3. 68% Showed what the client already knew about his or her condition

This indicates that when professionally evaluated, the workers' true perception and the evaluators' match most of the time. However, it is the strength of the FCE, not workers' perceptions, that brings out clients' true abilities in an objective way to allow release to work.

CLIENTS WHO HAD NOT RETURNED TO WORK

36 percent (32 of 89) of clients who participated in the telephone survey had not returned to work at six months post FCE.

These clients indicated the reason for not returning to work was:

1. 12% Litigation
2. 19% The client's old job was not available
3. 3% A new job was not available
4. 62% He/she was not physically capable of working
5. 3% Things seemed stalled—no progress in the process

Information related to use of the FCE test results was identified for all 89 clients participating in the study. For those 32 clients who had not returned to work:

Clients indicated use of the FCE by others to be:

1. 70% Used by others to understand how the client could best work
2. 30% Did not seem to be used by others

The FCE appropriately identifies both those clients who will and those who cannot function safely to perform their work tasks.

Client understanding of the test and test results was indicated to be:

1. 90% Helped the client understand physical function
2. 10% Confused the client

continues

Exhibit 40–9 (continued)

Regarding future exercise, the clients indicated that the FCE:

1. 68% Encouraged the client to want to get stronger
2. 3% Discouraged the client about exercise
3. 29% Didn't matter
4. 0% Offered no information regarding exercise

In two-thirds of these cases, the clients felt we had encouraged them to want to exercise. Even though the client may not be returning to work, it is important to assist and encourage the worker in self-responsibility to reinforce the progress of return to functional status.

From the clients' perspectives, the FCE:

1. 10% Verified clients were more capable than they thought
2. 16% Verified clients were less capable than they thought
3. 74% Verified what the clients already knew about their condition

This indicates that when professionally evaluated, the workers' true perceptions and the evaluators' match most of the time. Again, it is the strength of the FCE that verifies ability, or in some cases, inability due to physical status, to perform required work tasks.

Courtesy of Isernhagen and Associates, Inc., Duluth, Minnesota.

Exhibit 40–10 Work-Hardening Outcome Study, Center for Functional Performance, Hammond, Louisiana

OUTCOME STUDY
CENTER FOR FUNCTIONAL PERFORMANCE
DECEMBER 17, 1992

In order to assess the effectiveness of the Center for Functional Performance, the charts of 20 clients referred for work hardening were reviewed.

I. Client Characteristics
  A. Diagnosis
    1. Various lumbar complaints (strain, pain, sprain): 11 clients
    2. HNP: 5 clients
    3. Miscellaneous
      a. CX Strain—1 client
      b. Hip FX—1 client
      c. Pectoralis strain—1 client
      d. Rotator cuff injury—1 client
  B. Referral Time
    1. The mean time from date of onset to referral was 26 weeks.
    2. The median time from date of onset to referral was 17 weeks.
  C. Physicians
    1. Referrals were taken from 11 different physicians.

continues

Exhibit 40–10 (continued)

D. Employers
   1. Clients were employed at 14 different companies. Ten of 20 were in occupations that required continuous lifting.

II. Program Characteristics/Outcomes
   A. Six of the 20 patients either failed to complete more than 2 weeks of work hardening or were discharged for noncompliance (5 of 6 were recommended to be released to return to work.)
      1. Three of 6 have returned to work.
      2. Two have had benefits terminated.
      3. One is still receiving compensation.
   B. Fourteen of 20 clients "successfully" completed their work hardening program. (All 14 were recommended to be released to some form of gainful employment.)
      1. Six of the 14 have returned to their former position.
      2. Five of the 14 are still receiving compensation. (One of these was released to return to work but has not; another has been awarded disability.)
      3. Three of the 14 have been released to return to work but thus far have failed to do so.
   C. Characteristics of clients still receiving benefits (five workers' compensation cases). *Note:* All five were recommended for a return to work.
      1. Length of benefits:
         a. Four of 5 averaged 54 weeks of compensation payments.
         b. One client has been on compensation for 157 weeks.
      2. Continuing Care:
         a. Three of 5 are currently *not* participating in any type of rehab.
         b. One has been released to return to work.
         c. One is receiving passive modalities.

CONCLUSIONS

Of those clients who successfully completed their Work Hardening program, 100 percent were recommended to return to work, as it had been objectively noted that they had the functional capacity to physically tolerate a return to gainful employment, and 85 percent were recommended to return to their former position. However, only 64 percent were released to do so.

The amount of time between date of onset and the initiation of work hardening was the primary indicator of poor treatment outcome—not necessarily diagnosis, occupation, or referral source.

Source: Copyright © 1992, Center for Functional Performance, Hammond, Louisiana.

2. The Center for Functional Performance in Hammond, Louisiana, has undertaken standardized quality programming and measurement of its success. A copy of a work-hardening outcome study performed by this facility is noted in Exhibit 40–10.

3. The Rehabilitation Institute at Northeast Georgia Medical Center recently completed a return to work outcome study in patients with work-related low back pain.[21] A retrospective analysis on 72 workers' compensation patients with low

back pain injuries who had participated in a comprehensive work-hardening program was performed to establish the correlation between evaluation findings and subsequent return-to-work outcomes. Overall, 42 percent of the patients had successfully returned to work, 33 percent had settled their claims but had not returned to work, and 25 percent had neither settled nor returned to work. An analysis of the predictive value of evaluation findings on return-to-work outcome was undertaken. Minimal time out of work (under three months) showed the strongest independent correlation (92 percent). Among the subjects with time out of work under three months, negative Waddell signs combined with either valid functional capacities or trunk strength measurements were found to provide the best predictors of positive outcome (80 percent and 73 percent, respectively).

RESPONSE TO THE OUTCOMES MEASUREMENT CHALLENGE

The focus on outcomes is still new. Specialists engaged in work injury evaluation and work rehabilitation should recognize the benefits of quality management and outcomes measurement. Organizations will move forward to ensure their position as high-quality providers of industrial rehabilitation services by using outcome information.

Why should providers in industrial rehabilitation be interested in quality outcomes and the measurement process? Reasons to measure outcomes include:

- *Improving the quality of care:* As providers in work rehabilitation, we are in the business of returning injured workers to the worksite. Anything less than a quality performance by ourselves does a disservice to our clients and the employers and physicians who refer them to us. By focusing on the best outcome possible, we help clients achieve the best functional status possible.
- *Saving workers' compensation costs:* Effectiveness of care and treatment can lead to returning the worker to the job site, thereby reducing the number of lost workdays and increasing productivity. Appropriate cost and effective treatments attract customers—employers, physicians, and clients alike.
- *Helping you to remain competitive:* The organization that monitors outcomes and proves effectiveness in the care and treatment of the injured worker will survive in a very competitive industry. Through outcomes measurement and the sharing of data with referral sources, you will strengthen your reputation as a quality provider of work rehabilitation services. The purpose is to become the provider that "customers" choose when they need industrial rehabilitation services.

CONCLUSION

To what extent will attempts to improve quality and measure outcomes benefit work rehabilitation? When compared with traditional quality assurance, quality improvement in industrial rehabilitation implies simply that not only quality of care matters but also the *value* of service—effectiveness, reliability, cost-effectiveness, and the best in "customer" service.

Those successful in industrial rehabilitation will accept a leadership role in addressing a critical issue that will allow survival in the competitive marketplace: to provide assurance of value in the rehabilitative care received by disabled workers.

All of the evidence in the literature to date leads toward the belief that a shift is occurring, indicating a trend from simply reducing cost to increasing value. Bringing together quality management and outcomes measurement will lead us to value.

Solid business and marketplace reasons make it practical for industrial medicine professionals to move forward into the realm of outcomes measurement and management. Quality is at a critical point in the evolution of the work injury prevention and work rehabilitation system. Employers, employees, physicians, and specialists in industrial rehabilitation are finally aligned and dedicated to providing clinically effective care in an efficient manner. All involved in the work injury continuum of care realize that "industrial health" for employees translates directly to financial health of their organization, whether they be the purchaser, provider, or recipient of the care.

Quality outcomes in work rehabilitation will be the key to success and survival into the next century.

NOTES

1. G. Wiley, A Scramble for Facts, *Rehab Management*, June/July 1993, 165–167.
2. W.C. Deaton, Outcomes Measurement/Outcomes Management: Implications for Marketing, *Physical Therapy Today*, Spring 1993, 61–62.
3. P. King, Outcome Analysis of Work Hardening Programs, *American Journal of Occupational Therapy* 47, no. 7 (1993):595–693.
4. P. Ellwood et al., The Future: Clinical Outcomes Management, in *Health Care Quality Management for the 21st Century*, ed. The American College of Physician Executives (Tampa, Fla.: Hillsboro Printing Company, 1991), 465–483.
5. J. Zamduloff, A Conversation with Paul Ellwood, M.D.: Rehabilitation, Outcomes Management and Health Care Reform, *Rehab Management*, June/July 1993, 27–32.
6. P. Helm Williams, Industrial Rehabilitation: Developing Guidelines, *PT Magazine*, March 1993, 65–68.
7. L. Goldstein, Defining Quality: New Challenges for Rehabilitation, *Rehabilitation Today*, June 1993, 27–30.
8. For further information, contact the Agency for Health Care Policy and Research, 2101 E. Jefferson Street, Suite 502, Rockville, Md. 20852, (301) 594-1357.
9. For further information, contact the Uniform Data System for Medical Rehabilitation, State University of New York at Buffalo, (716) 829-2076.
10. Deaton, Outcomes Measurement, 61–62.

11. D. Berwick, Sounding Board: Continuous Improvement as an Ideal in Health Care, *New England Journal of Medicine* 320, no. 1 (1989):53–56.
12. Hospital Corporation of America, *FOCUS-PDCA. Hospital Quality Improvement Process: Strategy for Improvement* (Nashville, Tenn.: Hospital Corporation of America, Quality Resource Group, 1989).
13. D. Berwick et al., *Curing Health Care: New Strategies for Quality Improvement* (San Francisco: Jossey-Bass Publishers, 1991).
14. G. Wiley, An Interview with John Melvin, M.D., *Rehab Management*, June/July 1993, 32–34.
15. Contact National Back Injury Network, University Research Park, 8701 Mallard Creek Road, Charlotte, N.C. 28262, (704) 548-0661.
16. G. Steffen, Quality Medical Care: A Definition, *JAMA* 260 (1988):56–61.
17. C. Ellingham, University of Minnesota Physical Therapy Program Materials, P.O. Box 388, Minneapolis, Minn. 55455.
18. Deaton, Outcomes Measurement.
19. S. Abeln, A Different Look at the Workers' Compensation Environment, in *Starting and Managing a Successful Physical Therapy Practice* (Washington, D.C.: American Physical Therapy Association Private Practice Section).
20. Wiley, Scramble for Facts.
21. J. Alday and F. Fearon, Evaluative Factor Correlation in Return to Work Outcomes in Patients with Work Related Low Back Pain (Paper presented at a conference, "Achieving and Measuring Positive Outcomes in Rehabilitation," sponsored by the Rehabilitation Institute of Northeast Georgia Medical Center, Atlanta, October 1993).

Chapter 41

Reasonable Accommodation and the Americans with Disabilities Act

Barbara L. Kornblau and Melanie T. Ellexson

For many years rehabilitation professionals adapted environments to enhance the participation of individuals with disabilities in work, rest, and leisure activities. Through the years, therapists and individuals with disabilities battled the frustrations of inadequate funding, unreasonable employers, inaccessible transportation, and inadequate public facilities.

On July 26, 1990, Congress passed the Americans with Disabilities Act (ADA; Title I, 29 CFR § 1620.1 et seq.; Title III, 28 CFR § 36.1 et seq.) opening new vistas for employment, public access, public accommodation, communication, and transportation. One theme runs through all five sections of Public Law 101-336—"reasonable accommodations."

The ADA seeks to integrate society by prohibiting the exclusion of individuals with physical or mental disabilities from jobs, services, activities, or benefits. Reasonable accommodation constitutes the pivotal means to that end. Under the ADA, employers must now take assertive action to provide reasonable accommodations to enable individuals with disabilities to perform a job. No longer must individuals with disabilities rely on charitable companies willing to make accommodations. Now employers and employees will find that making reasonable accommodations is the law, not the exception.

Congress included reasonable accommodations in the ADA as more than a symbolic gesture. The reasonable accommodation must be an effective accommodation. It must provide an opportunity for a person with a disability to achieve the same level of performance and to have the same benefits and privileges as a nondisabled person in a similar situation.[1]

THE INJURED WORKER AND THE ADA: BASIC DEFINITIONS

The obligation to make reasonable accommodations extends to new employees as well as existing or current employees and employees who acquire on-the-job injuries. Work injury management has always included making reasonable accommodations, although they have not necessarily been called reasonable accommodations. Therapists often recommend changes in the workplace, with the ultimate goal of return to work.

Unfortunately, before the ADA, making these accommodations for injured workers fit into the "optional" category. For example, before the ADA, employers were free to not make accommodations. State workers' compensation laws provided little guidance about return to work. State worker's compensation laws lacked requirements for accommodations. For example, an employer could refuse to return an employee to his job if the employer did not want to buy a cart to transport materials the worker could no longer carry because of his on-the-job injury. Now, under the ADA, many injured workers face more opportunities to actually return to work through reasonable accommodation.

To understand how reasonable accommodations affect the injured worker, one must first understand some basic definitions found in the ADA and how they relate to those with work injuries. The ADA presents the reader with a basic question: Under the ADA, will injured workers be considered disabled and thereby entitled to the ADA's protection, specifically reasonable accommodations?

The ADA defines *disability* broadly in three categories. The first definition states that the term includes "one who has a physical or mental impairment that substantially limits one or more major life activities" (29 CFR § 1630.2(g)(1)). A physical or mental impairment includes "any physiological disorder or condition, cosmetic disfigurement or anatomical loss affecting one or more of the following body systems: neurological, musculoskeletal, special sense organs, respiratory (including speech organs), cardiovascular, reproductive, digestive, genito-urinary, hemic and lymphatic, skin, and endocrine" (29 CFR § 1630.2(h)(1)). Mental or psychological disorders include mental retardation, organic brain syndrome, emotional or mental illness, and specific learning disabilities (29 CFR § 1630.2(h)(2)).

Substantially limited means an individual is "unable to perform a major life activity that the average person in the general population can perform" or "significantly restricted as to the condition, manner or duration under which an individual can perform a particular major life activity as compared to the condition, manner or duration under which the average person in the general population can perform that same major life activity" (29 CFR §1630.2(j)(1)(i), (ii)). Major life activities include many activities of daily living functions such as caring for oneself, performing manual tasks, talking, seeing, hearing, speaking, learning, and working (29 CFR § 1630.2(i)). Under the ADA, with respect to the major life activity of working, "substantially limited" means one is significantly limited in the ability to perform either a class of jobs or a broad range of jobs in various classes as compared to the average person having comparable training, skills, and abilities (29 CFR § 1630.2(j)).

Individuals with work injuries that result in "obvious" or "visible" disabilities, such as paraplegia, quadriplegia, head injuries, amputations, severe depression, serious hand injuries, and blindness would fall under the umbrella of the first definition of disability.

The second definition of *disability* under the ADA includes one who has a record of such a substantially limiting impairment (29 CFR § 1630.2(g)(2)). The injured worker who has a history of a five-year-old back injury, a previous disk surgery, or a knee injury would fall into this category. This category also includes persons with a history of a disabling condition such as an individual who has had cancer that is now in remission, or who has a past psychiatric condition.

If an injured worker's impairment is not substantially limiting, but the employer perceives that the claimant has an impairment that substantially limits his or her ability to work, the claimant will probably find protection under the ADA's third definition. The third definition of *disability* includes "being regarded as having an impairment." Under this definition, someone perceives another as having an impairment, based upon myths, misperceptions, fears, and stereotypes about disabilities (29 CFR § 1630.2(g)(3); 29 CFR §1630 Appendix to Part 1630—Interpretive Guidance on Title I of the Americans With Disabilities Act). According to the Equal Employment Opportunity Commission (EEOC), the third definition "applies to decisions based upon unsubstantiated concerns about productivity, safety, insurance, liability, attendance, cost of accommodations, accessibility, workers' compensation costs, or acceptance by co-workers and customers" (EEOC § II-11).

This definition would include, for example, the workers' compensation claimant, whom others perceive as disabled because of a severe facial burn. This category might also include the workers' compensation claimant whom the employer perceives as disabled because of the employer's fear that a prior soft tissue or back injury will reoccur, cause poor attendance, and increase workers' compensation costs. In reality, this claimant does not have an impairment. However, the employer perceives him or her as disabled, because of myths, stereotypes, misperceptions, and fears about persons who have had back injuries.

Not every worker injured on the job will find himself or herself considered disabled under the ADA. An award of workers' compensation or the assignment of a high impairment rating will not guarantee an individual protection by the ADA (EEOC § IX-2). Certainly workers with medical-only cases will not find protection under the ADA, nor will workers with nonchronic, nonpermanent, rapidly healing injuries such as a broken leg. However, in situations where an individual's work-related injuries fail to be "disabling" and/or "substantially limiting" under the ADA, the worker still may find protection under the ADA.

Seriously injured workers who find themselves unable to work for six months because of their injuries would have a "record" of a substantially limiting impairment. An employer who refused to hire or promote them because of their record, despite their full recovery from injuries, would probably violate the ADA (EEOC, § IX-3). Many individuals with work-related injuries absent a disabling condition or a record of a substantially disabling condition will probably find themselves considered disabled because their employers regard them as having a disability. Anyone who has worked in the area of work injury rehabilitation or management knows that employers and coworkers often perceive workers injured on the job as disabled. The supervisor who refuses to place a qualified, recovered laborer with a healed low-back sprain, claiming, "once on comp, always on comp," is basing his or her conclusion on misperception and is violating the ADA.

The ADA lists specific exclusions to the definition of *disabled*. The exclusions from ADA protection include transvestites, homosexuals, pedophiles, exhibitionists, voyeurs, gender disorders not caused by physical impairments, and other sexual behavior disorders (29 CFR §1630.3(d)(1)). Also excluded are individuals who partake in the illegal use of drugs, compulsive gamblers, kleptomaniacs, pyromaniacs, and alcoholics whose alcohol use prevents them from performing their jobs (29 CFR §1630.3(d)(2),(3)). The ADA protects former illegal drug users as disabled as long as they are rehabilitated. Current drug users do not find any protection under the ADA.

The ADA only protects the "qualified individual with a disability." According to the ADA regulations, a "qualified individual with a disability" is an individual with a disability who satisfies the requisite skill, experience, education, and other job-related requirements of the employment position that that individual holds, or desires, and who, with or without reasonable accommodations, can perform the essential functions of the position that he or she desires to hold (29 CFR § 1630.2(m)). The concept of a "qualified individual with a disability" gives rise to the employer's responsibility to make a reasonable accommodation. Where the worker falls short of meeting the qualified individual with a disability standard, the employer has no duty to make a reasonable accommodation.

DEFINING REASONABLE ACCOMMODATION UNDER THE ADA

Title I of the ADA defines *reasonable accommodations* as any change in the work environment or the way work is customarily performed that enables an individual with a disability to enjoy equal employment opportunity (29 CFR §1630.2(o)). However, instead of pronouncing a definition limited strictly to the work itself, Title I reminds the reader to view this section in concert with Title III of the ADA. Title III requires places of public accommodation to take the necessary steps to ensure that no individual with a disability is excluded, denied services, segregated, or otherwise treated differently because of the absence of auxiliary aids or services (28 CFR § 36.1 et. seq.). Reasonable accommodations for places of public accommodations under Title III include auxiliary aids and services as well as modification in policies, practices, or procedures, removal of barriers, and alternative forms of service delivery (28 CFR § 36.1 et. seq.).

This chapter focuses on reasonable accommodation under Title I as it relates to employment. However, many of the conditions and accommodation examples given may also apply to other ADA provisions. Under the employment provisions of the ADA, employers must consider reasonable accommodation throughout all employment-related functions such as the application and hiring process, performance of essential job functions, and determination of benefits and privileges of employment. Employers must always consider reasonable accommodation requests and thoroughly investigate possible solutions to employment obstacles. These employer's responsibilities continue throughout the course of employment, beyond the time of hire.

The employee also bears some responsibility for reasonable accommodation. He or she must reveal the need for an accommodation and must assist in the process of finding a workable solution to the work-related problem. An employer need not accommodate an individual with a disability unless the individual requests an accommodation. Therapists should keep this in mind so they can prepare their clients to ask for accommodations. If the clients do not ask for the accommodation, the employer does not have to make the accommodation.

The ADA's definition of *reasonable accommodations* includes modifications or adjustments to the job application process to enable a qualified applicant with a disability to participate in the process and have a chance to acquire the position. For example, a reasonable accommodation might include providing a reader for an individual with a learning disability such as dyslexia or an interpreter for an individual who is deaf. As a reasonable accommodation, an employer could provide a job application printed in large format, or a recorded job application allowing for verbal response to application questions.

Modifications or adjustments to the work environment, or to the manner or circumstances under which the job is customarily performed, is the second category of reasonable accommodations under the ADA. Employers must make these modifications or adjustments upon request if they enable a qualified individual with a disability to perform the essential functions of the job and if they are reasonable. Therapists will find the possibilities under this category limited only by the creativity of those involved in the accommodation process. In this category, examples of reasonable accommodations include:

- allowing a person to work at home programming or entering data;
- providing a person with cerebral palsy a headset to allow for note taking while talking on the phone
- providing a clerk who uses a wheel chair with lateral files for easier access
- allowing a person who requires two-hour rest periods during the day to

change his or her schedule so that person may take breaks and still work a full day

- providing a pushcart for a mail carrier who has sustained a back injury, enabling him to keep his or her route[2]

In looking at reasonable accommodation in the work context, therapists must focus on the work that needs to be done and how a particular person could do the task.

The ADA's third category of reasonable accommodations includes modifications or adjustments to allow an employee with a disability to enjoy the same benefits and privileges of employment as his or her coworkers (EEOC § III-4). Benefits and privileges need not be identical but must be equal. For example, suppose an employer provides a break room with food and beverage machines on the second floor of a building that lacks an elevator. The employer might reasonably accommodate an employee with a disability by providing food, beverage machines, and space for the employee to eat with coworkers on the first floor. The facility does not have to be exactly the same but must allow for the same activity without segregating the person with a disability from coworkers (EEOC § III-4). If an employer sponsored a company bowling team, the employer would need to make certain the bowling alley was accessible. However, like other employees, the employee would be responsible for supplying any personal equipment necessary to allow him or her to participate in the sport, such as orthopedic bowling shoes.

Insurance, including health, life, and pension, is a consideration in reasonable accommodation. Insurance coverage must be equal for all employees and allow equal access. Employers may not provide different terms, conditions, or costs to employees. Employers may not refuse to hire an individual with a disability because a current insurance plan will not cover the individual or because cost of the plan will increase. Plans may continue to include preexisting condition clauses as long as they apply across the board to all employees. "The plan must limit its exclusion to the pre-existing condition, and may not completely deny coverage for an illness or injury unrelated to the condition."[3, p.34] The Equal Employment Opportunity Commission and case law will provide more guidance in this area in the future.

The employer's obligation to make reasonable accommodations extends only to accommodate physical or mental conditions resulting from a disability (EEOC § III-4). It does not apply to accommodations a person may request for another reason. Congress included reasonable accommodations in the ADA as a means to reduce barriers to employment. For this reason, employers may require proof of the need for a reasonable accommodation, and they may evaluate the specific needs of an employee to determine acceptable accommodation solutions (EEOC § III-6).

Although it is the employer's obligation to provide reasonable accommodations, in some instances the individual with a disability may provide his or her own reasonable accommodation. For example, an individual who uses a mouth stick for data entry may prefer to provide his or her own mouth stick since he or she already has one available. In certain instances, where a particular piece of equipment proves to be an undue hardship for the employer to provide, the individual may be able to obtain and provide the device. Organizations such as the Lions Clubs will often purchase equipment to enlarge print or otherwise assist an individual who is blind or has low vision if the equipment facilitates employment.

There are other alternatives if the high cost of an accommodation presents an impediment to the employer. The employer should give the employee an opportunity to pay the difference between the amount the employer can afford without creating an undue hardship, and the cost of the accommodation. The employer cannot require the person to provide his or her own accommodation. However, certain situations may lend themselves to this solution.

WHAT IS INCLUDED IN REASONABLE ACCOMMODATION?

The ADA provides employers with guidance as to which areas to consider when making reasonable accommodation. First, the ADA reminds employers that making facilities accessible is a reasonable accommodation. This often includes widening doorways, installing automatic door openers, cutting curbs into existing walkways, ramping stairs, providing adequate parking for employees with disabilities, and even installing elevators.

Reasonable accommodations under the ADA also include job restructuring. Job restructuring involves changing how or when a worker performs an essential function. Job restructuring may include job-sharing plans or reassigning work tasks. The ADA also cautions employers to consider part-time or modified work schedules when evaluating accommodation needs. For example, establishing flexible work hours to allow an individual requiring special transportation the ability to work within those transportation schedules would be a reasonable accommodation. Developing part-time positions may also be another way of accommodating other individuals with disabilities.

Employers and employees must understand that reasonable accommodation does not mean a lowering of quality or quantity standard. The ADA specifically excludes accommodations that require the employer to lower production rates. An accommodation that effectuates a lower production rate falls outside the definition of *reasonable* accommodation. Therefore, the employer need not make the accommodation. According to the ADA, quality and production can and must be maintained even if this means the employer cannot make an accommodation.

Reasonable accommodation may include transferring or reassigning the individual with a disability to a vacant position within the company. If an individual is no longer able to perform the essential functions of his or her job but can successfully complete the required tasks of another position, and if that other position has or will have a vacancy in a reasonable period of time, the employer should consider the transfer. Transferring or reassigning an individual implies the individual need not be subject to an interview and other formalities required for a new hire.

When the employer transfers or reassigns an individual to a vacant position, the employer need not maintain the previous level of salary if the new position pays less. Conversely, if the new position pays more, the employer must compensate the individual at the new or higher rate of pay (EEOC § III-10(5)).

Many employers have instituted light or restricted duty programs to allow workers with medical restrictions to return to work, thereby managing their workers' compensation costs. Sometimes the employer allows the employee to participate in this program for only a limited time period.

Employers need not create light duty positions unless the "heavy duty" tasks are marginal functions that the employer can reassign to another employee through job restructuring, a reasonable accommodation. Often, the light duty positions are totally different from the worker's regular job. The ADA does not require an employer to create this kind of light or restricted duty job. However, if the employer has a vacant light or restricted duty position, it should consider reassigning the individual to the open position as a reasonable accommodation. If the employer creates a temporary position, the reassignment to that position need only be temporary. The ADA does not require that the employer make a temporary position permanent.

The ADA does not obligate employers to create light/restricted duty jobs as a means of temporary accommodation for an individual with only a temporary disability. The ADA does not cover individuals with a temporary disability. A person who breaks his or her leg and whom the doctors expect to recover in six to eight weeks has a temporary disability. Should the person ultimately sustain permanent weakness in the leg, which affects job performance by limiting climbing or walking long distances, for example, the employer should consider accommodation.

Transferring or reassigning individuals who develop long-term or permanent disability will affect workers' compensation placements. Employers must afford the same consideration for reasonable accommodation under the ADA as they would to any other similarly situated person with a non-work-related disability.

Another key component of reasonable accommodation includes the acquisition or modification of equipment or devices to facilitate performance of essential work functions. This may include, for example, devising a jig that allows an individual with one hand to complete a work task, or purchasing a cart to transport items for an individual with a back injury and lifting restrictions.

Readers with experience in making workplace accommodations know there are often several ways to accommodate an individual with a disability to allow for work performance. Employers may face a range of devices or equipment from the one-hundred-dollar model to the several-thousand-dollar computerized version that accomplishes the same task for a lot more money. The equipment or device need not be the best or most sophisticated or state of the art; however, it must work. If the employer finds himself facing a choice between a hydraulic lift and a ramp, the employer may choose the less expensive ramp, providing the ramp allows the individual the necessary access.

In addition to equipment, the ADA requires employers to make appropriate adjustments or modifications of examinations, training materials, policies, and procedures as a reasonable accommodation. The employer should evaluate readers, interpreters, alternative testing formats, and policy or procedural allowances for variations in the way employees perform tasks. For example, if a grocery store has a policy that all cashiers must stand while working, the employer should look at changing the policy to allow a worker with a foot neuroma to sit while working.

Requirements for certain adjustments to meet reasonable accommodation go beyond those required by the ADA. For instance, supported employment programs that utilize job coaches or require restructuring of jobs beyond what may be necessary under ADA are encouraged by EEOC and in no way violate the ADA (EEOC § III-4). While employers need not provide job coaches, many employers choose to work with employees who use job coaches through programs such as the state vocational rehabilitation programs.

THE REASONABLE ACCOMMODATION PROCESS

Through its legislative mandate, the ADA stresses two premises basic to the reasonable accommodation process: communication and flexibility. To make reasonable accommodations, the employer must accept that "we've always done it that way" no longer justifies inaction when it comes to making accommodations for individuals with disabilities. The concept of reasonable accommodations demands a willingness to try new methods to accomplish the same task. Employers must not make accommodation decisions in a vacuum. To fulfill their obligation to make reasonable accommodations, they must participate in discussions with the employee or prospective employee needing or requesting the accommodation. The ADA seeks to avoid situations where the employer makes inappropriate accommodations for an employee or no accommodations at all because of its ignorance about accommodations.

The process of identifying appropriate reasonable accommodations involves an informal, interactive dialogue between the qualified individual with a disability who is in need of accommodations and the employer. The process should identify the precise limitations resulting from the disability and potential reasonable accommodations that could overcome the limitation (EEOC § III-8). When an individual requests an accommodation to assist him or her in the performance of a job, the employer must use a problem-solving approach to create or develop appropriate reasonable accommodations.

The first step in the reasonable accommodation process requires the employer to analyze the particular job involved to determine the job's purpose and essential functions. Next, the employer should consult with the individual with a disability to determine the specific job-related limitations imposed by the person's disability and how the individual could overcome the limitations. For example, if an employee has a back injury, rather than assume the employee cannot perform, the employer should discuss with the employee the specific limitations he or she might have as an information operator and strategies for overcoming those limitations.

During the next stage in the process, the employer, in conjunction with the employee, should identify potential accommodations and evaluate the effectiveness each would have in enabling the person to perform the job's essential functions. Suppose an employee has a carrying restriction resulting from a back injury. The employee and the employer can identify specific possible accommodations such as a roller conveyor or a cart. At this stage in the process, the employer and employee must also look at whether the items on the list of potential accommodations are effective accommodations.

As part of the process, the employer must consider the preference of the individual in need of accommodations; however, the employer's obligation is to select and implement the accommodation that is most appropriate for both the employee and employer. The employer need not choose the best accommodation, only an accommodation that works. For example, employers may find themselves facing a $10 accommodation solution and a $10,000 computerized version of the same accommodation solution. As long as both accommodations work, the employer need only provide the $10 accommodation.

The employee does have the right to refuse an accommodation. However, if the employee refuses an accommodation, he or she runs the risk of being no longer qualified for the job (EEOC § III-8). For example, Jim, an employee of Gold Bar Steel Co., developed a mobility problem due to progressive arthritis. As an essential function of his job, Jim walked one-half mile, four times a day, delivering work orders. His supervisor noticed that the trips took Jim longer and longer to complete. This curbed Jim's other work, making him less efficient and unable to complete all of his assignments. Jim's supervisor offered him the use of an electric cart to make these trips as an accommodation to his disability. Jim refused the accommodation, claiming he didn't need the cart and didn't want to draw attention to himself by using the cart.

Once Jim refused to use the cart, his supervisor was forced to tell management that Jim no longer qualified for the job since Jim's quantity and quality suffered as a result of Jim's inability to deliver the work orders on time. Jim's refusal of an accommodation put him in danger of termination for poor job performance.

If the employer and the person needing the accommodation fail to identify an appropriate accommodation, the employer may benefit from technical assistance. The ADA regulations and the EEOC's Technical Assistance Manual identify professionals and others able to provide technical assistance as physicians, psychologists, rehabilitation counselors, occupational and physical therapists, and others who have direct knowledge of the particular disability. Organizations such as the Job Accommodation Network, Arthritis Foundation, Easter Seal Association, United Cerebral Palsy Association, and other disability advocate groups fall under the umbrella of those with direct knowledge of the disability (EEOC § III-10(6)).

The reasonable accommodation decision-making process follows a natural progression. First the employer must determine whether the worker is an individual with a disability. To accomplish this step, the employer must ask three questions:

1. Does the worker have a physical or mental impairment that substantially limits one or more major life activities?
2. Does the worker have a history of a substantially limiting impairment?
3. Is the worker regarded as having such an impairment?

If the employer cannot answer yes to one of these three questions, the worker is not disabled and the employer does not have to accommodate.

During the second step in decision making, the employer determines whether the worker is a qualified individual with a disability. The worker must satisfy the requisite skills, experience, education, and other job-related requirements, and must be able to perform the essential functions of the job with or without reasonable accommodation. If the worker is not a *qualified* individual with a disability, the employer does not have to accommodate.

The third step involves determining whether the accommodation is reasonable. The accommodation must not be unduly costly, extensive, substantial, disruptive, or a fundamental alteration to the nature or operation of the business. If the accommodation fits one of the aforementioned categories, it may be considered an undue hardship to the company and thus not reasonable. If the accommodation is not reasonable, the employer does not have to make the accommodation.

The ADA defines *undue hardship* as any action requiring significant difficulty or expense for the company when considering the following factors:

- The nature and net cost of the accommodation, considering tax credits, deductions, and/or outside funding (EEOC § III-9). Only the net cost to the employer would be considered.
- The overall financial resources of the facility involved in providing the accommodation (EEOC § III-9).

- The number of persons employed at a particular site and the effect on the expenses and resources of that particular facility.
- The overall financial resources of the business and the type and location of all its facilities (EEOC § III-9). The EEOC's ADA Technical Assistance Manual for Title I gives an example of an independently owned fast food franchise that receives no money from the franchisor. Since the financial relationship between the franchisor and franchisee is limited to payment of an annual fee, only the resources of the franchise would be considered in determining whether an accommodation was an undue hardship.
- The composition, structure, and functions of the workforce (EEOC § III-9).
- Geographic separateness and administrative and/or fiscal relationships between the facility and the parent organization must be considered.
- The impact of an accommodation upon the operation of the facility, including the impact on the ability of other employees to perform their duties and the impact on the facility's ability to conduct business (EEOC § III-9). For example, it would fundamentally alter the nature of a temporary construction site and be unduly costly to expect the company to make the site accessible for someone in a wheelchair. An individual with a disability who required a very warm environment for comfort might negatively affect other employees' productivity if the employer maintained higher temperatures in the workplace.

  If a custom-redesigned workstation prevented employees on the other two shifts from using the workstation, the redesigned workstation would constitute an undue hardship. Where accommodations adversely affect other employees' performance, an employer could make a case for the undue hardship exception. However, should employees simply complain about working around or with a person with a disability, without any bearing on their performance, no undue hardship would exist. Such circumstances obligate the employer to inform, educate, and raise the sensitivity level of coworkers and other employees.
- The terms of collective bargaining agreements (EEOC § III-9). The employer's primary responsibility is to consult with the union about reasonable accommodation in each individual situation. The ADA suggests that employers and unions develop provisions in their labor agreements permitting employers to take necessary actions to comply with the law (EEOC § III-9). Labor unions fall under the same ADA mandates and obligations with which employers must comply (EEOC § III-9).

  Circumstances will arise where the terms of a collective bargaining agreement may help determine whether an accommodation poses an undue hardship. For example, a worker who sustains a back injury and can no longer perform heavy labor requests, as an accommodation, transfer to a vacant clerical position. If the new position fell under a separate collective bargaining agreement, with its own provisions and seniority lists, it might be an undue hardship to transfer the worker if others held seniority for the clerical job (EEOC § VII-11). Long Island Lighting and The International Brotherhood of Electrical Workers negotiated a partial solution to the seniority problem. Their agreement allows workers with 10 or more years' seniority assignment out of seniority order should they require a reasonable accommodation (EEOC § VII-11).

THE COST OF REASONABLE ACCOMMODATION

By far the most pervasive fear expressed by businesses about the ADA is the cost of reasonable accommodations. Numerous myths and misperceptions surround the reasonable accommodation obligation—probably the most popular is that "making reasonable accommodations for workers with disabilities is too expensive."

According to a 1987 study by the Lou Harris Group, three-fourths of all managers interviewed stated that the cost of employing individuals with disabilities equaled that of nondisabled workers. The EEOC estimates that employers will find the cost for making reasonable accommodations about $16 million annually.[4] However, these reasonable accommodations will provide productivity gains of more than $164 million annually and reduce government support payments and increase tax revenue by $222 million each year.[5]

The Job Accommodation Network (JAN) provides the following statistics regarding accommodations costs:

- 19 percent of all accommodations are between $1 and $50.
- 19 percent of all accommodations are between $50 and $500.
- 19 percent of all accommodations are between $500 and $1,000.
- 11 percent of all accommodations are between $1,000 and $5,000.
- More than two-thirds (69 percent) of accommodations cost less than $500.
- 50 percent of accommodations suggested cost less than $50.[6]

Another study of accommodations provided to handicapped employees by federal contractors and cited in the proposed regulations to Title I found:

- 51.1 percent of all accommodations are made at no cost.
- 18.5 percent cost between $1.00 and $99.
- 11.9 percent cost between $100 and $499.
- More than 80 percent cost less than $500.

- The average cost of an accommodation is $304.
- At least one-half of all individuals with disabilities require no accommodations in order to work.[7]

To assist with making reasonable accommodations, Congress provided several tax and other incentives for hiring individuals with disabilities. These include:

- The Architectural and Transportation Barriers Removal Deduction (Section 190 of the Internal Revenue Code), which allows businesses to deduct up to $15,000 for making existing facilities or public transportation vehicles accessible and usable by persons with disability (EEOC § III-11).
- The Targeted Jobs Tax Credit (Section 51 of the Internal Revenue Code), which allows businesses that hire individuals with disabilities a credit against their tax liability equal to 40 percent of the first year's wages. Individuals must be employed at least 90 days or have completed 120 hours (EEOC § III-11).
- The Disabled Access Credit (Section 44 of the Internal Revenue Code), which permits eligible small businesses to receive a tax credit for certain costs of complying with the ADA. Qualifying businesses may claim a credit of up to 50 percent of eligible expenditures that exceed $250 but do not exceed $10,250 (EEOC § III-11).
- The Rehabilitation Act of 1973, which authorizes annual funding to the State Vocational Rehabilitation Agencies for on-the-job training. The position must be full time and permanent and must pay more than minimum wage.
- The Job Training Partnership Act provides on-the-job training at a worksite and reimburses the employer 50 percent of the first six months' wages for each eligible employee (EEOC § III-11).

Individuals with disabilities will find other sources of assistance for reasonable accommodations. The Veterans Administration (VA) is one of the largest purchasers of assistive devices for persons with disabilities. The VA's systematic structure for providing assistive technology seeks to ensure cost-effective procurement of equipment that is both needed by and safe for eligible veterans.

Service clubs and organizations are another source of funding for assistive devices. Groups such as the Lions, Shriners, Kiwanis, Rotary, Sertoma, Knights of Columbus, and Elks historically fill many gaps that the human service delivery system leaves open.[8] Various loan programs also assist individuals with disabilities to obtain needed adaptive equipment. Examples of these programs include the Mitari/Canon Optacon II, a portable, tactile reading device for persons who are blind. The Easter Seals Society created the first national loan program designed to enable persons with disabilities to buy assistive technology.[9]

Subsidy programs also make products available to consumers with disabilities who may have difficulty paying for devices. These programs lower the cost of equipment or provide it at no cost. A subsidy may come in the form of a grant, discount, or rebate. Subsidies may target a specific product or may cover a variety of devices. Organizations offering subsidy programs include Associated Services for the Blind, located in Philadelphia, Digital Equipment Corporation, Tandy Corporation, IBM, Chrysler, and Volkswagen.[10]

Reasonable accommodations can run the gamut from simply increasing the awareness of the vast contributions available to society from people with disabilities to very tangible assistive devices. Reasonable accommodations can cost from nothing to thousands of dollars. Most cost very little. The concept of reasonable accommodation means allowing employers, employees, and others with knowledge of disability to enter into an interactive, problem-solving process. This process ultimately allows individuals with disability to enjoy the same privileges, benefits, acceptance, and recognition of accomplishment that nondisabled people enjoy. Most importantly, one must remember that reasonable accommodations provide opportunities for achievement and equality of benefits and privileges of employment.

HYPOTHETICAL ACCOMMODATION EXERCISES

Read the following hypothetical situations. Using the problem-solving approach outlined in this chapter and your knowledge of workplace accommodations, develop a course of action for accommodating these workers.

Remember, your first step should be to analyze the job and determine its purpose and essential functions. Next consult with the individual to determine the specific job-related limitations imposed by the disability and how the individual could overcome the limitations. In conjunction with the employee, identify potential accommodations and evaluate the effectiveness each would have in enabling the person to perform the job's essential functions. Consider the preference of the individual but select the accommodation that best suits both employer and employee.

The solutions outlined below show one possible solution. Readers will probably be able to develop numerous other solutions. Reasonable accommodations solutions are limited only by one's creativity.

Problem 1: Susan works as a cashier in a hospital cafeteria. Her hands have been tingling. The doctor informs her that she has carpal tunnel syndrome. The doctor states that her job has contributed to her condition. He also discovers that Susan has a neuroma in her right foot. Susan needs this job. Her employer likes her and wants to accommodate her.

Solution: The job analysis shows Susan must ring up items on the cash register, collect money, and make change. Susan's neuroma limits her ability to stand. Her physician prescribes bilateral splints that reduce wrist mobility. To accommodate Susan's limitations, the following accommodations are provided:

1. an adjustable bar stool height chair with arm supports so Susan can work while sitting instead of standing
2. adjustments to the height of the cash register to reduce the repetitive motions

Problem 2: The director of purchasing hires Bill as her administrative assistant. After two weeks, the director, unhappy with Bill's performance, tells him if he does not improve, she will have to fire him. He is not completing orders on time and he is omitting items from the orders. Bill tells his boss he has a psychiatric impairment that causes short-term memory loss and difficulty concentrating. He also has problems organizing his thoughts. The boss calls you as a consultant for advice.

Solutions: Once Bill discloses his disability, the boss must attempt to accommodate him. The job analysis reveals that Bill's desk is located in a noisy, open area with much traffic. Bill must review requisition forms and transfer the information onto purchase order forms. Occasionally, he must take down verbal orders. To accommodate Bill's limitations, the following accommodations are provided:

1. Move Bill's desk to a quiet work area with low traffic and few distractions.
2. Teach Bill a check-off system so he can compensate for the short-term memory deficits by checking off items as he records them and can check his work.
3. Encourage Bill to keep a small notepad with him so he can write down verbal orders.

Problem 3: Laura is a hairdresser. Her knee has been hurting her over the last three months. She goes to the doctor, who tells her she has rheumatoid arthritis and her condition will worsen if she doesn't avoid bending and standing for long periods of time. She works for a large chain of $10.00 hair cut salons. Her boss tells her she will have to quit her job. She calls an attorney and he calls you.

Solution: An analysis of Linda's job shows that hairdressers must stand for long periods of time and do a lot of repetitive bending. Linda can continue to perform her job if these tasks are eliminated. Linda is entitled to reasonable accommodations under the ADA because of her disability. To accommodate Laura's limitations, the following accommodations are provided:

1. a high chair on wheels or a lean-on type chair on wheels from which Laura can cut hair

2. a raised height storage cart for rollers and other equipment
3. moving Laura's workstation to the end of the row to allow more access for her accommodations

Problem 4: Bob is a dispatcher for a large trucking company. He sits at a large "L"-shaped console. He has developed a serious neck problem causing him chronic pain. He is having difficulty managing the computer screens and all the console indicators because of his neck problems. What would you do if called by the employer for assistance?

Solution: Bob's job contributes to his neck pain because of the design of the dispatcher console, his workstation. To accommodate Bob's neck problem, neck flexion and extension should be kept at a minimum. To accommodate Bob's limitations, the following accommodations are made:

1. The height of the computer screens is adjusted so they are at eye level.
2. Based upon the job analysis, the screens used most often are placed directly in front of Bob at eye level.
3. Bob is provided with an crgonomic chair with adjustable height and arms.

Problem 5: Mike is a garbage collector. He hurt his back and is ready to return to work. His doctor restricted him to lifting 60 pounds. The county, realizing it has an unusually high incidence of garbage collectors with back injuries, wants to return all of these workers to their jobs and to figure out a way to accommodate other workers to prevent further injuries.

Solutions: Many municipalities have turned to automated garbage collection systems. With these systems, customers use standard garbage cans that are wheeled to the garbage trucks, attached to the trucks, and mechanically lifted and emptied into the truck. With these automated systems, workers need not lift the garbage cans. These types of systems can allow workers with restorations to return to their jobs and can prevent further injuries in other workers.

NOTES

1. Equal Employment Opportunity Commission (EEOC), *A Technical Assistance Manual on the Employment Provision (Title I) of the Americans with Disabilities Act* (Washington, D.C.: U.S. Government Printing Office, 1992).
2. President's Committee on Employment of People with Disabilities, *Facts You Can Use* (Washington, D.C., 1992).
3. Bureau of National Affairs, *Americans with Disabilities Act (ADA) Manual Newsletter*, June 1992.
4. Bureau of National Affairs, *Americans with Disabilities Act (ADA) Manual Newsletter* 1, no. 3 (1992):13–18.

5. President's Committee, *Facts You Can Use*.
6. President's Committee on Employment of People with Disabilities, *Job Accommodation Ideas* (Washington, D.C., 1992).
7. Berkely Planning Associates for the Department of Labor, Employment Standards Administration, vol. 1, 17 June 1982, 29.
8. George Washington University Regional Rehabilitation Continuing Education Program and Electronics Industry Foundation Rehab Engineering Center, *Financing Assistive Technology* (Washington D.C., 1992).
9. Ibid.
10. Ibid.

Conclusion

Challenge to the Future

The contributors to the introductory chapter also provide focus to our challenges for the future. The actions they delineate can serve as an action plan for individual direction and teamwork goals.

ERGONOMICS

Don Chaffin: Cross-discipline ergonomics teams are needed for effectiveness.

Documentation of ergonomics when used to change work must be made. There must be outcome studies for ergonomics to prove what works.

There should be an emphasis on the blend of life and engineering sciences as a basis for guiding ergonomics. This will require graduate education and dedication.

Melanie Ellexson: Identify the supporting body of knowledge regarding injury prevention and workplace safety. The only philosophy we can support is one that looks at the work, the worker, and the workplace and utilizes the team approach to problem solving.

We must find better ways to help employers match the individual worker to the job.

PREVENTION

Gunnar Andersson: Emphasize prevention in the workplace and learn more about the etiological factors of work and work-related industries. This would include emphasis not only on the physical work environment, but on the psychological work environment and life factors as well.

Michael Oliveri: Bring safety and health education and prework screening into the countries and industries where these are not known. These measures will be preventative for work injury and will need to be introduced scientifically.

EARLY INTERVENTION

Gunnar Andersson: If injury does occur, treat the patient quickly and effectively and emphasize activity, activity, activity. Early return to work is desired.

Marilyn Peterson: Returning to meaningful work as early as possible is vital. Jobs should not be "made up," they should be real and they should bring the employee satisfaction.

OUTCOMES

Peter Towne: Study the effect of disciplines, not only the effect of services. For example, physical therapy's effectiveness in work injury management should be studied for its professional role as well as the treatment techniques such as modalities, exercise, etc. By studying both techniques and professions, outcomes should direct future services.

Glenn Carmen: Identify roadblocks in returning injured workers to work. By identifying the roadblocks, we can then divide them into those constraints that cannot be removed and those people or constraints that can be removed from the system.

Dennis Hart: All clinicians must be accountable and responsible for their clinical services.

All outcomes should be measured. The use of unsupported clinical opinions about level of capacity or return to work should no longer be in the hands of physicians or clinicians who do not have measurements to support their conclusions.

In the absence of quality outcomes measures in clinical effectiveness studies of industrial rehabilitation services, standards for clinical procedures should be used. Clinicians fortunately have developed and are using standards that should promote more consistent clinical measures and assessments of patients.

LIGHT DUTY

Melanie Ellexson: We need to educate employers about reasonable accommodation and alternative work. This requires medical practitioners to learn more about business operations, how companies work, and what their productivity requirements are. The more medical professionals learn, the better we can teach and help companies with prevention of injury.

PHILOSOPHY

Mark Rothstein: Companies should change their emphasis from the dollar to the worker and emphasize injury prevention and employee rehabilitation. This can be done in a nondiscriminatory way. Employers can use both OSHA and ADA to their own benefit by allowing better interface between the work and the worker.

Glenn Carmen: In a union environment, labor and management must work together as a team working cooperatively and for the same goal. That goal is restoring our employees to full recovery using the "state-of-the-art" medical rehabilitation techniques. Labor must be fully apprised of the treatment program and its objectives. The future will be for cooperation, not for adversarial relationships.

Michael Oliveri: Rehabilitation should also be available to the group of people with some chronic disability who still manage to keep on working. Deterioration and the development of a chronic handicap could then be prevented.

Jean Brisson: We must commit to teamwork and mutual respect.

MANAGEMENT

Glenn Carmen: In disability management, take a "seamless" approach. This means understanding that a disability is a disability regardless of cost. Employ strategies for managing occupational and nonoccupational disability that are identical. Emphasis must be away from challenging the validity of claims to changing the level of disability. This is achieved by a staff knowledgeable in the concepts of early intervention and thorough treatment of soft tissue and back injuries. Therefore the emphasis should be on reducing disability, not challenging claims.

Marilyn Peterson: All involved should recognize when it is obvious that the employee will never be able to meaningfully contribute to the company's obligation and address the eventuality that further vocational rehabilitation will be required.

Vert Mooney: We must try to influence state regulations for workers' compensation soft tissue injury compensation. Our goals should be to demand functional definitions of severity of injury by objective testing after six weeks. Indefinite care for soft tissue injury is not acceptable. Continued care only on the basis of documented progress should be allowed. Cases should be discontinued if function does not improve.

FINANCIAL

Dennis Hart: Referral for profit should be eliminated, as this erodes accountability and responsibility.

Glenn Carmen: Establish a baseline of hard disability costs. This baseline can then be compared both historically and to outside benchmarks.

Target areas for costs *can* be managed.

Set realistic targets for cost containment or improvement.

MEDICAL

Jed Downs: Injured tissues, especially chronically injured tissues, must be recognized as such and the injured worker should not be expected to return to the job that was the cause of the injury and function as though nothing had ever gone wrong. Job analysis and ergonomic modifications are important. The injured worker and the reason for the injury represent an opportunity for primary prevention when studied.

Vert Mooney: Buy the concept that soft tissue injury (employer responsible) that doesn't heal by a month means lack of effective healing and that the "weak link" must be specifically strengthened. This is using medical parameters similar to cardiac rehabilitation.

Extensive search for a diagnosis in individuals without neurological deficit is usually unwarranted. The weak link identification is adequate and should be explored further.

COMMUNICATION

Jean Brisson: Industry must look at the results other companies have accomplished by promoting their safety and ergonomics in the workplace. Employers can learn from one another to commit to taking control of their injuries and working with health care providers to promote safety.

CONCLUSION

Two thoughts broached the entire process of problems and solutions. They are quoted here since they are especially succinct.

Jed Downs: Treatment must be as active as possible. Function is enhanced by pain control, and function needs to be recognized as the goal we are working toward. Pain can only be measured subjectively. Functional capacities determine an individual's freedom to choose among vocational and avocational options.

Therefore the individual is seen as a whole person with both vocational and avocational interests. Rehabilitation of the vocational problems in a functional manner will help the whole worker.

The second comment relates to the individual.

Leonard Matheson: We must appreciate the power of the individual's urge toward competence. That can be tapped to strengthen and focus motivation. We must focus on improving function, competence, and self-reliance rather than on remediation of symptoms.

The authors of this book and the experts included in the introduction and conclusion have set the direction for our challenge for the future. Summarized, we can find:

- *Teamwork* is essential. Industry workers, employers, and union members must work closely with the medical community of physicians, therapists, psychologists, and chiropractors, who also must integrate vocational counselors and case managers. There must be a bonding with the legal system to work in a proactive way for fair case resolution.
- *Mutual respect* will allow the worker's needs to come forward and the outcomes to be measured and assessed for the good of the worker.
- *Function* must be emphasized over pain. Pain control is important, but function is the goal for the worker in his or her whole life as well as the work injury management system.

Working in work injury management is not only challenging but extremely rewarding. One only has to see the relief on injured workers' faces when they find out they truly can function again and return to work to recognize that our specialties have power for positive change. One only has to see employers finally understand that they do not have an "unsafe" workplace, but rather that there are some very effective things that can make the worksite safer and more productive to realize that our intervention can affect society proactively. These positive solutions bring a sense of safety and camaraderie to workers, employers, and providers.

As we progress from the date of this publication, we will promote:

- the integrity of and respect for the worker
- the empowerment of the employer and employee to continue to provide productive and safe work
- the scientific and medical arts rationale for treatments that enable professionals to be effective and meaningful in the work injury management system

The challenge for the future will be for us to continue to work together and bring out the best in each other. When we commit to the destination rather than the road, progress is accelerated and we are rewarded with new paths, views, and traveling companions.

—Susan Isernhagen, 1994

Index

B

C

D

E

N

O

Q

R

S

T

U

V

W

NOTES

NOTES

NOTES

NOTES

NOTES

NOTES

NOTES